Project 233

IUGS UNESCO IGCP

Terranes In The Circum-Atlantic Paleozoic Orogens

R.D. DALLMEYER, Athens, USA
J.D. KEPPIE, Halifax, Canada
(Directors)

Volumes already published in the series:

R. D. DALLMEYER and E. MARTÍNEZ GARCÍA (Eds.):
Pre-Mesozoic Geology of Iberia

R. D. DALLMEYER and J. P. LÉCORCHÉ (Eds.):
The West African Orogens and Circum-Atlantic Correlatives

R.D. Dallmeyer J.P. Lécorché (Eds.)

The West African Orogens and Circum-Atlantic Correlatives

With 196 Figures

Springer-Verlag
Berlin Heidelberg New York
London Paris Tokyo
Hong Kong Barcelona
Budapest

Dr. R. D. Dallmeyer
Department of Geology
University of Georgia
Athens, Ga 30602
USA

Dr. J. P. Lécorché
Laboratoire de Géophysique et Géodynamique,
case 422
Faculté des Sciences et
Techniques de St. Jérôme
Université d'Aix-Marseille 3
13397 Marseille Cedex 13
France

ISBN-13: 978-3-642-84155-2 e-ISBN-13: 978-3-642-84153-8
DOI: 10.1007/978-3-642-84153-8

Library of Congress Cataloging-in-Publication Data
The West African orogens and circum-Atlantic correlatives / R. D. Dallmeyer, J. P. Lécorché, eds. p.cm. Includes bibliographical references and index.
1. Geology--Africa, West. 2. Stratigraphic correlation. 3. Geology, Structural--Africa, West. I. Dallmeyer, R. D., 1944- . II. Lécorché, J. P. (Jean Paul), 1933- . QE339.W46W47 1990 551.7'00966--dc20

Softcover reprint of the hardcover 1st edition 1991

32/3145-543210 Printed on acid-free paper

Preface

In December, 1987, Project 233 of the International Geological Correlation Program hosted an international conference in Nouakchott, Mauritania. Discussions were focused on the "Tectonothermal Evolution Of The West African Orogens And Circum-Atlantic Terrane Linkages". A wealth of new information was presented during the conference, and it was apparent that the time was appropriate to prepare a systematic volume dealing with the geology of northwest Africa. The present volume is an outgrowth of this conference, but is not merely a compilation of the papers presented in Nouakchott. Instead, it represents a coordinated volume designed to present a balanced, comprehensive view of our present understanding of West African geology and potential correlations in other Circum-Atlantic Paleozoic orogens.

We sincerely appreciate the efforts of all the reviewers whose comments have greatly helped to improve the quality of this volume. We also thank the various contributors for their diligence and perseverance in manuscript preparation. The calibre of the present volume is a reflection of their labor.

Athens/Marseille, Autumn 1990

R. D. DALLMEYER
J. P. LÉCORCHÉ

Contents

Chapter 6
Circum-Atlantic Correlations

Contributors

You will find the addresses at the beginning of the respective contributions.

Chapter 1

Introduction

R. D. Dallmeyer[1] and J. P. Lécorché[2]

The geology of Africa is dominated by the Kalahari, Congo (Zaire) and West African cratons (Fig. 1). These are composed of late Archean and early-middle Proterozoic crystalline basement rocks which are nonconformably overlain by middle to late Proterozoic-earliest Paleozoic shelf sedimentary sequences and successions within younger Paleozoic basins (Fig. 1). Kennedy (1964) and Clifford (1968) pointed out that most regions surrounding the African cratons record affects of ca. 750–500 Ma tectonothermal activity (Pan-African). The history of these orogenic belts and the role of plate tectonic processes in their evolution has been uncertain and controversial (e. g. Kroner 1977, 1979). Results of recent, collaborative field, geochemical, geophysical and geochronological investigations in the West African orogenic terranes have significantly enhanced our understanding of the tectonothermal evolution of the enigmatic Pan-African orogens. These data are summarized in this volume.

[1] Department of Geology, University of Georgia, Athens, Georgia 30602, USA
[2] Faculté des Sciences et Techniques de St. Jérôme, Université d'Aix-Marseille 3, 13397 Marseille Cedex 13, France

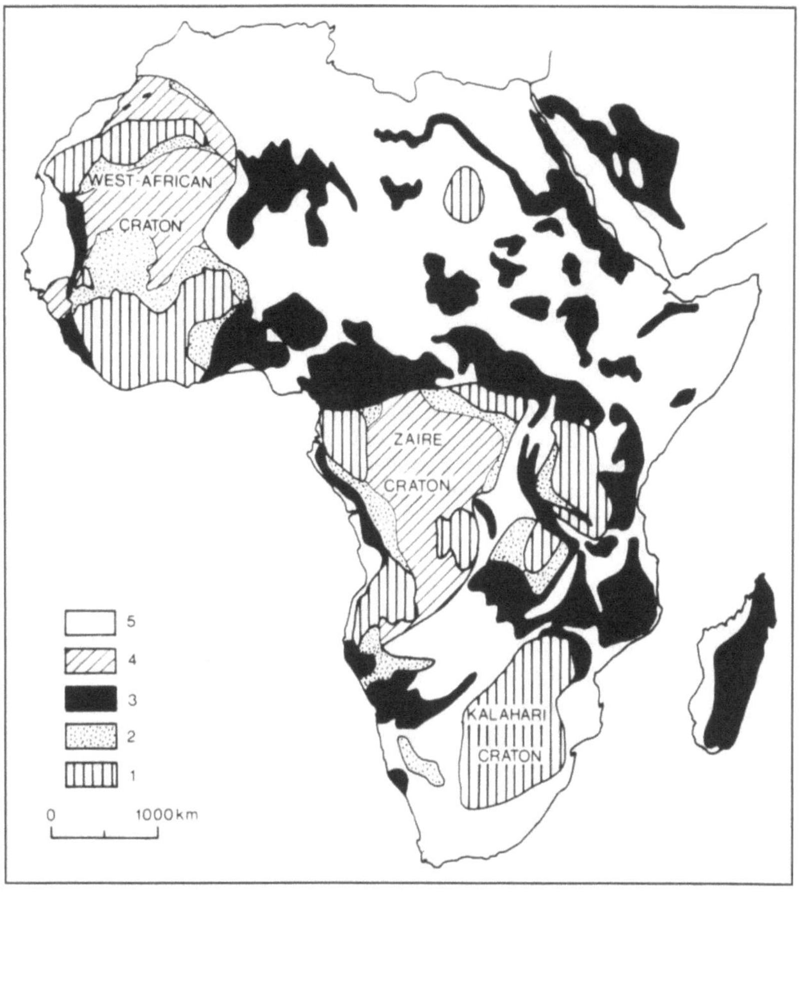

Fig. 1. The African cratons and distribution of late Proterozoic-earliest Paleozoic (Pan-African) orogenic belts: *1* Archean-early Proterozoic crystalline basement; *2* middle to late Proterozoic epicontinental clastic sedimentary sequences; *3* Pan-African orogens (including structural units of penetratively reworked basement); *4* Paleozoic basins; *5* Post-Paleozoic cover successions

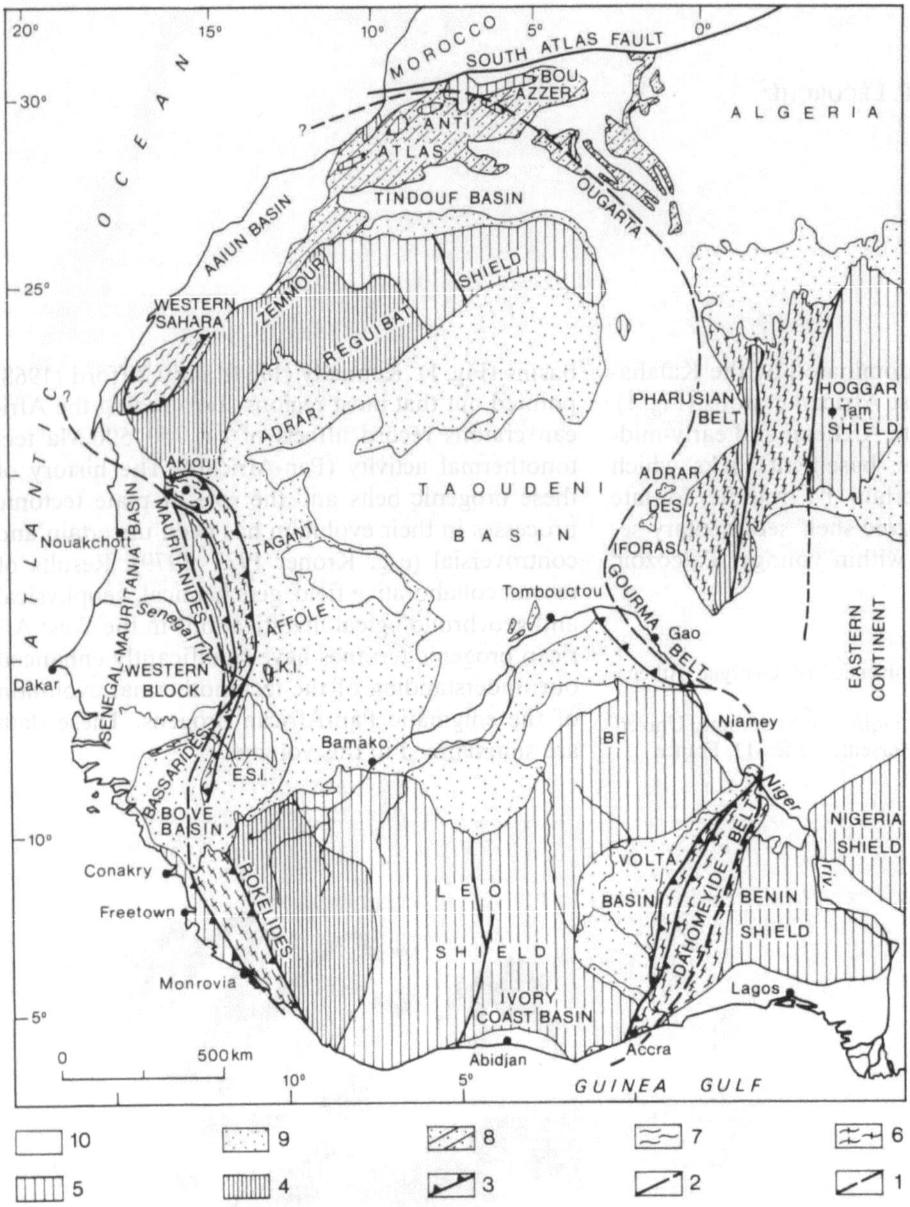

Fig. 2. Geologic setting of northwest Africa: *1* limit of the West African craton; *2* major high-angle fault; *3* major thrust fault; *4* late Archean crystalline basement; *5* early-middle Proterozoic crystalline basement (Birrimian/Eburnean); *6* late Proterozoic-earliest Paleozoic (Pan-African) orogenic belts; *7* structural units emplaced in the late Paleozoic; *8* sequences recording late Paleozoic penetrative strain and/or metamorphism; *9* late Proterozoic-late Paleozoic cover sequences; *10* post-Paleozoic cover sequences

The regional setting of northwest Africa is outlined in Fig. 2. Archean and early to middle Proterozoic crystalline rocks of the West African craton are exposed in the extensive Reguibat and Leo shields. Smaller tracts are exposed in the Kéniéba and Kayes inliers (K. I. and E. S. I. in Fig. 2). Crystalline basement of the West African craton is nonconformably overlain by several successions of sedimentary rocks ranging in age from middle-late Proterozoic to Devonian. Both basement and cover units have been affected by late Proterozoic-earliest Paleozoic tectonothermal events and comprise a variety of orogenic belts, including the (Fig. 2): Pharusian (Trans-Saharan), Dahomeyide, Rokelide, Bassaride and Mauritanide orogens. Variably penetrative late Paleozoic tectonothermal reworking occurred within the Bassaride and Mauritanide orogens during Carboniferous collision of Gondwana and Laurentia.

The initial three chapters of this volume discuss the overall crustal structure of northwest Africa, and outline characteristics of the basement and cover sequences of the West African craton. Systematic descriptions of the various West African orogenic belts are then presented. The concluding chapters of this volume explore potential extensions of West African structural elements within other Circum-Atlantic, Paleozoic orogens.

References

Clifford TN (1968) Radiometric dating and the pre-Silurian geology of Africa. In: Radiometric dating for geologists. Intersci, NY, pp 299-416

Kennedy NK (1964) The structural differentiation of African in Pan-African (500 million years) tectonic episode. Res Inst Afr Geol, Leeds Univ, GB, Eight Rep, pp 48-49

Kroner A (1977) The Precambrian geotectonic evolution of Africa: Plate accretion versus plate destruction. Precambrian Res 4: 163-213

Kroner A (1979) Pan-African plate tectonics and its repercussions on the crust of northeast Africa. Geol Rundsch 68: 565-583

Chapter 2

Geophysics and the Crustal Structure of West Africa

J. ROUSSEL[1] and A. LESQUER[2]

1 Introduction

West Africa, facing the Laurentian basement of North America and the Brazilian craton of South America in a Bullard-type predrift reconstruction, is largely identified to the West African craton (a Precambrian granitized basement) and its surrounding Upper Proterozoic (Pan-African) mobile belts. In contrast with relatively complete geological studies (Fabre 1983), West Africa did not benefit from a large number of extensive geophysical studies. Only gravity measurements have resulted in a nearly complete regional mapping and thereby are useful for investigation of the crustal structure and clarification of geological interpretation. In places, seismological studies such as the detection of teleseismic travel time anomalies or surface waves velocity dispersion measurements, magnetotelluric soundings, and heat flow measurements have been made. Considering the total absence of deep seismic profiles, all these data furnish a precious adjunct to the knowledge of the crustal and upper mantle structures of West Africa.

After a brief description of the regional geology, a review of existing geophysical data and proposed interpretations derived from geophysical-geological relationships are presented.

2 Major Geological Features in West Africa (Fig. 1)

West Africa is composed essentially of two major tectonic elements: the West African Precambrian craton with its sedimentary cover and the surrounding mobile belts, largely of Upper Proterozoic age but locally affected by Paleozoic overprinting.

2.1 The West African Craton Proper

This crops out in the Reguibat and Leo shields, in the Kayes and eastern Senegal inliers, and in the Anti-Atlas inliers south of the South Atlas main fault, which roughly forms the boundary between the Anti-Atlas and the Alpine Atlas Mountains. The craton is partly covered by Upper Proterozoic to Late Paleozoic sediments of the Taoudeni basin. Two large series are differentiated into the Precambrian cratonic basement (Bessoles 1977): Late Archean rocks dated at about 2500-3000 Ma are exposed in the southwestern part of the Reguibat shield (Barrère 1967) and in the western portion of the Leo shield. Early Proterozoic or Birrimian rocks mainly affected by the Eburnean orogeny (2000-1800 Ma) are the main parts of the cratonic basement in the Leo shield, the inliers of eastern Senegal and Kayes, and also make up the northeastern part of the Reguibat shield. In the Leo shield, especially in Burkina Faso (Upper Volta) and Ivory Coast, geochronological data suggest a Burkinian cycle (2400-2100 Ma) distinct from the Eburnean orogeny (Caen-Vachette 1988). Granites were emplaced between 2000 and 1700 Ma (Vachette et al. 1975). These intrusions mark the end of the crustal evolution of the West African craton, which has been stable since that time.

The epicontinental sedimentary cover of the craton is formed by generally thin and horizontal Middle and Upper Proterozoic to Late Paleozoic sediments of the Taoudeni basin in Mauritania and Mali. These series have been well studied in Mauritania Adrar (Trompette 1973; Bronner et al. 1980). However, the borders of the basin were tectonized during subsequent orogenesis in Pan-African time. In the Gourma basin on the eastern margin of the craton, strong subsidence occurred with a thickness

[1] Laboratoire de Géophysique Université d'Aix-Marseille 3, 13397 Marseille Cedex 13, France
[2] Centre Géologique et Géophysique Université des Sciences et Techniques du Languedoc, 34060 Montpellier Cedex, France

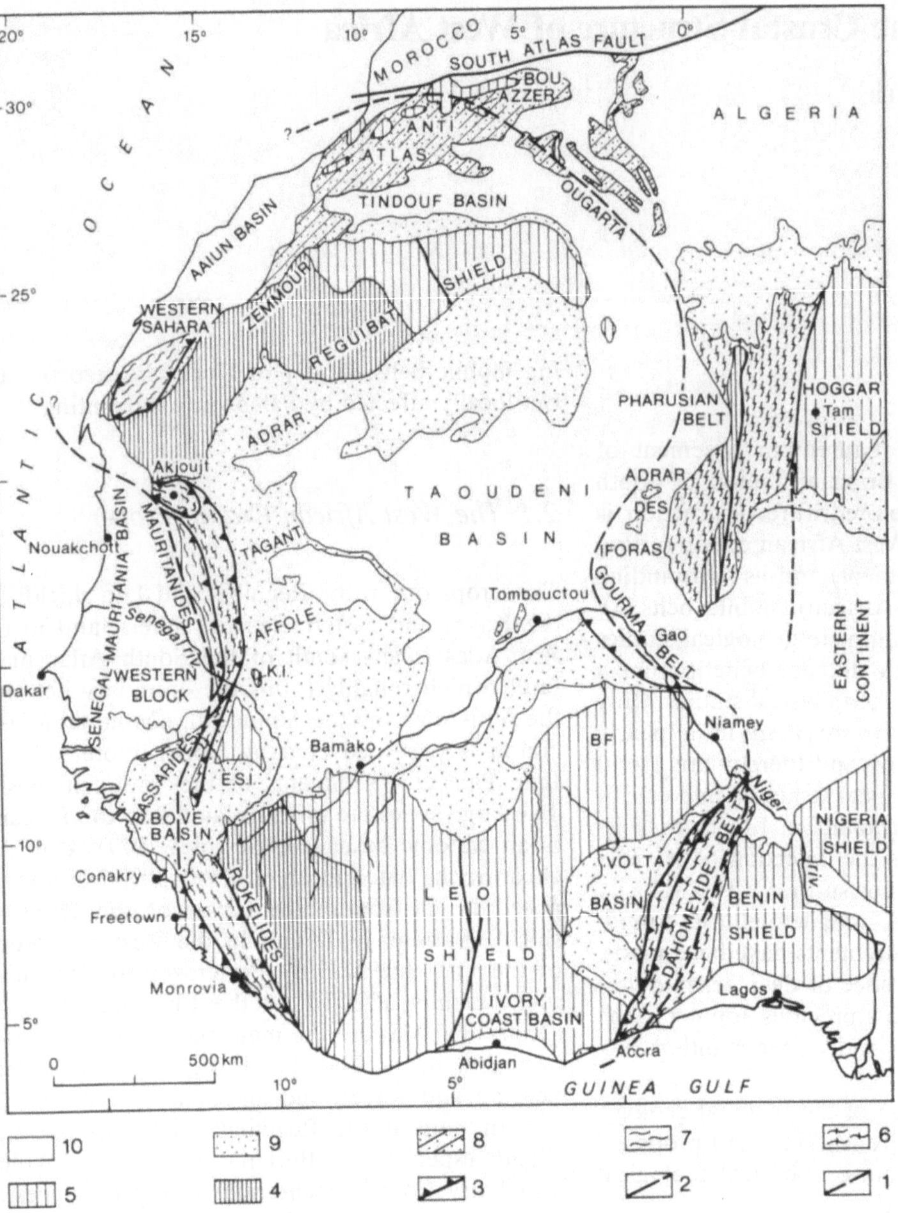

Fig. 1. Major structural elements of West Africa. *1* Limits of the West African craton; *2* faults; *3* thrusts; *4* Late Archean basement; *5* Early Proterozoic (Birrimian) basement; *6* Pan-African orogens; *7* Late Paleozoic orogens; *8* Late Proterozoic to Late Paleozoic folded cover; *9* Late Proterozoic to Late Paleozoic cover; *10* post-Paleozoic cover. *ESI* eastern Senegal inlier; *KI* Kayes inlier (Early Proterozoic basement); *BF* Burkina Faso; *Tam* Tamanrasset

of over 8000 m of sediments (Reichelt 1972; Moussine-Pouchkine and Bertrand-Sarfati 1978).

In addition to the Taoudeni basin, sediments of Upper Proterozoic to Upper Carboniferous age also have deposited in smaller basins. The Tindouf basin (Gevin 1960) lies between the northern margin of the Reguibat shield and the Anti-Atlas. Its northern border was folded during Paleozoic orogenesis. The Volta passive margin basin in Ghana and Burkina Faso formed the foreland of the Dahomeyide orogen (Affaton 1975; Affaton et al. 1980). Finally, the Bové basin in Guinea (Villeneuve 1984) contains Paleozoic sediments which mask the continuity of the West African orogens between Southern Senegal and Sierra Leone.

2.2 The Surrounding Pan-African Mobile Zone

This zone was first mentioned by Kennedy (1964) as a main tectonothermal episode which affected at 550±100 Ma the borders of the previously cratonized West African craton. At the present time, Pan-African belts (ca. 700-500 Ma) are found almost in continuity around the West African craton except along its southern margin, and a number of workers have clarified the geology of these belts.

To the east of the West African craton, the Pharusian-Dahomeyide orogens extend through Hoggar in Algeria (Bertrand and Caby 1977; Caby 1970), the Iforas and Gourma regions in Mali (Reichelt 1972; Caby 1978) and the Dahomeyide chain in Benin and Togo (Affaton 1975; Trompette 1979). The Dahomeyides are interrupted on the south at the continental margin of the Gulf of Guinea, but an extension has been postulated in South America (Brazilian belt, Lesquer et al. 1984).

The Pharusian-Dahomeyide orogens are interpreted as the result of an oceanic closure and collision around 600 Ma (Black et al. 1979) between the West African passive margin and an eastern continent including the Hoggar shield and the Benin and Nigeria shields. The suture can be traced from Morocco to Togo (Leblanc 1976; Bayer and Lesquer 1978; Black et al. 1979; Trompette 1979). In the Anti-Atlas (Morocco) on the northern edge of the craton obducted ophiolite has been found at Bou Azzer (Leblanc 1976, 1981). This northern margin extending to Ougarta behaved as an active zone during most of the Paleozoic (Black 1987).

Along the western margin of the West African craton, the Mauritanide-Bassaride-Rokelide orogens stretch out from southern Morocco to Liberia. The Mauritanides (Sougy 1962) and Bassarides (Villeneuve 1984) suffered a complex polyphase tectonothermal evolution (Chiron 1973; Lécorché 1980; Le Page 1983; Dia 1984; Villeneuve 1984). Evidence of Pan-African deformation has been reported (Chiron 1973) but the Pan-African belt is in part obscured by the overprinting of Paleozoic deformation.

In the light of new geochronological data, the history of the Mauritanide-Bassaride orogens is mainly characterized by two major Pan-African tectonothermal events at 650 Ma (Pan-African I) and around 550-500 Ma (Pan-African II), and, finally, by late Paleozoic (ca. 300 Ma) overprinting and thrusting (Dallmeyer and Villeneuve 1987; Dallmeyer et al. 1987; Villeneuve and Dallmeyer 1987; Lécorché et al. 1989; Dallmeyer and Lécorché 1989). The first orogenesis at 650 Ma is interpreted as the result of the collision of the passive margin of the West African craton with a western continental block previously rifted away from the craton but with a limited intervening oceanic domain. The second collision event at 550-500 Ma is marked by a tectonothermal intensity which increases from the north (central Mauritanides) through the Bassarides and is dominant in the Rokelides of Sierra Leone and Liberia where the Pan-African I orogenesis seems to be missing (Allen 1969; Thorman 1976; Culver and Williams 1979). The Pan-African II event may have been resulted from collision of the Guyana craton of South America with the southern part of the West African craton.

2.3 Late Paleozoic Event in Mauritanides and Bassarides

A late Paleozoic event at 300-280 Ma developed during collision of Gondwana and Laurentia and led to Hercynian overprinting and extensive tectonic transport toward the east in the northern and central Mauritanides (Dallmeyer and Lécorché 1989). This Hercynian event decreases southwards with local overprinting in the Bassarides (Dallmeyer and Villeneuve 1987; Villeneuve and Dallmeyer 1987) and is totally absent in the southern Bassarides and Rokelides (Lécorché et al. 1989).

2.4 The Atlantic Margin of West Africa

This margin results from the opening of the present North Atlantic ocean. The coastal plains, mainly including the Guinea, Senegal, Mauritania and Aaiun basins, occupy the whole area between the Atlantic and the Mauritanide-Bassaride-Rokelide orogens onto which they have transgressed. Their deep structure is known primarily from petroleum exploration works. First extensional episodes occurred in Triassic and resulted in rifting and associated volcanic activity and salt deposits (Guieu and Roussel 1984). In the Senegal basin, the synrift tectonics has been partly evidenced in the light of seismic sections which typically show thinned continental basement with north-south normal faulting and tilted blocks (Chanut and Micholet 1988). The subsequent subsided basin comprises thick Mesozoic-Cenozoic deposits with a thickness of 10000 m or more in the Dakar region (Roussel and Liger 1983). The post-rift subsidence history has been recently studied and documented in Senegal basin (Latil-Brun 1987; Flicoteaux et al. 1988). In contrast, in the Gulf of Guinea, the Ivory Coast basin, which has developed in a transform margin environment (Spengler and Delteil 1966), suffered moderate subsidence whose history seems to be related to successive shear tectonics episodes (Latil-Brun et al. 1988).

3 Lithospheric Structure of West Africa from Seismological, Magnetotelluric and Heat Flow Data

Few geophysical studies of the deep structure of the western part of African continent are presently available. Nevertheless, some information can be deduced from deep electromagnetic soundings, surface waves studies and heat flow density variations.

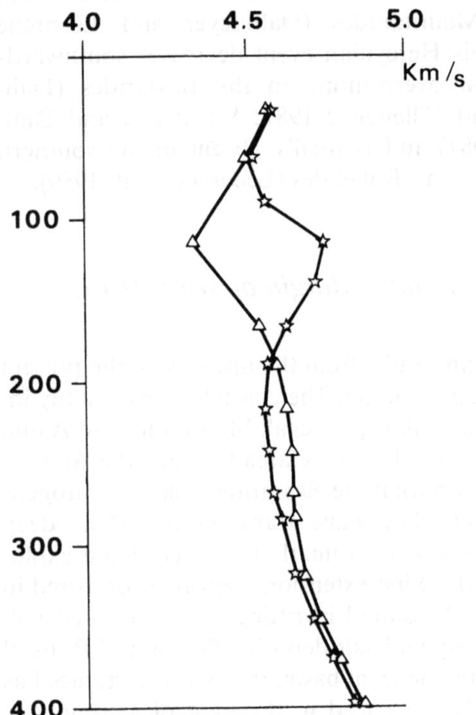

Fig. 2. Models of shear-velocity structure for the cratons and mobile zones (modified after Dorbath and Montagner 1983)

3.1 Seismological Data

P-wave studies of the West African craton margin in Senegal (Briden et al. 1981; Dorbath et al. 1983) revealed that a major seismic discontinuity, extending down to 150–200 km, separates the cratonic structural block (characterized by highest velocities) from the western mobile zone.

The study of surface wave velocities along paths crossing the African continent provides models for vertical distribution of lithospheric velocities and allows one to generalize the local P-wave results to the whole continent. The S-wave velocity model (Fig. 2) proposed by Dorbath and Montagner (1983) shows that the cratonic areas are characterized by higher shear-velocities than the mobile zones. A mean velocity difference of 0.22 km/s is shown at depths between 80 and 180 km. At greater depth this difference is reversed but weaker. Because only few distinct paths are considered and also because an a priori crude regionalization in cratons and mobile zones is used for data inversion, it is obvious that this model must be regarded as an average model which does not account for lateral inhomogeneities within the lithosphere. For instance, the delay time calculation for Tamanrasset observatory in South Algeria (5°E Longitude) shows that the Central Hoggar block (Pan-African belt) probably displays the same deep velocities distribution as the cratonic lithosphere of the Ivory Coast (Dorbath and Dorbath 1984).

A recent interpretation of long period records of surface waves by Hadiouche and Jobert (1988) without any a priori geological constraints improves the lateral resolution and provides a more accurate three dimensional image of the lithospheric structure beneath Africa.

The West African craton is characterized by a fast lithosphere down to 240 km, whereas the Pan-African eastern mobile belt is characterized by a slow lithospheric zone 150 km thick which trends NW-SE and extends from Sahara to Red Sea (Hadiouche and Jobert 1988). The cross-section of SV-velocity variation along parallel 20° N (Fig. 3) shows that the eastern and western limits of the fast lithosphere roughly correspond with inferred Pan-African suture zones. Eastward, this limit appears as a vertical zone situated between longitude 0° and longitude 6° E. It is noteworthy that between these two latitudes the Hoggar shield is divided by major meridian shear zones into units between which correlation is not always possible (Caby 1987). This observation may support the hypothesis that the structuration deeply extends within the lithosphere

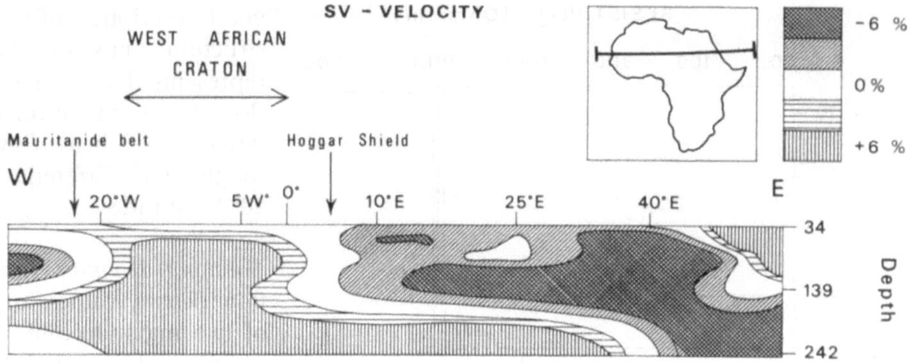

Fig. 3. Cross section at about 20° N of SV-velocity variations in West Africa. Reference value is 4.5 km·s^{-1} (Hadiouche and Jobert 1988)

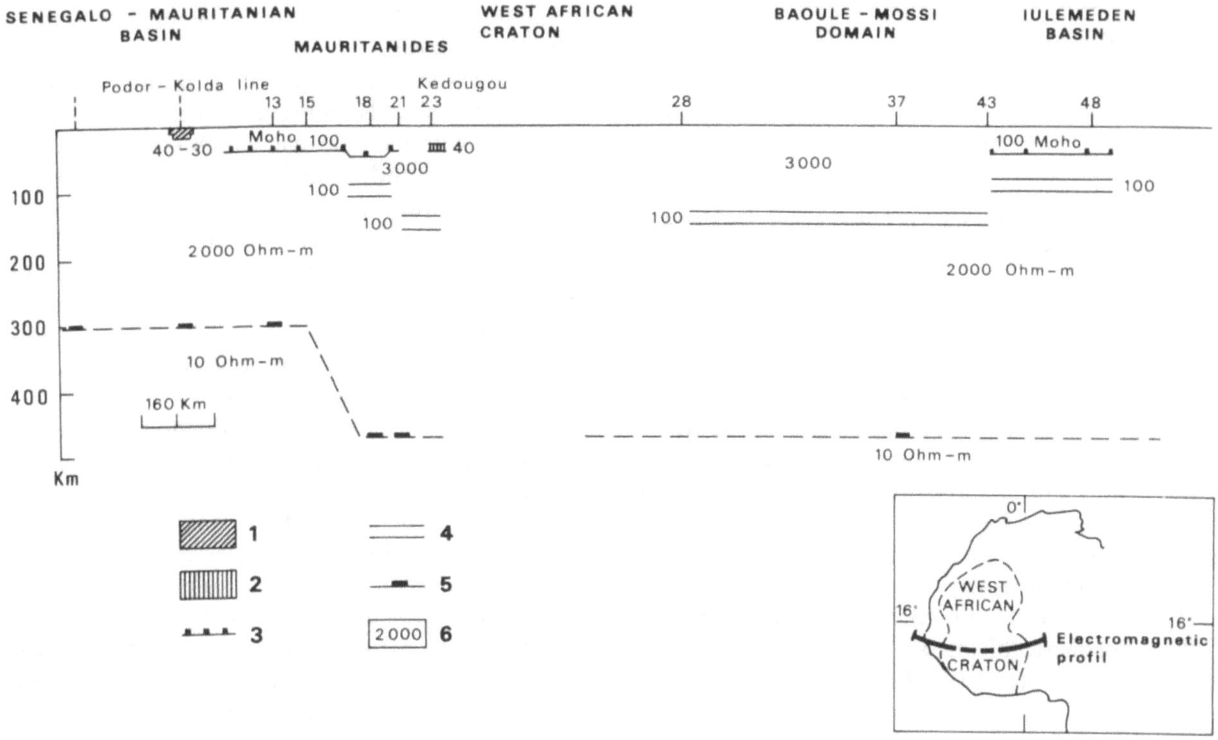

Fig. 4. Schematic model for conductive structures in the West African lithosphere. *1* Conductive body in the upper crust beneath the Senegal basin (20–30 ohm·m); *2* conductive body in the lower crust of the West African craton (40 ohm·m); *3* Moho; *4* conductive layer in the upper mantle (100 ohm·m); *5* depth of the lowest conductive layer; *6* resistivity in the upper mantle (2000 ohm·m) (after Ritz and Robineau 1986)

(that agrees with delay-time value at Tamanrasset). However, the slow lithospheric zone which extends from Sahara to Red Sea appears as oblique to Pan-African structures. This observation may indicate that this low velocity structure is not related to the Pan-African events and consequently may be younger. But lateral resolution of surface waves does not allow one to be more precise concerning the relationships with the geological framework (particularly the 0° latitude Pharusian suture zone).

3.2 Magnetotelluric Data

Magnetotelluric soundings (Ritz 1985) provide information on electrical conductivity inhomogeneities in the lithosphere. Fig. 4 from Ritz and Robineau (1986) shows that the deep geoelectrical structures of the craton and the mobile belt are strongly different (but the question is the degree to which the results of few magnetotelluric soundings are representative of all the West African craton):

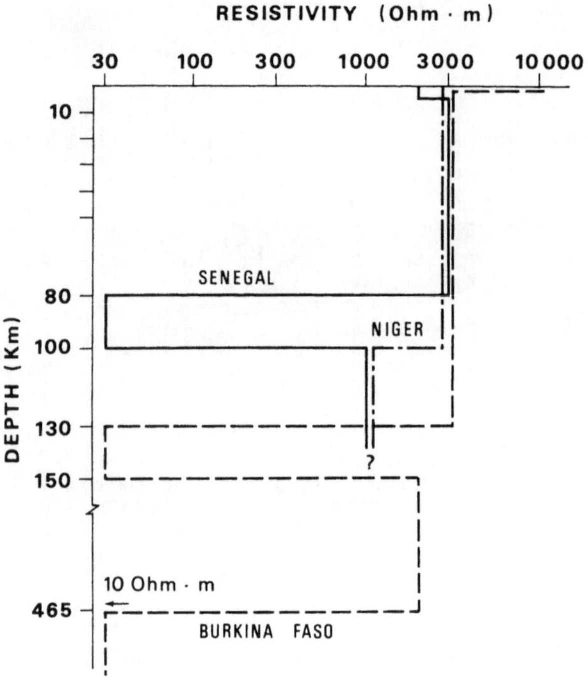

Fig. 5. Geoelectric model for the West African craton (Burkina Faso) and the mobile belts (Senegal, Niger) (modified from Ritz 1985)

(1) Low electrical resistivities (50–100 ohm·m) are found 130–150 km beneath the craton and 80–100 km beneath the mobile belts and (2) the resistivity falls to 10 ohm·m at 460 km depth beneath the craton and at 300 km depth beneath the Senegal-Mauritanian basin.

As pointed out by Ritz and Robineau (1986), it is still difficult to interpret one of these conductive layers as lithosphere boundary, and the problem of the lithospheric thickness is not yet resolved. Nevertheless, it is noteworthy that the low resistivity levels (Fig. 5), situated between 80–150 km depth, correlate with changes in shear velocities (Fig. 2).

3.3 Heat Flow Data

In 1982, only 12 pieces of heat flow density data were available in West Africa. As a result of the heat flow density program organized by the Centre Géologique et Géophysique de Montpellier (CNRS), 132 classical measurements and 540 oil exploration boreholes data (Bottom hole temperatures or Drill stem tests) are now available. They allow an initial insight into the thermal regime of West Africa (Brigaud et al. 1985; Lucazeau et al. in press), but they are not numerous enough to allow a good description of relationships with the different structural units. On the map in Fig. 6 (which only represents long wavelength components of heat flow density variations), changes in the pattern and intensity of heat flow density values delineate roughly two different domains, north and south of 20° N latitude.

Southern Domain. South of 20° N latitude, heat flow density values can be discussed in the context of two major tectonic units, i.e., the West African craton and the mobile belt. The Eburnean Leo shield is characterized by an average heat flow (30 ± 10 mW m^{-2}) lower than on the Pan-African basement of the Touareg (Hoggar) and Benin-Nigeria shields (51 ± 8 mW m^{-2}) or on the Senegal-Mauritania basin.

These values are compatible with those observed within other shields and reveal a general decreasing heat flow with tectonic age, which is the worldwide rule (Vitorello and Pollack 1980). The difference in average heat flow density between cratonic domain and surrounding mobile belts seems to be significant and may be interpreted as being due to greater crustal heat production within mobile belts than in craton, or alternatively due to differences in lithosphere deep structure as proposed by Ballard and Pollack (1987) for South Africa.

The present lack of heat production data and of information on the deep crust structure does not allow the interpretation of heat flow density in relation to crustal heterogeneities. However, using simplifying assumptions on the average crustal thickness and the mean heat production, we can propose some bounds for deep thermal structure, assuming that the West African craton and the mobile belts are homogeneous domains. If we accept a minimal value of 10 mW m^{-2} for the mantle heat flow beneath the craton (Pinet and Jaupart 1987), the total crustal contribution for the Eburnean Leo shield would be 20 ± 10 mW m^{-2}. For a 35 km crustal thickness, crustal heat production therefore would be 0.6 μW m^{-3}. This value agrees with the mean crustal heat production estimated to 0.4–1 μW m^{-3} by Ballard and Pollack (1987) and Jaupart and Provost (1985). Such a value significates that the heat-productive upper crust (granites, greenstone belts) 5–10 km thick (from gravity and magnetotelluric data) is underlain by a middle and lower crust depleted in radio-active elements (granulite?). For such a model of crust, a geotherm can be calculated: it shows 1200–1300 °C temperatures at 200–250 km. If we suppose that the contrast in surface heat flow between craton and mobile belt en-

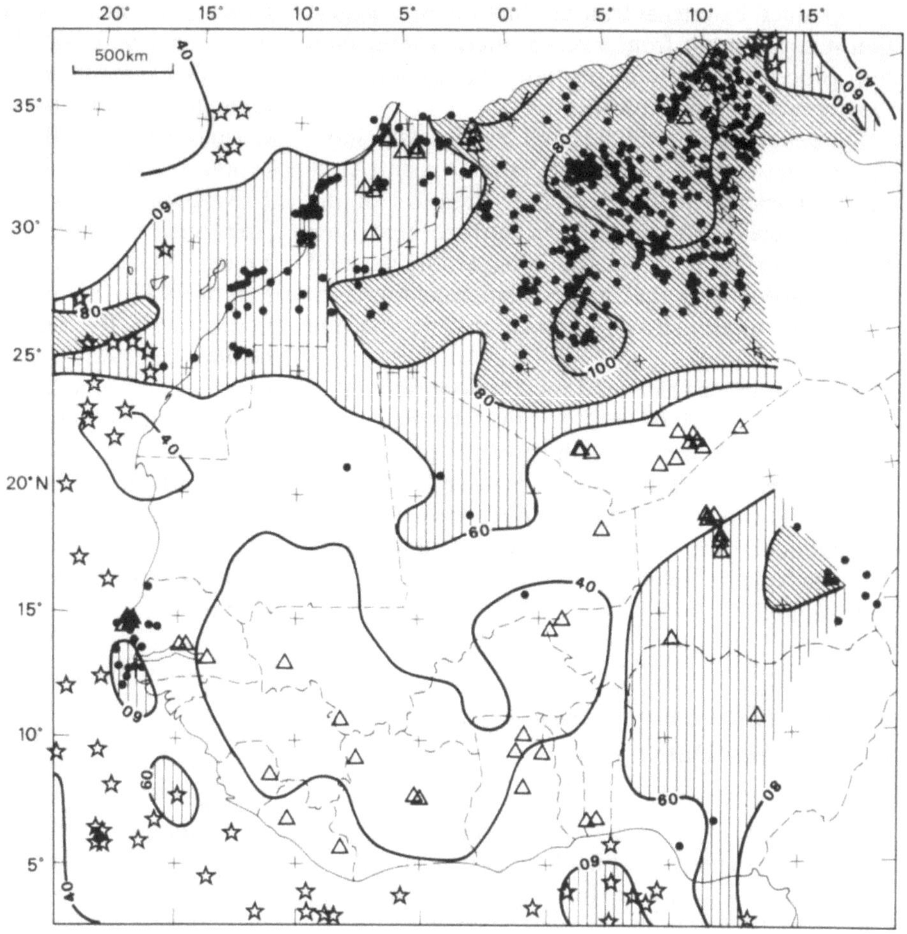

Fig. 6. Map of heat flow density regional variations. *Open triangles* Classical measurements. *Solid circles* oil well data: *BHT* (bottom hole temperature) or *DST* (drill stem test). *Stars* Off-shore published data (Lucazeau et al. 1988)

tirely arises from crustal heat production, no major differences in deep thermal structure are observed between the two domains, which could account for changes in electrical conductivities and seismic velocities. Alternatively, if we suppose that the mantle heat flow beneath the belt is higher than beneath the craton, deep thermal structure appears very different. If for example, we accept, that all of the contrast in surface heat flow arises from mantle heat flow, temperatures of 1200-1300 °C should be reached at 100-150 km depth beneath the belt. This result is in agreement with the magnetotelluric and seismic results. It leads to the proposal that the southern part of the West African craton is characterized by a high resistivity, high velocity, low-temperature thicken root. Some other results may be interpreted as evidence that the deep lithosphere of the Pan-African belt (east of the craton) is heterogeneous.

Northern Domain. North of 20° N, a roughly E-W trending zone characterized by high heat flow density (60-100 mW m^{-2}) extends from Canary islands to Lybia (Tackerist and Lesquer 1989; Lucazeau et al. in press). This zone crosses the northern part of the West African craton in the Tindouf basin where the average value of heat flow density is 90± 20 mW m^{-2}. Southward, the extension of this high flow density high is yet unknown, but it is noteworthy that the two heat flow determinations in the Taoudeni basin display higher heat flow density values than in the Leo uplift.

In Algeria, the heat flow density anomaly of Sahara (80-100 mW m^{-2}) is secant to the N-S Pan-African structures. As shown by Lesquer et al. (in press), no significant thermal disturbance is associated with the Hoggar domal uplift. From various observations, including petrographic and gravimetric constraints, the uplift is interpreted as being due to a now cooled altered lithosphere emplaced during Late Mesozoic to Early Cenozoic times (Lesquer et al. 1988).

It is likely that such a high heat flow density zone oblique to major crustal structures cannot be entirely explained by crustal heat production. We suggest that it could result from a recent thermal rejuvenation process, possibly in relation with Cenozoic to Quaternary alkaline volcanism.

Oil exploration boreholes and gravity data show that the basement beneath the Sahara basin (Algeria) corresponds to the northern extension of the Hoggar shield. The contrast in surface heat flow (about 48 mW m^{-2} in the Hoggar shield and 100-80 mW m^{-2} in the Sahara basin) may therefore be reliable to an increase of 50-30 mW m^{-2} of the mantle heat flow.

The geotherms calculated for this zone show that temperatures of 1200-1300 °C should be reached at 50-80 km depth, indicating the possible presence of zones of partial melting at shallow depth. This hypothesis is supported by the surface waves studies of Hadiouche and Jobert (1988), which suggest that low lithospheric velocities are associated with the Sahara area.

4 Crustal Structure of West Africa from Gravity Data

Deep seismic profiles are dramatically missing in West Africa. The few available data concerning the deep crustal structure are from magnetotelluric surveys (Ritz 1985). On the geoelectric cross section of Fig. 4 (after Ritz and Robineau 1986) a depth of 30-40 km is proposed for the Moho beneath the western margin of the West African craton, whereas the crustal thickness should be only 25-30 km for the Senegal-Mauritania basin. Beneath the eastern part of the craton in Burkina Faso, no pronounced conductivity change is evidenced at crust mantle boundary. Eastward, the Pan-African crust under the Iullemeden basin (Fig. 4) displays a thickness of 30-40 km.

Two poorly documented local seismological studies are available which indicate a depth of 28-42 km for the Moho beneath the Ivory Coast (Bard et al. 1976) and 30-40 km for Hoggar (Merlet 1962). Using surface wave dispersion, Hadiouche and Jobert (1988) have recently proposed a mean crustal thickness of 30 km for all of West Africa.

Aeromagnetic surveys exist too, but they are restricted to some sedimentary areas such as in the Volta basin (Ako and Wellman 1985), in the Senegal basin (BRP 1956) or in Liberia (Behrendt and Wotorson 1970). The correlation of the gravity and magnetic fields may aid in some regions to gain an insight into the deep anomaly sources; for example, in the Senegal basin (Liger 1979).

In fact, gravity is the only geophysical data allowing a global investigation of the West Africa crustal structure. But unfortunately, gravity alone cannot provide information on crustal thickness.

Available gravity data in West Africa are mainly from regional surveys conducted by ORSTOM, the French Office for Overseas Scientific Research (Crenn 1957; Crenn et al. 1962; Crenn and Rechenmann 1965; Rechenmann 1965; Louis 1971; Rechenmann 1971). Additional surveys have been more recently made to complete the gravity mapping over the eastern edge of the West African craton in the Tanezrouft region of Southern Algeria (Rechenmann 1973; Bourmatte 1977) and in the Adrar des Iforas region in northeastern Mali (Ly 1979; Ly et al. 1980).

Other gravity surveys are also available on the southeast margin of the West African craton in Ghana (Ako and Murray 1983; Hastings 1983) and on its southwestern margin in Guinea (Akhmetjanov et al. 1976), in Guinea Bissau (Amorin Ferreira 1986) and Liberia (Behrendt and Wotorson 1970; Behrendt et al. 1974).

Figure 7 presents a simplified Bouguer anomaly map compiled from the different surveys performed mainly by ORSTOM.

Mean Bouguer anomaly varies between 30-35 mgal and the anomalies range in wavelength from a few kilometers to several hundreds of kilometers. The short wavelength anomalies are attributed to upper crust features such as faults, sedimentary basins, intrusives bodies, etc. The long wavelength anomalies are caused by deeper or larger features such as changes in crustal thickness, and in composition of the crust or of the upper mantle. Except North Africa, Hoggar and Guinea uplifts, topography is very flat (300-500 m) and the variations of Moho depth would reflect essentially isostatic compensation of large scale crustal heterogeneities.

In Fig. 8, where the major gravity trends are shown together with main structural features, discordant trend patterns between West African craton and surrounding mobile belts are obvious. The belt of positive anomalies which outline the West African craton boundaries appears as one of the most striking features of the gravity map.

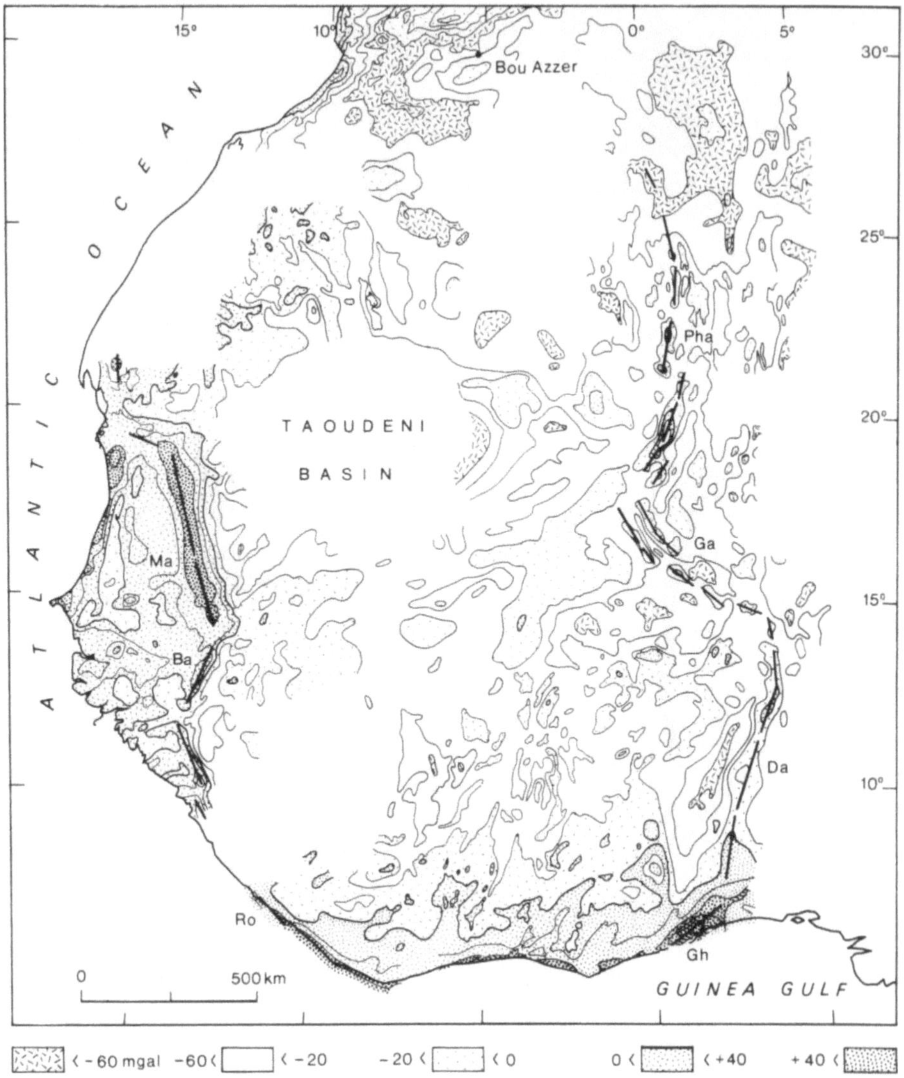

Fig. 7. Simplified Bouguer gravity anomaly map of West Africa. *Contour interval* 20 mgal; *heavy lines* Circum-West African craton belt of gravity highs; *Ma* Mauritanide anomaly; *Ba* Bassaride; *Ro* Rokelide; *Gh* Ghana; *Da* Dahomeyide; *Ga* Gourma; *Pha* Pharusian (modified after Lesquer et al. 1984)

4.1 Gravity Field and Relationships with the Major Structural Features of the West African Craton

The analysis of gravity field leads to the division of the West African craton into distinct provinces characterized by a consistent set of gravity and a structural trend pattern (Fig. 8).

In the Leo shield, the gravity field is not well defined over the Archean nucleus of Liberia-Guinea-Sierra Leone (Fig. 8, LN); but the N-E trending anomalies mapped by Behrendt and Wotorson (1970) show that this nucleus has a gravity signature different from the Eburnean (Birrimian) domain east of the Sassandra fault. On the other hand, two gravity provinces can be distinguished in the Eburnean basement of the Leo shield: the Birrimian province and the Ghanean nucleus (Lesquer et al. 1984). The narrow Birrimian domain is characterized by short wavelength high-intensity anomalies strongly elongated NNE-SSW. The gravity field within the Ghanean nucleus (Fig. 8, GN) is smoother, with less elongated short wavelength anomalies trending both in NNE-SSW and NW-SE directions.

The anomalies closely correlate with the upper crustal heterogeneities. The high Bouguer anomalies correspond to the greenstone belts (2–5 km thick) containing dense metasediments and metavolcanites. The low anomalies correspond to areas where granitoids predominate. Within the Birrimian province, the steepest gravity gradients correlate with the major N-S trending Eburnean shear zones.

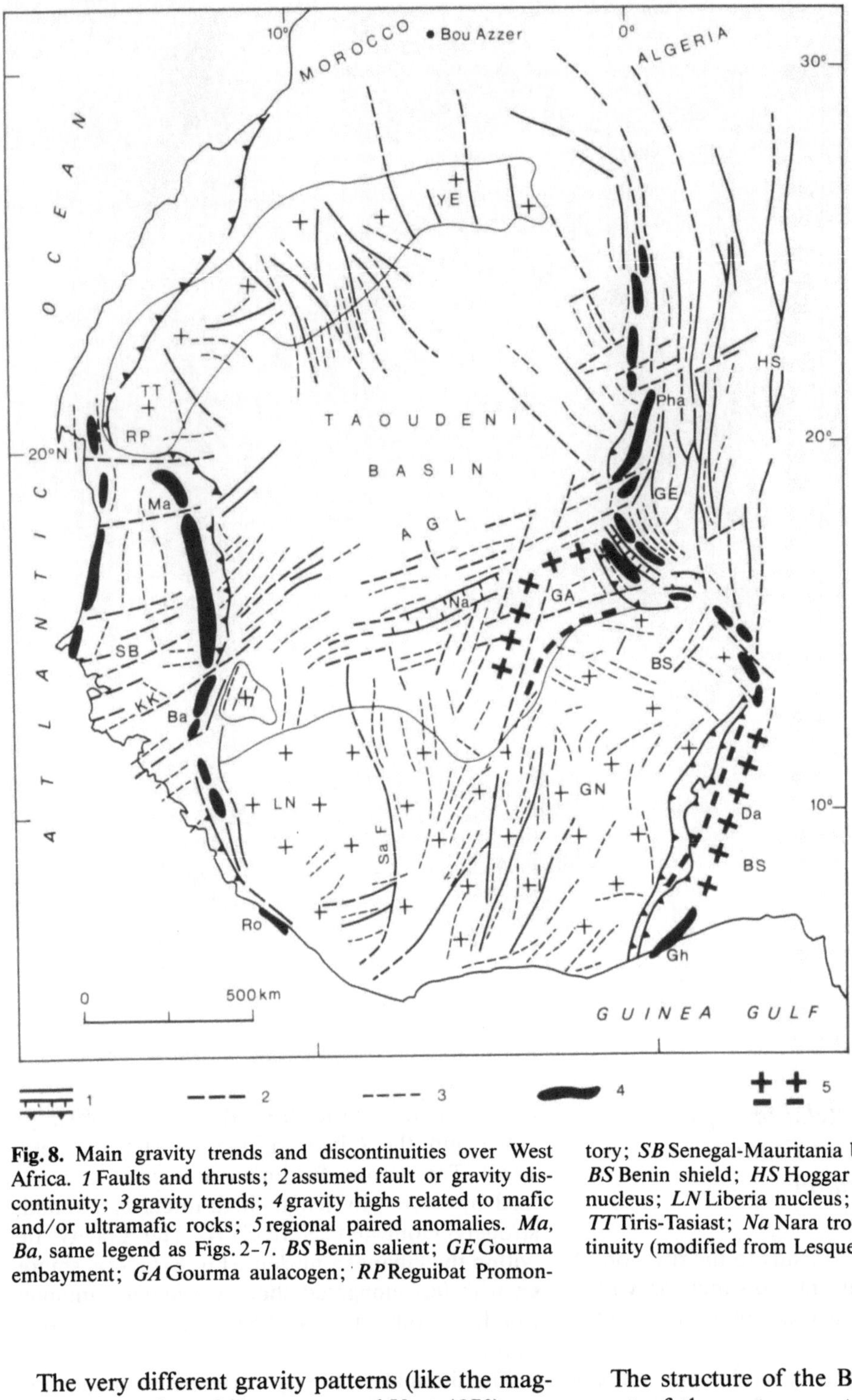

Fig. 8. Main gravity trends and discontinuities over West Africa. *1* Faults and thrusts; *2* assumed fault or gravity discontinuity; *3* gravity trends; *4* gravity highs related to mafic and/or ultramafic rocks; *5* regional paired anomalies. *Ma, Ba,* same legend as Figs. 2-7. *BS* Benin salient; *GE* Gourma embayment; *GA* Gourma aulacogen; *RP* Reguibat Promontory; *SB* Senegal-Mauritania basin; *AGL* Adrar-Guinea line; *BS* Benin shield; *HS* Hoggar or Touareg shield; *GN* Ghana nucleus; *LN* Liberia nucleus; *SaF* Sassandra fault; *YE* Yetti; *TT* Tiris-Tasiast; *Na* Nara trough; *KK* Kidira-Kolda discontinuity (modified from Lesquer et al. 1984)

The very different gravity patterns (like the magnetic observation of Strangway and Vogt 1970) suggest that these two areas may have suffered a different tectonic evolution. This assumption must be examined in relationship with geochronological determinations which show evidence for a Burkinian orogenic cycle independent of the Eburnean cycle (Caen-Vachette 1988).

The structure of the Birrimian (Eburnean) basement of the eastern part of Reguibat shield (Yetti area, Fig. 8, YE) is also revealed by the succession of NNW-SSE elongated gravity highs and lows. As for the Leo shield, the major faults which divide the basement into crustal blocks of different mean density are clearly seen (Bronner and Roussel 1978). They can be extrapolated southward under the sedimentary cover of the Taoudeni basin where they seem to have controlled the post Precambrian sedimentation. Westward, the granulitic Archean domain of Tiris-Tasiast (Fig. 8, TT) is characterized by higher gravity values and appears as a denser crustal block, less structurated, with gravity trends oblique to those of the Yetti area.

The southern part of the Taoudeni basin exhibits a very consistent set of WSW-ENE gravity trends oblique to the structural and gravity trends of Leo and Reguibat shields. This large (500 km wide, 1200 km long) transverse region extending from eastern Senegal to northeastern Mali has been called "the Adrar-Guinea line" (Bayer and Lesquer 1978; Fig. 8, AGL). It can be correlated with a variety of elongated geological features (faults, dolerite dykes, troughs, horsts) of variable age (Precambrian, Mesozoic and may be more recent according to Guiraud et al. 1987). Photogeological studies based on LANDSAT imagery (Simon et al. 1982) also reveal that two lineament bands parallel to gravity trends crosscut this area.

The most interesting gravimetric feature of this area is the broad arcuate positive anomaly (+30 mgal relative to the background) associated with the Gourma basin (Fig. 8, GA) and bounded southward by a coextensive and parallel negative anomaly (−40 mgal relative to the background; Fig. 8). Eastward these two anomalies are truncated by the anomalies associated with the Pan-African suture zone. The Gourma gravity high can be interpreted as a thick layer (5 km) of dense basic effusive rocks situated just under the sedimentary cover (in the deepest part of the basin) or as a deep and less dense crustal block (Lesquer and Moussine-Pouchkine 1980). North of the Niger river loop, a belt of short wavelength positive anomalies displaying a linear shape and an extension of nearly 800 km borders the Gourma gravity high and the 400-km-long negative anomaly associated with the Nara trough (CGG 1956). From gravity interpretation the fault-controlled Nara trough (Fig. 8, Na) is filled with more than 4000 m of sediments (Paleozoic?). The belt of positive anomalies correspond to shallow (3 km), 3- to 5-km-thick dense bodies (0.2 to 0.3 g cm^{-3}). Eastward, the continuation of the Nara faults correspond to the limit between the deep, dense Gourma structure and the belt of shallower dense bodies.

The shape of the basin, its evolution and the presence of a great number of basic rocks led to the interpretation of this structure as an aulacogen, related to the pre-Pan-African rifting that occurred 800–850 Ma ago along the eastern margin of the West African craton (Bayer and Lesquer 1978; Moussine-Pouchkine and Bertrand-Sarfati 1978).

From gravity data, the "failed arm" structure can be traced southward along more than 200 km. There, it appears to be controlled by the northward extension of the Leo shield shear zones, but there is no geological evidence for the coextension of the highly subsiding Gourma basin.

Surface geology fails to explain the long wavelength negative anomaly. It has been interpreted as crustal thickening due to downward deflection of the crust, partly caused by the weight of the inferred dense basic effusive rocks. Such an explanation would imply the existence of a symetrical gravity low on the northern side of the Gourma gravity high. But this long wavelength gravity low is not obvious on the gravity map. Perhaps it is hidden by the anomalies associated with the upper crustal structure. If it does not really exist, then we have a dipole anomaly characteristic of an assymetrical deep crustal structure inconsistent with the hypothesis of a crustal flexuration associated with the Gourma basin formation. The position of the Gourma paired anomaly in front of the Ghanean nucleus (Fig. 8, GN) as well as the general gravity pattern and tectonic trends around this nucleus are very similar to those found across structural province boundaries (Gibb et al. 1983; Lesquer and Louis 1982; Wellman 1978). Therefore, an alternative hypothesis may be tentatively proposed that assumes the Gourma paired anomaly may reflect the existence of an old major structural boundary (Burkinian, Eburnean?) between two geophysically distinct basement provinces, i. e. the Ghanean nucleus and a northern domain corresponding westward to the Birrimian gravity province. The Gourma high would correspond to a dense crustal structure (deep crust?) isostatically compensated for by a downward flexuration of the Ghanean nucleus northern margin.

The shape of the Gourma basin which overprints this structural boundary would have been controlled by pre-existing discontinuities. If the Gourma positive anomaly is not related to effusive rocks associated with rifting process but with a pre-existing deep crustal structure, the aulacogen model becomes doubtful and the Gourma may be considered as a passive margin basin (Bertrand-Sarfati et al. 1988).

As a consequence of this hypothesis, we may infer that contemporaneously with the emplacement of the dense Gourma structure the basement of the Taoudeni basin suffered intense deformations. Thus, the ENE-WSW transverse zone, well defined by geophysical and geological trends (Adrar Guinea line), may be regarded as evidence for the existence of a wide Proterozoic mobile belt reactivated throughout geological times.

This zone bounds southward the deeper part of the Paleozoic-Mesozoic Taoudeni basin which is roughly ENE-WSW elongated. In the northeastern part of the basin, where the gravity data are widely spaced, the gravity trends seem coherent with structural features of the Reguibat shield. Westward, in front of the Mauritanide belt (Fig. 8, Ma), we again observe oblique ENE-WSW trends. Unfortunately, because of the lack of gravity data we cannot know if the central part of the Taoudeni basin displays the same gravity signature. However, field observations and petroleum exploration data (Villemur 1967) reveal that the subsidence axis has been controlled by ENE-WSW structural trends since the Upper Precambrian times, with occurrence of fault bounded troughs like the noteworthy Tagant trough (Bronner et al. 1980), filled with more than 5000 m of sediments. These observations may support the assumption that the evolution of the whole Taoudeni basin was controlled by ENE-WSW crustal zones of weakness inherited from Proterozoic deformations.

To conclude, gravity data support the subdivision of the West African craton into a number of discrete crustal blocks (Liberian, Ghanean and Tiris-Tasiast nuclei) surrounded by highly deformed areas including belts of dense rocks. Delineation of these different areas must be improved and their relative age remain questionable. Meanwhile, we think they must be taken into account to understand the geodynamic evolution of the West African craton.

4.2 Gravity Signature Related to the Pharusian-Dahomeyide Orogens

From South Algeria to the Gulf of Guinea, the Pan-African belt is characterized by a consistent set of N-S elongated gravity anomalies (Figs. 7, 8). The eastern edge (generally hidden by sediment deposits) of the West African craton can be readily identified gravimetrically. It is marked by a string of elongated gravity highs with an amplitude of over 30 mgal locally attaining 80 mgal. The major N-S trending shear zones, which divide in longitudinal units the Pharusian belt, are seen gravimetrically to have northward and southward prolongations which appear to have played a role in the shape of the Paleozoic basins.

In the Hoggar shield, the alternation of gravity highs and lows closely correlate with the different terranes and basement blocks that compose the Pharusian belt (Rechenmann 1973; Bourmatte 1977). In the polycyclic central Hoggar and the eastern Hoggar the gravity field is less documented. It appears smoother and displays NNW-SSW trends with less steep gradients (Lesquer et al. 1988).

In the western branch of the Pharusian belt, the negative anomalies are associated with a thick flysch-like sequence of green volcanic graywackes (The "serie verte"), Middle Proterozoic quartzitic sequences or post-Pan-African molasse. In the eastern branch of the Pharusian belt and in the polycyclic central Hoggar synorogenic batholiths are characterized by large negative anomalies, whereas small gravity lows are associated with late or post-orogenic restricted granites (Bourmatte and Lesquer, unpubl work 1977). The positive anomalies are associated with a wide range of rock types from granulites to peridotites or gabbros. East of the 4° 50 E shear zone some gravity highs are proposed to be connected with dense magmatic bodies intruded in the deep crust in relation with a Late Mesozoic rifting process (Lesquer et al. 1988). In the Adrar des Iforas (Republic of Mali), deep dense bodies are associated with the post-orogenic alkaline ring-complexes province (Ly et al. 1984).

The existence of linear gravity anomalies along the edge of the West African craton was one of the main arguments which led to the proposal of a collision process for Pan-African orogeny (Black et al. 1979). A two-dimensional interpretation using inverse approach and linear programming (Bayer and Lesquer 1978; Ly et al. 1980) shows that the positive anomalies correspond to superficial unrooted structures (6 to 12 km thick). The subvertical to easterly dipping geometry of these structures is in agreement with the general tectonic transport, with thrusting towards the craton. The calculation of minimum density give values over 2.9 g cm^{-3}. Thus, these structures can be interpreted as basic or ultrabasic bodies. In the Iforas region they correspond to basic to ultrabasic rocks outcropping in an island-arc assemblage context. In the Gourma and Benin regions they correspond to masses of granulitic metagabbros (Caby 1987).

In Togo and Benin, the positive anomalies linked to the suture zone (Fig. 8, Da) are bounded westward by a long wavelength negative anomaly (– 40 mgal relative to the background) partly associated with the Atakora nappes (quartzites and phyllites) (Trompette 1979) overthrust on the passive margin Voltaian basin. This anomaly may be related either to (1) superficial low density materials corresponding to thickening (about 8 km) of the Volta sediments and/or of the Atakorian quartzite nappes under the gneissic internal nappes, or to (2) a deep thickened and isostatically equilibrated crust. This thickening (about 3 km) may correspond to isostatic response to the upthrusted dense lower crust (high-grade gneisses) of the Pan-African domain.

Schematic interpretative profiles across the Pharusian (Fig. 8, Pha), Gourma (Ga) and Dahomeyide (Da) anomalies are shown in Fig. 9.

Southward, in Ghana, a positive anomaly (Fig. 8, Gh) trending NE-SW (correlated with basic granulites) suggest the virgation toward the west of the suture zone, whereas structural pattern evidenced by gravity surveys in northeast Brazil suggest the extension of the N-S Pan-African structures under the Parnaiba basin (Lesquer et al. 1984).

In a Bullard pre-drift reconstruction the suture zone exhibits a shape with a large salient called the Benin salient (Fig. 8, BS), flanked northward by the Gourma embayment (Fig. 8, GE) and southward by the Parnaiba embayment (in Brazil). Gravity anomalies which define the suture have a different wavelength, therefore a different origin according to their situation along the suture. As observed in other Proterozoic belts (Gibb et al. 1983; Lesquer and Louis 1982; Gibb and Thomas 1976) the cratonic salient (Benin salient, Fig. 8, BS) is associated with a long wavelength anomaly which reflects a probable deep crustal origin. In contrast, shorter wavelength anomalies correspond to vertical structure along the salient flanks and the Gourma embayment (Fig. 8, GE) is characterized by several eastward dipping basic bodies.

Assuming that this geometry is inherited from the initial shape of the Precambrian craton, a model of evolution involving an indentation process (Molnar and Tapponnier 1975; Tapponnier 1977) of the eastern continent by the rigid Benin indenter has been proposed (Bayer and Lesquer 1978; Lesquer and Louis 1982). This explains the relationships between the suture geometry and both the gravity and tectonic patterns: (1) development of thrusting arcs around the promontory; (2) lateral extrusion tectonics along strike-slip shear zones; (3) dense unrooted bodies probably set during final stage of collision and thereby well preserved in the Gourma embayment; (4) strong crustal thickening and major lower crust upthrusting resulting from considerable shortening in front of the Benin salient.

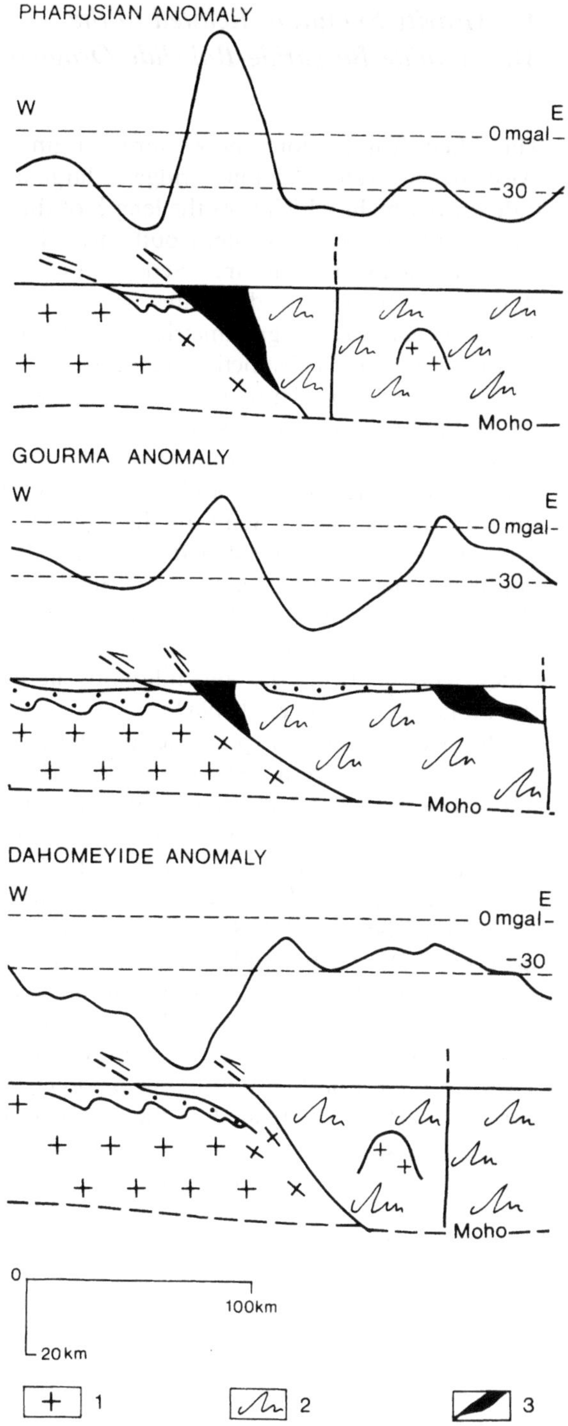

Fig. 9. Gravity profiles across the Pharusian-Dahomeyide orogens and proposed schematic crustal models. *1* West African craton crust; *2* accreted zone and/or Pan-African crust; *3* mafic and/or ultramafic bodies (adapted from Lesquer et al. 1984)

4.3 Gravity Signature Related to the Mauritanide-Bassaride-Rokelide Orogens

The salient feature along the western margin of the West African craton is a linear ridge of large-amplitude gravity highs which runs the length of the West African orogen in the western outboard of its exposed segments in Mauritania, Senegal and Guinea (Figs. 7, 8). This feature delineates two crustal provinces which may be gravimetrically well defined on the basis of their distinctive signature.

The eastern crustal province is dominated by a broad regional negative anomaly (ca. -30 mgal) within which occur minor troughs and culminations corresponding to NE-SW and ENE-WSW structures in the basement. Some of these gravity features, which are associated with the West African craton as mentioned in Chapter 3, Section 1, are clearly seen to continue to the west for about 50 km beyond the exposed front of the central Mauritanides. This confirms the allochthonous nature of the orogen (Lécorché et al. 1983).

The western crustal province (western block, Fig. 1) is marked by a generally positive Bouguer anomaly (ca. $+10$ mgal) which contrasts with the gravity low observed over the craton. It corresponds with the coastal plain and its basement. In this area, and particularly in the Senegal-Mauritania basin, two sets of NS and NE-SW trending features are observed: (1) Gravity trends with a NE-SW orientation are particularly well developed in the southern part of the basin. They are locally associated with gravity highs and reflect lineament-controlled dense intrusive bodies probably of basalt or dolerite (Liger 1979). These are related to the reactivation of old transverse fractures zones (Fig. 8, KK) during the opening of the Atlantic ocean. It is interesting that they are in the prolongation of the Adrar-Guinea lineament described above. (2) Gravity anomalies elongated in a NS direction predominate northward of 15° N. Some appear to be offset by transverse discontinuities. They are associated with magnetic lineations and reflect the probable existence of tensional features related to the Triassic rifting and subsequent opening of the Atlantic. This interpretation is supported by the rotational block faulting pattern of the pre-Mesozoic basement inferred from refraction sections (Chanut and Micholet 1988).

The positive anomaly which parallels the coast (Figs. 7, 8) has a steep seaward gradient that results in crustal thinning from continental to oceanic crust. The coastal gradient is associated with a series of superimposed local gravity highs and is nearly colinear with an important magnetic anomaly which appears to form an on-shore segment of the West African coastal magnetic anomaly (WACMA; Roussel and Liger 1983). This feature, which is in the prolongation of the Brunswick magnetic anomaly in a pre-Mesozoic reconstruction, could correspond to a north-south segment of the Late Paleozoic suture (Nelson et al. 1985) but Hatcher et al. (1987) think that this feature is far to the east of the suture.

Putting aside these features linked to the Atlantic break-up, it is evident that the regional positive gravity background over the western province suggests control by a basement thinner or denser than the West African craton. This interpretation is supported by a seismological study based on P-wave travel time residual (Briden et al. 1981) and magnetotelluric data (Ritz and Robineau 1986), as discussed in a previous section.

In the Bassaride orogen (southern Senegal and Guinea, Ba in Figs. 7, 8) the proposed two-dimensional models (Ponsard et al. 1982; Roussel et al. 1984) indicate a western crustal block thicker than the West African crustal block. The suture which marks the discontinuity in density separating the two blocks dips westward (~45°) beneath the western denser (reactivated) province (Fig. 10). This is consistent with the uppermost crustal part of the seismic models (Briden et al. 1981; Dorbath et al. 1983) and with other geological data (Villeneuve et al. 1984). The belt of gravity highs which delineates the two crustal provinces corresponds to dense unrooted bodies that are squeezed along the suture. These bodies may correspond either to remnants of oceanic crust skinned and upthrusted during the subduction (Ponsard 1984) or to mafic granulite facies metamorphic rocks derived from the lower crust (Ponsard et al. 1988) and similar to the granulites which are exposed in the Rokelide in Liberia (Behrendt et al. 1974).

In the Rokelide orogen of Liberia (Fig. 8, Ro), a positive Bouguer anomaly exists along the coast and continental shelf from Sierra Leone to Ivory Coast (Behrendt and Wotorson 1970). This anomaly parallels a thrust fault zone dipping seaward which marks the boundary of the Pan-African and Liberian age provinces and correlates with mafic granulites in the Monrovia region. According to Behrendt et al. (1974), this anomaly is attributed to a belt of granulite rocks that formed deep within the crust and were uplifted during Pan-African time (ca. 550 Ma) in relation to a closing oceanic domain. In contrast, field evidence for continental suturing in Sierra Leone is lacking although possible correlations with Rokelide rocks in Liberia and Guinea are suggested (Williams and Culver 1988). These authors prefer a model of rift structure with limited closure that differs greatly from that of Bassarides.

Along the Mauritanide orogen, the suture between the West African craton (negative gravity anomaly) and the western thinner or denser continental block (positive anomaly) is characterized by a very different gravity signature: in contrast with the sharp belt of gravity highs which mark the Pan-African suture in the Bassarides as mentioned above, the so-called Mauritanide anomaly (Guetat 1981; Roussel and Lécorché 1983; Ma in Figs. 7, 8) is a wide (about 100 km) positive ridge of long wavelength, with an amplitude greater than 50 mgal. Elongated for more than 700 km between the Reguibat shield and southeastern Senegal, it remains closely parallel to the trend of the Mauritanide orogen. In fact the anomaly has been related to the belt by several authors, but with various interpretations: mafic intrusion or folded layer of basaltic rocks (Blot et al. 1962); up-bulging of the upper mantle (Crenn and Rechenmann 1965); grano-dioritic ridge (Chiron 1973); remnants of oceanic crust (Dillon and Sougy 1974). The anomaly has been more recently interpretet (Guetat 1981) as a mantle rooted, mafic or ultramafic, elongate massif that dips significant westward. According to Guetat (1981) and Lécorché et al. (1983) thinner crust occurs beneath the coastal western block (Fig. 10, Model 1). This interpretation is consistent with magnetotelluric soundings (Ritz and Robineau 1986). For Lesquer et al. (1984) the reworked Pan-African western block is thicker and denser (Fig. 10, Model 2). These two alternative models, in the absence of other geophysical constraints, exemplify the fundamental ambiguity in gravity interpretation.

Two transverse discontinuities in the gravity pattern mark the northern (20° N trend) and southern (Kidira-Kolda lineament, Fig. 8, KK) structural boundaries respectively, of both the Mauritanide anomaly and the major element of the exposed Mauritanide orogen. These probably pre-existing structures were reactivated during the emplacement of the deep-seated mass (Lécorché et al. 1983) and its further reworking. Indeed, recent geochronological data indicate that a late Paleozoic tectonothermal event is superimposed to the Pan-African orogen and is concentrated precisely on this major segment of the belt (Dallmeyer and Lécorché 1989; Lécorché et al. 1989).

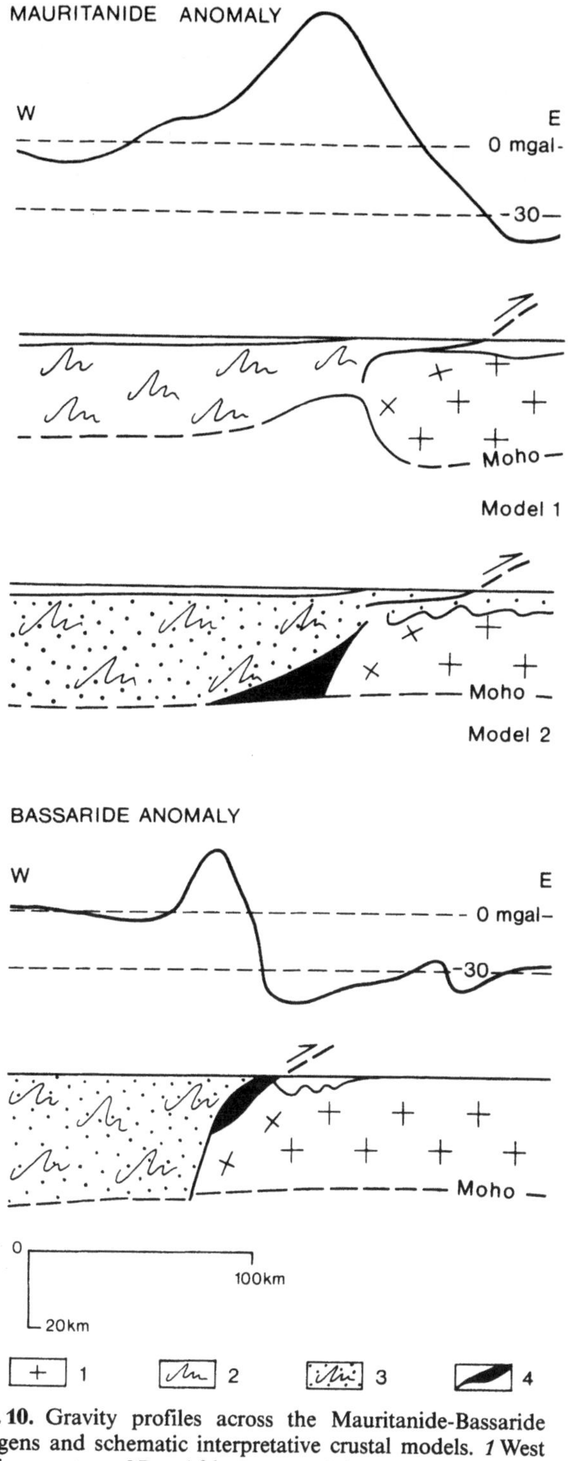

Fig. 10. Gravity profiles across the Mauritanide-Bassaride orogens and schematic interpretative crustal models. *1* West African craton; *2* Pan-African crust; *3* denser reworked Pan-African crust; *4* mafic and/or ultramafic bodies. Two alternative models are proposed for the Mauritanide anomaly: model 1, after Lécorché et al. (1983); model 2, after Lesquer et al. (1984). Bassaride anomaly: adapted from Ponsard (1984)

These coincidences are interpreted (Lécorché and Roussel 1987) as indicating that the specific geometry of the Mauritanide anomaly could be associated with compressional stresses related to a renewed significant eastward translation of the western coastal block, during collision of Gondwana and eastern North America between the northern and southern transverse boundaries, both probably acting as transform boundaries.

The existence of this western coastal block (Senegal microplate) is consistent with paleomagnetic data (Lefort and Van der Voo 1981) which indicate that a crustal block was probably located between the Mauritanide and Appalachian orogens during the Late Paleozoic amalgamation of Pangea.

In front of the westerly extension of the Reguibat shield, the Mauritanide anomaly appears to be deflected toward the west (Fig. 8), as are the outcrops of the chain (Fig. 1). Furthermore, the correlations of gravimetric and magnetic lineations around the Reguibat promontory (Fig. 8, RP) and in the coastal plains of eastern North America indicate that both eastern central Appalachian and Mauritanide terranes are bent. This leads the westerly promontory of the Reguibat shield to be considered as an indenter (Lefort 1984). The indentation of the central Appalachians in the Chesapeake Bay region probably resulted in major Carboniferous thrusting arcs and large-scale strike-slip displacements parallel to the trend of the southern Appalachian and central Mauritanides (Vauchez et al. 1987; Lefort et al. 1988).

5 Conclusion

This chapter used the available geophysical, mainly gravity, data to investigate the crustal structure and terranes boundaries associated with the West African craton and the surrounding Pan-African belts.

On the Precambrian basement, mainly on the Leo and Reguibat shields, gravity highs generally correspond to granulite rocks and greenstone belts, whereas gravity lows are associated with areas where granitoids predominate. Gravity trends clearly show that major faults affecting the shields' outcrops extend beneath the sedimentary cover of the Taoudeni basin. In the southern part of the basin, these structural trends are interrupted by a large transverse WSW-ENE trending gravimetric lineament, the Adrar-Guinea line, which has been correlated with a series of geological features of various age such as the Precambrian Gourma aulacogen or the Cretaceous Nara trough. Finally, the Bouguer anomaly pattern and gravimetric trends allow the recognition in the West African craton of distinct structural provinces: the Tiris-Tasiast nucleus in the Reguibat shield and the Liberia and Ghana nuclei in the Leo shield appear to be well individualized crustal blocks which are surrounded by distinct areas where high deformation occurred in the course of the Eburnean orogeny.

Along the eastern margin of the West African craton, the Pan-African province is dominated by north-south trending gravity anomalies which outline major shear zones associated with the belt. The boundary between the craton and the Pharusian-Dahomeide orogens is marked by a string of elongated gravity highs. Interpretative models of profiles across the boundary show that positive anomalies correspond to eastward dipping unrooted mafic and/or ultramafic bodies. This is consistent with a Pan-African continental collision between the West African craton and the active margin of an eastern continent including the Hoggar, Benin and Nigeria shields.

Along the western margin of the West African craton, the gravity highs which parallel the Mauritanide-Bassaride orogens also delineate two gravimetrically well defined crustal terranes: the craton is dominated by a broad gravity low; in contrast the western basement terrane corresponding to the coastal plain is characterized by a relatively positive Bouguer anomaly. This gravity background is consistent with a western terrane (coastal block) thinner or denser than the West African craton. Two-dimensional models indicate that the suture separating the two blocks dips westward beneath the coastal block. In the Bassarides, short wavelength gravity highs reflect dense unrooted bodies trapped along the Pan-African suture. The long wavelength positive anomaly which parallels the Mauritanide orogen is interpreted as a mantle rooted mafic and/or ultramafic structure. The specific geometry of this anomaly is tentitavely related to a renewed translation of the coastal block (Senegal microplate) during collision of Gondwana and eastern North America.

Thus, gravity data allow the structural provinces and terrane boundaries in West Africa to be defined and point at the collision model for the Pan-African geodynamic evolution in agreement with our geological understanding of the orogen.

References

Affaton P (1975) Etude géologique et structurale du Nord-Ouest Dahomey, du Nord Togo et du Sud-Est de la Haute-Volta. Trav Lab Sci Terre Saint-Jérôme, Marseille, B 10: 201

Affaton P, Sougy J, Trompette R (1980) The tectono-stratigraphic relationships between the Upper Precambrian and lower Paleozoic Volta basin and the Pan-African Dahomeyide orogenic belt (West Africa). Am J Sci, New Haven 230: 224-248

Akhmetjanov BS, Loutsenko VF, Diallo M (1976) Carte des anomalies de Bouguer et résultats des travaux géophysiques en Guinée occidentale. In: Notice explicative de la carte géologique de la République de Guinée, Feuille de Gaoual. Rapp Technoexport, Conakry, pp 184-205

Ako JA, Murray AS (1983) Bouguer gravity anomaly map of Ghana. 1:1000000. Geol Surv Ghana, Accra

Ako JA, Wellman P (1985) The margin of the West African craton: the Voltaian basin. J Geol Soc Lond 142: 625-632

Allen PM (1969) The geology of part of an orogenic belt in Western Sierra Leone. West Africa. Geol Rundsch 58: 588-620

Amorin Ferreira H (1966) Observacoes gravimetricas no territorio de Guinea Portuguese. Serv Meteorol Nac, Lisboa, 9 pp

Ballard S, Pollack HN (1987) Diversion of heat by Archean cratons: a model for southern Africa. Earth Planet Sci Lett 85: 253-264

Bard JP, Burlot R, Delaye F (1976) Etude préliminaire de la morphologie de la discontinuité de Mohorovicic dans la région de Toumodi (Côte d'Ivoire) et essai d'interprétation géologique. C R Acad Sci, Paris 283: 1153-1156

Barrère J (1967) Le groupe précambrien de l'Amsaga entre Atar et Akjoujt (Mauritanie). Etude d'un métamorphisme profond et de ses relations avec la migmatisation. Mém Bur Rech Géol Min 42: 275

Bayer R, Lesquer A (1978) Les anomalies gravimétriques de la bordure orientale du craton ouest-africain: géométrie d'une suture panafricaine. Bull Soc Géol Fr 6: 863-876

Behrendt JC, Wotorson CS (1970) Aeromagnetic and gravity investigations of the coastal area and continental shelf of Liberia. West Africa, and their relation continental drift. Geol Soc Am Bull 81: 3563-3574

Behrendt JC, Schlee J, Robr JM, Silverstein MK (1974) Structure of the continental margin of Liberia, West Africa. Geol Soc Am Bull 85: 1143-1158

Bertrand JM, Caby R (1977) Geodynamic evolution of the Pan-African orogenic belt: a new interpretation of the Hoggar shield. Geol Rundsch 67: 357-388

Bertrand-Sarfati J, Moussine-Pouchkine A, Caby R (1988) Les corrélations du Protérozoique au Cambrien en Afrique de l'Ouest: nouvelle interprétation géodynamique. Bull Soc Géol Fr, Paris 5: 855-865

Bessoles B (1977) Géologie de l'Afrique. 1, le Craton Ouest africain. Mém Bur Rech Géol Min, Paris 88: 402

Black R (1987) Introduction to West African orogens and circum-Atlantic terrane linkages, IGCP 233 Conf, Nouakchott, Mauritania, abstracts and program, Univ Georgia (USA), pp 27-37

Black R, Caby R, Moussine-Pouchkine A et al. (1979) Evidence for Late Precambrian plate tectonics in West Africa. Nature (Lond) 278: 223-227

Blot C, Crenn Y, Rechenmann J (1962) Eléments apportés par la gravimétrie à la connaissance de la tectonique profonde du Sénégal. C R Acad Sci, Paris 254: 1131-1133

Bourmatte A (1977) Etude gravimétrique du Tanezrouft (Algérie). Thèse 3e cycle, Univ Montpellier, Fr, 131 pp

Briden JC, Whitcombe DN, Stuart GW, Fairhead JD, Dorbath C, Dorbath L (1981) Depth of geological contrast across the West African craton margin. Nature (Lond) 292: 123-128

Brigaud F, Lucazeau F, Ly S, Sauvage J-F (1985) Heat flow from the West African shield. Geophys Res Lett 12: 549-552

Bronner G, Roussel J (1978) Données gravimétriques et structure de la dorsale Reguibat (Mauritanie). 6e Réun Annu Sci Terre, Soc Géol Fr (ed), p 72

Bronner G, Roussel J, Trompette R, Clauer N (1980) Genesis and geodynamic evolution of the Taoudeni basin (Upper Precambrian and Paleozoic), Western Africa. In: Bally AW, Bender PL, McGetchin TR, Walcott RT (eds) Dynamic of plate interiors. Geodyn Ser 1, Am Geophys Union, Washington, pp 81-90

Bureau de Recherches Pétrolières (BRP) (1956) Magnétisme aéroporté du Sénégal. Rapp inédit, Compagnie Gén Géophys (CGG), Paris, 28 pp

Caby R (1970) La chaîne pharusienne dans le Nord-Ouest de l'Ahaggar (Sahara central, Algérie): sa place dans l'orogenèse du Précambrien supérieur en Afrique. Thesis, Univ Montpellier, Fr, 336 pp

Caby R (1978) Paléogéodynamique d'une marge passive et d'une marge active au Précambrien supérieur: leur collision dans la chaîne panafricaine du Mali. Bull Soc Géol Fr 6: 857-861

Caby R (1987) The Pan-African belt of West Africa from the Sahara desert to the Gulf of Benin. In: Schaer JP, Rodgers J (eds) The anatomy of mountain ranges. Princeton Univ Press, USA

Caen-Vachette M (1988) Le craton ouest-africain et le bouclier guyanais: un seul craton au Protérozoique inférieur? J Afr Earth Sci 2: 479-488

Chanut C, Micholet J (1988) Le stade rift dans le bassin sédimentaire du Sénégal. J Afr Earth Sci 7: 397-400

Chiron JC (1973) Etude géologique de la chaîne des Mauritanides entre le parallèle de Moudjeria et le fleuve Sénégal (Mauritanie). Un exemple de ceinture plissée précambrienne reprise à l'Hercynien. Mém Bur Rech Géol Min 84: 284

Compagnie Générale de Géophysique (CGG) (1956) Etude hydrologique par prospection électrique et sismique dans la plaine du Gondo (Soudan). Rapp Mission, Paris

Crenn Y (1957) Mesures gravimétriques et magnétiques dans la partie centrale de l'AOF. Interprétations géologiques. Rapp ORSTOM, Paris, 39 pp

Crenn Y, Rechenmann J (1965) Mesures gravimétriques et magnétiques au Sénégal et en Mauritanie occidentale. Cah ORSTOM Sér Géophys, Paris 6: 59

Crenn Y, Blot C, Metzger J, Rechenmann J (1962) Mesures gravimétriques et magnétiques en Afrique occidentale de 1956 à 1958. Cah ORSTOM Sér Géophys, Paris 3: 45

Culver SJ, Williams HR (1979) The last Precambrian and Phanerozoic geology of Sierra Leone. J Geol Soc Lond 136: 605-618

Dallmeyer RD, Lécorché JP (1989) $^{40}Ar/^{39}Ar$ Polyorogenic mineral age record within the Central Mauritanide orogen, West Africa. Geol Soc Am Bull 101: 55-70

Dallmeyer RD, Villeneuve M (1987) $^{40}Ar/^{39}Ar$ mineral age record of polyphase tectonothermal evolution in the southern Mauritanide orogen, southeastern Senegal. Geol Soc Am Bull 98: 602-611

Dallmeyer RD, Caen-Vachette M, Villeneuve M (1987) Emplacement age of post-tectonic granites in southern Guinea (West Africa) and the peninsular Floride subsurface: implications for origins of southern Appalachian exotic terranes. Geol Soc Am Bull 99: 87-93

Dia O (1984) La chaine panafricaine et hercynienne des Mauritanides face au bassin protérozoique supérieur à dévonien de Taoudeni dans le secteur-clé de Mejeria (Taganet, Sud RIM): Lithostratigraphie et Tectonique. Un exemple de tectoniques tangentielles superposées. Thesis, Univ Aix-Marseille III, 516 pp

Dillon WP, Sougy JM (1974) Geology of West Africa and Canary and Cape Verde Islands. In: Nairn AE, Stehli FG (eds) The Ocean basins and margins. Plenum, NY 2: 315-390

Dorbath C, Dorbath L (1984) Approche sismologique de la structure de la lithosphère en Afrique de l'Ouest. Thèse, Univ Paris VI, 307 pp

Dorbath L, Montagner JP (1983) Upper mantle heterogeneities in Africa deduced from Rayleigh wave dispersion. Phys Earth Planet Inter 32: 218-225

Dorbath C, Dorbath L, Le Page A, Gaulon R (1983) The West African craton margin in eastern Senegal: a seismological study. Ann Geophys 1: 25-36

Fabre J (1983) Afrique de l'Ouest. Introduction géologique et termes stratigraphiques. In: Fabre J (ed) Lexique stratigraphique international. Pergamon Press, Oxford, 386 pp

Flicoteaux R, Latil-Brun MV, Michaud L (1988) Histoire de la subsidence post-rift du bassin côtier mauritano-sénégalo-guinéen. Relation avec l'amincissement crustal pendant la période jurassique à Crétacé inférieur. Comparaison avec l'évolution des marges péri-atlantiques au niveau de l'Atlantique Central et Equatorial (Côte est des USA, Sud-Sahara, Côte d'Ivoire et Plateau du Demerara). J Afr Earth Sci 7: 345-359

Gevin P (1960) Etudes et reconnaissances géologiques sur l'axe cristallin Yetti-Eglab et ses bordures sédimentaires. Bull Serv Carte Géol Alger 23: 328

Gibb RA, Thomas MD (1976) Gravity signatures of fossil plate boundary in the Canadian shield. Nature (Lond) 262: 199-200

Gibb RA, Thomas MD, Lapointe PL, Muckhopadhyay M (1983) Geophysics of proposed proterozoic sutures in Canada. Precambrian Res 19: 349-384

Guetat Z (1981) Etude gravimétrique de la bordure occidentale du craton ouest africain. Thèse 3e cycle, Univ Montpellier, 185 pp

Guieu G, Roussel J (1984) Une interprétation des bassins côtiers ouest-africains à partir de l'histoire des dépôts salifères, dans le cadre de l'ouverture de l'Atlantique. Bull Soc Géol Fr 26: 1149-1164

Guiraud R, Bellion Y, Benkhelil J, Moreau C (1987) Post-Hercynian tectonics in Northern and Western Africa. Geol J 22: 433-466

Hadiouche O, Jobert N (1988) Geographical distribution of surface waves velocities and 3-D upper mantle structure in Africa. Geophys J 95: 87-109

Hastings DA (1983) An updated Bouguer anomaly map of south-central West Africa. Geophysics 48: 1120-1128

Hatcher RC Jr, Zietz I, Litehiser J (1987) Crustal subdivisions of the eastern and central United States and a seismic boundary hypothesis for eastern seismicity. Geology 15: 528

Jaupart C, Provost A (1985) Heat focusing, granite genesis and inverted metamorphic gradients in continental collision zones. Earth Planet Sci Lett 73: 385-397

Kennedy W (1964) The structural differentiation of Africa in the Pan-African (±500 millions years) tectonic episode. Eighth Annu Rep Res Inst Afr Geol, Leeds Univ, UK, pp 48-49

Latil-Brun MV (1987) Subsidence et évolution thermique du bassin du Sénégal. Comparaison avec la marge divergente homologue est-américaine et les marges transformantes de Côte d'Ivoire et des Guyanes. Thesis, Univ Aix-Marseille III, 200 pp

Latil-Brun MV, Flicoteaux R, Blarez E, Mascle J, Bessis F, Brunet MF (1988) Variation de la subsidence le long d'un segment de marge transformante: l'exemple du plateau continental de Côte d'Ivoire. C R Acad Sci, Paris 306: 83-88

Leblanc M (1976) Proterozoic oceanic crust at Bou Azzer. Nature (Lond) 261: 34-35

Leblanc M (1981) The late Proterozoic ophiolites of Bou Azzer (Morocco): evidence for Pan-African plate tectonics. In: Kröner A (ed) Precambrian plate tectonics. Elsevier, Amsterdam, pp 435-451

Lécorché JP (1980) Les Mauritanides face au craton ouest-africain. Structure d'un secteur-clé: la région d'Ijibiten (Est d'Akjoujt, RI de Mauritanie). Thesis, Univ Aix-Marseille III, Fr, 446 pp

Lécorché JP, Roussel J (1987) Geophysically defined basement terranes and suture in the Mauritanides and Bassarides: Tectonic significance. In: Tectonothermal evolution of the West African orogens and Circum-Atlantic terrane linkages. Progr Abstr, Nouakchott, pp 135-137

Lécorché JP, Roussel J, Sougy J, Guetat Z (1983) An interpretation of the geology of the Mauritanides orogenic belt (West Africa) in the light of geophysical data. Geol Soc Am Mem 158: 131-147

Lécorché JP, Dallmeyer RD, Villeneuve M (1989) Definition of tectonostratigraphic terranes in the Mauritanide, Bassaride and Rokelide orogens, West Africa. In: Dallmeyer RD (ed) Terranes in the Circum-Atlantic Paleozoic orogens. Geol Soc Am Spec Pap 230: 131-144

Lefort JP (1984) Mise en évidence d'une virgation carbonifère induite par la dorsale Reguibat (Mauritanie) dans les Appalaches du Sud (USA). Arguments géophysiques. Bull Soc Géol Fr, Paris 26: 1293-1303

Lefort JP, Van der Voo R (1981) A kinematic model for the collision and complete suturing between Gondwanaland and Lanrussia in the Carboniferous. J Geol 89: 537-550

Lefort JP, Max MD, Roussel J (1988) Geophysical evidence for the location of the NW boundary of Gondwanaland and its relationship with two older satellite sutures. In: Harris AL, Fettes DJ (eds) The Caledonian-Appalachian orogen. Geol Soc Lond Spec Publ 38: 49-60

Le Page A (1983) Les grandes unités des Mauritanides aux confins du Sénégal et de la Mauritanie. L'évolution structurale de la chaîne du Précambrien supérieur au Dévonien. Thesis, Univ Aix-Marseille III, Fr, 518 pp

Lesquer A, Louis P (1982) Anomalies gravimétriques et collisions continentales au Précambrien. Geoexploration 20: 275-293

Lesquer A, Moussine-Pouchkine A (1980) Les anomalies gravimétriques de la boucle du Niger. Leur signification dans le cadre de l'orogenèse pan-africaine. Can J Earth Sci, Ottawa 17: 1538-1545

Lesquer A, Beltrao JF, de Abreu FAM (1984) Proterozoic links between northeastern Brazil and West Africa: a plate tectonic model based on gravity data. Tectonophysics, Amst 110: 9-26

Lesquer A, Bourmatte A, Dautria JM (1988) Deep structure of the Hoggar domal uplift (central Sahara, south Algeria)

from gravity, thermal and petrological data. Tectonophysics 152: 71-88

Lesquer A, Bourmatte A, Ly S, Dautria JM (in press) First heat flow determination from the central Sahara: relationship with the Pan-African belt and Hoggar domal uplift. J Afr Earth Sci, Oxford

Liger JL (1979) Structure profonde du bassin côtier sénégalo-mauritanien, interprétation de données gravimétriques et magnétiques. Thèse, Univ Marseille, Fr, 155 pp

Louis P (1971) Contribution géophysique à la connaissance géologique du bassin du Lac Tchad. Mém ORSTOM, Paris 42: 311

Lucazeau F, Lesquer A, Vasseur G (in press) Trends of heat flow density from West Africa. In: Cermak V, Rybach L, Decker E (eds) Terrestrial heat flow and the structure of the lithosphere. Blackwell

Ly S (1979) Etude gravimétrique de l'Adrar des Iforas (Nord-Est Mali). Thèse docteur-ingénieur, Univ Montpellier, Fr, 107 pp

Ly S, Albouy Y, Chauvin M, Foy R, Lachaud JC, Lesquer A (1980) Apport de la gravimétrie à la compréhension de la chaîne panafricaine dans l'Adrar des Iforas. Cah ORSTOM Sér Geophys 17: 37-57

Ly S, Lesquer A, Ba H, Black R (1984) Structure profonde du batholite occidental de l'Adrar des Iforas (Mali): une synthèse des données gravimétriques et géologiques. Rev Géol Dyn Géogr Phys 25: 33-44

Merlet J (1962) Note relative aux phases sismiques observées entre 100 et 200 km dans le massif du Hoggar. C R Acad Sci, Paris 255: 3441-3443

Molnar P, Tapponnier P (1975) Cenozoic tectonics of Asia: effects of a continental collision. Sciences 189: 419-426

Moussine-Pouchkine A, Bertrand-Sarfati J (1978) Le Gourma: un aulacogène du Précambrien supérieur? Bull Soc Géol Fr, Paris 6: 851-857

Nelson KD, Arnow JA, McBride JH et al. (1985) New COCORP profiling in the southeastern United States. Part I: Late Paleozoic suture and Mesozoic rift basin. Geology 13: 714-718

Pinet C, Jaupart C (1987) The vertical distribution of radiogenic heat production in the Precambrian crust of Norway and Sweden: geothermal implications. Geophys Res Lett 14: 260-263

Ponsard JF (1984) La marge du craton ouest-africain du Sénégal à la Sierra Leone: interprétation géophysique de la chaîne pan-africaine et des bassins du Protérozoique à l'Actuel. Thèse, Univ Aix-Marseille III, 198 pp

Ponsard JF, Lesquer A, Villeneuve M (1982) Une suture pan-africaine sur la bordure occidentale du craton ouest-africain? C R Acad Sci, Paris 295: 1161-1164

Ponsard JF, Roussel J, Villeneuve M, Lesquer A (1988) The Pan-African orogenic belt of southern Mauritanides and northern Rokelides (southern Senegal and Guinea, West Africa): gravity evidence for a collisional suture. J Afr Earth Sci 7: 463-472

Rechenmann J (1965) Mesures gravimétriques et magnétiques en Côte d'Ivoire, Haute Volta et Mali méridional. Cah ORSTOM Sér Géophys, Paris 5: 43

Rechenmann J (1971) Cartes gravimétriques et magnétiques du Nord Mauritanie et notice explicative n° 46. Rapp ORSTOM, Paris, 4 pp

Rechenmann J (1973) Mesures gravimétriques dans le Tanezrouft oriental (Algérie). CNRS-CRZA (ed), Paris 17: 70

Reichelt R (1972) Géologie du Gourma. Mém Bur Rech Géol Min, Paris 53: 213

Ritz M (1985) La distribution de la résistivité électrique dans la lithosphère de l'Afrique de l'Ouest suivant un profil ouest-est. Implications géodynamiques. Thèse, Univ Paris VI, 199 pp

Ritz M, Robineau B (1986) Crustal and upper mantle electrical conductivity structures in West Africa: geodynamics implications. Tectonophysics 124: 115-132

Roussel J, Lécorché JP (1983) Regional gravity trends associated with the Mauritanides orogen (West Africa). In: Schenk PE (ed) Regional trends in the geolgoy of the Appalachian-Caledonian-Hercynian-Mauritanide orogen. NATO ASI Ser, C 116, Reidel, Dordrecht, pp 63-71

Roussel J, Liger JL (1983) A review of deep structures and ocean-continent transition in the Senegal basin (West Africa). Tectonophysics 91: 183-211

Roussel J, Lécorché JP, Ponsard JF, Sougy J, Villeneuve M (1984) Panafrican to Hercynian deformations in the Mauritanides and tectonic significance of gravity anomalies. Tectonophysics 109: 41-59

Simon B, Brisset A, Roussel J, Sougy J (1982) Confrontation de la télédétection (analyse numérique et analogique, téléinterprétation à petite échelle) avec la cartographie géologique classique et les données gravimétriques du Mali sud-occidental (Afrique de l'Ouest). Bull Soc Géol Fr, Paris 24: 13-22

Sougy J (1962) West African fold belt. Geol Soc Am Bull 73: 871-876

Spengler A de, Delteil JR (1966) Le bassin secondaire-tertiaire de Côte d'Ivoire. In: Reyre D (ed) Bassins sédimentaires du littoral africain. 1ère partie: littoral atlantique. Assoc Serv Géol Afr, Paris, pp 99-113

Strangway DW, Vogt PR (1970) Aeromagnetic tests for continental drift in Africa and South America. Earth Planet Sci Lett 7: 429-435

Tackerist D, Lesquer A (1989) Mise en évidence d'importantes variations régionales du flux de chaleur en Algérie. Can J Earth Sci 26: 615-626

Tapponnier P (1977) Evolution tectonique du système alpin en Méditerranée, poinçonnement et écrasement rigide plastique. Bull Soc Géol Fr 19: 437-460

Thorman CH (1976) Implications of klippen and a new sedimentary unit at Gibi Mountain, Liberia, West Africa, in the problem of a Pan-African Liberian age province boundary. Geol Soc Am Bull 87: 851-856

Trompette R (1973) Le Précambrien supérieur et le Paléozoïque de l'Adrar de Mauritanie (bordure occidentale du bassin de Taoudeni, Afrique de l'Ouest). Un exemple de sédimentation de craton. Etude stratigraphique et sédimentologique. Thesis, Univ Aix-Marseille III, Fr. Trav Lab Sci Terre Saint-Jérôme, Marseille, B 7: 702

Trompette R (1979) Les Dahomeyides du Bénin, Togo et Ghana: une chaîne de collision d'âge pan-africain. Rev Géol Dyn Géogr Phys 21: 339-349

Vachette M, Rocci G, Sougy J et al. (1975) Ages radiométriques Rb/Sr, de 2000 à 1700 Ma de séries métamorphiques et granites intrusifs précambriens dans la partie N et NE de la dorsale Reguibat (Mauritanie septentrionale) 7ème Colloque international de Géologie Africaine, Florence. In: Trav Lab Sci Terre St-Jérôme, Marseille, B 11: 142-143

Vauchez A, Kessler SF, Lécorché JP, Villeneuve M (1987) Southward extrusion tectonics during the Carboniferous Africa-North America collision. Tectonophysics 142: 317-322

Villemur JR (1967) Reconnaissance géologique et structurale du nord du bassin de Taoudeni. Mém Bur Rech Géol Min, Paris 51: 172

Villeneuve M (1984) Etude géologique sur la bordure sud-ouest du craton ouest-africain. La suture panafricaine et l'évolution des bassins sédimentaires protérozoiques et paléozoiques de la marge NW du continent de Gondwana. Thesis, Univ Aix-Marseille III, Fr, 552 pp
Villeneuve M, Dallmeyer RD (1987) Geodynamic evolution of the Mauritanide, Bassaride, and Rokelide orogens (West Africa). Precambrian Res, Amst 37: 19-28
Villeneuve M, Lesquer A, Ponsard JF, Roussel J (1984) Géologie et Géophysique du Sud des Mauritanides et du Nord des Rokelides. Les témoins d'une suture panafricaine en bordure occidentale du craton ouest-africain. In: Klerks J, Michot J (eds) African Geology. Tervuren, Belg, pp 57-65
Vitorello I, Pollack HN (1980) On the variation of continental heat flow with age and the thermal evolution of continents. J Geophys Res 85: 983-995
Wellman P (1978) Gravity evidence for abrupt changes in mean crustal density at the junction of Australian crustal blocks. Bureau Mineral Resources. J Austr Geol Geophys 3: 153-162
Williams HR, Culver SJ (1988) Structural terranes and their relationships in Sierra Leone. J Afr Earth Sci 7: 473-477

Chapter 3

Crystalline Basement of the West African Craton

G. Rocci[1], G. Bronner[2] and M. Deschamps[1]

The West African craton is a very extensive portion of Precambrian crust (~4500000 km^2), stable since 1700 Ma ago, bounded on all sides by more recent mobile belts mainly of Pan-African age, such as the Mauritanide fold belt (Fig. 1) on the western edge.

More than half of the craton is covered by sediments, the main part of them constituting the huge Taoudeni Basin. Even the earliest sediment (1000 Ma) are generally undeformed, except within the Mauritanides fold belt.

The crystalline basement comprises a western domain consisting essentially of Archean rocks, and a central and eastern domain where Lower Proterozoic (Birimian and Burkinian) terranes are exposed.

Several synthetic studies on the West African craton have been published during the past fifteen years, so we shall present some new data on selected zones.

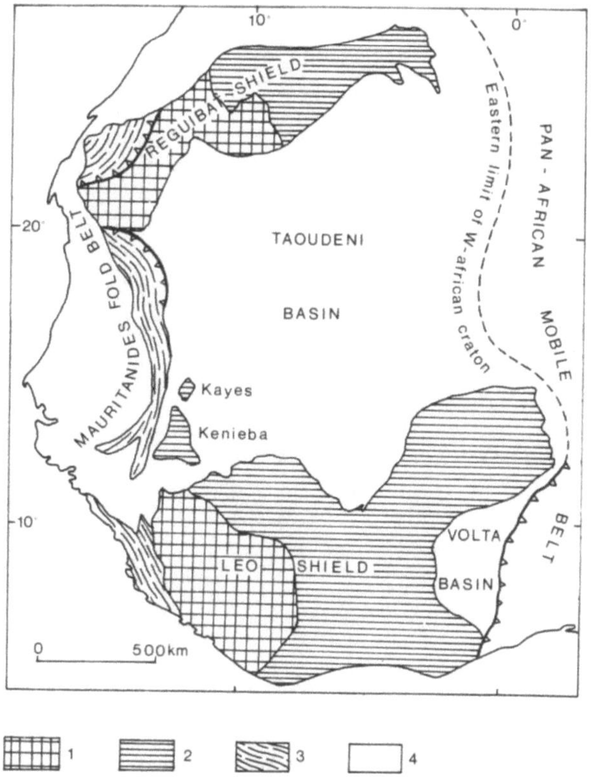

Fig. 1. The West African craton. *1* Archean domain; *2* Eburnean domain (mainly Birimian); *3* Hercynian-Pan-African western mobile belt; *4* sedimentary cover

1 Archean

1.1 Archean of the Reguibat Shield

1.1.1 General Setting and Structure

The West African craton crops out in two main uplifts: the Leo shield in the south and the Reguibat shield in the north. The latter occurs in Mauritania, western Sahara, Morocco and Algeria (Fig. 1).

The Reguibat shield is limited by the Paleozoic basin of Tindouf in the north, the Upper Proterozoic and Paleozoic basin of Taoudeni in the southeast and the Pan-African, Caledonian and Hercynian fold belt of the Mauritanides in the southwest and the west (Fig. 2).

About 1500 km long in the southwest/northeast direction, 250 to 400 km wide, it extends between the 20th and the 27th northern parallel, and the 3rd and the 16th western meridian. It has a desert climate, of Saharian type. The country is a flat desert sprinkled by many inselbergs ("guelb" in arabian dialect) which are the only massive outcrops, mainly in the central and southwestern part. Its average altitude increases regularly from 150 m in the west to 350 m in the east.

It is covered by several types of recent sediments such as desert gravels and pavement ("rag"), hamada and more or less fixed sand which constitutes

[1] Laboratoire de Pétrologie, Faculté des Sciences, Université de Nancy I, BP 239, 54506 Vandoeuvre-lès-Nancy cédex, France

[2] Laboratoire de Géophysique et Géodynamique, Faculté des Sciences et Techniques de St Jérôme, 13397 Marseille cédex 13, France

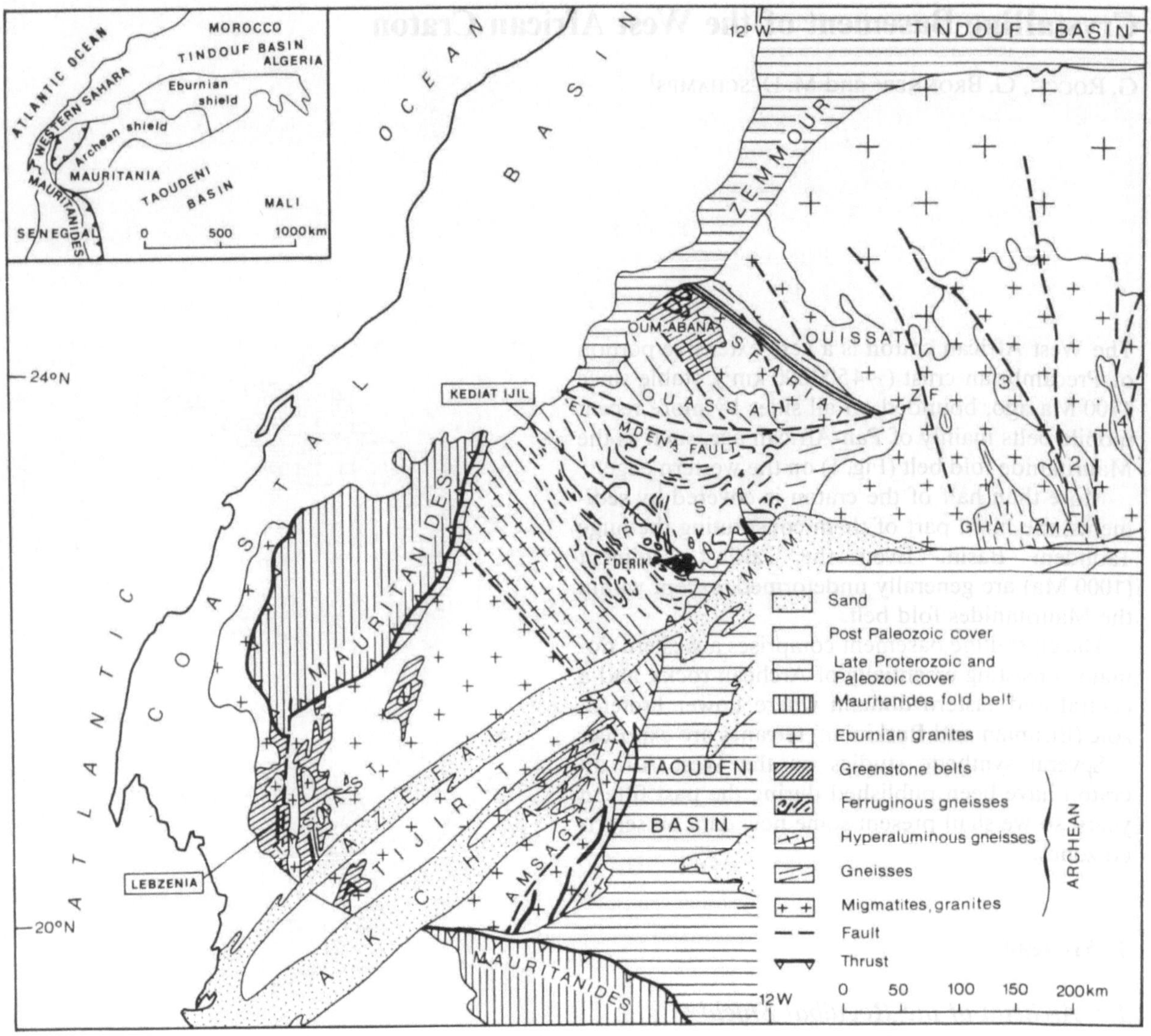

Fig. 2. General map of the Archean Reguibat shield. *1* Temimichat Tsabia; *2* Sebkha Oudian Kharroub; *3* Sebkha Ghallaman; *4* Karet; *ZF* Zednes Fault

extensive dune ridges northeast/southwest in direction. Through this thin skin mainly only steep dipping or vertical unweathered beds of the basement crop out. An aerial and satellite imagery shows a very tight network made of straight or curved lines, constituting structural maps of high resolution (Figs. 3, 4).

The Reguibat shield may be divided in two main parts:

1. The central and eastern shield or "Eburnean shield", mainly with ages not older than 2000 Ma, broken up by several faults, the number of which increases from west to east, with a predominance of intrusive granites in the west and of volcanic formations in the east;
2. The western shield or "Archean shield", where ages older than 2500 Ma are dominant and where metamorphic volcanic and volcanoclastic series and granites of Proterozoic age are absent. Late faults are scarce and the rejuvenation of the biotite at 1600 Ma is the only Eburnean event which affects the Archean shield.

The Archean shield contrasts with the Eburnean shield by exhibiting a major metamorphism of very high grade at 3000 Ma, and by the abundance of migmatites and ferruginous quartzites. The main structural feature is a migmatitic core 300 km in diameter where gneiss septa are still present. It is centered around longitude 14°30′ W and latitude 20°30′ N and is surrounded in the north and the east by concentric and continuous belts of gneisses. From core (SW) to margin (NE), aluminous (sillimanite gneisses), ferruginous (magnetite quartzites), siliceous and calcareous (marbles) formations occur in the gneissic and leptynitic background. Most of these rocks are considered as paragneisses while some of them (amphibolites, gabbros) have an indubitable volcanic origin (Barrère 1969). These belts are mainly exposed in Amsaga in the southeast and Tiris in the northeast. Towards the southwest, in the Tijirit and Tasiast, because of lesser erosion, preserved septa of greenstone belts increase in size. In the Tasiast district they constitute typical mantled gneiss-dome structures (Fig. 6). In the north, the Tiris is separated from the triangular block of the Ouassat-Sfariat district by the E-W trending El Mdena fault, which runs along 400 km. In the northwest extremity of this triangle, the Oum Abana Group crops out. Detected through photogeological analysis only, it appears similar to the Tasiast Group, exhibiting mantled gneiss dome features (Fig. 2).

The remaining old basement (Ouissat and Ghallaman districts), devoid of ferruginous quartzites, is composed of Archean septa which exhibit less metamorphism towards the east. It is invaded by the Ghallaman migmatitic granites and cross cut by N. S. trending mylonitic faults which outline the boundary between the two shields. In the Eburnean shield some outcrops of probable Archean age constitute the Chegga Group, at the Mauritania/Algeria border.

In the F'Derik district, a particular series is exposed: the Ijil Group which is intercalated between the Tiris Group and the Lower part of the Upper Proterozoic series of the Taoudeni basin (1000 Ma). It is constituted of allochthonous units less metamorphosed than the basement. The presently mined hematitic ore of the Kediat Ijil is located within this group, whose age is uncertain ($\geqslant$1720 Ma).

All available data concerning the Archean groups of the Reguibat shield are reported together in Table 1 and located in Fig. 2. The main events are listed in Table 4.

1.1.2 Geochronological Data

1.1.2.1 Main Events

The age of the main metamorphism obtained through isochron studies (Rb/Sr with $\lambda = 1{,}42.10^{-11}$) is Archean (Table 1). For the most part, the age of the Tiris Group appears younger than that of the Ghallaman and Amsaga Groups. Numerous analyses made on the Tasiast and Amsaga Groups (migmatites, migmatitic granites and pegmatites), on whole rocks or on minerals excepting biotites, yield conventional ages also younger than that of the main metamorphism, without significant differences between the different types of rocks (Table 2). Even these ages have to be taken carefully, as they mean that migmatization occurred clearly after the main metamorphism. The average age is 2617 Ma.

1.1.2.2 Eburnean Rejuvenation, Eburnean Event in the whole Reguibat Shield: the Problem of the Ijil and Chegga Groups

The data in Table 3 show that in the central and eastern Reguibat shield, the oldest ages are those of the metamorphism of the Eburnean groups, followed by those of biotite in the Archean groups and the Eburnean granites. The ages of these latter obtained by isochrons are younger than that of the biotite and seem to be the age of their intrusion. Thus the regional metamorphism and crustal reworking precede the granite intrusion. The phenomena are not synchronous. Furthermore, these data show that the average age of the biotites of the western

Table 2. Conventional ages (Ma) of migmatites, migmatitic granites, granites and pegmatites from the Tasiast and Amsaga Groups

Rocks	Conventional ages Rb/Sr		References
	Whole Rocks	Minerals	
Migmatites		2721 ± 56	Vachette (1964)
Migmatites	2681 ± 161		Giraudon and
Migmatitic granite	2394 ± 92		Vachette (1964)
Granite	2533 ± 122		Bonhomme (1962)
Pegmatite	2908 ± 243		Barrère (1967)
Pegmatite	2798 ± 78		
Pegmatite	2654 ± 74		
Pegmatite	2484 ± 83		
Pegmatite		2642 ± 27	
		2729 ± 45	
		2661 ± 76	
		2670 ± 93	
		2151 ± 168	

Table 1. Synoptic data table of the Archean groups of the Reguibat shield

AREA	TASIAST		AMSAGA	TIRIS	
GROUP	LEBZENIA Bronner et al. (1985)	TASIAST Bonnici and Giraudon (1963)	AMSAGA Barrère (1967)	TIRIS Bronner (1975b)	IJIL Bronner and Chauvel (1979)
AGE	Archean probable	2842 to 2340 Ma (conventional ages on whole rocks and minerals)	3016 ± 135 Ma (isochron)	2779 ± 84 Ma (isochron)	⩾ 1754 ± 38 Ma (biotite)
STRUC-TURE	Isoclinal fold phase Mantled gneiss dome	Isoclinal fold phase Mantled gneiss dome	2 Isoclinal fold phases 1 Kilometric fold phase	3 Isoclinal fold phases 1 Diapiric phase	2 Fold phases, boudinag[e] brecciation, thrust 7 Allochthonous units
LITHOSTRATIGRAPHY	Amphibolites (1000–2000 m) Ferruginous quartzites (50–500 m): - oxide facies - silicate, carbonate facies Ultramafic rocks	Gneiss (5000–10000 m) Migmatites Granites Pegmatites	Top Menerat-Rogouga (200–500 m): Amphibolites Ferruginous quartzites (6 beds): mesobanded, oxide facies, silicate facies interbedded with amphibolites and marble Aidzidine (1000 m): Biotite gneisses Garnet bearing leptynites Sillimanite bearing gneisses Amphibolo-pyroxenites Jamra (3000 m): Garnet-bearing leptynites Bou Rhzama (6000 m): Pyroxeno-amphibolites Hypersthene gneisses Bottom Rag el Abiod complex: migmatites, migmatitic granites, final granites, pegmatites anorthosite, gabbros (premetamorphic)	El Aouj Formation (> 8000 m): Ferruginous quartzites, 8 beds (50–500 m): mesobanded, isotropic, oxide facies interbedded with: - biotite gneisses - hypersthene gneisses - pyroxeno-amphibolites - leptynites - quartzites - marble - migmatite - migmatitic granite anorthosite (premetamorphic)	Ferruginous quartzites (~1000 m): - micro and mesobande[d] oxide facies - silicate, carbonate fac[ies] - red jaspers - breccia Quartzites (heavy minerals) Micaschists Amphibolites Marble Dolomite Garnatiferous and staurolite bearing micaschists Conglomerate Hematitic ore: - metasomatic process - leaching process
PARAGENESES NON-FERRUG.	Qtz + Bt + A Ep + Hbl/Act + Chl + Ep + Ab Qtz + Hbl + Fs + Ep + Bt + Ilm/Sil + Chl + Ab	Hbl + Pl + Ep ± Chl + Spn + Ap + Mag Qtz + Pl + Bt + Ep Qtz + Kfs + Ab + Bt + Mag + Ep	Hbl + Alm + Pl ± Ep ± Qtz C + Fo + Tr/Atg Qtz + Kfs + Opx + Di + Alm + Ap + Zrn Qtz + Sil + Alm/Ms + Bt + Qtz	Qtz + Hbl + Alm + Ap + Zrn/ Bt//Chl + Act Qtz + Kfs + Opx + Di + Alm + Ap + Zrn/Bt Qtz + Sil + Alm/Ms + Bt + Qtz	Qtz + Bt + Mag + Alm + Hbl + Pl + Ep + Grt ± Qtz/Act + Ab
PARAGENESES FERRUGINOUS	Qtz + Mag + (Bt)/Chl Qtz + Mag + Sil + A + Bt/Chl Mag + A/Stp + C		Qtz + Opx + Di + Mag	Qtz + Hem + Mag + Gru + Di Qtz + Hem + Mag + Bt + A ± Cpx/Bt//Chl Qtz + Hem + Mag + Ap + Chl Qtz + Hem + Mag + A ± Stp + Ank Qtz + Hem + Mag	Qtz + Hem + Mag Qtz + Hem + Act Qtz + Bt + Ank + Dol/ Tlc + Chl Qtz + Hem + Mag + Hb[l] Chl + Alm + Cld + Tur
META-MORPHIC FACIES	1 Amphibolite-albite-epidote 2 Greenschist	Granulite-amphibolite	Granulite-amphibolite	1 Granulite 2 Amphibolite-albite-epidote 3 Greenschist	1 Amphibolite-albite-epidote 2 Greenschist
REFERENCES	Bronner et al. (1985)	Bonnici and Giraudon (1963) Giraudon and Vachette (1964) Vachette (1964)	Blanchot (1955) Barrère (1967, 1969) Barrère and Blanchot (1964) Bronner et al. (1990) Vachette and Bronner (1975)	Arribas (1968) Besnus et al. (1969) Bronner (1969, 1972, 1974, 1975a, 1975b, 1976, 1978, 1988) Bronner et al. (1984a, b) Bronner et al. (1990) Cuney et al. (1975) Maurin (1975) Plote (1962) Vachette and Bronner (1975)	Besnus et al. (1969) Blanchot (1955) Bronner (1969, 1970, 1[...] 1976, 1978, 1988) Bronner and Chauvel (1979) Bronner and Fourno (1988) Bronner and Lécorché (1970) Destombes (1963) Huvelin (1962) Spindler (1958) Vachette and Bronner (1975)

OUASSAT		GHALLAMAN		EASTERN REGUIBAT SHIELD		SYMBOLS
OUASSAT Rocci (1957)	OUM ABANA Marchand et al. (1985)	GHALLAMAN Rocci (1957) WEST	EAST	CHEGGA WEST (MAURITANIA) Sougy (1966)	EAST (ALGERIA) Gévin (1951)	
Archean probable	Archean probable	Archean probable	3288 ± 162 (isochron)	Archean probable		$\lambda = 1.42 \times 10^{-11}$
…iat monoclinal range: …clinal fold phases, flat- …ng, boudinage, thrust	Mantled gneiss dome	2 Isoclinal fold phases	?		Isoclinal fold phase	
Ferruginous quartzites: - massive - micro- and meso-banded - oxide or silicate, carbonate facies - breccia - migmatite - ferruginous quartzites and amphibolites - leptynites/migmatites - ferruginous quartzites (3 beds) interbedded with marble, amphibolites - migmatites - ferruginous quartzites, marble - migmatites …atitic ore	Top (~3000 m) - leptynites/migmatites - ferruginous quartzites (100–200 m) - quartzites - amphibolites - migmatites/migmatitic granites Bottom	Leptynites Quartzites Ferruginous quartzites Marble Amphibolites Pyroxene gneisses Migmatites/migmatitic granites Calc-alkaline granite (muscovite, biotite) Granodiorite (biotite and apatite)	WEST A Sebkha Ghallaman: Gneisses, Leptynites, Amphibolites B Sebkha Oudian Kharroub: Leptynites, Hyperaluminous gneisses, Feldspathic micaschists C Karet: Leptynites, Amphiboloschists, Quartzites, Marble EAST	Leptynites Gneisses Amphibolites Migmatites Calc-alkaline granites	Gabbros Diorite Migmatites Granites (Chenachane-Erg Chech)	
…Grt + Stp + Ox/ + Chl + Pmp An + Di Mp + Alm + Opx + …b + Tlc + Pl + A/Ab + Chl …lc + Fo/Srp …Tr + Fo + C + Do/ …l + Tlc …Cpx + Fo + Do + …z/Chl + Tlc	?	Qtz + Kfs + Pl + Grt + Ap + Zrn Qtz + Pl + Hbl + Di + Opx + Mag C + Di + Pl + Spn C + Fo ± Spl Pl + Hbl + Di + Opx + Grt + Mag		?/Ab + Ep + Chl		Rock-forming minerals after: Kretz (1983) and: A Amphibole C Carbonate Cpx Clinopyroxene Fs Feldspar MP Microperthite Ox Oxide
Hem + Mag + Ap Mag + Hem + A + …+ C + Do + Ap + C + …z + Fo + Ap + Qtz + Do + + Act/Tr Hem ± Mag	?					Parageneses: Parag. 1 / Parag. 2 // Parag. 3
…nulite-amphibolite …enschist	?	Granulite-amphibolite	A Amphibolite B Epidote-amphibolite C Epidote-amphibolite	1 Amphibolite 2 Greenschist		
…uat and Coulombeau …72) …hand et al. (1985) … (1957) … (1952b)	Marchand et al. (1985)	Barbey (1974a, b) Bornuat and Coulombeau (1972) Marchand et al. (1971) Rocci (1957) Vachette (unpubl. data)		Buffière et al. (1965a, b) Cancé et al. (1960) Caron et al. (1971) Gévin (1951) Lasserre and Lameyre (1969) Sabaté and Lameyre (1973) Sougy (1960)		Kretz (1983)

Archean groups are much younger (approximately by 200 Ma) than those of the central and eastern Reguibat shield. This age difference may be interpreted as thermal inertia and suggests the existence of a wave of heat moving from the main occurrence of the Eburnean granites towards the western part of the Reguibat shield where these granites are absent. Finally, the reason why no Archean ages have been obtained in the Chegga Group, if Archaean actually occurs, may be due to the proximity of the Eburnean granites, which produced a strong reworking erasing Archean ages.

The conventional ages of the Ijil Group obtained on whole rocks (micaschists consisting only of quartz and biotite) and on biotite, are Eburnean. Therefore, they may represent either the low-grade regional metamorphism which is a specific characteristic of this group or the Eburnean rejuvenation indicated by biotites of the Tiris Group. This problem has not yet been solved.

Much later on, around 1600 Ma, crustal extension allowed formation of a very dense network of basic dykes. By 200 Ma after the main Eburnean events, the Reguibat shield was already sufficiently rigid to fracture on a regional scale (Table 3).

1.1.3 Lithostratigraphy

The Archean Reguibat shield has been studied for more than 50 years. Subdivided by natural and political borders, sometimes impassable, the different parts or districts have therefore been described independently. They have been defined as geological entities under several local terms without using an homogeneous vocabulary.

In order to simplify the exposition, three main lithostratigraphic domains may be pointed out:

1. The Tasiast-Lebzenia domain, to which can be related the Oum Abana Group respectively in the southwestern and northwestern parts of the shield, characterized by greenstone belts and a mantled gneiss dome feature;

Table 3. Eburnean geochronological events through the whole Reguibat Shield

Localization			Ages Rb/Sr in Ma. ($\lambda = 1.42 \times 10^{-11}$)		
			Isochron	Conventional	References
Central and Eastern Reguibat Shield	Chegga Groups (Tilemsi)		2023 ± 52	1982 ± 47 1982 ± 104 1987 ± 104 (biotites)	Sabaté and Lameyre (1973) Lasserre and Lameyre (1973)
	Archean Groups			2015 average age (biotite) 5 meas.	
	Birimian II Groups	Iguidi	2056 ± 110		Vachette et al. (1973)
		Aïoun	2043 ± 19		
		Abd el Malek	2032 ± 71		
		Imourene			
	Eburnean Granites		1958 average age 8 meas.	1993 average age (biotite) 10 meas.	Lameyre and Lasserre (1967) Rocci (1966) Bonhomme (1962) Vachette et al. (1973) Lasserre et al. (1970)
	Dykes			1438 1514 K/Ar	Dosso et al. (1979)
Western Reguibat Shield	Archean Groups	Tiris Amsaga Tasiast		1811 average age (biotite) 14 meas.	Vachette (1964) Giraudon and Vachette (1964) Bonhomme (1962) Barrère (1967) Vachette and Bronner (1975) Dallmeyer and Lécorché (1990)
	Dykes	N-S ENE-WSW N20	1609 ± 45	1610 1757 K/Ar	Dosso et al. (1979)

2. The Amsaga-Tiris-Ouassat domain, strongly metamorphosed with very extensive iron formations;
3. The Ghallaman domain, with a bimodal association consisting of dominant granulites and pyroxene bearing gneisses, interpreted as volcanic and volcanoclastic rocks which are well developed in the eastern part of the Reguibat shield.

Between these domains, correlations are not easy to establish (Table 1). This difficulty is due to the very large distances from one domain to another and to the presence of regional faults which cut the shield into many megarhomboclasts. Nevertheless, some particular lithostratigraphic features may be pointed out:

1. Abundance of iron formations, especially in the Tiris and Ouassat Groups;
2. Association of iron formations and marble in the Amsaga and Ouassat Groups;
3. Predominance of felsic rocks (~80–90% of the volume) in the western and central Reguibat shield;
4. Abundance of volcanic and volcano-clastic horizons in the eastern Reguibat shield;
5. Omnipresence of amphibolites throughout the Archean;
6. Ubiquitous migmatites, migmatitic granites, late orogenic granites and pegmatites.

1.1.3.1 Iron Formations: an Archean Signature. The Red Jasper of the Ijil Group

Iron formations are one of the most typical lithostratigraphic elements of the Archean shield. They occur in most of the Archean groups of the western part of the shield, are scarce in the centre and are absent in the eastern part. Three main types have been defined (Bronner et al., 1990):

1. The *Lebzenia type* occurs in the Tasiast greenstone belts (Bronner et al. 1985). It is characterized

Fig. 3. Map of the Tiris district showing structural trending drawn by iron quartzites. *1* Sand; *2* Late Proterozoic cover (Taoudeni basin); *3* Ijil Group; *4* Archean basement; *A* Well-oriented megascopic folds, often of elongated ellipsoidal shape; *B* fold interference features; *5* fault; *6* thrust; *7* dome (After Bronner 1978)

by a meso- and microbanding with alternation of quartz oxide facies with some remnants of original detrital grains and silicate-carbonate facies where the layers have a variable composition. The primary sediment seems to have been a mixture of Fe-smectite glauconite and Fe-hydroxide interbedded with centimetric-scale silica beds (Bronner et al., 1990). Therefore, it may be interpreted as a clastic accumulation devoid of volcanic activity, even if it is close to volcanic formation. It can be compared with the Algoma type (Gross 1972, 1980).

2. The *Tiris type* presents two main facies: facies A is isotropic of granular tendency with grains of 0.1 to more than 10 mm; facies B is anisotropic with an obvious metamorphic foliation parallel to a clear bedding made of millimeter to centimeter alternation of quartz-rich beds, silicate- and magnetite-rich beds. The original sediments were probably a mixture of Al-poor smectites and Fe-hydroxides (Bronner et al., 1990). Intense metamorphic processes (Table 1) produced the segregation of SiO_2 which differentiated in separate layers. This type can be related both with the Hamersley type in average composition and with the Lake Superior type for its trace element content.

3. The *Ijil type* is characterized by both mesobanding, with alternation of siliceous and ferruginous layers whose thickness varies from several millimeters to 1 or 2 cm and microbanding (<0.5 mm), and ironstones with red jaspers (Bronner and Chauvel 1979). It is usually fine-grained and contains quartz layers with rims of replaced clastic quartz grains, some layers with heavy minerals (zircon, apatite and tourmaline) or chert layers with allochems. The ironstones are associated with siliceous clastic and mica schists. They are devoid of volcanic or plutonic rocks. The original sediments were grits, sandstones and thinly bedded sediments such as thin argillaceous and silty sequences with an iron-rich matrix and some oolitic and intraclastic horizons. Mesobanding is interpreted as the result of a segregation through diagenesis and metamorphism. The very low amounts of major elements except silica and iron allow a comparison with the Hamersley type, but, as for the Tiris type, it may be also compared with the Lake Superior type with regard to trace element content.

The Ijil type shows a secondary evolution which does not exist in the other types. Hard hematitic ore occurs in extension zones through a metasomatic process followed by an intense leaching which causes sporadic complete desilicification.

Among the several ironstones, the Ijil type appears to be a particular lithology and is the only one which contains red jasper.

1.1.3.2 Migmatites, Migmatitic Granites, Granites and Pegmatites

Migmatization is developed in all the Archean rocks. Nevertheless, some districts were only weakly affected by anatexis, but none were totally free from this phenomenon which represents, on a continental scale, the final stage of a single major metamorphic event.

The migmatization is more intense in the southern and western parts of the shield than in the eastern one. During the melting process, some very refractory lithologies, such as ironstone layers and, to a lesser extent, basic rocks, remained competent, forming isolated boudins during ductile deformation of the migmatitic melt. There is a gradation from migmatites to migmatitic granites and pegmatites. They are mainly formed by melting of granulites and leucogneisses, and show either diffuse, gradational (in situ migmatites) or sharp (intrusive migmatites) contacts. The evolution from a migmatite to a migmatitic granite is characterized by progressive loss of the foliation sharpness. At this stage, the granites may form individual stocks with gradational or sharp contacts such as granodiorites in the Tasiast Group (Timmimichat-Tioulit granites) which are clearly related to the migmatitic complex. Some granites cross cut the gneisses but show gradational contacts with the migmatites (Touijenjert and Modreïgue granites, Amsaga). They are often porphyroblastic, biotite- and hornblende-bearing granites locally grading into charnockites (quartz + microcline + plagioclase + hornblende + orthopyroxene) revealing that they formed under granulite-facies conditions.

The so-called final granites (Barrère, 1967) are muscovite or biotite (the most common), bearing calc-alkaline granites. They generally cross cut the gneisses.

Pegmatites seem to be the latest stage of the migmatization process. They are widespread and clearly spatially associated with the migmatites. In the Tasiast area, they are located in the upper part of the gneiss domes where they constitute diffuse stocks. Two generations of pegmatites may be distinguished. The first, usually pink, is made of quartzo-feldspathic material only. These pegmatites invade open fractures formed during the second isoclinal fold phase in the ironstones of the Tiris where disharmonic structures appear. The second generation is white and composed of several mineral phases in addition to a quartzo-feldspathic matrix: black tourmaline, biotite, muscovite, beryl. In the Tasiast some pegmatites are of sodi-lithic type (quartz + microcline + albite + lepidolite + columbo-tantalite + spodumene + beryl + Li-tourmaline).

1.1.4 Metamorphism

Metamorphic data are summarized in Table 1. They show the following features:

1. Granulite/amphibolite facies conditions existed everywhere in the western Reguibat shield whereas the metamorphic intensity decreased eastward;
2. Over almost the whole shield this main metamorphic event is followed by a retromorphism to greenschist facies;
3. In the Tasiast, two metamorphic facies exist together. This fact confirms the existence of two distinct groups: early basement (Tasiast s. s. Group) with greenstone belts (Lebzenia Group) constituting mantled gneiss dome structures. No unconformity has yet been observed between the two groups;
4. In the Tiris district, the superposition of two groups which underwent very different metamorphic facies is in accordance with the existence of a thrustplane between them. The Ijil Group clearly overthrusted the Tiris Group in the Kediat Ijil, whereas the El M'Haoudat Range, which constitutes an isolated slice of Ijil Group between two blocks of basement, overthrusted the western block but was overthrusted by the eastern one. In these two rootless terranes unconformity between the two groups is unequivocal (Bronner, 1988).

1.1.5 Tectonics

The main geometric marker is a very well-developed foliation generally parallel to the bedding. The main trend of the most visible folding, following several isoclinal fold phases, illustrates the megascopic structure disposed around the migmatitic core situated in the southwestern part of the Reguibat shield (Fig. 2). In general, folding is characterized by kilometer-sized structures with well-developed antiforms, sometimes of ellipsoidal shape, showing steeply plunging closures and vertical axial planes. The synforms are narrow and complicated by many folds of several wavelengths and interference patterns are common.

1.1.5.1 Diapiric Features

Two regions illustrate megascopic tectonic features which constitute an evolution peculiar to the general fold phase: the Tiris district in its northern and northwestern part and the Tasiast district.

The first region (Fig. 3), studied in detail (Bronner 1976, 1981), is characterized by an abundance of ferruginous quartzites. It shows catazonal facies and weak migmatization. The main tectonic phase produced folds of about 2 to 3 km in wavelength. The folds developed locally into antiformal swellings to become spherical or ellipsoidal bulbs outlining a system of large polygonal cells with vertical walls. The most common dimensions measured in the horizontal plane are comprised between 8 × 8 and 12 × 30 km. The domes have few folds of lower wavelength suggesting extension (Fig. 4), whereas synforms developed into very narrow gutters with many folds. Where three domes are neighbours, the triangle they bound is centered by a navel structure, exhibiting folds with radial axial planes and constrictional features, suggesting sinking (Fig. 5). Dur-

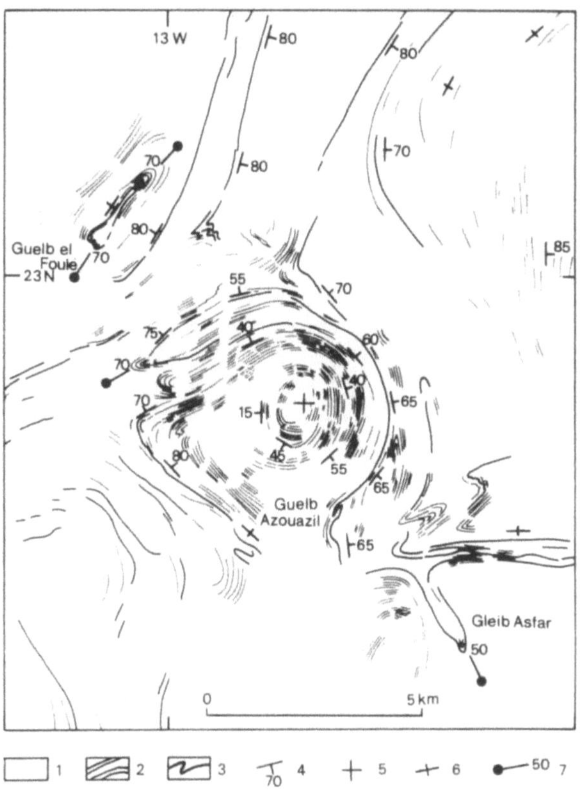

Fig. 4. Detailed map of the Azouazil dome. *1* Reg (desert pavement) and sand; *2* quartzo-feldspathic and basic rocks; *3* iron quartzites; *4* beds or foliation: dip in degrees; *5* beds or foliation: horizontal; *6* beds or foliation: vertical; *7* fold axis (dip in degrees) (After Bronner 1978)

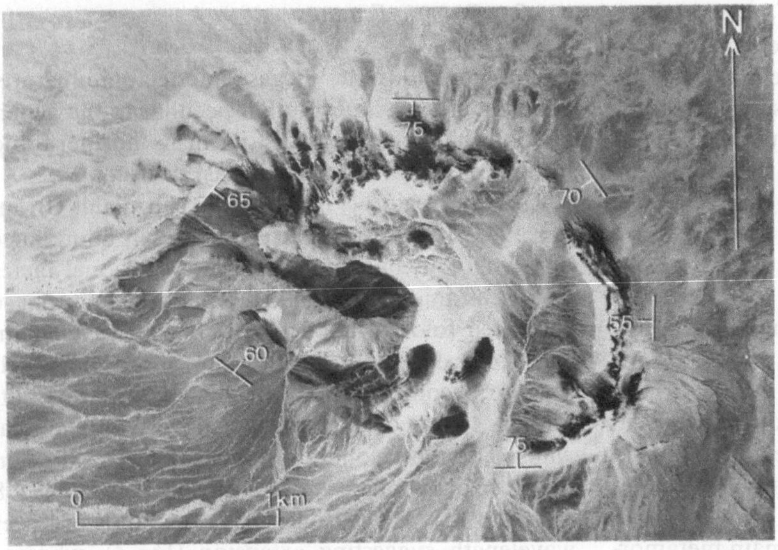

Fig. 5. Aerial photograph of the Guelb Tintekrat el Beida: a navel-shaped sinking feature at the junction of three neighbouring domes

ing doming and sinking, buckling was predominant with very important thinning and thickening, stretching and deflection. Ferruginous quartzites show a very ductile behaviour, whereas the flow deformation of the gneissic material suggests a very low viscosity, which may occur only under conditions of high temperature and hydrostatic pressure.

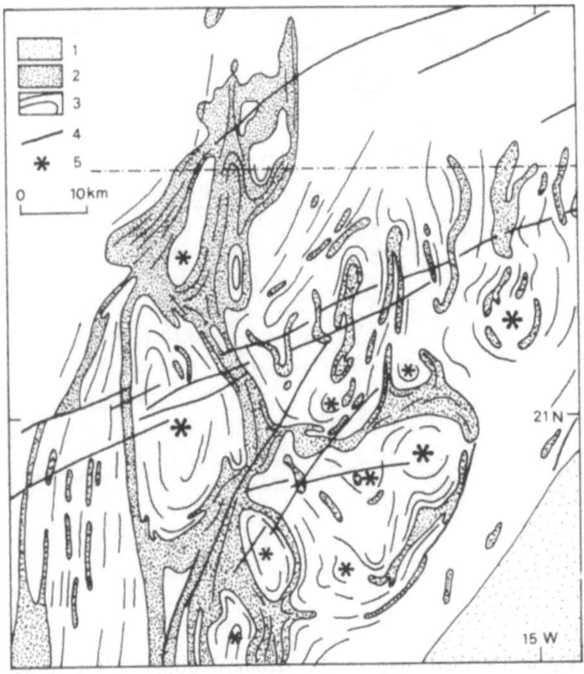

Fig. 6. Structural map of the Tasiast district showing the mantled gneiss dome feature drawn by the greenstone belt network. *1* Sand; *2* greenstone belt; *3* quartzo-feldspathic basement; *4* fault; *5* dome

The density map established from sample measurements and mining data shows that bulbs are made of rocks with granite density (2.63 g/cm^3), whereas the sinking bodies have higher density (3.1-3.3 g/cm^3). These data lead to the consideration of gravity as an active element in the genesis of such deformation.

The second region was not mapped in detail but is known by geological and field compilation. Nevertheless, the study revealed typical mantled gneiss dome structures (Fig. 6). The dome shapes are between 16×16 and 18×36 km in size; one dome measures 20×40 km with an apex at more than 10 km above the topographic plane and another composed of four smaller domes, is about 40×45 km (Fig. 7).

The density contrast is about 0.4 to 0.6 g/cm^3 between the granitic rocks and the basic and ultrabasic rocks which overlie them and form narrow rim synforms surrounding the domes. From the boundaries to the center, some domes show evident petrographic differentiation, from simple gneisses and migmatites to granitic rocks rich in pegmatites. This suggests mobilization during doming. These kind of structures have been observed also in the northwestern part of the Ouassat district where the Oum Abana Group has been defined (Fig. 2).

In these two examples, the shapes and dimensions of the domes are different and so are the rocks involved, although the density contrast is the same. In both examples the main direction of predomal large folds may still be observed. Possible antiformal fold intersection might have favoured the rise of some domes. In the first example, high

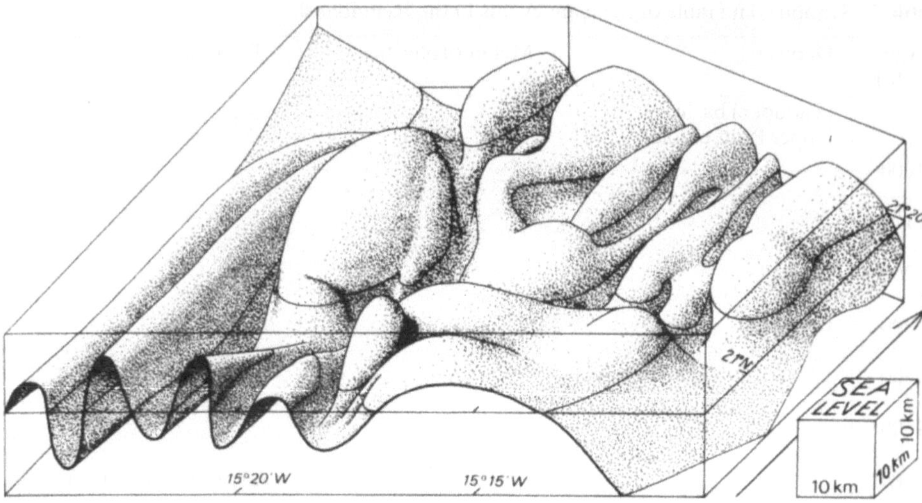

Fig. 7. Three-dimensional reconstruction of the geometry of the greenstone belts

density material, concentrated after early fold superposition, should have produced the gravity instability which generated the sinking and doming.

In both examples, wavelengths of predomal antiforms might have controlled the eventual dome dimensions. Although the kilometer-sized fold phase exists everywhere in the Archean shield, a high density contrast seems to be necessary to allow development of predomal antiforms. This is favoured by decreasing viscosity which may be related to migmatization and ductile behaviour of ferruginous quartzites.

Furthermore, these deformations illustrate that the slice of crust involved is more than 20 km thick (Fig. 7) and that the primary vertical line has not moved very much. This is in accordance with the fact that most of the domes show a quite vertical axis of rotation symmetry, as do the related fold axis, that doming-sinking stopped suddenly and that the isograd surfaces were quite horizontal over several hundred kilometers (Table 4).

1.1.5.2 Fracturing

Mapping by field and remote sensing photogeological approaches of the Reguibat shield shows it to be composed of several megarhomboclasts. These latter are limited by main fracture zones which played repeatedly as shear, extension or thrust planes. The last important movements were normal faults which took place during the deposition of lowermost Upper Proterozoic layers (1000 Ma).

All these brittle events occurred mainly after the end of the Archean period.

An intracratonic extensive event marked by intrusion of basic dykes took place around 1600 Ma, confirming that the Reguibat shield was probably cratonized before that time.

This extension is quite different from one district to the other, and seems to be linked to the petrographic constitution and the structure of the rocks involved. The most important swarms of dykes, usually NNE-SSW trending, occur in migmatized basement where the foliation has quite disappeared, such as in the felsic basement of the Tasiast district, the migmatized core, the south-western part of the Amsaga and the eastern part of the Ouassat. The cumulative width of the basic intrusions reaches locally 50% of the total width. In the other regions, the anisotropic architecture is probably too tight to allow brittle deformations on a wide scale. This view is confirmed by the brittle feature in domes (Tiris district) where the dyke network is of a radial shape, and by refraction phenomena observed where dykes cross over an oblique competent bed.

These basic intrusions generally occur through several episodes. In the Tasiast and the Amsaga, three occurrences have been defined (Barrère 1967). The first one, the most important, is made of gabbros oriented NNE-SSW; the second is made of dolerite oriented NE-SW and the latter of gabbros, N-S and E-W to ENE-WSW are less abundant; the last one (ENE-WSW) may be followed for several hundred kilometers throughout the Western Reguibat shield with regular frequency, every 20 or 30 km one of them (Touyerma dyke, Amsaga) has provided an age of 1500 Ma (Dosso et al. 1979).

Table 4. Synoptic data table of the main events in the Reguibat shield

Age (Ma)	Deposit	Metamorphism	Plutonism	Structure
	(Taoudeni basin) Upper Proterozoic			
1000				Fracturation NW-SE
1100				
1200				
1300	Erosion			Up lift
1400				
1500				Intracratonic distension, dykes (W Archean shield)
1600			Syenite (Tabatanat)	Shear zone (Zednes)
1700			Last Eburnean granites (Bir Moghrein)	Thrust NE→SW (Chegga/Aguelt Nebka, Ijil Group/basement)
1800		Biotite: W Archean groups (av. age)		
1900		First anorogenic granites	Eburnean granites	End of the orogenic stage
2000		Biotite: central + eastern Archean groups (av. age)	Biotite: Eburnean granites (av. age)	Fracturation NW-SE
2100	Birimian II groups Erosion	Retromorphosis (Chegga Group)	Yetti granite + hornfels	
2200	Birimian I groups			
2300	Erosion	Regional metamorphism: Birimian II groups		Up lift
2400				
2500	Banded iron formations (Ijil Gr.)			
2600				
2700		Pegmatites ↑		
2800		Granitization		↑
2900		Migmatization		Diapiric features Mantled gneiss domes
3000		Main granulite/amphibolite grade metamorphism		Isoclinal fold phases
3100	Iron formations			
3200	Mantle intrusions (Lebzenia Group)		Anorthosite and gabbros intrusions	Fracturation NE-SW
3300		First granulite grade metamorph.		
3400	Protolite (Tasiast Group)			

1.2 Archean of the Leo Shield

This part of the West African craton will be described very briefly, simply to establish a comparison with the Reguibat shield, especially on lithostratigraphical and structural topics.

1.2.1 General Setting

Political division of this poorly exposed region of limited accessibility makes it difficult to produce a coherent view of the main structure of the Leo shield. Recently several syntheses (Bessoles 1977; Cahen et al. 1984 and Wright et al. 1985) have given a general view of the largest portion of the West African craton. They provide most of the available data and an exhaustive bibliography.

The Leo shield (Fig. 8), like the Reguibat shield, can be roughly divided into two parts: the western Archean shield within Sierra Leone, Liberia and the western extremity of the Ivory Coast, and the eastern Eburnean shield or Baoule-Mossi domain, within the Ivory Coast, Burkina Faso and parts of Ghana and Niger. The boundary between these two shields is marked by a major shear zone (the Sassandra-Trou faults) which may be equivalent to the Zednes fault in the Reguibat shield. This boundary is more obviously delimited in the Leo shield, but the distribution of the two Precambrian zones is hard to establish. In the central and eastern parts of this shield many small isolated zones of Archean migmatites remain among the Birimian formation, while in its western part the Archean basement is variably imprinted with the Eburnean events.

Along its southwestern border, the Archean shield or Kenema-Man domain (Fig. 9), is overthrust northwestward by the Kasila belt. An important shear zone, the Todi shear zone, which may be

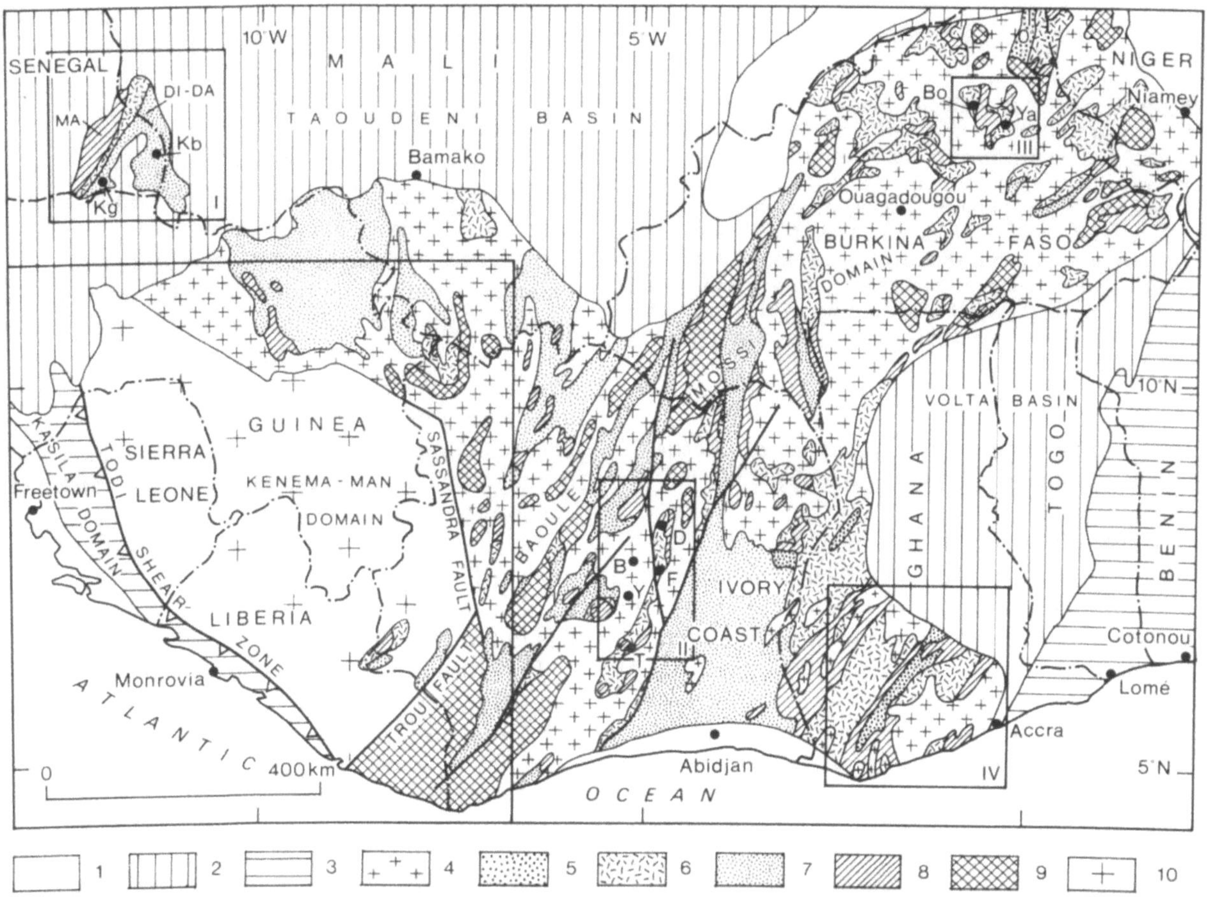

Fig. 8. General map of the Leo Shield. *1* Phanerozoic cover; *2* Upper Proterozoic cover; *3* Pan-African mobile belts; *4* undifferentiated basement and Eburnean granites; *5* Tarkwaian molasse; *6* volcano-sedimentary rocks; *7* flysch; *8* basic rocks; *9* reactivated basement and/or Burkinian; *10* Kenema-Man domain - Archean of western domain. *I* Kenieba - Kedougou inlier; *Kg* Kedougou, *Kb* Kenieba, *MA* Mako series, *DI-DA* Diale - Dalema series; *II* Yaoure unit - Toumodi - Fetekro unit; *Y* Yamoussoukro, *B* Bouaké, *F* Fetekro, *T* Toumodi, *D* Dabakala; *III* Bouroum - Yalogo unit; *Bo* Bouroum, *Ya* Yalogo; *IV* Ghana. *1* Trou shear zone; *2* Sassandra shear zone; *3* Todi shear zone

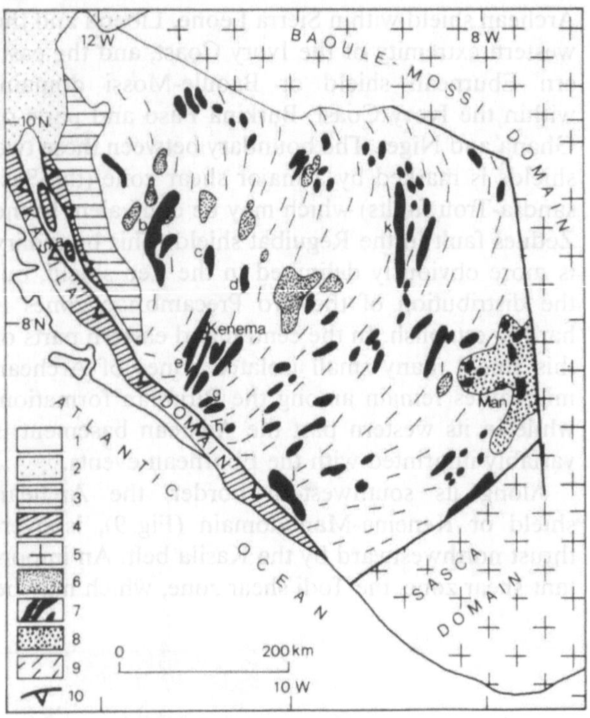

Fig. 9. General map of the Kenema-Man domain (after Rollinson and Cliff, 1982 and wright et al., 1985). *1* Post Palaeozoic; *2* Late Proterozoic to Palaeozoic cover; *3* Pan-African Rokelide belt; *4* Supracrustal belts of the Marampa (a) and Kasila domains; *5* Eburnian domain; *6* Liberian granites; *7* Supracrustal belts: (b) Sula Mountains, (c) Nimini Hills, (d) Gori Hills, (e) Kambui Hills, (f) Bagla Hills, (g) Mano River Mine, (h) Bomi Hills, (i) Bong Range, (j) Goe Range, (k) Simandou belts, (1) Nimba Mountains; *8* Man charnockite complex; *9* Gneiss-migmatite-granulite basement (with foliation trends); *10* Thrust

traced about 1700 km through Guinea, Sierra Leone and Liberia, marks the limit between the two domains.

The Kasila group and the neighbouring Marampa group, consisting also of allochthonous terranes, are considered as Archean terranes (Williams and Culver 1988) according to radiometric and stratigraphic data (Hurley et al. 1971; Hedge et al. 1975). Nevertheless, in these two groups, Eburnean and Pan-African imprints are important (main structures at right angles to those of the Archean basement, mainly Eburnean and Pan-African ages, metamorphic grade contrast with the basement). Therefore, only the Kenema-Man domain will be compared to the Reguibat shield.

1.2.2 Geochronological Data (Table 5)

In the Baoule-Mossi domain the Eburnean ages yielded by presumably Archean rocks indicate a total rejuvenation similar to that in the eastern Reguibat shield. In the Kenema-Man domain the ages are mainly Archean, except those given by biotite which underwent an Eburnean resetting.

In the Kenema-Man domain, especially in the Ivory Coast, the geochronological data (Camil and Tempier 1982; Camil et al. 1983) indicate, more clearly than in the Reguibat shield, the existence of two distinct events: the first at 3125±131 Ma or Leonian event, reaching granulite-facies conditions, the second at 2850±70 Ma or Liberian event, of hornblende-granulite and amphibolite facies conditions.

As in the Reguibat shield, the biotite resetting presents an evident diachronism through Guinea, north and south Sierra Leone and, less clearly, between the Baoule-Mossi and Sasca domains. From northeast to southwest, over about 300 km, the average age decrease from 1980 to 1760 Ma. Here too, a wave of heat seems to start where the Eburnean granites are most abundant and to propagate southwestward through the Archean basement.

1.2.3 Lithostratigraphy (Table 5)

Two main components can be distinguished: a migmatite and gneiss formation which constitutes the basement complex and greenstone and metasediments belts considered as supracrustal relics. The basement complex is dominated by quartzo-feldspathic biotite and hornblende-bearing migmatites and granitic gneisses. They may locally contain remnants of an early sedimentary cycle comprising quartzites, ironstones, quartzo-feldspathic gneisses and schists, graphitic schists and calc-schists. It is intruded by basic and ultrabasic igneous rocks and contains autochthonous granitoids.

The supracrustal formations occur as linear belts and small bodies. Their thickness varies from several kilometers in the large belts to a few hundred meters in the small bodies.

The typical lithostratigraphical sequence (Rollinson 1978) may be given by the Gori Hills or the Sula Mountains (Sierra Leone) outcrops:

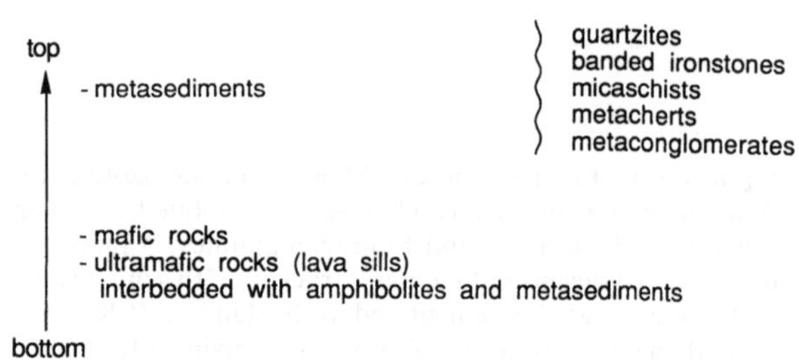

Table 5. Synoptic data table of the main events in the Leo shield compared with those of the Reguibat shield

AGE (Ma)	Sierra Leone, Guinea	Liberia	Ivory Coast: Man Domain	Ivory Coast: Baoulé-Mossi Domain	Reguibat Shield
1600	South (Biotite (av. age))		Sassandra mylonite		Zednes mylonite
			Biotite (av. age)		
1800	North (Biotite (av. age))	Biotites (av. age)		Biotite (av. age)	Biotite (west) (av. age)
		Granites, pegmatites Diorite			First anorogenic gr.
2000	Guinea (Biotite (av. age))	Norite, ultramafic rocks, Quartzites, iron formation, greenstones			Biotite (center and east) (av. age)
2200	Burkinian event				
2400					
			Pegmatites		
2600	Intrusive granites	Granites	Migmatization		↑ Pegmatites
	N-S folds Thermal event	Fold phase Thermal event	Charnockite intrusion	Thermo tectonic event	
2800	Iron formations Greenstone belts Ultramafic rocks				Granitization Migmatization
3000	Pegmatites Granitoids, E Thermal event	Granulites Foliated granites	Iron formation + mantle magmatism		Main granulite amphibolite grade metamorphism
		Gneisses Migmatites	Thermo-tectonic ev.		
3200	Iron formation Early greenstone belts		Migmatitic gneisses (grey gneisses)		
3400			(Protolite)		(Protolite)
References	Allen et al. (1967) Beckinsale et al. (1980) Bonhomme (1962) Hurley et al. (1971) Macfarlane et al. (1981) Rollinson and Cliff (1982) Vachette (1964)	Hedge et al. (1975) Hurley et al. (1971)	Bonhomme (1962) Camil (1984) Camil et al. (1984) Papon (1973)		

From northwest to southeast the thickness of the greenstone belts decreases while the proportion of iron formations increases, as in the western Reguibat shield. However in this latter domain, typical supracrustal belts have only been defined in the Tasiast (Lebzenia Group) and in the Ouassat districts (Oum Abana Group) where greenstones are well developed. Similar greenstone belts are probably present elsewhere as relics (Menerat-Rogouga Formation, Amsaga Group).

Detailed studies, in northeastern Sierra Leone by Macfarlane et al. (1981) distinguished two separate suites of greenstone belts. The older (Loko Group) is composed of amphibolites, serpentines, quartzites and banded ironstones. The younger (Kambui Supergroup) consists of a lower volcanic formation (massive and pillow lavas of basic and ultrabasic composition) overlain by tuffs, psammites, pelites and banded ironstones. A similar sequence has been recognized in the Gola forest of southwestern Sierra Leone (Wilson 1965). Thus, more clearly than in the Reguibat shield, the Archean crust of Kenema-Man domain reveals three distinct orogenic cycles: the "Preleonian cycle", known only as an old basement in relation to the later cycles, the Leonian and the Liberian cycles. Both of the latter are characterized by their own greenstone belts with metasediments and related basic and ultrabasic intrusions, and each is followed by a thermo-tectonic event giving rise to migmatites, granitoids and pegmatites. Before each cycle the former basement is hardened and eroded then reactivated with generation of granitic magmas by crustal melting.

However, geologists who have worked in several

countries of the Kenema-Man domain do not all agree with this point of view. Nevertheless, they recognized everywhere the two main elements: the crystalline basement and greenstone belts.

As in the Reguibat shield, iron formations are one of the most characteristic facies of the Kenema-Man domain. They are usually made up of banded layers of quartz/iron oxides (magnetite and/or hematite, with 35% Fe reaching 65% Fe through metasomatic and leaching enrichment). They are interbedded with greenstones and mafic rocks but, as in the Lebzenia Group, the main depositional processes seem to be independent of volcanic activity. The banded iron formations of the Kenema-Man domain occur in several places such as the Sula Mountains, South Nimini Hills, Gori Hills, Bagla Hill, Mano River mine, Bomi Hills, Bong Range, Goe Range, Nimba Mountains, Simandou belts and Man district. They occur also in the Marampa Group where they constitute a probable Archean signature. They can be linked to those of the Lebzenia and Oum Abana Groups and Menerat-Rogouga Formation, but seem to be very different from those of the Tiris Group where greenstone belts are not obvious. Some of them are very similar to the red jasper-bearing banded iron formations of the Ijil Group. They show the same enrichment processes. Nevertheless, banded iron formations reveal special physico-chemical depositional conditions and constitute an important Archean signature.

1.2.4 Metamorphism

Metamorphic conditions are very similar to those defined in the Reguibat shield. The main metamorphic grade varies from the northwest to the southeast: from greenschist to epidote-amphibolite facies conditions (600 °C/5 kb) and then to granulite-facies conditions (750 °C/7 kb). In the Man district, Camil (1984) defined an earlier thermal or Leonian event, of higher grade granulite facies than the Liberian one. This fact was also pointed out by Macfarlane et al. (1981) for Sierra Leone: a first amphibolite-facies thermal event (Leonian event, Loko Group) was followed by a low-pressure greenschist-amphibolite facies (Liberian event, Kambui Supergroup). According to Williams (1978), the evolution of the Kenema-Man domain may also be defined in terms of one single continuous Liberian event rather than two distinct events.

In both shields, steep metamorphic gradients are commonly observed between basement gneisses and migmatites of amphibolite to granulite facies on one side, and supracrustal belts of greenschist to amphibolite facies on the other side.

1.2.5 Structural Features

The regional trend of the tectonic grain (Fig. 9) swings from northeasterly in the South to northerly and northwesterly in the North, giving a curve quite similar to that of the western Reguibat shield (Fig. 2). Locally an early E-W structural trend has been recognized in Sierra Leone. Tightly folded belts lie parallel to the regional trend but some of them, such as the Simandou Group, are clearly discordant on a regional scale with the basement complex. Except for the Ijil Group, this fact has never been observed in the Reguibat shield. On the other hand, there is no evidence of mantled-gneiss-dome structures here, even if greenstone belts are more developed than in the Reguibat shield.

In the Man district, Camil (1984) distinguished two successive fold phases, both post-anatectic. A first isoclinal phase is followed by a northeast-southwest orientated fold phase, of hectometric to kilometric wavelength, which deforms an earlier foliation. The second phase seems to be very similar to the main fold phase of the Reguibat shield. It controls, as in the Tasiast district, the shape and occurrence of the greenstone relics.

The early E-W fold phase, illustrated locally as tight minor folds, is associated with an amphibolite-facies mineral assemblage. It characterizes the Leonian tectonothermal event.

The Leo shield is cut by many faults of NE-SW direction, parallel to the general trend in the southwestern part of the shield, but their network is less precisely known than in the Reguibat shield. Many mylonites and shear zones occur with probable southeastward dip but of unknown age (earlier than the Todi shear zone and probably than the Sassandra fault). This main fault direction is only known in the southern part of the Reguibat shield.

2 Birimian

Except for the N-W and S-W segments which are composed of Archean rocks, the main part of the West African craton is of Lower Proterozoic age: 2000 ± 200 Ma. The formations are generally gently metamorphosed to the greenschist facies, but intruded by extensive and various granitoids. Referring to the Birim chain in Ghana, whose terranes were called Birimian, more than half of the outcrops are presently covered by platform sediments of Taoudeni Basin (Taoudeni syneclise), the western margin of which is included into the Mauritanides. The Birimian is exposed to the north and south of this huge basin, respectively in the Reguibat uplift and the Leo shield. Birimian basement is exposed in two inliers in the southwestern part of the Taoudeni basin, at Kayes (Mali) and Kedougou (Senegal), Kenieba (Mali). A synthesis of all the data prior to 1975 was made methodically and with many details, region by region by Bessoles (1977). The reader is referred to this work in order to obtain an historical account of the Birimian. We intend to present here a sketch on Birimian stratigraphy and geochronology of the West African craton, then to gather some of the more significant new data disseminated in the literature, or even unpublished, relating to four zones:

1. Reguibat uplift,
2. Kedougou/Kenieba (K-K) inlier,
3. Central part of Ivory Coast,
4. Ghana and northeast of Burkina Faso.

The Birimian formations are exposed as narrow basinal structures separated by ridges of granitoid massifs. This is the chief structural style in the Leo shield, also called the Man shield. The belts are made of quite vertical strata, characterized by rectilinear strikes for many kilometers.

The stratigraphic succession of the metamorphosed strata composing the main part of these belts is therefore very difficult to establish. Excluding the Reguibat uplift, controversy about the superimposition of two main components, sometimes called Lower Birimian and Upper Birimian, is going on. For Junner (1954) the Lower Birimian of Ghana was volcano-sedimentary and detrital, the Upper Birimian being essentially volcanic. Milesi et al. (1986) have the same opinion for Birimian formations of the K-K inlier, and for Mali and Ivory Coast (Milesi et al. 1987). On the contrary, Tagini (1960, 1971) considers that basaltic formations, wrongly interpreted as ophiolites, occur at the base. Recent works, some of which are presented in a later section, favour a basal succession of basaltic volcanics followed by volcanically-derived sediments and differentiated volcanics.

Recently, some authors (Lemoine et al. 1985) demonstrate that some terranes, now called Burkinian, and previously attributed to an undated basement or to the Archean in Burkina Faso and Ivory Coast, are older than the Birimian supracrustals but clearly Lower Proterozoic.

The available radiometric data are numerous but uneven in quality (cf. Bessoles 1977; Cahen et al. 1984). Early Rb/Sr age determinations on the granitoids of Ivory Coast led Bonhomme (1962) to define the Eburnean orogeny near 2000 ± 150 Ma. These dates were confirmed by U/Pb on zircons (Cahen et al. 1984). At present, new measurements, especially by Sm/Nd method on metavolcanics, confirm that the oldest ages are between 2.2 and 2.1 Ga (Dia 1988).

2.1 *Reguibat Uplift* (Fig. 10)

The Birimian formations of the Reguibat uplift are exclusively located in the so-called Eburnean Shield which extends to the east of the 11° W meridian (east of the Zednes fault). They are represented by volcanic and sedimentary rocks frequently exposed in narrow north-trending strips lying unconformably on the Archean basement. The general scheme of the Reguibat uplift, mainly Archean in the western part and Birimian in the eastern one, is similar to the organization in the Leo Shield.

2.1.1 Lithostratigraphy and Geochronology

The Birimian formations of the Reguibat Shield belong to two units that have been considered by several authors (Sabaté and Lameyre 1973; Barbey 1974b) as representative of two orogenic cycles. Each of these units is represented by series which differ by all their characters (lithology, volcanism, structure, metamorphism) and field observations suggest that the lower unit was folded, metamorphosed and eroded before the deposition of the upper one.

The lower unit (Birimian I) is mainly composed of four volcano-sedimentary series all including conglomerate, greywacke, arkose and schist associated with basic to felsic lavas and pyroclastic rocks. These series are affected by polyphase folding and metamorphism and are intruded by calc-alkaline granites prior to the deposition of the upper unit.

The main representatives of this unit are the Yetti series in Algeria (Gevin 1951) and the Aguelt Nebkha, Tsalabia el Khadra and Adam Talha series in Mauritania (Sougy 1952a, 1960; Deschamps and Rocci 1975; Barbey 1974b).

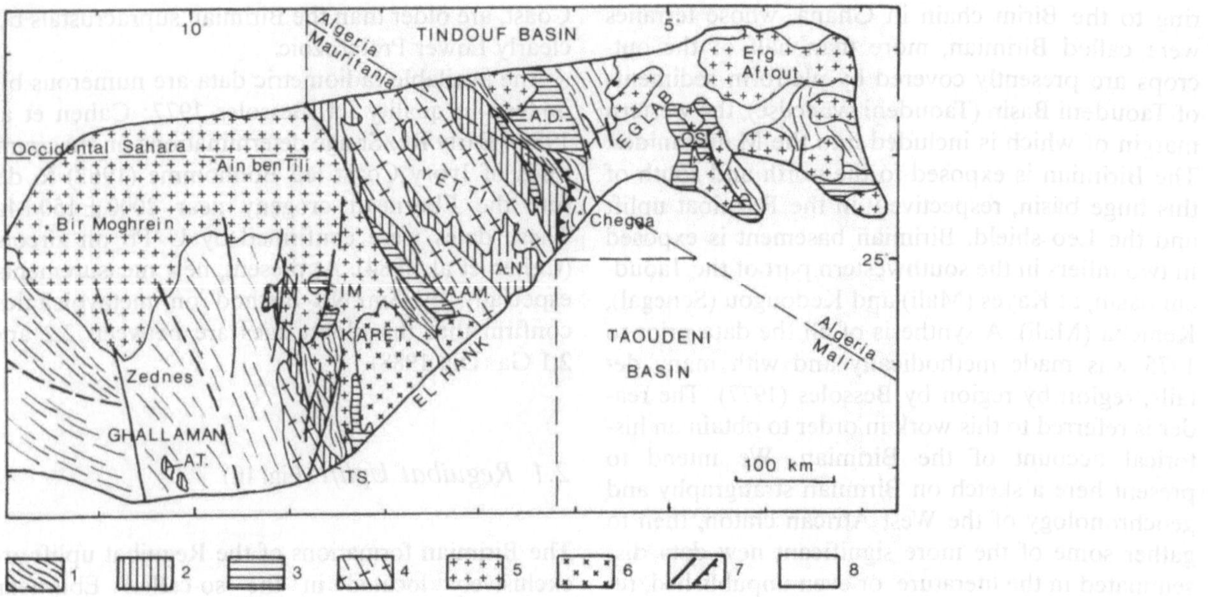

Fig. 10. General map of the Birimian of the central and eastern Reguibat Shield. *1* Pre-Birimian gneisses and granites (Ghallaman and Chegga series); *2* Birimian I; *3* Birimian II; *4* Birimian I granites (Yetti, Karet); *5* Birimian II granites (Aïn ben Tili, Aftout); *6* rapakivi granites; *7* major faults and thrusts; *8* post-Birimian formations. Birimian series: *Y* Yetti, *AN* Aguelt Nebkha; *TS* Tsalabia el Khadra; *AT* Adam Tahla; *OS* Oued Souss; *AD* Akilet Deilel; *AAM* Aioun abd el Malek; *IM* Imourene

The upper unit (Birimian II) is composed of non- or slightly metamorphosed rocks which are mainly detrital and include sandstone, arkose and conglomerate with, in their upper part, mainly felsic lava and pyroclastics. These are all intruded by huge granitoid bodies which are often closely associated with microgranites and even rhyolites attesting to their high level of emplacement.

The Algerian Oued Souss and Akilet Deilel series (Buffière et al. 1965 a, b; Sabaté 1973) and the Mauritanian Aïoun-abd-el-Malek and Imourène series (Sougy 1951, 1960) are highly representative of this upper unit.

The Rb/Sr radiometric data (see also Table 3) are not numerous and are themselves rather old, but confirm the relative chronology as deduced from field observations.

No data is yet available concerning the Birimian I series. However, the Yetti granite which intrudes and metamorphoses the Yetti and Aguelt Nebkha series, yields an age of 2164 ± 100 Ma (Lasserre et al. 1970). This age thus indicates an upper limit for the Birimian I.

Concerning the Birimian II, the Oued Souss and Akilet Deilel series have been dated to 2030 ± 35 Ma (Sabaté 1978). The Imourene and Aïoun-abd-el-Malek series gave an age of ca. 2040 Ma (Vachette et al. 1973). According to these authors, this age represents a tectono-metamorphic event. The granites intrusive into the Birimian II have been emplaced between 1970 and 1755 Ma.

2.1.2 Folding and Metamorphism

Lower Group. The polyphase character of the folding and of the metamorphism is well established in several series of the lower group (Birimian I). This is illustrated by the Yetti series which successively underwent an isoclinal phase P_1 with a schistosity S_1 and a second phase P_2 with N 150 symmetric folds and slaty-cleavage S_2 oblique on S_1 (Sabaté 1972).

Likewise, the Adam Talha series shows the superposition of two metamorphic and folding phases. The first one is isoclinal with foliation S_1 parallel to axial planes. The second is represented by N 170 hectometric similar folds generally coaxial with the first phase and folding the S_1 foliation. Large poikiloblastic tremolite form along the S_2 cleavage and contain contorted S_1 inclusion fabric (Barbey 1974 b).

The two phases took place in the P, T conditions of the greenschist facies with albite + epidote + actinolite or tremolite + quartz in metabasic rocks and quartz + chlorite + hematite ± muscovite in siliceous rocks.

Throughout Birimian I, the metamorphism rarely exceeds greenschist facies; epidote-hornblende parageneses have been locally observed in the Tsalabia el Khadra series (Deschamps and Rocci 1975).

Upper Group. The structural style is generally of one fold phase. Metamorphism is of low grade, sometimes absent. Local high-grade metamorphism, with incipient anatexis, is related to the emplacement of large post-tectonic granitic bodies. The whole area has been affected by an important period of thrust tectonics after the deposition of the Birimian II series. This style is particularly well developed in the eastern part of the Shield. Marchand et al. (1971) pointed out the overthrust nature of the old basement on the Aguelt Nebkha series near Chegga. Likewise, Sabaté (1973) has demonstrated overthrusting of the Akilet Deilel series (Birimian II) onto the Yetti series (Birimian I). Some other series (Aïoun-abd-el-Malek, Tsalabia el Khadra) appear also to have been deformed by thrusting. Important strike-slip faults cut up the whole Birimian architecture, with horizontal displacements of several tens of kilometers. The mylonites associated with this tectonic style along the Zednes fault yield an age of 1670 Ma (Caen-Vachette 1988).

2.1.3 Igneous Activity

Volcanism. The volcanic formations, represented either by lava-flows or dykes but also by pyroclastics, represent an important volume throughout the Birimian series. These formations provide an additional reason to distinguish the two Birimian units: basic and intermediate volcanic rocks are predominant in the lower unit while felsic rocks occur mainly in the upper unit. Geochemical data are still incomplete. Units closely studied are the Tsalabia el Khadra series for Birimian I (Deschamps et al. 1986) and the Akilet Deilel series (Sabaté 1978) for Birimian II.

Tsalabia el Khadra

This series outcrops in the southern Karet in a narrow synclinal structure dismembered into small fragments by north-trending strike-slip faults.

The lower part of the sequence is detrital, including metaconglomerates and metagreywackes with intercalations of basic metavolcanics and rare carbonates. The upper part is fine grained and quartzo-pelitic with intermediate to felsic volcanics.

The metabasic rocks are geochemically similar to present-day ocean-floor tholeiites with particular REE contents moderately exceeding ten times the C_1 (Evensen et al. 1978) chondrites and with $(La/Yb)_N$ ratios very close to 1. Intermediate and felsic volcanics exhibit well-documented calc-alkaline features, showing an LREE enrichment with $(La/Yb)_N$ ratios higher than 3.

The succession of lower tholeiitic and upper calc-alkaline volcanism has been observed in some other Birimian series of West Africa (Burkina-Faso, Senegal). This fact led Deschamps et al. (1986) and Zonou (1987) to propose a geodynamic model of evolution rather similar to the model of Kröner (1984).

Akilet Deilel

The rocks are mainly sedimentary and detrital (sandstone, arkose, conglomerate) in its lower part and become predominantly volcanic at the top. Volcanic rocks belong to a continuous suite ranging from typically calc-alkaline basaltic andesites to rhyolites. The geochemistry of major and trace elements led Sabaté (1978) to propose for this suite an origin by fractional crystallization of a basic magma derived by partial melting of a mantle source in a destructive continental margin environment.

Plutonism. The Birimian formations are cross-cut by several generations of essentially granitic rocks.

The first generation is represented by Karet and Yetti granites which intrude the Birimian I series. These rocks are biotite and amphibole granodiorites to granites. They are associated with a dense network of aplitic and pegmatitic veins but have no microgranular facies. This fact differentiates them from later granites.

Isotopic data about Yetti granite shows a relative heterogeneity with initial $^{87}Sr/^{86}Sr$ ratio within a large range, between 0.703 and 0.714 (Lasserre et al. 1970; Vachette et al. 1973), suggesting contributions from both crustal and mantle sources.

The Karet granite (Rocci 1955) has a high K/Rb ratio (Baouch 1981) indicating the possibility of at least in part a mantle origin. In Pearce diagrams (1984) it lies in the domain of the granites associated with volcanic arcs (VAG).

The second generation consists of a variety of post-Birimian II granites dated from 1950 to 1750 Ma. The oldest ones are calc-alkaline and form an exceptionally extensive complex (> 100 000 km^2) followed by some more limited yet extensive intrusions of rapakivi and alkali granites.

Calc-alkaline granites are represented by the Aïn ben Tili Massif in Mauritania and the Aftout pluton in Algeria. In spite of their large size, these granites are very homogeneous. They are pink-coloured, coarse-grained and unfoliated biotite granites. They contain no metamorphic enclaves, but decimetric xenoliths of amphibole microdiorites are common and may sometimes constitute some important massifs especially in the Aftout domain (Algeria). The common transition to microgranites (Sougy 1951) constitute another remarkable feature of these granites. Their wide distribution is therefore not to be related to a particularly deep level of erosion (Rocci 1975). The age of these post-tectonic granites is about 1970 Ma (Vachette et al. 1973).

In the southern Karet, a large massif of rapakivi granite (Rocci 1955) of subalkaline composition was studied by Daval (1982) who demonstrated a magmatic origin of its diagnostic mantled K-feldspar megacrysts. As with other rapakivi worldwide, this massif dated at ca. 1870 Ma cannot be related to any yet known orogenic event. Neither gabbro nor anorthosite have been as yet observed in association with these rocks, which show common transition to microgranite and rhyolites, indicating a shallow level of intrusion. The relatively high $^{87}Sr/^{86}Sr$ initial ratio (0.7098) precludes a direct mantle origin and suggests that melting of the lower crust may have played a significant role in the genesis of these rocks as demonstrated by several authors (e. g. Anderson 1983).

Alkaline granites constitute a great number of generally small-sized intrusions as isolated masses, e. g. Djebel Drissa massif (Algeria) with its volcanic cover preserved (Kahoui 1988) or grouped into beautiful inselbergs in the Bir Moghrein area where Rocci (1955) has shown the existence of 12 successive intrusions. Their compositions are monzonitic to peralkaline and their ages are between 1900 and 1750 Ma (Vachette et al. 1973). A few occurrences of alkali syenites are also known; the youngest one (Tabatanat massif, studied by Rocci 1955), with an age of ca. 1500 Ma, represents the latest event in the Precambrian history of the Reguibat Shield.

In conclusion, all post-Birimian II granitoids have in common a post-tectonic or anorogenic setting and a shallow level of intrusion. In Pearce's typology, all of them are in the geodynamic setting of whithin plate granites (WPG).

2.2 Birimian of the Leo Shield (Fig. 8)

From the Mauritanide orogen in the west to the Dahomeyide orogen in the east, three distinct domains will be briefly presented.

2.2.1 Kedougou - Kenieba Inlier

This inlier covering about 16000 km² is triangular in shape, its eastern side limited by the Plateau Mandingue, beyond the Falemé river, its southern side by Upper Proterozoic rocks of the Segou-Madina - Kouta series, and its western limit, the southern branch of the Mauritanide orogen. The precambrian formations are distributed in parallel stripes, with SSW-NNE strike, named by Bassot (1966, 1969) from the west to the east: Mako series, Dialé series and Daléma series. These formations, either chiefly volcanic or volcano-sedimentary or sedimentary, are cross-cut by granitoids in elongated complexes of variable composition and age.

According to Bassot (1969) basaltic volcanics, mainly exposed in Mako series, are overlain by detrital sedimentary flysch-type and carbonates series developed in Dialé and Daléma series. This succession remains, however, hypothetic, a structural relationship being very difficult to observe.

2.2.1.1 Mako Series

The Mako series (Debat et al. 1982, 1983; Seyler et al. 1984; N'Gom 1985; Dioh 1986; Dia et al. 1987; Dia 1988) is exposed in a 20- to 40-km-wide north-trending band within the inlier. At its base are basic metavolcanics: pyroxene-basalts with amygdules, metabasalts in thick, quite fresh, pillowed flows, with epiclastic intercalations associated with mega-sills and dykes of metadiorites. Several large sills of peridotite-gabbro with cumulate textures are intrusive in these metabasalts, although their precise stratigraphic position is unknown.

Good cross-sections occur in the vicinity of Mako, along the Gambia River. The pillowed flows are visible as a continuous sequence from the Mako locality through to Medina-Foulbé near the Falemé River. In the central region of Sabodala where some gold mineralization is concentrated, N'Gom (1985) described for the first time pillows with very well preserved and deformed "variolas" determining a strong lineation.

Massive basalt flows are also widespread and predominate in the South (Debat et al. 1982). They are tholeiites with MORB affinities (N'Gom 1985; Deschamps et al. 1986). Metasediments are not scarce but are more common in the South: schists, quartzites and some limestones. Tuffs and volcanic breccias are also present.

Meta-andesites and metarhyolites are exposed, mainly in dykes cross-cutting the former rocks.

Their calc-alkaline geochemistry has been established by N'Gom (1985). The Mako series has been highly deformed. Beds, often vertical or steeply dipping to the NW, have been isoclinally folded with a strong schistosity parallel to the stratification. The emplacement of generally NW trending calc-alkaline dykes has been driven by subsequent strike-slip tectonics.

A slight dynamothermal metamorphism affected the Mako series, forming low-grade parageneses which, in basic rocks, are ab + chl + act + ep + titanite ± qz ± calc, typical of temperatures above 400 °C for pressure of 3 kb (Liou 1971). Locally, especially in the northern part of the belt, amphibole-bearing gneisses are present. These are considered by Dia (1988) as variably migmatized slices of a formerly deep-seated layered plutonic complex. The age (Rb/Sr and Sm/Nd) of these particularly high-grade formations is the same as the Mako low-grade series.

According to Dia (1988), the characteristics of metabasic rocks (Ti poor tholeiites slightly enriched in LREE and without significant Nb anomaly) are transitional between those of N-MORB and those of IAT and are typical of an immature intra-oceanic arc environment. The intermediate to acidic calc-alkaline lavas would correspond to a more mature stage in the arc building.

2.2.1.2 *Dialé Series*

To the east of the Mako series and cropping out nearly over half of the inlier, the Dialé series (Bassot 1969; Diallo 1983), is composed of weakly metamorphosed sedimentary rocks mainly composed of marble, schist, greywacke, sandstone and conglomerate with some metavolcanics. According to Chtocolov (1973), this series is 8000 m thick and may be subdivided into four groups where detrital sediments are predominant. As in the Mako series, these interpretations are tentative because deformation obliterates primary structures. A gravimetric study (Ponsard 1984) assigns only a maximum thickness of 3500 m to entire Birimian terranes. From a geodynamic point of view the Dialé series corresponds to an epicontinental domain with platform carbonates, in full contrast to the island arc character of Mako series.

2.2.1.3 *Daléma Series*

This series is very similar to Dialé and differs in having a calc-alkaline volcano-sedimentary sequence (Bassot 1987). It lies to the East of the great Saraya plutonic complex which caused long-lived thermal metamorphism (Witschard 1965; N'Diaye 1986). The volcanic rocks are exposed to the south and east of a north-trending volcano-plutonic axis. Pyroclastics such as tuffs, agglomerates and breccias, are scarce. Effusive volcanics are represented by opx and ol - free andesites and dacites. The most widespread formations, so-called Boboti granites, are microdiorites and granitoids cross-cutting this complex.

2.2.1.4 *Magmatic Intrusions*

As in other Birimian area, numerous plutonic intrusions of intermediate to felsic composition occur, which have been recently dated by Bassot and Vachette (1984). The oldest rocks are represented by the Badon-Kakadian batholith (Bassot 1966), of great size, heterogeneous and orientated conformably with the major structures of the enclosing Birimian. It is made of diorite, granodiorite and granite with calc-alkaline trondjhemite chemistries. In the northern part of the massif, in the Falemé River, Dia (1985, 1988) described the plutonic complex of Laminia, a magmatic suite of gabbroic cumulates to adamellites evolving by fractional crystallization.

This complex is probably cogenetic with the Kakadian massif but is moderately deformed and lacks microcline phenoblasts characteristic of the Kakadian complex. Moreover, the so-called migmatites mapped by Witschard (1965) in the Sonfara-Sandikounda region are in fact a huge layered differentiated complex of ultramafic to trondjhemitic composition (Dioh 1986; Dia 1988). Variably mobilized amphibolites of unknown origin have been locally observed in the complex; they could represent fragments of a pre-Mako material or slices of this series metamorphosed at a deeper structural level (see Sect. 2.2.1.1).

Elsewhere, muscovite leucogranites and biotite granites are intrusive into the Dialé and Daléma series. The largest body is the Saraya granite (Bassot 1966) which contains uranium mineralization. Valero et al. (1985) show that the Saraya massif is a composite of several plutons emplaced diapirically during the regional eburnean deformation which is represented by WNW-ESE shortening and slight overthrusting to the SE. The intrusion is surrounded by staurolite-biotite-muscovite micaschist, garnet-two mica micaschist and amphibole-scapolite marble (N'Diaye 1986).

Finally, Birimian plutonic complexes, including small intrusive plutons, have been emplaced since 2045 to 1977 Ma ago (Bassot and Vachette 1984).

2.2.1.5 Conclusion

The most striking feature of the K-K inlier is the absence of Archean. The oldest rocks are the tholeiitic metabasalts not older than 2200 Ma. The K-K inlier is set between two main Archean domains: Reguibat Uplift to the North and Man-Leo Shield to the South. According to geophysicists (Ponsard 1984) this paradox could be explained by the fact that K-K Birimian is apparently rootless. But this hypothesis is inconsistent with isotopic data (Dia 1988; Abouchamy et al. 1989), implying the absence of any previous sialic crust.

2.2.2 Central Part of Leo Shield

To the east of the Man Archean Shield, across a zone more than 1000 km wide, are exposed Birimian terranes. They are distributed along a set of belts generally orientated SW-NE, highly variable in width. The belts are composed of sedimentary, volcano-sedimentary or volcanic sequences, in greenschist (locally amphibolite) facies, separated by granites and sometimes by older variably migmatitic rocks considered as reactivated basement.

In an exhaustive study of the Birimian of Ivory coast, Tagini (1971), influenced by the geosynclinal concept, explained the structure of the region as a set of eugeosynclinal, narrow, elongate zones of subsidence with an important "initial" volcanic component. The corresponding miogesynclines being composed of volcanosediments, flysch and molasses. These concepts are now discredited but applicability of modern plate tectonics model is very difficult because regional structures are often obliterated by large granitic intrusions and reliable trace-element data are still very scarce.

Papon in Bessoles (1977) subdivided the belts into two types; type I: deep troughs; type II: shallow basins. According to his map (Fig. 71, Bessoles's memoir) type I is well represented in central part, especially in the Ivory Coast.

To illustrate the structure and composition of the central part of the West African craton, we shall describe in the Ivory Coast the Yaouré and Toumodi-Fetekro units.

2.2.2.1 Yaouré Unit

Situated to the west of Yamoussoukro, administrative capital of the Ivory Coast, the Yaouré Unit covers 4000 km^2 of a peneplain with scattered hills composed of basic rocks and granites. A detailed study of this region (Jaujou 1970), demonstrated that the folded series has at its base metabasic rocks overlain by intermediate to felsic metavolcanics containing breccias and tuffs. These are intercalated with a few quartzite horizons and manganiferous quartzite sediments: conglomerate, greywacke, siliceous shale and sandstone. Granitoids of eburnean age are widely exposed. Biotite-granites make up the Bouaké plateau as foliated bodies such as the Ferkessedougou muscovite leucogranites. Smaller intrusive plutons, called Bondoukou-type, made of granodiorite, granite and alkaline granite, occur ubiquitously. More recent studies (Fabre 1985a, b 1987; Fabre et al. 1987; Fabre and Matheis 1987) establish a detailed succession of Yaouré formations which are essentially volcanic and not sedimentary as previously stated (Jaujou 1970). They lie on an inferred remobilized Archean basement. At the base, the Lower Birimian group is represented by metabasalts and orthoamphibolites, probably of subaqueous origin, overlain by an Upper Birimian pyroclastic group composed of 10% sediments, 25% pyroclastic flows, 45% tuffs and 20% lavas of a total thickness of more than 1000 m. The pyroclastics are divisible into a dacitic subgroup overlain by an andesitic subgroup. Finally, a complex of ultrabasic to basic cumulate rocks is intrusive into the volcanic suite.

New geochemical data (Fabre et al. 1987) indicate a tholeiitic character (MORB) for the lower basalts and ultrabasic/basic complexes and a tholeiitic and calc-alkaline island arc trend for the explosive unit (andesite and dacite).

The whole unit of Yaouré underwent a greenschist metamorphism locally reaching amphibolite facies. The deformation was high and produced a set of anticlines and synclines open in the south and tight in the north (Fig. 11). Large subvertical north-trending strike-slip fractures with sinistral displacement affected the unit and a later conjugate fault system formed in response to an E-W compression stage (Fabre 1985a).

2.2.2.2 Toumodi - Fetekro (T-F) Unit

This well-exposed narrow elongate unit is 280 km long. Its structural situation is interesting, with foliated granites and granodiorites exposed on its western side and, on its eastern side, the great granitic complex named Dabakala-Sideradougou uplift complex. Within the belt, volcanics and hypabyssal basic formations of the Yaouré unit are present. They were previously interpreted to be "initial" metavolcanics (Vitali 1970; Yacé 1976).

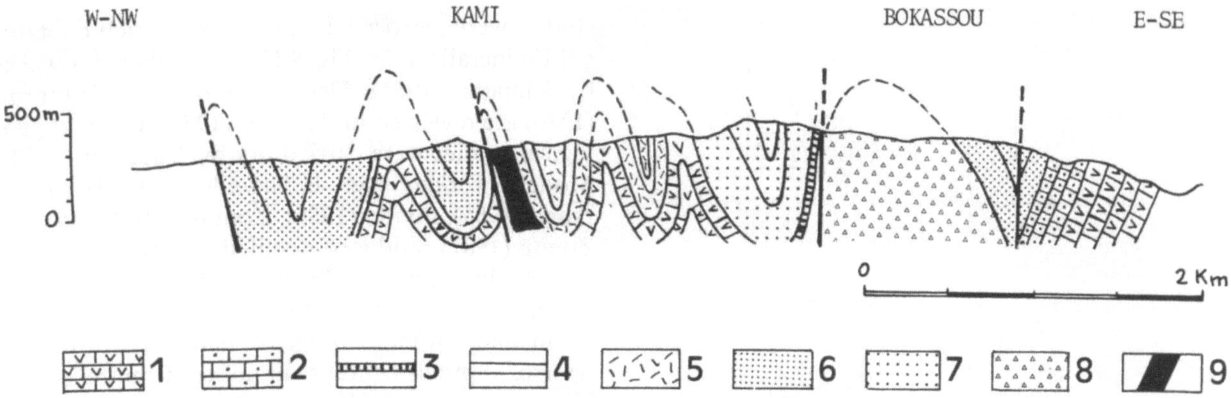

Fig. 11. Geological sections in the folded Birimian volcanic series of Yaouré (after Fabre 1987, simplified). *1* Basalts; *2* andesites; *3* porphyric dacites; *4* rhyodacitic tuffs; *5* sediments; *6* fine, undifferentiated pyroclastites; *7* andesitic pyroclastic flows; *8* gabbros; *9* ultrabasites

Many authors regarded these rocks as ophiolites because pillowed lavas are exposed. However, neither ultramafics nor gabbro and its differentiates are observed in association with the basalts, and these rocks cannot then be considered as true ophiolites with their diversity and geodynamic significance (Rocci 1977, 1980).

In the T-F unit, those metavolcanics are underlain by a metamorphic bi-modal basement known as the ante-birimian, which is described, further to the east, in the Dabakala uplift, by Lemoine (1985) as migmatitic gneisses. These basic formations are probably overlain southwards by more felsic horizons with dacitic and rhyolitic tuffs (Yacé 1976).

Predominant in the volcano-sedimentary unit are highly explosive pyroclastics, felsic lava and derived volcano-detrital sediments. The upper part of the unit is represented by monotonous flyshoid rocks unconformably overlying these formations and exhibiting a rhythmic sequence of conglomeratic rocks containing pebbles of the metavolcanics.

The T-F unit was folded by tight NNE-SSW folds and metamorphosed to the greenschist facies during the eburnean orogeny.

Plutonic complexes, mainly felsic in composition, were studied in detail in the NE part and in the Dabakala uplift (Lemoine 1985). Only a few of the synkinematic tonalitic to granodioritic intrusions are of Burkinian age. For the most part, granitoid of the Dabakala region are eburnean. Lemoine (1985) distinguished older granites transformed into orthogneisses during regional deformation. A second metamorphic phase imprinted on the lower Birimian volcano-sedimentary rocks is dated at 2056 ± 35 Ma (Caen-Vachette and Lemoine 1983). During this second episode or later, porphyroid biotite granodiorite and biotite-amphibole granodiorite were emplaced. Along the north-trending Brobo ductile transcurrent fault an early leucogranite (Brobo granite) is connected with the reworking of an old burkinian thrust zone (Lemoine 1982). A slightly younger muscovite granite was emplaced after the second metamorphic episode. During a post-tectonic phase, numerous small granitoid plutons were intruded into the Birimian terranes, forming an hypovolcanic tonalite-trondjhemite suite including some syenites (Bondoukou-type granite), dated at 1990 ± 20 Ma.

2.2.3 Eastern Part of the Leo Shield

The West African craton is bounded to the SW by a Pan-African mobile belt from Tilemsi (Mali) through Gourma, in the borderland of Mali and Niger, through to the Gulf of Guinea along a narrow zone in Togo and Benin. A substantial portion of the basement is covered by the Volta Basin. Two sectors will be described, one north of the Volta Basin, i. e. NE of Burkina Faso, and the second, to the south, in Ghana, which for a long time was the type region.

2.2.3.1 The Birimian of Burkina-Faso

In his tentative classification of the Birimian belts, Papon (in Bessoles 1977) erected Burkina-Faso formations in Group II (shallow basins). In this region, volcanic and sedimentary formations do not occur in regular belts but are coalescent segments enclosed in a granitic mass.

The region selected is the Bouroum-Yalogo unit, situated 150 km NE of Ouagadougou. It is mapped as a Y-shape belt of 3000 km^2. Detailed geological

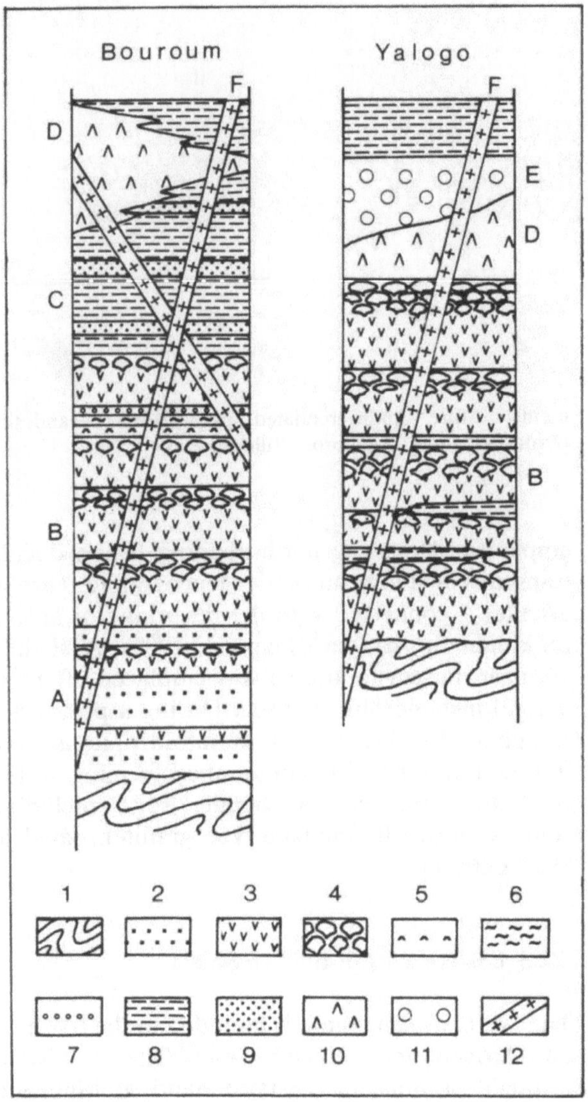

Fig. 12. Simplified stratigraphical column of the Bouroum - Yalogo proterozoic trough (*left* Bouroum; *right* Yalogo). *1* Metamorphic basement (amphibolites, leptynites, quartzites) of undetermined age (Burkinian or Lower Birimian); *2* detritic immature sediments (greywackes, quartzose sandstones); *3* tholeiitic subdoleritic gabbros and dolerites (sills and flows); *4* tholeiitic pillow lavas, mini-pillows and pillow fragment breccias; *5* tholeiitic welded tuffs of intermediate composition; *6* epiclastics; *7* calc-alkalic intermediate to felsic pyroclastic rocks; *8* pelitic deposits; *9* banded iron-quartzites; *10* calc-alkaline pyroclastic deposits (breccias, vitroclastic and welded tuffs); *11* conglomerates bearing calc-alkalic and rare tholeiitic fragments; *12* late calc-alkaline andesitic and dacitic dykes. *A* Predominant immature sediments sequence with minor sills and flows; *B* predominant tholeiitic flow group with associated welded tuffs and epiclastics; *C* sedimentary sequence with pelites, shales and BIF; *D* volcaniclastic sequence; *E* coastal conglomerate; *F* calc-alkalic dykes

studies were performed on this unit, which contains gold mineralization (Fig. 8 III). Unpublished works by Samama (1963), Delfour (1965) and Jeanbrun (1970) are reviewed in Bessoles (1977). General geological studies were carried out by Ducellier (1963). Modern workers such as Sawadogo (1983), Ouedraogo (1985), Sondo (1986), Zonou et al. (1985) and Zonou (1987) produced detailed reports.

For a long time, in this particular region, the occurrence of highly metamorphosed rocks, attributed to an ante-Birimian orogeny, had been observed. Sparse, unpublished data do not confirm the occurrence of Archean rocks in the Bamba and Bouroum series. However, it is highly probable that this basement to the overlying Birimian is linked to the Burkinian (Lemoine et al. 1985), also called Dabakalian. An age of 2386 ± 35 Ma was obtained on a deformed microgranite cross-cutting a magmatically-derived leucogneiss (Wenmenga 1986); amphibolites and leptynites collected by Zonou yield an age of 2200 to 2100 Ma (Boher 1987).

The Birimian formations occur in parallel north-trending belts composed of alternating volcanic and volcano-sedimentary complexes and a sedimentary suite. According to Zonou (1987), the Birimian of Bouroum can be divided into four main units, the lithostratigraphic succession of which is from the base to the top (Fig. 12):

Unit A: detrital (sandstone, greywacke, siliceous pelite ...) with intercalation of basic volcanic flows and sills;
Unit B: tholeiitic (basal massive flows with overlying pillow-lavas);
Unit C: pelites and banded iron quartzites in very thick strata;
Unit D: essentially pyroclastic (breccias, welded or vitroclastic tuffs) of calc-alkaline character.

In this complex, the volcanic formations are represented by basalts, icelandites and rare thin horizons of garnet-dacites. The hypabyssal rocks are dolerites and gabbros with variable Fe-Ti content, and associated granophyres and trondjhemites. Pyroclastic formations are abundant and exceptionally fresh; two types occur: icelandite pyroclastic flows associated with pillow-lavas and andesitic to dacitic tuffs and breccias.

Geochemically, the basalts and associated rocks are tholeiites relatively impoverished in Nb and Ti with reference to N-MORB (Le Roex et al. 1983). Some of them exhibit REE patterns depleted in LREE. Icelandite and garnet-dacites are also tholeiitic but somewhat different. According to Zonou

(1987), this fact is probably due to limited crustal contamination not apparent in the basalts. The pyroclastic and associated rocks are calc-alkaline and probably formed by partial melting of a mantle source previously contaminated by crustal recycling.

Meta sedimentary rocks are either of chemical or detrital origin. They are generally immature at the base (coarse-grained greywacke, conglomerate, sandstone) of shallow basin deposit, and more evolved (quartzopelites and BIF) to the top.

The Birimian formations underwent a low-grade metamorphism into the greenschist facies. Two metamorphic episodes transformed the lavas: an initial static hydrothermal metamorphism to the amphibolite grade, and a subsequent low-grade, low-pressure metamorphism. Volcano sedimentary units have suffered only a very weak metamorphism to anchizonal grade.

The Birimian of Bouroum is only moderately deformed; structures are rarely penetrative and the schistosity is poorly defined. Many zones of ductile North-trending strike-slip deformation cut the region and are the characteristic style of Birimian (Eburnean) tectonics. Compressive phases later built a closely spaced set of conjugate fractures alternatively sinistral and dextral.

As in other parts of the Leo Shield, granitoids are widespread but are not different from those described for the Ivory Coast.

In conclusion, the structural organization is complex and it was not possible to confirm stratigraphic relations between volcanic and sedimentary units. They are probably pene-contemporaneous and correspond, at least in part, to contrasting geodynamic environments. Some of the sediments are probably younger than the volcanics, representing a basin fill environment.

2.2.3.2 *Birimian of Ghana*

In Bessoles' (1977) synthesis for Southern Ghana, four elongate basins of SSW-NNE strike contain sedimentary deposits such as phyllite, tuff and greywacke overlain by conglomerate and sandstone. According to Ghana geologists, these rocks are representative of the Lower Birimian (cf. Kesse 1986). The Upper Birimian is characterized by important volcanic activity, producing lavas, pyroclastics and tuffs. The Eburnean orogeny strongly folded the rocks in these basins, with concomitant metamorphism. At the same time, granitoids were emplaced: older granites of the so-called Cape coast type, and younger granites of Dixcove type. Erosion of the Birimian range was responsible for a very thick (4000 m) molasse known as Tarkwaian. It consists of coarse, poorly sorted, immature sediments possibly derived of intermontane environment (Sestini 1971). Hirdes et al. (1987) describe in Ghana five parallel, evenly-spaced, several hundred kilometers-long volcanic belts consisting mainly of low-grade metamorphic tholeiitic lavas (N-MORB). These are separated by basins containing chiefly dacitic, volcaniclastic and argillitic sediments of various lithofacies, isoclinally folded (Fig. 8 IV).

In the transition zone between volcanic belts and sedimentary basins, crop out cherts, manganiferous horizons and carbonates and C-rich sediments, considered as typical exhalative facies.

The Cape coast type granitoids occur as foliated batholiths in the central parts of the sedimentary basins. They are peraluminous to granodioritic in composition. The Dixcove type are small unfoliated intrusions, metaluminous in character and often tonalitic, emplaced into the volcanic belts.

The model proposed by these authors supposed the existence of a previously differentiated sialic crust. Volcanics, granitoids and sediments have been emplaced pene-contemporaneously between 2.2 and 2.1 Ga. The Tarkwaian is restricted to this region and its northern extension. The deposits are real molasses coming out of the destruction of Birimian ranges, but their age remains Lower Proterozoic. According to recent geochronometry (Agyei et al. 1987) the Cape coast granite complex (2051 ± 45 Ma) has a low $^{87}Sr/^{86}Sr$ initial ratio, and a reworked Archean (?) basement may be represented by poorly dated gneisses (ca. 2800 Ma).

2.3 *Summary and Discussion*

One of the most striking features of the Birimian formations of the West African craton is the absence of well-defined orogenic belts similar in scale and length to those of the late-proterozoic (e.g. Pan-African) and phanerozoic. Birimian deposits and Eburnean events are scattered throughout a very wide area (> 1000 km) with relatively constant characteristics but without regional polarity except in some areas (e.g. K-K inlier). In mobile zones where Pan-African belts are well defined, very strong metamorphic events of Eburnean (ca. 2000 Ma) age have also been recognized, especially in the eastern margin of the craton (Bertrand 1974; Latouche 1978).

2.3.1 Comparison Between Leo and Reguibat Shield

In these two blocks, Birimian terranes are located in their eastern parts and are separated from the western Archean domains by large faults: the Sassandra (Leo) and Zednes (Reguibat) mylonite zones.

In these two domains, the Birimian terranes are commonly displayed along narrow north-trending basins separated by gneissic and granite ridges. Similar relationships are observed in the Guiana Shield (Gibbs and Barron 1983). Moreover, paleomagnetic and radiometric data (Onstott and Hargraves 1981; Caen-Vachette 1988) suggest that these domains formed a single craton during Archean and lower Proterozoic times. Opening of large oceanic areas within this crustal block during the lower Proterozoic is therefore unlikely, but stretching with crustal thinning and in some cases true but limited oceanization is not to be excluded.

2.3.1.1 Lithostratigraphy and Volcanism

The Birimian series of West Africa show a distinct volcano-sedimentary character. Doubts remain, however, about the true chronological sequence, but the lower Birimian is essentially volcanic and the upper is more sedimentary in character.

The nature and the age of the "basement" of these series still remain problematic although sedimentary features of some of them sometimes suggest the existence of an older sialic crust. Radiometric data on rocks considered representative of this crust were obtained prior to 1975 by Rb/Sr method and did not give any reliable Archean age. For long time this was interpreted as being due to isotopic rejuvenation related to crustal reworking. However, new determinations (e.g. Boher 1987; Dia 1988; Abouchamy et al. 1989) by Sm/Nd, Pb/Pb and Rb/Sr confirm the lack of Archean age. Moreover, the $^{87}Sr/^{86}Sr$ ratios are very close to the mantle values inferred for this period and suggest that this "basement" is in fact juvenile mantle-derived material of Birimian age.

The lower-Birimian volcanism is bimodal, mainly tholeiitic and poorly differentiated. The lavas are interbedded with coarsely detrital and immature sediments. It is important to observe that, although tholeiites are rather similar to N-type MORB and sometimes pillowed, no evidence of true layered oceanic crust has been discovered to date (no Birimian ophiolites).

The upper-Birimian volcanism, also bimodal but predominantly felsic and mainly calc-alkaline, is associated with fine-grained quartzo-pelitic sediments. However, the relative scarcity of true andesites in these series is an important character which differentiates them from modern, subduction-related, calc-alkaline series.

2.3.1.2 Metamorphism and Folding

The most striking features of the metamorphism is its moderate intensity, rarely exceeding low grade in lower Birimian and even very low grade in upper Birimian. This fact reminds one the metamorphism of Archean greenstone-belts, which cannot be explained without assuming the existence of a relatively cool crust that isolated the series from a hotter than present-day mantle (Kröner 1984).

The metamorphism that affected the lower Birimian is contemporaneous with at least one phase of isoclinal folding. The upper Birimian folding is less intense and not associated with penetrative metamorphic cleavage.

Finally, the whole Birimian is affected by a strong thrust tectonics and by important strike-slip faulting prior to the emplacement of large post-tectonic granitic bodies.

2.3.1.3 Plutonism

Throughout the West African craton, granitoid of intermediate to felsic composition are abundant. They display a great structural, textural and geochemical variety, and for the most part their ages are eburnean (2100–1850 Ma). Some of them are older and Burkinian (2300–2100 Ma). In the Leo Shield, they were classified as Baoulé-type granites (mainly syntectonic and foliated granites) and Bondoukou-type (mainly smaller sub-circular intrusive and discordant granites) according to Bodin's (1951) terminology. This classification should be discarded; Bonhomme (1962) defined the eburnean orogeny on the base of the first geochronological data on these two types of granites and locally the timing is not consistent. For example, the granite of Bondoukou locality has been recently dated to 2170 Ma, i.e. of Burkinian age (Touré et al. 1987).

On the whole, these Birimian granitoids yield low initial strontium ratios; but, especially in the Reguibat Shield, big massifs such as the Yetti and Archeouat, yield relatively high ratios (Vachette et al. 1973).

Finally, many granitoid bodies were emplaced at a high level into their sub-volcanic cover and include basic co-magmatic remnants.

2.3.2 Geodynamic Evolution and Final Remarks

The geodynamic evolution of the West African craton during the Lower Proterozoic still remains problematic. Previous hypotheses point to the existence of an older (Archean) sialic crust whose thinning and stretching should have induced the formation of narrow basins with subsequent volcanism and sedimentary filling. Succession of tholeiitic and calc-alkaline volcanism has been interpreted, after Kröner (1984), as reflecting crustal stretching followed by convergent motions with delamination of subcrustal lithosphere (Deschamps et al. 1986; Zonou 1987).

This model is now inconsistent with recent isotopic data (see Sect. 2.3.1.1): materials believed to be Archean basement yield in most cases Birimian ages and initial ratios (Sr, Nd) close to mantle values at 2 Ga suggest that they are juvenile and mantle-derived. Moreover, Birimian volcanics do not show any evidence of crustal contamination.

Another remarkable feature pointed out by these new data is the lack of any polarity throughout the whole, 1000-km wide Birimian domain.

The Lower Proterozoic of West Africa appears therefore as a period of rapid crustal growth. According to Abouchamy et al. (1989) this process may have occurred through accretion of immature arcs and back-arc basins. Subsequent crustal thickening may have induced the generation of late calc-alkaline magmas (Dia 1988) through partial melting of a lower, amphibolitic, part of this new crust.

Terranes previously regarded as Archean basement are similar to grey gneisses or so-called TTG series of other regions, with especially low Sr and Nd initial ratios. They could have been derived from partial melting of short-lived basaltic juvenile crust of early Proterozoic age, according to Martin (1987) model.

These two stages in the development of Proterozoic crust should be respectively represented by Burkinian (2.3 to 2.1 Ga) and Birimian (2.1 to 1.9 Ga).

A nearly identical sketch has been proposed by Gruau et al. (1985) for lower Proterozoic granite-greenstone terranes of French Guiana (South America) which were probably linked to West Africa 2 Ga ago. Likewise, Cordani et al. (1988) have shown that crustal accretion in the Brasilian Shield during the Proterozoic was more important than during the Archean. Generally speaking, it is now admitted, on the basis of isotopic data (Patchett and Arndt 1986), that precambrian crustal growth is very much contrasted during the Lower Proterozoic between two types of regions: the first one is essentially reworked Archean crust, the second one is more than 80% newly differentiated materials such as the western and mid-continent regions of the United States (Nelson and DePaolo 1985; Bennet and DePaolo 1987).

References

Abouchami W, Michard A, Albarede F, Rocci G, Fabre R, Dia A (1989) The contribution of mafic magmas to crustal growth in Early Proterozoic West Africa. 5th EUG Meet, Strasbourg, Terra Cognita 1 (1); p 363

Agyei EK, van Landewijk JE, Armstrong RL, Harakal JE, Scott KL (1987) Rb-Sr and K-Ar geochronometry of Southeastern Ghana. J Afr Earth Sci 6: 153-161

Allen PM, Snelling NJ, Rex DC (1967) Age determination from Sierra Leone. In: Hurley PM et al. (eds) Variations in isotopics abundances of strontium, calcium and argon and related topics. Rept 1381-15, 15th Ann Progr Rpt, Massach Inst Technol, Cambridge, pp 17-22

Anderson JL (1983) Proterozoic anorogenic plutonism of North America. Geol Soc Am Mem 161: 133-154

Arribas A (1968) El Precambrio del Sahara español y sus relaciones con la series sedimentarias mas modernas. Bol Geol Min Madrid, 79 (5): 445-480

Baouch S (1981) Contribution à l'étude géochimique et géochronologique des formations précambriennes de la Dorsale Reguibat septentrionale (Mauritanie). DEA Univ Nancy I, 31 pp

Barbey P (1974a) Etude pétrologique d'une série volcanosédimentaire catazonale archéenne: le groupe de Temmimichate Tsabya (Mauritanie du Nord) et de sa couverture éburnéenne. Thèse 3ème cycle, Univ Nancy I, 105 pp

Barbey P (1974b) The Adam Talha epizonal sequence and a general scheme for the eburnean orogenic belt of West Africa. Precambrian Res 2: 255-262

Barrère J (1967) Le Groupe précambrien de l'Amsaga entre Atar et Akjoujt (Mauritanie). Etude d'un métamorphisme profond et de ses relations avec la migmatisation. Thèse Sci Clermont-Ferrand Mem BRGM, 42, 278 pp

Barrère J (1969) Aperçu sur le métamorphisme et la migmatisation dans le Précambrien de l'Amsaga (Mauritanie sudoccidentale). Bull Soc Géol Fr 7 (11): 150-159

Barrère J, Blanchot A (1964) Données nouvelles sur le Groupe de l'Amsaga (Mauritanie). Rept 22nd Session Int Geol Congr, New Delhi, 10: 210-220

Bassot JP (1966) Etude géologique du Sénégal oriental et de ses confins guinéo-maliens. Thèse Clermont 1963, Mém BRGM, 40, 322 pp

Bassot JP (1969) Aperçu sur les formations précambriennes et paléozoïques du Sénégal oriental. Bull Soc Géol Fr, 7 (11): 160-169

Bassot JP (1987) Le complexe volcano-plutonique calcoalcalin de la rivière Daléma (Est Sénégal): discussion de sa signification géodynamique dans le cadre de l'orogène éburnéenne (Protérozoïque inférieur). J Afr Earth Sci 6: 505-519

Bassot JP, Vachette M (1984) Données géochronologiques et géochimiques nouvelles sur les granitoïdes de l'Est Sénégal: implication sur l'histoire géologique du Birrimien de cette région. Afr Geol, Klerkx J, Michot J (eds), Tervuren, pp 191-209

Beckinsale RD, Gale NH, Pankhurst RJ et al. (1980) Discordant Rb-Sr and Pb-Pb whole rock isochron ages for the Archaean basement of Sierra Leone. Precambrian Res 13: 63-76

Bennet VC, DePaolo DJ (1987) Proterozoic crustal history of the western United States as determined by neodymium isotopic mapping. Geol Soc Am Bull 99: 674-685

Bertrand JML (1974) Evolution polycyclique des gneiss précambriens de l'Aleksod (Hoggar central, Sahara algérien). Aspects structuraux, petrologiques, géochimiques et géochronologiques. Thèse Doct Etat, Univ Montpellier, CNRS-CRZA, 370 pp

Besnus Y, Bronner G, Mosser C, Oksengorn S (1969) Etudes géochimiques et minéralogiques sur la province ferrifère du Tiris (Précambrien de la Dorsale Reguibat, Fort-Gouraud, Mauritanie). Bull Serv Carte Géol Als Lorr, 22 (4): 311-328

Bessoles B (1977) Géologie de l'Afrique. Le craton Ouest Africain. Mém BRGM 88, 403 pp

Blanchot A (1955) Le Précambrien de Mauritanie occidentale (esquisse géologique). Thèse Sci Nancy Bull Dir Féd Min Géol AOF, Dakar, 17, 308 pp

Blanchot A, Dumas JP, Papon A (1972) Carte géologique de la partie méridionale de l'Afrique de l'Ouest. BRGM, Paris

Bodin L (1951) Contribution à l'étude des granites birrimiens dans le précambrien de l'AOF. Bull Dir Mines AOF, Dakar, 12, 113 pp

Boher M (1987) Pétrographie et géochimie des formations de la Daléma (Sénégal Oriental). Les formations métamorphiques du sillon de Yalogo (Burkina Faso). Premières datations Sm-Nd et Rb-Sr. DEA Univ Nancy I, 40 pp

Bonhomme M (1962) Contribution à l'étude géochronologique de la plate-forme de l'Ouest Africain. Thèse Clermont, Ann Fac Sci Univ Clermont Ferrand, 5, 62 pp

Bonnici PJ, Giraudon R (1963) Le Groupe de Tasiast, nouvelle unité lithostratigraphique du socle antécambrien de la Mauritanie occidentale. Bull Soc Géol Fr, 7: 1119-1123

Bornuat M, Coulombeau C (1972) Prospection de la dorsale Reguibat centrale. Rapp Bur Rech Géol Min Dakar, 153 pp (unpubl)

Bronner G (1969) Esquisse géologique et structurale de guelbs à magnétite du Précambrien de la région de F'dérik (Fort-Gouraud, Mauritanie). Trav Lab Sci Terre, Marseille, Saint Jérôme, 10 (3), 11 pp

Bronner G (1970) Plissement, cisaillement, boudinage et laminage: mécanismes essentiels de la formation de la brèche d'Idjil (Précambrien de la dorsale Reguibat, Fort-Gouraud, Mauritanie septentrionale). Bull Serv Carte Géol Als Lorr 23 (3), 20 pp

Bronner G (1972) Etude géologique et structurale d'El Mherizat (Précambrien du Tiris), dorsale Reguibat, Mauritanie. Trav Lab Sci Terre, St-Jérôme, Marseille, 10 (14), 14 pp

Bronner G (1974) Structure du Précambrien du Tiris (dorsale Reguibat, Mauritanie). Rôle des quartzites ferrugineux dans la genèse des structures à caractère diapirique. 2ème Réun Ann Sci Terre, Pont-à-Mousson, pp 85

Bronner G (1975a) Structure du Précambrien du Tiris, Dorsale Reguibat (Précambrien). Trav Lab Sci Terre, St-Jérôme, Marseille, (B) 11: 147-148

Bronner G (1975b) Le Précambrien du Tiris, région de Fort-Gouraud. In: Notice de la carte géologique de la République Islamique de Mauritanie à 1/1000000. Ed BRGM, Paris, pp 103-108

Bronner G (1976) Tectonique et reconcentrations: exemple des quartzites ferrugineux précambriens de Mauritanie. Application de l'analyse structurale. Mém H Sér Soc Géol Fr, 7: 167-174

Bronner G (1978) Carte géologique de F'Derik-Oum Dférat à 1/100000. Trav Lab Sci Terre, St-Jérôme, Marseille, 6 feuilles

Bronner G (1981) Diapiric structures of the Archean basement of the South-West Reguibat shield (Mauritania). J Struct Geol 3 (1): 91

Bronner G (1988) La chaine de M'Haoudat (Mauritanie). Rapp Inédit SNIM, 33 pp, 3 cartes

Bronner G, Chauvel JJ (1979) Precambrian banded iron formations of the Ijil Group (Kediat Ijil, Reguibat Shield, Mauritania). Econ Geol 74: 77-94

Bronner G, Fourno JP (1988) Campagne audio-magnétotellurique sur la chaine de M'Haoudat (Mauritanie). Rapp Inédit SNIM, 22 pp

Bronner G, Lécorché JP (1970) Etude pétrographique et microstructurale des formations du mur de la mine de Tazadit (Kediat-Ijil, Mauritanie). Trav Lab Sci Terre, St-Jérôme, Marseille, 10 (3), 11 pp

Bronner G, Marchand J, Sougy J (1984a) Carte géologique provisoire du Maroc à l'échelle du 200000°. Notice explicative de la feuille Mijak (NF28-XXIV). Serv Géol Maroc, Notes Mém 325: 9. Trav Lab Sci Terre, St-Jérôme, Marseille, 10 (64), 9 pp

Bronner G, Marchand J, Sougy J (1984b) Carte géologique provisoire du Maroc à l'échelle du 200000°. Notice explicative de la feuille Twizarfaten (NF28-XXIII). Serv Géol Maroc, Notes Mém 333: 13. Trav Lab Sci Terre, St-Jérôme, Marseille, 10 (67), 13 pp

Bronner G, Chauvel JJ, Triboulet C (1985) Archean metamorphic iron formations and associated basic rocks of Lebzenia (Tasiast, South-Western Reguibat Shield, Mauritania). Sci Géol Bull, Strasbourg, 38 (4): 337-357

Bronner G, Chauvel JJ, Triboulet C (1990) Geochemistry and knowledge of banded iron formations: the West African Shield, an example. In: Ancient banded iron-formations and deposits. Theophrastus, Athens, Greece: 135-156

Buffière JM, Fahy JC, Petey J (1965a) Etude géologique de la partie orientale de la dorsale Reguibat. Région des Eglab et secteur nord du Yetti. Rapp Inéd SERMI, Paris, ALG 63-09-IV, 230 pp

Buffière JM, Fahy JC, Petey J (1965b) Notice explicative de la carte géologique au 1/500000° de la région de l'Eglab et de la bordure nord du Yetti. SERMI, Paris, 38 pp, 1 carte

Caen-Vachette M (1988) Le craton ouest-africain et le bouclier guyanais: un seul craton au Protérozoïque inférieur? J Afr Earth Sci 7 (2): 479-488

Caen-Vachette M, Lemoine G (1983) Chronologie des événements éburnéens dans la région Dabakala (Côte d'Ivoire). 12th Coll Géol Afr, Bruxelles (abstr), p 17

Cahen L, Snelling MJ, Delhal J, Vail JR, Bonhomme M, Ledent D (1984) The geochronology and evolution of Africa. Clarendon Press, Oxford, 512 pp

Camil J (1984) Pétrographie, chronologie des ensembles granulitiques archéens et formations associées de la région de Man (Côte d'Ivoire). Implications pour l'histoire géologique du craton ouest-africain. Thèse Univ Abidjan, Côte d'Ivoire, 79, 306 pp

Camil J, Tempier P (1982) Age à 2850 Ma déterminé par la méthode Rb-Sr sur roche totale pour la série des gneiss à hyperstène et granulites roses associées dans la région de Man (Côte d'Ivoire). Signification de l'âge archéen. CR Acad Sci Paris D 294: 131-133

Camil J, Tempier P, Caen-Vachette M (1983) Schéma chro-

nologique et structural des formations de la région de Man (Côte d'Ivoire). 12th Coll Afr Géol (abstr), p 18
Camil J, Tempier P, Caen-Vachette M (1984) Schéma chronologique et structural des formations de la région de Man (Côte d'Ivoire). In: Géologie africaine. Mus R Afr Cent, Tervuren, pp 1-10
Cancé E, Lameyre J, Perolle H, Rivet P (1960) Mission Mauritanie 1959-1960. Rapport de fin de mission. Rapp BIA, Paris 743, 76 pp
Caron JP, Deschamps M, Marchand J, Rocci G, Simon B, Sougy J (1971) Rapport scientifique des tournées sur le socle de la dorsale Reguibat entre Tenoumer et Chegga (Nord de la Mauritanie). Trav Lab Sci Terre, St Jérôme, Marseille, (E) 7: 73-79
Chtocolov LP (1973) Projet de recherches minières sénégalo-soviétiques (contrat 2000). Rapport final sur les travaux de prospection pour la recherche de l'or filonien et alluvionnaire au Sénégal oriental en 1971. Geological Survey of Senegal, 278 pp (unpubl)
Cordani UG, Teixeira W, Tassinari CCG, Kawashita K, Sato K (1988) The growth of the Brasilian Shield. Episodes, 11 (3): 163-166
Cuney M, Bronner G, Barbey P (1975) Les paragenèses catazonales des quartzites à magnétite de la province ferrifère du Tiris (Précambrien de la Dorsale Reguibat, Mauritanie). Pétrologie, Paris, 1 (2): 103-120
Dallmeyer RD, Lécorché JP (1990) 40Ar/39Ar polyorogenic mineral age record within the northern Mauritanides orogen, West Africa. In: Matte P (ed) Variscan terranes. Tectonophysics, Spec Iss, V. 177, 81-107
Daval D (1982) Sur quelques granites anorogéniques du Protérozoïque moyen de la Dorsale Reguibat (Mauritanie septentrionale). Confirmation de la présence de granites Rapakivi. DEA Univ Nancy I, 46 pp
Debat P, Diallo DP, Rollet M, Seyler M (1982) Mise en évidence d'une série magmatique basique dans les formations birrimiennes du Sénégal oriental. CR Acad Sci Paris 294: 211-213
Debat P, Diallo DP, N'Gom PM, Rollet M, Seyler M (1984) La Série de Mako dans ses parties centrale et méridionale (Sénégal Oriental, Afrique de l'Ouest). Précisions sur l'évolution de la série volcano-sédimentaire et données géochimiques préliminaires sur les formations magmatiques post-tectoniques. J Afr Earth Sci, Oxford, GB, 2 (1): 71-79
Deschamps M, Rocci G (1975) Un nouveau témoin du Protérozoïque inférieur dans la Dorsale Reguibat: la Série de Tsalabia el Khadra (Mauritanie). CR Acad Sci Paris 280: 943-946
Deschamps M, Karche JP, Lapierre H, N'Gom PM, Rossi M, Zonou S (1986) Diversité du magmatisme birimien dans le Craton Ouest-Africain (Burkina-Faso, Sénégal oriental, Mauritanie). In: Les Formations birimiennes d'Afrique de l'Ouest. Occas Publ, CIFEG, Paris, pp 45-64
Destombes JP (1963) Sur la structure géologique de la Kédiat Idjil près de Fort-Gouraud (République Islamique de Mauritanie). CR Acad Sci Paris, 256: 219-221
Dia A (1985) Le complexe plutonique de Laminia: un exemple d'association acide-basique dans le massif de granitoïdes de Kakadian (Sénégal Est). 13th Colloq Afr Geol, St Andrews (abstr), pp 66-67
Dia A (1988) Caractères et significations des complexes magmatiques et métamorphiques du secteur de Sandikounda-Laminia (Nord de la boutonnière de Kedougou, Est du Sénégal). Un modèle géodynamique du Birimien de l'Afrique de l'Ouest. Thèse d'Etat, Univ Dakar, 350 pp
Dia A, Rocci G, Bassot JP (1987) Lithostratigraphie des complexes magmatiques de Sandikounda-Laminia (Sénégal oriental). Données nouvelles sur le Protérozoïque inférieur de la Fenêtre de Kédougou. 14th Colloq Afr Geol, Berlin (abstr), p 46
Diallo DP (1983) Contribution à l'étude géologique de la Série du Dialé (Birrimien) dans les Monts Bassaris (Sénégal Oriental). Thèse 3ème cycle, Univ Dakar, 185 pp
Dioh E (1986) Etude des roches magmatiques birimiennes de la région de Sonfara-Laminia - Médina Foulbé (Sénégal Oriental). Thèse 3ème cycle, Univ Nancy I, 144 pp
Dosso L, Vidal P, Sichler B, Bonifay A (1979) Age précambrien de dolérites de la Dorsale Reguibat (Mauritanie). CR Acad Sci, Paris, 288: 739-742
Ducellier J (1963) Contribution à l'étude des formations cristallines et métamorphiques du Centre et du Nord de la Haute-Volta. Mém BRGM 10, 320 pp
Eaton GP (1984) The Miocene Great Basin of Western North America as an extending back-arc region. Tectonophysics 102: 275-295
Evensen NM, Hamilton PJ, O'Nions RK (1978) Rare earth abundances in chondritic meteorites. Geochim Cosmochim Acta 42: 1199-1212
Fabre R (1985a) Les effets tectoniques des décrochements N-S senestres dans les formations volcaniques birrimiennes de la région du Yaouré, centre de la Côte d'Ivoire. CR Acad Sci, Paris 300: 955-960
Fabre R (1985b) Le volcanisme ignimbritique birrimien et ses mécanismes de mise en place dans le bassin du Yaouré, centre de la Côte d'Ivoire. CR Acad Sci, Paris 301: 1355-1360
Fabre R (1987) Lithostratigraphie du Birimien (Précambrien moyen) dans le centre de la Côte d'Ivoire (Afrique de l'Ouest), région du Yaouré. Bull Soc Géol Fr 8/III (4): 657-663
Fabre R, Matheis G (1987) Caractérisation géochimique du magmatisme birrimien dans le centre de la Côte d'Ivoire (Afrique de l'Ouest): ses implications géodynamiques. 14th Colloq Afr Geol, Berlin (abstr), p 47
Fabre R, Matheis G, Utke AW (1987) Caractérisation géochimique du magmatisme birimien dans le centre de la Côte d'Ivoire (Afrique de l'Ouest). Ses implications géodynamiques. In: Matheis G, Schandelmeier H (eds) Current Research in African Earth Sciences, Extended Abstract. 14th Colloq Afr Geol, Berlin, pp 21-24
Gevin P (1951) Sur la structure du Massif cristallin Eglab-Yetti (Sahara occidental). CR Acad Sci, Paris 233: 1129-1130
Gibbs AK, Barron CN (1983) The Guiana shield reviewed. Episodes 2: 7-14
Giraudon R, Vachette M (1964) Mesures d'âges absolus sur des formations de Mauritanie. CR Acad Sci, Paris 258: 3520-3523
Gole MJ, Klein C (1981) Banded iron formations through much of Precambrian times. J Geol 89: 169-183
Gross GA (1972) Nature et répartition géographique et géologique des gisements de minerai de fer. In: Les ressources mondiales en minerai de fer, publ des Nations Unies, pp 13-33
Gross GA (1980) A classification of iron formations based on depositional environments. Can Min 18: 215-222
Gruau G, Martin H, Leveque B, Capdevila R (1985) Rb-Sr and Sm-Nd geochronology of Lower Proterozoic granite-greenstone terrains in French Guiana, South America. Precambrian Res 30: 63-80
Hamilton W (1979) Tectonophysics of the Indonesian regions. US Geol Surv Prof Pap 1078, 345 pp

Hedge CE, Marvin RF, Naeser CW (1975) Age Provinces in the basement rocks of Liberia. J Res US Geol Surv 3: 425-429

Hirdes W, Leube A, Mauer R, Moorbath S (1987) The evolution of the Birimian Supergroup in Ghana West Africa. 14th Colloq Afr Geol, Berlin (abstr), p 48

Hurley PM, Leo GW, White RN, Fairbairn HW (1971) Liberian age province (about 2700 Ma) and adjacent provinces in Liberia and Sierra Leone. Geol Soc Am Bull 82: 3483-3490

Huvelin P (1962) Précision sur la genèse de la brèche d'Idjil (Fort-Gouraud, Mauritanie). Chronologie de la formation des minerais de fer et la brèche. Bull Soc Géol Fr, 5ème Sér 4: 322-328

Jaujou M (1970) Levé géologique du degré carré de Bouaké. Dir Min Serv Géol Abidjan, 151 pp

Junner NR (1954) Notes on the classification of the Precambrian of West Africa. Congr Geol Int, 19ème Session, Alger 20: 114-117

Kahoui M (1988) Etude d'un complexe granitique différencié et de sa couverture volcanique. Incidence métallogénique - Cas du Djebel Drissa (Massif des Eglab, Algérie). Thèse Univ Nancy I, 258 pp

Kesse GO (1986) The Birimian system in Ghana. CIFEG, Paris, Occas Publ 10: 85-97

Kretz R (1983) Symbols for rock-forming minerals. Am Miner 68: 277-279

Kröner A (1983) Proterozoic mobile belts compatible with the plate tectonic concept. In: Medaris GL Jr (ed) Proterozoic geology. Geol Soc Am Mem 161: 59-74

Kröner A (1984) Changes in plate tectonics style and crustal growth during the Precambrian. Bull Soc Geol Fr 7/26 (2): 297-319

Lameyre J, Lasserre M (1967) Etude géochronologique des syénites alcalines et néphéliniques du massif annulaire de Hassi el Fogra (Mauritanie du Nord). CR Acad Sci Fr, D 265: 733-736

Lasserre M, Lameyre J (1969) Données stratigraphiques et géochronologiques sur la dorsale précambrienne du Pays Reguibat (Mauritanie du Nord, Algérie). Rés Comm 5ème Coll Géol Afr, Clermont-Ferrand, 9-11 avril Ann Fac Sci Univ Clermont-Ferrand, n° 41, Géol Minér 19: 48-49

Lasserre M, Lameyre J (1973) Les leptynites du groupe de Tilemsi (SW algérien), pétrographie, géochronologie. VIIème Coll Int Géol Afr, Florence. Trav Lab Sci Terre, St-Jérôme, Marseille, 1975 (B) 11: 97-98

Lasserre M, Lameyre J, Buffière JM (1970) Données géochronologiques sur l'axe précambrien Yetti-Eglab en Algérie et en Mauritanie du Nord. Bull BRGM, 2ème Sér Sect IV, 2: 5-13

Latouche L (1978) Etude pétrographique et structurale du Précambrien de la région des Gour Oumelalen (Nord-Est de l'Ahaggar, Algérie). Thèse Doct Etat, Univ Paris VII, 255 pp

Lemoine S (1982) Le décrochement ductile de Brobo, un linéament éburnéen majeur, son rôle possible dans l'orogenèse éburnéenne en Côte d'Ivoire. CR Acad Sci, Paris 295: 601-606

Lemoine S (1985) Quelques aspects nouveaux de l'orogenèse éburnéenne en Côte d'Ivoire. Actes 110ème Congr Nat Soc Sav, Montpellier, VI: 53-64

Lemoine S, Tempier P, Bassot JP et al. (1985) The Burkinian, an orogenic cycle precursor of the Eburnean of West Africa. 13th Colloq Afr Geol, St Andrews (abstr), p 27

Le Roex AP, Dick HJB, Erlank AJ, Reid AM, Frey FA, Hart SR (1983) Geochemistry, mineralogy and petrogenesis of lavas erupted along the southern Indian ridge between the Bouvet triple junction and 11 degrees east. J Petrol 24 (3): 267-318

Liou JG (1971) P-T stabilities of laumontite, wairakite, lawsonite, and related minerals in the system $CaAl_2Si_2O_8$-SiO_2-H_2O. J Petrol 12 (2): 379-411

Macfarlane A, Crow MJ, Arthur JW, Wilkinson AF, Aucott JW (1981) The geology and mineral resources of northern Sierra Leone. Overseas Mem 7, Inst Geol Sci, UK London MMSO, 134 pp

Marchand J, Sougy J, Rocci G et al. (1971) Etude photogéologique de la partie orientale de la dorsale Reguibat et de sa couverture sud (Mauritanie). Trav Lab Sci Terre, vol 10, No 11, St-Jérôme, Marseille, 167 pp

Marchand J, Bronner G, Sougy J (1985) Carte géologique provisoire du Maroc à l'échelle du 200000°. Notice explicative de la feuille Aghzoumal (NG 28 VI). Trav Lab Sci Terre, St-Jérôme, Marseille 10 (69), 28 pp

Martin H (1987) Petrogenesis of Archean trondjhemites, tonalites and granodiorites from eastern Finland: major and trace element geochemistry. J Petrol 28 (5): 921-953

Maurin G (1975) Etude structurale de la région Nord-Ouest de F'Dérik. Précambrien inférieur ferrifère du Tiris (dorsale Reguibat, Mauritanie). Application à la recherche du minerai de fer. Thèse 3ème cycle, Univ Aix-Marseille, 117 pp

Milesi JP, Diallo M, Dommanget A et al. (1986) Caractérisation litho-structurale de deux ensembles successifs dans les séries birimiennes de la boutonnière de Kedougou (Sénégal - Mali) et du Niandan (Guinée): implications gîtologiques. Occas Publ 1986/10 CIFEG, Paris, pp 113-121

Milesi JP, Diallo M, Dommanget A et al. (1987) Quelques questions sur les séries du Protérozoïque inférieur (Birimien) de l'Afrique de l'Ouest. 14th Colloq Afr Géol, Berl (abstr), p 51

N'Diaye PM (1986) Etude géologique et métallogénique de la partie septentrionale du granite de Saraya (Sénégal oriental). Thèse Univ Dakar, 109 pp

Nelson BK, DePaolo DJ (1985) Rapid production of continental crust 1.7 to 1.9 b.y. ago: Nd isotopic evidence from the basement of the North American mid-continent. Geol Soc Am Bull 96: 746-754

N'Gom PM (1985) Contribution à l'étude de la Série birrimienne de Mako dans le secteur aurifère de Sabodala (Sénégal oriental). Thèse Univ Nancy I, 135 pp

Onstott TC, Hargraves RB (1981) Proterozoic transcurrent tectonics: paleomagnetic evidence from Venezuela and Africa. Nature (Lond) 289: 131-136

Ouedraogo A (1985) Etude de quelques unités plutoniques basiques éburnéennes dans le sillon de Bouroum-Yalogo au Nord-Est du Burkina-Faso. Thèse 3ème cycle, Univ Nancy I, 2, 64 pp

Papon A (1973) Géologie et minéralisation du Sud-Ouest de la Côte d'Ivoire. Mém Bur Rech Géol Min, Paris, 80, 284 pp

Patchett PJ, Arndt NT (1986) Nd isotopes and tectonics of 1.9-1.7 Ga crustal genesis. Earth Planet Sci Lett 78: 329-338

Plote H (1962) Carte hydrogéologique du Tiris occidental entre Fort-Gouraud et le Rio de Oro (Campagne 1961-62). Rapp BRGM, Dakar, DAK 62-A36, 128 pp

Ponsard JF (1984) La marge du craton Ouest africain du Sénégal à la Sierra Leone. Interprétation géophysique de la chaîne panafricaine et des bassins du Protérozoïque à l'Actuel. Thèse Univ Aix-Marseille III, Trav Lab Sci Terre, Marseille 25, 198 pp

Rocci G (1955) Formations métamorphiques et granitiques de la partie occidentale du Pays Reguibat (Mauritanie du Nord). Thèse Univ Nancy 1955, Mém Dir Féder Min Géol AOF, Dakar, 484 pp

Rocci G (1966) Signification pétrogénétique des minéraux caractéristiques et accessoires du Massif granitique de Fort Trinquet (Mauritanie). IIIème Coll Géol Afr, Tervuren

Rocci G (1975) Le Pays cristallin Reguibat: description des terrains et commentaires. In: Notice explicative de la carte géologique à 1/1000000° de la Mauritanie. BRGM, Paris, pp 77-101

Rocci G (1977) Les ophiolites précambriennes: mythes et réalités. Bull Soc Géol Fr 7 (19): 947-952

Rocci G (1980) Ophiolites 80: une formation géologique élevée au rang de marqueur géodynamique. Mém H Sér Soc Géol Fr 10: 39-50

Rollinson HR (1978) Zonation of supracrustal relics in the Archean of Sierra Leone, Liberia, Guinea and Ivory Coast. Nature (Lond) 272: 440

Rollinson HR, Cliff RA (1982) New Rb-Sr age determination of the Archaean basement of eastern Sierra Leone. Precambrian Res 17: 63-72

Sabaté P (1972) Structure de la Série du Yetti (Sahara occidental algérien). CR Acad Sci, Paris 275: 2591-2594

Sabaté P (1973) La jointure Yetti-Eglab dans la Dorsale précambrienne du Pays Reguibat (Sahara occidental algérien). CR Acad Sci, Paris 276: 2237-2239

Sabaté P (1978) Données géochimiques et radiométriques sur les volcanites calco-alcalines précambriennes de l'Eglab (Sahara occidental algérien). Esquisse de leur évolution géotectonique. Bull Soc Géol Fr 7/20 (1): 81-90

Sabaté P, Lameyre J (1973) Magmatism and metamorphism in the Yetti-Eglab precambrian formations of the Reguibat Dorsale (occidental Algerian Sahara). 7ème Coll Afr Géol, Florence. Trav Lab Sci Terre St-Jérôme, Marseille, 1975, 11: 131-132

Sawadogo J (1983) Etude géologique du sillon birrimien de Yalogo dans la région de Ganguol (Nord de la Haute-Volta). Thèse Doct Géol App Fac Sci Tech Univ Besançon, 165 pp

Sestini G (1971) Paleocurrents and exploration for gold in the Tarkwaian of Ghana. 15th Ann Rep Res Inst Afr Geol Univ Leeds, pp 23-25

Seyler M, Debat P, Diallo DP, N'Gom PM, Rollet M (1984) La Série de Mako dans ses parties centrale et méridionale (Sénégal oriental, Afrique de l'Ouest). Précisions sur l'évolution de la série volcano-sédimentaire et données géochimiques préliminaires sur les formations magmatiques post-tectoniques. J Afr Earth Sci 2 (1): 71-79

Sondo D (1986) Les minéralisations en or de Bouroum-Sud (NE Burkina-Faso, Afrique occidentale) dans leur contexte géologique et structural: approche métallogénique. Thèse Doct 3ème cycle, Univ Pierre & Marie Curie, Paris, 185 pp

Sougy J (1951) Rapport Annuel. Serv Géol AOF, Dakar, pp 55-57

Sougy J (1952a) Rapport Annuel. Serv Géol AOF, Dakar, pp 22-26

Sougy J (1952b) Rapport de reconnaissance sur le minerai de fer des Sfariats (Mauritanie). Rapp Dir Fédér Géol Afr Occ Fr, Dakar, 55 pp (unpubl)

Sougy J (1960) Les séries précambriennes de la Mauritanie nord-orientale (AOF). Rept 21th Int Géol Congr Norden, Copenhague, part IX, sect 9, pp 59-68

Spindler JP (1958) Exploration géologique des bordures N et E de la Kediat Idjil. Rapp inédit, 39 pp

Tagini B (1960) Hypothèses nouvelles pour une esquisse structurale du SE de la Côte d'Ivoire. Rapport Sodemi, Abidjan, 180 pp (unpubl)

Tagini B (1971) Esquisse structurale de la Côte d'Ivoire. Essai de géotectonique régionale. Thèse Univ Lausanne, Sodemi, 302 pp

Touré S, Caen-Vachette M, Tempier P (1987) Nouvelles données pétrographiques, géochimiques et géochronologiques du massif granitique de Boundoukou (Côte d'Ivoire). Mise en évidence d'un âge Burkinien, par isochrone Rb/Sr sur roches totales. J Afr Earth Sci 6: 269-274

Vachette M (1964) Essai de synthèses des déterminations d'âges radiométriques de formations cristallines de l'Ouest Africain (Côte d'Ivoire), Mauritanie, Niger). Ann Fac Sci Univ Clermont-Ferrand 25: 7-29

Vachette M, Bronner G (1975) Ages radiométriques Rb/Sr de 2900 et 2700 Ma des séries précambriennes de l'Amsaga et du Tiris, dorsale Reguibat (Mauritanie). 7ème Coll Géol Afr, Florence (abstr). Trav Lab Sci Terre St-Jérôme, Marseille, (B) 11: 147-148

Vachette M, Rocci G, Sougy J, Caron JP, Marchand J, Tempier C (1973) Ages radiométriques Rb/Sr, de 2000 à 1700 Ma, des séries métamorphiques et granites intrusifs précambriens de la partie N et NE de la Dorsale Reguibat (Mauritanie). 7ème Coll Géol Afr, Florence. Trav Lab Sci Terre St-Jérôme, Marseille 11: 142-143

Valero J, Bouchez JL, Debat P et al. (1985) Mise en place et évolution du granite de Saraya au sein des formations birrimiennes de la fenêtre de Kédougou-Kayes. 13ème Coll Géol Afr, St Andrews, pp 30-31

Vitali G (1970) Contribution à l'étude géologique du Birrimien. Stratigraphie et structure du complexe volcano-sédimentaire de Fetekro. Note BRGM, Paris, 26 pp

Wenmenga V (1986) Pétrologie des ensembles lithologiques du Protérozoïque Inférieur au Nord-Est de Ouagadougou (Burkina-Faso, craton Ouest-Africain). Thèse Doct Univ Clermont-Ferrand, 275 pp

Williams HR (1978) The Archaean geology of Sierra Leone. Precambrian Res 6: 251-268

Williams HR, Culver SJ (1988) Structural terranes and their relationships in Sierra Leone. J Afr Earth Sci 7 (2): 473-477

Wilson NW (1965) Geology and minerals reserves of part of the Gola Forest, south eastern Sierra Leone. Sierra Leone Geol Surv Bull 4, 102 pp

Witschard F (1965) Contribution à l'étude géologique, pétrographique et métallogénique des massifs granitiques du Sénégal oriental. Thèse Univ Lausanne, Mém BRGM, Fr 44, 172 pp

Wright JB, Hastings DA, Jones WB, Williams HR (1985) Geology and mineral resources of West Africa. Allen and Unwin, Lond, 187 pp

Yacé I (1976) Le volcanisme éburnéen dans les parties centrale et méridionale de la chaîne précambrienne de Fettekro en Côte d'Ivoire. Thèse Univ Abidjan, 141 pp, Annexes

Zonou S (1987) Les formations leptyno-amphiboliques et le complexe volcanique et volcano-sédimentaire du Protérozoïque inférieur de Bouroum-Nord (Burkina-Faso, Afrique de l'Ouest). Thèse Univ Nancy I, 294 pp

Zonou S, Karche JP, Lapierre H, Lemoine S, Rossy M (1985) Volcanismes tholéïtique et calco-alcalin dans les formations du Birrimien supérieur de Bouroum (NE du Burkina Faso). CR Acad Sci, Paris 301: 309-314

Chapter 4

Cover Sequences of the West African Craton

Janine Bertrand-Sarfati,[1] Alexis Moussine-Pouchkine,[1] Pascal Affaton,[2] Roland Trompette,[3] Yves Bellion[4]

The West African craton, as defined by Kennedy (1964), represents a very large area stabilized after the Eburnean orogeny, 2150 ± 150 Ma ago (Black and Fabre 1983, Bessoles 1983). These old rocks outcrop today in two separate shields: the Reguibat Shield and the Man-Leo Shield. The cover sequences, beginning after a long period of erosion, appear in roughly concentric belts, in the Taoudenni basin, between the two shields. Younger deposits are found in the eastern part of the craton toward the Pan-African suture zone.

Two distinct methods are used to subdivide the cover sequences: (1) structural, distinguishing cratonic sedimentation areas from pericratonic basins and (2) stratigraphical.

Specific sedimentation features have been described in the Proterozoic cover of the Taoudenni basin (Bertrand-Sarfati and Moussine-Pouchkine 1988), defined as a cratonic sedimentation area. The area is characterized by a wide surface, almost flat, involving a rapid response of the whole area to eustatic movements. The sedimentary sequences indicate a very great lateral continuity of the facies, together with generally low sedimentation rates. These conditions have been more or less recognized throughout the geological history of the craton (Bertrand-Sarfati et al. 1977). The cratonic cover is subdivided according to stratigraphical features.

Toward the edges of the cratonic area there are several pericratonic basins, related to orogenic events occurring outside the craton itself. In contrast to cratonic areas, they are characterized by a smaller size, a thicker sediment sequence and higher subsidence rates. They are: the Gourma basin and the Volta basin, relative to the Pan-African orogeny, and the Tindouf basin, relative to the Hercynian orogeny.

1 Cratonic Sedimentation on the Taoudenni Basin

The Taoudenni cover has been subdivided into five main sedimentary sequences separated by discontinuity surfaces. These discontinuities correspond to important tectonic events, correlating with major tectonic events beyond the craton.

Sequence 1. This sequence is the first to be deposited over the post-Eburnean surface. It is considered to be Middle to Late Proterozoic, according to biostratigraphy and isotopic dating.

Sequence 2. This sequence follows a discontinuity surface which corresponds to the Pan-African collision, during which the cratonic area was tilted, slightly folded, faulted and eroded. Sedimentation began with a widespread glacigenic event. The age of these rocks is poorly known; brachiopods of late Cambrian-Ordovician affinities are found in the uppermost beds. Shelly fossils newly found in the southwestern part of the Taoudenni basin have been referred to as Early Cambrian.

Sequence 3. The basal discontinuity corresponds probably to Taconic movements. Sedimentation (Ordovician-Silurian) began with widespread glacigenic deposits of Ordovician age and continued with Late Ordovician and Silurian deposits.

Sequence 4. The disconformity surface is evidenced by the fact that part of the Silurian and Early Devonian rocks are lacking. It may correspond to Caledonian movements elsewhere. The marine sequence (Devonian-Carboniferous) begins with Emsian transgression, continuing up to the Namurian. Con-

[1] Centre Géologique et Géophysique, CNRS U. P. 361, USTL, 34060, Montpellier, France
[2] Fac. des Sciences, Département de Géologie, Université du Bénin, Lomé, Togo
[3] CNRS U. A. 132, Univ. de St Jérôme, AIX-Marseille III, 13397, Marseille, France
[4] Fac. des Sciences, Département de Géologie, Fac. Sc. 84000 Avignon-France.

tinental red beds characterize the upper part of the Paleozoic.

Sequence 5. Unconformable surfaces resulted from the Hercynian orogeny when folding and faulting took place, especially along the western, northern and eastern borders of the Taoudenni basin (Mauritanides, Morocco and Ougarta). Meso-Cenozoic continental sedimentation continued on the craton, with very thin deposits, except when graben and trough are filled with thick Mesozoic deposits, in the Timétrine-Tilemsi and Néma-Nara areas.

Sequence 1: Middle to Late Proterozoic Sequences

The Proterozoic sequence crops out as a continuous belt along the northern border of the Taoudenni basin, resting on the Reguibat Shield from Atar-Mauritania to Grizim-Algeria (Fig. 1). Along the southern border, from south of Senegal to Burkina Faso it lies on the Man Shield. Toward the east and the west, only scattered outcrops are known. Sismic prospection and drill holes in the central Taoudenni basin provide data about the thickness and extension of the sequences (Beicip 1980, 1981).

Eburnean orogeny has made up the cratonic area. Between the last orogenic events (2150 ± 150 Ma) and the beginning of the sedimentation (before 1000 ± 100 Ma), a major erosional phase peneplaned the whole area, which was probably more extant toward the West and the East.

Description of the Sequence (Supergroup 1). It has been subdivided into three groups according to li-

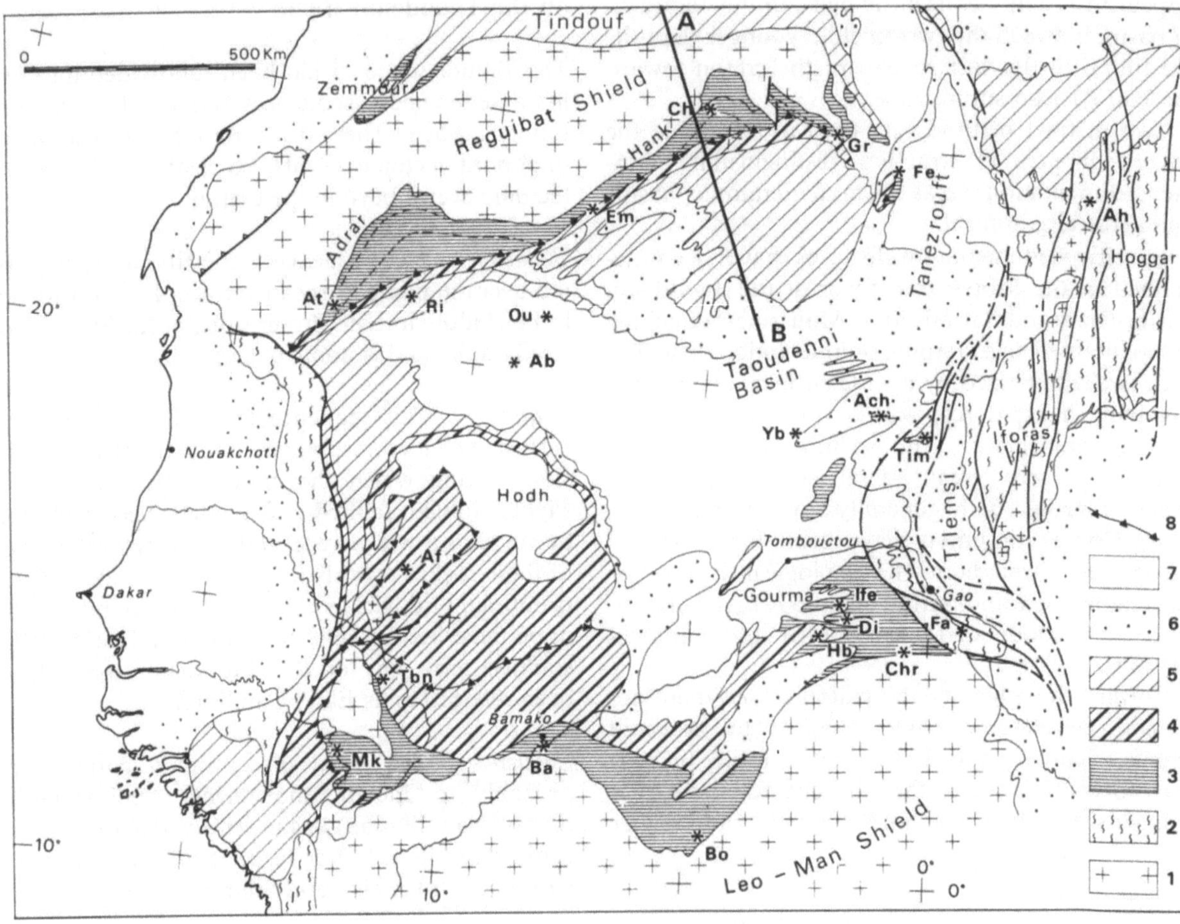

Fig. 1. Geological map of western Africa: *1* Eburnean basement; *2* mobile belts (Pan-African); *3* Sequence 1, Middle-late Proterozoic; *4* Sequence 2, Cambrian-Ordovician; *5* Sequence 3 and 4, Late Ordovician to Silurian and Devonian to Carboniferous; *6* Sequence 5, Mesozoic; *7* Quaternary; *8* post Pan-African tillite (from Bertrand-Sarfati et al. 1987). Sections: north and east of Taoudenni basin: *At* Atar; *Ri* Richat; *Em* El Mreiti; *Ch* Chegga; *Gr* Grizim; *Fe* Fersiga; *Ab* Abolag; *Ou* Ouassa; *Yb* Yarba; *Ach* Achaïkar. South of Taoudenni basin: *Af* Afollé; *Mk* Madina Kouta; *Tbn* Tambaoura nord; *Ba* Bamako; *Bo* Bobo Dioulasso. Gourma pericratonic basin: *Hb* Hombori; *Di* Dimamou; *Ife* In Fereren; *Chr* Christine; *Fa* Fafa; *Tim* Timetrine. Hoggar Pan-African fold belt: *Ah* inlier which displays a cratonic section, with stromatolites

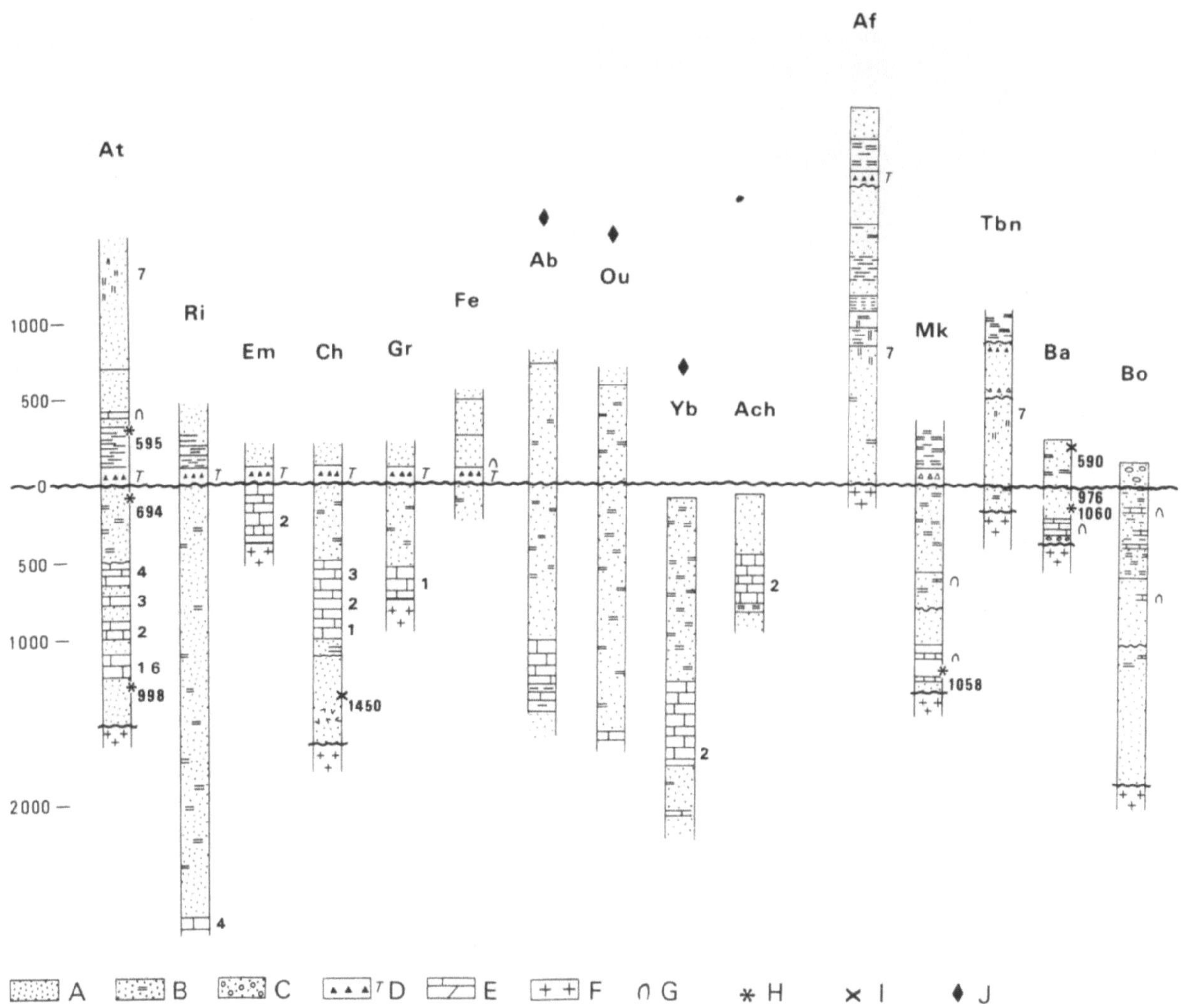

Fig. 2. Stratigraphic chart of Sequences 1 and 2, separated by the Pan-African discontinuity surface; north and east of the Taoudenni basin and south of the Taoudenni basin (see Fig. 1 for section names; from Bertrand-Sarfati et al. 1987). *A* sandstones; *B* shales-siltstones; *C* red molasses; *D* glacigenic deposits; *E* carbonates; *F* basement; *G* stromatolites; *H* Rb/Sr ages on clay; *I* K/Ar on glauconite; *J* boreholes. Paleontology: Middle-Late Proterozoic stromatolites: *1 Conophyton; 2 Inzeria* and *Serizia; 3 Jurusania; 4 Tifounkeia;* others *5* carbonate algae (Vendian); *6* Acritarches; *7* Skolithos

thology (Trompette 1973). The groups described in Atar have been traced with slight variations over the northern edge of the Taoudenni basin, photogeologically and in the field (Gevin 1960; Buffières et al. 1965, 1966; Villemur 1967; Deynoux 1971; Deynoux et al. 1971; Bertrand-Sarfati 1972; Marchand et al. 1972; Simon et al. 1979). Along the southwestern edge, the sedimentary cover is mainly siliciclastic and the different groups defined along the northern edge of Taoudenni, cannot be recognized (Fig. 2). Correlations have recently been proposed, between the cratonic cover sequences and the sedimentary and tectonic sequences within the Pan-African orogenic belt (Bertrand-Sarfati et al. 1987). Tentative paleogeographic maps of the Taoudenni cratonic area (Bertrand-Sarfati and Moussine-Pouchine 1988; Fig. 3) have been proposed, using sedimentological subdivisions based on sequence boundaries as time-lines (Christie-Blick et al. 1988).

1. The Char Group, marine, is mainly siliciclastic, sandstones, siltstones and mudstones, with very rare laminated stromatolitic carbonates with evaporites. An isotopic age, Rb/Sr on clay minerals, is evaluated about 998 ± 32 Ma, in Atar (Clauer 1976). The Middle Proterozoic pole, reported from the Char Group (I-2) "has no correlative in the African paleomagnetic data set" (Perrin 1987; Perrin et al. 1988). Toward the East, this group is recorded as

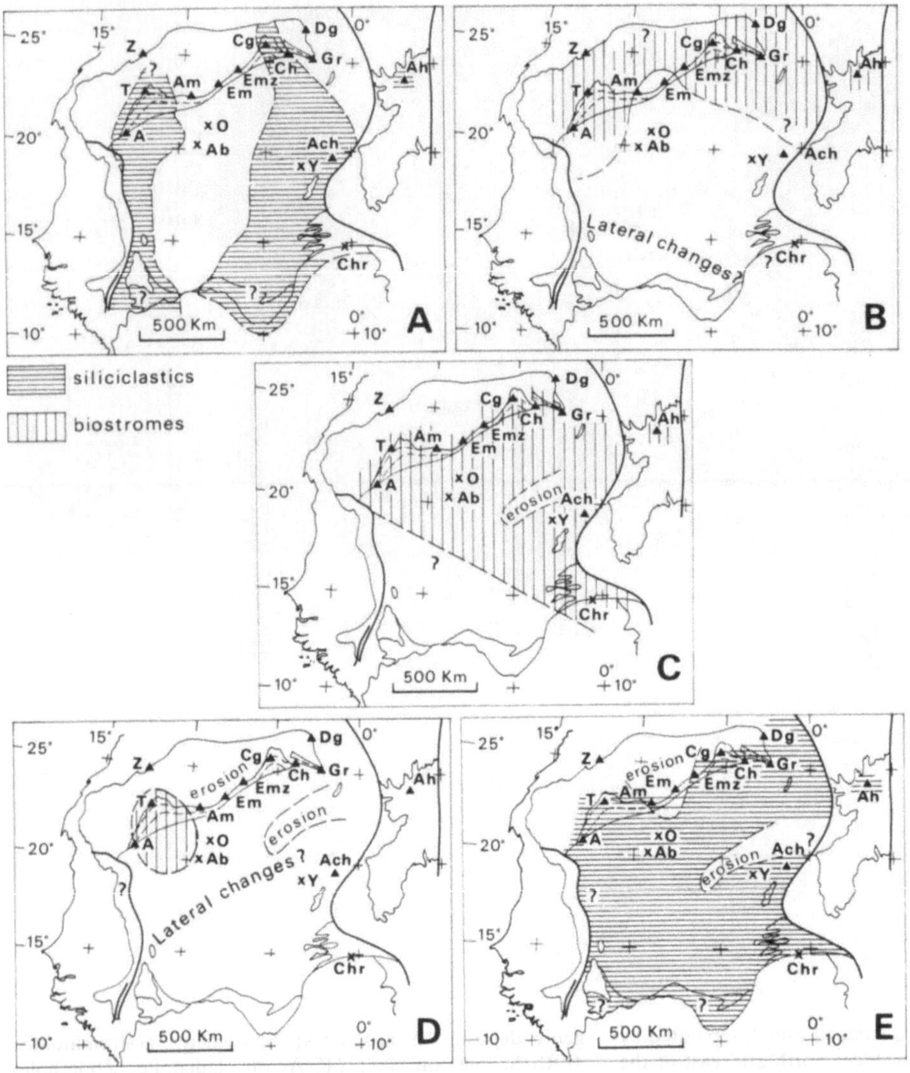

Fig. 3. Paleogeographic maps of Sequence 1, on the Taoudenni basin. *A* Char Group; *B, C, D,* three main sequences of the Atar Group; *E* Assabet el Hassiane Group. *Ah* Ahnet figures the cratonic-type sequence found enclosed in the Pan-African domain (Bertrand-Sarfati and Moussine-Pouchkine 1988)

Groupe de Douik (Chegga) and disappears toward Chenachane. In the West, siliciclastics are found at the base of the Achaïkar Group but correlations are tentative. Toward the southwest of the Taoudenni craton, the entire cover is siliciclastic (Bassot 1966; Bassot et al. 1981; Ouedraogo 1981; Villeneuve 1984; Keita 1984) and the correlation with the Char Group relays on some isotopic datings: 1058 Ma, Rb/Sr ages on clay (Bassot et al. 1963), 990 Ma, K/Ar on glauconite (Bozhko et al. 1971) and 1060 Ma, also on clay minerals (Clauer 1976). The central part of the basin according to subsurface data is free of sediment (Beicip 1980).

2. The Atar Group comprises shales and siltstones and abundant carbonates with columnar stromatolites in biostromes. Radiometric data, Rb/Sr on clay minerals (Clauer 1976; Clauer et al. 1982), give ages ranging from 890±35 to 775±52 Ma. In Atar, paleomagnetic data (Perrin 1987) indicate a low latitude position for the West African craton. The stromatolite biostromes indicative of Upper Riphean (Bertrand-Sarfati and Raaben 1970; Amard 1986) have been traced all along the northern Taoudenni (El Mreiti Group), up to Grizim, on the eastern margin in the Gourma region, they are known from bore-holes and from outcrop (Achaïkar Group; Bertrand-Sarfati and Moussine-Pouchkine 1978 a). These biostromes have been interpret-

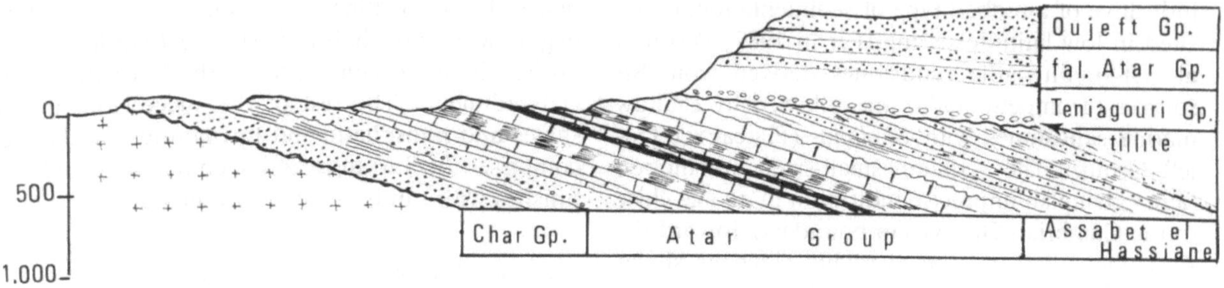

Fig. 4. Schematic stratigraphical cross-section of Sequences 1 and 2 in the Atar area, illustrating the erosional surface underlying the glacigenic (tillite) deposits (Perrin et al. 1988)

ed as corresponding to eustatic events which innundated the area, inducing the widespread growth of stromatolite biostromes, thus usable as time-lines. Therefore, three paleogeographic maps are proposed. The southwestern Taoudenni basin is not correlated on the maps, because of the lack of comparable stromatolite biostromes. The first map begins with a Conophyton biostrome. Except for a promontory in the El Mzereb- El Mreiti area, the northern part of the craton was covered, comprising the Reguibat Shield up to Zemmour (Fig. 3b). The second sequence starts with a biostrome of columnar stromatolites with a tussocky microstructure. It corresponds to the maximum extent of the depositional area (Fig. 3c). The third sequence is restricted to the Atar area (Fig. 3d). The same succession has been recorded from inliers within the Pan-African orogenic belt of Hoggar (Amasine Group, Ahnet area; Moussine-Pouchkine et al. 1988; Fig. 1).

3. The Assabet el Hassiane Group, marine siliciclastic, lying slightly unconformably on the Atar Group, is relatively much thicker (Bronner et al. 1980). It reaches 3000 m in the Richat area. These rocks can be traced, toward the east of the northern Taoudenni border, except where they have been eroded (Cheikhia-Bir Amrane Groups). No correlations are drawn with the mainly siliciclastic formations of the south of the Taoudenni basin (Fig. 3e). It is possible, however, that the Assabet el Hassiane Group and equivalents covered a maximum surface on the Taoudenni craton. It had been suggested (Bertrand-Sarfati et al. 1987) that these rocks may correspond to deposition of orogenic volcano-clastic rocks in the Pan-African fold belt of the Hoggar. These rocks are dated between 836±4 and 635±5 Ma (Caby et al. 1982)

Sequence 2: Cambro-Ordovician

Pan-African orogeny affected the craton during the collision that formed the Pan-African fold belt. There was diverse thrusting and folding in front of the suture situated east of the craton. On the cratonic area, these tectonic events resulted in tilting, faulting and folding of Sequence 1, followed by an erosion that partly or totally removed the cover (Fig. 4). The best examples are Foum el Alba, (Beicip 1980), Chenachane area (Buffières et al. 1965), Adrar of Mauritanie, (Trompette 1973) and southern Mali (Bassot et al. 1981). This surface results from the Pan-African collisional phase between 620 and 580 Ma (Caby 1978; Black et al. 1979; Caby and Andréopoulos-Renaud 1989).

Description of Sequence 2. Sequence 2 is well known because of the presence of widespread glacigenic deposits. The age of this major glacial event has been bracketed between 630 and 595 Ma, in the Atar area (Clauer and Deynoux 1987). North of the Taoudenni basin these rocks, overlying the post Pan-African erosional surface, (Zimmerman 1960; Biju-Duval and Gariel 1969; Deynoux 1980; Deynoux et al. 1985) are represented by a layer (0 to 50 m) of argilaceous glacigenic sediments, the tillite s. s., overlain by a thin veneer of carbonates with baryte (Djbeliat Group). In the south and southwest, these deposits rest over a sedimentary unit (150 m or more) of sandy-argillaceous subglacial sediments (Bakoye Group). In the Affolé-Hodh area, the sediments are mainly periglacial and eolian sandstones (Deynoux et al. 1989; Proust et al. 1990), reaching 1000 m (Fig. 2). Transportation of the glacigenic sediments was from north to south; the ice sheet was centered in the north of the Taoudenni basin. In the Madina Kouta area, SW of the basin, (Figs. 1 and 2) the tillite is overlain by carbonates with shelly fossils of Early Cambrian age (Culver et al. 1988a, b).

In the Atar area, on top of the tillite s. s., green marine siliceous shales (Teniagouri Group), have given a Rb/Sr age on clay minerals of 595±43 Ma (Clauer 1976). Above these, the Falaise d'Atar Group comprises mainly sandstones and red shales indicative of a redbed type of sedimentation, deposited in low latitude setting (Perrin 1987; Perrin et al. 1988). Fluviatile sandstones derived from SE sources (Trompette 1973) are followed by shallow marine sandstones with *Skolithos* (Plateaux d'Oujeft Group). They contain small-sized brachiopod fauna giving an imprecise indication of an age close to the Cambrian-Ordovician boundary. Toward the East, the tillite and the overlying siliceous shales (Azlaf Group) are followed by 500 to 1000 m of sandstones (Kreb en Naga Group). Higher up, a conglomeratic episode, the Sobti Conglomerate, may be interpreted as a sequence boundary at the base of the Erg Chech Group (Villemur 1967; Fabre 1983, 1988). The Sobti Conglomerate disappears toward the West; under recent cover but toward the East, it displays a remarkable continuity over 200 km (Beicip 1981). It overlies a very planar erosional surface with dreikanters which can be traced toward the nearby Hoggar, where it meets the infra-tassilian discontinuity surface (Beuf et al. 1971). The basal unit of the Erg Chech Group, sandstones with *Skolithos* can be correlated with the Plateaux d'Oujeft Group. Southwest of Taoudenni, glacigenic deposits are overlain by the Nara shales and the shales and siltstones of the Mali Group.

Sequence 3: Ordovician and Silurian

A disconformity is present toward the centre of the Taoudenni basin, grading to an angular unconformity toward the West (Dia et al. 1969), close to the Mauritanide mobile zone. The cratonic shield is uplifted by epeirogenic movements that may correspond to the Taconic phase. The same situation is envisaged by Beuf et al. (1971) for the discontinuity within the nearby cratonic area of the Hoggar.

Description of Sequence 3. Following these movements a wide ice sheet was installated with its center located on the south of the Hoggar, Pan-African fold belt. The glacigenic deposits are sandy (100 to 200 m thick), mostly glacial and periglacial (Beuf et al. 1961, Deynoux 1980). Sediments were transported from south to north or southeast to northwest.

Silurian rocks are known only in the region situated to the north and west of the Taoudenni basin. Eastward, the Silurian rocks are missing, either by non-deposition or by erosion (Beicip 1980). The best studied sections are in the Adrar of Mauritania (Monod 1952; Trompette 1973). There, a graptolitic fauna permitted recognition of all stages of the Silurian. Transgression, representing the glacio-eustatic phase due to melting of the Ordovician ice sheet, begins with silty shales of Middle Llandovery, deposited in sharp contact upon the Ordovician glacial sediments. The rest of the succession, mainly shales and phosphatic and ferruginous sandstones, comprises the Tarranon and Wenlock and probably finishes with sandy shales of Ludlovian.

Sequence 4: Devono-Carboniferous

The discontinuity surface reflects diverse stratigraphical lacunas. Silurian beds are thinner towards the central part of the Taoudenni basin. The Early Devonian rocks are partially absent and the first deposits correspond to the eustatic transgression of Emsian strata. The significance of that discontinuity is not well understood; it may be due to Caledonian movements.

Description of Sequence 4. Devonian rocks are known from the northern edge of the Taoudenni basin, and the petroleum exploration drill-holes have shown their presence in a WSW-ENE, elongated area, parallel to the southern border of the Reguibat Shield. Within this area they lie unconformably over Silurian beds to the West and Ordovician beds to the East (Villemur 1967; Trompette 1973; Fabre et al. 1976; Dia 1984). The oldest strata are Late Siegenian to Late Emsian. Although the Famennian is not recognized paleontologically, the Devonian strata seem to grade upward progressively to Carboniferous. The best known sections of Devonian rocks on the West African craton are from the Taoudenni region (Villemur 1967; Beicip 1980). Here, resting directly upon the glacial sandstones of Late Ordovician, one can see about 10 m of fine sandstones alternating with black shales, with an Emsian brachiopod fauna, grading to carbonate layers with Late Eifelian-Givetian faunas. The sequence continues with about 130 m of silty shales with upward-increasing sandstones layers. The rare brachiopod fauna gives only inacurate late Devonian ages. By drawing correlations with the Devonian basins to the East and to the North, the sandstone layers of this upper part are thought to be equivalent to the Strunian sandstones and correspond to a probable sea level fall (Fabre and Moussine-Pouchkine 1971). This assumption is reinforced by the fact that the overlying rocks are Early Carboniferous in age. The overall thickness of the Devonian shales is about 170 m. A paleomagnetic study of the Devonian rocks in Mauritania indicates that, for that period, the West African craton was situated under lower latitudes than was previously thought (Kent et al. 1984).

Carboniferous beds are restricted to the NE part of the Taoudenni basin, where they form a round-shaped basin, the southern part of which is covered by post-Carboniferous continental deposits (Villemur 1967). Conformable on the Devonian rocks, the sequence began by shales, interbedded with immature siltstones and sandstones (Bir en Naharat Formation, about 130 m) desposited in a shallow marine environment. Continental deposits are reported from the West. This formation is Late Tournaisian to Early Visean (Legrand-Blain 1985). After a hiatus corresponding to the Middle Visean, an important sea level rise leads to the deposition of Late Visean to Early Namurian carbonates (Safia and Aïcha Formations, 150 m). They contain a rich fauna of brachiopods, corals, pelecypods and foraminifers. They are followed by a thick sequence of continental red shales and channelized sandstones with lacustrine carbonates and evaporite layers, corresponding to the installation of a continental environment on the West African craton. The base of that formation contains numerous plant remains of Namurian age (Fabre 1967). Its upper part is thought to be Westphalian to Permian (Fabre et al. 1976). These red beds are 400 m thick in outcrops but they reach more than 1000 m in the centre of the basin, south of Taoudenni (subsurface data, Beicip 1980). Such a great thickness, especially when compared to that (800 m) which represents the whole sequence from Early Ordovician to Lower Namurian, is equivalent to that measured within the pericratonic Tindouf basin (Fig. 11).

Sequence 5: Meso-Cenozoic

The Hercynian orogeny strongly affected the Taoudenni area mainly during Permian (Fabre 1988). The Reguibat Shield merges out, giving rise to the present features of the so-called Taoudenni basin. On the craton the very last sediments affected by this orogeny are Late Carboniferous. From that time, the West African craton has remained emerged and only thin veneers of continental sediments, mainly white sandstones with silicified woods, are recorded, testifying to a total lack of subsidence. These deposits are not found east of the 3° W meridian. Except for pericratonic graben (Timétrine-Tilemsi) along the Pan-African suture and an intracratonic trough (Néma-Nara), the sediments on the craton will not be detailed.

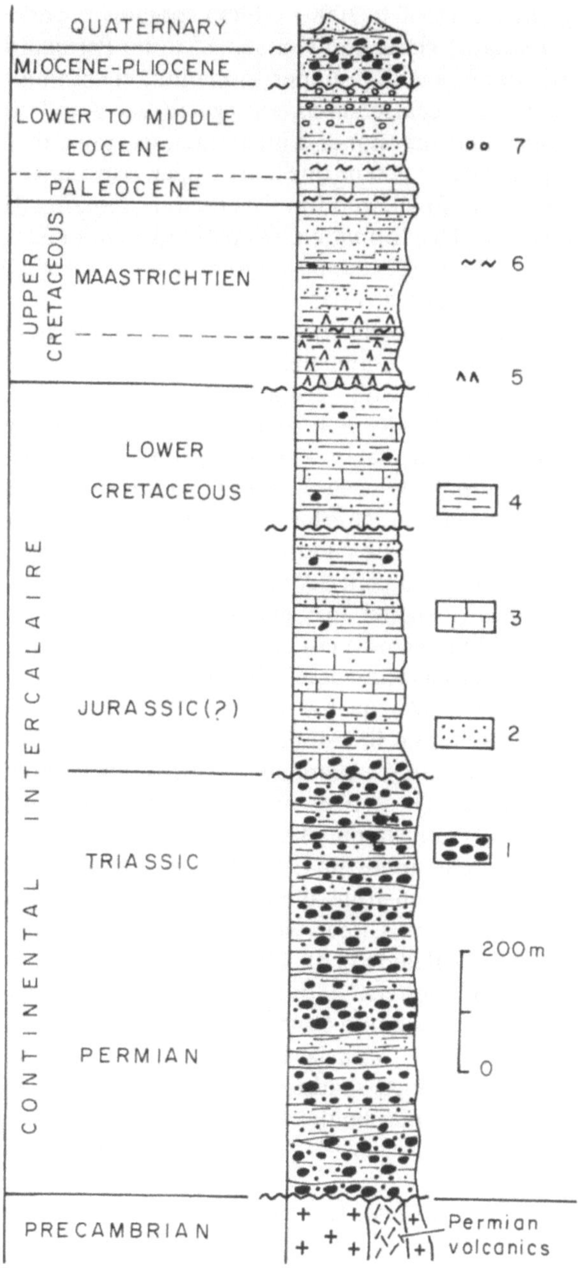

Fig. 5. Synthetic lithostratigraphic column of the eastern edge of the Taoudenni Basin (Timétrine-Tilemsi area). *1* Pebble, conglomerate; *2* sand, sandstone; *3* limestone, carbonates; *4* clay, shale; *5* gypsum; *6* Pelecypods; *7* oolith

The Timétrine-Tilemsi zone

In this zone (Fabre et al. 1982; Bellion 1987; Fig. 1), on the eastern edge of the basin, post-Carboniferous rocks outcrop from Tanezrouft to Gao, around a Proterozoic inlier which is intruded by late Permian (Liégeois et al. 1983) or perhaps more recent (Sauvage and Savard 1985) anorogenic syenite and carbonatite complexes.

The "Continental Intercalaire" (Kilian 1931) is composed of two formations separated by an unconformity (Fig. 5). The lower one, Tesoffi conglomerates (500 to 1000 m thick) contains reworked carbonatite clasts and is assumed to be Permian to Triassic in age. The upper formation, consisting of alternating coarse sands, variegated clays and siltstones and limestones with intraclasts, toward the top, is 500 to 600 m thick. Its upper part contains numerous fossil wood and vertebrate remains (dinosaurs). This formation, in comparison with Saharian sequences, is supposed to be Early Cretaceous in age and perhaps partly Jurassic.

A lagoonal-marine sequence (500 to 600 m thick) overlies the "Continental intercalaire". The lower part, begining with several gypsum-rich layers, is composed of silty and gypsiferous clays and fine sands with thin layers containing abundant Maastrichtian pollen, located about 60 m above the base of the sequence. Northward, intercalations of thin layers of limestone and gypsum increase in number (Ayme and Hubault 1955). This lower part of the sequence assigned to Cenomanian-Turonian may be of Campanian-Maastrichtian age, in comparison with Saharian sequences. The upper part of the lagoonal-marine sequence (Cornet 1948) consists of two fossiliferous limestones banks (Terrecht 1 and 2 of Monod 1939) separated by shelly clays, marls and sands. These beds are very widespread from north to south (Radier 1959). They are also present in boreholes more than 500 km to the southeast. East of the Timétrine, these strata have been dated from the Late Upper Maastrichtian to the Late Paleocene (Bellion et al. 1989). The Late Paleocene limestone (Terrecht 2) is overlain by shales with phosphatic sand and bone beds at the top, dated as Early Eocene (Monciardini 1966). The sequence ends with clayed sandstones and variegated clays with ferruginous oolitic layers (dated as Early to Middle Eocene, in Niger, Boudouresque et al. 1982).

The "Continental terminal" (Kilian 1931) overlies unconformably all the older beds. It comprises 20 to 40 m of non-folded clastics, red sandy clays, pebbles and quartz gravels with some remains of silicified fossil wood. It is probably Mio-Pliocene in age.

In this Timétrine-Tilemsi region, the Permian to Mid-Eocene sedimentary cover is folded and faulted. Dips of beds are generally low, less than 20°, except near the basement outcrops or the great lineaments, where they steepen abruptly (Cornet 1948; Karpoff 1960). In these areas, the beds are commonly overturned and the fractures are reverse faults. Such faults limit the western edge of the Timétrine Proterozoic basement, which rests above the overturned Mesozoic. Near the Algeria-Mali border, sediments are folded with N 20° to N 60° E-trending, fold axes. All these structures are the result of a post-Middle Eocene compressional tectonic event. A microtectonic analysis shows that the direction of shortening is oriented N 145° E (Bellion 1987; Bellion and Guiraud 1988).

The Néma-Nara Trough

In this area, toward the south, post Paleozoic continental sediments overlie the unfolded "Nara shales" (Chamley et al. 1980) and intrusive dolerites. These dolerites have been dated as Early to Middle-Lias (about 190-180 Ma, K/Ar; Bellion 1987). The 20- to 30-m-thick beds of fine siliciclastic found in the north thicken abruptly toward the south, as shown by geophysical data (Huot 1957). The trough, limited by N 70° E and N-S trending faults, is located on the northern edge of a great tectonic lineament cutting accross Africa from Guinea to the Red Sea (Guiraud et al. 1985). This diamond-shaped tectonic trough is filled with about 400 m of coarse clastics and dolerites at the base. These clastics are of inferred Middle-Late Jurassic age according to the fossil wood *Brachyoxylon* (Dars 1957). Therefore, the opening of the Nara pull-apart basin is interpreted as being the result of right-hand lateral movement of Africa, north of the Guinea-Nubia lineament, and related to the fast Central Atlantic spreading, begining after the Bathonian (Olivet et al. 1984). Between the Upper Cretaceous and the Present, several periods of ferruginization affected the deposits.

These ferruginized continental sediments are folded and faulted during two compressive tectonic events, responsible for reverse faults, strike-slip faults and associated folds (Bellion et al. 1984). Structural analysis has shown that the oldest shortening direction averages N 140° E, while the youngest trends are roughly N 30° E. In comparison with the microtectonic data of the Timétrine-Tilemsi area, the oldest tectonic episode corresponds to the post-Middle Eocene phase, well known in both North and West Africa (Bellion 1987; Guiraud et al. 1987). The second episode has not yet been dated.

The numerous Liassic dolerite sills and dykes in the northern and southern parts of the Taoudenni basin and the deformations in Timétrine-Tilemsi and Néma-Nara areas, are evidences of important post-Hercynian intraplate tectonics. Furthermore, the most deformed zones are located along a large tectonic lineament (Nema-Nara) or along the unstable Pan-African suture zone (Timétrine-Tilemsi).

2 The Pericratonic Basins

Situated along the outer edges of the cratonic area, the pericratonic basins are characterized by a smaller size than the cratonic basin, a more active subsidence and a thick sedimentary sequence.

The Gourma basin

The Gourma region (Fig. 1), located southwest of the Taoudenni basin and west of the Pan-African orogenic belt of Iforas, comprises two distinct structural units: an autochthonous part and an allochthonous part, thrust over the first one (NW-SE) and composed of successive thrust belts with distinct sediment sequences. The autochthonous Gourma, folded with increasing intensity toward the allochthonous belt, is the only part considered below. The overall thickness of the sedimentary sequence, filling the basin, is estimated to be more than 8000 m.

Sequence 1. The basal unit consists of a thin veneer of quartz sandstones, outcropping along the borders of the Man and Leo Shield (Grès de Base, Grès de Firgoun, Grès de Gassa), sometimes overlain by a thin carbonate unit. The Irma formation, overlying these basal sandstones, (Reichelt 1972; Fig. 6) contains, in the Christine borehole, stromatolites which are comparable to those from the nearby Achaïkar Group (Fig. 1) and are correlated to the Atar Group (Fig. 3c). These dolomites are overlain by the shaly-carbonate unit of Koro (Keita 1984), but their relationships with the overlying dolomites of Sarnyéré are not recognized.

The upper unit represents the filling of the Gourma trough. Toward the centre of the basin where the substratum is not known, Formation I comprises shales, coarse sandstones and conglomerates (Fig. 6). In the eastern, allochthonous part of the Gourma, in the Fafa area, a similar unit is disconformable over the Eburnean basement (Caby and Moussine-Pouchkine 1978; Sacko 1985). Formation II is essentially composed of red turbiditic shales and siltstones. The carbonate-rich Formation III displays a platform-to-basin facies evolution (Bertrand-Sarfati and Moussine-Pouchkine 1983a) from turbiditic limestones in the NE toward Dimamou, to the transgressive dolomites of Sarnyéré, corresponding to a tidal flat setting, in which diagenetic features testify to repeated emersion (Bertrand-Sarfati and Moussine-Pouchkine 1983b; Fig. 7). Higher up, marine sedimentation extends largely over the whole Gourma basin, with the carbonates of Massi Formation, the sandstones of Hombori Formation and the shales of Oualo Formation.

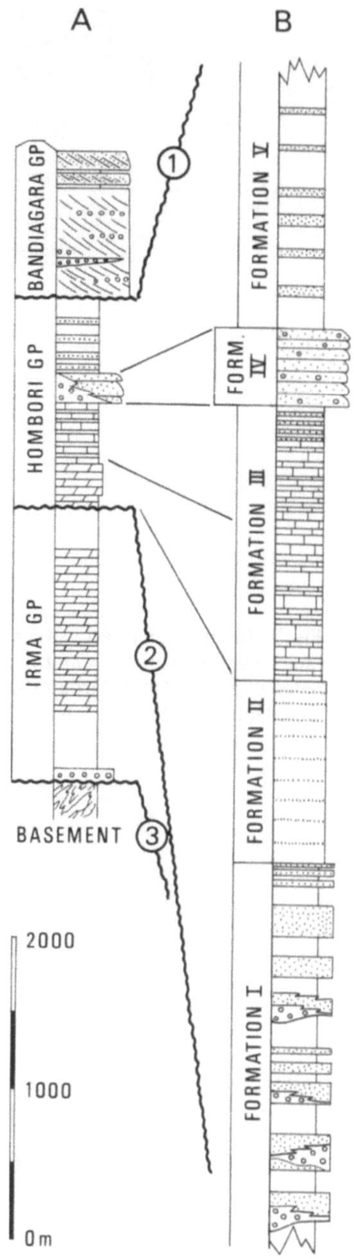

Fig. 6. Lithostratigraphic sections in the Gourma basin area. *A* Southern border (Hombori and Christine; Fig. 1); *B* center of the basin (in Fereren, Fig. 1). Discontinuities: *1* post-Pan-African discontinuity separating Sequence 1 and Sequence 2 (Bandiagara Group); *2* disconformable surface separating within sequence 1, the cratonic sediments from the sediments of the Gourma s. s.; *3* major disconformity on the Eburnean basement

In the Sarnyéré Formation, a carbonate Algae of uncertain affinities (Bertrand-Sarfati 1979) suggests a late Vendian to very Early Cambrian age. This

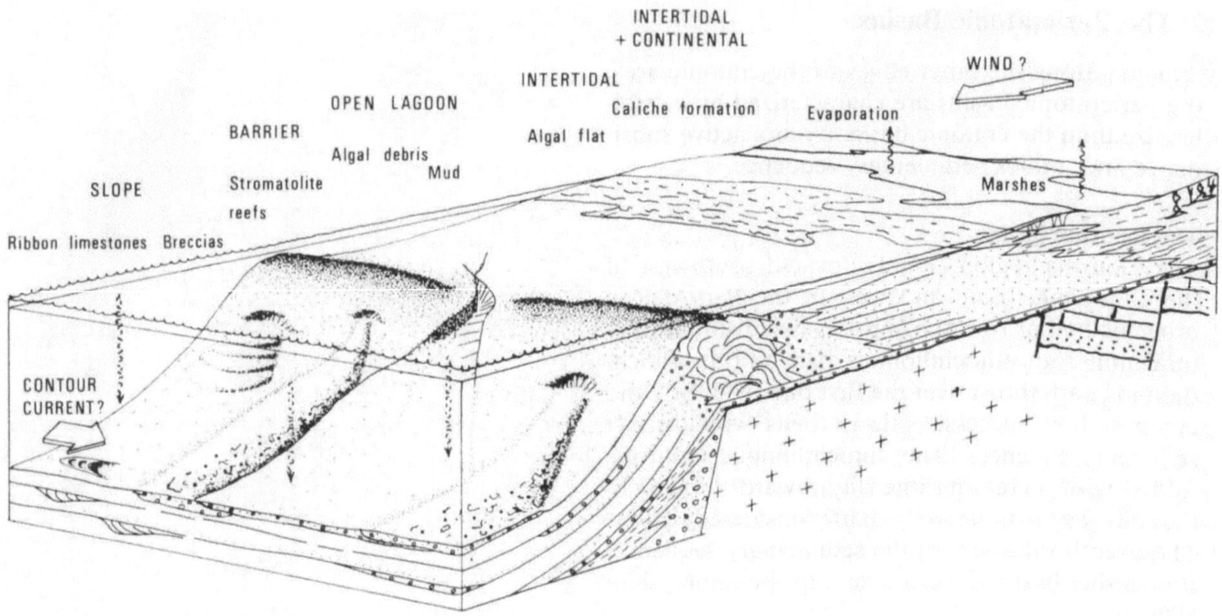

Fig. 7. Interpretative reconstruction of the platform-to-basin facies evolution in the Gourma: Formation III and carbonates of the Hombori Group (Sarnyéré formation) in Dimamou-Hombori area (see Fig. 6). Tilted and faulted blocks of cratonic-type sediments are inferred from bore-holes data (see Figs. 1, 2; Christine; from Bertrand-Sarfati and Moussine-Pouchkine 1983 a)

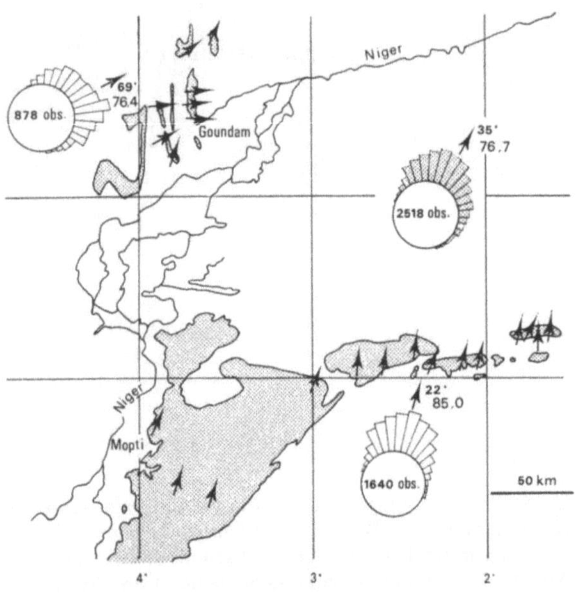

Fig. 8. Map of the paleocurrents in the sandstones of the Bandiagara Group (*arrows* are stations representing about 100 measures; from Moussine-Pouchkine et Bertrand-Sarfati 1978)

age could indicate that sedimentation in the Gourma basin and folding which affected the sediments may be late Pan-African.

Sequence 2. The discontinuity, corresponding to the post-Pan-African disconformity and recognized on the West African craton, separated, in the Gourma, the strongly folded sediments of the upper unit from the slightly folded Bandiagara Group. This discontinuity surface corresponds to that in the nearby Bamako area, which separates the Sadiola and Koutiala formations from the underlaying Sotuba Group (Bassot et al. 1981; Fig. 2).

The Bandiagara Group represents Sequence 2. It is composed of thick coarse fluviatile sandstones that are marine, with *Skolithos* in the upper part. Very thick in the north, 1000 m at Hombori, they are thinner toward the south (100 m). Paleocurrents directions indicate a definite slope to the NE (Moussine-Pouchkine and Bertrand-Sarfati 1978; Fig. 8). It is evident that this depression was related to the existence of the Gourma basin. The fluviatile character, the lack of glacial features and the south-north current directions, converging toward a narrow depression, prevent the drawing of a precise correlation with the nearby formations (especially the tillite s. s.). The northeastern part of the Bandiagara sandstones is slightly folded by a late Pan-African phase.

Structure. The relationships of the Gourma basin with the Pan-African orogeny have been variously interpreted, but always recognized (Reichelt 1972; Lesquer and Moussine-Pouchkine 1980; Moussine-Pouchkine and Bertrand-Sarfati 1978; Caby 1979; Sacko 1985).

Two geodynamic hypothesis can be used to explain the development of the Gourma basin, perpendicular to the Pan-African trends. The first one, argued by Lesquer and Moussine-Pouchkine (1980), considers Gourma as an aulacogen, linked to a failed arm of a triple junction, in relation to the rifting of the Pan-African domain, about 800 to 870 Ma (Black et al. 1979). The positive gravimetric anomaly found under the Gourma can, however, also be interpreted as a relict of ante-Pan-African history. The second hypothesis presents Gourma as a foreland basin (Caby 1987, 1989) developed during the Pan-African collision (around 620–580 Ma). Despite the indication of a possible Vendian age for the upper unit, the fact that sediments were not clearly derived from the nearby thrust zone and were turbiditic, but not molassic, suggests that further studies are necessary to confirm or reject the foreland basin hypothesis.

The Volta Basin

The Volta basin, which forms the western border of the Dahomeyide Pan-African belt, stretches from Ghana, Togo, Benin to Burkina Faso and Niger (Fig. 9). It rests with a major unconformity on the Eburnean basement (2000 Ma), the Man-Leo Shield.

Following the pioneering works of Junner and Hirst (1946), the studies of Saunders (1970), Sougy (1971) and Affaton (1975, 1987), have led to the recognition of the supergroups of Bombouaka at the base, Pendjari or Oti and Talamé at the top (Fig. 10), which are different from the earlier subdivisions: Lower, Middle and Upper Voltaian.

The Bombouaka Supergroup (about 1000 m) comprises the Dapaong and Panabako groups, which are dominantly sandstones and encased the clayey siliceous group of the "Fosse aux Lions" (Fig. 10). The fine and mature detritic sediments are deposited mostly on epicontinental marine platform (Drouet 1986). The erosional and cartographic unconformities observed between and within the groups, and numerous indices of emersion, suggest a waverer transgression and a slow and irregular subsidence. The upward increase in subsidence intensity is interpreted as increasing, though limited, instability. An Rb/Sr isotopic age on clay minerals of 993 ± 65 Ma have been calculated for the "Fosse aux Lions" formation (Clauer 1976; Clauer et al. 1982).

The Pendjari (or Oti) Supergroup (about 2500 m) rests with an erosional unconformity, partly of glacial origin, on different units of the Bombouaka Supergroup, and on the basement. The lower group, Kodjari Group, comprises the triade (tillite, s. s., carbonates with baryte and silexites), which is well known in the Taoudenni basin. The upper group, the Porga Group is composed by rhythmical shales and siltstones with phosphorite lenses at the base. The Pendjari Supergroup is made up of fine immature material (graywackes), deposited in relatively deep marine environment. An Rb/Sr isotopic age on clays, of 660 ± 9 Ma, has been given (Clauer 1976; Clauer et al. 1982). Biostratigraphic data, stromatolites and Acritarchs *(Chuaria)* have been published (Bozhko et al. 1974; Amard and Affaton 1984).

The Tamalé Supergroup (500 m) known only in Ghana, rests unconformably on the Pendjari Supergroup (Fig. 10). It compromises shales and siltstones of the Yendi Group and the coarse molassic sandstones of the Tamalé Group, beginning with a diamictite and partly glacial conglomerates. Sediment sources were in the East, i. e. in the Dahomeyides and the deposition centers migrate westward later on.

The Volta basin appears as composite, formed by superposition of a cratonic basin, a probable passive margin and a foreland basin. The first one was probably formed between 1000 and 660 Ma, the second between 660 and 600 Ma. The age of the third one is still a matter of controversy; Late Proterozoic for some people, Early Paleozoic for others.

Structure. The present-day Volta basin is only part of a larger basin, incorporated to the east to the Dahomeyides. The synthesis of the available geophysical data (Affaton 1987) gives information about the paleotopography and the structure of the basement. The paleosurface (Ako and Murray 1983; Ako and Wellman 1985) is located at less than 3 km depth, in the West and North and 5 to 7 km in the southeastern parts of the Volta basin. The structure is that of a disymmetrical basin with a gentle eastward slope and a steep westward slope. Paleoreliefs and paleodepressions have their axis subparallel to the Eburnean structures of the nearby Man-Leo Shield. Basic sills and intrusions have been recognized together with normal faults (reverse faults are rare) resulting in horst and graben which induce a progressive thickening of the Bombouaka and Pendjari Supergroups toward the east. Some of

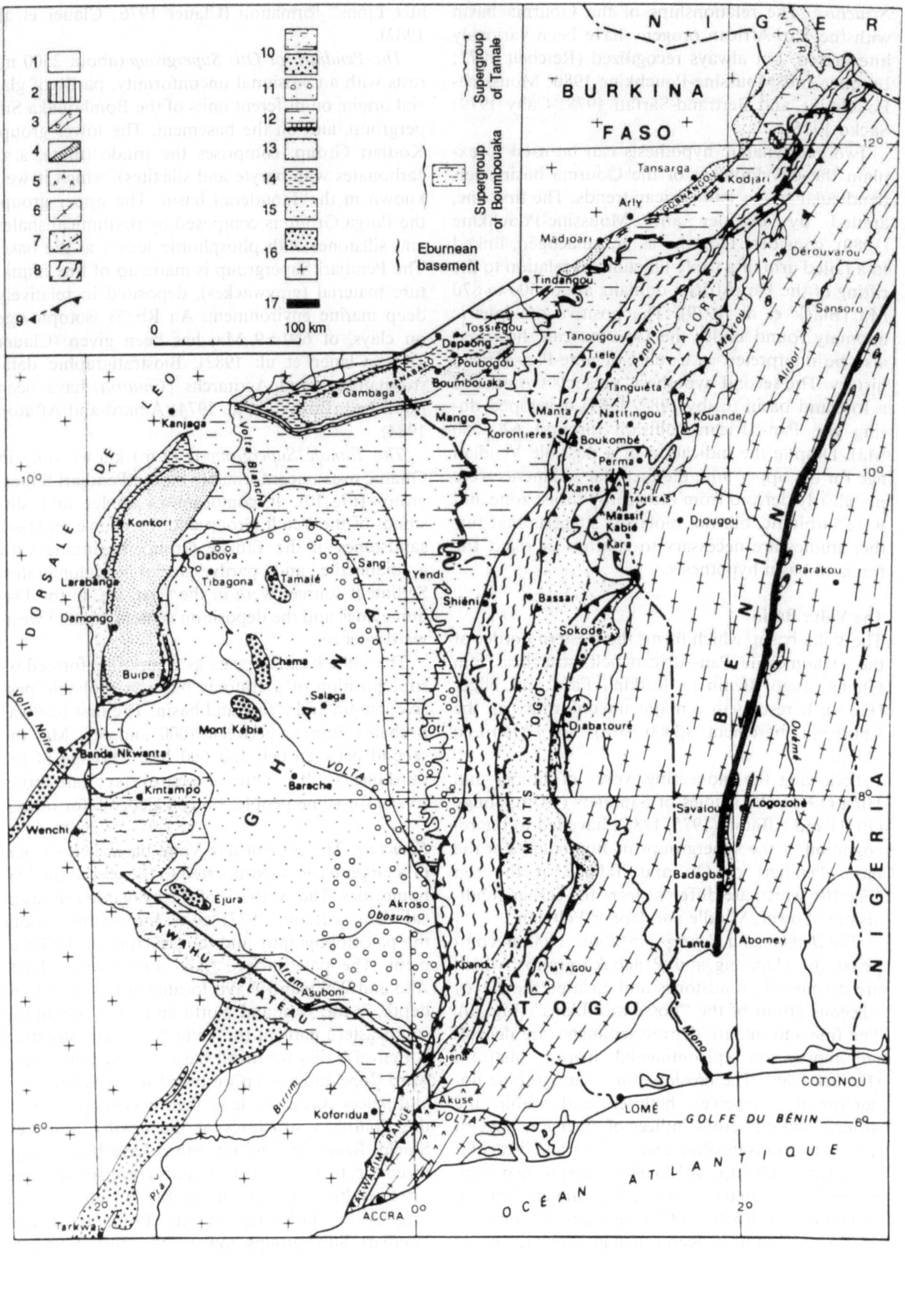
1
2
3
4
5
6
7
8
9
10
11
12
13
14
15
16
17
Supergroup of Tamale
Supergroup of Boumbouaka
Eburnean basement
0 100 km
NIGER
BURKINA FASO
BENIN
NIGERIA
GHANA
TOGO
DORSALE DE LÉO
MONTS TOGO
ATACORA
KWAHU PLATEAU
AKWAPIM RANGE
GOLFE DU BÉNIN
OCÉAN ATLANTIQUE
Tansarga
Kodjari
Arly
Madjoari
Dérouvarou
Tindangou
Sansoro
Tossiégou
Dapaong
Poubogou
Boumbouaka
Tanougou
Tiele
Tanguiéta
Natitingou
Kouandé
Gambaga
Mango
Manta
Korontières
Boukombé
Perma
Kanjaga
Kanté
TANEKAS
Massif Kabié
Djougou
Kara
Konkori
Daboya
Tibagona
Sang
Tamalé
Yendi
Larabanga
Damongo
Shiéni
Bassar
Parakou
Sokodé
Chama
Buipe
Salaga
Mont Kébia
Djabatouré
Banda Nkwanta
Kintampo
Barache
VOLTA
Oti
Volta Noire
Wenchi
Savalou
Logozohé
Badagba
Ejura
Akroso
Obosum
Kpandi
Agou
Asuboni
Lanta
Abomey
Ajena
Akuse
Koforidua
LOMÉ
COTONOU
Mono
Ouémé
Birrim
Pra
ACCRA
Tarkwa
12°
10°
8°
6°
0°
2°

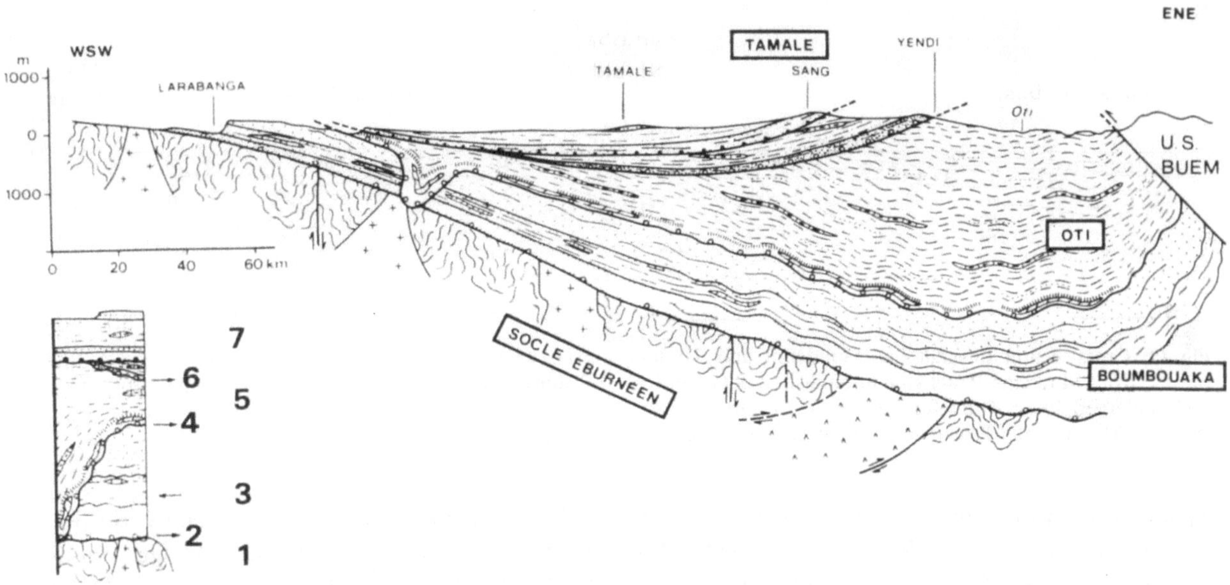

Fig. 10. Schematic cross-section of the Volta basin and generalized stratigraphical section. *1* Eburnean basement; *2* major discontinuity; *3* Bombouaka Supergroup, 993 ± 65 Ma; *4* disconformable surface, *Pro parte* glacial; *5* Pendjari (Oti) Supergroup, 660 ± 9 Ma; *6* disconformable surface; *7* Tamalé Supergroup

these fractures are restricted to the basement and the base of the sedimentary sequences; they are probably contemporaneous to the first sediment deposition and the late structuration of the basin. The basin is folded in the east toward the Dahomeyide.

Geophysical and stratigraphical data permit the distinguishing of three stages in the geodynamic evolution of the basin. First the deposition, on the eastern margin of the West African plate, of an epicontinental siliciclastic sequence, thickening eastward. Then a sedimentary wedge, also thickening eastward toward a Pan-African paleo-ocean which has entirely deasappeared to-day, but is recognized further north in the Hoggar. Tholeiitic-to-transitional (Affaton 1987) subaqueous volcanism occurred, however, during the ocean opening phase, in the eastern part of the basin, folded and metamorphosed. It constitutes the relicts of this Pan-African oceanization. The third stage is the constitution of the Pan-African Dahomeyide fold belt, around 600 Ma, and its conjunction with the eastern part of the basin. The erosion of the uplifted Pan-African belt supplies sediment to the foreland basin, the depocenter of which migrates toward the West African craton. Therefore, the Volta basin records precisely the history of the adjacent Dahomeyide belt which consists in a cratonic sedimentation, oceanization and subduction phase followed by a collision. The history of the Volta basin, up to the glacial event at the end of the Proterozoic, can be correlated with that of the cratonic basin of Taoudenni (Trompette 1973; Deynoux et al. 1985), whereas its late history is dominated by the Dahomeyide orogenic events.

◁

Fig. 9. Geological map of the Volta basin and the external units of the Pan-African belt of the Dahomeyides, according to Sougy (1971) and Affaton (1987). *Inset*: Internal units of Dahomeyides: *1* Post-Paleozoic sediments; *2* Devonian-Carboniferous; *3* quartzites with kyanite; *4* volcano-sedimentary formations; *5* basic and ultrabasic series; *6* schist, gneiss, migmatite, amphibolite, granite; *7* structural unit of the Atacora; *8* structural unit of the Buem; *9* thrust; *10* Keba Group; *11* Yendi Group; *12* Pendjari Supergroup; *13* Panabako Group; *14* Fosse-aux-lions Group; *15* Dapaong Group; *16* Eburnean basement; *17* Ultramylonites

The Tindouf basin

Tindouf basin lies north of the Reguibat shield, the northern part of the Taoudenni basin (Fig. 11). Toward the east it grades to the Reggane basin, now under a Quaternary cover of sand dunes. The present structuration of the Tindouf basin results from the Hercynian tectonic, stronger on its north and western edges, along the Anti-Atlas orogenic belt (Morocco). The north and south flank of the basin are very different: sedimentation began much earlier in the north, whereas the first recorded sediments in the southern flank were of Tremadoc-Arenig ages (Destombes et al. 1969).

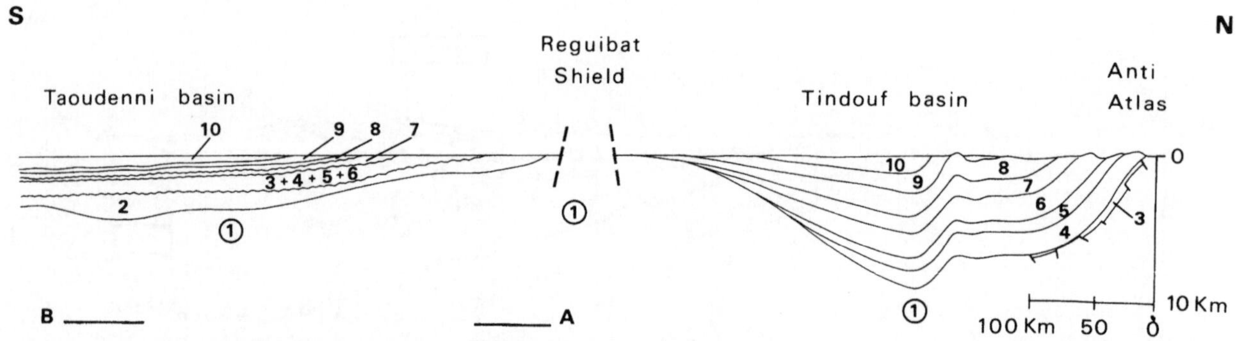

Fig. 11. Cross-section through the Taoudenni and Tindouf basins; from Moussine-Pouchkine unpublished data (vertical scale enlarged 10×; *A–B* refers to Fig. 1). *1* Eburnean basement; *2* Middle-Late Proterozoic; *3* Latest Proterozoic (P. III); *4* Early Cambrian; *5* Middle Cambrian; *6* Ordovician; *7* Silurian and Early Devonian; *8* Middle-Late Devonian; *9* Early Carboniferous; *10* Late Carboniferous

Sequence 1. In the Anti-Atlas area, the sequence is strongly affected by the Pan-African orogeny and will not be described here. Very limited outcrops (50 m) of Sequence 1 remain in the Zemmour area (El Thlethyate Group, Sougy 1969). There, stromatolites have been correlated to those of the Atar Group (Fig. 3b).

Sequence 2. In the Anti-Atlas the cover begins with the "Precambrian III", mainly volcanics and red conglomerates and sandstones giving an U/Pb age of 563±20 and 578±15 (Juery et al. 1974). It is followed by a thick sequence of carbonates with thrombolites and stromatolites (Adoudounian) intruded by volcanics indicative of an U/Pb age of 530±10 Ma (Ducrot and Lancelot 1977). Shales and sandstones formations are superposed in a continuous thick pile (3000 to 4000 m) up to the Late Ordovician. There were no equivalents of these deposits on the south flank of the Tindouf basin except for some very limited areas (shales and sandstones, Gara Sayada formation; Gevin 1960; Menaked formation, Rognon et al. 1972).

Sequences 3 and 4. Late Ordovician epeirogenic movements resulted in strong erosion of the Tindouf basin before the deposition of the Ordovician tillite. Reaching 100 m in the north, the glacigenic deposits rest disconformably on the basement in the south. In contrast with the Taoudenni basin there are no discontinuities, on the northern Tindouf basin, in the marine sedimentation from the Silurian deposits up to the Middle Namurian (about 5000 m). However, the sedimentary sequence is thinner in the south. The Reguibat Shield was under water and the Tindouf and Taoudenni basin communicate. Since the Middle Namurian, sedimentation has been essentially continental red beds.

The sequence comprises, especially during the Paleozoic, a thick sedimentary wedge, deposited from south to north, along the northern edge of the West African craton. The total thickness reaches 10000 m, however, the sediments remain deposited under shallow water marine conditions.

3 Conclusion

The West African craton was achieved about 2000 Ma ago, after the Eburnean orogeny. It formed a huge area extending over most of present-day West Africa and probably further east and west, beyond the Pan-African suture of Hoggar-Iforas and the Mauritanides thrust. The uplift of the cratonic block has lasted about 1000 Ma, resulting in a practically flat erosional surface, on which no sediments dated between 2000 and 1000 Ma have been yet recorded. The global event that caused sedimentation to take place again on the West African craton, about 1000–1100 Ma, has not been identified.

In that Middle-Late Proterozoic sedimentary pile, diverse eustatic events have been recognized, which can be referred to as sequence-boundaries and therefore considered as time-lines. These sea-level rises resulted in the covering of the whole region, except for some elevated areas. In northern, eastern and central Taoudenni, deposition started with mature siliciclastic marine sediments. In the Atar group, and equivalents, three stromatolite-bearing sequences have been described (the third being restricted in size). Directions of sediment transport are uncertain, but it is possible to imagine an open sea toward the northwest of the craton.

The last sequence, Assabet el Hassiane marine siliciclastic, and its equivalents, was the most widespread. It may follow the distensive Pan-African events, perhaps related also to the appearance of Gourma and Volta pericratonic basins. The sequences, west to southwest of the Taoudenni basin, as well as in the Gourma and Volta basins, are, however, not comparable to those defined above, and precise correlations cannot be drawn.

A conspicuous discontinuity can be traced on the whole cratonic area, represented by an erosional surface caused by a slight but definite tectonic event affecting the Proterozoic cover. The tectonic event is attributed to the Pan-African collision, dated between 620–580 Ma. This discontinuity surface can be used as a time-line and correlations can be made on that basis. Except in the southwest of the craton, the first post-Pan-African sediments deposited were glacigenic beds, covering the whole cratonic area, including the Volta basin. However, tillite is not recorded in Tindouf and Gourma basins. The ice sheet was centred in the north of the craton, and the transport directions were N–S. The interpretation of the tillite as being post-Pan-African in age is supported by the new discovery of Early Cambrian shelly fossils in the overlaying beds, and new information about the Rb/Sr ages in Atar. This interpretation of a post-Pan-African position of the tillite is challenged in the Volta basin (Affaton and Trompette, this work). In the opinion of these authors, the Pendjari Supergroup, with an age of 660 ± 9 Ma and its basal glacigenic beds, similar to the Djbeliat Group, was incorporated to the Dahomeyide Pan-African domain. Therefore, it is considered as ante Pan-African. According to that interpretation, tillite have to be considered as diachronous, from south to north (Deynoux et al. 1978). In contrast with that hypothesis, Caby (1987, 1989) argues that Pendjari Group may have been an early molasse of the Pan-African orogeny, the isotopic age of the Pendjari Group being influenced by detrital white micas.

Above these layers, deposition of red beds (Early Cambrian-Ordovician) which lack a marine fauna, testifies to an uplift of the Reguibat Shield, separating the Taoudenni from the Tindouf basin. In the latter, the relatively thick succession (in relation to that in cratonic regions standard) can be attributed to the uplift and weathering of the nearby Pan-African belt. Higher up, sedimentation appears to be concordant in the Taoudenni basin, up to the glacial beds of the Late Ordovician. The Sobti Conglomerate, however, was probably deposited over an erosional surface that appears to correlate with the infra-tassilian erosional surface of the nearby Hoggar-Iforas area. That surface represents the ultimate peneplanation which terminated the Pan-African orogeny, uplift and erosion. From that time, the West African craton presents a common history with this newly formed craton.

The craton history since deposition of the Late Ordovician glacigenic deposits (ice sheet was south of Iforas, indicating tilting of the craton toward the north) and up to the Namurian, reflects diverse epeirogenic and eustatic movements. In post-Namurian times, the major change is the graditional passage to continental red beds sedimentation. Like the Cambrian red beds, these are thicker than usual cratonic sequences, as a consequence of the uplift and erosion of the Hercynian belt. From that period the craton remained as a continental environment, up to the present time. The Permo-Jurassic distension phases are recorded by extensive dolerite intrusions throughout the Taoudenni basin. No lavas corresponding to these sills and dykes have been recorded yet. Important post-Hercynian tectonics are also evidenced by the location of graben filled by Mesozoic sediments along the unstable Pan-African suture zone (Timétrine-Tilemsi) or large tectonic lineament (Néma-Nara).

On the craton borders are located diverse pericratonic basins of various ages, Gourma, Volta, Tindouf. They are characterized by an asymmetrical sedimentary wedge, a high sediment thickness reflecting high subsidence of the basins. However, these basins have various geodynamic settings, related to the adjacent, tectonically active areas, and not always clearly understood: foreland basin, aulacogene or passive margins.

References

Ako JA, Murray AS (1983) Bouguer gravity anomaly map of Ghana 1/1000000. Publ Ghana Geol Surv Dept Accra

Ako JA, Wellman P (1985) The margin of the West African craton: the Voltaian basin. J Geol Soc Lond 142: 625–632, 9 fig

Affaton P (1975) Etude géologique et structurale du Nord-Ouest Dahomey, du Nord-Togo et du Sud-Est de la Haute-Volta. Trav Lab Sci Terre, St Jérome, Marseille, B 10: 201

Affaton P (1987) Le bassin des Volta (Afrique de l'Ouest): une marge passive, d'âge protérozoïque supérieur, tectonisée au Panafricain (600 ± 50 Ma). Thèse Doct Sc Fac Sci, St Jérôme, Marseille, 462 pp

Amard B (1986) Microfossiles (Acritarches) du Proterozoïque supérieur dans les shales de la formation d'Atar (Mauritanie). Precambrian Res 31: 69–95

Amard B, Affaton P (1984) Découverte de *Chuaria circularis* (Acritarche) dans le bassin des Volta (Haute Volta et Bénin, Afrique de l'Ouest). Age protérozoique terminal de la formation de la Pendjari et de la tillite sous-jacente. C R Acad Sci, Paris, 299 II (14): 975–980

Ayme JM, Hubault C (1955) Mission de préreconnaissance

Pétrolière en AOF, rapport de fin de mission 1954-1955, Soudan Rapp Inéd Bur Rech Pétrol Dept Explo, Dakar

Bassot JP (1966) Etude géologique du Sénégal oriental et de ses confins Guinéo-Maliens. Thèse Clermont, 1963, Mém Bur Géol Min, Paris, 40: 322

Bassot JP, Bonhomme M, Roques M, Vachette M (1963) Mesures d'âges absolus sur les séries précambriennes et paléozoïques du Sénégal oriental. Bull Soc Géol Fr (5) (3): 401-405

Bassot JP, Meloux J, Traore H (1981) Notice explicative de la carte géologique au 1/1500000 de la république du Mali. Minist Dév Ind Min Mali, Dir Géol Min, Bamako

Beicip (1980) Synthèse géologique et pétrolière du bassin de Taoudeni. Dir Gener Min Mali, Dir Min Géol, Mauritanie, 31 PL, Notice 139

Beicip (1981) Nouvelles observations géologiques dans le bassin de Taoudeni. Rapp 87

Bellion Y (1987) Histoire géodynamique post-paléozoïque de l'Afrique de l'Ouest d'après l'étude de quelques bassins sédimentaires (Sénégal, Taoudeni, Iullemmeden, Tchad). Thèse Sci, Avignon, Pub occ 1989/17 CIFEG 302 pp

Bellion Y, Guiraud R (1988) Déformation d'origine compressive d'âge intra-éocène à l'Ouest de l'Adrar des Iforas (Mali). C R Acad Sci, Paris, 307 II: 529

Bellion Y, Benkhelil J, Guiraud R (1984) Mise en évidence de déformations d'origine compressive dans le Continental intercalaire de la partie méridionale du bassin de Taoudeni (Hodh oriental, confins mauritano-maliens). Bull Soc Géol Fr XXVI 6: 1137-1147

Bellion Y, Saint-Marc P, Damotte R (1989) Contribution à la connaissance des dépôts marins au passage crétacé-tertiaire dans la vallée du Tilemsi (Nord- du Mali). J Afr Earth Sci 9: 187-194

Bertrand-Sarfati J (1972) Stromatolites columnaires du Précambrien supérieur, Sahara Nord-occidental. Inventaire, morphologie et microstructure des laminations, corrélations stratigraphiques. Thèse Montpellier, CNRS (CRZA), Sér Géol 14: 242

Bertrand-Sarfati J (1979) Une algue inhabituelle verte, rouge ou bleue dans une formation dolomitique présumée d'âge Précambrien supérieur. Bull Centr Rech Explor Prod, Elf-Aquitaine 3: 453-641

Bertrand-Sarfati J, Moussine-Pouchkine A (1978a) Le groupe d'Achaïkar à l'Ouest du Timetrine (Mali): un témoin de l'aire sédimentaire cratonique saharienne au Précambrien supérieur. C R Somm Soc Géol Fr 2: 62-66

Bertrand-Sarfati J, Moussine-Pouchkine A (1978b) Mise en évidence d'une discordance du groupe de Bandiagara sur la formation sédimentaire du Précambrien supérieur (Gourma, Mali). C R Somm Soc Géol Fr 2: 59-61

Bertrand-Sarfati J, Moussine-Pouchkine A (1983a) Platform-to-basin facies evolution: the carbonates of late Proterozoïc (Vendian), Gourma, West Africa. J Sediment Petrol 53 (1): 275-293

Bertrand-Sarfati J, Moussine-Pouchkine A (1983b) Pedogenetic and diagenetic fabrics in the upper Proterozoïc Sarnyéré formation (Gourma, Mali). Precambrian Res 20: 225-242

Bertrand-Sarfati J, Moussine-Pouchkine A (1988) Is cratonic sedimentation consistent with actualistic models? An example from Upper Proterozoic of the West African Craton. In: Jackson MJ (ed) Aspects of Proterozoic sedimentary geology. Sediment Geol 58: 255-276

Bertrand-Sarfati J, Raaben ME (1970) Comparaison des ensembles stromatolitiques du Précambrien supérieur du Sahara occidental et de l'Oural. Bull Soc Géol Fr 12: 346-371

Bertrand-Sarfati J, Fabre J, Moussine-Pouchkine A (1977) Géodynamique des aires sédimentaires cratoniques: quelques exemples sahariens. Bull Centr Rech Explor Prod, Elf-Aquitaine 1: 217-231

Bertrand-Sarfati J, Moussine-Pouchkine A, Caby R (1987) Les corrélations du Protérozoïque au Cambrien en Afrique de l'Ouest: nouvelle interprétation géodynamique. Bull Soc Géol Fr III (5): 855-865

Bessoles B (1983) Le craton Ouest-africain. Lex Stratig Int, Fabre Edit, Pergamon Press, Oxford, pp 27-34

Beuf S, Biju-Duval B, de Charpal O, Rognon P, Gariel O, Bennacef A (1971) Les grès du Paléozoïque inférieur au Sahara. Edit, Technip, Paris, 480 pp

Biju-Duval B, Gariel O (1969) Nouvelles observations sur les phénomènes glaciaires "eocambriens" de la bordure nord de la synéclise de Taoudeni, entre le Hank et le Tanezrouft (Sahara occidental). Paléogéogr Paléoclimato Paléoécol 6: 283-315

Black R, Fabre J (1983) A brief outline of the geology of West Africa. Lex Stratig Int, Fabre Edit, Pergamon Press, Oxford, pp 17-26

Black R, Caby R, Moussine-Pouchkine A et al. (1979) Evidence for late Precambrian plate tectonics in West Africa. Nature (Lond) 278: 223-227

Boudouresque L, Dubois D, Lang J, Trichet J (1982) Contribution à la stratigraphie et à la paléogéographie de la bordure occidentale du basin des Iullemmeden au Crétacé supérieur et au Cénozoïque (Niger et Mali, Afrique de l'Ouest). Bull Soc Géol Fr 4: 685-695

Bozhko NA (1969) Stratigraphy and Tectonics of the Voltaian basin. Ann Fac Sci Univ Clermont-Ferrand, Geol Min Fr 41 (19): 3-4

Bozhko NA, Kasakov GA, Trofimov DM, Knorre KG, Gatinski JA (1971) Nouvelles déterminations de l'âge absolu des glauconies d'Afrique occidentale. Dokl Akad Nauk SSSR 198

Bozhko NA, Pykhova NG, Raaben MY (1974) Upper Precambrian Biostratigraphy of Africa. (Biostratigrafii verkhnego dokembriya Afriki). Dokl Akad Nauk SSSR 214: 643-646

Bronner G, Roussel J, Trompette R, Clauer N (1980) Genesis and geodynamic evolution of the Taoudenni cratonic basin (Upper Precambrian and Paleozoic, Western Africa). In: Bally AN, Bender PN, McGetchin TR, Walcott RI (eds) Dynamics of Plate interiors. Geodyn Ser 1, Am Geophys Un, pp 81-90

Buffieres JM, Fahy JC, Petey J (1965) Etude géologique de la partie orientale de la dorsale Réguibat. Région des Eglab et secteur Nord du Yetti. Rapp Inéd Soc Et Réal Min Industr (SERMI). Paris, Déc n°ALG 63-09-IV, 230 pp, Carte Géologique

Buffieres JM, Fahy JC, Petey J (1966) Sur l'ensemble précambrien Yetti-Eglab et sur sa couverture infratillitique en territoire algérien. C R Acad Sci, Paris, D 262: 1513-1516

Caby R (1978) Paléogéodynamique d'une marge passive et d'une marge active au Précambrien Supérieur: leur collisions dans la chaîne pan africaine du Mali. Bull Soc Geol Fr XX (5): 857-861

Caby R (1979) Les nappes précambriennes du Gourma dans la chaîne pan africaine du Mali. Rev Géol Dyn Géogr Phys 21 (5): 365-376

Caby R (1987) The pan-African belt of West Africa from the Sahara desert to the gulf of Benin. In: Schaer JP, Rodgers J (eds) The anatomy of mountain ranges. Princeton Univ Press, pp 129-170

Caby R (1989) Precambrian terranes of Benin-Nigeria and

Northeast Brazil and the Late Proterozoic South Atlantic fit. Geol Soc Am Spec Pap 230: 145-158

Caby R, Andreopoulos-Renaud U (1989) Age U/Pb à 620 Ma, d'un pluton synorogénique de l'Adrar des Iforas (Mali). Conséquences pour l'âge de la phase majeure de l'orogenèse pan africaine. C R Acad Sci, Paris, 308: 307-310

Caby R, Moussine-Pouchkine A (1978) Le horst birrimien de Bourré (Gourma oriental, République du Mali): nature et comportement au cours de l'orogénèse pan-africaine. C R Acad Sci, Paris, 287: 5-8

Caby R, Andreopoulos-Renaud U, Gravelle M (1982) Cadre géologique et géochronologique U/Pb sur zircon des batholites précoces dans le segment pan africain du Hoggar central (Algérie). Bull Soc Géol Fr 24 (4): 677-684

Chamley H, Deynoux M, Robert C, Simon B (1980) La sédimentation argileuse du Précambrien terminal au Dévonien dans la région du Hodh (bassin cratonique de Taoudeni, Sud-Est mauritanien). Ann Soc Géol Nord, pp 73-82

Christie-Blick N, Grotzinger JP, von der Borsh CC (1988) Sequence stratigraphy in Proterozoic succession. Geology 16: 100-104

Clauer N (1976) Géochimie isotopique du Strontium des milieux sédimentaires. Application à la géochronologie de la couverture du craton Ouest-africain. Mem Sci Géol, Strasbourg, 45: 256

Clauer N, Deynoux M (1987) New information on the probable isotopic age of the Late Proterozoic glaciation in West Africa. Precambrian Res 37: 89-94

Clauer N, Caby R, Jeannette D, Trompette R (1982) Geochronology of sedimentary and metasedimentary Precambrian rocks of the west African craton. Precambrian Res 18 (1/2): 53-72

Cornet A (1948) Sur la réalité des mouvements post-crétacés au Sahara. Trav Inst Rech Sahar 5: 33-46

Culver J, Pojeta J, Repetski JE, Robineau B (1988) Signification stratigraphique de microfossiles du Cambrien inférieur dans le bassin de Taoudéni, à la frontière Guinée-Sénégal. C R Acad Sci, Paris, t 307, Sér II, pp 651-656

Culver J, Pojeta J, Repetski JE (1988) First record of Early Cambrian shelly microfossil from West Africa. Geology 16: 596-599

Dars R (1957) Sur l'existence du Continental intercalaire au NE de Nara (Afrique occidentale française). C R Somm Soc Géol Fr, p 248

Destombes J, Sougy J, Willefert S (1969) Révisions et découvertes paléontologiques (Brachiopodes, Trilobites et Graptolites) dans le Cambro-Ordovicien du Zemmour (Mauritanie septentrionale). Bull Soc Géol Fr 7 (11): 185-206

Deynoux M (1971) Essai de synthèse stratigraphique du bassin de Taoudeni. Trav Lab Sci Terre, St-Jérôme, Marseille B 3

Deynoux M (1980) Les formations glaciaires du Précambrien terminal et de la fin de l'Ordovicien en Afrique de l'Ouest. Deux exemples de glaciation d'inlandsis sur une plate-forme stable. Thèse Doct Etat, Univ Aix-Marseille, 554 pp

Deynoux M, Tempier C, Trompette C (1971) Etude photogéologique de la partie orientale de la dorsale Reguibat et de sa couverture sud (Mauritanie). Tome 1: synthèse géologique. Trav Lab Sci Terre, St Jérôme, Marseille XI (11): 167

Deynoux M, Trompette R, Clauer N, Sougy J (1978) Upper Precambrian and lowermost Paleozoïc correlations in West Africa in the western part of central Africa. Probable diachronism of the Late Precambrian tillite. Geol Rundsch 67: 615-630

Deynoux M, Sougy J, Trompette R (1985) Lower Paleozoic rocks of West Africa and the western part of central Africa. In: Holland CH (ed) Lower Paleozoic of Northwestern and West Central Africa. Wiley, N Y, pp 337-495

Deynoux M, Kocurek GA, Proust JM (1989) Late Proterozoic periglacial eolian deposits on the West african platform, in the Taoudenni basin in Western Mali. Sedimentology, 36: 531-549

Dia O (1984) La chaine panafricaine et hercynienne des Mauritanides, face au bassin, protérozoique à dévonien, de Taoudenni, dans le secteur de Mejeria (Taganet, Sud RIM): lithostratigraphie et tectonique. Thèse, Univ St Jérôme, Marseille, 516 pp

Dia O, Sougy J, Trompette R (1969) Discordance de ravinement et discordance angulaire dans le Cambro-Ordovicien de la région de Méjéria (Taganet occidental, Mauritanie). Bull Soc Géol Fr 7 (11): 207-221

Drouet JJ (1986) Le cycle sédimentaire et les milieux de dépôt du Précambrien supérieur infratillitique, dans le bassin des Voltas au Nord Togo. J Afr Earth Sci 5: 455-464

Ducrot J, Lancelot JR (1977) Problème de la limite Précambrien-Cambrien: étude radio-chronologique par la méthode U/Pb sur zircon du volcan du Jbel Boho. Can J Earth Sci 14: 2771-2777

Fabre J (1961) Les formations continentales post-viséennes de Taoudenni (Sahara occidental). C R Acad Sci, Paris, 253 (2): 293-295

Fabre J (1967) Le Paléozoïque terminal à faciès grès rouge au Sahara central et occidental. 6th Congr Géol Carbonifère, Sheffield, II: 737-744

Fabre J (1983) Afrique de l'Ouest. Introduction géologique et termes stratigraphiques. Lex Stratig Int, Fabre Edit, Pergamon Press, Oxford, 386 pp

Fabre J (1988) Les séries du Paléozoïque d'Afrique: une approche. J Afr Earth Sci 7 (1): 1-40

Fabre J, Moussine-Pouchkine A (1971) Régressions et transgressions permo-carbonifères sur le Nord-Ouest de la Plate-forme africaine; épirogenèse ou variation eustatique? Bull Soc Géol Fr 7 XIII (162): 140-145

Fabre J, Caby R, Girod M, Moussine-Pouchkine A (1976) Introduction à la Géologie du Sahara algérien, et des régions voisines I. la couverture Paléozoïque. SNED Edit, Alger, p 420

Fabre J, Ba H, Black R et al. (1982) La chaine pan-africaine, son avant pays et la zone de suture au Mali. Notice explicative de la carte géologique et gravimétrique de l'Adrar des Iforas au 1/500000. Dir Nat Géol Min Mali, DNGM Bamako 85, 1 carte coul.

Gevin P (1960) Etudes et reconnaissances géologiques sur l'axe cristallin Yetti-Eglab et ses bordures sédimentaires. 1ère partie: bordures sédimentaires. Pub Serv Carte Géol Algérie, Nlle Sér 23: 328

Guiraud R, Issawi B, Bellion Y (1985) Les linéaments guinéonubiens, un trait structural majeur à l'échelle de la plaque africaine. C R Acad Sci, Paris, 300 II: 17

Guiraud R, Bellion Y, Benkhelil J, Moreau C (1987) Post-hercynian tectonics in Northern and Western Africa. Geol J 22: 433-466

Huot G (1957) Etude par prospection électrique et sismique réfraction de la région de Goundam-Nara-Néma en 1955-56. Arch Serv Hydr AOF, Rapp CGG Inéd

Juery A, Lancelot JR, Hamet J, Proust F, Allegre CJ (1974) L'âge des rhyolites du Précambrien III du Haut Atlas et le problème de la limite Précambrien-Cambrien. 2° Réun Ann Sci Terre, Nancy, 230

Junner NR, Hirst T (1946) The geology and hydrology of the Voltaian Basin. Mem Gold Coast Geol Surv, Accra, 8: 51

Karpoff R (1960) La géologie de l'Adrar des Iforas. Thèse Sci, Paris, 1958. Bull Serv Géol Prosp Min Afr Occident 30: 265 pp

Keita ND (1984) Etude géologique des formations sédimentaires de la partie Sud orientale du bassin précambrien supérieur et paléozoïque de Taoudenni au Mali (région du plateau de Bandiagara). Thèse, 3ème cycle, Marseille, 209 pp

Kennedy WQ (1964) The structural differentiation of Africa in the Pan African 500 M.y) tectonic episode. Leeds Univ Inst Afr Geol Ann Rep Sci Result, (abstr), pp 48-49

Kent DV, Dia O, Sougy JMA (1984) Paleomagnetism of lower-middle Devonian and upper-Proterozoic-Cambrian (?) rocks from Mejeria (Mauritania, West Africa). Geodyn Ser AGV 12: 99-115

Kilian C (1931) Les principaux complexes continentaux du Sahara. C R Somm Soc Géol Fr, p 109

Legrand-Blain M (1985) Taoudenni basin. In: Wagner RH, Winkler Prins CF, Granados LF (eds) The Carboniferous of the world. vol 2, pp 319-322

Lesquer A, Moussine-Pouchkine A (1980) Les anomalies gravimétriques de la boucle du Niger. Leur signification dans le cadre de l'orogenèse pan-africaine. Can J Earth Sci 17: 1538-1545

Liegeois JP, Bertrand JM, Black R, Caby R, Fabre J (1983) Permian alkaline undersaturated and carbonatite province, and rifting along the West african craton. Nature (Lond) 305 (5929): 42

Marchand J, Trompette C, Sougy J (1972) Etude photogéologique de la région El Mreiti-Mejahouda-Agaraktem. Trav Lab Sci Terre, St Jérôme, Marseille X 21: 30

Monciardini C (1966) La sédimentation éocène au Sénégal. Mem BRGM 43: 65, 34 pl

Monod Th (1939) Notes géologiques sur les confins sahariens du Soudan français. Rev Geogr Phys Géol Dyn XII (4): 461-478

Monod Th (1952) L'Adrar mauritanien. Pub Dir Mines, AOF, Dakar, 286 pp

Moussine-Pouchkine A, Bertrand-Sarfati J (1978) Le Gourma: un aulacogène du Précambrien supérieur? Bull Soc Géol Fr 7 XX (6): 851-857

Moussine-Pouchkine A, Bertrand-Sarfati J, Ball E, Caby R (1988) Les séries sédimentaires et volcaniques protérozoïques impliquées dans la chaîne pan africaine: la région de l'Adrar Ahnet (NW Hoggar, Algérie). J Afr Earth Sci 7: 57-75

Olivet JL, Bonin J, Beuzart P, Auzende JM (1984) Cinématique de l'Atlantique nord et central. Rapp Sci Tech 54: 108, 8 pl, CNEXO

Ouedraogo C (1981) Etude géologique des formations sédimentaires du bassin précambrien supérieur et paléozoïque de Taoudenni en Haute Volta. Trav Lab Sci Terre, St Jérôme, Marseille, Sér X 45: 51

Perrin M (1987) Paléomagnétisme des séries rouges à aimantations multiples d'âge protérozoïque, d'Afrique et d'Amérique du Nord. Doc Trav CGG 11: 365

Perrin M, Elston D, Moussine-Pouchkine A (1988) Paleomagnetism of Proterozoic and Cambrian strata, Adrar de Mauritanie, Cratonic West Africa. J Geophys Res 93: 2159-2178

Proust JN, Deynoux M, Guillocheau F (1990) Effets conjugués de l'eustatisme et de l'isostasie sur les plate-formes stables en période glaciaire. Exemple des dépôts glaciaires du Proterozoïque supérieur de l'Afrique de l'Ouest au Mali occidental. Bull Soc Géol Fr., 8, VI, 673-682

Radier H (1959) Contribution à l'étude géologique du Soudan oriental (AOF). I. Le Précambrien saharien du Sud de l'Adrar des Iforas. II. Le bassin crétacé et tertiaire de Gao. Thèse Sci, Strasbourg, 1959. Bull Serv Géol Prosp Min 26: 550

Reichelt R (1972) Géologie du Gourma (Afrique occidentale), un seuil et un bassin du Précambrien supérieur, stratigraphie, tectonique, métamorphisme. Mem Bur Rech Géol Min, Paris, 53: 213

Rognon P, Biju-Duval B, de Charpal O (1972) Modelé glaciaire dans l'Ordovicien supérieur saharien: phases d'érosion et glaciotectonique sur la bordure nord des Eglab. Rev Géogr Phys Géol Dyn 2 XIV (5): 507-528

Sacko S (1985) Contribution à l'étude structurale du Gourma oriental (chaine pan-africaine du Mali). Thèse 3° cycle, USTL, Montpellier, 130 pp

Saunders RS (1970) Early Paleozoic Orogeny in Ghana: Foreland stratigraphy and structure. Geol Soc Am Bull 81: 233-240

Sauvage JF, Savard R (1985) Les compositions sous-saturés à carbonatite de la région d'In Imanal (Sahara Malien); une présentation. J Afr Earth Sci 3 (1/2): 143-149

Simon B, Deynoux M, Keita N, Marchand J, Rossi P, Trompette R (1979) Le Précambrien supérieur et la base du Paléozoïque de la partie Sud-Ouest du bassin de Taoudenni. Essai de synthèse. 10ème Colloq Géol Afr, Montpellier, (abstr), pp 108-109

Sougy J (1969) Grandes lignes structurales de la chaîne des Mauritanides et de son avant-pays (Socle précambrien et sa couverture infracambrienne et paléozoïque), Afrique de l'Ouest. Bull Soc Géol Fr 11: 133-149

Sougy J (1971) Remarques sur la stratigraphie du Protérozoïque supérieur du Bassin Voltaïen; influence de la paléosurface d'érosion glaçiaire de la base du group de l'Oti sur le tracé sinueux des Voltas et de certains affluents. C R Acad Sci, Paris, 272: 800-803

Trompette R (1973) Le Précambrien supérieur et le Paléozoïque inférieur de l'Adrar de Mauritanie (bordure occidentale du bassin de Taoudenni, Afrique de l'Ouest). Un exemple de sédimentation de craton. Etude stratigraphique et sédimentotogique. Trav Lab Sci Terre, St Jérôme, Marseille, Sér B, 7: 573

Villemur JR (1967) Reconnaissance géologique et structurale du Nord du bassin de Taoudenni. Mem Bur Rech Géol Min, Paris, 51: 152

Villeneuve M (1984) Etude géologique sur la bordure Sud-Ouest du craton Ouest africain et l'évolution des bassins sédimentaires proterozoïques et paleozoïques de la marge Nord-Ouest du continent de Gondwana. Thèse Fac Sci St Jérôme, Marseille, 550 pp

Zimmermann M (1960) Nouvelle subdivision des séries antégothlandiennes de l'Afrique occidentale (Mauritanie, Soudan Sénégal). Int Geol Congr, Copenhagen, 8: 26-36

Chapter 5

The West African Orogenes

The Pan-African Trans-Saharan Belt in the Hoggar Shield (Algeria, Mali, Niger): A Review

Anne-Marie Boullier[1]

1 Introduction

The Pan-African belt of central northwest Africa is part of the Transaharan belt (Cahen et al. 1984) and has been interpreted by Black et al. (1979) and Caby et al. (1981) as a collision belt formed during a Wilson cycle. The Tilemsi suture zone separates the stable West African Craton (passive margin) from the mobile belt (active margin). Pan-African deformation and metamorphism affected the whole Hoggar shield from the Tilemsi region at least to the 8°30′E lineament and is recognized on a 1000-km-wide area (Fig. 1). Actually, the Transaharan belt appears as a complex assemblage of NS trending geological domains which have undergone different evolutions. Large areas are still poorly known and several interpretations have been proposed for some others. In this review paper, I will present the geological evolution of each domain and discuss these different interpretations. Correlations of these domains with the Togo-Benin belt and with the Nigerian shield have been proposed by several authors (Bessoles and Trompette 1980; Caby et al. 1981; Ajibade et al. 1987; Caby, in press).

Bertrand and Caby (1978) subdivided the Hoggar shield into three main structural domains, respectively from east to west: the eastern Hoggar, the central polycyclic Hoggar and the Pharusian belt. I will keep the main first-order domains but will disthinguish some additional subdivisions in the Pharusian belt in the light of the recent studies in the Adrar des Iforas: the eastern Pharusian belt, the western Pharusian belt, the central Iforas batholith, the Tilemsi island-arc domain and the Gourma-Timetrine foreland nappes.

For each domain I will present the lithology, the tectono-metamorphic evolution and the Pan-African magmatism.

[1] Centre de Recherches Pétrographiques et Géochimiques. BP 20, 54501 Vandoeuvre-lès-Nancy Cedex, France

2 The Eastern Hoggar: The Tiririne-Aouzeguer Belt

2.1 *In Algeria*

2.1.1 The Djanet-Tafassasset Domain

The Tiririne Pan-African belt (Bertrand et al. 1978) is parallel to the 8°30′ meridian and separates the polycyclic Central Hoggar to the west from the Djanet-Tafassasset domain to the east. The latter is a segment of a pre-Pan-African fold belt trending NW-SE which was subdivided into three domains separated by vertical shear zones (Caby and Andreopoulos-Renaud 1987). The stratigraphic relationships between these three domains are unknown. The Tiririne Formation lies unconformably on the basement of the western domain which is composed mainly of granitoids, remnants of a shelf-type carbonate sequence, felsic lavas and related intrusives, and terrigeneous and volcanic-derived flysh-type deposits; ultramafic and mafic rocks have also been observed by the same authors. A syn- to late-tectonic batholith crosscutting the ultramafics has been dated at 729 ± 8 Ma by U/Pb on zircons (Caby and Andreopoulos-Renaud 1987).

2.1.2 The Tiririne Formation

The Tiririne Formation is exposed in variably complex synclines in which it reaches a thickness of up to 8000 m (Fig. 2; Blaise 1957, 1961; Bertrand et al. 1978). The base (Unit I) is characterized by clays and limestones. The overlying Unit II is almost entirely clastic with sequences of polymictic conglomerates (of possible glaciogenic origin), arkoses, greywackes and silts; some limestones occur towards the south. The detrital source seems to be exclusively located to the east. Unit III has a more regular lateral development of clastic sediments with more mature rock types including red beds on top.

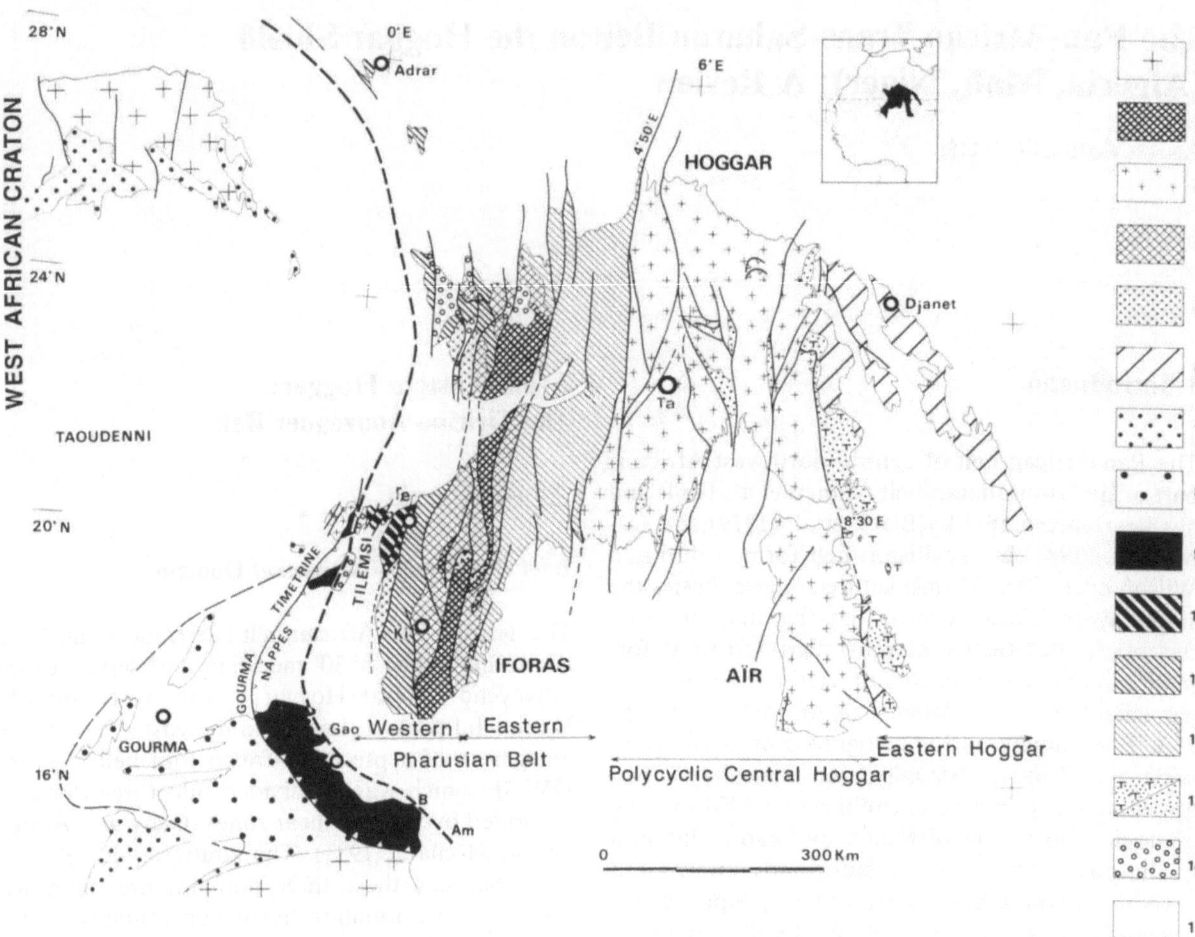

Fig. 1. Simplified geological map of the Touareg shield and adjacent areas (modified from Caby et al. 1981). *1* Reguibat and Leo shields; *2* slightly reactivated Eburnean granulites within the Pharusian belt; *3* reactivated pre-Pan-African gneisses; *4* undifferentiated gneisses highly reactivated during the Pan-African; *5* gneisses affected by late high-temperature - low-pressure metamorphism; *6* undifferentiated rocks of the eastern Hoggar (metamorphism at ca. 730 Ma); *7* Upper Proterozoic shelf sediments; *8* slope-basin sediments of the Gourma aulacogen; *9* Gourma and Timetrine nappes; *10–12* Late Upper Proterozoic greywackes and magmatic rocks of the Tilemsi island arc *(10)*, of the western Pharusian belt *(11)*, of the eastern Pharusian belt *(12)*; *13* Late Upper Proterozoic volcanodetritic schist belts in the central and eastern Hoggar with *T* Tiririne and *Proche Ténéré* Groups; *14* molassic *Série Pourprée* or Nigritian; *15* Palaeozoic and Mesozoic cover. *Ta* Tamanrasset, *Ki* Kidal, *Te* Tessalit

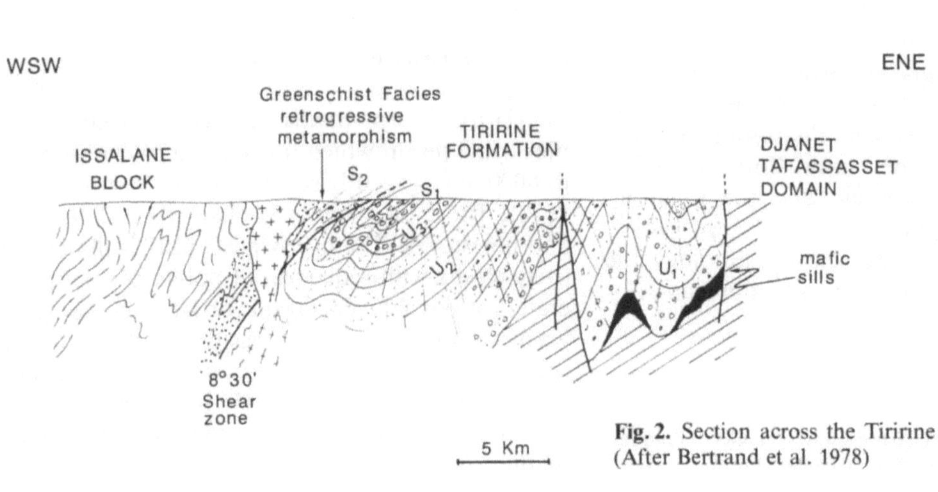

Fig. 2. Section across the Tiririne linear fold belt in Algeria (After Bertrand et al. 1978)

Some pretectonic igneous rocks have intruded the Tiririne Formation: olivine gabbros, layered sills of mafic biotite diorites, dykes of diabase and by dykes and sills of rhyolites to rhyodacites and of granodiorites which have been dated at 660±5 Ma by U/Pb on zircon (Bertrand et al. 1978).

2.1.3 The Tectono-Metamorphic Evolution of the Tiririne Belt

The structure of this linear fold belt is controlled by the 8°30′E dextral shear zone which grades progressively to the north into a thrust (Bertrand et al. 1978). Mylonites and ultramylonites outline the shear zone and exhibit a horizontal NS-trending stretching lineation. A metamorphic and strain gradient is described by the authors from south to north along the belt and from east to west in the Tiririne Formation across the belt:

Zone A: very low grade to lower greenschist facies with incipient E-W shortening to the south of 22°30′N.

Zone B: intense deformation under greenschist-facies conditions to the north of 22°30′N. In the western Issalane block, tight folds with westward dipping N-S striking axial planes deform an earlier pre-Pan-African foliation. In the Tiririne Formation along the shear zone, folds are tight to isoclinal with steep fold axes and vertical axial-plane cleavage. Strain decreases eastward.

Zone C: very high strain under middle to upper amphibolite-facies conditions reaching migmatization. In this zone a thrusting component of the Issalane block onto the highly deformed Tiririne Formation has been recognized. A synkinematic adamellite has been dated at 604±13 Ma and a late kinematic granitic pluton yielded 585±14 Ma by U/Pb on zircon (Bertrand et al. 1978).

2.2 *In Niger (Aïr)*

2.2.1 The Eastern Domain

In the Aouzeguer region (SE Aïr, Fig. 3), an eastern basement (Eberjeghi gneisses and granodiorite) is overlain by the molassic Proche Ténéré Formation which has been described by Raulais (1959), Black at al. (1967), Kehrer et al. (1975) and Cosson et al. (1987). This formation has been correlated with the Tiririne Formation with which it is in continuity across the border and similar in lithology: siltstones, sandstones, arkoses with conglomeratic intercalations, fan-conglomerates, breccias and calcitic greywackes with some siliceous tuffs.

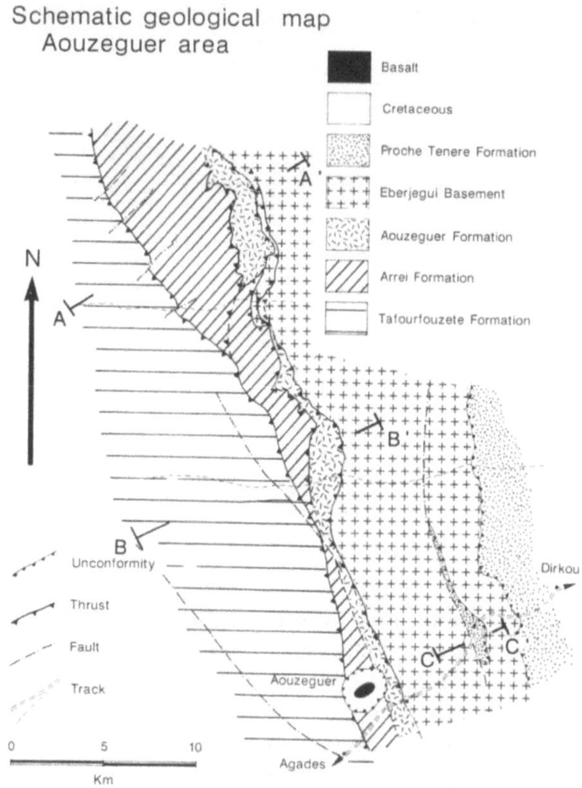

Fig. 3. Geological map of the Aouzeguer area modified after Kehrer et al. (1975). Cross-sections of Fig. 4 are located

2.2.2 The Western Domain

The western basement is composed of three main lithological and structural units which are separated by westward dipping thrusts (Cosson et al. 1987). These units are, from west to east and from top to bottom (Fig. 4):

1. The Tafourfouzète Unit, mainly composed of granitic gneisses with amphibolitic layers and of a platform sedimentary sequence (quartzites, aluminous schists and marbles). No basement-cover relationship has been observed between the sediments and the gneisses.
2. The Arrei Unit, made up of quartzofeldspathic, arenitic and volcanoclastic sediments with tholei-

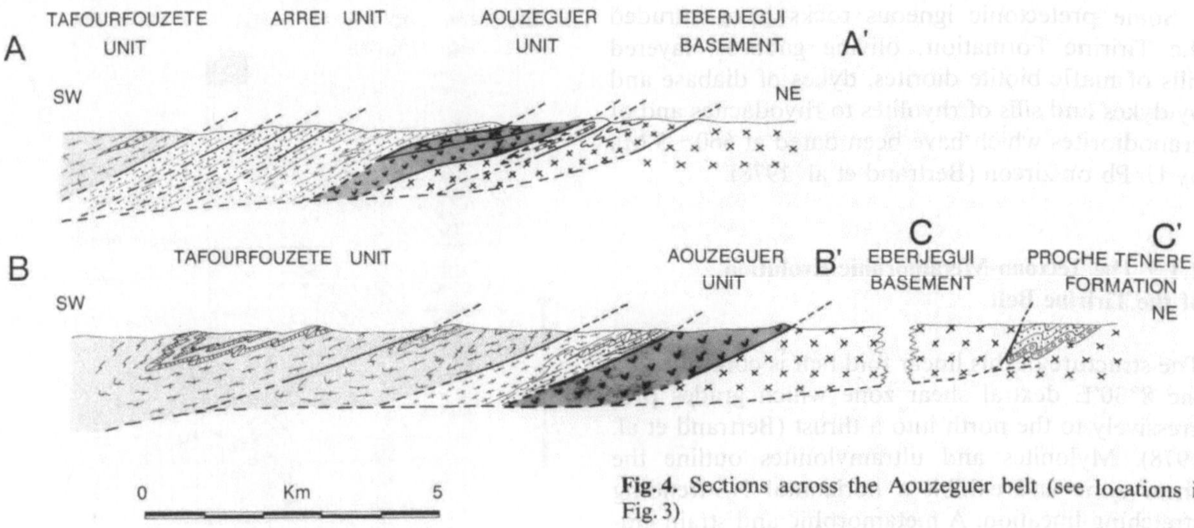

Fig. 4. Sections across the Aouzeguer belt (see locations in Fig. 3)

itic basaltic sills. Granitic bodies were emplaced into the sediments before the deformation.

3. The Aouzeguer Unit, composed of serpentinized ultramafics and gabbroic cumulates, diorites and granophyres. Some rodingites have been observed and suggest an oceanic metamorphism.

2.2.3 Tectono-Metamorphic Evolution

The three western units were thrust onto the Eberjeghi basement and the Proche Ténéré Formation is involved in this deformation. The metamorphic grade decreases across the tectonic pile from top (amphibolite facies) to bottom (greenschist facies) with breaks along the thrust planes. All the structural units show a westward-dipping foliation or schistosity with a well-expressed stretching lineation: elongated pebbles, fibres in extension veins, quartz and feldspar ribbons. This stretching lineation trends N 100° to N 80° in the uppermost Tafourfouzète Unit, N 60° to N 40° in the intermediate Arrei Unit and N 40° to N 20° in the lowermost Aouzeguer mafic and ultramafic schists Unit, and in the foliated Proche Ténéré molasse and Eberjeghi basement (Cosson et al. 1987).

This change in the stretching lineation direction could be interpreted either as a rotation in time or as due to contemporaneous combined wrenching and thrusting during shortening (Burg et al. 1987). In the latter case, the wrenching component would be compatible with the dextral sense of shear along the Tiririne belt (Bertrand et al. 1978).

2.3 Comparison and Discussion

Does the Aouzeguer-Tiririne belt correspond to a Pan-African suture zone? To the north, Bertrand et al. (1978) have shown that the Tiririne belt separates two different continental basements: the eastern Djanet-Tafassasset basement and the western Issalane basement, the latter being compared to the Central Hoggar. However, the Aouzeguer mountains are the only known occurrence of ultramafics along the belt. Pre-730 Ma ultramafic bodies have been described by Caby and Andreopoulos-Renaud (1987) in the Tiririne basement. Moreover, Kehrer et al. (1975) describe pebbles of ultramafic schists in the Proche Ténéré conglomerates. Consequently, the Aouzeguer Formation could represent earlier oceanic rocks that may represent a pre-730 Ma suture subsequently reworked during a Pan-African (s. s.) shortening event involving the Tiririne and Proche Ténéré, Upper Proterozoic, molassic Formations.

3 The Polycyclic Central Hoggar

3.1 The Main Lithological Associations

Independent of any stratigraphic interpretation, three main lithological associations may be defined in Central Hoggar (Fig. 5): (1) quartzo-feldspathic gneisses and granites, (2) banded high-grade metasediments including quartzites, marbles and metapelites, and (3) low-grade metavolcanics and greywackes.

The two former associations represent the Suggarian and the latter the Pharusian, as defined by Kilian (1932) and Lelubre (1952). Among the high-grade metasediments two groups were formerly distinguished: (1) those intimately associated with quartzo-feldspathic gneisses, the Arechchoum Series and (2) the Aleksod Series which lies in structural unconformity upon group 1. The two groups are very similar in lithology but a set of mafic dykes was emplaced in the older group prior to the main deformation (Bertrand 1974). In places, statigraphic unconformities are preserved, for example in the Gour Oumelalen area (Latouche 1978) and in the Tazat area (Blaise 1967; Bertrand et al. 1968), but tectonic contacts are more usual. From the distinction of these two groups of metasediments, together with a non-critical acceptance of a classical Pharusian unconformity below the low-grade schists (which was considered as giving a lower limit for the Upper Proterozoic), the existence of a Kibaran event at about 1000 Ma was previously proposed in Central Hoggar (Bertrand 1974). Some geochronological data seemed at that time to support this assumption (Bertrand and Lasserre 1976). In fact, recent structural and geochronological investigations do not confirm the existence of a Kibaran event in Central Hoggar (Bertrand et al. 1986a; Barbey et al. 1989).

On the basis of geochronological data (Picciotto et al. 1965; Latouche and Vidal 1974; Bertrand and Lasserre 1976; Bertrand et al. 1978, 1984b, 1986a; Vialette and Vitel 1981), the Eburnean event dated at about 2000 Ma is seen in quartzo-feldspathic gneisses including orthogneisses and charnockites and from the older metasediments, whereas the Pan-African event corresponds to nappe tectonics and granitoid emplacement ca. 600 Ma ago. No age older than Eburnean is well constrained in Central Hoggar but Rb/Sr whole-rock results (Bertrand and Lasserre 1976) and a Pb/Pb whole-rock isochron (3480 ± 90 Ma) quoted by Latouche (1978) suggest that Archean crust is involved.

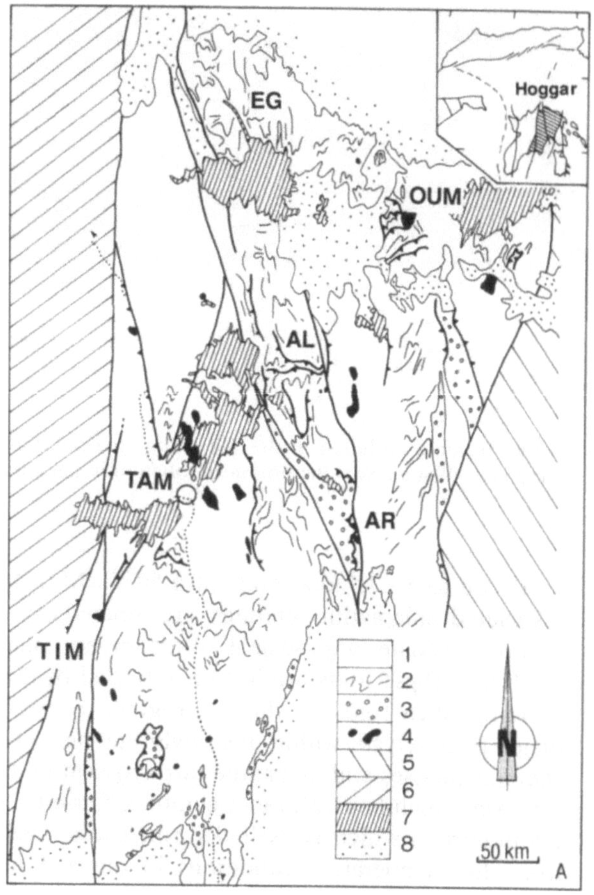

Fig. 5. Polycyclic Central Hoggar (modified after Bertrand et al. 1986a). *1* Granites and quartzofeldspathic gneisses; *2* high-grade metasedimentary rocks (*1* + *2* Suggarian); *3* low-grade schists (Pharusian); *4* Taourirt granites; *5* Issalane domain; *6* eastern Pharusian belt; *7* and *8* Phanerozoic lavas and sediments. Localities on map are Aleksod *(AL)*, Arefsa *(AR)*, Egere *(EG)*, Oumelalen *(OUM)*, Tamanrasset *(TAM)* and Timgaouine *(TIM)*

3.2 The Main Pan-African Tectono-Metamorphic Evolution

The metamorphic evolution of Central Hoggar reflects the dual influences of the Eburnean basement and of the Pan-African imprint (Latouche 1978; Vitel 1979; Ouzegane 1981). Detailed structural, metamorphic and geochronological work of a case study area situated to the SW of Tamanrasset (Boullier et Bertrand 1981; Ouzegane 1981; Bertrand et al. 1986a) has shown that the tectono-thermal history of the Central Hoggar may be interpreted as resulting from Pan-African thrust tectonics and crustal thickening. This area is characterized by a strong contrast between two completely different tectonic units separated by a mylonite belt gently dipping toward the NE (Fig. 6). The tectonic units are as follows:

Lower Tectonic Unit: the Iherane Gneisses and Migmatites. This comprises granulite-facies metasediments which are invaded by migmatites and nebulitic granites. Two successive metamorphic assemblages were defined by Ouzegane (1981) and correspond respectively to more than 8 kbar and 800 °C

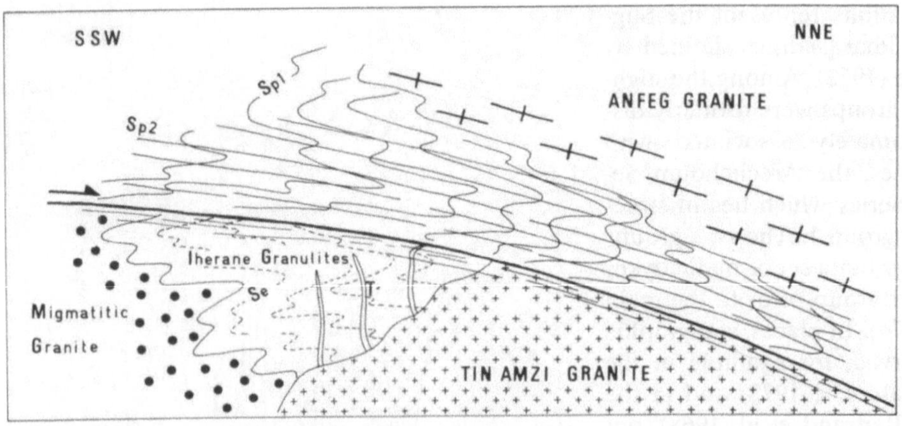

Fig. 6. Tectonic evolution in the Tin Amzi area (after Bertrand et al. 1986a). *Se* is the Eburnean foliation in the granulites; *Sp1* is the main Pan-African refoliation in the gneisses; and *Sp2* is the superimposed lower-grade foliation.

(M1) and to 5±1.5 kbar and 600±50 °C (M2). M2 associated migmatitic granites are related to or slightly postdate large-scale EW recumbent isoclinal folds. These folds are in turn refolded by NE-trending open folds. A third retrogressive assemblage is observed within the mylonite belt. The Iherane gneisses and migmatites are crosscut by the Tin Amzi high-level plutonic complex (granitic to granodioritic) which is itself deformed under relatively low-temperature conditions in the mylonite belt.

The Upper Tectonic Unit. In this unit, gneisses have a gently northeastward dipping foliation Sp1 and show a conspicuous NE stretching lineation. This foliation corresponds to upper amphibolite-facies conditions and to a complete tectonic and metamorphic reworking of ancient gneisses similar to the Iherane granulitic gneisses (lower unit). This Sp1 foliation is deformed by tight overturned folds with NW-trending axes, developing locally a new axial planar cleavage Sp2 at low angle to the first foliation (less than 30° toward the northeast). The Anfeg and Amsel plutons crosscut Sp1, but Sp2 affects their outer zone. They are considered as broadly syntectonic and emplaced under amphibolite-facies conditions.

The Mylonitic Belt. This separates the two tectonic units described above and is 50 to 100 m thick. The mylonitic surface is parallel to Sp2 and reflects epidote-amphibolite facies conditions, it bears a N30° stretching lineation and shows shear sense criteria which indicate a NNE sense of movement for the upper unit. The overall parallelism between Sp1, Sp2 and the mylonitic foliation suggests a continuum between these deformation phases: they are interpreted as the result of progressive tangential deformation occurring while metamorphic conditions were waning (Bertrand et al. 1986a).

U/Pb studies on zircon on both lower and upper unit rocks (Bertrand et al. 1986a) have shown that: (1) the granulitic M2 metamorphism and the associated migmatites are 2075±30 Ma old; (2) the emplacement age of the Tin Amzi plutonic complex in the lower unit is not known precisely (612+50/−20 Ma), but sphenes indicate cooling ages of 578±6 Ma, and (3) the emplacement of the Anfeg pluton in the upper unit took place at 615±5 Ma.

These results confirm the existence of important Pan-African thrust tectonics in Central Hoggar (Boullier and Bertrand 1981; Bertrand et al. 1986a) which partly reworked the continental Eburnean crust 615 Ma ago. Similar thrust tectonics have also been described in other places in Central Hoggar (see Bertrand et al. 1986a for a review) and in the Timgaouine area (Lapique et al. 1986), where horizontal foliation and associated stretching lineation are well expressed in the Aouilene orthogneisses (Fig. 7) and dated between 629 Ma (zircons) and 614 Ma (sphenes; Bertrand et al. 1986b). Farther west, thrusting grades to wrenching along the Tin Di-Tin Eifei lineament, which may be considered as a major decoupling zone outlining the true limit of the so-called Central Polycyclic Hoggar (Lapique et al. 1986).

In Aleksod, occurrence of eclogites in one of the major mylonitic zones stresses the importance of these thrusts. Sautter (1985), from a petrological study, indicated that the eclogites suffered high-pressure hydrous eclogite-facies conditions at 750 °C and more than 15 kbar, and subsequently

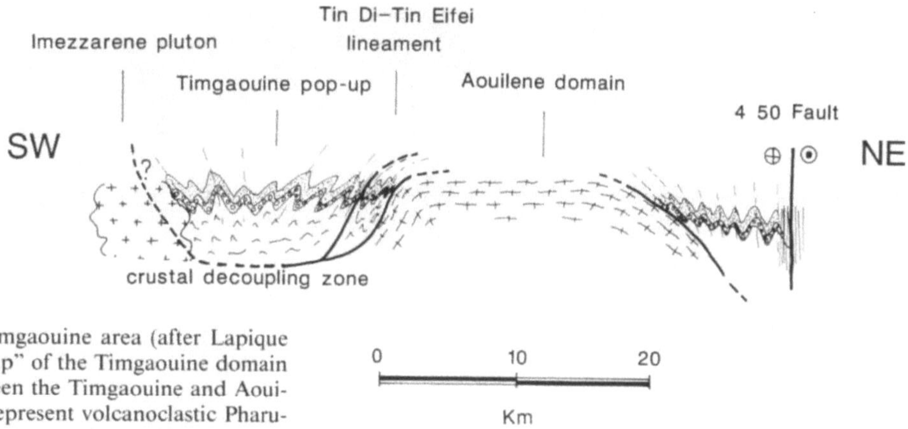

Fig. 7. Interpretation of the Timgaouine area (after Lapique et al. 1986) showing the "pop-up" of the Timgaouine domain and the decoupling zone between the Timgaouine and Aouilène domains. The sediments represent volcanoclastic Pharusian II Group

have been retrogressed with the host rocks (metasediments, orthogneisses) under upper amphibolite-facies conditions.

As can be seen in the Timgaouine area and in the Arefsa, the Pan-African thrust tectonics involved not only the Eburnean basement and high grade metasediments, but also the so-called Pharusian sediments of Upper Proterozoic age (Pharusian II in Timgaouine, Lapique et al. 1986; volcanoclastics of the Arefsa in the classical definition area of Pharusian, unpublished observations by Briedj and Bertrand).

3.3 *Late Pan-African Tectonic Evolution*

After the tangential tectonics, shield-scale strike-slip ductile shear zones and brittle faults were pruduced, which were long considered the main tectonic manifestation of the Pan-African event in Central Hoggar (Bertrand and Caby 1978; Vitel 1975). These vertical shear zones did not operate at the same time; some of them were related to the thrust tectonics, such as the Tin Di-Tin Eifei lineament (Lapique et al. 1986), while others clearly postdate this event, such as the 4° 50′ dextral fault. The Taourirt granites (Boissonnas 1973), which are post-tectonic relative to the thrusting event, crosscut some of these shear zones (In Amguel) but are also deformed by some others (4° 50′). There seems to be an overall control of the Taourirt granites by the latest wrench faults.

3.4 *Pan-African Magmatism*

The Pan-African granitic plutonism of Central Hoggar has been studied recently (Bouabsa 1987; Moulahoum 1988; Bertrand et al. 1987). Syntectonic plutons (Anfeg, Tefferkit) show subalkaline (or monzonitic calc-alkaline) affinities with variable mantle contamination (Anfeg: 615 Ma, Sri = 0.705). The post tectonic "Taourirt" granites are high-level intrusive plutons with subvolcanic apical differentiations; they are peraluminous in character suggesting a crustal origin (In Tounine granite: 521 Ma, Sri = 0.723, Moulahoum 1988). The Sn-W mineralizations are related to a younger magmatic group of unknown age: the albite-topaz-bearing fine-grained granites and porphyry dykes which crosscut the Taourirt granites. Earliest magmas have been interpreted as formed at the crust-mantle interface during the thrusting event. The peraluminous "Taourirt" magmatism is controlled by local extensional structures induced by the wrench fault network.

4 The Eastern Pharusian Belt

In the eastern branch of the Pharusian belt, large areas remain unknown, especially in the N, NE and SW, on both sides of the traverse mapped by Gravelle (1969) between 22° N and 23° N.

In the Silet area (Fig. 8) studied by Gravelle (1969), two distinct tectono-metamorphic cycles were defined, Pharusian I and Pharusian II, separated by a major unconformity (Bertrand et al. 1966).

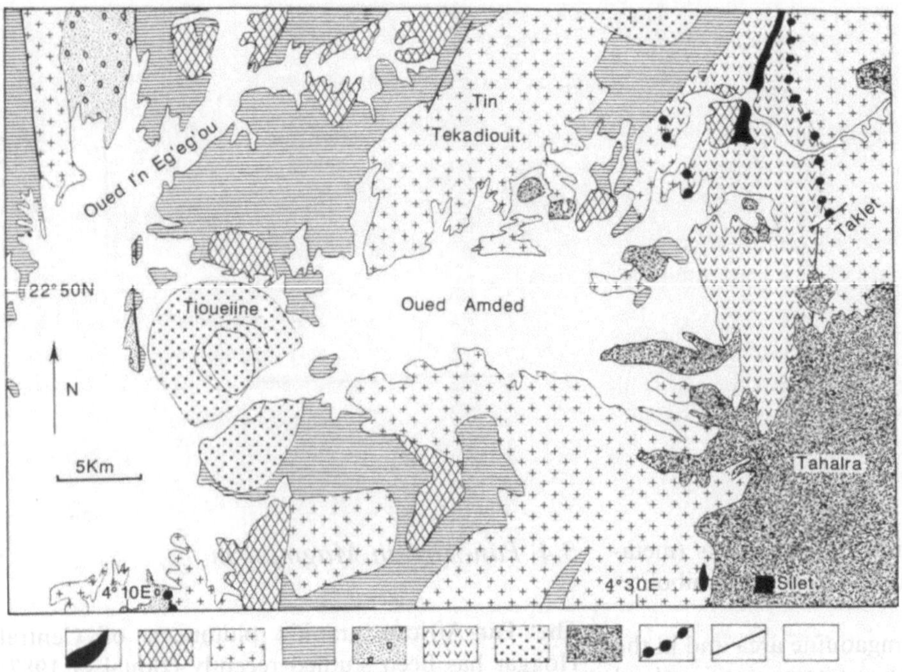

Fig. 8. The Silet area (after Gravelle 1969). *1* Serpentinized ultramafics; *2* metamorphic gabbros and diorites; *3* calc-alkaline Pharusian I granitoids; *4* volcanic and volcanoclastic Pharusian I Group; *5* gresopelitic Pharusian II Group; *6* Irrelouchem Pharusian II volcanics; *7* post-tectonic Taourirt granites; *8* Phanerozoic basalts; *9* unconformity; *10* fault; *11* alluvial sands

4.1 Pharusian I

The oldest cycle began with the deposition of a carbonate platform formation (Timesselarsine Series) on a presumed Eburnean basement: the unconformity occurs near Timgaouine where Conophyton stromatolites are well preserved (Gravelle and Lelubre 1957). Some pillow or massive basalts with transitional to weakly alkaline affinities indicate a continental-rift environment (Dupont et al. 1987; Dupont 1987).

Lenses of ultramafic to mafic material are also present in the Timesselarsine Series (Gravelle 1969). A preliminary geochemical and petrographical study precludes an oceanic tholeiitic affinity as proposed by Abed (1983), but suggests an island-arc or marginal-basin environment (Dupont 1987).

The major magmatic event occurring during the Pharusian I cycle is represented by early calc-alkaline quartz diorites and tonalites and by the post-tectonic emplacement of the Taklet batholith: granodiorites, diorites, calc-alkaline granites described by Gravelle (1969), Fabriès and Gravelle (1977), Dupont (1987) and Dupont et al. (1987). Caby et al. (1982) have obtained ages of $868 \pm 8 - 5$ Ma for the older group and 839 ± 4 Ma for the Taklet batholith (U/Pb on zircons). The Timesselarsine Series was deformed by isoclinal recumbent folds under greenschist to amphibolite-facies metamorphic conditions (Lapique et al. 1986; Boullier 1982) prior to the emplacement of the older calc-alkaline magmatism ($868 + 8 - 5$ Ma). Thus the Pharusian I cycle has been interpreted by Dupont (1987) as a rifting event followed by a subduction.

4.2. Pharusian II

The Pharusian II cycle began with the deposition of the Amded Series which contains detrital elements from the Pharusian I Formation described above. The upper part of the Amded Series is intruded by dolerites, basalt and rare tholeiitic andesites in an island-arc site (Dupont et al. 1987; Dupont 1987; Lapierre et al. 1987). The calc-alkaline Irrelouchem Volcanic Series overlies the Amded Series and has been dated at 680 ± 36 Ma (Liégeois, in Dupont 1987) by Rb/Sr isochron (Sri = 0.7031).

Finally, granodiorites and rare granites were emplaced which display calc-alkaline geochemical characteristics similar to those of the Irrelouchem Series, suggesting an active margin site.

All the Pharusian II magmatic episodes have island-arc (Dupont 1987) or active continental margin (Chikhaoui 1981) characteristics. They predate the D2 Pan-African (s. s.) event which is expressed by vertical schistosity, open folds and, in the Timgaouine area, by wrenching grading to thrusting (Fig. 7, Lapique et al. 1986) which occurred between 629 and 614 Ma (Bertrand et al. 1986b), (Sect. 4.3 and Fig. 7 in this volume).

The Immezarene pluton is late tectonic relative to D2; it is a porphyritic granite which has been dated at 583 ± 7 Ma (Bertrand et al. 1986b).

5 The Western Pharusian Belt

The two branches of the Pharusian belt are very different in lithology and are separated by the Adrar Fault. If the eastern branch is not well known near the Adrar Fault, more data are available on the western branch, especially in its southern part (Central Iforas), where extensive work has been done since 1976.

5.1 *The 3.5–2 Ga Old Basements*

5.1.1 The Granulitic Units

Two major granulitic units (In Ouzzal and Iforas granulitic units) and two smaller units (Adrar Bezzeg and Tin Essako, in Mali, Fig. 9) are involved in the Pharusian belt. Another may exist north of Adrar Bezzeg in Algeria (Kiniakine, pers. commun.). They form narrow, several hundred km-long NS-trending domains with mylonitic margins. Although the two units show some slight differences in the rock-type ratio, metamorphic grade and intensity of Pan-African reworking, they present similar Archaean and Eburnean evolutions.

Both units comprise granitic gneisses, charnockitic intrusives, ultramafics and mafics and metasediments (aluminous, Al-Mg, quartzofeldspathic gneisses, magnetite bearing quartzites and marbles). Norites are often intrusive into the metasediments. Granitic gneisses and charnockitic intrusives often have subalkaline (or monzonitic calc-alkaline) affinities. The Archaean age of the material is indicated by Rb/Sr data (Ferrara and Gravelle 1966) and by U/Pb data on zircon (upper intercepts, Lancelot et al. 1976). However, the lower U/Pb intercept (zircon data, same authors) shows that the age of the granulitic metamorphism is Eburnean (~2.1 Ga).

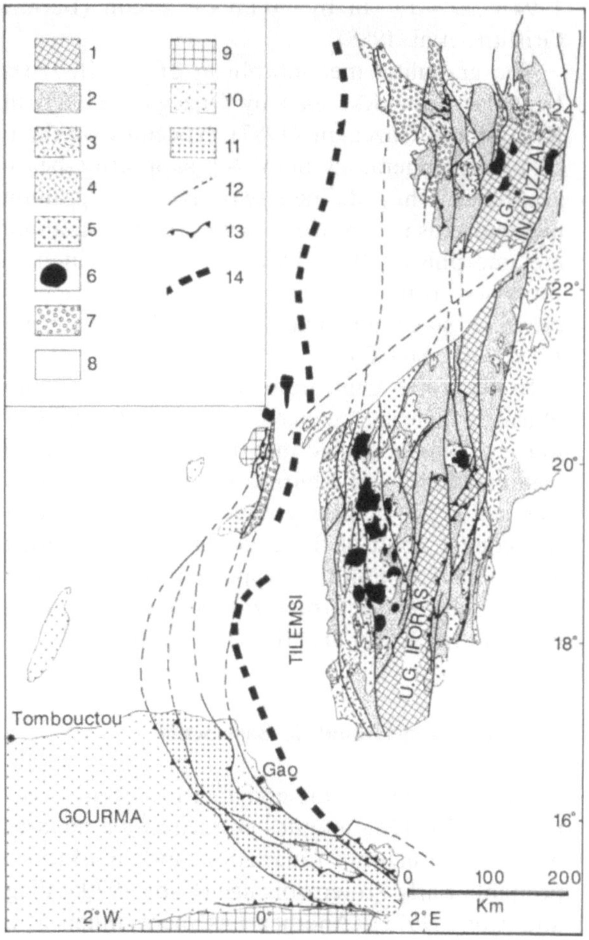

Fig. 9. Simplified geological map of the western Pharusian belt, the central Iforas batholith, the Tilemsi magmatic island arc and the foreland nappes (after Boullier 1982, and Fabre et al. 1982). *1* 2 Ga granulitic units *(U. G. Iforas* and *U. G. In Ouzzal); 2* 2 Ga basement and Middle to Upper Proterozoic sediments reworked during Pan-African tectonics; *3* volcanoclastic Late Upper Proterozoic sediments; *4* Tilemsi Island-Arc Group; *5* Pan African granitoids (mainly in the central Iforas batholith); *6* post-tectonic alkaline granitoids and rhyolites (east of the suture zone); undersaturated Permian alkaline ring-complexes (west of the suture zone); *7 Série Pourprée* and Nigritian molasses; *8* Phanerozoic cover; *9* 2 Ga basement of the West African craton; *10* passive margin sediments; *11 (10)* involved in Gourma and Timetrine nappes; *12* fault; *13* thrust; *14* suture zone as defined by gravity anomalies (Bayer and Lesquer 1978)

Just after the granulitic metamorphic event, several episodes of Eburnean magmatic activity can be distinguished: (1) intrusion of syenites, (2) intrusion of an early generation of carbonatites and (3) intrusion of late carbonatites. These late carbonatites show uncommon mineralogy and high REE concentrations (Ouzegane et al. 1988). The carbonatitic events have been dated at

1994+22/−17 Ma by U/Pb on zircon (Bernard-Griffiths et al. 1988).

The granulitic metamorphism of the In Ouzzal Unit has been described by Ouzegane (1987) and Kienast and Ouzegane (1987), especially on the basis of the mineralogy of Al-Mg sapphirine-bearing granulites. They defined two stages of granulitic metamorphism: an early high-pressure granulitic metamorphism (10 ± 1.5 kb, 800-900 °C) which they supposed to be Archaean in age, and a later low-pressure stage attributed to the Eburnean (5 ± 1.5 kb, 700-800 °C).

In the Iforas Unit, from structural and petrographical evidence, Boullier (1982) and Boullier and Barbey (1988) assumed that both stages of granulite metamorphism are Eburnean in age and reflect a classical (P, T, t) history related to thickening and subsequent uplift of the crust during the Eburnean event. Such an uplift is supported by the existence of unconformable Proterozoic sediments lying on both the In Ouzzal and Iforas granulitic units.

5.1.2 The Non-Granulitic Basements

In NW Hoggar the Tassendjanet basement is composed of 2090 Ma granites crosscutting a metasedimentary formation (quartzites and marbles) of probable Early Proterozoic age (Caby 1970; Allegre and Caby 1972). Another basement inlier is exposed at the border between Algeria and Mali (Tisangenine inlier, Caby 1973) and presents many similarities to the Tassendjanet basement.

5.2 The Middle to Upper Proterozoic Formations

Post-Eburnean sedimentary of volcanoclastic formations are known in the NW Hoggar and Iforas regions.

5.2.1 Middle Proterozoic Quartzites

In NW Hoggar where these formations were deformed and metamorphosed during Pan-African events, their age has been determined by dating intrusive rhyolitic sills at 1755 ± 10 Ma (Caby and Andreopoulos-Renaud 1983). An unmetamorphosed and well-preserved equivalent is known in the Ahnet mountains north of the In Ouzzal Granulitic Unit (Moussine-Pouchkine et al. 1988).

In the Iforas region, Middle Proterozoic quartzitic formations are abundant in the Tin Essako area, where many alkaline rhyolitic sills or lavas occur as in NW Hoggar. Their basement is not known but is suspected to be similar to the Eburnean granulitic units (Davison 1980; Boullier 1982).

5.2.2 Upper Proterozoic Platform Sequence

In the Tassendjanet area, the non-granulitic basement is unconformably overlain by a platform sequence, the so-called Stromatolite Series (Caby 1970; Moussine-Pouchkine et al. 1988). It is composed of quartzites with a basal microconglomerate, black and purple pelites, marbles and dolomites with Conophyton. The same formation overlies the Adrar Bezzeg granulitic unit in the Ourdjan hills in the Iforas region (Fabre et al. 1982). This formation has been correlated with the platform cover of the West African Craton in the Taoudeni Basin (Atar Group, Bertrand-Sarfati 1972). In the Tassendjanet area, pretectonic mafic sills emplaced in the Stromatolites Series have been dated at 793 ± 32 Ma (Clauer 1976).

5.2.3 Upper Proterozoic Volcanoclastic Formations

Volcanoclastic formations are well known in NW Hoggar where they are called *Série Verte* (Green Series) by Caby (1970). An undeformed and unmetamorphosed equivalent of the *Série Verte* outcrops in the northern part of the In Ouzzal unit but the unconformity has never been observed. It is mainly composed of unmetamorphosed andesitic lava flows (Chikhaoui et al. 1978; Chikhaoui 1981). Porphyritic andesitic dykes which crosscut the granulites may be related to this volcanic episode. The Stromatolite Series and the *Série Verte* are generally separated by tectonic contacts.

The volcanoclastic formations are rarely well dated except for the Tafeliant Group in the Iforas region: it unconformably overlies a 696-Ma quartz diorite (Caby and Andreopoulos-Renaud 1985) and is intruded by 634-Ma rhyolitic dykes (Liégeois 1987). In the northern Iforas region, the Oumassene volcanoclastics are also of assumed Upper Proterozoic age and are composed of andesitic lavas, breccias and rare sediments (tuffs, conglomerates and arkoses, Chikhaoui 1981). Many other volcanoclastic formations of the Iforas region also show similarities to the Tafeliant Group or the Oumassene formation (see for example the Ibedouyen volcaniclastics, Boullier 1982). Correlations with the Pharusian I and Pharusian II groups in the eastern Pharusian belt are impossible, because of the difficulty of comparing terranes on both sides of the Adrar Fault and the lack of geochronological data.

5.3 *The Kidal Gneissic Assemblage*

The Pan-African evolution of the Iforas granulitic unit and of the whole central Iforas in the western Pharusian belt cannot be dissociated from the evolution of the Kidal Assemblage, which has been defined by Boullier et al. (1978) as "resulting from the common high-grade tectono-metamorphic evolution during the major deformation phase of a granulitic basement, metasediments and pre- to syn-tectonic intrusives".

The basement is probably made of Eburnean granulites identical to the main In Ouzzal and Iforas granulitic units and to the smaller Adrar Bezzeg granulitic unit (see Fig. 11 for location), as the same rock types have been observed (spinel clinopyroxenites, norites, magnetite-bearing quartzites, etc.; Boullier 1982).

North and east of the Iforas Granulitic Unit, part of the metasediments is attributed to the Middle Proterozoic; thick aluminous quartzites are associated with alkaline metavolcanic or meta-igneous gneisses dated at 1837 ± 179 Ma (Caby and Andreopoulos-Renaud 1983).

West of the Iforas Unit, platform-type metasediments from the Kidal Assemblage (quartzites, schists and marbles) are overlain by basaltic flows with tholeiitic affinities (Leterrier and Bertrand 1986). Such metasediments are assumed to represent a Upper Proterozoic shelf-type formation similar to the Stromatolite Series.

Pretectonic intrusives range from ultrabasites to gabbros and anorthosites. They show the same chemical affinity as the basaltic flows cited above and have been attributed to a back-arc spreading center (Leterrier and Bertrand 1986). The large (60% of the volume of the Kidal Assemblage) pretectonic metadiorite-metatonalite plutons suggest a significant change in the geotectonic environment; a continental setting related to a subsidiary back-arc spreading developed during a subduction stage is the most likely interpretation (Leterrier and Bertrand 1986).

5.4 *The Pan-African Tectono-Metamorphic Evolution*

In Central Iforas, three successive ductile deformation phases (D1, D2, D3) have been defined by Boullier (1979, 1982) and a fourth brittle phase (D4) has been described by Ball (1980). Correlations between the D2, D3 and D4 phases and the main magmatic episodes of the Iforas Batholith have been discussed by Boullier et al. (1986) and Liégeois et al. (1987) (see Sect. 6.3 in this volume).

Figure 10 gives a schematic interpretation of the central Adrar des Iforas and its tectonic pile comprising three main structural units: from bottom to top, the Kidal gneissic assemblage, the Iforas granulitic Unit and the allochthonous metasediments.

The main deformation phase in Central Iforas is D1, which can be subdivided into two subsidiary phases interpreted as a continuum (Boullier 1982; Champenois et al. 1987). D1 corresponds also to a continuous metamorphic evolution (10 ± 1 kbar and 700 °C ± 50 °C, to 7–8 kbar and 650–750 °C); migmatization took place in the latest stage. The S1 foliation, where not affected by the later deformation events, is flat-lying and bears a roughly N-S trending stretching lineation. The S1 foliation in the Kidal Assemblage and the mylonitic foliation of the northern and eastern borders of the Iforas granulitic Unit and of the southern tail of the In Ouzzal Unit, are parallel. The Kidal Assemblage lies structurally below the Eburnean granulites (Fig. 10). Assumed Upper Proterozoic platform sediments (equivalent to the Stromatolites Series) have also been involved in the D1 thrust tectonics and form an uppermost nappe above the Iforas granulites (Ibedouyen, Boullier et al. 1978; Boullier 1982). This D1 event has been interpreted as resulting from the thrusting of the granulites onto the Kidal Assemblage in a roughly northward direction (Boullier et al. 1978; Wright 1979; Davison 1980; Boullier 1982; Champenois et al. 1987). Caby (1987a) favoured an alternative interpretation, contending that the Eburnean granulites are rooted in and outlined by vertical strike-slip lithospheric shear-zones. Two ages have also been suggested for the D1 tectonics considering the relationship between the D1 and D2 structures relative to a 696 Ma old intrusive in the southermost part of the central Iforas; D1 is considered to be younger than 696 Ma by Ball and Caby (1984) and Caby and Andreopoulos-Renaud (1985) and older than 696 Ma by Boullier et al. (1978).

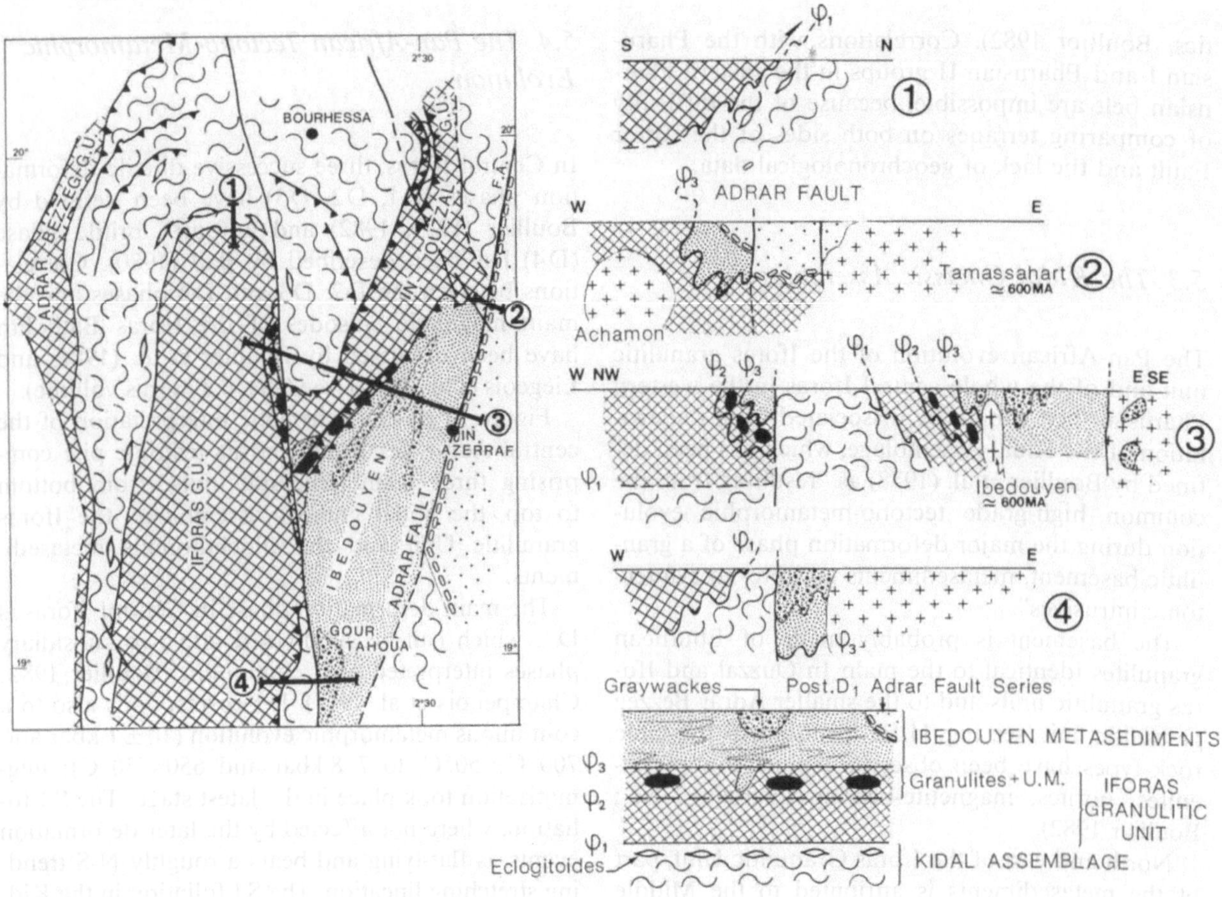

Fig. 10. Interpretation of the Central Iforas domain. The superposition of the different structural units is schematized in the lower right section

The D2 phase refolds the S1 foliation and thrust planes and corresponds to a NS-SE to E-W shortening, which is expressed by upright to overturned folds and by some sinistral NS to NNE-SSW wrenching movements (Boullier 1982). In the Ibedouyen hills (Fig. 10), a volcanoclastic formation is also involved in the D2 deformation and may be correlated to the Tafeliant Group described by Ball and Caby (1984). The age of the D2 phase is constrained between 620 Ma and 590 Ma (Ball and Caby 1984; Bertrand et al. 1984a; Liégeois and Black 1984; Liégeois et al. 1987; Caby et al. 1985; Caby and Andreopoulos-Renaud 1989). Metamorphic conditions were lower in the Kidal assemblage during D2 but still in amphibolite facies.

The D3 phase is characterized by NS to N 20° dextral shear zones, one of which bounds the Iforas granulitic Unit on the west (Abeibara-Rahrous shear zone) and was active between 566 and 535 Ma (Lancelot et al. 1983). A 7 km offset of the Abeibara-Rahrous shear zone has been estimated by Boullier (1986). The D3 phase corresponds to NE-SW compression.

The Iforas granulitic Unit has recorded the whole Pan-African evolution: deformation, metamorphism and magmatism (Boullier 1979, 1982; Bertrand et al. 1984a; Boullier and Barbey 1988). The deformation is expressed within the Iforas Granulitic Unit by curved shear zones, mainly attributed to D2, which controlled gabbroic to granitic intrusives. Prograde metamorphism superimposed on Eburnean resorption coronitic reactions indicates that the granulites suffered P, T conditions as high as 620 ± 50 °C and 5 ± 1 kbar; Boullier and Barbey (1988) integrated this metamorphic evolution in the Pan-African structural thrusting scheme proposed by previous authors (Boullier et al. 1978; Wright 1979; Davison 1980; Boullier 1982; Champenois et al. 1987).

5.5 The Molassic Formations

Molassic sediments were named Nigritian by Karpoff (1960) in the Adrar des Iforas region and are equivalent to the *Série Pourprée* (Purple Series) of NW Hoggar (Caby and Moussu 1967). In the Adrar des Iforas, the molasse series begins with ignimbrites, felsic lavas, breccias and rhyolites which are intruded by the latest alkaline ring-complexes and which could have been fed by the NS dykes dated at 543 ± 9 Ma (Liégeois and Black 1984). The molassic sediments show important sedimentological variations across the Touareg shield (they are, for example, much more developed in the Pharusian belt than in the polycyclic central Hoggar) and in the western pharusian itself, where they were deposited in isolated basins (Fabre et al. 1988). They probably suffered the D3 tectonic event (NS dextral shear zones).

Along the Adrar Fault, a clastic sedimentary series lying unconformably on the deformed and metamorphosed Upper Proterozoic marbles has been attributed by Caby et al. (1985) to the *Série Pourprée* molasse. It is intruded by granites at $581 + 7/-6$ Ma, very similar to the Immezzarene pluton in the eastern Pharusian belt (583 ± 7 Ma, Bertrand et al. 1986b). These granites are considered as pre-D3 by Caby et al. (1985) but as syn- to late-tectonic relative to a Pan-African thrusting event by Boullier et al. (1978) and Lapique et al. (1986). These conflicting interpretations of the structural evolution of the sediments along the Adrar Fault raise one of the most important remaining questions about the structural and geodynamic evolution of the whole Hoggar Shield. No modern structural data and no detailed maps are available in the 100 km-wide area between the Adrar fault and the Timgaouine region. This area is crucial to our understanding of the relationship between the 2-Ga-old reworked basement of both the western Pharusian belt and the Polycyclic Central Hoggar, the 850-Ma-old Pharusian I terrane lying in between. Is there another suture here?

6 The Central Iforas Batholith

A large (250×100 km^2) composite calc-alkaline batholith occurs close to the Pan-African suture (Fig. 9) between the Tilemsi island-arc terrane and the reworked basement of the western Pharusian belt (Kidal Assemblage), affected by the D1 nappe tectonics. Three main magmatic stages have been distinguished (Liégeois et al. 1987).

6.1 The Cordilleran Stage

The cordillera (by anology with the Andes) is older than 620 Ma and is essentially composed of volcanosedimentary sequences (Chikhaoui 1981) representing marine, aerial or subaerial environments (Fabre 1982). Some pretectonic plutons are also observed (Ball and Caby 1984). In the Adrar des Iforas region, the Tafeliant Series lies unconformably on platform sediments and on a $696 + 8 - 3$ Ma old granodiorite (U/Pb on zircon, Caby and Andreopoulos-Renaud 1985) and contains glaciogenic conglomerates (Caby and Fabre 1981) and rhyolitic sills which have been dated at 634 Ma (Rb/Sr whole rock isochron, Liégeois 1987). This series is considered to be older than the D1 event by Ball and Caby (1984) and Caby and Andreopoulos-Renaud (1985), whereas it is considered by Boullier (1982) and Liégeois et al. (1987) to lie on folded platform sediments and thus to post-date this D1 tectonic event. Large cordilleran-related plutons are rare. To the north, the Oumassene formation is probably a time-equivalent of the Tafeliant formation, but it comprises a large volume of andesite flows and tuffs in addition to restricted basal clastics and agglomerates (Chikhaoui 1981). Similar, more felsitic formations crop out along the eastern edge of the central Iforas Batholith; they rest unconformably upon deformed (D1) assumed Upper Proterozoic quartzites and marbles (Ourdjan area, J. M. Bertrand, pers. commun., 1990).

6.2 The Collision Stage

The Iforas batholith is essentially late- to post-tectonic. Its emplacement occurred after the closure of the western ocean and during the collision of the western Pharusian belt with the West African Craton (620–580 Ma; Bertrand and Davison 1981; Liégeois and Black 1984, 1987; Liégeois et al. 1987; Caby and Andreopoulos-Renaud 1989). Although diachronism possibly exists between different parts of the batholith, the same evolution of intrusive phases has been observed from mafic to more felsic compositions: quartz monzodiorites, granodiorites, porphyritic or fine-grained granites. All the late-tectonic plutons display a well-defined high-K calc-alkaline trend (Liégeois et al. 1987) and similar ages (Rb/Sr isochrons 595 ± 25 to 581 ± 15 Ma) with similar Sri ratios (0.7033 to 0.7053).

6.3 Post-Collision Magmatism

6.3.1 The Calc-alkaline to Alkaline Transition

This transition occurred in post-tectonic conditions relative to D2 and after considerable uplift of the belt. The relationships with D3 (NS to N20° dextral strike-slip shear zones or faults) and D4 tectonics (brittle sinistral NNW-SSE and dextral ENE-WSW conjugate faults, Ball 1980) have been discussed in detail by Boullier et al. (1986), while the petrographic, petrogenetic and geochronological point of view has been given in Liégeois and Black (1984, 1987).

The latest calc-alkaline magmatic stage is expressed by spectacular E-W acid dyke swarms (Fig. 11) which are clearly younger than D3. They have yielded Rb/Sr isochrons in the range 565 ± 14 Ma and 544 ± 12 Ma (Liégeois and Black 1984). Unroofing began during or just after the emplacement of the E-W dykes, leading to cooling of the crust.

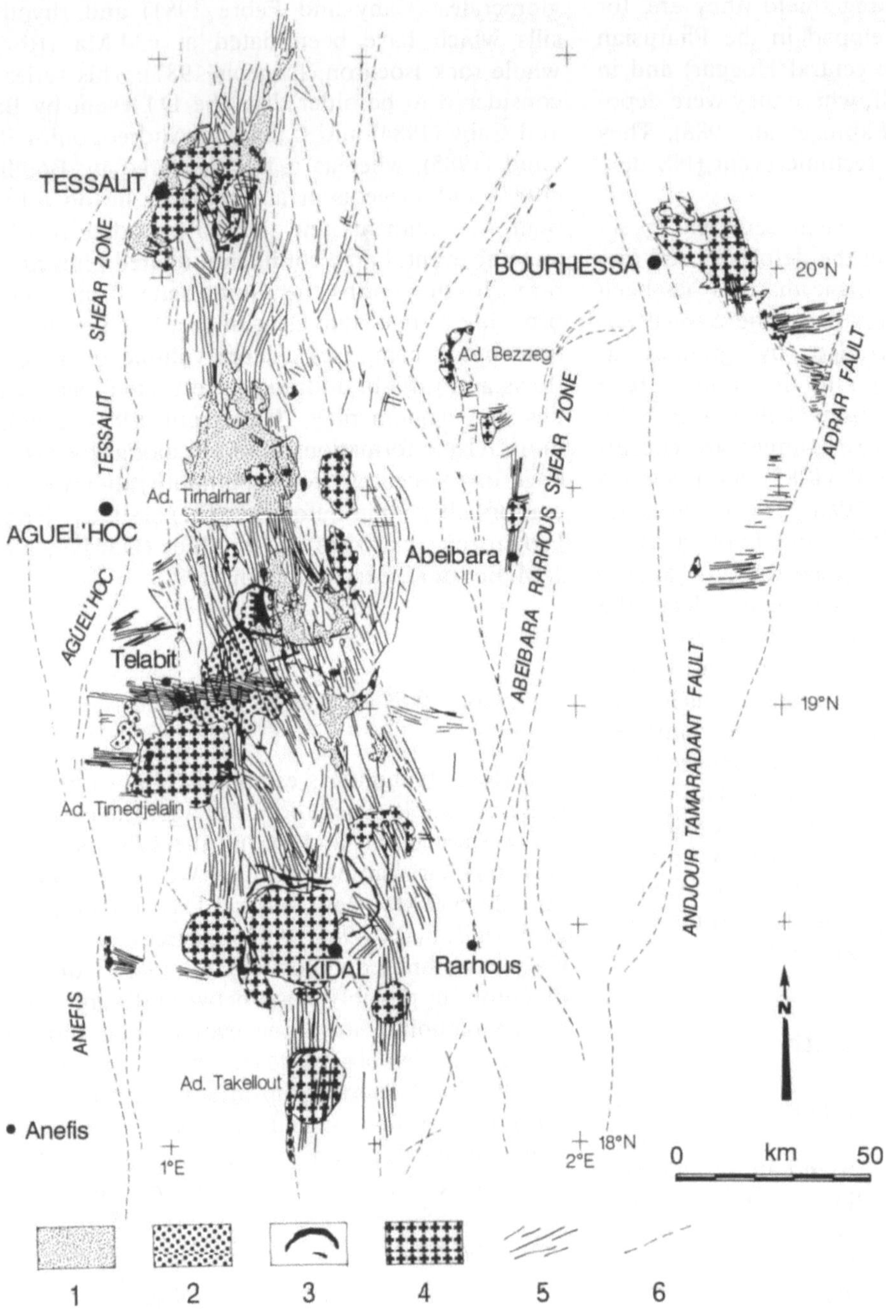

Fig. 11. Geological map of the alkaline province of the Adrar des Iforas after Fabre et al. (1982) and Boullier et al. (1986). *1* rhyolites; *2* alkaline granite of Tarhmert; *3* ring dykes; *4* alkaline ring complexes; *5* dykes; *6* faults

6.3.2 The Alkaline Stage

Subsequently, NS felsic alkaline dyke swarms (Fig. 11) were intruded (543 ± 9 Ma, Rb/Sr isochron, Liégeois and Black 1984) along the entire length of the batholith; they predated and were partly contemporaneous with the emplacement of the alkaline ring-complexes which are typical A-type granitoids (Ba et al. 1985). The slightly arcuate shape of the NS dykes swarms also seems to outline the large positive gravity anomalies described by Ly et al. (1984). These anomalies have been interpreted as resulting from the presence, at depth (12 km), of dense intrusive mafic bodies related to the alkaline magmatism. Two ring-complexes have been dated by Liégeois and Black (1984) by Rb/Sr isochrons: Kidal at 561 ± 7 Ma and Timedjelalen at 549 ± 6 Ma. In the Iforas batholith, the transition from subduction-related calc-alkaline magmatism to within-plate alkaline magmatism in a very short time-span (560–540 Ma) corresponds also to a reversal of the regional stress field during and after collision ($\sigma 1$: N 135–110° (D2); N 50–60° (D3); N 105° (D4); Boullier et al. 1986). Liégeois and Black (1987) suggested that the first reversal (D2–D3) is responsible for the slicing of the eastward-dipping subducted plate (lithosphere beneath the West African Craton), thereby allowing the rise of asthenospheric mantle believed to be the source of the alkaline magmatism. The tapping of this new source occurred between D3 and D4 events.

7 The Tilemsi Island-Arc

7.1 *Lithology*

The 100 km-wide Tilemsi magmatic arc preserved along the suture zone (Figs. 9, 12) has been studied by Caby et al. 1989).

Bimodal volcanics overlying a marble-dolomite sequence represent the oldest unit of the Tilemsi Group. Pillowed metabasalts interfingered with dacitic breccias represent arc tholeiites. They are overlain by a 3-km-thick unit of volcanic greywackes and conglomerates showing turbiditic sedimentary features and cut by various dyke swarms.

Near Tessalit, these formations are overlain by a metapelitic unit which was interpreted as a glaciomarine deposit and correlated with the Tafeliant formation (cordilleran stage, Sect. 6.1).

7.2 *Magmatism*

Several generations of plutons and associated dykes were emplaced in the Tilemsi group (Caby et al. 1989):

1. laccoliths of layered gabbro intruded the oldest carbonate formation. Geochemical data suggest a tholeiitic character and a MORB affinity (Liégeois 1987; Caby et al. 1986);
2. gabbro-norite lopoliths induced an early high-temperature metamorphism in the greywackes which were transformed in grey gneisses displaying granulite-facies assemblages. A mobilisate has yielded a 710 + 6 − 3 Ma age (U/Pb on zircon) for the metamorphism of the grey gneisses;
3. metaquartz-dioritic sheets intruded the grey gneisses and the high-grade metagreywackes at 726 + 7 − 3 Ma;
4. granodiorites with a calc-alkaline affinity were emplaced within the volcano-sedimentary formations at 635 ± 5 Ma to the east of the arc.

7.3 *Deformation*

Pan-African deformation and metamorphism affected the whole Tilemsi arc (Caby et al. 1989). In the north, upright folds are associated with steep, axial-plane, slaty cleavage generated in a flattening regime. Strain progressively increases toward steep, N-trending sinuous shear zones developed in greenschist-facies conditions.

Towards Aguel'hoc to the south, low-dipping and recumbent folds are associated with amphibolite-facies metamorphism. The metamorphic grade increases toward the Aguel'hoc gneisses, dominantly metasedimentary in origin (cordierite-bearing gneisses, marbles, quartzites) but including also the same meta-igneous suite as in the Tilemsi arc. The Aguel'hoc gneisses are characterized by upper amphibolite to granulite-facies assemblages and interlayered with synmetamorph trondjhemite, norite and troctolite bodies. A model Rb-Sr age on biotite at 575 ± 10 Ma indicates the age of cooling under the retention temperature for biotite in this province (Bertrand and Davison 1981).

7.4 *Interpretation*

The Tilemsi arc is characterized by an abundance of pre-, syn- und post-tectonic mafic to intermediate plutons and by the lack of granites.

Sr, Nd and Pb isotope studies on metaquartz-diorites and on metagreywackes are in agreement with a depleted mantle source, excluding any significant derivation from an older sialic source and supporting an ensimatic origin for the magmatic arc (Liégeois 1987; Caby et al. 1989). On the other hand, the Sr and Nd isotopic compositions of a pretectonic granodiorite suggests the presence of a crustal reservoir. This change in magmatic source is interpreted by Caby et al. (1989) as the result of accretion of the ensimatic arc to the western Pharusian belt, before the continent-continent collision. From the similarity of the Pb and Nd isotopic characteristics of the magmatic rocks of the island-arc and of the batholith, Liégeois (1987) suggested that the same mantle source was available during the subduction stage and the collision stage, but with greater crustal contamination during the latest stage.

8 The Gourma-Timetrine Foreland Nappes

These nappes were thrust onto the West African craton (Timetrine) or onto its passive margin (Gourma). They are exposed west of the gravity high (Bayer and Lesquer 1978; Ly 1979) and interpreted as outlining the suture.

8.1 Timetrine

Several nappes are exposed in the Timetrine area and were extensively described by Caby et al. (1981) and Caby (1987a). The highest nappe is made of oceanic material represented by serpentinites, metagabbros and diabases (Karpoff 1960; Leblanc 1976; Caby et al. 1981). The lowest unit is composed of sericite quartzites, chlorite-albite schists and pillowed basalts, which probably represent the sediments and lavas deposited on the passive margin (Caby et al. 1981).

8.2 Gourma

This area was studied by Reichelt (1972), Caby (1979), Moussine-Pouchkine and Bertrand-Sarfati (1978), Lesquer and Moussine-Pouchkine (1980) and Davison (1980). The Gourma basin s. s. represents a passive margin basin with flysh-type terrigeneous sediments which are affected by upright folds with a steep axial-planar slaty cleavage parallel to the Gourma arc. The Gourma basin has been interpreted as an aulacogen (Moussine-Pouchkine and Bertrand-Sarfati 1978; Lesquer and Moussine-Pouchkine 1980).

The foreland nappes (Fig. 12) consist mostly of sericite-chlorite phyllites and of a series with quartzites, marbles and siliceous dolomites. A nappe of rooted turbiditic terrigeneous sediments was also emplaced as one of the highest nappes close to the Bourré window which is part of the Eburnean basement (de la Boisse and Lancelot 1977).

The internal nappes (Fig. 12) consist of aluminous quartzites and schists which have undergone high-pressure metamorphism with P, T conditions estimated at 15 kb and 550 °C (de la Boisse 1981). These nappes are rooted beneath the Amalaoulaou granulitic mafic body which was dated ca. 800 Ma by U/Pb on zircon (de la Boisse 1979) and was partially overprinted by eclogitic metamorphism. These mafic rocks are located on the site of the gravity high (Bayer and Lesquer 1978).

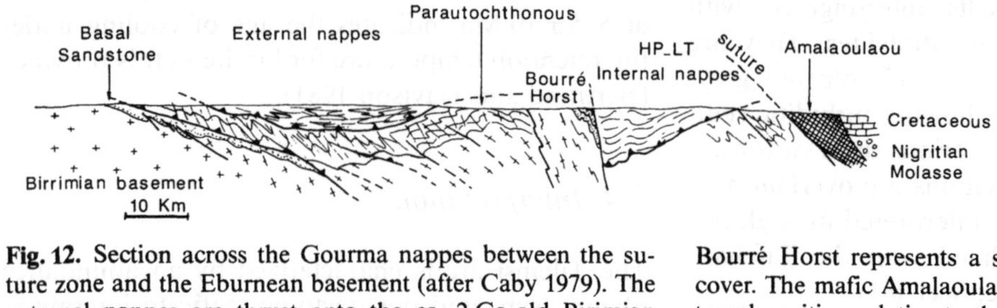

Fig. 12. Section across the Gourma nappes between the suture zone and the Eburnean basement (after Caby 1979). The external nappes are thrust onto the ca. 2 Ga-old Birimian basement and its autochthonous sedimentary cover. The Bourré Horst represents a slice of the same basement and cover. The mafic Amalaoulaou complex is exposed in an internal position relative to the HP-LT nappes and is a lateral equivalent to the Tilemsi magmatic island arc

9 Conclusion

The Hoggar shield is a complex assemblage of NS trending domains whose stratigraphic and tectonometamorphic evolution is not always easy to correlate. To summarize, several domains may be distinguished from E to W:

1. The Djanet-Tafassasset domain in eastern Hoggar and its Nigeran equivalent, the Eberjegui domain, was involved in a magmatic event at ca. 730 Ma (Caby and Andreopoulos-Renaud 1987). This domain is probably not very wide as Pan-African (s. s.) granites have been recognized in the Tibesti Massif by Pegram et al. (1976);
2. The polycyclic central Hoggar, where two main tectonic, metamorphic and magmatic events at 2 Ga and 600 Ma have been recognized;
3. The Pharusian I domain, which is only known in a narrow NS-trending belt west of the Tin Di-Tin Effei lineament and of the 4° 50′E fault, and which was involved in two main orogenic events at around 850 Ma and 600 Ma;
4. The western Pharusian belt of the Iforas and of NW Hoggar, in which an Eburnean basement and its Middle to Upper Proterozoic assumed sedimentary cover was involved in an early Pan-African event (pre-696 Ma) and in the Pan-African (s. s., ca. 600 Ma) event (Boullier et al. 1978), or only in the Pan-African (s. s.) event (Ball and Caby 1984). Volcanoclastic sediments of a cordillera were deformed before and during the emplacement of the Pan-African (s. s.) Central Iforas Batholith;
5. The Tilemsi island-arc domain;
6. The Timetrine and Gourma foreland nappes.

Except for the two westernmost domains, the whole shield presents as a common feature the various sedimentary formations, often volcanoclastic and glaciogenic, deposited between 730 and 600 Ma: the Tiririne and *Proche Ténéré* Formations, the low-grade Pharusian sediments in central Hoggar, the Pharusian II Formation, the *Série Verte* in NW Hoggar, the Tafeliant Group and the Ibedouyen greywackes in the Adrar des Iforas. All these sedimentary formations have suffered Pan-African deformation and metamorphism at ca. 600 Ma; consequently, even if these Upper Proterozoic sedimentary formations (and particularly the associated volcanics) show different depositional environments and tectonic evolution from one domain to another, it can be assumed that the Hoggar shield was an entity just before the deposition of these series and that the major accretion event took place before that time (Caby 1987b).

What are the traces of this accretion event? They could be seen in the basement of the Late Proterozoic series, as in the Eberjegui or Djanet-Tafassasset domains, or in the Pharusian I domain where pre-730 Ma or pre-850 Ma respectively, and post-2 Ga deformation is clearly expressed. In contrast, no deformation has been observed in Polycyclic Central Hoggar between 2 Ga and 615 Ma. In the western Pharusian belt, two interpretations have been proposed and supplementary geochronological data are necessary to solve the uncertainty. Boullier et al. (1978) and Boullier (1982) suggested that the thrust tectonics involving the 2 Ga granulites represent an early Pan-African event, which could be related to the main accretion event. Ball and Caby (1984) interpreted the thrust tectonics as strictly equivalent to the shortening deformation of the Late Proterozoic Tafeliant Group; in the latter case the central Iforas behaved passively during the accretion stage as did the Polycyclic Central Hoggar. However, it should be stressed that the continental blocks represented are very different: in Central Polycyclic Hoggar, Archaean rocks are rare, Lower Proterozoic sediments are abundant and Pharusian sediments and volcanics correspond probably to uppermost Proterozoic, while in the western Pharusian belt, Archaean material is dominant, the Middle Proterozoic is represented by thick fluviatile quartzites and the Upper Proterozoic by a carbonate platform.

The Upper Proterozoic volcanic and volcanoclastics sediments emplaced or deposited later than ca. 800 or 700 Ma, have different chemical characteristics throughout the Hoggar shield. Tholeiitic basalts were observed in the Arrei unit (Aouzeguer belt, Cosson et al. 1987) and calc-alkaline andesitic basalts in Arefsa (Central Hoggar, Leterrier and Briedj, pers. commun.); calc-alkaline series have been described in Pharusian II (Dupont 1987) and in the Adrar des Iforas region (Chikhaoui 1981). This suggests that one or several subduction zones operated during the Upper Proterozoic and were accompanied by some local extensional tectonics.

During the Pan-African deformation (s. s.) the accretion scars were reworked, mostly inducing strike-slip movement during the continent-continent collision along the Tilemsi suture on the west: Tiririne-Aouzeguer belt, Tin Di-Tin Effei lineament, Adrar Fault (?). These strike-slip movements could well be explained as the result of an oblique collision with the West African Craton margin.

Many questions remain to be answered before an understanding of the global evolution of the Hoggar shield is possible: what is the exact age of D1 in the Adrar des Iforas? Why were some large sialic domains deformed by thrusting during the collision and others by wrenching? What were the relative importance, the kinematics and the succession in time of the Pan-African strike-slip shear-zones? Is there any space and time migration of the deformation? The Pan-African Trans-Saharan belt of West Africa has been interpreted as the result of a Wilson Cycle ending around 600 Ma with collision and suturing between the West African Craton and the Pan-African mobile belt by Black et al. (1979) and Caby et al. (1981). If the latest stages of the Pan-African history of the belt are now well known, the first stages and especially the accretion stage of this history need to be better understood.

Acknowledgements
The author is very grateful to J.M. Bertrand for fruitful discussions and help during the preparation of the manuscript, and to R. Caby and J. Macaudière for comments on the first draft. A. Vauchez is acknowledged for his very careful and interesting review of the manuscript. W.L. Brown kindly improved the English draft. CRPG contribution number 792.

References

Abed H (1983) Paléosuture à caractères ophiolitiques du Protérozoïque supérieur dans la région de Silet, Hoggar occidental (Algérie). Thèse 3ème cycle Univ Nancy I, p 172

Ajibade AC, Woakes M, Rahaman MA (1987) Proterozoic crustal development in the Pan-African regime of Nigeria. In: Kröner A (ed) Proterozoic lithosphere evolution. AGU Geodyn Ser 17: 259-271

Allegre CJ, Caby R (1972) Chronologie absolue du Précambrien de l'Ahaggar occidental. C R Acad Sci, Paris, 275 D: 2095-2098

Ba H, Black R, Benziane B et al. (1985) La province des complexes alcalins saturés de l'Adrar des Iforas, Mali. J Afr Earth Sci 3: 123-142

Ball E (1980) An example of very consistant brittle deformation of a wide intracontinental zone: late Pan-African fracture system of the Tuareg and Nigeria shield; structural implications. Tectonophysics 61: 363-379

Ball E, Caby R (1984) Open folding and wrench movement, their relationships with horizontal tectonics in the Pan-African belt of northern Mali. In: Klerkx J, Michot J (eds) African Geology. MRAC Tervuren, pp 75-89

Barbey P, Bertrand JM, Angoua SE, Dautel D (1989) Petrology and U/Pb geochronology of the Telohat migmatites (Aleksod, central Hoggar, Algeria). Contrib Mineral Petrol 101: 207-219

Bayer R, Lesquer A (1978) Les anomalies gravimétriques de la bordure orientale du craton Ouest African: géométrie d'une suture pan-africaine. Bull Soc Géol Fr XX: 863-876

Bernard-Griffiths J, Peucat JJ, Fourcade S, Kienast JR, Ouzegane K (1988) Origin and evolution of 2 Ga-old carbonatite complex (Ihouaouène, Hoggar, Algeria): Nd and Sr isotopic evidence. Contrib Mineral Petrol 100: 339-348

Bertrand JM (1974) Evolution polycyclique des gneiss du Précambrien de l'Aleksod (Hoggar Central, Sahara algérien). Aspects structuraux, pétrologiques, géochimiques et geochronologiques. In: CNRS CRZA (ed) Sér Géol 19: 350

Bertrand JM, Caby R (1978) Geodynamic evolution of the Pan-African orogenic belt: a new interpretation of the Hoggar shield (Algerian Sahara). Geol Rundsch 67: 357-388

Bertrand JM, Davison I (1981) Pan-African granitoid emplacement in the Adrar des Iforas mobile belt (Mali): a Rb-Sr isotope study. Precambrian Res 14: 333-361

Bertrand JM, Lasserre M (1976) Pan-African and pre-Pan-African history of the Hoggar (Algerian Sahara) in the light of new geochronological data from the Aleksod area. Precambrian Res 3: 343-362

Bertrand JM, Boissonnas J, Caby R, Gravelle M, Lelubre M (1966) Existence d'une discordance dans l'Antécambrien du "Fossé Pharusien" de l'Ahaggar occidental (Sahara central). C R Acad Sci, Paris, 262 D: 2197-2200

Bertrand JM, Caby R, Fabriès J, Vitel G (1968) Sur la structure et l'évolution orogénique du Précambrien du Tazat (Ahaggar oriental). C R Somm Soc Géol Fr 8: 257-258

Bertrand JM, Caby R, Ducrot J, Lancelot J, Moussine-Pouchkine A, Saadallah A (1978) The late Pan-African ensialic linear fold belt of eastern Hoggar (Algeria): geology, structural development, U/Pb geochronology, tectonic implications for the Hoggar shield. Precambrian Res 7: 349-376

Bertrand JM, Michard A, Carpena J, Boullier AM, Dautel D, Ploquin A (1984a) Pan-African granitic and related rocks in the Iforas granulites (Mali). Structure, geochemistry and geochronology. In: Klerkx J, Michot J (eds) African Geology. MRAC Tervuren, pp 147-165

Bertrand JM, Michard A, Dautel D, Pillot M (1984b) Ages U/Pb éburnéens et pan-africains au Hoggar central (Algérie). Conséquences géodynamiques. C R Acad Sci, Paris, 298 D: 643-646

Bertrand JM, Michard A, Boullier AM, Dautel D (1986a) Structure and U/Pb geochronology of the Central Hoggar (Algeria). A reappraisal of its Pan-African evolution. Tectonics 5: 955-972

Bertrand JM, Meriem D, Lapique F, Michard A, Dautel D, Gravelle M (1986b) Nouvelles données sur l'age de la tectonique pan-africaine dans le rameau oriental de la chaine Pharusienne (région de Timgaouine, Hoggar, Algérie). C R Acad Sci, Paris, 302: 437-440

Bertrand JM, Moulahoum O, Bouabsa L, Cheilletz A, Charoy B, Boullier AM (1987) Pan-African younger granites of central Hoggar (Algeria): structure, petrology, geochronology and mineralization. 14th Coll Afr Geol, Berl, CIFEG, Paris, (abstr), pp 70-71

Bertrand-Sarfati J (1972) Stromatolites columnaires du Précambrien supérieur du Sahara nord-occidental. CNRS CRZA (ed) Sér Géol 14: 82

Bessoles B, Trompette R (1980) Géologie de l'Afrique. La chaine Pan-Africaine "zone mobile d'Afrique centrale (partie sud) et zone mobile soundanaise". Mém BRGM 92: 396

Black R, Jaujou M, Pellaton C (1967) Carte géologique du

Massif de l'Aïr (1/500000). BRGM, Notice explicative, p 57

Black R, Caby R, Moussine-Pouchkine A et al. (1979) Evidence for late Precambrian plate tectonics in West Africa. Nature (Lond) 278: 223-227

Blaise J (1957) Une "Série Intermédiaire" au Sahara central: la série de Tiririne (Ahaggar oriental). In: Les Relations entre Précambrien et Cambrien. Coll Int CNRS, Paris, pp 217-222

Blaise J (1961) Sur la stratigraphie des séries anté-cambriennes dans la région du Tafassasset moyen (Ahaggar oriental). Bull Soc Géol Fr III: 184

Blaise J (1967) Le Précambrien du Tazat. Sa place dans les structures du Hoggar oriental. CNRS, Sér Géol 7: 197

Boissonnas J (1973) Les granites à structures concentriques et quelques autres granites tardifs de la chaine Pan-Africaine en Ahaggar (Sahara Central, Algérie). CNRS CRZA, Sér Géol 16: 662

Bouabsa L (1987) Intrusions granitiques à albite-topaze: minéralisations stannowolframifères et altérations hydrothermales associées. L'exemple du Hoggar central, Algérie. Thèse 3ème cycle Univ Nancy I, p 193

Boullier AM (1979) Charriage et déformations de l'Unité Granulitique des Iforas au cours de l'orogénèse Pan-Africane. Rev Géogr Phys Géol Dyn, Paris, 21: 377-382

Boullier AM (1982) Etude structurale du centre de l'Adrar des Iforas (Mali). Mylonites et tectogénèse. Thèse d'Etat INPL, Nancy, p 327

Boullier AM (1986) Sense of shear and displacement estimates in the Abeibara-Rharous late Pan-African shear zone (Adrar des Iforas, Mali). J Struct Geol 8: 47-58

Boullier AM, Barbey P (1988) A polycyclic two-stage corona growth in the Iforas Granulitic Unit (Mali). J Metamorphic Geol 6: 235-254

Boullier AM, Bertrand JM (1981) Tectonique tangentielle profonde et couloirs mylonitiques dans le Hoggar central polycyclique (Algérie). Bull Soc Géol Fr XXIII: 17-22

Boullier AM, Davison I, Bertrand JM, Coward MP (1978) L'unité granulitique des Iforas: une nappe de socle d'âge Pan-Africain précoce. Bull Soc Géol Fr XX: 877-882

Boullier AM, Liégeois JP, Black R, Fabre J, Sauvage M, Bertrand JM (1986) Late Pan-African tectonics marking the transition from subduction-related calc-alkaline magmatism to within-plate alkaline granitoids (Adrar des Iforas, Mali). Tectonophysics 132: 233-246

Burg JP, Balé P, Brun JP, Girardeau J (1987) Stretching lineation and transport direction in the Ibero-Armorican arc during the Siluro-Devonian collision. Geodinamica Acta 1: 71-87

Caby R (1970) La chaine Pharusienne dans le Nord-Ouest de l'Ahaggar (Sahara central, Algérie); sa place dans l'orogénèse du Précambrien supérieur en Afrique. Thèse Etat, Univ Montpellier, p 336

Caby R (1973) Les complexes précambriens du Nord de l'Adrar des Iforas: leurs relations chronologiques et structurales et leur place dans la chaine panafricaine du Sahara central. C R Acad Sci, Paris, 277 D: 2301-2304

Caby R (1979) Les nappes précambriennes du Gourma dans la chaine pan-africaine du Mali. Comparaison avec les Alpes occidentales. Rev Géogr Phys Géol Dyn 21: 365-376

Caby R (1987a) The Pan-African belt of West Africa from the Sahara desert to the gulf of Benin. In: Schaer JP, Rodgers J (eds) The anatomy of Mountain Ranges. Princeton Univ Press, USA, pp 129-169

Caby R (1987b) Upper Proterozoic continental accretion and collisions in northwest Africa. Terra Cognita 7: 150

Caby R (in press) Precambrian terranes of Benin-Nigeria and Northeast Brazil and the late Proterozoic south Atlantic fit. Geol Soc Am Mem

Caby R, Andreopoulos-Renaud U (1983) Age à 1800 Ma du magmatisme calco-alcalin associé aux métasédiments monocycliques dans la chaine Pan-Africaine du Sahara Central. J Afr Earth Sci 1: 193-197

Caby R, Andreopoulos-Renaud U (1985) Etude pétrostructurale et géochronologique d'une métadiorite quartzique de la chaine Pan-Africaine de l'Adrar des Iforas (Mali). Bull Soc Géol Fr 8: 899-903

Caby R, Andreopoulos-Renaud U (1987) Le Hoggar oriental, bloc exotique cratonisé à 730 Ma dans la chaine pan-africaine du Nord du continent africain. Precambrian Res 36: 335-344

Caby R, Andreopoulos-Renaud U (1989) Age U-Pb à 620 Ma d'un pluton synorogénique de l'Adrar des Iforas (Mali). Conséquences pour l'âge de la phase majeure de l'orogénèse Pan-Africaine. C R Acad Sci, Paris 308, II: 307-313

Caby R, Fabre J (1981) Tillites in the latest Precambrian strata of the Touareg Shield (Central Sahara). In: Hambrey MJ, Harland WB (eds) Earth's pre-Pleistocene glacial record. Cambridge Univ Press, Cambridge, UK, pp 146-149

Caby R, Moussu H (1967) Une grande série détritique du Sahara: stratigraphie, paléogéographie et évolution structurale de la "Série Pourprée" dans l'Aseg'rad et le Tanezrouft oriental (Sahara algérien). Bull Soc Géol Fr IX: 876-882

Caby R, Bertrand JM, Black R (1981) Pan-African closure and continental collision in the Hoggar-Iforas segment, central Sahara. In: Kröner A (ed) Precambrian Plate Tectonics. Elsevier, Amst, pp 407-434

Caby R, Andreopoulos-Renaud U, Gravelle M (1982) Cadre géologique et géochronologie U/Pb sur zircon des batholites précoces dans le segment pan-africain du Hoggar central (Algérie). Bull Soc Géol Fr XXIV: 677-684

Caby R, Andreopoulos-Renaud U, Lancelot JR (1985) Les phases tardives de l'orogénèse Pan-Africaine dans l'Adrar des Iforas oriental (Mali): lithostratigraphie des formations molassiques et géochronologie U/Pb sur zircon de deux massifs intrusifs. Precambrian Res 28: 187-199

Caby R, Liégeois JP, Dupuy C, Andreopoulos-Renaud U (1986) The Tilemsi magmatic arc and the Pan-African suture zone in northern Mali. Int Field Conf Proterozoic Geol Geochem, Central Colorado, USA, (abstr), p 88

Caby R, Andreopoulos-Renaud U, Pin C (1989) Late Proterozoic arc/continent and continent/continent collision in the Pan-African Trans-Saharan belt of Mali. Can J Earth Sci 26: 1136-1146

Cahen L, Snelling NJ, Delhal J, Vail JR (1984) The geochronology and evolution of Africa. Clarendon Press, Oxford, p 512

Champenois M, Boullier AM, Sautter V, Wright LI, Barbey P (1987) Tectonometamorphic evolution of the gneissic Kidal Assemblage related to the Pan-African thrust tectonics (Adrar des Iforas, Mali). J Afr Earth Sci 6: 19-28

Chikhaoui M (1981) Les roches volcaniques du Protérozoique supérieur de la chaine Pan-Africaine (Hoggar, Anti-Atlas, Adrar des Iforas). Thèse Doct Etat Univ Montpellier, p 183

Chikhaoui M, Dupuy C, Dostal J (1978) Geochemistry of Late Proterozoic volcanic rocks from Tassendjanet (NW Hoggar, Algeria). Contrib Mineral Petrol 66: 157-164

Clauer N (1976) Géochimie isotopique du strontium des milieux sédimentaires: applications à la géochronologie de la couverture du Craton Ouest Africain. Univ Strasbourg Sci Géol Mém 45: 256

Cosson Y, Boullier AM, Rocci G (1987) Pan-African deformation at a plate boundary outlined by ophiolites in Aîr (Niger). Int Conf: Tectonothermal evolution of the west african orogens and circum atlantic terrane linkages. Nouakchott, (abstr), pp 61-64

Davison I (1980) A tectonic, petrographical and geochronological study of a Pan-African belt in the Adrar des Iforas and Gourma (Mali). Ph D Thesis, Univ Leeds and Centre Géol Géophys, Montpellier, 360 pp

de la Boisse H (1979) Pétrologie et géochronologie de roches cristallophylliennes du bassin du Gourma (Mali). Conséquences géodynamiques. Thèse 3ème cycle Univ Montpellier, p 54

de la Boisse H (1981) Sur le métamorphisme du micaschiste éclogitique de Takamba (Mali) et ses conséquences paléogéodynamiques au Précambrien supérieur. C R Somm Soc Géol Fr 3: 97-100

de la Boisse H, Lancelot JR (1977) A propos de l'événement à 1000 Ma en Afrique occidentale: 1: le "Granite de Bourré". C R Somm Soc Géol Fr 4: 223-226

Dupont PL (1987) Pétrologie et géochimie des ensembles magmatiques Parusien I et II, dans le rameau oriental de la chaine pharusienne (Hoggar, Algérie). Implications géodynamiques pour l'évolution d'une chaine mobile au Protérozoïque Supérieur. Thèse Doct Univ Nancy I, p 283

Dupont PL, Lapierre H, Gravelle M, Bertrand JM (1987) Caractérisation du magmatisme Protérozoique supérieur en Afrique de l'Ouest et implications géodynamiques: des rifts intracratoniques au Pan-Africain? Can J Earth Sci 24: 96-109

Fabre J (1982) Pan-African volcano-sedimentary formations in the Adrar des Iforas (Mali). Precambrian Res 19: 201-214

Fabre J, Ba H, Black R, Caby R, Leblanc M, Lesquer A (1982) Carte géologique et gravimétrique de l'Adrar des Iforas au 1/500000ème. Direction Géol Min Bamako, Mali

Fabre J, Ait Kaci A, Bouima T, Moussine-Pouchkine A (1988) Le cycle molassique dans le rameau trans-saharien de la chaine panafricaine. J Afr Earth Sci 7: 41-55

Fabriès J, Gravelle M (1977) Modalités du magmatisme au Protérozoique supérieur dans la région de Silet (Hoggar Central). Bull Soc Géol Fr XIX: 995-1004

Ferrarra G, Gravelle M (1966) Radiometric ages from western Ahaggar (Sahara) suggesting an eastern limit for the West African craton. Earth Planet Sci Lett 1: 319-324

Gravelle M (1969) Recherches sur la géologie du socle précambrien de l'Ahaggar centro-occidental dans la région de Silet-Tibehaouine. Thèse Doct Etat Univ Paris, p 298

Gravelle M, Lelubre M (1957) Découverte de Stromatolites du groupe des Conophytons dans le Pharusien de l'Ahaggar occidental. Bull Soc Géol Fr 6: 435-442

Karpoff R (1960) La géologie de l'Adrar des Iforas (Sahara central). Thèse Doct Etat, Univ Paris, Publ BRGM, Dakar

Kehrer P, Hollinger J, Merkt J, Tessensohn F (1975) Rapport de fin de mission 1971-1973 sur la feuille de NE 32 X, Barghot (Massif de l'Aïr). Bundesanst Geowiss Rohstoffe, Hannover, RFA

Kienast JR, Ouzegane K (1987) Polymetamorphic Al, Mg-rich granulites with orthopyroxene-sillimanite and sapphirine parageneses in Archaean rocks from Hoggar, Algeria. Geol J 22, Thematic Issue, and In: Bowden P, Kinnaird J (eds) African Geology Reviews. Wiley, pp 57-79

Kilian C (1932) Sur des conglomérats précambriens du Sahara central: le Pharusien et le Suggarien: C R Somm Soc Géol Fr 4: 87

Lancelot JR, Vitrac A, Allegre CJ (1976) Uranium and lead isotopic dating with grain-by-grain zircon analysis: a study of complex geological history with a single rock. Earth Planet Sci Lett 29: 357-358

Lancelot JR, Boullier AM, Maluski H, Ducrot J (1983) Deformation and related radiochronology in a late Pan-African mylonitic shear zone. Contrib Mineral Petrol 82: 312-326

Lapierre H, Bendali M, Dupont PL, Gravelle M (1987) Nouvelles données stratigraphiques et structurales sur le rameau oriental de la chaine pharusienne (région de Silet, Hoggar, Algérie). C R Acad Sci, Paris, 303: 1731-1736

Lapique F, Bertrand JM, Meriem D (1986) A major Pan-African decoupling zone in the Timgaouine area (western Hoggar, Algeria). J Afr Earth Sci 5: 617-625

Latouche L (1978) Le Précambrien de la région des Gour Oumelalen (NE de l'Ahaggar, Algérie). Thèse Doct Etat, Univ Paris VII, p 225

Latouche L, Vidal Ph (1974) Géochronologie du Précambrien de la région des Gour Oumelalen (NE de l'Ahaggar, Algérie). Un exemple de mobilisation du strontium radiogénique. Bull Soc Géol Fr XVI: 195-203

Leblanc M (1976) Proterozoic oceanic crust at Bou Azzer. Nature (Lond) 261: 34-35

Lelubre M (1952) Recherches sur la géologie de l'Ahaggar central et occidental (Sahara central). Bull Serv Carte Géol Algérie, Alger, 2ème Sér, 22, 2 vols

Lesquer A, Moussine-Pouchkine A (1980) Les anomalies gravimétriques de la boucle du Niger. Leur signification dans le cadre de l'orogenèse panafricaine. Can J Earth Sci 17: 1538-1545

Leterrier J, Bertrand JM (1986) Pretectonic tholeiitic volcanism and related transitional plutonism in the Kidal assemblage (Iforas Pan-African belt, Mali). J Afr Earth Sci 5: 607-615

Liègeois JP (1987) Le batholite composite de l'Adrar des Iforas (Mali). Géochimie et géochronologie d'une succession magmatique du calco-alcalin à l'alcalin dans la cadre de l'orogénèse pan-africaine. Thèse Doct Etat, Univ Libre Bruxelles, Belgique, p 335

Liégeois JP, Black R (1984) Pétrographie et géochronologie Rb-Sr de la transition calcoalcalin-alcalin fini-Pan-Africaine dans l'Adrar des Iforas (Mali): accrétion crustale au Précambrien supérieur. In: Klerkx J, Michot J (eds) African Geology. MRAC, Tervuren, pp 115-145

Liégeois JP, Black R (1987) Alkaline magmatism subsequent to collision in the Pan-African belt of Iforas (Mali). In: Fitton JG and Upton BGJ (eds) Alkaline igneous rocks. Geol Soc Spec Publ 30: 381-401

Liégeois JP, Bertrand JM, Black R (1987) The subduction- and collision-related Pan-African composite batholith of the Adrar des Iforas (Mali): a review. Geol J 22, Thematic Issue and In: Kinnaird J, Bowden P (eds) African geology reviews. Wiley, Chichester, pp 185-211

Ly S (1979) Etude gravimétrique de l'Adrar des Iforas (NE Mali). Thèse Doct Ing Univ Montpellier

Ly S, Lesquer A, Ba H, Black R (1984) Structure profonde du batholite occidental de l'Adrar des Iforas (Mali): une synthèse des données gravimétriques et géologiques. Rev Géol Dyn Géogr Phys 25: 33-44

Moulahoum O (1988) Dualité du magmatisme d'âge pan-africain: aspects structuraux et pétrologiques des granites subalcalins et alumineux de la région de Tamanrasset (Hoggar Central, Algérie). Thèse Doct Ing Univ Nancy I, p 145

Moussine-Pouchkine A, Bertrand-Sarfati J (1978) Le Gourma: un aulacogène du Précambrien supérieur. Bull Soc Géol Fr XX: 851-857

Moussine-Pouchkine A, Bertrand-Sarfati J, Ball E, Caby R (1988) Les séries sédimentaires et volcaniques anorogéniques protérozoiques impliquées dans la chaîne pan-africaine: la région de l'Adrar Ahnet (NW Hoggar, Algérie). J Afr Earth Sci 7: 57-75

Ouzegane K (1981) Le métamorphisme polyphasé granulitique de la région de Tamanrasset (Hoggar central). Thèse 3ème cycle Univ Paris VII, p 171

Ouzegane K (1987) Les granulites Al-Mg et les carbonatites dans la série de l'In Ouzzal, NW Hoggar, Algérie. Thèse Doct Etat Univ Paris VI, p 433

Ouzegane K, Fourcade S, Kienast JR, Javoy M (1988) New carbonatite complexes in the Archaean In'Ouzzal nucleus (Ahaggar, Algeria): mineralogical and geochemical data. Contrib Mineral Petrol 98: 277-292

Pegram WJ, Register JL, Fullagar PD, Ghuma MA, Rodgers JJW (1976) Pan-African ages from a Tibesti massif batholith, southern Lybia. Earth Planet Sci Lett 30: 123-128

Picciotto E, Ledent D, Lay C (1965) Etude géochronologique de quelques roches du socle cristallophyllien du Hoggar. Proc 15th Coll Int CNRS Géochronol, Nancy, (abstr), pp 277-289

Raulais M (1959) Esquisse géologique sur le massif cristallin de l'Air (Niger). Bull Soc Géol Fr I: 786-813

Reichelt R (1972) Géologie du Gourma (Afrique occidentale). Un "seuil" et un bassin du Précambrien supérieur. Stratigraphie, tectonique, métamorphisme. Mém BRGM 53: 213

Sautter V (1985) An eclogite paragenesis from the Aleksod basement, Central Hoggar, South Algeria. Chem Geol 50: 331-347

Vialette Y, Vitel G (1981) Geochronological data on the Amsinassene-Tefedest block (Central Hoggar) and evidence for its polycyclic evolution. Precambrian Res 9: 241-254

Vitel G (1975) Mylonitisation, tectonique cassante et linéamentaire du Hoggar Central: leur rapports avec les granites d'age Pharusien. Rev Géogr Phys Géol Dyn XVII: 413-426

Vitel G (1979) La région Tefedest-Atakor du Hoggar Central (Sahara). Evolution d'un complexe granulitique précambrien. Thèse Doct Etat Univ Paris VII, p 324

Wright L (1979) The pattern of movement and deformation during the Pan-African, in the Adrar des Iforas. X Coll Geol Afr, Montpellier, (abstr), p 77

The Dahomeyide Orogen: Tectonothermal Evolution and Relationships with the Volta Basin

P. AFFATON[1], M. A. RAHAMAN[2], R. TROMPETTE[3], and J. SOUGY[4]

1 Indroduction

The Dahomeyide orogen is located along the southeastern margin of the West African craton, and is exposed throughout eastern Ghana, Togo, Benin, Nigeria and Cameroon (Fig. 1). The West African craton has remained tectonically stable since ca. 2000 Ma (Black 1985a, b; Camil 1984). Western sectors of the craton are characterized by an Archean domain represented by grey gneisses and migmatitic complexes (amphibolite and granulite metamorphic facies), acid to basic granulites and charnockitic complexes, together with subordinate greenstone rocks, supracrustal metasedimentary and metavolcanoclastic rocks and granites. These rocks were affected by orogenic events at ca. 3000 Ma (Leonian) and ca. 2800 Ma (Liberian). Southeasternmost segments of the Archean domain were partially remobilized at ca. 1800 Ma during Eburnean orogenesis. Western sectors of this domain were variably affected by late Proterozoic (Pan-African) orogenic events.

Eastern portions of the West African craton are underlain by crystalline rocks of the Birrimian system. These are comprised of greenstone rocks (metabasalts), metasedimentary and metavolcanoclastic units and flysch-type deposits. The volcanic rocks vary from basic to acid in composition. Traditionally, the Birrimian system has been subdivided into lower and upper series. It is overlain by an extensive and gently folded molassic sequence known as the Tarkwaian. During the Eburnean orogeny, the Birrimian system was extensively intruded by syn- to post-tectonic granites which vary in composition

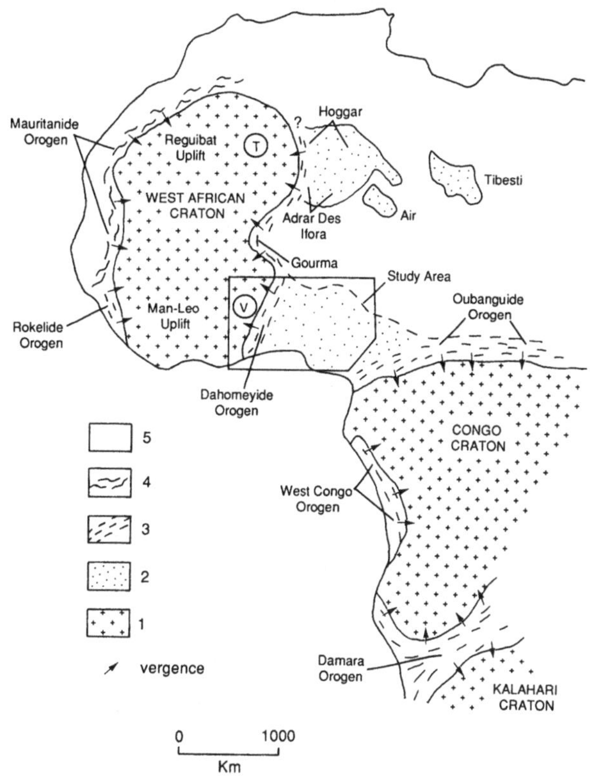

Fig. 1. Location of the Dahomeyide orogen. *T* Taoudeni basin; *V* Volta basin; *1* Archean to Eburnean (Lower Proterozoic) domains (their cover is not represented); *2* Pan-African domains; *3* well-known Pan-African orogens; *4* Mauritanide orogen (Pan-African and Hercynian orogen); *5* Paleozoic to recent cover

from diorites to two-mica granites and alkaline plutons. Tectonothermal activity in this area is represented by the Burkinian event (2400–2100 Ma), which affected lower formations (the Dabakalian) and the Eburnean event (2100–1800 Ma) which affected especially the upper formations (to which the stratigraphic term Birrimian is applied in a rather restricted sense).

Large areas of the West African craton are covered by undeformed sedimentary rocks of Upper Proterozoic to Quaternary age. These sedi-

[1] Département de Géologie, Faculté des Sciences, Université du Bénin, BP 30389, Lomé, Togo
[2] Department of Geology, Obafemi Awolowo University, Ile-Ife, Nigeria
[3] Institut de Géodynamique, U. R. A.-C. N. R. S., Sophia Antipolis 1, Av. A. Einstein, 06560 Valbonne, France
[4] Département de Géologie, Faculté des Sciences de Saint-Jérôme, 13397 Marseille Cedex 13, France

mentary rocks comprise the Taoudeni (central) and Volta (southeastern) basins.

Most of the Congo craton has remained tectonically quiescent since ca. 2600 Ma. It consists of an Archean system and lower to upper Proterozoic cover (Black 1985b; Cahen et al. 1984; Weber 1969; Cosson 1957; Weber and Bonhomme 1975; Klerkx 1985). The Archean is represented by high-grade (upper amphibolite-granulite facies) gneiss complexes, granitic and migmatitic gneisses, greenstone belts, metasedimentary and metavolcanosedimentary belts older than 2400 Ma. These rocks were affected by tectonothermal events at ca. 2950 Ma and 2700 Ma.

The Francevillian sequence constitutes lower portions of the cover succession. It is 400-2000 m thick and comprised of sandstones, pelites, jaspers, dolomitic limestones and volcanosedimentary rocks. The Ogoue system, exposed on the western margin of the Congo craton, is a metamorphosed equivalent of the Francevillian sequence. It was affected by the Tadilian (Eburnean) orogeny dated at 2088 ± 60 Ma (Weber and Bonhomme 1975) which also partially remobilized older cratonic sequences. The upper part of the sedimentary cover is represented by Lindian and west-Congolian successions to the North and the Katangian and Bushimay sequences in the south and southeast. They are mainly of upper Proterozoic age with a few middle Proterozoic sequences (the Bushimay and Chela Groups in south Angola). Easternmost segments of the Congo craton were reactivated during the Kibaran orogeny at ca. 1200 Ma (Klerkx 1985).

The Oubanguide fold belt (Fig. 1) is made of two main lithological units (Poidevin 1985; Poidevin et al. 1981): (1) slightly metamorphosed, mainly clastic sedimentary rocks; and (2) gneisses and granitoids. The metasedimentary rocks are roughly time-equivalent to sedimentary cover rocks within the adjacent Congo craton. The Upper sequence starting with a diamictite horizon, locally interpreted as glaciogenic in origin, can be correlated with the West Congo sequence (Upper Proterozoic) and the Lower sequences correlate with the Kibaran sequence (Middle Proterozoic). Among gneisses and granitoids, it is possible to distinguish large tracts of polycyclic basement with a few Archean remnants and late Proterozoic (Pan-African) migmatites and gneisses such as the Yaounde sequence in southern Cameroon (Nzenti et al 1984, 1988; Nedelec et al. 1986).

The Dahomeyide fold belt is a portion of a more extensive orogen (Fig. 1), sometimes called the Transaharian fold belt (Cahen et al. 1984; Black 1985b), which extends from the Hoggar (Algeria) to the Dahomeyides (Ghana, Togo, Benin, Nigeria and Cameroon), through the Adrar des Iforas and Gourma (Mali). It possibly continues northwards into southern Morocco. Southwards, on the other side of the Atlantic Ocean, it may correlate with the Pan-African sequence exposed in northeastern Brazil. From Hoggar to the Benin Gulf, the Transaharian fold belt is thrust onto the West African craton. The Gourma and east-Dahomeyide orogens are thrust on a cratonic cover represented by southeastern margins of the Taoudeni and the Volta basins. From west to east, monoclinal flat-lying sedimentary sequences within the Volta basin are progressively folded and metamorphosed. In the easternmost part, inverted and faulted units constitute external structural units of the Dahomeyide orogen. This area therefore provides the opportunity to relate the chronology and geodynamic evolution of a Pan-African orogen with the depositional history within an adjacent and partly synchronous basin (Affaton 1975, 1987).

2 Lithostratigraphy and Structure Within the Volta Basin

Sedimentary rocks within the Volta basin rest with major unconformity on the Eburnean Man-Leo uplift (Fig. 1). Following the early work of Junner and Hirst (1946), the studies of Saunders (1970), Leprun and Trompette (1969), Sougy (1971) and Affaton (1975, 1987) have led to the recognition of three supergroups (Fig. 2) which, in ascending stratigraphic order, are the Bombouaka Supergroup, the Pendjari (or Oti) Supergroup, and the Tamale Supergroup. These represent slightly different stratigraphic divisions than the lower, middle, and upper Voltaian subdivisions previously used in Ghana.

The Bombouaka Supergroup (average thickness of 1000 m) comprises two groups which are made up dominantly of sandstones that sandwich a third soft clay-siliceous group. These detritic, fine and mature sediments were deposited mostly on an epicontinental marine platform along the edge of a peneplained basement. Erosional and cartographic unconformities observed between and within the groups, as well as the numerous indices of emersion, suggest an irregular transgression and a slow and variable subsidence.

The Pendjari (or Oti) Supergroup rests with an erosional unconformity, partly of glacial origin, on different formations of the Bombouaka Supergroup

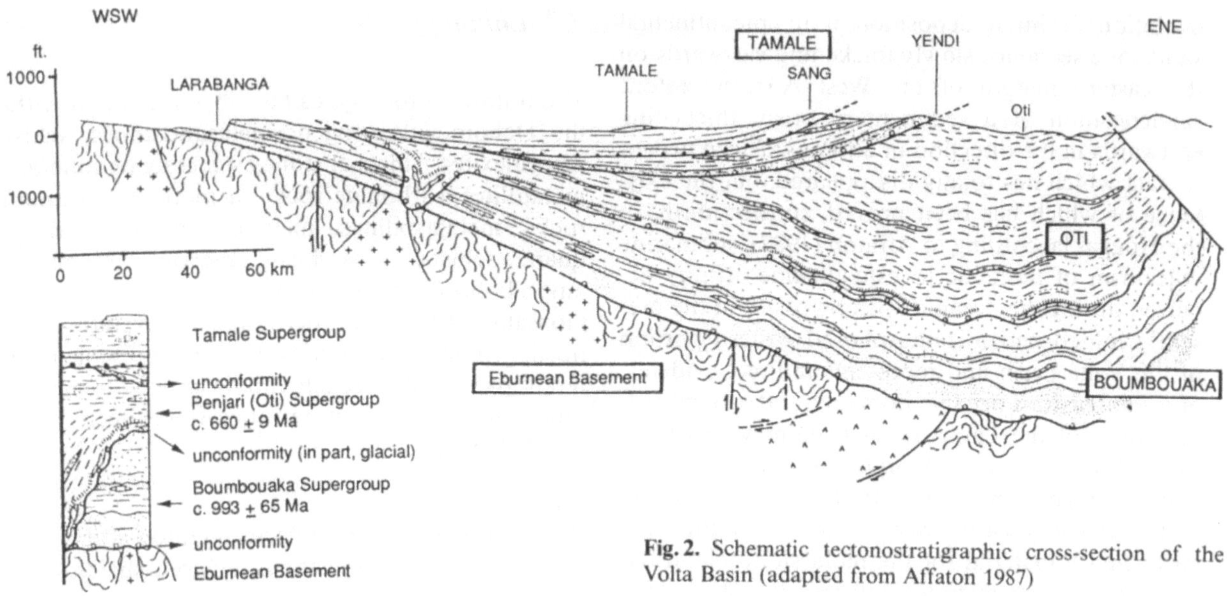

Fig. 2. Schematic tectonostratigraphic cross-section of the Volta Basin (adapted from Affaton 1987)

and, locally, directly on crystalline basement units. It has an average thickness of 2500 m. It comprises the lower Kodjari group formed by an association of tillite (s. s.)-dolomitic limestone with baryte-silexite (this association is well known in the Taoudeni basin) and the upper Porga group with rhythmic argilites and siltstones with phosphorite lentils locally found at the base (Trompette et al. 1980). As a whole, the Pendjari Supergroup is made up of fine and immature materials with notably greywackes, deposited in a relatively deep and more subsident marine environment, probably on a passive margin developed on the edge of the Man-Leo uplift, and prograding eastward towards a Pan-African oceanic domain.

The Tamale Supergroup is only known in Ghana where it unconformably overlies the Pendjari Supergroup. Its thickness is about 500 m. It is made up of the lower Yendi Group which contains mainly clay and siltstones and the sandy, coarser continental and molassic upper Kebia Group characterized by basal members of diamictites and conglomerates partly of glacial origin. The materials were easterly derived (i. e. from the Dahomeyides) and sedimentary deposition centers migrated westwards (Fig. 2). This supergroup represents a typical foreland molassic basin.

The Volta basin appears to be a composite basin which formed by the filling of a cratonic epicontinental basin, a probable passive margin, and a foreland basin. It comprises three supergroups. Probably the first one formed between 1100 and 700 Ma, the second one between 700 and 600 Ma, a period during which a Pan-African ocean of probable reduced dimension developed. The age of the third one is controversial. It is upper Proterozoic to basal Cambrian for some people, and lower Paleozoic for others, and thus could be time equivalent with the "Sekondian and Accraian Series" developed along the Atlantic coast of Ghana.

The present Volta basin is nothing but a portion of a wider basin whose eastern sectors have been incorporated in the Dahomeyide orogen. Synthesis of available geophysical data (Affaton 1987) shows that the basement unconformity is less than 3 km deep under western and northern parts and between 5 and 7 km deep in southeastern segments. Roughly, the Volta basin has an assymetric shape, with an eastward descending gentle slope and a steeper, west-sloping eastern margin. It shows in detail very important paleorelicts and paleodepressions whose axes are subparallel to Eburnean structures within Man- Leo basement uplift. Large basic sills and intrusions have been found in this substratum, as well as normal and rarely reversed faults delimiting horsts and grabens. These suggest a progressive eastward thickening of the Bombouaka and Pendjari (Oti) Supergroups. Some of these numerous fractures affect only the substratum and the lower part of the sedimentary sequence; they are probably penecontemporaneous with the first deposits and the differenciation of the basin. Moreover, this basin is folded in its eastern part as the Dahomeyide belt is approached.

The geophysical data combined with stratigraphic characteristics suggest a three phase geodynamic evolution: (1) initial deposition of an epicontinental sandstone sequence slowly thickening eastwards on the eastern margin of the West African craton; (2) deposition of a sedimentary prism thickening eastwards in the direction of an Pan-African oceanic tract which has totally disappeared, but the existence of which has been proved, further north, in the Hoggar [according to Affaton (1987), submarine volcanism of tholeiitic to calc-alkaline affinity occurred during the oceanic opening phase in the presently folded and metamorphosed eastern portion of the Volta basin and represents indirect evidence of a Pan-African oceanic domain]; and, (3) individualization of the Pan-African Dahomeyide belt at ca. 600 Ma, and the incorporation of the whole eastern fringe of the basin into this orogen. Erosion of the elevated Pan-African domain supplied the sedimentary material for the molassic foreland basin. Depocenters migrated towards the West African craton. The Volta basin therefore recorded the three-phase tectonothermal evolution of the adjacent Dahomeyide belt. Thus, the first part of the history of the Volta basin (until the uppermost Proterozoic glaciation) can be correlated with that of the other extensive West African cratonic basins (Taoudeni type: Trompette 1973; Deynoux et al. 1985). The second and third parts of its evolution are very different, and entirely dominated by the tectonothermal events involved in the development of the Dahomeyide orogen.

3 The Dahomeyide Fold Belt

3.1 Introduction

The Dahomeyide fold belt comprises two different lithologies. To the west, external structural units are made up of metasedimentary rocks which represent tectonized and metamorphosed equivalents of the Bombouaka and Pendjari (Oti) Supergroups. To the east, the dominant facies are schists, gneisses and granitoids which can be subdivided into three lithotectonic categories: (1) polycyclic basement; (2) metamorphic equivalents of the Volta basin formations (including molasse deposits); and (3) highly metamorphosed metasedimentary units which can represent either basement or unrecognizable Pan-African material.

3.2 Lithology

The following rock types have been mapped within the Dahomeyide orogen: metasandstones, schists, metasilexites, metasiltstones, metaconglomerates, metatillites, carbonate rocks or marbles, ultramafic rocks, metajaspillites, metahematitites, quartzites, quartz schists, micaschists, gneisses, migmatites, amphibolites, granites, charnockites, eclogites, metabasalts, calc-silicate rocks, and pyroxenites. Contrasting lithological associations are present in different parts of the orogen and, from west to east, six lithotectonic zones or terranes have been recognized (Fig. 3).

Zone I

Zone I borders the West African craton and is composed mainly of sandstones, schists, shales, silexites, ultramafic rocks, quartzites and quartz schists, micaschists, marbles, metahematitites, metajaspillites and gneisses. These rocks, with the exception of the gneisses, constitute the external units of the Dahomeyide orogen and have been termed the Buem and Atacora structural units. Affaton (1975, 1987) demonstrated that the materials of these structural units are equivalent to rocks of Bombouaka and Pendjari (Oti) Supergroups within the Volta basin. The gneisses (such as the 2000 Ma Kara orthogneiss; Caen-Vachette et al. 1979) represent remobilized basement of the West African craton. This zone is characterized by an absence of Pan-African granites and migmatites.

Zone II

Zone II is a narrow tract characterized by a high positive gravity anomaly. It is underlain by quartzites, micaschists, gneisses, amphibolites, pyroxenites, ultramafic rocks, eclogites, and noritic leucogabbros. The basic and ultrabasic rocks forming the Dérouvarou, Kabyè, Djabatouré, Lato, Agou, and Akuse complexes are believed to represent either mantle diapirs or remnants of a paleo-oceanic crust with characteristics of an ophiolitic complex. The Pan-African suture at the eastern margin of the West African craton has thus been defined by rocks of this zone (Crenn 1957; Sagbohan 1972; Trompette 1979 and in Bessoles et Trompette 1980; Affaton 1987).

Zone III

Zone III is made up predominantly of migmatites, gneisses, syenites, granites, micaschists, marbles, quartzites, and amphibolites. The granites account for less than 15% of the surface area of this zone. In

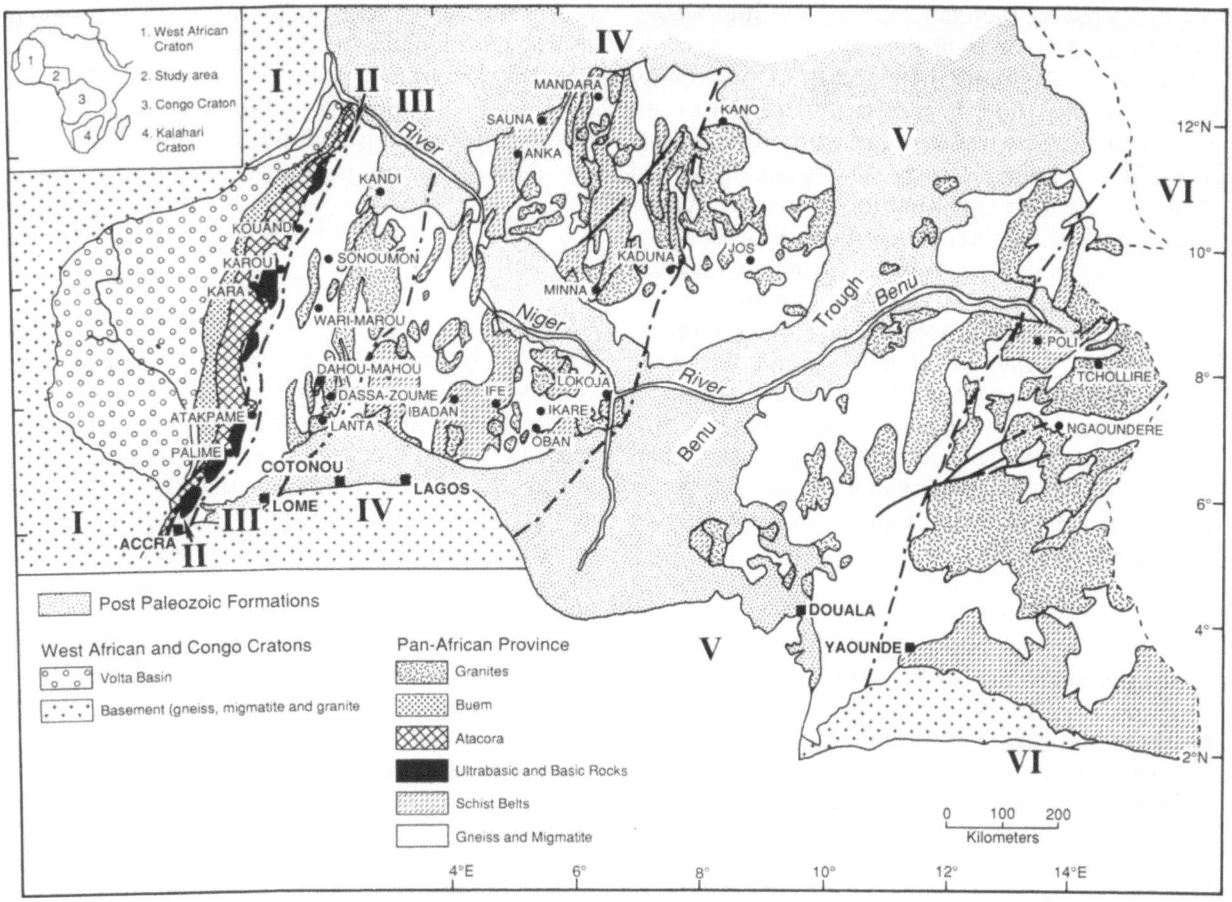

Fig. 3. Generalized geologic map of the Dahomeyide orogen outlining the six lithotectonic zones discussed in this chapter *(I-VI)*

general they are absent near the boundary with Zone II and become increasingly important in volume eastwards. In Zone III, the orthogneisses and amphibolites forming the Sonoumon complex display calc-alkaline geochemical affinities (Affaton et al. 1978). In southeastern parts of this zone, a very narrow belt, known as the Dahou Mahou basin, exposes shales, sandstones, polymict conglomerates, basalts, rhyolites and rhyodacites which uncomformably overlie gneisses and migmatites. It represents a Pan-African molasse deposited in a graben associated with post-tectonic uplift of the high-grade terrains.

Zone IV

Zone IV is characterized by the presence of several north-northeast and south-southwest trending schist belts surrounded by a gneiss-migmatite complex. Both the schist belts and the gneiss-migmatite complex are intruded by granite and charnockite plutons. The schist belts are composed predominantly of pelitic to semi-pelitic rocks, quartzites, polymict metaconglomerates, banded iron formations, marbles, calc-silicate rocks, metagreywackes, amphibolites, ultramafic and acid to intermediate metavolcanic units. Contrasting lithological associations occur within the different schist belts. The gneiss complex comprises quartzo-feldspathic gneisses and migmatites, granitic gneisses, mafic schists, and amphibolites. The granites fall into two broad age groups. Most are associated with the late Proterozoic Pan-African orogeny and include pre-, syn-, late- and post-tectonic granites. A less well-known group is associated with the Eburnean orogen and is now represented by penetratively foliated orthogneisses (mylonitic metagranites). The Pan-African granites vary in composition from tonalites to granites and syenites. The volume of Pan-African granites in this zone increases significantly from ca. 20% in the west to ca. 50% in the east. Charnockites form a distinct petrographic assemblage within this zone. Compositionally, they show two geochemical trends including plutonic calc-alkaline and Fe-enriched fractionation trends.

The Ife-Ilesha schist belt occurs in the central portion of this zone. It consists of two lithostratigraphic units: (1) a western unit, composed mainly of mafic to ultramafic rocks and metapelites; and (2) an eastern unit composed of quartzites, quartz-schists, and metapelites. These two units are separated by a major fault zone: Geochemical studies (Ajayi 1981; Klemm et al. 1984; Rahaman et al. 1988) suggest that the western unit evolved as an ocean basin. Other schist belts within this zone (Ogezi 1988; Bafor 1982) have also been shown to have evolved in similar oceanic environments. Rahaman et al. (1988) suggested that the distribution of rocks in this schist belt can be explained by the closure of a small ocean along a westerly dipping subduction zone.

Zone V

Zone V is characterized by two main features: (1) the preponderance of Pan-African granitic rocks which account for more than 80% of the exposed bedrock; and (2) a general absence of schist belts. However, minor occurrences of rocks ascribed to the schist belts have been described from the Oban Massif (Rahman et al. 1981; Ekwueme and Onyeagocha 1986).

Charnockites, gneisses, and migmatites, also of possible Pan-African age, account for the remainder of the outcrops. Granitic high-level, anorogenic ring-complexes of Carboniferous-Cretaceous age (Rahaman et al. 1984), commonly refferred to as the "Younger Granites" of Nigeria, are emplaced along the western margin of this zone (mainly in the Jos Plateau). Similar magmatic ring complexes, but of Cenozoic age, are concentrated along the Cameroon volcanic line. Cretaceous-Recent sediments of the Benue Trough, Anambra, Chad, Bornu and Niger basins cover ca. 50% of this zone and separate it into contrasting western and eastern parts. Calc-alkaline plutonic suites comprising gabbros, biotite-diorites, granodiorites and granites have been recognized in some parts of this zone (e.g. the Mandara massif: Rahaman 1988b).

Zone VI

Zone VI is lithologically similar to zone IV (Toteu et al. 1988). It comprises three assemblages: (1) a Lower Proterozoic basement of migmatites, amphibolites, and para- and orthogneisses reactivated during the Pan-African tectonothermal activity; (2) schist belts made up predominantly of quartzites, micaschists, metaconglomerates, calc-silicate rocks, metagreywackes with intercalated metavolcanics of tholeiitic to alkaline affinity; and (3) a Pan-African granitic suite of mainly calc-alkaline affinity; less important volumes of syn-late and post-tectonic plutons and dikes of subalkaline-alkaline affinity also occur locally.

3.3 Structure

Foliation

The Dahomeyide orogen is characterized by large scale lithological banding which is responsible for the geometric relations of the different rocks present in the area. It occurs as a steep foliation, trending north-south to northeast-southwest. Near the eastern margin of the orogen in Cameroon, it trends northeast-southwest but along the western border it strikes north-northeast to south-southwest. In some local areas, this trend has an east-west orientation. Within each lithological unit, a main foliation can be defined. In the schists and gneisses it is a mineralogical banding (S_1), generally associated with tight to isoclinal folds. This foliation trend is often parallel to the large scale litholigical banding described above. This feature corresponds to the first recognizable tangential deformation (D_1) of Pan-African age within the orogen. Commonly, during a second deformation (D_2), this banding is folded with the production of an axial planar schistosity which is well developed in schists and gneisses. Often, this schistosity (S_2) is parallel to S_1.

A third deformation (D_3), recognized within the Atacora and Ife-Ilesha schist belts, produced large-scale open folds, with rarely developed axial plane fracture cleavage. In the Atacora structural unit, D_1, D_2, and D_3 are in general coaxial. In some of the schist belts, a D_4 folding phase is recognized. It is represented by large scale virgations. The mean axes of these are parallel to D1-D3 structures. In general, it has not been possible to separate in the field pre-Pan-African structural elements, which probably developed during the Eburnean orogeny (2000 ± 200 Ma) and possibly earlier orogenic events, from Pan-African structures. However, two extreme points of view exist. Caby (1987, 1988) suggests that most rocks within the Dahomeyide orogen did not undergo any deformation prior to the Pan-African orogeny. On the other hand, Affaton (1987, 1988) and Ajibade (1988) suggest that pre-Pan-African rocks underwent several deformational events prior to the Pan-African. The fine scale banding seen in some grey gneisses and orthogneisses of Eburnean age may represent relicts of these structures (e.g. in the Ibadan orthogneiss). Generally, old foliations or traces of deformation are transposed or obliterated during the Pan-African tectogenesis.

Lineation
Lineation is commonly defined by the alignment of minerals. A gently plunging stretching lineation, trending north-south and probably perpendicular to the direction of transport of D_1 structures, is found throughout the orogen. A crenulation or intersection lineation is generally well developed during D_2 deformation; it is often subparallel to the first lineation generation.

Nappes
D_1 deformation is believed to correspond to an originally flat lying foliation. On the western margin of the orogen, Affaton (1987) has shown that D_1 deformation began with isoclinal folds and ended with generation of large nappes. The large-scale isoclinal folds with shallow dips of foliation, observed in the inner parts of the orogen (e.g. western side of the Ife-Ilesha area) are believed to be associated with the production of D_1 nappes.

Shear Zones
Several major transcurrent faults have been described from the orogen. They have orientation varying from north-south (Kandi, Zungeru, Ifewara fault zones), NE-SW (Kalanga fault zone) to east-northeast to west-southwest (Ngaoundere, Benue fault zones). Until recently, the steep, north-south trending faults which separate different rock groups in the Hoggar Shield were unknown in the Dahomeyide orogen. Affaton et al. (1978) described the Kandi fault as a major dislocation in Benin Republic. It is a prolongation to the south of the 4° 50′ fault in the Hoggar (Caby et al. 1981) and of the Sobral fault in Brazil.

Some of these faults have been reactivated during the Phanerozoic. They are marked by steeply dipping zones of fault breccia and mylonites. Various kinematic criteria suggest that they show mainly a dextral sense of shearing. The timing of the initiation of the first major period of mylonitization was probably diachronous. Available evidence from the Pan-African granites, however, suggests that most of the mylonitisation is associated with D_2 deformation. The bulk of these granites appears to have been deformed in a subsolidus state. Generally, they have a steep foliation and a horizontal lineation and are thought to have been emplaced in the framework of a transcurrent regime. However, Ajibade et al. (1979) suggest that the north-south trending Zungeru mylonite zone was associated with D_1 structures and subsequently folded by D_2 deformation. The available mineralogical evidence (Boesse and Ocan 1988; Rahaman 1988b) suggest that mylonitization occurred in a high temperature regime and was subsequently retrogressed.

3.4 Metamorphism

Mineral assemblages reported from each of the six terranes or zones which comprise the Dahomeyide orogen indicate contrasting metamorphic histories during the Pan-African time. In Zone I, within the external units of the orogen, Affaton (1987) described high pressure-low temperature paragenesis (which include phengitic micas, pumpellyite, chlorite, piedmontite, and aragonite) from the Buem structural unit, and medium/high pressure-medium temperature paragenesis (characterized by green-blue-actinolite, zoisite, biotite, ripidolite in basic rocks) from the Atacora structural unit.

From Zone II, Menot and Seddoh (1985) described chloritoid, kyanite, garnet, phengite, omphacite, clinopyroxene and orthopyroxene assemblages in basic rocks from Lato Hill in southern Togo. They showed that these rocks crystallized at about 16 kb and 700 °C, which indicates a high pressure-high temperature type of metamorphism in this area. Most of the quartzites within this zone are kyanite-chloritoid bearing and indicate a high pressure-medium temperature metamorphism. In Zone III, biotite, garnet, and kyanite have been described and indicate medium high pressure-medium temperature type metamorphism. Mineral assemblages reported from several parts of Zone IV point to a bimodal distribution of metamorphic types (Rahaman 1988b). Widespread occurrences of chlorite, biotite, almandine garnet, staurolite, and sillimanite in rocks of appropriate composition indicate a medium pressure-medium temperature type of metamorphism. On a more restricted scale, assemblages which include cordierite, sillimanite, and anthophyllite have been documented (Rahaman and Ocan 1988; Ekwueme 1983; McCurry 1976; Sacchi 1968) and point to a low pressure-medium temperature type of metamorphism. Biotite, garnet, kyanite, sillimanite have been described from Zone V (Ekwueme and Onyeagocha 1986) and indicate a medium pressure-high temperature type of metamorphism.

Mineral assemblages from Zone VI suggest a bimodal distribution of metamorphic types (Toteu et al. 1987). A medium high pressure-medium temperature type of metamorphism is defined by the presence of rutile, garnet and kyanite in pelitic rocks. Cordierite bearing assemblages define a medium/low pressure-high temperature type of metamorphism.

Metamorphic grade

Mineral parageneses reported within the Dahomeyide orogen indicate that metamorphic grade varies from lowermost greenschist or zeolite facies to granulite facies. The Buem and western parts of the Atacora structural units record an anchizonal grade metamorphic assemblage (Affaton 1987). The Atacora structural unit and most of the schist belts of Nigeria and Cameroon have been metamorphosed to a greenschist or lower amphibolite facies grade. Locally, they may reach the upper amphibolite facies as in the Uwet area where kyanite and sillimanite have been described (Ekwueme and Onyeagocha 1986).

In general, the migmatitic quartzo-feldspathic gneisses lack diagnostic index mineral and it is often not possible to define precisely their metamorphic grade. The mineral assemblages formed in most place indicate amphibolite facies grade.

Granulite facies metamorphism has been documented in several places within the orogen. A north-south trending belt of granulite facies rocks has been mapped along the Savalou meridian within Zone III. A less well-defined belt of granulite facies rocks occurs north of the Cretaceous-Recent sediments in southern Benin Republic (Vedogbeton 1988). Rahaman and Ocan (1988) described granulite facies rocks exposed in more than 200 sq km within the migmatite-gneiss complex of Ikare area in Zone IV. Granulite facies rocks have been described in Zone VI from the Oubanguide fold belt (north of the Congo craton) around Yaounde where they form the core of a Pan-African nappe, and from the Nyong region at the Congo border. They also occur as relicts of Eburnean age in Pan-African amphibolite facies rocks around Geri and west of Tchollire (Macaudière et al 1988). These granulitic sectors of Zone VI belong to the Oubanguide Fold Belt which has been thrust onto the Congo Craton.

The time relation between metamorphism and deformation is poorly known and has been documented from only a few widely scattered areas. In general, the evidence, mainly from Zone IV, points to a polymetamorphic history. McCurry (1976) and Rahaman (1976) recognized two periods of syntectonic metamorphism (M_1 and M_3) separated by a period of static metamorphism (M_2). Rahaman and Ocan (1988) described four phases of metamorphism, M_1–M_4, from the Ikare area. M_1 and M_2 were progressive. M_2 in general reached amphibolite facies and locally attained granulite facies grade. The peak of M_2 metamorphism postdated the D_2 transcurrent deformation. M_3 and M_4 were retrogressive and reached lower amphibolite-greenschist facies grade. In the Ife-Ilesha area, Ocan et al. (1988) described three phases of metamorphism (M_1–M_3) and two phases of deformation. The first phase was associated with D_1 and led to development of garnet, staurolite, and sillimanite in rocks of appropriate composition and possibly eclogite facies assemblage, now retrogressed, in basic rocks. M_2, syntectonic and associated with D_2 transcurrent deformation, reached middle greenschist facies, with abundant development of biotite. M_3 was retrogressive. Correlations between the various metamorphic episodes described above are uncertain and the peaks of metamorphism are in the main diachronous.

In Zone III, M_1 metamorphism, which in general reached amphibolite-granulite facies grade, is thought to be associated with D_1 deformation. M_2 metamorphism was syntectonic and associated with D_2 deformation. It was responsible for the retrogression of the granulite facies rocks and the widespread development of amphibolite facies grade metamorphism in this zone (Affaton 1987).

In Zone I, M_1 and M_2 metamorphism were progressive and reached greenschist facies. They were associated with D_1 and D_2 deformations. In Zone II, M_1 metamorphism, produced eclogite facies rocks and is associated with D_1 deformation. M_1 assemblages were retrogressed in amphibolite facies conditions (M_2) during D_2 deformation (Menot and Seddoh 1985; Affaton 1987).

In Zone VI, M_1 metamorphism reached amphibolite facies grade and is associated with D_1. M_2 produced abundant migmatites and granulite facies assemblages between D_1 and D_2 or during D_2 deformation (Macaudière et al. 1988; Toteu et al. 1987).

The metamorphic imprint in the Dahomeyide orogen is now believed to be of Pan-African age. Older interpretations of the geology of the area ascribed Archean and Eburnean ages to the granulite facies rocks (i.e., for example Hubbard 1975). An important question that needs clarification in understanding the metamorphic history of this orogen is whether or not a widespread granulite facies metamorphism affected the area prior to the Pan-African event. Earlier interpretations (Hubbard 1968, 1975) suggested that the granulite facies rocks represent relicts of an earlier Archean or Eburnean more widespread granulite facies metamorphism. Field and petrographic evidence from the Ikare area (Zone IV) and around Savalou (Zone III) indicate that production of granulite facies mineral assemblage was of restricted occurrence and post-dated transcurrent D_2 deformation which affected the Pan-African granites. It is therefore doubtful that most of the area ever reached granulite facies grade metamorphism. However, Macaudière et al. (1988) have described from northern Cameroon (Zone VI) relict granulite facies grade rocks which have yielded U-Pb zircon ages of ca. 2100 Ma within a predominantly Pan-African domain. These suggest that granulite facies metamorphism during earlier orogenic events was, as in Pan-African times, of restricted occurrence.

3.5 *Geochronology*

There is an uneven distribution of available radiometric ages within the orogen. A relatively large amount of radiometric data is available from Zones IV-VI and these data must serve as a reference for the whole orogen. The dates or ages of the rocks have been obtained using a variety of techniques which include K-Ar and Rh-Sr on whole rocks and minerals, and U-Pb on zircon, sphene and monazite. Nd-Sm technique has been used to study the eclogites from Zone II.

Zone I

Caen-Vachette (1979) and Caen-Vachette et al. (1979) obtained a three-point Rb-Sr whole-rock isochron age of 2064±90 Ma (initial ratio of 0.7133±0.0039) for the orthogneiss of Kara. The Kara orthogneiss of Eburnean age thus represents the basement on which the Atacora rocks were deposited. A whole-rock mineral (muscovite, k-felspar, plagioclase) isochron from the orthogneiss yielded Rb-Sr age of 608±17 Ma, and records a signature of the Pan-African tectonothermal activity in this zone.

Zone II

Bernard-Griffiths et al. (1985) studied the eclogite facies rocks of this zone using the Nd-Sm technique. Two of the three samples gave TDM ages between 700 and 900 Ma, suggesting that the protoliths were of Pan-African age. The third sample, a mylonitized eclogite, yielded TDM age of 510 Ma. This is thought to closely date the mylonitization event which appears to have followed culmination of eclogitic metamorphism. Breda (1982) obtained a Rb-Sr whole-rock muscovite date of 1195±33 Ma on kyanite and chloritoid-bearing quartzited from Gorobani in northwestern Benin Republic. This result was thought to indicate the existence of the Kibaran orogeny in this country.

Zone III

Available Rb-Sr geochronological data by Bonhomme (1962), Caen-Vachette et al. (1979) and Breda (1985) on granites, orthogneisses, and migmatites and granulite facies rocks from this zone suggest that the rocks have been affected by three orogenic events, including phases of Eburnean, Kibaran, and Pan-African activities. The evidence for the Eburnean and Kibaran events is not convincing. Bonhomme (1962) reported model Rb-Sr whole-rock dates of 1708±220 Ma and 1925±90 Ma for the Kouandé and Karou gneisses, respectively. Caen-Vachette et al. (1979) obtained a six-point whole-rock Rb-Sr isochron age of 1708±169 Ma from orthogneisses from southwestern Togo. They considered this age as a minimum and the initial $^{87}Sr/^{86}Sr$ ratio of 0.7085 as a maximum. Moreover, the granulite facies rocks of Savalou in Benin was correlated with the Iforas' granulite block which records a U-Pb zircon age of 2100 Ma (Lancelot et al. 1976). Breda (1985) reported a four-point, Rb-Sr whole-rock isochron age of 1214±117 Ma from granulites of this zone III (initial $^{87}Sr/^{86}Sr$ ratio is 0.70200±0.00073). This result was thought to suggest that the granulite facies metamorphism was of Kibaran age. Unfortunately, the MSWD was not given but, on the diagram, only one of the four points lies on the isochron. A two-point Rb-Sr age of 916±69 Ma was also obtained on migmatitic gneisses from south of Dassa-Zoumé. The data reported are very poor, and the Kibaran dates are considered to be geologically meaningless. All supposed Eburnean ages appear relatively too young to be reasonably associated with the Eburnean orogeny.

The Pan-African event is geochronologically well-documented. Ages on migmatites and granitoids vary from ca. 640 to 560 Ma. Caen-Vachette et al. (1979) obtained whole-rock Rb-Sr ages of 609± 30 Ma ($^{87}Sr/^{86}Sr = 0.7069 \pm 0.0007$; MSWD = 1.7) and 610 ± 18 Ma ($^{87}Sr/^{86}Sr = 0.7108 \pm 0.0005$; MSWD = 1) from Nikki and Agramarou migmatites. The syntectonic syenitic Wari-Marou Kpessou complex yielded a six-point Rb-Sr whole-rock isochron of 638±16 Ma ($^{87}Sr/^{86}Sr = 0.7067 \pm 0.0003$; MSWD = 0.5; Caen-Vachette et al. 1979). Caen-Vachette (1975) reported ages of 568, 598, and 576 Ma for the granites of Sinendé, Savè, and Fita, respectively.

Zone IV and V

Rahaman (1988a) and Ajibade et al. (1988) summarized the available radiometric dates on these zones. U-Pb data on zircons from the early grey gneisses, which are the oldest recognizable rock types in the basement complex, indicate ages of ca. 2400 Ma (Pidgeon et al. 1976; Rahaman and Lancelot, unpubl). Whole-rock Rb-Sr ages of 2283±70 Ma (Grant 1970) and 1904±18 Ma (Rahaman et al. 1983) were obtained on the Ibadan and Igbetti granite gneisses, respectively. The initial $^{87}Sr/^{86}Sr$ ratios are low (0.700-0.702) and are similar to values expected from upper mantle material. Therefore, it seems that the ca. 2400-2100 Ma Dabakalian system (Lemoine et al. 1985; Tempier 1986), or Lower Birrimian Supergroup (Hirdes et al. 1987), is not underlain by pre-existing continental crust but, instead, represents juvenile material. The ca. 2400 Ma event is thus seen as a crust-forming event. This crust underwent its first major tectonothermal event during the Eburnean orogeny.

Several dates which fall in the Kibaran spectrum (Grant et al. 1972; Ogezi 1977; Ajibade 1980; Caen-Vachette and Ekwueme 1988) have been reported. These dates are considered here to be geologically meaningless. Zircons from the Ile-Ife granite gneiss dated at 1172±142 Ma by a poorly-defined, whole-rock Rb-Sr isochron by Grant et al. (1972) have yielded U-Pb ages of 1850 Ma (Rahaman 1988a).

Ogezi (1977) and Ajibade (1980) obtained a ten-point, whole-rock, Rb-Sr isochron age of 1098±65 Ma ($^{87}Sr/^{86}Sr$ ratio=0.703-0.704) on phyllite from the main schist belt. Phyllitic rocks from the southern extension of this schist belt, however, yielded a 5-point, Rb-Sr whole-rock isochron age of 566±12 Ma ($^{87}Sr/^{86}Sr$ ratio=0.7094±0.0005) and a 3-point, Rb-Sr whole-rock isochron age of 681±36 Ma ($^{87}Sr/^{86}Sr$ ratio=0.7054±0.0008). The whole-rock Rb-Sr isochron results on the similar rocks with a similar tectonic history are thus seen to be widely different.

Caen-Vachette and Ekwueme (1988) published a whole-rock Rb-Sr isochron date of 1315±72 Ma ($^{87}Sr/^{86}Sr$ ratio=0.7016 and MSWD=10.7) on metasedimentary schists from Lokoja area in central Nigeria, on the eastern border of Zone IV. These rocks, like the phyllites dated by Ogezi (1977) and Ajibade (1980), are characterized by a low initial $^{87}Sr/^{86}Sr$ ratio which suggests that these sediments were either mantle-derived or volcanic and immature sediments with a short crustal residence time. In fact, these phyllites are derived from fine-grained mature sediments and the Kibaran dates are considered to be geologically meaningless. The Pan-African orogen is well documented by U-Pb data on zircons from charnockites and associated granites (Tubosun 1983; Tubosun et al. 1984; Van Breemen et al. 1977), whole-rock Rb-Sr isochron ages on granites and charnockites (Van Breemen et al. 1977; Rahaman et al. 1983; Grant 1978) and conventional Rb-Sr and K-Ar ages on minerals.

Ages obtained from granite whole-rock Rb-Sr isochrons and U-Pb studies on zircons range from ca. 677 to 518 Ma. Rahaman (1988a) divided granitic activity into three broad groups: older than ca. 600 Ma; 560±40 Ma; and, ca. 500 Ma. The peak of Pan-African magmatism is dated by U-Pb on zircons at ca. 610±40 Ma. The 580 Ma reported by Tubosun et al. (1984) is thought to date granulite facies metamorphism in Idanre area. The ca. 500 Ma granites have ages which are similar to the K-Ar and Rb-Sr ages on biotites. They characteristically have metamorphic aureoles developed around them. They were therefore emplaced after the country rocks had cooled to below the closure temperature for biotite.

Zone VI

The geological evolution of Zone VI is well constrained by a relatively large number of whole-rock Rb-Sr results (Bessoles and Lasserre 1977; Nzenti et al. 1988) which have been supported by more recent data obtained from U-Pb dating of zircons.

An Archean domain, the Ntem complex, is present between the Nyong region and the border of Cameroon with Gabon and Congo. It represents the northern part of the Congo craton and is composed of granulite facies gneisses. Whole-rock Rb-Sr isochron ages on these rocks (Lasserre and Soba 1976) have yielded ages older than ca. 2800 Ma. Mineral isochrons and conventional ages on biotites from Ntem complex, by K-Ar and Rb-Sr techniques, yielded ages between 1840±75 Ma and 2440±50 Ma. Lasserre (1964, 1967) and Lasserre and Soba (1976) show that the Archean Ntem complex was reactivated during the Eburnean orogeny. Further evidence for this Eburnean orogenic event has been obtained from granulitic remnants within the Pan-African domain, north of the Ngaoundéré fault line, in the region around Gezi and west of Tcholliré. Penaye (1988) and Toteu et al. (1989) obtained U-Pb ages of ca. 2100 Ma on zircons from these granulites.

U-Pb ages on zircons, whole-rock Rb-Sr isochron ages, and conventional Rb-Sr and K-Ar ages on minerals in the range ca. 800-500 Ma define clearly the signature of Pan-African orogenesis. Toteu et al. (1987) obtained an upper intercept U-Pb age of

830 ± 12 Ma on metarhyolite interlayered within the Poli schist belt and interpreted this as volcanic activity related to the initial stage of the Pan-African evolution. Nearly concordant U-Pb zircon isochrons on rock units from a basic to intermediate plutonic suite in northern Cameroon, which was previously considered to be part of an Archean basement, yielded upper intercept ages of ca. 630 Ma (Toteu et al. 1987). These results, together with low initial $^{87}Sr/^{86}Sr$ ratios and calc-alkaline geochemical characteristics of the rocks, were used to argue in favor of large amounts of juvenile crustal additions in this part of the Dahomeyide orogen during the Pan-African orogeny. Earlier, Lasserre and Soba (1979) dated the migmatization in southern Cameroon by whole rock Rb-Sr isochron at 565 ± 22 Ma.

4 Discussion

Most workers in the Precambrian orogens of West Africa appear to agree that plate tectonic models can be applied to explain the evolution of the Pan-African belt immediately east of the West African craton (Fig. 4; Burke and Dewey 1972, 1973; Burke et al. 1977; Bertrand and Caby 1978; Leblanc 1976, 1981; Black et al. 1979a, b; Affaton 1987; Trompette 1979 and in Bessoles and Trompette 1980; Wright and Ajibade 1987; Bertrand et al. 1986; Fitches et al. 1985; Rahaman et al. 1988).

Burke and Dewey (1972, 1973) suggested that the Pan-African belt east of the West African craton resulted from the closure of an Atlantic-type ocean by continent/continent (Himalayan-type) collision between the western continent of Birrimia and the eastern continent of Dahomea. According to these authors, the Buem structural unit should be considered as a vestigial fragment of an oceanic crust and represents the Pan-African suture zone. More recent detailed work (Trompette 1979 and in Bessoles and Trompette 1980; Affaton 1987) demonstrated that this suture zone is not located within the Buem, but is situated east of the Atacora structural unit where it is represented by the ultramafic/mafic rocks of Zone II (Fig. 3). Black et al. (1979b), on the other hand, based on their work in Mali, suggested that the closure of an ocean between an active eastern continental margin (Andean type) and a passive western continental margin along the eastern side of the West African craton was responsible for the development of the Pan-African belt. This view has been amply supported by recent works done in Togo, Benin, Nigeria, and Cameroon (Affaton et al. 1980; Trompette in Bessoles and Trompette 1980; Affaton 1987; Rahaman 1988a, b; Caby 1987; Toteu 1986).

Continental fragmentation appears to have been initiated ca. 1100 Ma ago and led to the formation of a Pan-African proto-ocean east of the West African Craton at ca. 900 ± 100 Ma. A passive western continental margin developed (i. e., Volta basin). Closure of the ocean began with subduction processes operating along an eastward dipping Benioff zone. The upper Proterozoic Sonoumon complex within Zone III (a Pan-African arc?), which consists largely of volcano-detritic rocks accompanied by calc-alkaline volcanism and plutonism, and the widespread calc-alkaline to subalkaline granites were established as a result of this process. These rocks have many characteristics which are comparable with modern island arcs and active continental margins (Affaton et al. 1978).

Behind this possible Pan-African arc (?), several, often fault-bounded, Pan-African volcano-sedimentary basins (evolving later into schist belts) were developed. The basins, near the craton margin, were small and contained more volcanic rocks than sedimentary units (e. g. the Buem), whereas basins further to the east were more extensive and tended to contain far more fine-grained sedimentary materials than volcanic rocks. In the Ife-Ilesha area and the Sado-Anka region (southwestern and northwestern Nigeria respectively) and in the Poli area of northern Cameroon, the crust thinned considerably and led to the emplacement of ultramafic and mafic rocks with ocean floor affinities, but with some amount of crustal contamination (Rahaman et al. 1988; Ogezi 1988; Bafor 1988; Toteu 1986; Ngako 1986).

Closure occurred along the eastern margin of the West African craton. This produced the parautochthonous Buem structural unit and the Atacora nappes, both made of materials from the passive continental margin. Because of thrusting and development of nappes, thickening of the crust occurred and this led to widespread rejuvenation of pre-existing crust and production and emplacement of granites. Most of the basins behind the Pan-African arc mentioned above probably also closed during this phase. Renewed thinning and further extension in a few zones led to the creation of grabens filled with Pan-African molassic materials and mafic to acid volcanic rocks (e. g., Dahou-Mahou, Lanta, Kandi, and Sauna basins). In some areas, these volcano-detritic sequences were folded and metamorphosed into zeolitic to greenschist facies, and in-

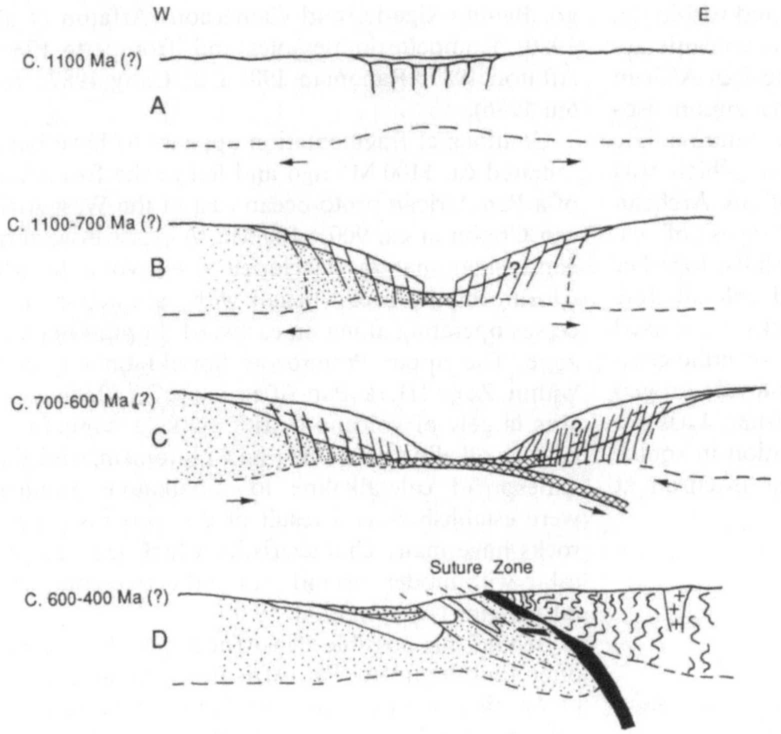

Fig. 4. Geodynamic evolution of western portions *(I–III)* of the Dahomeyide orogen (adapted from Affaton 1987). *A* Distention and first stage of rifting; *B* major stage of rifting leading to the creation of the Pan-African proto-ocean; *C* convergence and subduction of the Pan-African oceanic crust; *D* collision, Pan-African ocean closure, folding, thrusting, erosion and development of a molassic basin

truded by late granites and pegmatites during the terminal stages of the Pan-African orogeny.

The eastward dipping subduction process, the partial melting in the thickened crust and the last event of the Pan-African tectonothermal activity produced granitic magmas which were subsequently emplaced as pre-, syn-, late- and post-tectonic Pan-African granitoids. The large amount of these granitoids within the whole Dahomeyide orogen and their relative importance in some areas (Zone III–VI), as well as their association with ultramafic/mafic rocks with ocean floor affinities, indicate that a more complex geodynamic model must be considered (Fig. 5). For example, Wright and Ajibade (1987) considered the Pan-African domain as an aggregation or mosaic of island arcs, interarc (ensimatic) basins, and continental fragments. They suggested that all of the ultramafic/mafic assemblages found in the Pan-African "basement" of the Togo-Benin-Nigeria shield represent vestigial fragments of ophiolite suites, dismembered by collision and then largely removed by erosion. The sutures, along which these assumed residual ophiolites are located, are further assumed to have provided the loci for subsequent transcurrent faulting during the various collision events that built up the shield. According to these authors, ultramafic/mafic bodies associated with transcurrent shearing in the northwest (Sado-Anka region) and with low angle thrusting in the south (Ife-Ilesha area) are consistent with this model.

Alternatively, Rahaman (1988b) suggested that, besides the major Pan-African suture (Zone II), a smaller ocean was created in the Ife-Ilesha and Sado-Anka area. The closure of this ocean involved initial subduction along a westward dipping Benioff zone, producing calc-alkaline volcanic and plutonic rocks (e. g., Iwo plutonic rocks and Maiinchi batholith) forming a new arc. Final closure was eventually affected by continent/continent type collision, which led to a considerable thickening of the crust west of the Ife-Ilesha schist belt, dismembering of the oceanic crust, rejuvenation of pre-existing basement, and emplacement of subsequent granitoids. This hypothesis (Rahaman 1988b) implies a slight diachronism for Pan-African thermotectonic events across the Dahomeyide orogen (as has been shown in the Hoggar shield by Bertrand et al. 1986).

To better understand the Dahomeyide orogen, one must take into account the presence of a Pan-African possible suture zone along the northern margins of the Congo craton, within the Oubanguide orogen (Poidevin et al. 1981; Poidevin 1983, 1985; Nzenti et al. 1984; Nedelec et al. 1986). Such a suture, which is neither well-defined nor well-located, implies the possible closure of an ocean of unknown dimension, with subduction processes operating along a northward dipping Benioff zone.

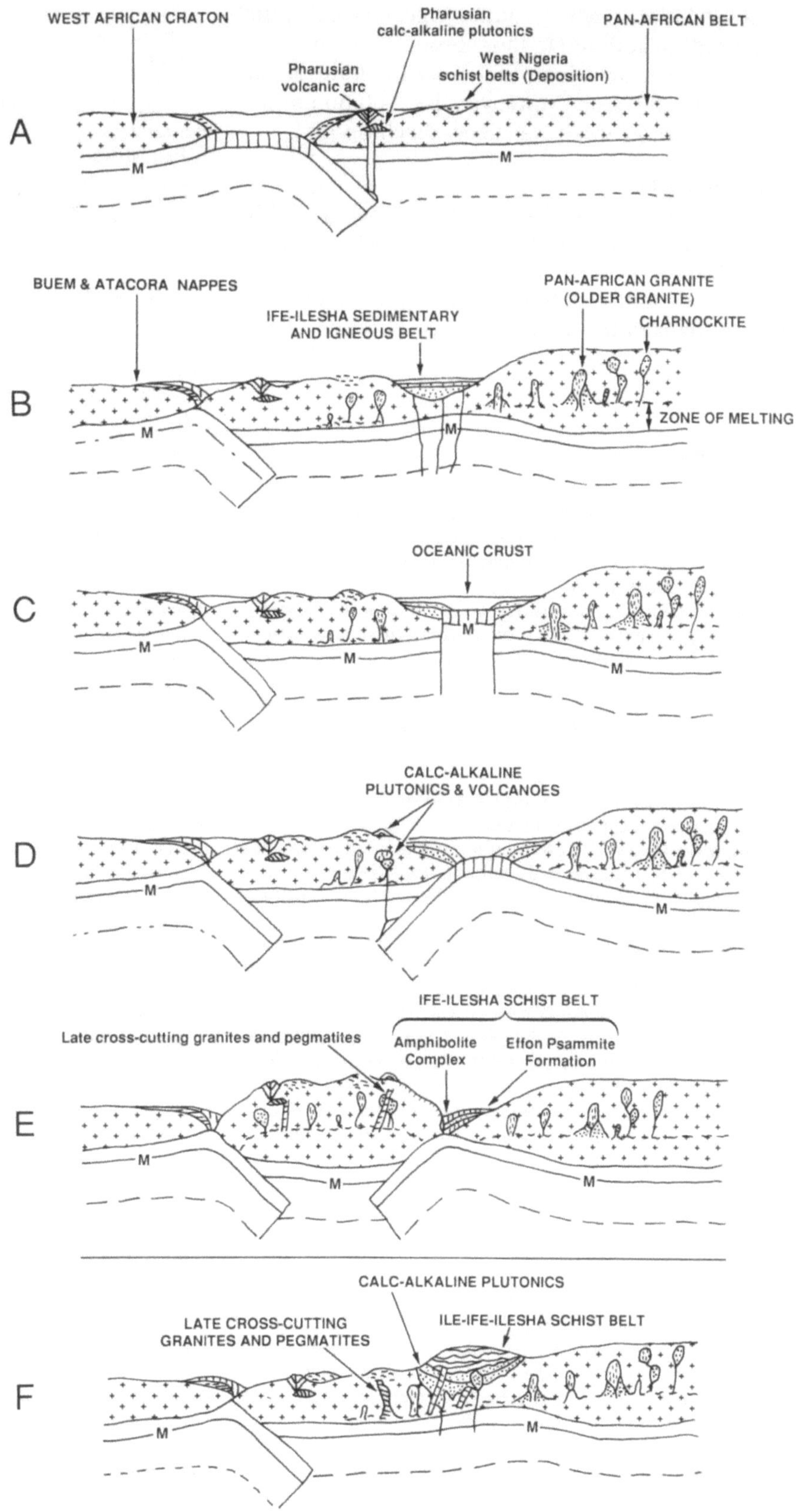

Fig. 5. Possible geodynamic evolution of the Dahomeyide orogen: *A* Subduction processes along margin of West African craton associated with east-dipping subduction; *B* oceanic closure along craton/Pan-African orogen margin with back-arc extension; *C* rifting with formation of oceanic lithosphere; *D* subduction processes along west-dipping subduction zone in Ife-Ilesha area; *E* closure of oceanic tract and formation of post-kinematic granites; *F* alternative interpretation (following stage *B*)

This evolution would produce volcanic rocks and granitoids in addition to those produced by the subduction processes along the eastern margin of the West African craton. Moreover, the evolution of the two subduction zones was likely diachronous.

Acknowledgments

This contribution would not have been possible without the encouragement and patience of Dr. R. D. Dallmeyer. Original versions of the manuscript were improved through his critical review. He is also thanked for assistance in preparation of the figures.

References

Affaton P (1975) Etude géologique et structurale du Nord-Ouest Dahomey, du Nord-Togo et du Sud-Est de la Haute-Volta. Thèse 3e cycle, Trav Lab Sci Terre St Jérôme, Marseille, Fr, Ser B, 10: 201

Affaton P (1987) Le bassin des Volta (Afrique de l'Ouest): une marge passive, d'âge protérozoïque supérieur, tectonisée au Panafricain (600 ± 50 Ma). Thèse Doct d'Etat, Fac Sci St Jérôme, Univ Aix-Marseille III, Fr, 462 pp

Affaton P (1988) L'unité structurale de l'Atacora et ses relations avec les unites structurales adjacentes. Int Meet Proterozoic Geol Tectonics High Grade Terrains, Univ Ife, Nigeria, 20 pp

Affaton P, Lasserre JL, Lawson LT, Vincent PL (1978) Notice explicative des cartes géologiques au 1/200000 de la République du Togo et de la République populaire du Bénin entre les 9e et 10e degrés de latitude nord (feuille Bassari-Djougou et feuille Parakou-Nikki). Rapp BRGM n° 78 RDM 055 AF, Orleans, Fr, 70 pp

Affaton P, Sougy J, Trompette R (1980) The tectonostratigraphic relationships between the Upper Precambrian and Lower Paleozoic Volta Basin and the Pan-African Dahomeyide orogenic Belt (West Africa). Am J Sci 280: 224-248

Ajayi TR (1981) On the geochemistry and origin of the amphibolite in Ife-Ilesha area, SW Nigeria. Nigerian J Min Geol 17: 179-196

Ajibade AC (1980) Geotectonic evolution of the Zungeru region, Nigeria. Ph D Thesis, Univ Wales, Aberystwyth, GB, 303 pp (unpubl)

Ajibade AC (1988) Evolution structurale et tectonique du socle nigérian avec spéciale référence au Nord-Ouest Nigéria. Int Meet Proterozoic Geol Tectonics High Grade Terrains, Univ Ife, Nigeria, 22 pp

Ajibade AC, Fitches WR, Wright JB (1979) The Zungeru mylonites, Nigeria: recognition of a major tectonic unit. Rev Geol Dyn Géogr Phys 21 (5): 359-363

Ajibade AC, Rahaman MA, Woakes M (1988) Proterozoic crustal development in the Pan-African regime of Nigeria. In: Kröner A (ed) Proterozoic lithospheric evolution. Am Geophys Union Spec Publ, pp 259-271

Bafor BE (1982) The Zungeru metavolcanics of north-western Nigeria: their geochemistry and tectonic setting. J Mining Geol 18 (23): 42-52

Bafor BE (1988) Some geochemical considerations in the evolution of the Nigerian basement in the Egbe area of southwestern Nigeria. In: Oluyide PO et al. (eds) Precambrian geology of Nigeria. Publ Geol Surv Nigeria, Kaduna, pp 277-288

Bernard-Griffiths J, Peucat JJ, Menot R-P, Seddoh KF, Lawson LT (1985) Sm-Nd study of some eclogites from Togo (West Africa). Terra Cognita 5 (4): 434

Bertrand JML, Caby R (1978) Geodynamic Evolution of the Pan-African Orogenic Belt: A new interpretation of the Hoggar Shield (Algerian Sahara). Geol Rundsch 67 (2): 357-388, 8 fig

Bertrand J-M, Michard A, Boullier AM, Dautel D (1986) Structure and U-Pb geochronology of Central Hoggar (Algeria): A reappraisal of its Pan-African evolution. Tectonics 5 (7): 955-972

Bessoles B, Lasserre M (1977) Le complexe de base du Cameroun. Bull Soc Geol Fr 7/XIX (5): 1085-1092

Bessoles B, Trompette R (1980) Géologie de l'Afrique - La chaîne Pan-Africaine (zone mobile d'Afrique centrale (partie sud) et zone mobile soudanaise). Mém Bur Rech Géol Min Orléans, Fr 92: 397

Black R (1985a) Bref apercu sur l'Archéen de l'Afrique centrale et de l'Ouest. CIFEG, Paris, Publ Occas 1985/4: 47-54

Black R (1985b) Le Protérozoïque inférieur: Afrique du Sud, Gabon, Afrique occidentale. CIFEG, Paris, Publ Occas 1985/4: 71-83

Black R, Ba H, Ball E et al. (1979a) Outline of the Pan-African geology of Adrar des Iforas. Geol Rundsch 68 (2): 543-564

Black R, Caby R, Moussine-Pouchkine A et al. (1979b) Evidence for late Precambrian plate tectonics in West Africa. Nature (Lond) 278 (5701): 223-227

Boesse JM, Ocan OO (1988) Géologie et évolution de la ceinture schisteuse d'Ifé-Ilésha (Sud-Ouest Nigéria). Int Meet Proterozoic Geol Tectonics High Grade Terrains, Univ Ife, Nigeria, 24 pp

Bonhomme M (1962) Contribution à l'étude géochronologique de la plateforme de l'Ouest africain. Ann Fac Sci Univ Clermont-Ferrand, n° 5, Géol Minéral, fasc 5: 62 pp

Breda IR (1982) Etude de cartographie géologique et de prospection minière de reconnaissance au Nord du 11e parallèle (Bénin). Rapp final Projet FED n° 4105-011-13-20, Geomineraria Italiana, Borgo S Dalmazzo, Italie (unpubl)

Breda IR (1985) Etude de cartographie géologique et de prospection minière de reconnaissance au Sud du 9e parallèle (Bénin). Rapp première phase Projet FED n° 5100-11-13-015, Geominieraria Italiana, Borgo S Dalmazzo, Italie (unpubl)

Burke KC, Dewey JF (1972) Orogeny in Africa. In: Dessauvagie TFJ, Whiteman AJ (eds) African Geology. Dept Geol Univ Ibadan, Nigeria, pp 583-608

Burke KC, Dewey JF (1973) An outline of Precambrian plate development. In: Tarling DM, Runcorm SK (eds) Implications of continental drift to the Earth sciences. Academic Press, Lond 2: 1035-1045

Burke K, Dewey JF, Kidd WSF (1977) World distribution of sutures. The sites of former oceans. Tectonophysics 40: 60-99

Caby R (1987) The Pan-African Belt of West Africa from the Sahara-Desert to the Gulf of Benin. In: Schaer JP, Rodgers J (eds) The anatomy of mountain ranges. Princeton Univ Press, USA, pp 129-170

Caby R (1988) Les terrains précambriens du Bénin, Nigéria, et Nord-Est du Brésil et les connections sud-atlantiques au Protérozoïque supérieur. Int Meet Proterozoic Geol Tectonics High Grade Terrains, Univ Ife, Nigeria, 23 pp

Caby R, Bertrand JML, Black R (1981) Pan-African ocean

closure and continental collision in the Hoggar-Iforas segment, central Sahara. In: Kröner A (ed) Precambrian plate tectonics. Elsevier, Amst, pp 407-434

Caen-Vachette M (1975) Age Pan-Africain des granites de Sinendé, Savé et Fita (Dahomey). C R Acad Sci, Paris, 281 (23) Ser D: 1793-1795

Caen-Vachette M (1979) Orthogneiss birrimiens et migmatisation panafricaine au Togo-Bénin. Résumés 10e Coll Géol Afr, Montpellier, p 22

Caen-Vachette M, Ekwueme BN (1988) Rb-Sr ages of schists in the metasedimentary belts in southeast Lokoja and their implications for the Precambrian evolution of central Nigeria. J Afr Earth Sci 7 (1): 127-131

Caen-Vachette M, Pinto KJM, Roques M (1979) Plutons éburnéens et métamorphisme dans le socle cristallin de la chaîne pan-africaine au Togo et au Bénin. Rev Géol Dyn Géogr Phys 21 (5): 351

Cahen L, Snelling NJ, Delhal J, Vail JR (1984) The Geochronology and Evolution of Africa. Clarendon Press, Oxford, 512 p

Camil J (1984) Pétrographie, chronologie des ensembles granulitiques archéens et formations associées de la région de Man (Côte d'Ivoire). Implications pour l'histoire géologique de craton ouest-africain. Thèse d'Etat, Univ Abidjan, Côte d'Ivoire, 306 pp

Cosson J (1957) Données nouvelles sur le système de l'Ogoué et les roches cristallines associées. Bull Dir Min Géol Afr Equat Fr 8: 21-32

Crenn Y (1957) Mesures gravimétriques et magnétiques dans la partie centrale de l'A. O. F. Interprétations géologiques. ORSTOM, Paris, 39 pp

Deynoux M, Sougy J, Trompette R (1985) Lower Palaeozoic rocks of West Africa and the western part of Central Africa. In: Holland CH (ed) Lower Palaeozoic of North-western and West-central Africa. Wiley, London, pp 337-495

Ekwueme BM (1983) Regional metamorphism of pelitic rocks, south-east of Lokoja, Nigeria. Nigerian J Min Geol 20 (1/2): 71-77

Ekwueme BN, Onyeagocha AC (1986) Geochemistry of metasedimentary rocks of Uwet area, Oban Massif, southeastern Nigeria. Geol Rundsch 75/2: 411-420

Fitches WR, Ajibade AC, Egbuniwe IG, Holt RW, Wright JB (1985) Late proterozoic schist belts and plutonism in NW Nigeria. J Geol Soc Lond 142: 319-337

Grant NK (1970) Geochronology of Precambrian basement rocks from Ibadan, southwestern Nigeria. Earth Planet Sci Lett 10: 29-38

Grant NK (1978) Structural distinction between a metasedimentary cover and an underlying basement in the 600 m. y. old Pan-African domain of northwestern Nigeria. Geol Soc Am Bull 89: 50-58

Grant NK, Hickman M, Burkholder FR, Powell JL (1972) Kibaran metamorphic belt in Pan-African domain of West Africa. Nature (Lond) (Phys Sci) 238: 90-91

Hirdes W, Leube A, Mauer R, Moorbath S (1987) The evolution of the Birimian supergroup in Ghana/West Africa. CIFEG, Paris, Publ Occas 1987/12: 48-49

Hubbard FH (1968) The association charnockite-older granite in southwestern Nigeria. Nigerian J Min Geol 3: 25-32

Hubbard FH (1975) Precambrian crustal development in western Nigeria: indications from the Iwo region. Geol Soc Am Bull 86: 548-554

Junner NR, Hirst T (1946) The geology and hydrogeology of the Voltaian basin. Mem Gold Coast Geol Surv, Accra, 8: 50

Klemm DD, Schneider W, Wagner B (1984) The Precambrian sedimentary sequence east of Ife and Ilesha/SW Nigeria. A Nigerian "greenstone belt"? J Afr Earth Sci 2 (2): 161-176

Klerkx J (1985) Le Protérozoïque inférieur en Afrique orientale. CIFEG, Paris, Publ Occas 1985/4: 85-90

Lancelot JR, Vitrac A, Allègre CJ (1976) Uranium and lead isotopic dating with grain-by-grain zircon analysis: a study of complex geological history with a single rock. Earth Planet Sci Lett 29: 357-366

Lasserre M (1964) Etude géochronologique par la méthode Strontium-Rubidium de quelques échantillons en provenance du Cameroun (Afrique équatoriale). Ann Fac Sci Univ Clermont-Ferrand n° 25, Géol Mineral n° 8, Etud Géochronol I: 53-67

Lasserre M (1967) Données nouvelles acquises, en géochronologie, par la méthode au strontium appliquée a l'étude des massifs cristallins du Cameroun. C R Somm Soc Géol Fr 3: 89-90

Lasserre M, Soba D (1976) Age libérien des granodiorites et des gneiss à pyroxène du Cameroun méridional. Bull BRGM, Paris, 2e Sér, IV (1): 17-32

Lasserre M, Soba D (1979) Migmatisation d'âge panafricain au sein des formations camerounaises appartenant à la zone mobile d'Afrique Centrale. C R Somm Soc Géol Fr 2: 64-68

Leblanc M (1976) Proterozoic oceanic crust at Bou Azzer. Nature (Lond) 261: 34-35

Leblanc M (1981) The late Proterozoic ophiolites of Bou Azzer (Morocco): evidence for Pan-African plate tectonics. In: Kroner A (ed) Precambrian plate tectonics. Elsevier, Amst, pp 435-451

Lemoine S, Tempier P, Bassot JP, Caen-Vachette M, Vialette Y, Wenmenga V, Joure S (1985) Le Burkinien, cycle orogénique précurseur de l'Eburnéen en Afrique de l'Ouest. 13th Coll Afr Geol, St Andrews, G B, pp 26

Leprun, J-C, Trompette R (1969) Subdivision du Voltaien du massif de Gobnangou (République de Haute-Volta) en deux séries discordantes séparées par une tillite d'âge éocambrien probable. C R Acad Sci, Paris, 269: 2187-2190

Macaudière J, Penaye J, Barbey P, Nzenti JP, Bertrand J-M (1988) Comparison of the evolution of the granulitic domains in the Pan-African orogen of northern and southern Cameroon. Int Meet Proterozoic Geol Tectonics High Grade Terrains, Univ Ife, Nigeria, 9 pp

McCurry P (1976) The geology of the Precambrian to Lower Palaeozoic rocks of northern Nigeria. A review. In: Kogbe CA (ed) Geology of Nigeria. Elizabethan, Lagos, Nigeria, pp 15-39

Menot RP, Seddoh KF (1985) The eclogites of the Lato Hills, south Togo, West Africa: Relics from the early tectonometamorphic evolution of the Pan-African orogeny. In: Smith DC, Franz G, Gebauer D (eds) Chemistry and petrology of eclogites. Chem Geol 50: 313-330

Nedelec A, Macaudière J, Nzenti J-P, Barbey P (1986) Evolution structurale et métamorphique des schistes de Mbalmayo (Cameroun). Implications pour la structure de la zone mobile pan-africaine d'Afrique centrale, au contact du craton du Congo. C R Acad Sci, Paris, II (303): 75-80

Ngako V (1986) Evolution métamorphique et structurale de la bordure Sud-Ouest de la "Série de Poli". Segment camerounais de la chaîne Panafricaine. Mém Doc, Centr Armoricain Etud Struct Socles, Rennes, 5: 185

Nzenti JP, Barbey P, Jegouzo P, Moreau C (1984) Un nouvel exemple de ceinture granulitique dans une chaîne protérozoïque de collision: les migmatites de Yaounde au Cameroun. C R Acad Sci, Paris, II (299): 1197-1199

Nzenti JP, Barbey P, Macaudière J, Soba D (1988) Origin and evolution of the late Precambrian high grade Yaounde gneisses, Cameroon. Precambrian Res 38: 91-109

Ocan OO, Rahaman MA, Boesse FM (1988) Caractéristiques pétrologiques du Schist belt d'Ife-Ilesha (Sud-Ouest Nigeria). Int Meet Proterozoic Geol Tectonics High Grade Terrains, Univ Ife, Nigeria, 26 pp

Ogezi AEO (1977) Geochemistry and geochronology of basement rocks from northwestern Nigeria. Ph D Thesis, Dept Earth Sci, Univ Leeds, UK, 298 pp (unpubl)

Ogezi AEO (1988) Geochemistry and origin of ensialic Alpine-type serpentinite associations from Mallam Tanko (Shemi) and Rabah (Wasagu), northwestern Nigeria. In: Oluyide PO et al. (eds) Precambrian geology of Nigeria. Publ Geol Surv Nigeria, Kaduna, pp 257-276

Penaye J (1988) Pétrologie et structure des ensembles métamorphiques du Sud-Est de Poli (Nord-Cameroun). Thèse Doct Inst Nation Polytech Lorraine, Nancy, 196 pp

Pidgeon RT, Van Breemen O, Oyawoye MO (1976) Pan-African and earlier events in the Basement Complex of Nigeria. 25th Congr Geol Int, Sydney, Australia (abstr), p 667

Poidevin JL (1983) La tectonique pan-africaine à la bordure nord du craton congolais: l'orogenèse des "Oubanguides". 12th Coll Afr Geol, Tervuren, B (abstr), p 75

Poidevin J-L (1985) Le Protérozoique supérieur de la République centrafricaine. An Sér 8°, Sci Geol, Mus R, Afr Centr Tervuren, Belgique, 91: 75

Poidevin JL, Alabert J, Miauton J-D (1981) Géologie des séries du Précambrien supérieur de la région de Bakouma (République centrafricaine). Bull Bur Rech Géol Min (2e Ser), 4(4) (1980-81): 311-318

Rahaman MA (1976) Review of the basement geology of southwestern Nigeria. In: Kogbe CA (ed) Geology of Nigeria. Elizabethan, Lagos, Nigeria, pp 41-58

Rahaman MA (1988a) Recent advances in the study of the basement complex of Nigeria. In: Oluyide et al (eds) Precambrian geology of Nigeria. Publ Geol Surv Nigeria, Kaduna, pp 11-43

Rahaman MA (1988b) Precambrian geology of Nigeria. Int Meet Proterozoic Geol Tectonics High Grade Terrains, Univ Ife, Nigeria, 34 pp

Rahaman MA, Ocan O (1988) The nature of granulite facies metamorphism in Ikare area, southwestern Nigeria. In: Oluyide et al. (eds) Precambrian geology of Nigeria. Publ Geol Surv Nigeria, Kaduna, pp 157-163

Rahaman MA, Emofurieta WO, Caen-Vachette M (1983) The potassic granite of the Igbeti area: Further evidence of the polycyclic evolution of the Pan-African belt in southwestern Nigeria. Precambrian Res 22: 75-92

Rahaman MA, Van Breemen O, Bowden P, Bennett JN (1984) Age migration of anorogenic ring complexes in northern Nigeria. J Geol 92: 173-184

Rahaman MA, Ajayi TR, Oshin IO, Asubiojo FOI (1988) Trace-element geochemistry and geotectonic setting of Ile-Ife schist belt. In: Oluyide et al. (eds) Precambrian geology of Nigeria. Publ Geol Surv Nigeria, Kaduna, pp 241-256

Rahman AAMS, Ukpong EE, Azmatullah M (1988) Geology of parts of the Oban Massif, Southeastern Nigeria. J Min Geol 15 (1): 38-56

Sacchi R (1968) The geology of the region around Bena in northern Nigeria. Mem Inst Geol Min Univ Padova, XXVI: 48

Sagbohan W (1972) Contribution à la géologie du Dahomey par l'utilisation des mesures gravimétriques et magnétiques. Thèse 3e cycle, Univ Strasbourg, Fr, 107 pp

Saunders RS (1970) Early palaeozoic orogeny in Ghana: foreland stratigraphy and structure. Geol Soc Am Bull 81: 233-240

Sougy J (1971) Remarques sur la stratigraphie du Protérozoique supérieur du basin voltaïen; influence de la paléosurface d'érosion glaciaire de la base du groupe de l'Oti sur le tracé sinueux des Volta et de certains affluents. C R Acad Sci, Paris, D, 272: 800-803

Tempier P (1986) Le Burkinien: cycle orogénique majeur du Protérozoïque inférieur en Afrique de l'Ouest. CIFEG, Publ Occasi 1986/10: 17-23

Toteu SF (1986) Chronologie des grands ensembles structuraux de la région de Poli. Accrétion crustale dans la chaine panafricaine du Nord Cameroun. Thèse Doct d'Etat, Univ Nancy, Fr, 197 pp

Toteu SF, Michard A, Bertrand JM, Rocci G (1987) U-Pb dating of precambrian rocks from northern Cameroon, orogenic evolution and chronology of the Pan-African belt of Central Africa. Precambrian Res 37: 71-87

Toteu SF, Bertrand JM, Penaye J, Macaudière J, Angoua S, Barbey P (1988) Cameroon, a tectonic keystone in the Pan-African network. Geol Assoc Can Spec Pap Trans-Hudson Orogen. J Lawry and Strauffer, Toronto (in press)

Toteu SF, Macaudière J, Bertrand M-M, Dautel D (1989) Metamorphic zircons from North Cameroon. Implications for the Pan-African evolution of Central Africa. Geol Rundsch (submitted)

Trompette R (1973) Le Précambrien supérieur et le Paléozoïque inférieur de l'Adrar de Mauritanie (bordure occidentale du bassin de Taoudeni, Afrique de l'Ouest). Un exemple de sédimentation de craton. Etude stratigraphique et sédimentologique. Thèse d'Etat, Trav Lab Sci Terre, Marseille St Jérôme, B 7: 702

Trompette R (1979) Les Dahomeyides au Bénin, Togo et Ghana: une chaîne de collision d'âge pan-africain. Rev Géogr Phys Géol Dyn 21 (5): 339-349

Trompette R, Affaton P, Joulia F, Marchand J (1980) Stratigraphic and structural controls of Late Precambrian phosphate deposits of the Northern Volta Basin in Upper Volta, Nigeria and Benin, West Africa. Econ Geol 75: 62-70

Tubosun IA (1983) Géochronologie U-Pb du socle précambrien du Nigéria. Thèse 3e cycle, Univ Sci Tech Languedoc, Montpellier, Fr, 50 pp

Tubosun IA, Lancelot IR, Rahaman MA, Ocan O (1984) U-Pb Pan-African ages of two charnockite-granite associations from southwestern Nigeria. Contrib Miner Petrol 88: 188-195

Van Breemen O, Pidgeon RT, Bowden P (1977) Age and isotopic studies of some Pan-African granites from north central Nigeria. Precambrian Res 4: 307-319

Vedogbeton N (1988) Synthèse préliminaire sur le Précambrien du Bénin. Int Meet Proterozoic Geol Tectonics High Grade Terrains, Univ Ife, Nigeria, 31 pp

Weber F (1968) Une série précambrienne du Gabon: le Francevillien. Sédimentologie, Géochimie, Relations avec les gites mineraux associés. Mém Serv Carte Géol Alsace-Lorraine, 28: 328 pp

Weber F, Bonhomme M (1975) Données radiochronologiques nouvelles sur le Francevillien et son environnement, "le phénomène d'Oklo". Int Atomic Energy Agency, Vienne, Occas Pap, pp 17-35

Wright JB, Ajibade AC (1987) The Togo-Benin-Nigeria shield - a story of Pan-African aggregation? 14th Coll Afr Geol, CIFEG, Paris, Publ Occas 1987/12, (abstr), p 93

The Rokelide Orogen

S. J. Culver[1], H. R. Williams[2], and R. Venkatakrishnan[1]

1 Introduction

The term "Rokelide" (Allen et al. 1967) was proposed for the orogenic belt in Sierra Leone that was deformed during the Pan-African tectonothermal event, about 550 Ma (Kennedy 1964). Upper Archean basement gneisses and upper Proterozoic to lower Cambrian clastics were deformed and metamorphosed during this orogenic event.

The Pan-African Rokelide orogen generally parallels the coastline of West Africa south of the Paleozoic Bové Basin (Fig. 1). It extends some 600 km from western Guinea along the length of Sierra Leone and into Liberia. The geologic connection of the Rokelide and Mauritanide orogens is hidden beneath the younger sedimentary rocks of the Bové Basin, whereas the southern continuation of the orogen extends offshore in Liberia and possibly offshore of the Ivory Coast.

The low relief terrane characterizing the orogen results in generally poor exposure. In both Sierra Leone and Liberia, most exposures are along rivers or streams, although quarries in Sierra Leone and coastal situations in Liberia afford additional outcrops. A cover of dense equatorial forest in the southern half of the orogen further hampers field investigations.

Although Rokelide rocks have been mentioned in many publications, their orogenic context has been addressed in detail and directly only for Sierra Leone by Allen (1969) and Williams and Culver (1982). The tectonic setting of lateral correlatives in Guinea and Liberia has been discussed briefly by Ponsard et al. (1988) and Thorman (1976). The complete lack of a paleontologic framework along the whole orogen and the paucity of geophysical data for the Sierra Leonean section further result in our restricted understanding of the Rokelides.

[1] Department of Geological Sciences, Old Dominion University, Norfolk, Virginia 23529-0496, USA
[2] Ontario Geological Survey, 77 Grenville Street, Toronto, M7A 1W4, Ontario, Canada

2 Paleontology and Stratigraphy

"Apart from a medusoid of unknown affinity collected by Dr. P. C. Reid (pers. comm.) from the southern extremity of the Rokel River Group outcrop ..." (Macfarlane et al. 1981, p. 30) no fossils are known from the Rokelide orogen. Thus the paleontology of the Rokelides can be succinctly summarized.

Rocks involved in the orogen range from upper Archean gneisses to lower Cambrian clastics. For ease of discussion, these rocks will be discussed by country.

2.1 Sierra Leone

Two Pan-African orogenic phases have been recognized in the Mauritanides, Pan-African I (ca. 650 Ma) and Pan-African II (ca. 550 Ma; Dallmeyer and Villeneuve 1987). The Pan-African orogenic event in Sierra Leone is dated by K/Ar method on schists and late tectonic granitoids at about 550 Ma (Kennedy 1964; Hurley et al. 1971; Beckinsale et al. 1981). Dallmeyer (1989) has recently reported post-metamorphic cooling ages in Sierra Leone ranging from 546 to 505 Ma. All dates indicate a Pan-African II event in Sierra Leone. No Pan-African I dates have been recorded. Pan-African II affects are restricted to a 100 km wide coastal zone underlain by rocks of Archean, Proterozoic and probable Cambrian age (Figs. 2, 3).

Two distinct basement units are recognized in the Rokelide orogen (Williams and Culver 1988): (1) the low-grade Kenema Assemblage granite-greenstone terrane, part of the West African craton, and (2) a linear belt of high grade supracrustal rocks, the Kasila Group (Fig. 3). Both units are Archean in age and were reworked during the Pan-African orogeny (Hurley et al. 1971; Williams and Culver 1988). They are separated by a zone of intense ductile shear deformation (thrust symbol in Figs. 1-3; Williams 1988).

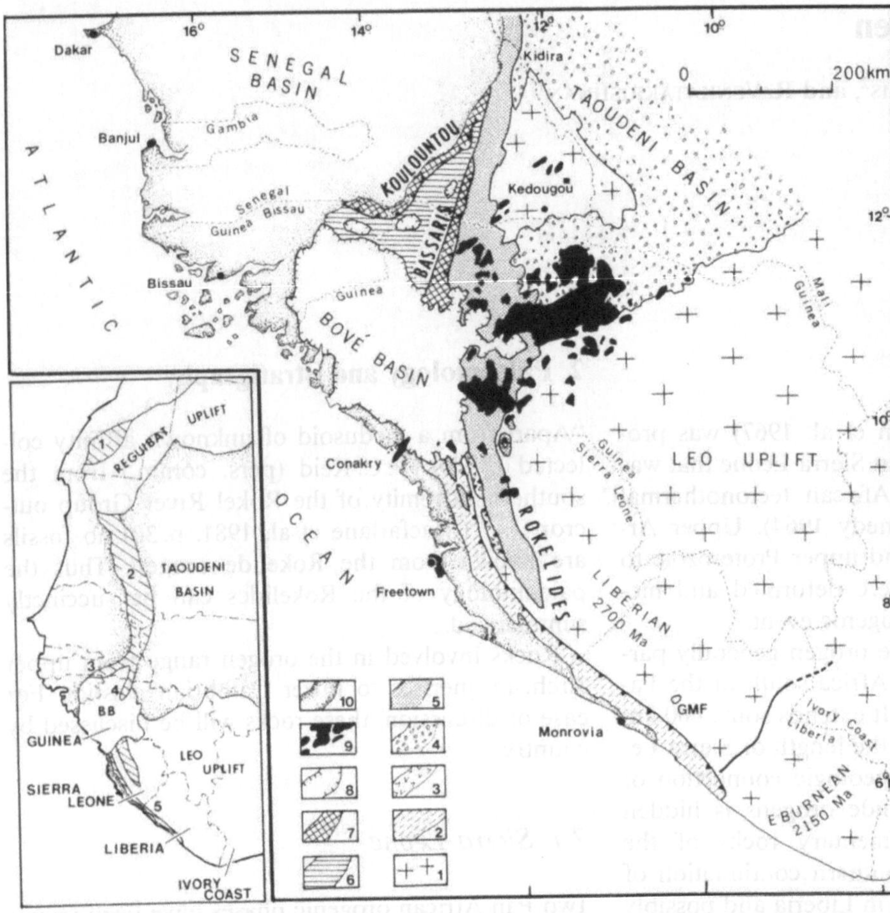

Fig. 1 Location of Rokelide orogen and general geology of West Africa. *1* Archean and lower Proterozoic basement of West African craton; *2* Archean basement reactivated in Pan-African; *3* Archean-lower Proterozoic klippen reactivated in Pan-African; *4* upper Proterozoic to mid-Ordovician undeformed sedimentary rocks of the cratonic Taoudeni Basin; *5* upper Proterozoic to mid-Cambrian sedimentary and volcanic rocks of the Rokelide and Mauritanide orogens; *6* mid-Cambrian to mid-Ordovician sedimentary rocks (molasse) of Mauritanide orogen; *7* Pan-African folded metamorphics of Mauritanide orogen with late Paleozoic Hercynian overprint along the Kouluntou; *8* upper Ordovician to Devonian mildly deformed sedimentary rocks; *9* Permian to Jurassic basic intrusives; *10* Mesozoic-Cenozoic coastal plain basins; *GMF* Gibi Mountain Formation (correlative to unit 5) overlain by klippen of unit 3. *BB* (inset map) Bové Basin. Compiled from Allen (1969), Williams (1978), Culver and Williams (1979a), Black (1980), Deynoux (1980), Tysdal and Thorman (1983) and Villeneuve (1984)

2.1.1 Kenema Assemblage

The heterogeneous granitic migmatites and gneisses of the Kenema Assemblage exhibit northwest-southeast trending structures in the area affected by the Pan-African orogeny (Figs. 2, 3). This is in sharp contrast to typical north-south to northeast-southwest trends of the granite-greenstone terrane in eastern Sierra Leone (Williams 1978, 1988). Rb/Sr whole rock radiometric ages of Kenema Assemblage rocks cluster around 2700 Ma (Fig. 2; Rollinson and Cliff 1982) and are referred to the Liberian tectonothermal event. Within the Rokelide orogen, reworked Kenema Assemblage gneisses give Pan-African, K/Ar cooling ages of around 550 Ma (Figs. 2, 3; see Beckinsale et al. 1981 and Cahen et al. 1984 for reviews).

2.1.2 Kasila Group

The Kasila Group (Fig. 3), exhibiting west-dipping planar fabrics parallel to the northwest-southeast outcrop, is composed of highly deformed mafic granulite, metaleucogabbroic intrusions, metasedimentary granulites and migmatites, and banded magnetite quartzite (Williams and Williams 1976; Williams 1988). The general stratigraphy of the Ka-

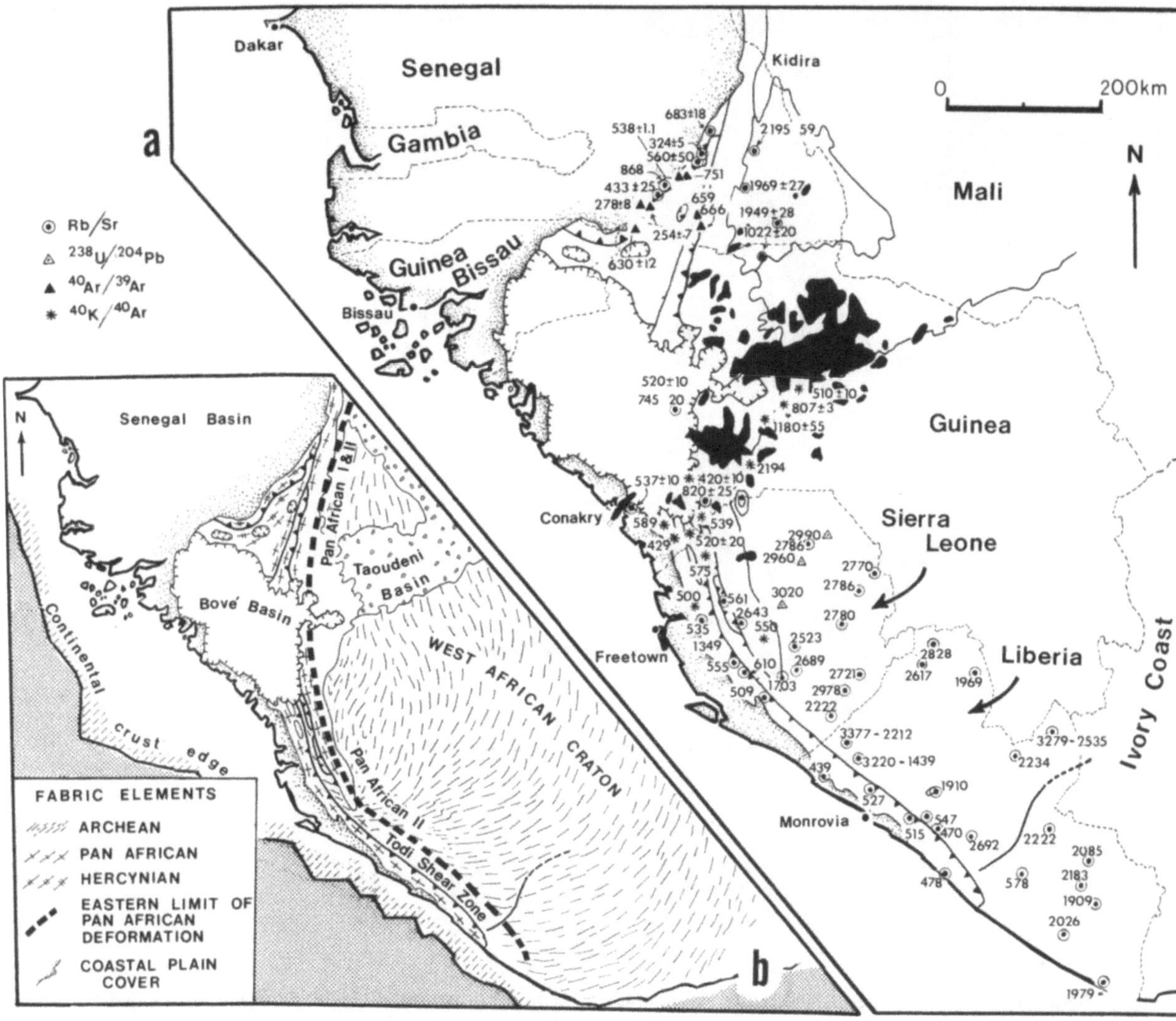

Fig. 2. Compilation of radiometric age determinations in the Koulountou and Bassaris branches of the Southern Mauritanides (in southern Senegal and eastern Guinea), and the Rokelide orogen in southern Guinea, Sierra Leone and Liberia. Basement ages in the Leo uplift are also given. A few age determinations have been omitted where the scale of map precluded their inclusion

sila Group is summarized in Fig. 4. On the basis of geological simplicity, the higher grade eastern mafic granulites, lying with a west-dipping thrust contact on Liberian granitic terrane, is probably stratigraphically lower than the ampholite facies western part that is dominated by metasediments but which contains minor mafic volcanics (Williams 1988). The lower, tectonic contact of the Kasila is one of intense ductile shear which displays overthrust sense of motion, and is coincident with highly deformed gabbroic and quartzitic rocks.

Allen et al. (1967) showed that Kasila Group gneisses are characterized by K/Ar cooling ages of around 550 Ma. Amphibole cooling ages were reported by Dallmeyer (1989) to be as young as 505 Ma. Unpublished U/Pb discordia ages on zircon reported in Williams and Williams (1976) suggested an Archean protolith for the Kasila that was deformed and metamorphosed under granulite facies conditions around 2700 Ma when Rb/Sr systems were reset (Hurley et al. 1971). Williams (1988) reported temperature and pressure conditions of metamorphism near the lower tectonic contact during ductile deformation to be 850 °C and ca. 9 kb.

The Kasila Group granulites are cut by zones of retrogressive metamorphism and greenschist facies chloritic shear zones that are signs of renewed, eastward directed thrusting. The occurrence in upper Proterozoic conglomerates (basal Rokel River

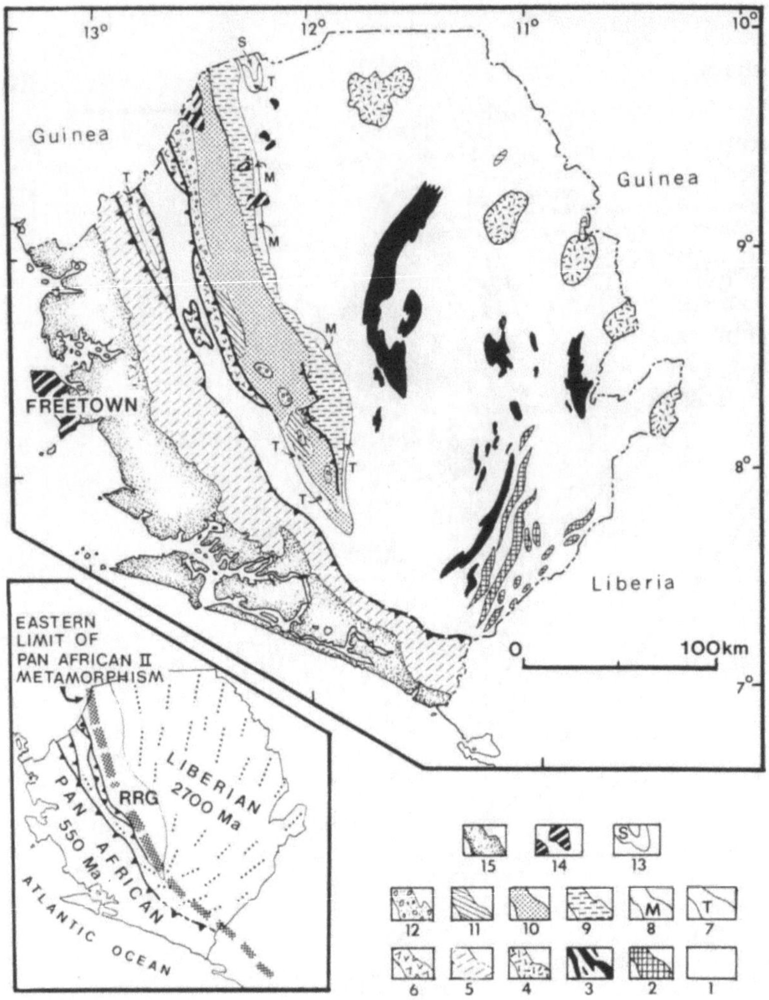

Fig. 3. Geology of Sierra Leone. *1–4* Kenema Assemblage; *1* Archean granitic basement; *2* granulite facies Archean supracrustals (Kambui Supergroup); *3* amphibolite facies Archean supracrustals (Kambui Supergroup); *4* Archean supracrustal granitoids; *5* Archean Kasila Group; *6* Archean (?lower Proterozoic) Marampa Group; *7–12* Rokel River Group; *7* Tabe Formation (T); *8* Makani Formation (M); *9* Teye Formation; *10* Mabole Formation; *11* Taia Formation; *12* Kasewe Hills Formation; *13* Saionia Scarp Group (S); *14* Mesozoic basic intrusives; *15* Bullom Group. *Inset* shows tectonic elements and age provinces; *RRG* Rokel River Group. Compiled from Williams (1978) and Culver and Williams (1979 a)

Group) of cobbles of both granulite and schist, identical to Kasila Group lithologies, argues for both high-grade deformation and metamorphism, and at least some of the retrogressive metamorphism, to have developed prior to the Pan-African II event. In addition, U/Pb and Rb/Sr geochronological data (Hurley et al. 1971; Hedge et al. 1975) suggest that protoliths for the Kasila Group are much older than Pan-African age, and are probably Archean. These data allow the implication that granulite and greenschist facies rocks were available in the immediate vicinity as source rocks for the Rokel River Group prior to the Pan-African II orogeny. As a result of eastward-directed thrusting, Pan-African planar fabrics developed parallel to and along existing Archean fabrics.

2.1.3 Marampa Group

To the east of the Kasila Group, low-grade supracrustal rocks of the recumbently folded Marampa Group tectonically overlie granitic terrane (Fig. 3). Although field relationships are unclear, Marampa Group rocks also probably are in fault contact with Rokel River Group strata. The Marampa Group contains ironstone, and mafic to felsic volcanic rocks and derived volcanogenic sediments that are similar to greenstone belt lithologies seen to the east

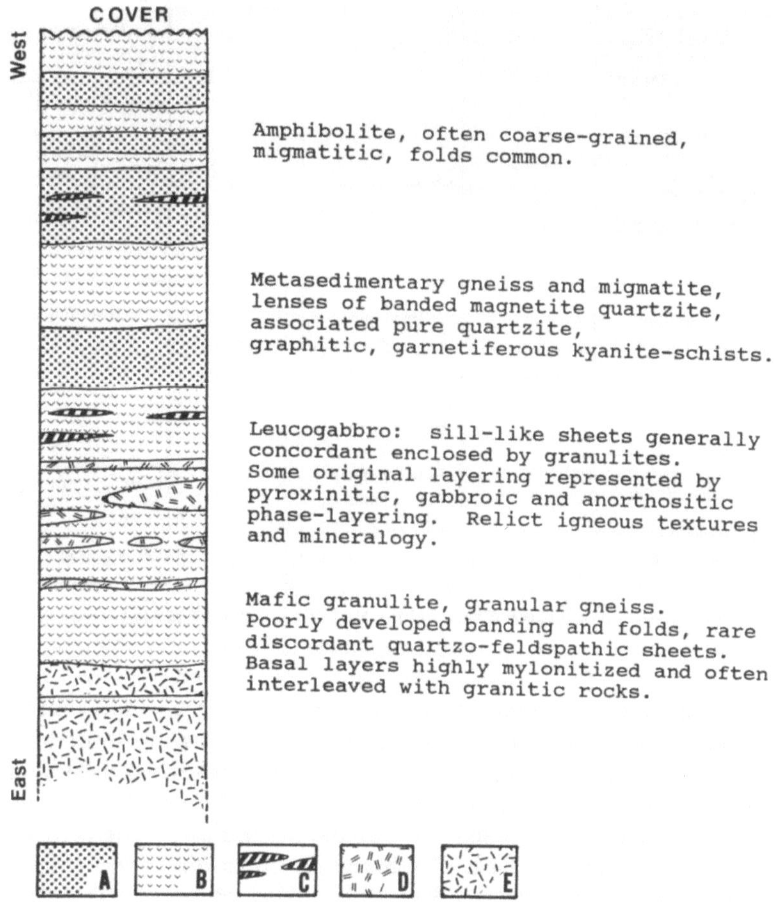

Fig. 4. Generalized stratigraphy of Kasila Group ignoring any tectonic duplication. The higher grade eastern part is probably stratigraphically lower than the lower grade western part. *A* metasedimentary gneisses/granulites; *B* basaltic amphibolites/granulites; *C* banded magnetite quartzite; *D* leucogabbro; *E* granitic terrane (Modified from Williams 1988)

within the Kenema Assemblage (Williams 1978). The Marampa Group is late Archean (Liberian) or early Proterozoic in age (Hurley et al. 1971).

The Marampa Group is divided into the basal metavolcanic Matoto Formation (Macfarlane et al. 1981), composed of serpentinite pods and mafic and intermediate, locally pillowed volcanics, and the overlying metasedimentary Rotokolon Formation (Macfarlane et al. 1981) composed of semipelites, psammites and banded ironstone. Williams (1988) emphasized the stratigraphic and lithologic similarities of the Marampa with supracrustal belts (Kambui Supergroup) in the adjacent Archean granite-greenstone terrane but pointed out corresponding sharp structural and metamorphic contrasts. Whereas typical Archean greenstone supracrustal successions of basaltic amphibolites and overlying greywackes, pelites and banded ironstone (Rollinson 1978; Williams 1978) occur in isolated elongate bodies engulfed by and infolded into the migmatitic granitoids (Williams 1988), the Marampa Group lies in tectonic contact on granitic terrane, is dominated by recumbent folds, and is characterized by greenschist facies metamorphism in the east, and amphibolite facies in the west (Macfarlane et al. 1981; Williams 1988).

Cahen et al. (1984) summarized the sparse geochronologic data from the Marampa Group. K:Ar determinations on muscovite from mica schists gave Pan-African ages of around 560 Ma (Fig. 2; Beckinsale et al. 1981). Whole-rock determinations on albite-mica schists gave model ages of 2075 and 2154 Ma (Hurley et al. 1971). Cahen et al. (1984) stated that these results were not conclusive but might suggest a long history before the Pan-African, a suggestion supported on lithologic grounds (Macfarlane et al. 1981; Williams 1988).

2.1.4 Rokel River Group

The Rokel River Group comprises the easternmost domain in the Rokelide orogenic belt in Sierra Leone. It occupies a belt some 30 km wide and 225 km long (Fig. 3). The Rokel River Series was defined and described by Dixey (1920, 1925) and

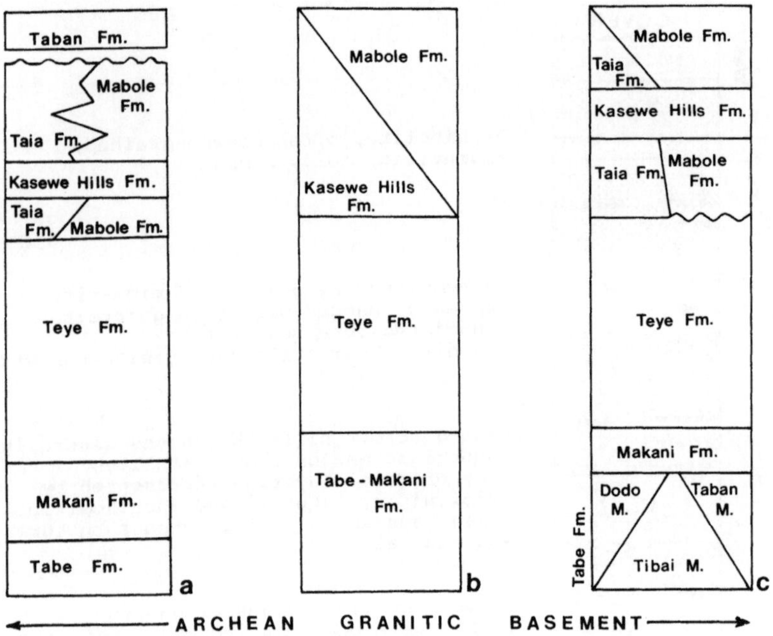

Fig. 5. Stratigraphy of the Rokel River Group. **a** after Allen (1969); **b** after Macfarlane et al. (1981); **c** after Culver et al. (1980)

subsequently was redefined by Allen (1968) as the Rokel River Group and was divided into six formations. Further redefinition of parts of the Rokel River Group were proposed by Culver et al. (1978, 1980), Culver and Williams (1979a) and Macfarlane et al. (1981). On the basis of lithologic correlations and limited radiometric data, the Rokel River Group is considered to be late Proterozoic to early Cambrian age (Culver and Williams 1979a).

Allen (1968) recognized a basal Tabe Formation, overlain by the Makani and Teye Formations, and succeeded by the laterally equivalent Taia, Mabole and Kasewe Hills Formations (Fig. 5). Allen (1968) also defined the Taban Formation which crops out to the west of the main Rokel River Group outcrop, which he considered to be younger than the Rokel River Group.

Culver et al. (1978) defined two members, the Tibai and Dodo, in the basal Tabe Formation and, subsequently, Culver and Williams (1979a, 1980) and Culver et al. (1980) assigned the Taban Formation to the Tabe Formation, renaming it the Taban Member (Fig. 5). Independently, Macfarlane et al. (1981) concluded that Taban strata should be considered as basal Rokel River Group and considered the Tabe and Makani formations to comprise a single lithostratigraphic unit. In this review, we follow the lithostratigraphic scheme of Culver et al. (1980) but discuss the points of view of Allen (1969, 1980) and Macfarlane et al. (1981).

The Tabe Formation rests unconformably on granitic basement and is exposed along the eastern and southwestern margins of the main Rokel River Group outcrop belt where it comprises the Tibai and Dodo members. A smaller elongate outcrop occurs 50 km to the west of the main belt in northern Sierra Leone (Fig. 3) and is represented by the Tibai and Taban members.

The Tibai Member (Fig. 5) consists of poorly sorted conglomerates originally interpreted as a marine transgressive deposit (Allen 1968), laminated siltstones and fine sandstones with isolated granitic clasts, and lenticular, graded feldspathic sandstones (Culver et al. 1980). The conglomerates have been interpreted as glacio-marine tillites (Culver et al. 1978; Culver and Magee 1987), the laminated siltstones as glacio-marine deposits containing ice-rafted dropstones (Vallance 1974; Culver et al. 1980), and the feldspathic sandstones as channel turbidites (Culver et al. 1980).

The trough cross-bedded feldspathic sandstones of the Taban Member have been interpreted as post-orogenic molasse (Allen 1968). However, Pollett (1951), Culver et al. (1980) and Macfarlane et al. (1981) noted Rokelide schistosity, folding and greenschist metamorphism within the Taban and suggested a basal position for the Taban within the Rokel River Group. The fluvial nature of these sandstones is agreed upon by all workers and Culver et al. (1980) further suggested a fluvio-glacial origin for these deposits.

Hence, the basal deposits of the Rokel River Group record glacial conditions over Sierra Leone in the late Proterozoic. These strata can be correlated with similar glacial deposits to the north in Senegal and Mauritania (Fig. 1; Deynoux 1980; Deynoux and Trompette 1981; Deynoux et al. 1985; Culver and Magee 1987) which are generally considered to record the worldwide late Proterozoic glaciation (Deynoux and Trompette 1981) and which are suggested to have been deposited within the radiometrically constrained interval of 630 to 595 Ma (Clauer and Deynoux 1987).

Overlying and laterally equivalent to the Tibai Member along the southeastern margin of the main Rokel River Group outcrop are the cross-bedded and ripple-marked sandstones of the Dodo Member (Fig. 5). Allen (1968), Vallance (1974) and Culver et al. (1980) concurred on the shallow marine, perhaps intertidal origin for these rocks. Macfarlane et al. (1981) grouped these deposits with the overlying Makani Formation.

Allen (1968) interpreted the interbedded silty claystones, orthoquartzites and subarkoses of the Makani Formation as probable neritic deposits. Macfarlane et al. (1981) agreed with this interpretation for strata that they placed in their Tabe-Makani Formation (Fig. 5), which includes all of the Makani Formation as used by Allen (1968) plus the underlying glacial strata.

The overlying Teye Formation is composed of a thick sequence of interbedded shales and sandstones interpreted by Allen (1968) as turbidites. The sandstones exhibit grading, sole structures, cross-lamination and mud clasts: some sequences consist entirely of siltstone and shale and may be distal turbidites (Culver and Williams 1979a).

The laterally equivalent Mabole and Taia formations overlie the Teye Formation (Fig. 5). Interbedded with them is the Kasewe Hills Formation (Allen 1968). The shales, siltstones, orthoquartzites and arkoses of the Mabole Formation were interpreted by Allen (1968) as shallow water deltaic deposits, whereas the mudstones and shales of the Taia Formation represent prodelta deposits.

The Kasewe Hills Formation is composed of tuffs and lavas of andesitic and dacitic composition (Allen 1968, 1969) that occur as lenses up to 2 km thick within the Mabole and Taia Formations. Allen (1968, 1969) considered these pillowed subaqueous lavas and air-borne pyroclastics to have been erupted from several volcanic centers within the Mabole and Taia Formation delta complex. Macfarlane et al. (1981) concurred with this interpretation on the basis of mapped isolated volcanic centers.

Although there are no direct geochronological data for the Rokel River Group, the basal glacigenic deposits can be correlated with similar strata in Senegal and Mauritania which are generally accepted to be late Proterozoic in age (Deynoux et al. 1985) and which are suggested to have been deposited within the radiometrically constrained interval of 630 to 595 Ma (Rb/Sr determinations on clays; Clauer and Deynoux 1987). Further, as pointed out by Allen (1969) and reiterated by Cahen et al. (1984), there is a parallelism of west-dipping planar fabrics between the Rokel River, Marampa and Kasila groups resulting from metamorphism and deformation during the Pan-African II Rokelide orogeny about 500–550 Ma (Allen et al. 1967; Allen 1969; Dallmeyer 1989).

Sedimentary rocks exposed in the Saionia Scarp in northern Sierra Leone (Fig. 3) are mentioned here although Culver and Williams (1979a) do not consider them to be part of the Rokelide orogen. These strata, described by Dixey (1925) as unconformably overlying Rokel River Group strata, have been considered as lateral equivalents of the late Ordovician basal "Supergroup 3" strata of the Taoudeni Basin to the north (Trompette 1973; Deynoux et al. 1985). Indeed, the laminated mudstones with isolated clasts within the Saionia Scarp Group have been suggested as indicating glacio-marine conditions of the late Ordovician glaciation that characterized much of West Africa (Reid and Tucker 1972; Tucker and Reid 1973; Bull and Culver 1979). Lithologic similarities of these strata with basal Rokel River Group rocks simply reflect the repetition of similar depositional environments.

This interpretation is in strong contrast with that of Macfarlane et al. (1981) who consider the entire sedimentary sequence at Saionia Scarp to belong to the Rokel River Group. They do not dispute the glacio-marine nature of the laminated mudstones but place these deposits within the Teye Formation, which does not, however, exhibit glacial characteristics in the main Rokel River Group outcrop belt. Although the field data of Macfarlane et al. (1974, 1981) suggest an angular unconformity between basal conglomerates (presumably tillites) of the Rokel River Group (10° dip to SW) and the overlying mudstones with isolated clasts (8 to 12° dip to E), they stated (Macfarlane et al. 1974, p. 42) that they "... have been unable to substantiate Dixey's (1925) unconformity between ... [the Saionia Scarp Group] ... and his Rokel River Series conglomerates ..."

We follow here the work of Dixey (1925) and Reid and Tucker (1972) in considering the Saionia Scarp Group as late Ordovician, as unconformably overlying the basal tillite of the Rokel River Group, and, therefore, as a sequence of strata deposited some 100 Ma after the deformation of the Pan-African Rokelide orogen. However, Villeneuve (1981, 1984) independently arrived at the same conclusions as Macfarlane et al. (1974, 1981) for the Saionia Scarp sequence exposed in Guinea. Clearly, more work is necessary to clarify this confusing stratigraphic situation.

2.2 *Guinea*

The geology of Guinea (Fig. 6) has been summarized by Villeneuve (1981, 1984). The northernmost part of the Rokelide orogen extends into Guinea and is overlain to the north by Paleozoic sedimentary rocks of the Bové Basin. Although map units (informally defined in Guinea) are broadly similar in the Sierra Leonean and Guinean Rokelides (Culver 1984), the geologic context of these units has been interpreted in very different ways.

2.2.1 Forécariah Group and Ouankifondi Group

Crystalline basement rocks in the western part of the Rokelide orogen are placed in the Forécariah group (Boufeev 1968; Torchine 1969; Villeneuve 1984) which is considered to be late Archean (2700 Ma) or early Proterozoic (Birrimian, 2000 Ma) in age (Villeneuve 1984). The Forécariah group is composed of a variety of gneisses, schists, migmatites and mylonites metamorphosed to amphibolite and granulite facies (Fig. 6). It is subdivided into the Kissi Kissi formation, the Forécariah formation, the Mahera formation, the Kounsouta formation and the "Ouankifondi complex" (Boufeev 1968; Torchine 1969). Based on structural contrasts, Villeneuve (1984) considered the mica-schists of the "Ouankifondi complex" to be distinct from the Forécariah group; the former was termed the Ouankifondi group in Villeneuve (1981) and Dallmeyer et al. (1987).

2.2.2 Bania Group and Kolenté Group

To the east of the Ouankifondi group is a sequence of volcanic-sedimentary rocks termed the Bania group (Fig. 6). It is composed of andesite, diabase, spilite and diorite (Villeneuve 1981) and exhibits pillow lavas at Mount Binia (Torchine 1969). Dallmeyer et al. (1987) inferred that the Bania group is laterally equivalent to the Ouankifondi group when

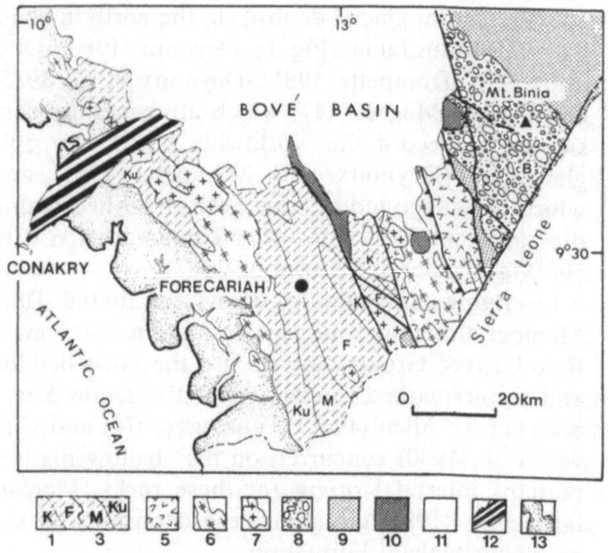

Fig. 6. Geology of southern Guinea. *1–4* Forécariah group; *K* Kissi Kissi formation; *F* Forécariah formation; *M* Mahera formation; *Ku* Kounsouta formation; *5* Ouankifondi group; *6* deformed granites; *7* undeformed Pan African granites; *8* Bania group; *9* Kolenté group; *10* Taban group; *11* Paleozoic Bové Basin sedimentary rocks (including Pita group); *12* Mesozoic basic intrusives; *13* Mesozoic-Cenozoic coastal plain cover. See text for discussion of age relations (Modified from Villeneuve 1984)

they stated (1987, p. 88) that the Bania group "... is progressively deformed and metamorphosed [to the west] where it is termed the Ouankifondi Group". This interpretation contrasts with data presented in Villeneuve (1984) which suggests a closer geologic relationship between the Bania group and the Kolenté group.

The Kolenté group is the equivalent of the upper Proterozoic to lower early Paleozoic Walidiala group and Mali group of northern Guinea and Senegal. The Walidiala group (Deynoux et al. 1985) commences with a basal tillite of late Proterozoic age (Bassot 1966; Deynoux et al. 1985; Culver and Magee 1987). Microfossils recovered from dolomite immediately overlying the tillite in northern Guinea place the upper part of the Walidiala group and the Mali group in the Cambrian (Culver et al. 1988 a, b). Hence, the Kolenté group, exposed in southern Guinea and characterized by a basal tillite, probably straddles the Precambrian-Cambrian boundary as does the Rokel River Group to the south in Sierra Leone.

The Kolenté group is composed of greenish clastics, sandy at the base of the sequence becoming finer grained towards the top (Bufeyev 1972; Villeneuve 1981, 1984). Dallmeyer et al. (1987) stated that the Kolenté unconformably overlies the me-

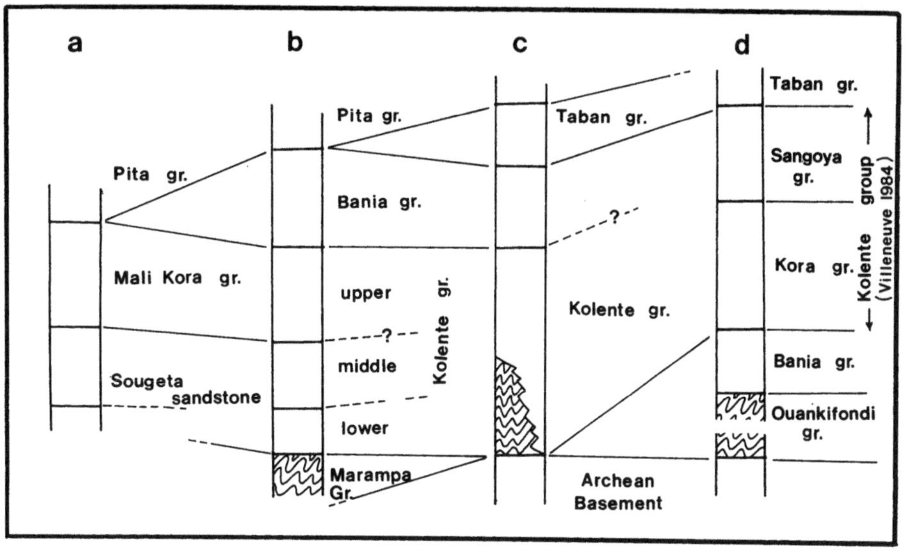

Fig. 7. Sections illustrating various authors' opinions on the stratigraphic position of the Bania group and Kolenté group. **a** Aymé et al. (1958); **b** Petrovsky (1973); **c** Torchine (1969); **d** Villeneuve (1981) (Modified from Villeneuve 1984)

tamorphic complex of the Bania and Ouankifondi groups. This interpretation contrasts with those of earlier workers (Fig. 7). Torchine (1969) and Petrovsky (1973) considered the Bania group to overlie the Kolenté group: a comparable interpretation applies in Sierra Leone where the volcanics of the Kasewe Hills Formation overlie most of the clastics of the Rokel River Group (Allen 1968; Culver and Williams 1979a). Villeneuve (1981, 1984), however, placed the Bania group stratigraphically beneath the Kolenté group even though he stated (p. 156) that "... the absence of observations of a contact between the two does not permit a choice between the two hypotheses".

To complicate matters further, Villeneuve (1984) correlated the glacigenic strata of the Kolenté group with those of the Saionia Scarp Group (Tucker and Reid 1973) in Sierra Leone. Tucker and Reid (1973) and Culver and Williams (1979a) consider the Saionia Scarp Group to be late Ordovician in age. Hence, a correlation of the Kolenté group with the Rokel River Group of Sierra Leone appears more likely.

2.2.3 Taban Group

Red sandstone and conglomerates of the Taban group outcrop in several small basins in Guinea immediately south of the Bové Basin (Fig. 6). These strata lie unconformably on metamorphic rocks of the Forécariah and Ouankifondi groups (Villeneuve 1984).

The Taban group is considered to be of fluviatile origin and is interpreted to be post-orogenic molasse (Villeneuve 1984). This interpretation agrees with Allen's (1968) for his post-Rokelide Taban of northern Sierra Leone but contrasts with the glacigenic, basal Rokel River Group interpretation of the Taban suggested by Culver and Williams (1979a) and Macfarlane et al. (1981). Villeneuve (1984) considered the Taban group to be the southernmost extent of the post-orogenic molasse of the Youkounkoun group of Guinea north of the Bové Basin.

2.3 Liberia

The outcrop of the Rokelide orogen terminates in south-central Liberia where it swings to the south onto the continental shelf (Figs. 1, 8). With the exception of one small outcrop, at Gibi Mountain (Fig. 8), the Pan-African orogen is represented by remobilized Archean basement rocks. This probably reflects the deeper erosional level of the Rokelide orogen southward, which is supported by resetting of all K/Ar ages.

2.3.1 Basement Affected by the Rokelide Orogeny

Approximately 90% of Liberia is underlain by Archean and lower Proterozoic granitic rocks (Liberian Geological Survey 1983). They form part of the Leo or Man Shield of the West African craton. The basement rocks can be divided into three major units on the basis of their radiometric age (Figs. 2, 8; Hurley et al. 1970, 1971; Hedge et al.

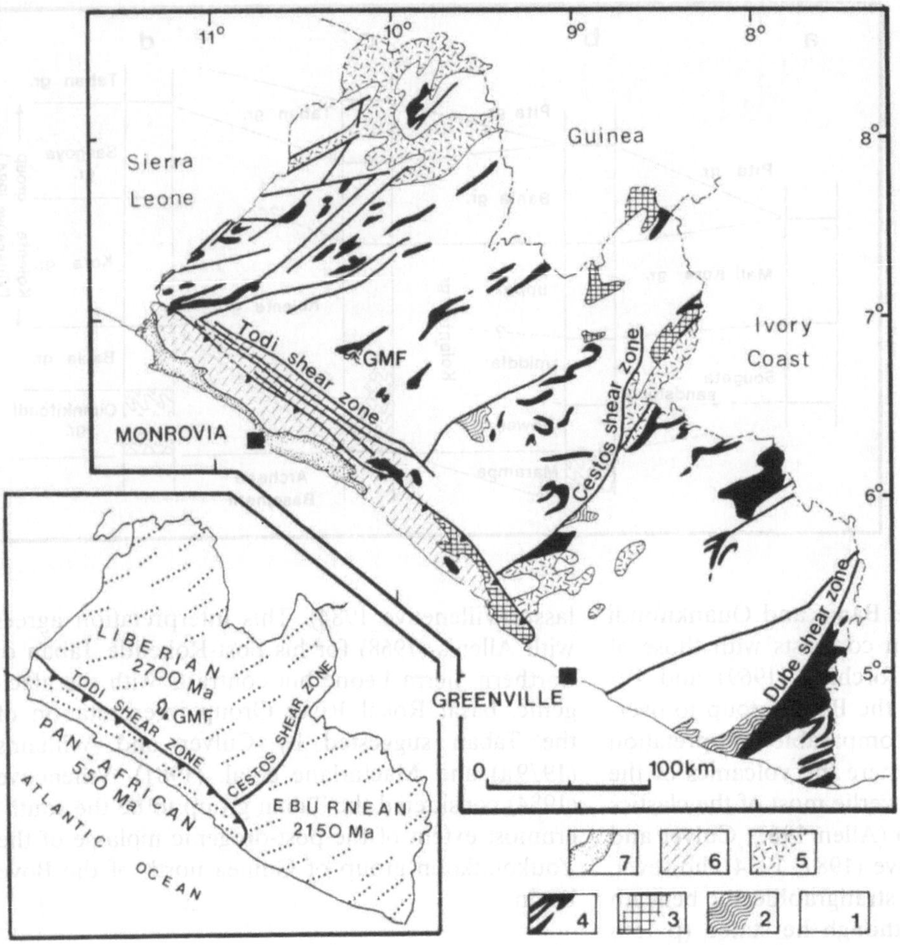

Fig. 8. Geology of Liberia. *1* Archean and Lower Proterozoic granitic basement; *2* Archean migmatites; *3* Archean granulites; *4* Archean amphibolitic supracrustals; *5* Archean granitoids and basic intrusives; *6* Archean gneisses reactivated in Pan-African; *7* Cenozoic coastal plain cover with small inliers of Devonian Paynesville Sandstone and Cretaceous Farmington River Formation east of Monrovia. *Inset map* shows tectonic elements and age provinces; *GMF* Gibi Mountain Formation. Modified from Tysdal and Thorman (1983) and Caen-Vachette (1988)

1975). The Liberian age province, composed of granite-greenstone terrane (Rollinson 1978), developed about 2700 Ma: the Eburnean age province was metamorphosed about 2150 Ma: the Pan-African age province was metamorphosed and intruded about 550 Ma. The boundary between the Liberian and Eburnean provinces is not well defined but the Liberian and Pan-African provinces are separated by a series of faults comprising the Todi Shear Zone (Tysdal 1977, 1978; Thorman 1976).

Gneisses of the Liberian and part of the Pan-African age provinces are metamorphosed to amphibolite grade. Granulite facies rock, however, are restricted to the Pan-African age province (Thorman 1974) and are considered by Tysdal and Thorman (1983) to have been originally part of the Liberian province.

2.3.2 Gibi Mountain Formation

Two small outliers (Fig. 9) of clastic sedimentary rocks, the Gibi Mountain Formation (Thorman 1977), form heavily forested hills 32 km northeast of the Todi Shear Zone (Fig. 8). They lie disconformably on Archean gneisses and are overlain by klippen of Precambrian itabirite-bearing quartzite (Thorman 1976).

The age of the Gibi Mountain Formation was suggested as late Precambrian to early Cambrian (Thorman 1976), based on lithologic similarities to basal Rokel River Group units in Sierra Leone. Culver et al. (1978, 1980) concurred with this suggestion which was further supported by later interpretations of depositional environments for the Gibi Mountain Formation (Magee and Culver 1986; Culver and Magee 1987).

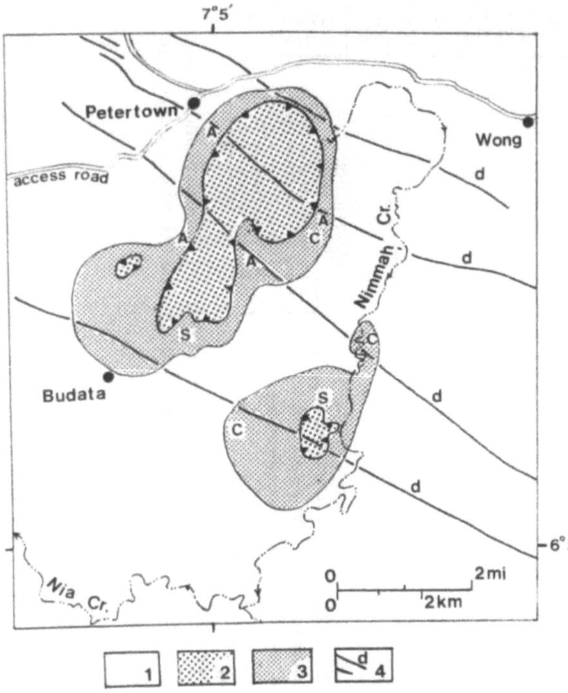

Fig. 9. Geology of Gibi Mountain area. *1* Archean gneiss; *2* Precambrian itabirite-bearing quartzite; *3* upper Proterozoic Gibi Mountain Formation; *C* conglomerate; *A* arkosic siltstones and sandstones; *S* shale; *4* diabase dikes. (Modified from Thorman 1976) Gibi Mountain Formation lithologic information form Thorman (unpubl.) and Magee and Culver (unpubl.)

The Gibi Mountain Formation consists of a basal conglomerate overlain by arkosic siltstones and sandstones which, in turn, are overlain by a shale (Fig. 9). The conglomerate is massive, matrix-supported, and contains angular to subrounded clasts of Archean country rock. The siltstones are irregularly laminated and contain isolated granitic clasts and arkosic clots. In thin section, isolated granitic fragments and large quartz grains depress underlying laminations.

These characteristics, plus the lithologic and stratigraphic similarities to the Tabe Formation of the Rokel River Group in Sierra Leone, led Magee and Culver (1986) to suggest a glacial origin for the conglomerates and siltstones of the Gibi Mountain Formation. A late Precambrian (ca. 620 Ma) age for these strata, corresponding with late Precambrian glacigenic deposits elsewhere in West Africa (Deynoux et al. 1985), thus seems likely. If so, the overlying shale, which is poorly exposed, may be laterally equivalent to the Makani or Teye formations of Sierra Leone on lithostratigraphic grounds.

2.4 Correlations Between Guinea, Sierra Leone and Liberia

The stratigraphic position of various units involved in the Rokelide orogen is the subject of much disagreement. Some of these problems have been briefly discussed above. We present here a correlation chart which expresses the latest opinions of Culver and Williams. The contrasting views of Villeneuve (1984) for Guinea north and south of the Bové Basin are also indicated (Fig. 10).

In northern Guinea and southeastern Senegal (Fig. 10, column A), two episodes of Pan-African orogenesis are recognized in the southern Mauritanide orogen (Villeneuve 1984), Pan-African I at 655-665 Ma and Pan-African II at 550-560 Ma (Dallmeyer and Villeneuve 1987). Only Pan-African II dates are recorded in the Rokelides. If the Pan-African I event did occur in western Sierra Leone, only the retrogressive effects at the eastern tectonic margin of the Kasila Group can be ascribed to it. No radiometric cooling ages or resetting can be directly attributed to the Pan-African I event, but the possibility that it was completely reset by the subsequent Pan-African II event cannot be discounted on structural and lithostratigraphic criteria. Based on lithostratigraphic correlations with the southern Mauritanides (Bassarides) Villeneuve (1984) and Dallmeyer et al. (1987) inferred the effects of Pan-African I orogenesis (Dallmeyer and Villeneuve 1987) in the Rokelides. The result of this extrapolation is shown in column B of Fig. 10.

If one accepts at face value the evidence to date that Pan-African II effects only are recorded in the Rokelides, then, based mainly upon fieldwork by Allen, Williams and Culver in Sierra Leone, a significantly different stratigraphic interpretation results (Fig. 10, columns C, D). As we shall see, based on the differing stratigraphies, the geodynamic evolution of the Rokelides is also subject to very different interpretations.

3 Structure

Rokelide tectonics in Guinea, Sierra Leone and Liberia resulted in the deformation of upper Proterozoic to Cambrian supracrustal rocks (the Rokel River Group and its lateral correlatives) and retrogression to amphibolite facies of the Archean infrastructure (Kenema Assemblage and Kasila Group and their lateral correlatives) to the west

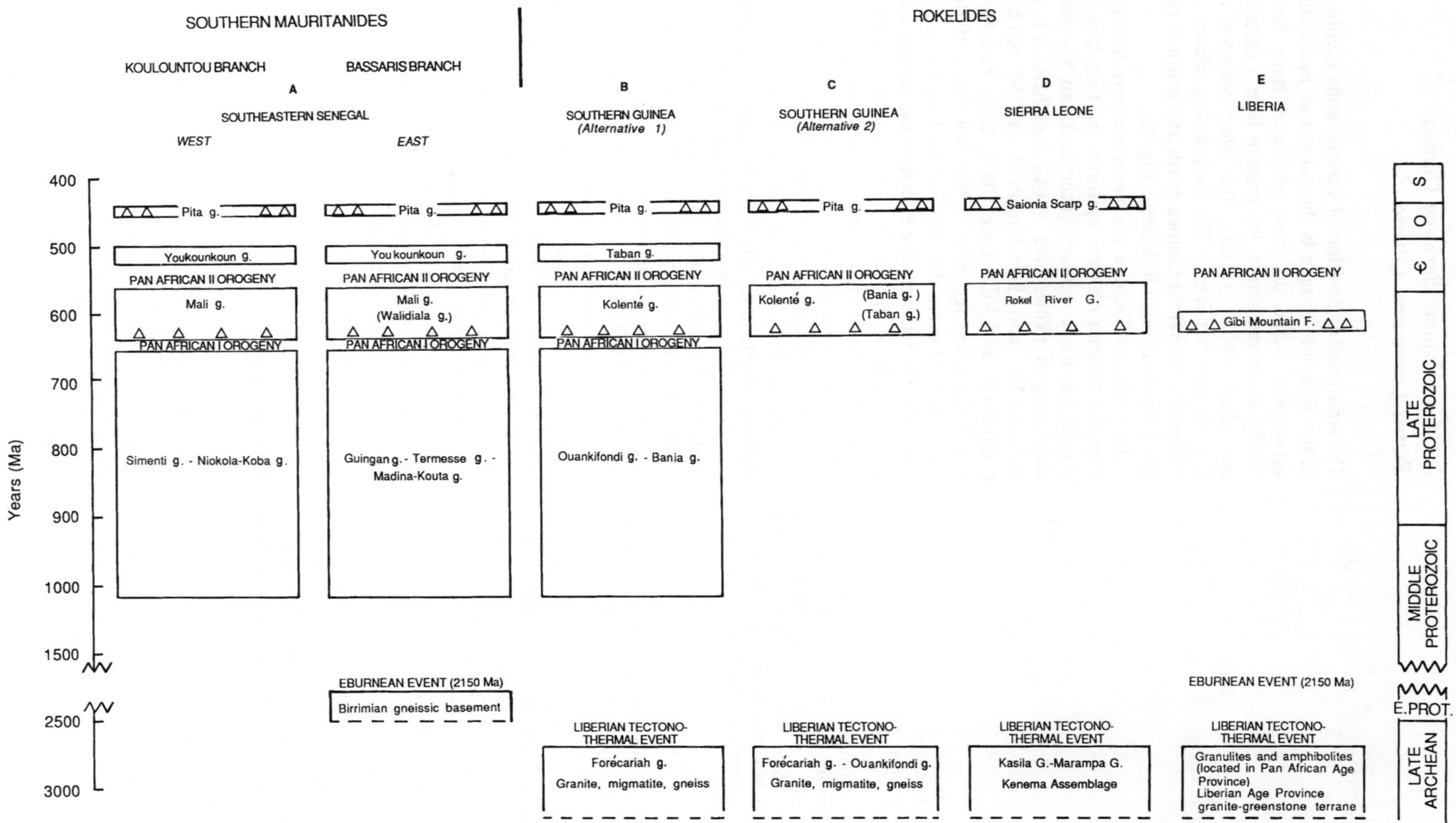

Fig. 10. Correlation chart for Liberia, Guinea, Sierra Leone and southeast Senegal. Note scale change at 1000 Ma. *A* and *B* after Villeneuve (1984) and Dallmeyer and Villeneuve (1987); *C* this report; *D* after Culver and Williams (1979a); *E* after Tysdal and Thorman (1983) and Magee and Culver (1986). *Triangles* indicate glacial deposits. See Figs. 2 and 4 for detail of Rokel River Group, and text for full discussion

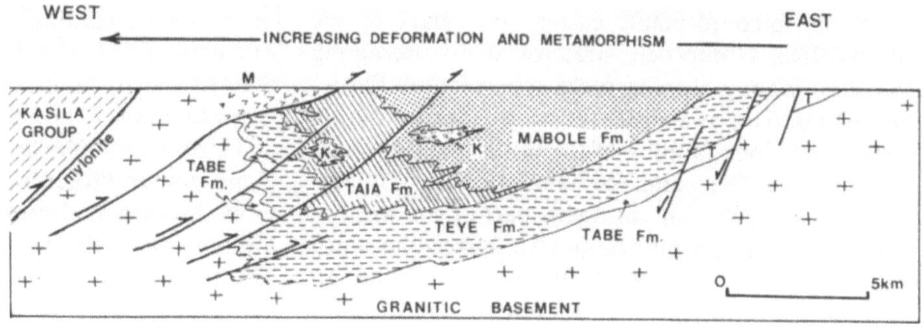

Fig. 11. Generalized cross-section of Rokel River Group synclinorium in northern Sierra Leone. *M* Marampa Group in tectonic contact on Rokel River Group; *T* Tabe Formation; *K* Kasewe Hills Formation

(Williams and Culver 1982). Archean structural trends thus have a controlling influence on Pan-African tectonics.

3.1 Sierra Leone

Foliation in the Liberian granite-greenstone terrane swings from a NE to NW trend in the vicinity of the boundary with the Kasila Group (Figs. 2, 3). Mylonite zones characterize the granitic terrane in the region between outcrops of the Rokel River Group and Kasila Group (Fig. 11). The boundary with the Kasila Group is a zone of grain-size reduction under granulite facies conditions (Williams 1988). Fault kinematics along shallow westward dipping planes were essentially ductile simple shear accompanied by dominant overthrust motion with minor dextral components (Williams and Culver 1988). This mylonitic belt, approximately 5 km in width (Williams 1988), is composed of very finely striped quartzo-feldspathic gneiss (Macfarlane et al. 1981). It is well developed in northern Sierra Leone but becomes much less distinct to the south.

The Kasila Group (Figs. 3, 4) is composed of penetratively foliated mafic granulites and amphibolites, metaleucogabbros, and metasedimentary migmatites and gneisses. Foliation trends are N-NW with westerly dips averaging 60°. Folds are extremely rare and isoclinal and of metre-scale. They are coplanar with the foliation and their axes plunge parallel with a mineral extension lineation on foliation surfaces (Williams 1978, 1988). The eastern tectonic contact with the Kenema Assemblage rocks consists of quartz-rich mylonites that have not been annealed by the Pan-African thermal event, and so are presumably coincident with, or later than it. However, the fabrics may only be recording the last motions along these highly strained rocks. Earlier movement history may be hidden by later events.

The Marampa Group (Fig. 3) is exposed as structural units with tectonic contacts, east of the Kasila Group. Stratigraphically and lithologically, Marampa Group rocks are similar to those of greenstone belts in the adjacent Kenema Assemblage granite-greenstone terrane (Williams 1978; Macfarlane et al. 1981), but their structure and metamorphism are in sharp contrast. Marampa Group rocks are deformed into east-facing recumbent folds, refolded by north-south trending upright structures (Macfarlane et al. 1981). The isolated outcrops of Marampa Group strata have been interpreted as klippen derived from the Kasila Group to the west (Williams 1978, 1979, 1988).

The overall structure of the Rokel River Group belt is an inclined synclinorium plunging to the north (Figs. 3, 11). Allen (1969) described the deformation of the Rokel River Group (Fig. 3) and recognized three phases of folding and a period of thrusting. The first phase (F_1) produced stratal extension (broken bed deformation) along bedding-parallel thrust faults. Asymmetric east-verging chevron folds with strong axial planar cleavage formed in the strongly deformed western part of the outcrop belt. F_1 folding appears to have been localized into distinct domains surrounded by relatively undeformed rock. F_1 axial traces trend NNW parallel with the Rokel River Group outcrop trend. We suggest that low-angle westerly dipping thrust faults parallel to these axial surfaces may be interpreted as late stage structures that preceded the "locking-up" of chevron fold geometries (e. g. Ramsay 1974).

Two subsequent, minor fold phases recognized by Allen (1969) are developed locally. They are characterized by vertical axial planes (F_2, north-south trend; F_3, east-west trend) and horizontal axes and did not alter the structural pattern of the first deformation phase.

At the unconformable eastern boundary of the Rokel River Group belt, strata occur in alternating zones of steeply dipping rocks separated by flat-lying strata. This is considered to be drape folding caused by block-faulting of the underlying Archean basement (Culver and Williams 1979a; Williams and Culver 1982). The southern part of the western boundary is an unconformity also, but to the north the western margin is an F_1 thrust fault. Although the contact is not exposed, Marampa Group rocks probably lie with tectonic contact on Rokel River Group strata (Fig. 11; Allen 1968, 1969).

3.2 Guinea

Archean (Liberian) and Pan-African structures are clearly recognizable in Guinea in rocks affected by the Rokelide orogeny and are similar to those in Sierra Leone (Figs. 2, 6). The mylonite belt and zone of cataclastic deformation continue to the north (e. g. Taban-Kafou fault) where they disappear under the Paleozoic strata of the Bové Basin (Villeneuve 1984). Deformation styles (Villeneuve 1981, 1984) of the Forécariah, Ouankifondi and Kolenté groups (equivalent to the Kasila, Marampa and Rokel River groups respectively) are the same as those recorded in Sierra Leone. The western boundary of the Kolenté group is mapped as a thrust whereas the eastern margin is an unconformity (Villeneuve 1981, 1984). Thus, different interpretations of the tectonic development of the Rokelides by workers in Guinea and Sierra Leone result, not from contrasting structural characteristics, but from different interpretations of stratigraphic position of the various units and their timing relative to the two Pan-African tectonic events.

3.3 Liberia

The 400 km-long shear zone separating the Kasila Group from Kenema Assemblage rocks in Sierra Leone continues into Liberia as the Todi Shear Zone (Thorman 1977; Tysdal 1977, 1978) (Figs. 2, 8) and extends an additional 300 km to the southeast until it makes a sharp bend to the south continuing offshore (Tysdal and Thorman 1983). It separates amphibolites and granulites of the Pan-African age province (equivalent to the Kasila Group) from the Liberian age province granite-greenstone terrane. The indistinct nature of this boundary in southern Sierra Leone continues into Liberia but parallelism of trends on either side of the shear zone soon give way to sharp discordance of structural and stratigraphic trends (Thorman 1974).

Thirty-two kilometers northeast of the Todi Shear Zone, upper Proterozoic sedimentary rocks of the Gibi Mountain Formation are preserved under klippen of itabirite-bearing quartzite (Figs. 8, 9; Thorman 1976). Two fold sets are found within these rocks; a first phase occurs throughout the allochthonous quartzites but only in the uppermost Gibi Mountain Formation close to the overlying thrust. A range of folding styles from shear folds to flexural and passive folds have been refolded about the axes of the second system which consists of large scale asymmetric to overturned folds whose axes trend NNE (Thorman 1976).

The allochthonous quartzites came from the southwest and most resemble rocks in the Goe-Fantro Range of the Pan-African age province (Fig. 8; Thorman 1976). Williams and Culver (1988) noted their likely correlation with quartzites in the Marampa Group in Sierra Leone, which also lies with probable tectonic contact on upper Proterozoic to Cambrian sedimentary rocks, the Rokel River Group. Thorman (1976) noted that thrust emplacement of the Gibi Mountain quartzites must have occurred during the Pan-African thermo-tectonic event about 550 Ma (equivalent to Pan-African II of the southern Mauritanides).

4 Metamorphism

Pan-African II age metamorphism of the Rokelide orogeny overprints Liberian age metamorphism along the western margins of the West African craton in Liberia, Sierra Leone and Guinea.

4.1 Sierra Leone

The bulk of the granite-greenstone terrane in Sierra Leone exhibits amphibolite facies conditions (Williams 1988). An increase in metamorphic grade occurs close to the Kasila Group both in the north (Williams 1978) and south (Rollinson 1978) of Sierra Leone.

The Kasila Group has been divided into a central zone of granulite facies rocks, flanked to the west with a zone of amphibolite facies, largely obscured by Cenozoic coastal deposits of the Bullom Group.

To the east, a zone of retrogression to amphibolite and greenschist facies is associated with eastwards directed overthrusting of the Kasila Group (see Fig. 4; Williams 1988).

Marampa Group klippen, composed of volcanics, pelites, psammites and hematite quartzite, are metamorphosed to the greenschist facies (Macfarlane et al. 1981). Metamorphic grade increases to the west towards the probable source of these nappe-like structures, the Kasila Group (Williams 1979, 1988).

Metamorphism in the Rokel River Group is confined to the small western outlier and western part of the main outcrop belt (Fig. 11). Greenschist metamorphism is indicated by the growth of sericite, actinolite, muscovite and biotite in pelites, sandstones and volcanics in the west (Allen 1969; Culver and Williams 1979a).

4.2 Guinea

Patterns of metamorphism in the Rokelides of Guinea appear similar to those recognized in Sierra Leone (Villeneuve 1984). The granite-greenstone terrane (Série de Dabola) east of the Kolenté group exhibits amphibolite facies conditions, the Forécariah group varies from amphibolite to granulite facies conditions, and the Ouankifondi group exhibits greenschist to amphibolite facies conditions (Torchine 1969; Petrovsky 1973; Villeneuve 1984).

The Bania group and the western part of the Kolenté group exhibit low grades of metamorphism whereas the eastern part of the Kolenté group in unmetamorphosed (Villeneuve 1981, 1984). An identical situation has been noted for the Rokel River Group in Sierra Leone (Allen 1969).

4.3 Liberia

Metamorphic trends in Sierra Leone continue into Liberia. The granite-greenstone terrane of the Liberian age province is metamorphosed to amphibolite facies, whereas Pan-African age province rocks (equivalent to the Kasila Group of Sierra Leone) exhibit both amphibolite and granulite facies conditions (Fig. 8).

The quartzites of the Gibi Mountain klippen rest on the Gibi Mountain Formation and exhibit dynamic metamorphism (Thorman 1976). Most rocks are protomylonite and individual quartz grains are elongated as much as 7:1 within the planar fabric (Thorman 1974). The underlying Gibi Mountain Formation is unmetamorphosed although parting planes of shales immediately underlying the thrust plane have a distinct sheen (Thorman 1976).

5 Igneous Activity

The story of Pan-African igneous activity in the Rokelide orogen is confusing given the various stratigraphic interpretations. The volcanics of the Marampa Group, which Williams and Culver (1988) consider to be Archean in age are described here because other workers suggest they are Pan-African in origin.

5.1 Volcanics of the Marampa Group

The Marampa Group of Sierra Leone (Fig. 3) has been divided into two formations, the metavolcanic Matoto Formation and the overlying metasedimentary Rotokolon Formation (Macfarlane et al. 1974, 1981). The Matoto Formation is composed of basaltic lavas with pillowed zones up to 25 m thick, interbedded serpentinites and flow-banded andesites. Thickness estimates of the Matoto Formation vary from 200 m (Macfarlane et al. 1981) to 750 m (Williams 1978). The overlying Rotokolon Formation is composed of semipelites and psammites with subordinate tuffs and andesites (Williams 1978; Macfarlane et al. 1981).

Allen (1968, 1969) suggested that the Marampa Group was the metamorphosed equivalent of the Rokel River Group and hence late Precambrian in age. However, Hawkes (1972), Williams and Williams (1976), Williams (1978, 1988) and Macfarlane et al. (1981) all consider that on the basis of radiometric ages, lithology and structural grounds, the Marampa is much older and probably equivalent in age to the upper Archean Kasila Group. Cahen et al. (1984) concurred with this conclusion.

A Pan-African age for the Marampa Group volcanics has been inferred by several workers since Allen's original suggestion (1968, 1969). Gass (1977) illustrated but did not discuss a large Pan-African mafic-ultramafic mass in Sierra Leone which probably represented the Marampa Group volcanics: Culver and Williams (1979b) disputed this interpretation. More recently, Dallmeyer et al. (1987) have inferred that the Ouankifondi group of the Rokelides of Guinea, which is the mapped continuation of the Marampa Group of Sierra Leone (Culver 1984), is the metamorphosed equivalent of the upper Proterozoic Bania group volcanics. On the basis of lithologic, structural, and radiometric age data (see pp. 132–133), Williams and Culver (1988) consider the Marampa Group volcanics to have been erupted, metamorphosed, and deformed in the late Archean and further affected, both tectonically and thermally, during the Pan-African Rokelide orogeny.

5.2 Volcanics of the Bania Group and Rokel River Group

The Bania group of Guinea south of the Bové Basin (Fig. 6) is composed of andesites, diabases, spilites, keratophyres and diorites (Villeneuve 1981). Exposures are extremely poor except at Mount Binia where pillowed basaltic lava has been noted (Torchine 1969; Villeneuve 1984).

Across the border in Sierra Leone (Fig. 3), volcanic and volcano-sedimentary rocks continue as the Kasewe Hills Formation of the Rokel River Group (Allen 1968, 1969). The volcanics occur in lenses up to 2 km thick within the Mabole Formation and the Taia Formation; the continuous belt of volcanics in northern Sierra Leone (Fig. 3) changes to isolated volcanic centers in the southern part of the Rokel River Group belt. Lavas and pyroclastics are interbedded with Mabole Formation deltaic rocks whereas only air-borne pyroclastics are interbedded with Taia Formation prodeltaic sedimentary rocks (Allen 1969).

The Kasewe Hills Formation is composed dominantly of andesites, but spilites, basalts and tuffs of dacitic composition also occur. Lavas are either massive or occur in layers (up to 35 m thick) exhibiting pillow structures (Allen 1968; Macfarlane et al. 1981). Allen (1968) suggested that many of the tuffs were erupted subaerially but were deposited in a subaqueous environment. Some banded tuffs, however, probably resulted from either subaqueous pyroclastic eruptions or from deposition through turbidity flows or slumps (Allen 1968).

5.3 Pan-African Intrusives

Macfarlane et al. (1981, p. 53) noted the "... conspicuous absence of igneous plutonic activity ..." in the Rokelide orogen of Sierra Leone. Minor quartz diorites (Williams and Williams 1976) and a trondhjemite (Hurley et al. 1971) were intruded into the Kasila Group of northern Sierra Leone around 500 Ma. In Liberia, Hedge et al. (1975) identified a lava and three intrusives of Pan-African age, a basalt in the Liberian age province, a syenite stock in the Eburnean age province, and a norite body and a diorite dike in the Pan-African age province.

The Rokelide orogen in Guinea contains several undeformed granite stocks intruded into the high grade metamorphic rocks of the Forécariah group (Fig. 6; Villeneuve 1984). One of these, the Coyah granite, has been dated at 533 ± 7 Ma (biotite-muscovite plus whole-rock Rb-Sr isochron age) (Dallmeyer et al. 1987). Field and petrographic evidence suggest the granite was emplaced post-tectonically at shallow crustal levels; the 533 Ma age, therefore, likely dates crystallization (Dallmeyer et al. 1987).

6 Local Geophysical Characteristics

Until recently geophysical investigations in Sierra Leone have been restricted to several small radiometric and electromagnetic surveys carried out as prospecting or mapping tools (Macfarlane et al. 1981). In contrast, Guinea and Liberia have been the subject of extensive geophysical investigations.

6.1 Sierra Leone

Magnetic and seismic reflection surveys have been conducted on the continental shelf (McMaster et al. 1973, 1975; Jones and Mgbatogu 1982) but the geophysical character of Sierra Leone has been unknown until a 1987 ORSTOM gravimetric survey (Bonvalot et al. 1988).

The Bouguer gravity anomaly patterns in Sierra Leone can be divided into two domains (Fig. 12). The larger of the two, the eastern domain, is underlain by the upper Archean Liberian granite-greenstone terrane and is characterized by broad wavelength, low-amplitude, generally negative anomalies that trend NE-SW. These patterns are similar to those observed in the northern part of Liberia (Fig. 15). The western domain is characterized by NNW-SSE trending, small wavelength, large amplitude, generally positive anomalies (up to +60 Mgals) that reflect the underlying Pan-African tectonic fabric. The two domains are separated by a steep gradient that can be traced south into Liberia.

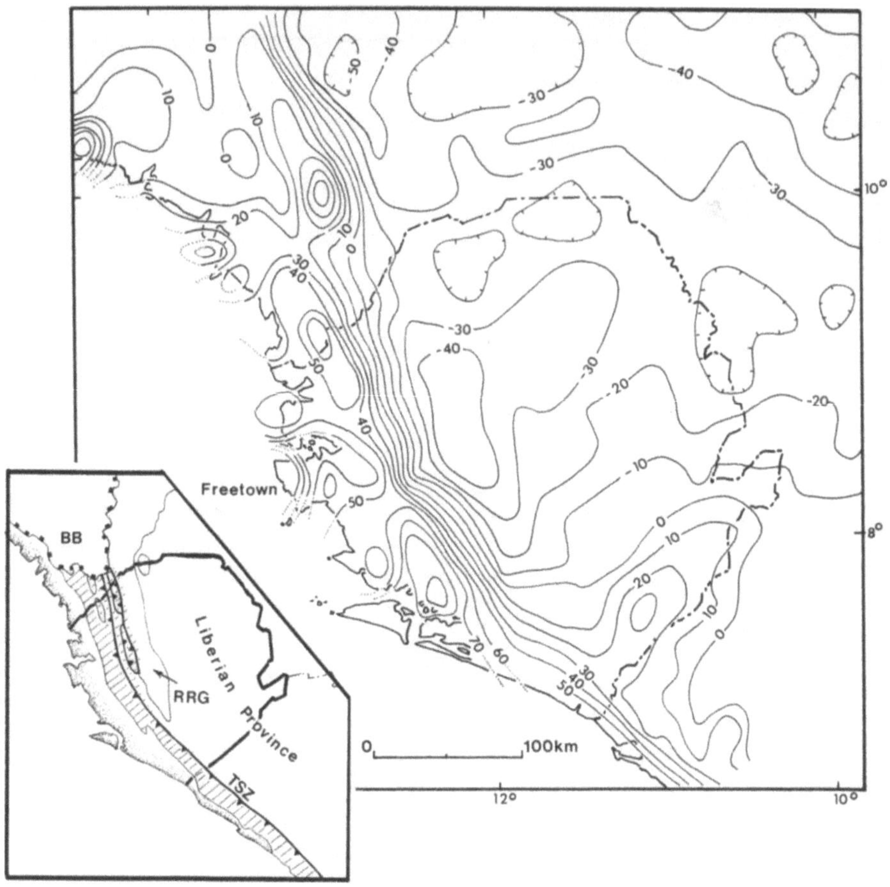

Fig. 12. Bouguer gravity map of Sierra Leone and parts of Liberia and Guinea (from Bonvalot et al. 1988). Contour interval is 10 milligals. The Liberian granite-greenstone domain is separated from the Pan-African domain (diagonally ruled) of the Rokelides (see *inset map*) by a NNW-SSE to NW-SE trending gradient that continues south to straddle the Todi Shear Zone (TSZ) of Liberia. In Sierra Leone the gradient lies just west of the Rokel River Group (RRG). *BB* Paleozoic Bové Basin

To the north the gradient swings inland to lie just west of the Bassaris (Fig. 13). At the coastline of Sierra Leone, the Mesozoic Freetown mafic igneous complex is marked by a large annular positive anomaly.

6.2 Guinea

The primary data of Amorin Ferreira (1966) and Akhmetjanov et al. (1976) are the basis of the Bouguer gravity anomaly map of Guinea and Guinea Bissau discussed in Ponsard (1984), Ponsard et al. (1982, 1988), Roussel et al. (1984) and Villeneuve (1984).

Ponsard et al. (1988) recently summarized the gravity anomalies associated with the Pan-African belt from the southern Mauritanides, across the later Bové Basin sedimentary rocks, to the northern Rokelides in southern Guinea (Fig. 13). Three principal domains are recognized: (1) an eastern domain characterized by a broad gravity low, corresponding to the Precambrian craton (Leo Uplift and Kenieba Inlier) and its upper Proterozoic cover; (2) a western domain characterized by a generally positive anomaly field, corresponding to the Mesozoic-Cenozoic Senegal coastal basin and the western part of the Paleozoic Bové Basin; and (3) a central domain represented by a generally north-south zone of gravity gradient and corresponding to the Pan-African orogenic belt (Bassarides in southern Senegal and northern Guinea and Rokelides in southern Guinea). The central domain exhibits a narrow linear belt of strong gravity highs flanked to the east by short wavelength gravity lows (Ponsard et al. 1988).

The crustal model interpreted from the regional gravity anomaly (Ponsard 1984; Ponsard et al. 1988) consists of two crustal blocks of contrasting density and thickness in isostatic equilibrium. The

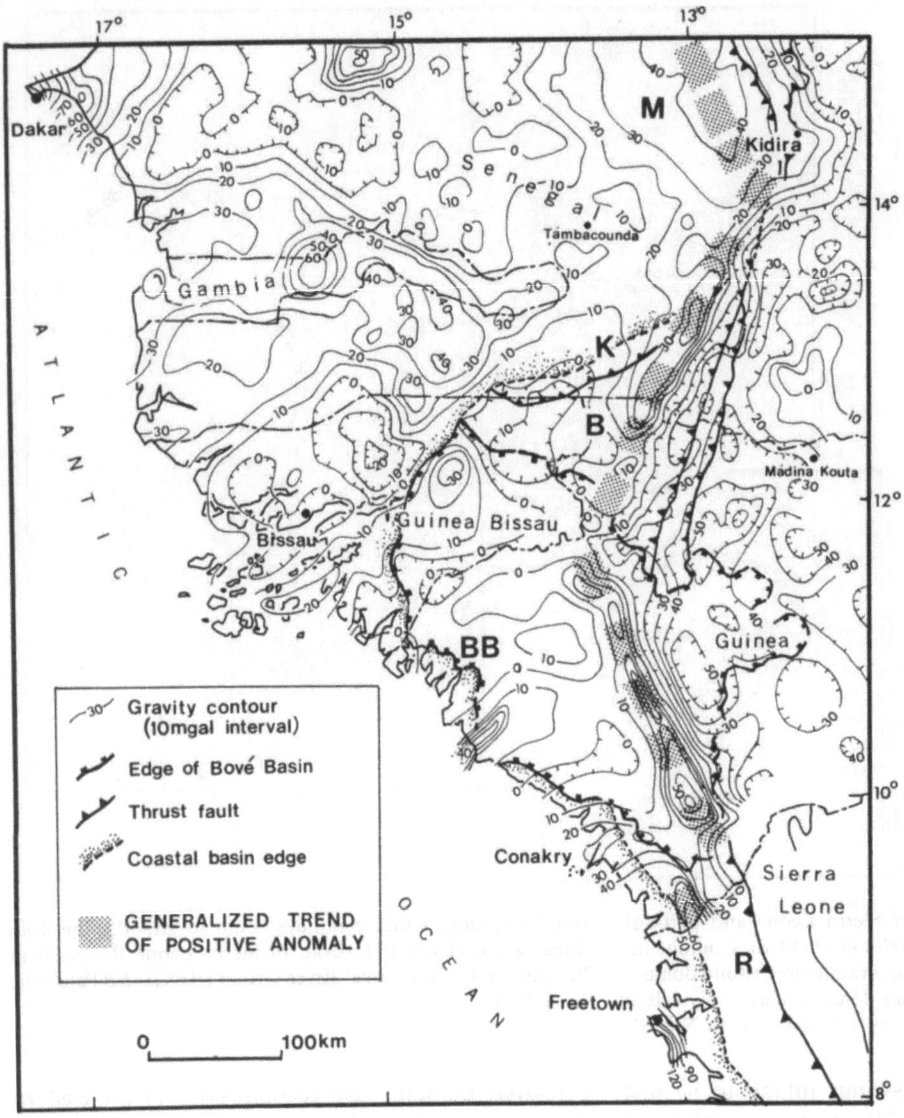

Fig. 13. Bouguer gravity anomaly map of Guinea, Guinea Bissau and southern Senegal. Modified from Ponsard et al. (1988). *M* Southern Mauritanide positive anomaly; *B* Bassaride positive anomaly; *K* Koulountou branch of southern Mauritanides; *BB* Bové Basin; *R* Rokelides

junction between them, represented by the gravity gradient, dips towards the west.

The eastern gravity lows within the central domain coincide directly with outcrops of volcanic and sedimentary rocks (Termesse group and Mali group in northern Guinea). The belt of gravity highs within the central domain is not related to outcropping strata but to a deeper source. Ponsard et al. (1988) suggested three hypotheses to account for the gravity highs but preferred the hypothesis that the dense bodies consist of granulite rocks that originated in the lower crust. These dense bodies become less deep from north to south which may reflect deeper levels of erosion southwards along the length of the Rokelides. Indeed, granulites are present in the Forécariah group of southern Guinea, the Kasila Group of Sierra Leone, and the Pan-African age province of Liberia. The outcrop pattern of the Rokel River Group of Sierra Leone also suggests deeper levels of erosion to the south.

6.3 Liberia

Aeromagnetic, aeroradioactivity and gravity surveys of Liberia have been reported by Behrendt and Wotorson (1970, 1974) and Behrendt et al. (1974). The magnetic data (Fig. 14) show long wavelength northeast-southwest trending anomalies that extend onto the continental margin to about the 3000 m submarine contour. Behrendt et al. (1974) suggested that continental basement extends to this contour. Simi-

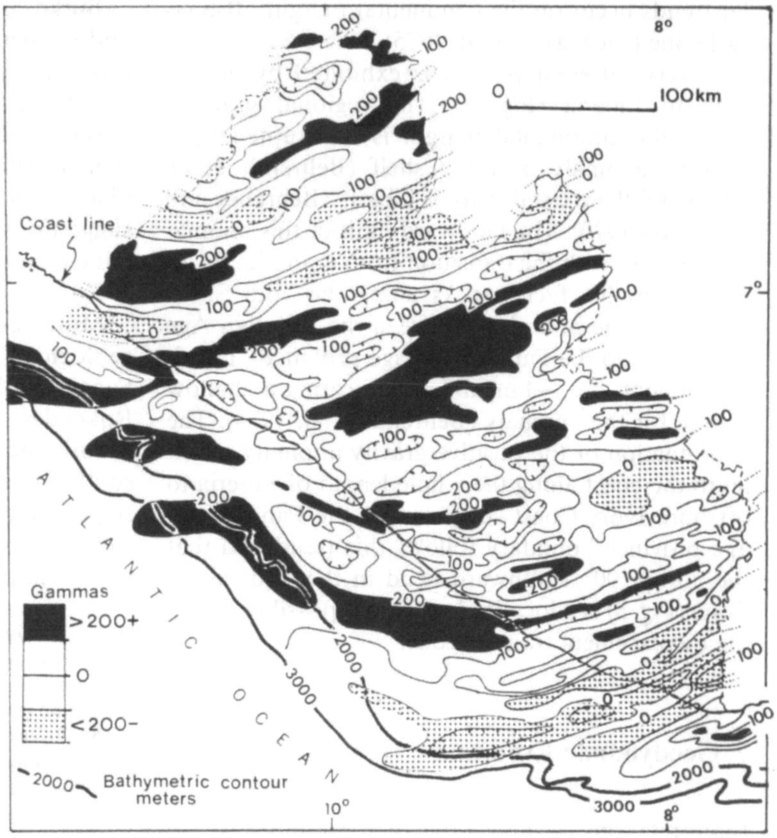

Fig. 14. Aeromagnetic anomaly map of Liberia. NE-SW trending anomalies reflect basement structural fabric (see Fig. 8) and extend onto the continental margin where they become parallel with ocean fracture zone trends. The Pan-African-Liberian age province boundary (see Fig. 8) is not readily discernable (Modified from Behrendt et al. 1974)

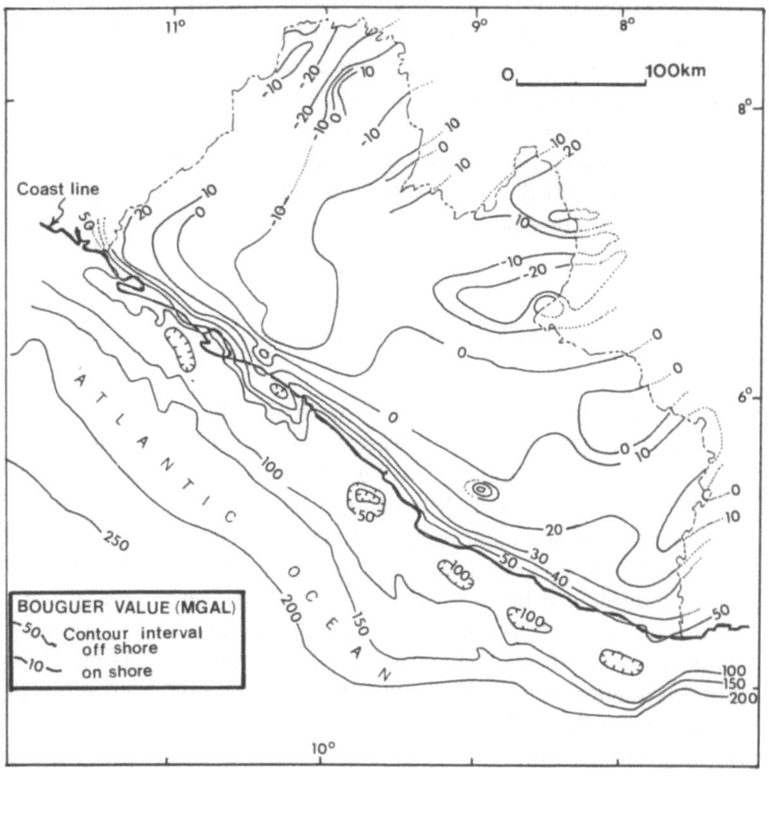

Fig. 15. Bouguer gravity anomaly map of Liberia. Note that the Archean basement fabric is broadly reflected in the NE-SW alignments of gravity contour patterns which merge with the steep NW-SE gradient, parallel with the coastline, marking the boundary between Pan-African and Liberian age provinces (Modified from Behrendt et al. 1974)

lar trends occur on the continental margin off Sierra Leone (McMaster et al. 1975).

A very different pattern is exhibited by gravity data in Liberia (Fig. 15). The regional gradient across the continental margin is interrupted by a positive anomaly over the shelf (Behrendt et al. 1974) and the coastal strip of Liberia (Behrendt and Wotorson 1970). The landward limit of this anomaly coincides with the transition from granulite facies to amphibolite facies rocks along the Todi Shear Zone (i. e. the boundary between the Liberian Age Province and the Pan-African age province). Locally, gradients are ~4 mgal/km, indicating a shallow source for the anomaly (Behrendt et al. 1974). The continuation of the positive gravity anomaly parallel to the coast along the whole length of Liberia to offshore Ivory Coast suggests that the Pan-African age province boundary follows this trend and that the abrupt 90° southward bend in the Todi Shear Zone (Fig. 8) is a localized variation, possibly a ductile displacement transfer zone.

7 Geodynamic Evolution

Interpretations of the geodynamic evolution of the Rokelide orogen depend on the assumed stratigraphic position of the various units. As noted above, considerable disagreement exists on this topic. Three major schools of thought have evolved with regard to two fundamental questions: (1) the nature and age of the boundary between the Archean granite-greenstone terrane of the Leo Shield and the granulite facies Archean supracrustals to the west; and (2) the age of the Marampa Group and its lateral correlatives.

7.1 The Pan-African Suture Hypothesis

Allen et al. (1967) proposed the term "Rokelide" to describe the orogenic belt in western Sierra Leone that was deformed during the Pan-African, about 550 Ma (Kennedy 1964). Allen (1969) recognized a basement sequence composed of the Kasila Group and a series of gneisses in the extreme western part of the granite-greenstone terrane; both units were deformed and metamorphosed during Rokelide orogenesis. Overlying this basement was a geosynclinal sequence composed of the sediments and volcanics of the Rokel River Group and the schists of the Marampa Group.

Burke and Dewey (1972) accepted this scheme and further proposed the existence of a cryptic suture, termed the Marampa Suture, extending southeast from the southern termination of the Rokel River Group belt (Burke and Dewey 1972, their Fig. 6). The idea that the Rokelides represent a plate-collision orogen was further supported by suggestions that the Sierra Leonean Rokelides contained possible ophiolites (Shackleton 1976; Gass 1977).

More recently, geophysical (Ponsard 1984) and geologic (Villeneuve 1984) data from Guinea and Senegal, led Ponsard et al. (1982) to suggest that the Rokelides of southern Guinea and the southern Mauritanides (Bassarides) of northern Guinea and southern Senegal exhibit a Pan-African suture. Five stages of evolution of this orogenic belt (Fig. 16) were recognized for the late Proterozoic (Ponsard et al. 1988, pp. 470–471):

"(a) distension and ocean opening (1600–850 Ma) with the formation of the volcanic-sedimentary terranes of the Termesse group in the eastern rifted passive margin and of the Guinguan group in the western margin; (b) convergence and ocean closure (850–700 Ma) marked by the subduction of the oceanic domain beneath the western crustal block acting as an active margin with calc-alkaline magmas (Niokolo Koba group) and accretion wedge involving the Guinguan group terranes; (c) collision and suturing (700–650 Ma) with overthrusting of the western continental block on the West African craton. At surface the volcanic-sedimentary material of the Guinguan group which has been largely remobilized during the subduction is thrust over the Termesse terranes which also were remobilized and folded during the colliding. The dense bodies revealed by the belt of gravity highs probably have been squeezed at depth along the suture during this stage which corresponds to the Pan-African orogeny sensu stricto; (d) isostatic recovery and erosion (650–550 Ma) marked by the deposits of the Mali and Youkounkoun groups; (e) late stage of Pan-African orogeny (550–500 Ma) expressed by folding and locally shallow small decollement in the Mali group terranes. Finally these terranes are capped with the Youkounkoun molasse."

Although this description is for the Bassaris section of the southern Mauritanides, Ponsard et al. (1988) considered that the same sequence of events characterizes the northernmost Rokelides in Guinea. Figure 10 indicates stratigraphic correlations but, most importantly, it should be emphasized that the Termesse group (Villeneuve 1980) of the southern Mauritanides is lithologically equivalent to the Bania group of the northern Rokelides of Guinea which can be traced southwards into the Kasewe Hills Formation of Sierra Leone. The Guingan group of the southern Mauritanides is lithologically equivalent to the Ouankifondi group of the northern Rokelides of Guinea which can be traced

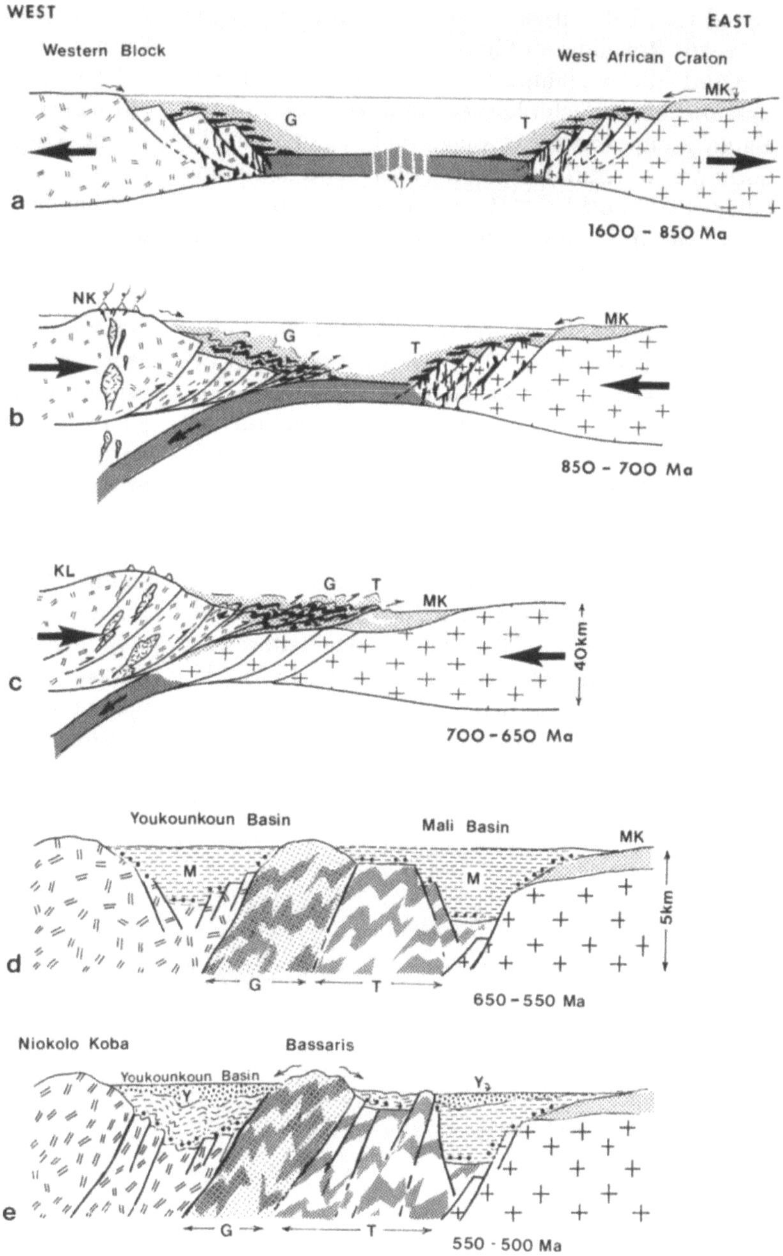

Fig. 16. Geodynamic evolution of the Pan-African belt in the southern Mauritanides and northern Rokelides according to Ponsard et al. (1988). See text for detailed explanation. *G* Guingan group; *T* Termesse group; *MK* Madina Kouta group; *NK* Niokola Koba group; *KL* Kouluntou group; *M* Mali goup; *Y* Youkounkoun group. Sections located across Koulountou and Bassaris branches of the southern Mauritanides (see Fig. 1). Note scale change between **c** and **d**. Redrawn from Ponsard at al. (1988)

southwards into the Marampa Group of Sierra Leone.

Although this model may well be the best scenario for the southern Mauritanides, considerable problems arise when attempting to apply it to the Rokelides.

First, the cryptic suture proposed by Burke and Dewey (1972) at the southern end of the Rokel River Group outcrop does not exist. Careful field work in this region reveals no indication of a structure that could be construed as a suture (Williams and Culver 1982).

Second, all geochronological data from the Rokelides (Fig. 2; see Beckinsale et al. 1981; Cahen et al. 1984) suggest a single orogenic event at approximately 550 Ma. This is the age of Pan-African II deformation in the southern Mauritanide orogen (Dallmeyer and Villeneuve 1987; Dallmeyer 1989). No Pan-African I dates of ca. 650 Ma (Dallmeyer and Villeneuve 1987) have been recorded in the Rokelides. The lack of traces of Pan-African I metamorphism and deformation was noted by Villeneuve and Dallmeyer (1987) even though Dallmeyer et al. (1987) and Ponsard et al. (1988) inferred that Pan-African I effects are present in the northern Rokelides of Guinea.

Third, the Marampa Group of Sierra Leone is demonstrably much older than the volcanics of the Bania group and the Kasewe Hills Formation. In addition, Macfarlane et al. (1981) and Williams (1978, 1988) noted that the stratigraphy of the Marampa Group is most similar to the Archean greenstones of the Kambui Supergroup (part of the Kenema Assemblage) to the east. Geochronologic data suggest a probable Liberian age for the Marampa Group and subsequent Pan-African resetting of radiometric systems (Beckinsale et al. 1981; Cahen et al. 1984). Two whole-rock Rb/Sr determinations on albite-mica schists gave model ages of 2075 and 2154 Ma (Hurley et al. 1971). Cahen et al. (1984) suggested that the sensitivity of the calculation to the assumed initial ratio of 0.705 is too great for these results to be accepted at face value. They do, however, indicate an extended history (Cahen et al. 1984) before the Pan-African event. The latter is recorded by ages of 560 ± 10 Ma and 562 ± 15 Ma derived from two K:Ar determinations on muscovite from Marampa Group mica-schists (Beckinsale et al. 1981). In summary, all recent workers in Sierra Leone consider the Marampa Group to be of late Archean (or possibly early Proterozoic age), deformed in the Liberian orogenic event (ca. 2700 Ma) or soon after, and reworked in the Pan-African (ca. 550 Ma).

The Marampa Group outcrop can be traced directly across the border into the Ouankifondi group of Guinea. It seems logical, therefore, to consider that the Ouankifondi group is of the same age and has the same history. Thus, the Ouankifondi group, at our present state of knowledge, cannot be the metamorphosed equivalent of the volcanic Bania group as suggested by Dallmeyer et al. (1987) and it cannot fulfill the geologic and tectonic role suggested in the model of Ponsard et al. (1988) for the northern Rokelides. The problem remains whether or not the Marampa and Ouankifondi groups are correlatives of the Guingan group of the southern Mauritanides. Recent $^{40}Ar/^{39}Ar$ mineral ages from southeastern Senegal suggest a much younger age for the Guingan group. Dallmeyer and Villeneuve (1987) recorded plateau ages of about 660 Ma (Pan-African I) which were interpreted as dating post-metamorphic cooling through argon retention temperatures.

Thus, the geodynamic history based on geologic, geophysical and geochronological data from the

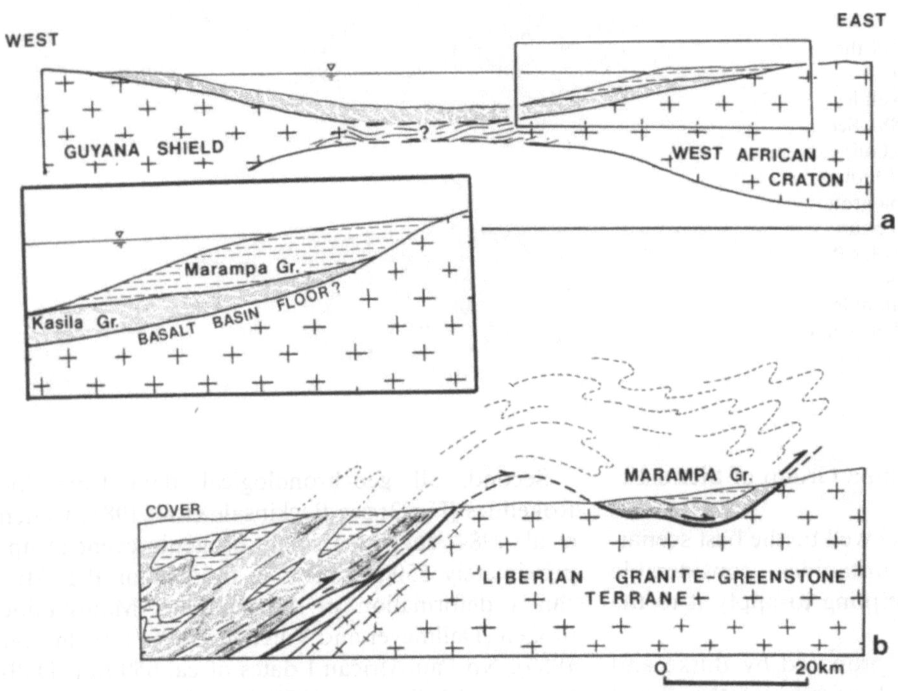

Fig. 17. Archean tectonic development of the supracrustal Kasila and Marampa groups in Sierra Leone according to Williams (1988). See text for full explanation. **a** Accumulation of Kasila and Marampa groups on thinned sialic crust of West African craton, *inset* illustrates stratigraphic relationships **b** structural telescoping of supracrustals resulting in interleaving of Kasila Group with Liberian basement and klippen of recumbently folded Marampa Group. Section drawn at present-day erosion level (Modified from Williams 1988)

southern Mauritanides of southeastern Senegal and northern Guinea does not fit observations from the Rokelides. The removal of the schists of the Marampa Group and Ouankifondi group from the geodynamic model proposed in Ponsard et al. (1988) most certainly weakens the argument for a Pan-African suture in the Rokelides.

7.2 The Archean Suture Hypothesis

Williams (1979, 1988) has suggested that the shear zone boundary between the Kasila Group and the granite-greenstone terrane is an Archean suture or terrane boundary. In this scenario, rocks of the Kasila Group and Marampa Group represent a su-

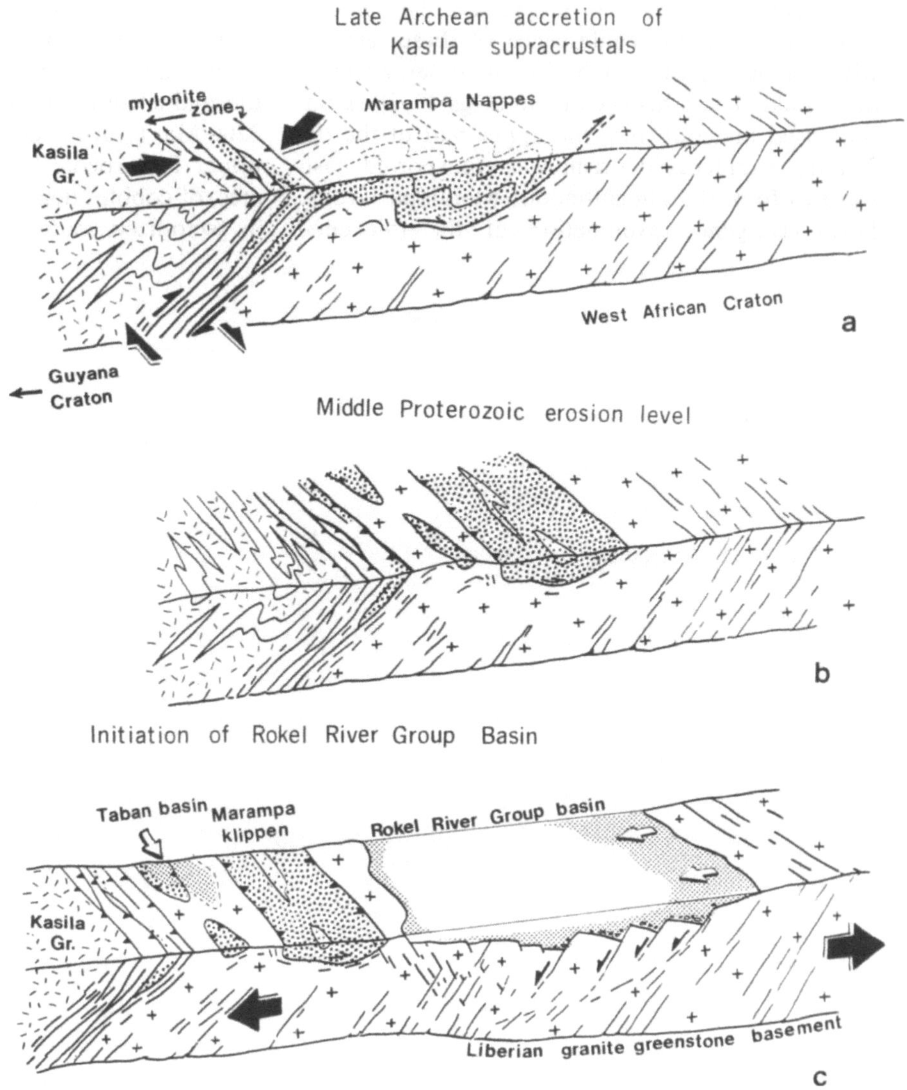

Fig. 18. Geologic development of the Rokelide orogen and its relationship to the Archean basement. **a** Eastward emplacement of Marampa nappes (coarse stipple) onto Liberian granite-greenstone basement during Late Archean collision of the Guyana and West Africa cratons; **b** Early to Middle Proterozoic erosion producing Marampa klippen. Note fracturing of West African craton east of the klippen; **c** Late Proterozoic (ca. 620 Ma) intracratonic basin infilling (Rokel River Group: *fine stipple*) begins east of Marampa klippen. Outlying Taban basin straddles Liberian basement and Marampa klippen. Deposition of Upper Proterozoic glacial strata (Tabe Formation). *Open arrows* indicate sediment transport directions; **d** basin deepens westward during deposition of Teye Formation turbidites; **e** Mabole Formation deltas prograde westward and interfinger with Taia Formation prodelta deposits in west. Several volcanic centers, concentrated in the western deeper part of the intracratonic basin, produce Kasewe Hills Formation (*black*); **f** eastward directed thrusting reworks Archean tectonic elements during Pan African II orogeny (ca. 550 Ma.). Retrograde metamorphism near eastern margin of Kasila Group and in mylonite zone. Greenschist facies metamorphism confined to western half of Rokel River Group and Taban outlier

pracrustal belt penecontemporaneously developed at and accreted onto the West African craton during or soon after aggregation of the granite-greenstone terrane (Fig. 17; Williams 1979). The Kasila Group and Marampa Group accumulated in a basin between the recently cratonized West Africa and Guyana shields. Williams (1988) further suggested that if a supracrustal sequence developed marginal to discrete granite-greenstone terranes, late Archean convergence as a result of large-scale plate motion would result in the coalescence of these terranes with one attempting to override the other, trapping the Kasila Group between them (Figs. 17, 18a). Supracrustal material from the upper part of the deforming supracrustal prism was transported eastward to form the allochthonous Marampa Group. Post-convergence exhumation of the thickened prism and its sheared contacts accounts for the high grade nature and zonation of the Kasila Group, the low-grade, recumbently folded Marampa Group, and the 5-km-wide zone of mylonites and highly deformed gneisses at the eastern margin of the Kasila Group (Fig. 18b).

Williams (1988) noted the U/Pb isotopic data that suggests that the Kasila Group is of late Archean age (Hedge et al. 1975). Rb/Sr whole-rock data on the Marampa Group, and on Liberian granitoids close to the contact between the Kasila Group and Kenema Assemblage, indicate that a thermal event or isotopic resetting occurred around 2200 Ma (Hurley et al. 1971). The presence, in upper Proterozoic Rokel River Group conglomerates, of cobbles derived from the Kasila Group and its tectonized contact zone, argues strongly against

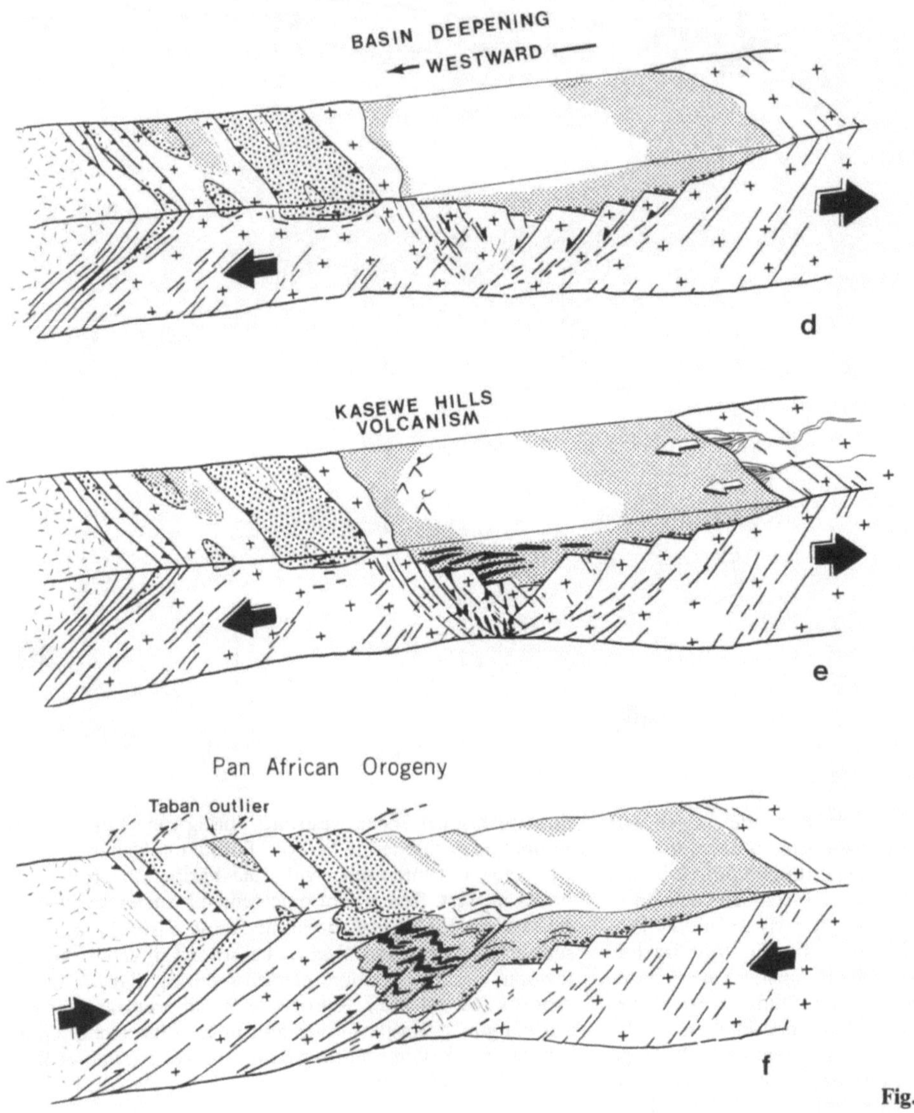

Fig. 18. d-f

a docking of the Kasila adjacent to the West African craton during the Pan-African, as inferred by Ponsard et al. (1982, 1988).

In the late Precambrian, limited rifting (Williams and Culver 1982) took place to the east of the Archean suture in the fractured crust of the overridden block (Fig. 18c). Rokel River Group deposits were laid down within this elongate basin that may have opened wider to the north in the southern Mauritanides (Bassarides; Williams and Culver 1982) perhaps to involve oceanic crust (compare Villeneuve and Dallmeyer 1987). Further to the north in the central Mauritanides (Fig. 1), oceanic crust formed between the West African craton and the Senegal microplate to the west (Lécorché et al. 1983; Dia 1984; Villeneuve and Dallmeyer 1987).

Ritz and Robineau (1986) independently suggested that the central Mauritanides may have had a different geodynamic evolution from the Bassarides and Rokelides. They suggested that the Bassarides, separated from the central Mauritanides by the Bissau-Kidira fracture zone (Fig. 2), developed as an intracontinental basin on stretched crust. This concurs with the model of limited rifting for the Rokelides suggested by Williams and Culver (1982).

Subsequent to accumulation of Rokel River Group sediments and volcanics (Figs. 18c, d, e), Pan-African II (550 Ma) closure of the Rokelide trough folded and induced metamorphism of western-most Rokel River Group deposits, caused retrograde metamorphism in the eastern part of the Kasila Group, reset mineral and rock isotopic systems (Hurley et al. 1971), and reactivated westerly dipping Archean thrust faults (Fig. 18f). The Marampa Group, at least in part, was tectonically retransported towards the east to lie with thrust contact on the Rokel River Group in Sierra Leone and the Gibi Mountain Formation in Liberia (Fig. 18f). Rokelide deformation was mild, and consisted of reactivation of existing thrusts and folding mainly in the western part of the Rokel River Group basin (Fig. 11). The thermal event that accompanied deformation partially reset Archean isotopic systems. The Pan-African "front" of metamorphism does not extend to the eastern part of the Rokel River Group (Fig. 3). Thus, these rocks can best be considered as slightly deformed strata on the margins of the foreland to the Rokelide orogen.

The considerable evidence for isotopic resetting in the western portion of the Rokelides during Pan-African II in Sierra Leone and Liberia cannot be related to the development of high grade metamorphism at that time (see pp. 113, 114, 136). The data at present are, however, permissive of the Kasila Group granulites being tectonically raised to their present position during the postulated Pan-African I event. Further geochronological work on both the Marampa and Kasila groups, using a number of minerals in order to determine cooling histories, will constrain this argument.

A similar developmental history for the Rokelide orogen in Liberia was suggested by Williams and Culver (1988). However, the geologists most familiar with Liberian geology do not concur.

7.3 The No Suture Hypothesis

Thorman (1976) argued strongly that the Todi Shear Zone, which forms the boundary between the Pan-African and Liberian age provinces in Liberia (Fig. 8), cannot be a suture. In noting that the shear zone represents a deep crustal environment, Thorman (1976) pointed out that stratigraphic units can be traced across the Todi Shear Zone in northern Liberia whereas major stratigraphic units and structural features are truncated by it further to the south (Fig. 8, inset). In southern Sierra Leone granulite facies rocks also occur on both sides of the indistinct boundary between the Kasila Group and the granite-greenstone terrane (Rollinson 1978; Williams 1978). Thorman (1976) suggested that field evidence indicates that displacement along the Todi Shear Zone is variable and complex and that the Pan-African age province rocks may be reworked rocks of the Liberian age province.

We believe that the field data in Liberia could be interpreted in an alternative fashion. In northern Sierra Leone the boundary between the Kasila Group and the granite-greenstone terrane consists of a complex interleaving of slivers from both terranes (Williams 1988). The structural grain on both sides of the boundary is parallel. We suggest that, in southern Sierra Leone and northern Liberia where the boundary is indistinct, a similarly deep level of crust is exposed on both sides of the sheared boundary. Granulites thus occur at the present erosion surface of both terranes and the sheared nature of the boundary results in the parallelism of structural trends on either side.

7.4 Discussion of Geodynamic Evolution

The Rokelide orogen recently has been the subject of field investigations by three groups of geologists, one in each of Guinea, Sierra Leone and Liberia.

Three fundamentally different models of geodynamic evolution have resulted. Two major factors have influenced these geologists; their concept of tectonostratigraphic position of various stratigraphic units, and interpretations of the nature and age of the sheared boundary between two Archean terranes.

Clearly, additional geochronological work is necessary, particularly in the northern Rokelides of southern Guinea and in Sierra Leone. Additional fieldwork in southern Guinea and Sierra Leone is also necessary to clarify the stratigraphic position of the Bania group and the Kasewe Hills Formation. Clarification of the nature of the sheared boundary between Archean and Pan-African age provinces in southern Sierra Leone and northern Liberia is fundamental to the understanding of the evolution of the southwestern margin of the West African craton. Although differences of interpretation will still likely result, those differences would surely be of lesser degree than those that presently exist.

References

Akhmetjanov BS, Loutsenko VF, Dialo M (1976) Carte des anomalies de Bouguer (1/500,000) et résultats des travaux géophysiques en Guinée occidentale. In: Notice explicative de la carte géologique de la Republique de Guinée, Feuille de Gaoual: Conakry, Guinea. Rapp Technoexport, pp 184-205

Allen PM (1968) The stratigraphy of a geosynclinal succession in western Sierra Leone. Geol Mag 105: 62-73

Allen PM (1969) Geology of part of an orogenic belt in western Sierra Leone. Geol Rundsch 58: 588-619

Allen PM (1980) Discussion on the late Precambrian and Phanerozoic geology of Sierra Leone. J Geol Soc Lond 137: 511

Allen PM, Snelling NJ, Rex DC (1967) Age determinations from Sierra Leone. 15th Annu Rep Dept Geol Geophys, Massachusetts Inst Technol, 23: 17-22

Amorin Ferreira H (1966) Observacoes gravimetricas no territoria da Guinea Portuguese: Lisboa. Serv Meteorol Nacl, pp 9

Aymé JM, Arnould M, Guillaume R, Liouville M, Spence J (1958) Rapport géologique sur la mission Soudan, Sénégal, Guinée. Comp Fr Petr, R G n° 42, Dakar, pp 139

Bassot JP (1966) Etude Géologique du Sénégal oriental et de ses confins Guinéo-maliens. Mem Bur Rech Géol Min, Paris, 40: 1-322

Beckinsale RD, Pankhurst RJ, Snelling NJ (1981) Appendix 2. The geochronology of Sierra Leone. In: Macfarlane A, Crow MJ, Arthurs JW, Wilkinson AF, Aucott JW. The geology and mineral resources of northern Sierra Leone. Overseas Mem, Inst Geol Sci, 7: 89-96

Behrendt JC, Wotorson CS (1970) Aeromagnetic and gravity investigations of the coastal area and continental shelf of Liberia, West Africa, and their relation to continental drift. Geol Soc Am Bull 81: 3563-3574

Behrendt JC, Wotorson CS (1974) Geophysical surveys of Liberia with tectonic and geologic interpretations. US Geol Surv Prof Pap 810: 33

Behrendt JC, Schlee J, Robb JM, Silverstein MK (1974) Structure of the continental margin of Liberia, West Africa. Geol Soc Am Bull 85: 1143-1158

Black R (1980) Precambrian of West Africa. Episodes 4: 3-8

Bonvalot S, Villeneuve M, Legeley A, Albouy Y (1988) Levé gravimetrique du sud-ouest du craton Quest-Africain. C R Acad Sci, Paris, 307: 1863-1868

Boufeev YL (1968) Notice explicative de la carte géologique de la République de Guinée au 1/200,000. Feuille C-28-XXIII et Nord de la feuille C-28-XXIX (Conakry). Conakry, pp 182

Bufeyev YU V (1972) The Late Proterozoic Rokel trough, West Africa, and its structure. Geotectonics 2: 102-104

Bull PA, Culver SJ (1979) An application of scanning electron microscopy to the study of ancient sedimentary rocks from the Saionia Scarp, Sierra Leone. Palaeogeogr Palaeoclimatol Palaeoecol 26: 159-172

Burke KC, Dewey JF (1972) Orogeny in Africa. In: Dessauvagie TF, Whiteman AJ (eds) African geology. Univ Ibadan, Ibadan, Nigeria, pp 583-608

Caen-Vachette M (1988) Le craton ouest-africain et le bouclier guyanais: un seul craton au Protérozoique inferieur? J Afr Earth Sci 7: 479-488

Cahen L, Snelling NJ, Delhal J, Vail JR, Bonhomme M, Ledent D (1984) The geochronology and evolution of Africa. Clarendon Press, Oxford, UK

Clauer N, Deynoux M (1987) New information on the probable istopic age of the late Proterozoic glaciation in West Africa. Precambrian Res 37: 89-94

Culver, SJ (1984) The Pan-African Rokelides of Sierra Leone: problems and possible correlations with Liberia and Guinea. Trav Lab Sciences Terre, St Jerome, Marseille, Ser B, 24: 50

Culver SJ, Magee AW (1987) Late Precambrian glacial deposits from Liberia, Sierra Leone, and Senegal, West Africa. Nat Geogr Res 3: 69-81

Culver SJ, Williams HR (1979a) Late Precambrian and Phanerozoic geology of Sierra Leone. J Geol Soc Lond 136: 605-618

Culver SJ, Williams HR (1979b) Discussion of the evolution of the Pan-African crystalline basement in NE Africa and Arabia. J Geol Soc Lond 136: 123

Culver SJ, Williams HR (1980) Discussion on the late Precambrian and Phanerozoic geology of Sierra Leone. J Geol Soc Lond 137: 511-512

Culver SJ, Williams HR, Bull PA (1978) Infracambrian glaciogenic sediments from Sierra Leone. Nature (Lond) 247: 49-51

Culver SJ, Williams HR, Bull PA (1980) Late Precambrian glacial deposits from the Rokelide fold belt, Sierra Leone. Palaeogeogr Palaeoclimatol Palaeoecol 30: 65-81

Culver SJ, Pojeta J Jr, Repetski J (1988a) First record of Early Cambrian shelly microfossils from West Africa. Geology 16: 596-599

Culver SJ, Pojeta J Jr, Repetski J, Robineau B (1988b) Signification stratigraphique de microfossiles du Cambrien inferieur dans le bassin de Taoudéni, a la frontiere Guinée-Sénégal. C R Acad Sci, Paris, 307: 651-656

Dallmeyer RD (1989) A tectonic linkage between the Rokelide orogen (Sierra Leone) and the St. Lucie metamorphic complex in the Florida subsurface. J Geol 89: 183-195

Dallmeyer RD, Villeneuve M (1987) $^{40}Ar/^{39}Ar$ mineral age record of polyphase tectonothermal evolution in the southern Mauritanide orogen, southeastern Senegal. Geol Soc Am Bull 98: 602-611

Dallmeyer RD, Caen-Vachette M, Villeneuve M (1987) Emplacement age of post-tectonic granites in southern Guinea (West Africa) and the peninsular Florida subsurface: implications for origins of southern Appalachian exotic terranes. Geol Soc Am Bull 99: 87-93

Deynoux M (1980) Les Formations glaciaires du Précambrien terminal et de la fin de l'Ordovicien en Afrique de l'Ouest. Deux exemples de glaciation d'inlandsis sur une plateforme stable. Thesis, Univ Aix-Marseille, pp 553

Deynoux M, Trompette R (1981) Late Precambrian tillites of the Taoudeni Basin, West Africa. In Hambrey MJ, Harland WB (eds) Earth's pre-Pleistocene glacial Record. Cambridge Univ Press, Cambridge, pp 123-131

Deynoux M, Sougy J, Trompette R (1985) Lower Paleozoic rocks of West Africa and the western part of Central Africa. In: Holland CH (ed) Lower Palaeozoic of north-western and west-central Africa. Wiley, NY, pp 337-495

Dia O (1984) La chaine panafricaine et hercynienne des Mauritanides face au bassin protérozoique superiéur à Devonien de Taoudeni dans le secteur clef de Mejeria (Taganet, Sud R. I. M.). Lithostratigraphie et tectonique. Un exemple de tectonique tangentielle superposée. Thesis, Univ Aix-Marseille, pp 516

Dixey F (1920) Report of the geological survey of Sierra Leone for 1920. Freetown, Sierra Leone (unpubl.)

Dixey F (1925) The geology of Sierra Leone. Q J Geol Soc Lond 81: 195-222

Gass IG (1977) The evolution of the Pan-African crystalline basement in northeast Africa and Arabia. J Geol Soc Lond 134: 129-138

Hawkes DD (1972) The geology of Sierra Leone. In: Dessauvagie TF, Whiteman AJ (eds) African geology. Univ Ibadan, Ibadan, Nigeria, pp 471-482

Hedge CE, Marvin RG, Naeser CW (1975) Age provices in the basement rocks of Liberia. US Geol Surv J Res 3: 425-429

Hurley PM, Leo GW, White RW, Fairbairn HW (1970) The Liberian age province (ca. 2700 m. y.) and adjacent provinces in Liberia and Sierra Leone. In: Hurley PM (ed) Variations in isotopic abundances of strontium, calcium and argon and related topics. Mass Inst Technol 18th Annu Progr Rep US Atomic Energy Comm 1381-18, pp 17-37

Hurley PM, Leo GW, White RW, Fairbairn HW (1971) Liberian age province (about 2700 Ma) and adjacent provinces in Liberia and Sierra Leone. Geol Soc Am Bull 82: 3483-3490

Jones EJW, Mgbatogu CCS (1982) The structure and evolution of the West African continental margin off Guinea Bissau, Guinea and Sierra Leone. In: Scrutton RA, Talwani M (eds) The ocean floor. Wiley, NY, pp 165-202

Kennedy WQ (1964) The structural differentiation of Africa in the Pan-African (±500 My) tectonic episode. 8th Annu Rep Res Inst Afr Geol, Leeds, pp 48-49

Lécorché JP, Roussel J, Sougy J, Guetat Z (1983) An interpretation of the geology of the Mauritanides orogenic belt (West Africa) in the light of geophysical data. Geol Soc Am Mem 158: 131-147

Liberian Geological Survey (1983) Summary of the Precambrian Geology of Liberia - Contribution to IGCP Project No. 108/144. Liberian Geol Surv, Monrovia, Liberia, pp 27 (unpubl.)

Macfarlane A, Crow MJ, Arthurs JW, Wilkinson AF (1974) The geology and mineral resources of northern Sierra Leone. Inst Geol Sci, Overseas Division, Int Rep 34: 203

Macfarlane A, Crow MJ, Arthurs JW, Wilkinson AF, Aucott JW (1981) The geology and mineral resources of northern Sierra Leone. Overseas Mem Inst Geol Sci 7: 103

Magee AW, Culver SJ (1986) Recognition of late Precambrian glaciogenic sediments in Liberia. Geology 14: 920-922

McMaster RL, Ashraf A, De Boer J (1973) Transverse continental margin fracture zone off Sierra Leone. Nature (Lond) 244: 93-94

McMaster RL, Christofferson E, Ashraf A (1975) Structural framework of continental shelf and slope off southwestern Sierra Leone, West Africa. Am Assoc Petrol Geol Bull 49: 2161-2171

Petrovsky A (1973) Notice explicative de la carte géologique de la république de Guinée au 1/200,000. Feuilles C-28-XVIII (Kindia). Conakry, pp 250

Pollett JD (1951) The geology and mineral resources of Sierra Leone. Colon Geol Miner Resour 1: 37-41

Ponsard JF (1984) La marge du craton ouest-africain du Sénégal à la Sierra Leone: interprétation géophysique de la chaîne panafricaine et des bassins du Protérozoique à l'Actuel. Thesis, Univ Aix-Marseille, pp 198

Ponsard JF, Lesquer A, Villeneuve M (1982) Une suture panafricaine sur la bordure occidentale du craton ouest-africain? CR Hebd Séances Acad Sci, Paris, Sér II, 295: 1161-1164

Ponsard JF, Roussel J, Villeneuve M (1988) The Pan-African orogenic belt of southern Mauritanides and northern Rokelides (southern Senegal and Guinea, West Africa): gravity evidence for a collisional suture. J Afr Earth Sci 7: 463-472

Ramsay JG (1974) Development of chevron folds. Geol Soc Am Bull 85: 1741-1754

Reid PC, Tucker ME (1972) Probable late Ordovician glacial marine sediments from nothern Sierra Leone. Nature (Lond) 238: 38-40

Ritz M, Robineau B (1986) Crustal and upper mantle electrical conductivity structures in West Africa: geodynamic implications. Tectonophysis 124: 115-132

Rollinson HR (1978) Zonation of supracrustal relics in the Archaean of Sierra Leone, Liberia, Guinea and Ivory coast. Nature (Lond) 272: 440-442

Rollinson HR, Cliff RA (1982) New Rb-Sr age determinations of the Archaean basement of eastern Sierra Leone. Precambrian Res 17: 63-72

Roussel J, Lécorché J-P, Ponsard JF, Sougy J, Villeneuve M (1984) Panafrican to Hercynian deformations in the Mauritanides and tectonic significance of gravity anomalies. Tectonophysics 109: 41-57

Shackleton RM (1976) Possible late-Precambrian ophiolites in Africa and Brazil. 20th Ann Rep Res Inst Afr Geol, Leeds, pp 3-7

Thorman CH (1974) Geology of the Monrovia Quadrangle, Liberia. US Geol Surv Int Rep IR-LI-66B, pp 18

Thorman CH (1976) Implications of klippen and a new sedimentary unit at Gibi Mountain, Liberia, West Africa, in the problem of the Pan-African - Liberian age province boundary. Geol Soc Am Bull 87: 851-856

Thorman CH (1977) Geologic map of the Monrovia Quadrangle, Liberia. US Geol Surv Misc Invest Ser Map I-775-D, Scale 1:250,000

Torchine NS (1969) Notice explicative de la carte géologique de la république de Guinée au 1/200,000. Feuille C-28-XII (Sierumba). Conakry, pp 180

Trompette R (1973) Le Précambrien supérieur et le Paléozoique inferieur de l'Adrar de Mauritanie (bordure occidentale du bassin de Taoudeni, Afrique de l'Ouest). Un exemple de sédimentation de craton. Etude stratigraphique et sédimentologique. Thesis, Univ Aix-Marseille, pp 702

Tucker ME, Reid PC (1973) The sedimentology and context of late Ordovician glacial marine sediments from Sierra Leone, West Africa. Palaeogeogr Palaeoclimatol Palaeoecol 13: 289-307

Tysdal RG (1977) Geologic map of the Buchanan Quadrangle, Liberia. US Geol Surv Misc Invest Ser Map I-778-D, Scale 1:250,000

Tysdal RG (1978) Geology of the Buchanan Quadrangle, Liberia. US Geol Surv Bull 1449: 31

Tysdal RG, Thorman CH (1983) Geologic Map of Liberia. US Geol Surv Misc Invest Ser Map I-1480, Scale 1:-1,000,000

Vallance G (1974) Geology of the country between Moyamba and Bo. Geol Surv Sierra Leone, pp 56 (unpubl.)

Villeneuve M (1980) Schéma géologique de Nord de la Guinée (Afrique de l'Ouest). CR Somm Soc Geol Fr 2: 54-57

Villeneuve M (1981) Résultats préliminaires d'une étude géologique du sud du Fouta Djalon (Guinée). CR Somm Soc Geol Fr 2: 55-59

Villeneuve M (1984) Etude géologique sur la bondure sud-ouest du craton ouest-africain. La suture panafricaine et l'evolution des bassins sédimentaires proterozoiques de la marge nord-ouest du continent de Gondwana. Thesis, Univ Aix-Marseille, pp 552

Villeneuve M, Dallmeyer RD (1987) Geodynamic evolution of the Mauritanide, Bassaride, and Rokelide orogens (West Africa). Precambrian Res 37: 19-28

Williams HR (1978) The Archaean geology of Sierra Leone. Precambrian Res 6: 251-268

Williams HR (1979) An Archean suture in Sierra Leone? Nature (Lond) 282: 608-609

Williams HR (1988) The Archean Kasila Group of western Sierra Leone: geology and relations with adjacent granite-greenstone terrane. Precambrian Res 38: 201-213

Williams HR, Culver SJ (1982) The Rokelides of West Africa-Pan-African aulacogen or back-arc basin? Precambrian Res 18: 261-273

Williams HR, Culver SJ (1988) Structural terranes and their relationships in Sierra Leone. J Afr Earth Sci 7: 473-477

Williams HR, Williams RA (1976) The Kasila Group, Sierra Leone, an interpretation of new data. Precambrian Res 3: 505-508

The Bassaride Orogen

M. Villeneuve,[1] J. P. Bassot,[2] B. Robineau,[3] R. D. Dallmeyer[4] and J. F. Ponsard[5]

1 Introduction

The Bassarides of eastern Senegal and northern Guinea constitute a segment of a continuous orogenic terrane which borders the western edge of the West African craton. The name, given by Villeneuve (1984), refers to the NNE-SSW Bassaris mountain ridge that runs across eastern Senegal and northern Guinea. We distinguish two tectonic branches. The eastern one, called the Bassaris branch, has a NNE-SSW trend, while the eastern one, called the Koulountou branch, has a NE-SW trend. These two branches were separated by the triangular Youkounkoun Basin (Fig. 1). The Bassaris branch is separated from the West African craton, represented here by the Kedougou inlier and the Madina-Kouta Basin, and by the Faleme and the Komba Basins. All of these geological elements are overlain by the Bové Basin which represents the Paleozoic cover of the Bassaride orogen.

The first description of the geology of this area was given by de Chételat (1935, 1939, 1946) who distinguished basement formations from a sandstone and shale cover sequence containing Silurian and Devonian fossils. In the first attempt at a regional syntheses, Bassot (1966) distinguished between crystalline basement units exposed in the Kedougou inlier from younger metavolcanic and metasedimentary sequences. He assigned the younger sequences of the Late Proterozoic and Cambrian and suggested that they had been deformed during the Caledonian orogeny. Bassot also provided a detailed stratigraphy for foreland sedimentary formations. Following Bassot, only local mapping and/or a prospecting survey by BRGM and Soviet-Guinean field parties were carried out. Villeneuve (1984) provided a recent synthesis of the geology of the Bassarides. This, combined with the geophysical survey of Ponsard (1984) and the geochronological work of Dallmeyer and Villeneuve (1987), forms the foundation for the present review.

[1] Institut de Géodynamique, Sophia-Antipolis I, Av. Albert Einstein, 06560, Valbonne (France)
[2] Département de Géologie et Minéralogie, Université de Clermont Ferrand II, 5, rue Kessler, 63038, Clermont Ferrand Cedex (France)
[3] Département de Géologie, Faculté des Sciences, Université cheikh Anta Diop, Dakar Fann, Sénégal
[4] Department of Geology, University of Georgia, Athens, Georgia, 30602, USA
[5] Laboratoire de Geophysique et Géodynamique, case 422, Faculté des Sciences et Techniques de St. Jerôme, Université d'Aix-Marseille 3, 13397 Marseille Cedex 13, France

2 Lithostratigraphy and Paleontology

The main geological units of the Bassaride orogen are shown on the map (Fig. 1) in an area restricted to southeasten Senegal and northwestern Guinea (Fig. 1 b).

A schematic cross-section (Fig. 2) based on geophysical investigations shows the vertical assemblage of these units. The basement is broadly covered by the volcano-sedimentary rocks of the Bassaris and Koulountou branches, and they are also covered by the Komba and Youkounkoun Basins. The general lithostratigraphic column (Fig. 3) shows the Birimian basement covered by the sedimentary groups of the Madina-Kouta Basin, on top of which we find the rocks of the Bassaride and Koulountou branches. These metasedimentary units are capped by the sedimentary rocks of the Komba and Youkounkoun Basins, which were covered by the clastics sediments of the Bové Basin. This column shows the unconformities between these differents units.

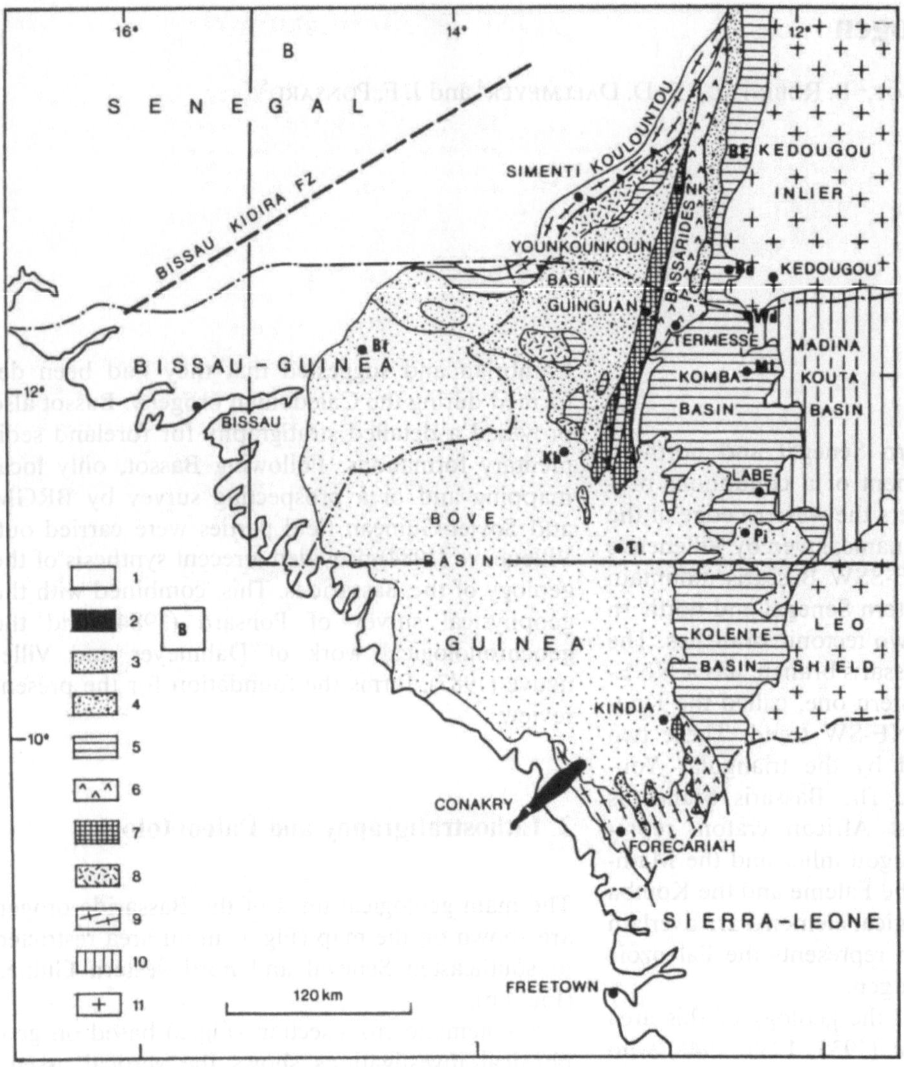

Fig. 1. Geological sketch map of the southwestern part of the West African margin. *1* Meso-Cenozoic basin sediments; *2* Mesozoic plutonic intrusion; *3* Paleozoic sediments of the Bové Basin; *4* Youkounkoun Group; *5* Mali and Batapa Groups; *6* Termesse Group; *7* Guinguan Group; *8* Niokolo-Koba Group; *9* Kouloutou Group; *10* Madina-Kouta Basin sediments; *11* Birimian basement of the Kedougou inlier and Archean basement of the Leo shield; *B* area of study for the Bassaride orogen

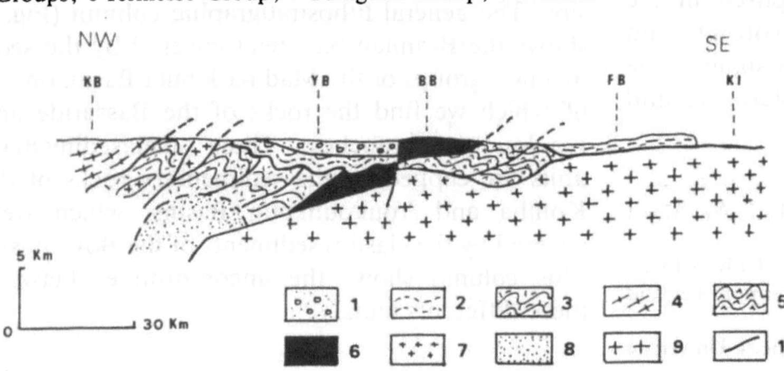

Fig. 2. Interpretative geological cross-section of Bassarides between Simenti and Kédougou in southern Senegal. *1* Youkounkoun Group; *2* Mali Group; *3* Niokolo-Koba Group; *4* Kouloutou Group; *5* Termesse Group; *6* Guiguan Group; *7* Pan-African age granitoids intruding the Niokolo-Koba Group; *8* high density body supposed to be granulitic rock type; *9* West African Craton basement; *10* thrust faults; *KB* Kouloutou branch; *YB* Youkounkoun Basin; *BB* Bassaris branch; *FB* Falémé Basin; *K1* Kedougou inlier

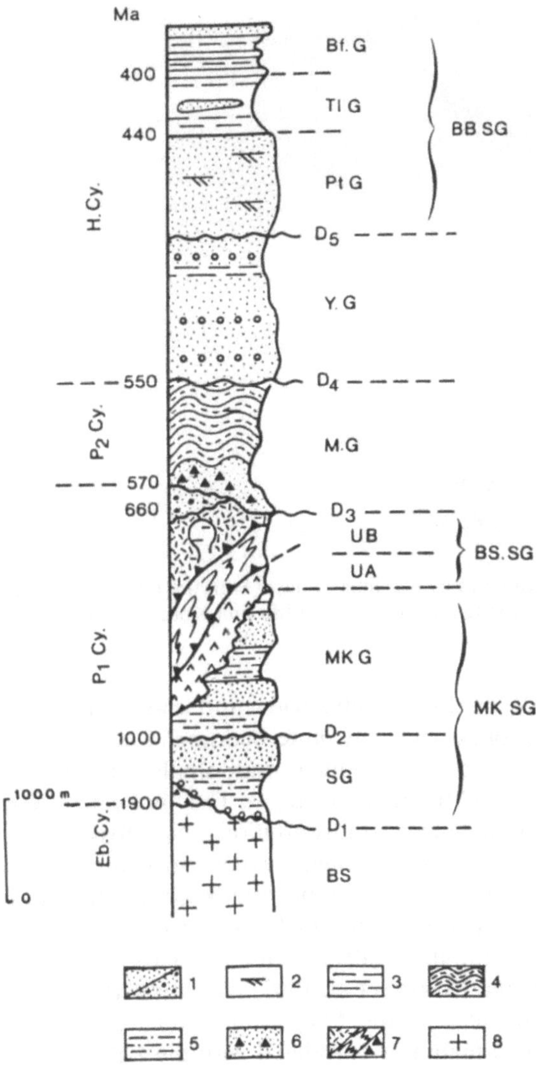

Fig. 3. Schematic lithological sequence of the Bassaris area. *1* Sandstones and conglomerates; *2* cross-bedding structures; *3* argillites of the Bové Basin; *4* argillites and siltstones of the Mali and Batapa Group; *5* siltstones and argillaceous sandstones of the Madina-Kouta Basin; *6* tillite of the Mali and Batapa Groups; *7* volcanoclastic sediments of the Bassaris and Koulountou branches; *8* crystalline basement of the West African Craton; *BfG* Bafata Group; *TIG* Télimélé Group; *PtG* Pita Group; *BBSG* Bové Basin sediments; *YG* Youkounkoun Group; *MG* Mali Group; *UA* Termesse Group; *UB* Guinguan Group; *BsSG* Bassaris and Koulountou branch rocks; *MKG* Madina Kouta Group; *SG* Ségou Group; *MKSG* Madina Kouta sediments; *Bs* basement; *Eb.Cy* Eburnean cycle; *PICy* Pan-African I cycle; *P2Cy* Pan-African 2 cycle; *H.Cy* Hercynian Cycle; *D* unconformities

2.1 The Older Geological Units

They are found only in the foreland of the Bassarides. They consists in a metamorphic and granitic basement overlain by the Upper Proterozoic sediments of the Madina Kouta Basin.

2.1.1 The Kedougou Inlier

In the Kedougou inlier (eastern Senegal), the Birimian volcano-sedimentary rocks have been metamorphosed and intruded by granitic plutons during the 2000-1800 May Eburnean orogeny (Bassot 1966). In these Birimian formations, Bassot (1966) distinguished three differents series with a SW-NE trend, following successively from NW to SE:

1. *The Mako series* with metabasic rocks; metaandesitic breccias, tuffs and greywackes;
2. *The Dialé series* with metabasites, red jaspers, conglomerates, marbles, schists and micaschists;
3. *The Daléma series* with schists, greywackes and conglomerates.

No lithostratigraphic sequence has been proposed for these series which were folded and more or less metamorphosed. The schistosity has generally a SW-NE direction and a strong dipping to the NW. Bassot recognized two different plutonic intrusions in these rocks units. In the western part (Mako serie) calc-alkaline granites were formed between 2250 and 2150 Ma, while in the eastern part (Diale and Dalema series) plutons occur at ca. 1987 Ma.

2.1.2 The Upper Proterozoic Cover

The Upper Proterozoic cover is represented by alternating shale and sandstone layers of the Madina-Kouta Basin which is located between the Kedougou inlier and the Leo shield (Fig. 1). This basin lengthening in an east-west direction seems to be a half graben basin with an important fault on its southern border (Fig. 4). It was studied by Arnould et al. (1960), Bassot (1966), Torchine (1976), Aguedelou and Fall (1979) and more recently by Villeneuve (1984). Most of these authors divided the sediments of the Madina-Kouta basin into a lower Segou Group (500 m thick) and an upper Madina-Kouta Group (700-1000 m), separated by a cartographic unconformity.

The Segou Group (500 m). This is composed of argilites, sandstones with mud cracks and salt traces, limestones and sandy limestones with onco-

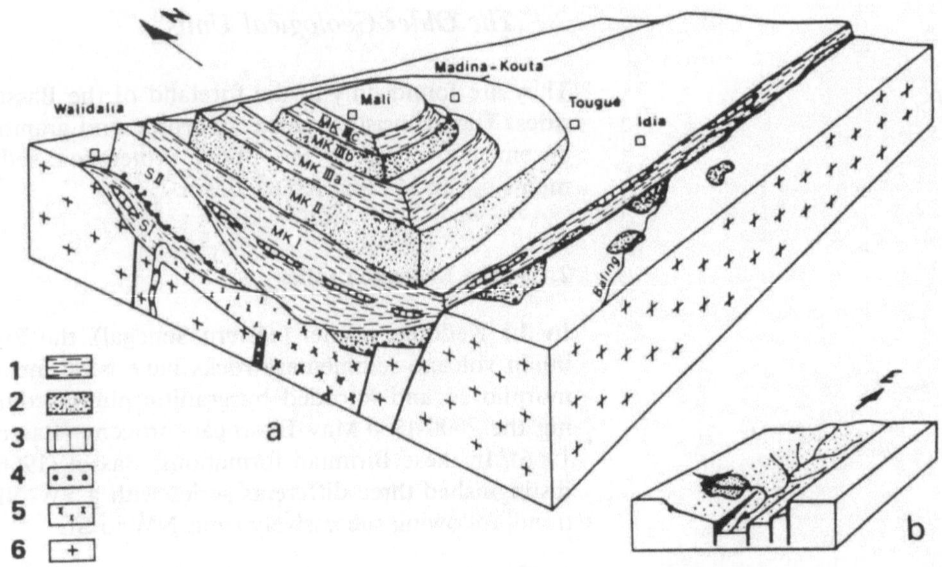

Fig. 4. a. Block diagram of the Madina-Kouta Basin. *1* Argillites; *2* sandstones; *3* limestones; *4* conglomerates; *5* rhyolites as supposed by gravimetric data (Ponsard 1984); *6* basement. **b** Interpretative model of the Madina-Kouta Basin

lites and oolites on the basis (Pellele formation, S I) and sandstones with glauconite on top (Dindefello formation, S II). The depositional environment of the Pellele formation could be a bordering platform or a flooded intracontinental platform with a "tidal flat" type of sedimentation (Villeneuve 1984). The depositional environment of the Dindefello formation seems to be a marine one, as it is underlined by the presence of glauconite.

The Madina-Kouta Group (700 to 800 m) was subdivided in to three different formations, which are, from the upper to the lower part:

The Fongolambi formation (MK I), made of an alternation of red sandstones with ripple marks, red or violet argillites and yellow or green limestones. Arnould et al. (1960) found stromatolites in these limestones. The cartographic unconformity between the Fongolambi formation and the Segou group is understood as a sequel of an important transgression underlined by a yellow limestone level with copper impregnation. The depositional environment seems to be intertidal. Stromatolites point out a quiet environment and mud cracks are consistent with frequent emersions.

The Kanta formation (MK II), which consists of pink or red feldspathic sandstones with glauconite and limonite. These sandstones are characterized by numerous current structures on top of the strata. The Kanta formation shows the same environment as the Dindefello sandstones.

The Dira formation (MK III), which consist of an alternation of sandstones with cross-bedding and ripple marks, argillites with mud cracks and sandy limestones with convolute bedding. Current marks point to an environment with shallow marine deposits and frequent emersion. They indicate a westward sediment transportation. No fossils were found in these sediments except stromatolites and oncolites, which gave no indication of the age of sediments deposition. Datation by Rb/Sr on argillites from the Pellele formation gave an age of 1020 Ma (Bassot et al. 1963). We know that the tillite on base of the Mali group which overlies the Dira formation is correlated with the Late Precambrian tillite of the Taoudeni basin, deposited at about 650 or 620 Ma (Deynoux 1980). The Madina-Kouta basin is considered (Villeneuve 1984) as an intracontinental basin on the northwestern margin of the Leo shield. This basin was probably opened on a sea located on its west side. The transgression observed at the base of the Fongolambi formation had a westward direction.

2.2 The Bassaride and Koulountou Branches

2.2.1 The Bassaride Branch

Trending roughly N-S, two parallel units named from east to west: the Termesse and the Guinguan Groups (lithologic groups) by Villeneuve (1984), consist of volcanic (basalt), volcano-sedimentary (lapilli, tuff, breccia) and sedimentary (sandstone, quartzite, jasper, argillite) rocks. Volcanic material of flows, dykes, sills and agglomerates are closely imbricated with sediments. Figure 5 shows a geological section across the Termesse group near the Senegal/Guinea boundary. This section shows the unconformity between the argillites of the Komba basin (Mali Group) above and volcano-sedimentary rocks of the Termesse Group below. No lithostratigraphic sequence has been determined because of the very intense deformation of this group, also called unit A by Villeneuve (1984). Figure 6 shows the depositional environment of the Termesse group rocks. Figure 7 presents a field section across the Guiguan Group (or unit B). Volcanic rocks are more important than in the Termesse group and the deformation is also stronger. Micaschists occur to the east of Tiaguissi and amphibolites in the western part of this village. Serpentines have been described by Chiron (1964) in the south of Senegal. For the same reasons as above we do not have a lithostratigraphic sequence of this group. We observed a tectonic contact between these Bassaris branch groups (Fig. 5). The submarine character of the volcanic activity has been discussed by Bassot (1966) and confirmed by a petrographic study (Dupont 1983). Pillow lavas have been found as well in the Termesse group (Villeneuve 1984) as in the Guinguan Group (Robineau, pers. commun.). Ages of these rocks are still problematic because they are not in contact in the field with those of the Kedougou inlier or with those of the Madina-Kouta basin. This contact is concealed by the Komba and the Faleme basins.

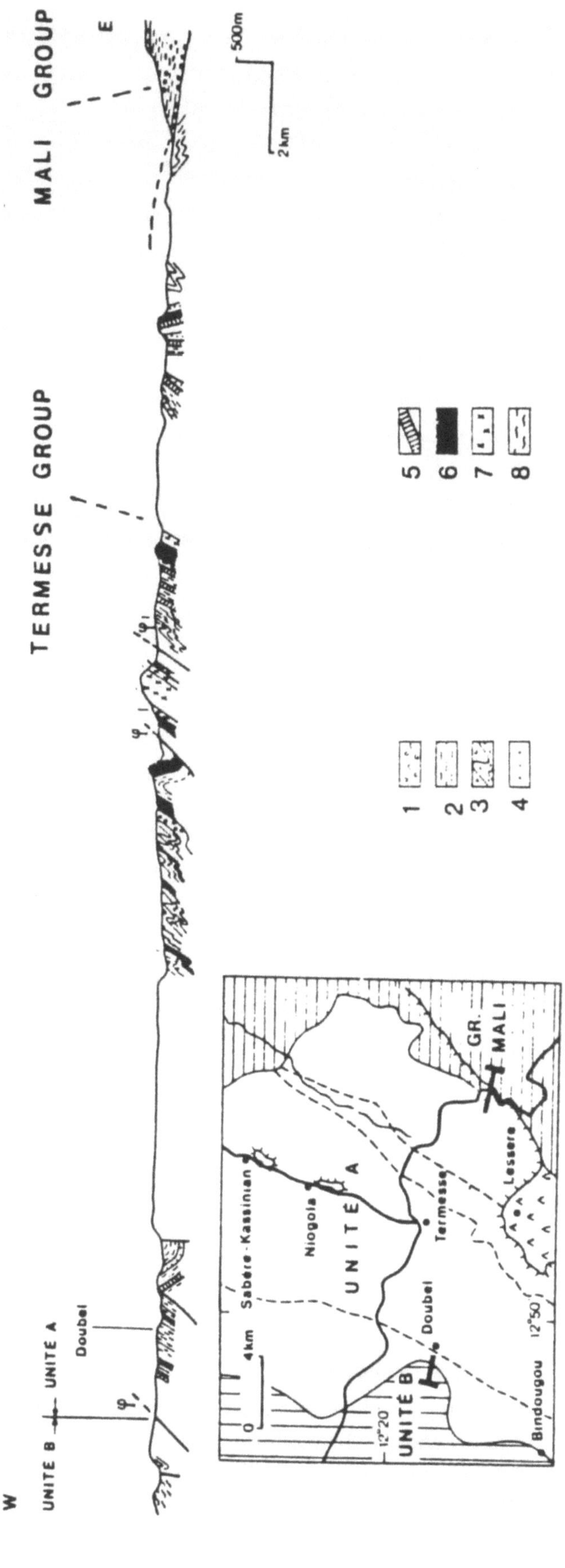

Fig. 5. Geological section across the eastern part of the Bassaris branch (Termesse Group). *1* Conglomerates; *2* argillites; *3* tuffs; *4* sandstones and greywackes; *5* jaspers, silexites; *6* spilites and basalts; *7* volcanic breccias; *8* schists and micaschists

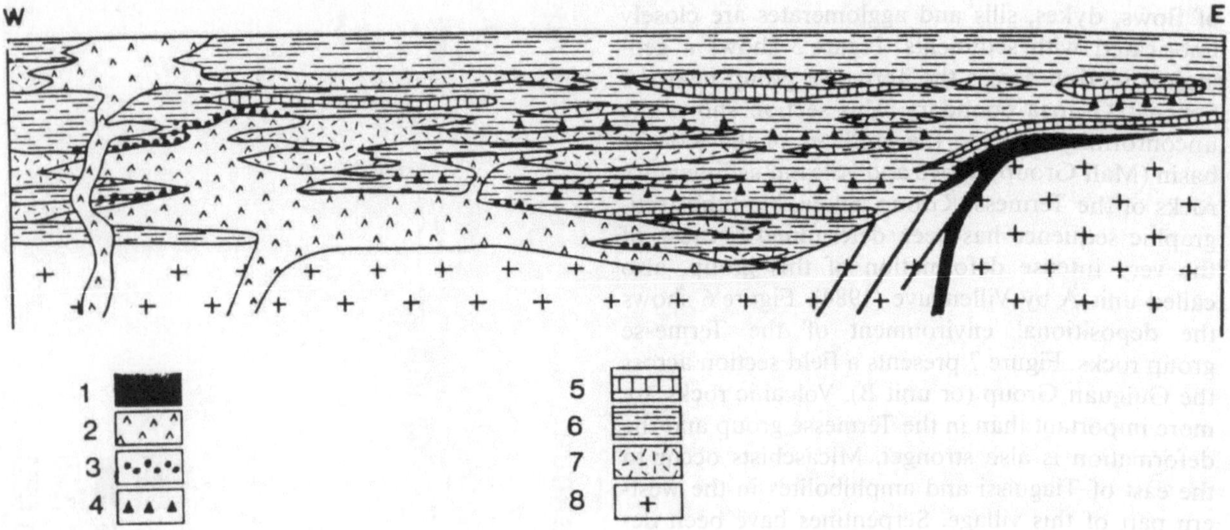

Fig. 6. Sedimentary environment of the Termesse Group. *1* Acid lavas; *2* basaltic lavas; *3* volcanic breccias; *4* conglomerates; *5* limestones and dolomites; *6* argillites and siltstones; *7* cinerites, greywackes and silexites (jaspers); *8* crystalline basement

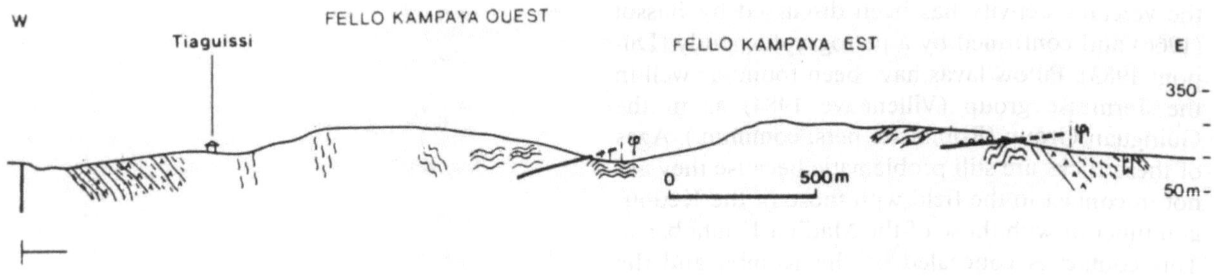

Fig. 7. Geological section across the western Bassaris branch (Guinguan Group). *1* Permo-Triassic dolerites; *2* conglomerates; *3* sandstones and quartzites; *4* micaschists; *5* argillites, schists and jaspers; *6* volcanic tuffs; *7* basalts and spilites

2.2.2 The Koulountou Branch

This branch is limited on its western side by the Meso-Cenozoic Senegalo-Mauritanian basin and on its eastern side by the Proterozoic to Paleozoic Youkounkoun basin. We distinguished (Villeneuve 1984) two lithological groups: the Niokolo-Koba Group to the east and the Koulountou Group to the west (Fig. 8).

The Niokolo Koba Group. Villeneuve (1984) distinguished three main types of rocks: granitoid, basic

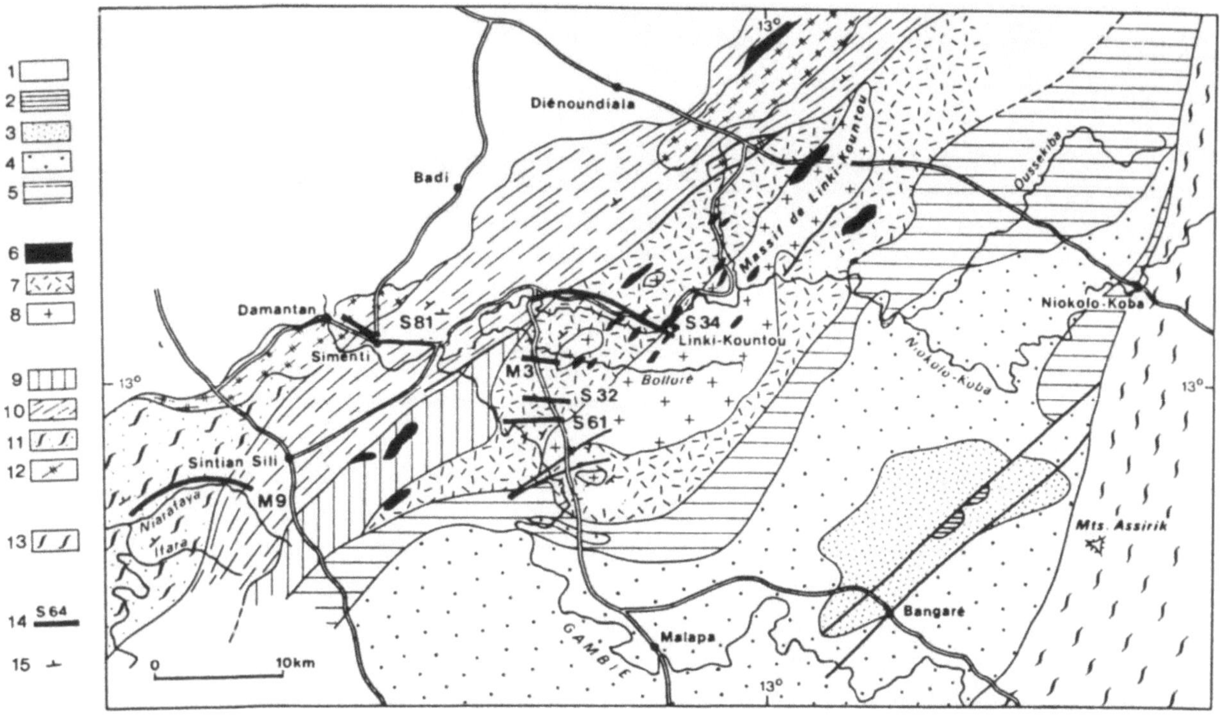

Fig. 8. Sketch map of the Koulountou area. *1* Meso-Cenozoic sediments; *2* Telimele Group (Bové Basin); *3* Pita Group (Bové Basin); *4* Youkounkoun Group (Youkounkoun Basin); *5* Mali and Batapa Groups (Komba and Youkounkoun Basin); *6* aluminous basalts and andesites (Niokolo-Koba Group); *7* eruptive complex of the Niokolo-Koba Group (dacites, rhyodacites, rhyolites and ignimbrites); *8* granites of the Niokolo-Koba Group; *9* Pantié formation (Koulountou Group); *10* Simenti formation (Koulountou Group); *11* Damantan formation (Koulountou Group); *12* orthogneiss of Damantan; *13* Bassaris branch; *14* location of sections in Figs. 9 and 10; *15* structural elements (direction and depth)

and acidic lavas. The granitic rocks constitute a broad batholith, the so-called "Linki-Kountou" massif, and some small outcrops along the Bolloré river. The basic lavas are represented by aluminous basalts and andesites but these rocks are few in number. Acidic lavas represent the greatest part of the volcanic rocks. We found dacites, rhyodacites, rhyolites and ignimbrites associated with tuffs, cinerites, grey, red and green jaspers, breccias and conglomerates. Figure 9 gives some examples of these associations. We see some hectometric to kilometric gentle folds but the many reverse faults do not allow us to set up a valuable lithostratigraphic succession of this group.

The Koulountou Group. Three formations are distinguished in this lithologic group, which also includes the Damantan gneiss:

1. The Pantié formation, consisting of dacites and rhyolites with a very strong slaty cleavage.
2. The Simenti formation, consisting of rhyolitic and rhyodacitic tuffs associated with chloritoschists, muscovitoschists and cataclastic quartzites.
3. The Damantan formation, which includes rhyolitic tuffs and white phylladic rocks. We mapped two orthogneissic massifs: one in the Simenti formation and the other around the Damantan bridge. Cross-sections in Fig. 10 show the great depth of the strata. We were unable to draw a lithostratigraphic succession here.

The contact between these two groups seems to be tectonic according to the observations along cross-sections presented in Fig. 8. No fossils occur in these rocks, but we observed an angular unconformity between the Niokolo Koba group on the basis and the argillites of the Youkounkoun basin on top (Villeneuve 1987 a). Thus, these rocks are more ancient than those of the Youkounkoun basin. In addition, Bassot and Caen-Vachette (1983 a) gave an age of 685 ± 15 Ma for the intrusion of the Niokolo-Koba granite.

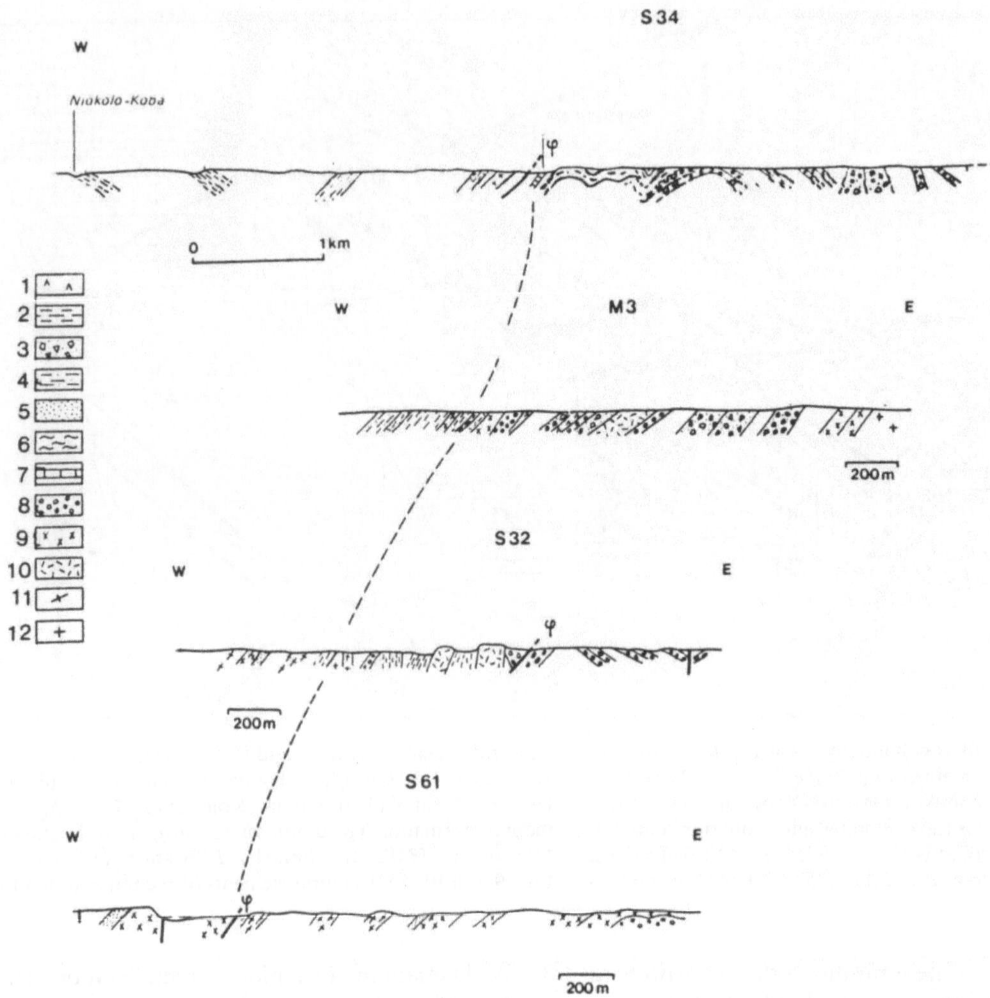

Fig. 9. Geological sections across the eastern part of the Koulountou branch (Niokolo-Koba Group). The *dashed line* corresponds to the limit between the Niokolo-Koba Group (on the *right*) and the Koulountou Group (on the *left*). *1* Permo-Triassic dolerites; *2* acid volcanic tuffs; *3* volcanic breccias; *4* conglomeratic tuffs; *5* sandstones; *6* schists and micaschists; *7* volcanic pouddingues; *8* rhyolites; *9* cinerites and jaspers; *10* gneiss; *11* granites (location of the sections in Fig. 8)

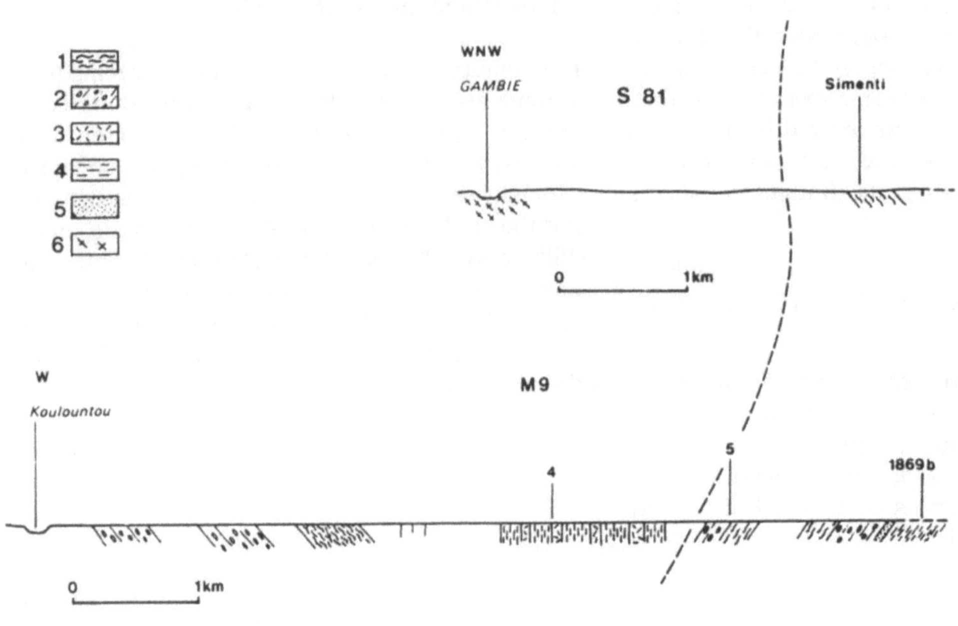

Fig. 10. Geological sections across the western part of the Koulountou branch (Koulountou Group). The *dashed line* corresponds to the limit between the Simenti and the Damantan formations. *1* Schists and micaschists; *2* breccia tuffs; *3* cinerites; *4* tuffs; *5* sandstones; *6* orthogneiss (location of sections in Fig. 8)

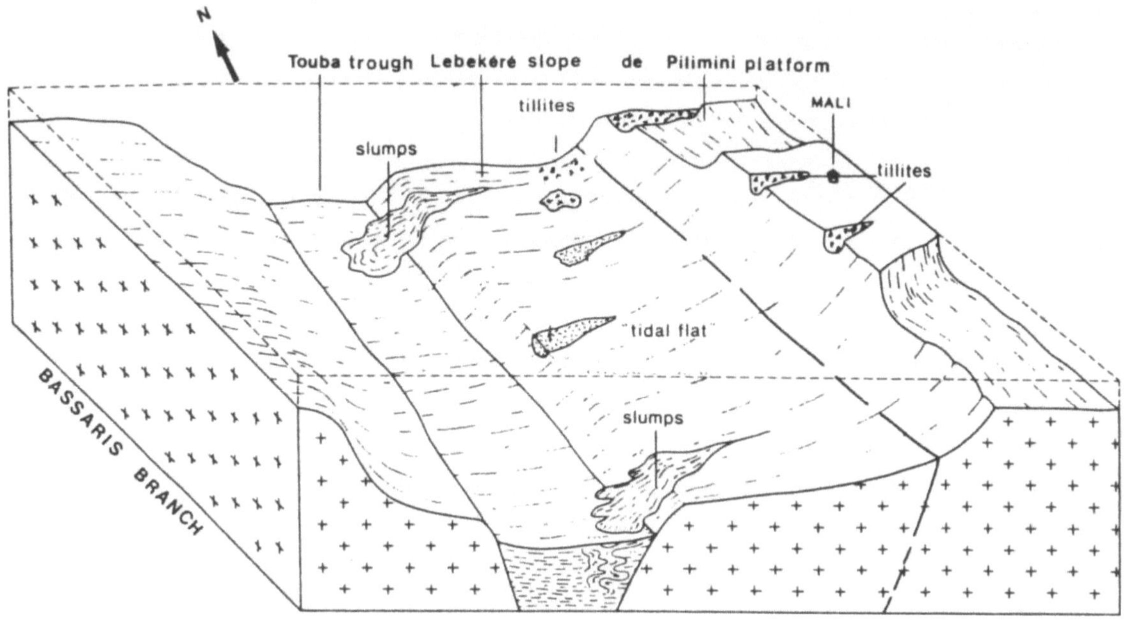

Fig. 11. Block diagram of the Komba Basin

2.3 The Komba, Faleme and Youkounkoun Basins

2.3.1 The Komba Basin

This basin is limited on its west sidy by the Kedougou inlier and the Madina-Kouta basin and on its eastern side by the Bassaris branch. Sediments of this basin are almost horizontal except on the western side. Hence, the angular unconformity between the sediments of the Komba basin and the metamorphic formations of the Birimian basement in the Kedougou inlier is well documented (Bassot 1966; Diallo 1983; Villeneuve 1980, 1984, 1987 a). However, the unconformity between the sediments of the Komba basin and the volcano-sedimentary rocks of the Bassaris branch has been observed in one only place, 10 km east of Termesse (Villeneuve 1984, 1987 a). This unconformity allows us to distinguish the Komba basin and the Bassaris branch. The block diagram of Fig. 11 shows the main structural elements of the Komba basin. From east to west, we distinguish: the "Pilimini platform", the "Lebekere slope" and the "Touba trough".

Figure 12 shows the lithostratigraphic sequences in these differents parts. On the "Pilimini platform", the succession begins with glacial deposits. These sediments are located mainly in paleovalleys of 150 to 200 m depth. The depositional environment of these glacial sediments is considered terrestrial in the Mali area and marine or fjord like the glacial environment in the Senegal area (Villeneuve 1988 a). These glacial deposits are overlain by argillites, siltstones and broad levels of sandstones. On the "Lebekere slope", the lithostratigraphic sequence above a thin level of marine glacial deposits (diamictites and argillites with scarce pebbles) is composed mainly of argillites and thin levels of sandstones. In the middle of the sequence we found 10 to 15 m of limestones and brecciated limestones. In several places we found thin levels of cherts. On the Touba trough, the sediments are composed mainly of green or grey argillites. Between the Lebekere slope and the Touba trough, many slumps capped by turbidites are described (Villeneuve 1984) and interpreted as the activity of a synsedimentary fault.

Locally, the glacial sediments are overlain by dolomites and cherts. But the association of tillites, cherts and dolomites, called the "triad" and considered as a reference level in the Taoudeni basin (Deynoux 1980), has been observed in only one place, at Walidiala near the Senegalese/Guinean boundary (location in Fig. 1). Except for some terrestrial glacial sediments on the Pilimini platform, all of these sediments were considered as marine deposits. After Villeneuve (1984), all of these sediments, including the glacial deposits on the basis, constitute the Mali Group.

No fossils have been found in sediments of the Komba basin except in the dolomites and argillites which capped the tillite at Walidiala. These shelly

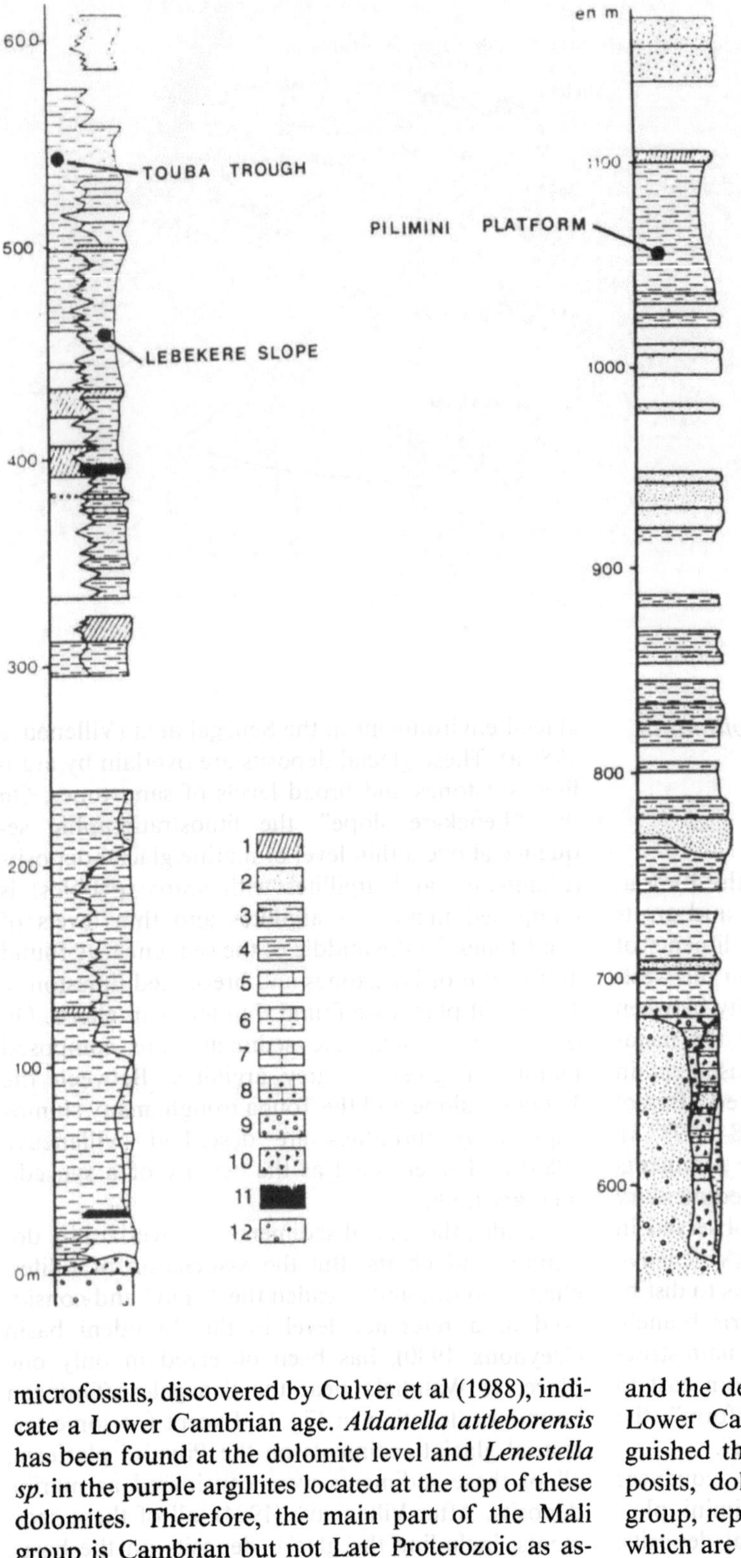

Fig. 12. Lithostratigraphic sequences on the Pilimini platform, the Lebekere slope and the Touba trough. *1* Hornfels from Permo-Triassic dolerites; *2* sandstones; *3* argillaceous sandstones; *4* siltstones; *5* limestones; *6* argillaceous limestones; *7* calcareous sandstones; *8* breccia limestones; *9* conglomerates; *10* glacial deposits (tillites); *11* cherts; *12* Madina-Kouta Basin sediments (basement of the Komba Basin)

microfossils, discovered by Culver et al (1988), indicate a Lower Cambrian age. *Aldanella attleborensis* has been found at the dolomite level and *Lenestella sp.* in the purple argillites located at the top of these dolomites. Therefore, the main part of the Mali group is Cambrian but not Late Proterozoic as assumed by several authors. Culver et al. (1988) supposed a hiatus of sedimentation between the deposition of glacial levels assumed to be Late Proterozoic in many places (Deynoux 1980; Affaton 1987), and the deposition of dolomites where they found Lower Cambrian shelly microfossils. They distinguished the Walidiala group, including glacial deposits, dolomites and cherts as well as the Mali group, represented by the argillites and sandstones which are on top of cherts. We agree with the existence of a hiatus but it seems to be too early to discuss stratigraphical correlations with the other basins of West Africa. In fact, we do not know exactly the time of deposition of the tillites in West Africa.

In the southern part of the Taoudeni basin, several levels of glacial deposits have been evidenced by Marchand et al. (1987) in the Bakoye group. Villeneuve (1988 c) suggests a correlation between the Bakoye group of the southern Taoudeni basin and part of the glacial deposits of the Komba basin. However, Culver et al. (1988) assume that the "triad", used till now as a reference level for a distant correlation, is not a good stratigraphical reference level. In fact, we see in this chapter that cherts and dolomites or limestones can occur at several levels in the lithostratigraphic sequence and their association could be an effect of local depositional conditions. For these reasons, we consider all rocks infilling the Komba basin to be one group only, called the Mali group. More investigations are needed to obtain a good stratigraphical view of this basin.

2.1.2 The Faleme Basin

The Faleme basin is the northward extension of the Komba basin. It constitutes a north-south trough between the Kedougou inlier and the Bassaris branch. In Fig. 1, we see two distinct lithologic units. The east unit is mainly argillaceous and the west unit is composed of sandstones and conglomerates. The cross-section shown in Fig. 13 presents the shape of the unconformity between the Birimian basement of the Kedougou inlier and the horizontal sedimentary cover of the Faleme basin on the eastern side of this basin. On the western side, the sediments are folded and the contact with the Bassaris branch is probably tectonic but it has never been observed. In the northern part of the Faleme basin, sediments are folded very close to the Kedougou inlier. Structurally, we distinguish the Bandafassi platform and the Namel trough. The Namel trough is limited by important faults, and the lithostratigraphic sequence is different from the Bandafassi platform to the Namel trough. On the Bandafassi platform, basal diamictite could be thick and associated with argillites en bloc and arkosic conglomerates, but on the Namel trough, glacial sediments are associated with limestones, phosphatic breccias and cherts. The upper part of the sequence consists mainly of argillites and siltstones. In reference to the Komba basin, we grouped these sediments in the Mali Group. The sandstones and conglomerates located on the western part of the Faleme basin could not be correlated with the Mali

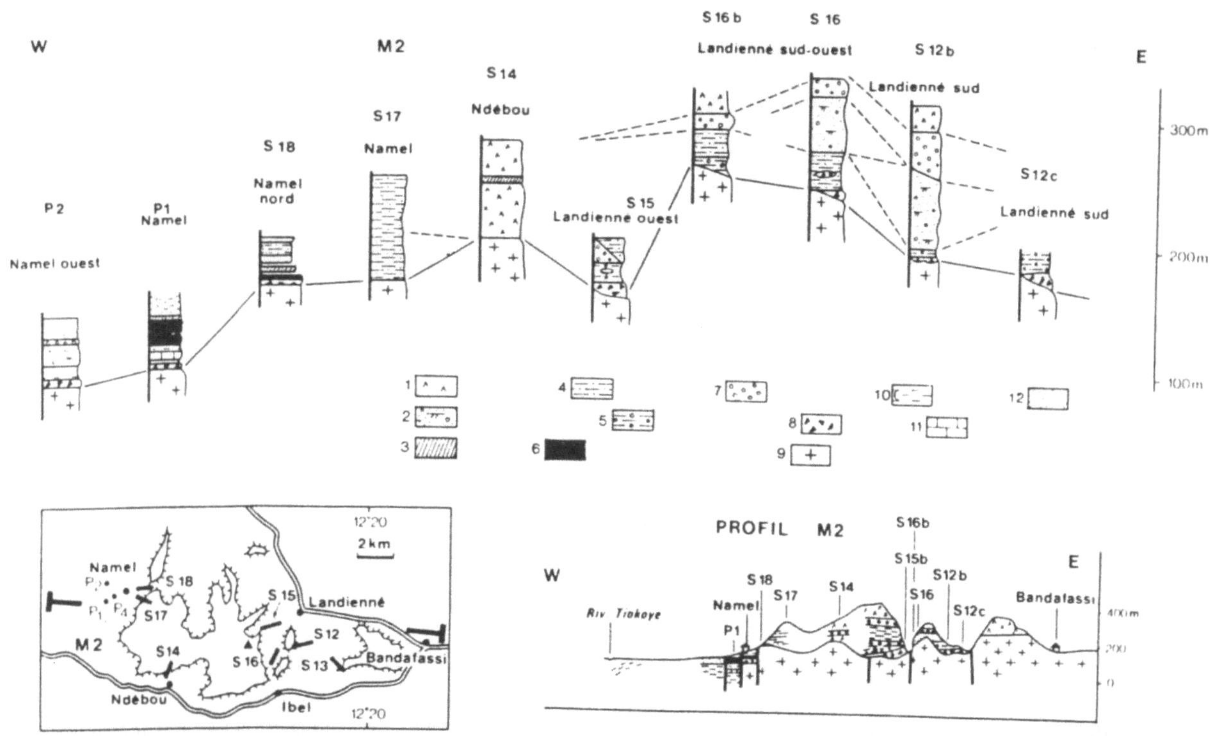

Fig. 13. Geological sections on the eastern part of the Feleme Basin. Bandafassi platform is on the *right* and Namel trough is on the *left*. *1* Permo-Triassic dolerites; *2* arkosic conglomerates; *3* cherts; *4* siltstones; *5* "argiles en bloc"; *6* phosphatic breccias; *7* conglomerates; *8* glacial sediments; *9* Birimian basement; *10* argillites; *11* limestones; *12* sandstones

Group due to a lack of observations between these two outcrops. The significance of these sandstones is still enigmatic. Bassot (1966) connected them to the post-Mali Group sediments.

2.1.3 The Youkounkoun Basin

The Youkounkoun basin, which occurs between the Bassaris branch, the Koulountou branch and the Bové basin, corresponds to an assemblage of two basins separated by the Kayan and Tominé inliers (Fig.14). Each of them may reach 3000 m in depth. Villeneuve (1984) distinguished two lithostratigraphic units separated by an angular unconformity.

The lower one is composed of conglomerates considered as mixtites and overlain by argillites, siltstones and limestones in thin levels (Fig. 15). These sediments, which occur also in the Bassaris branch (Figs. 14, 15), constitute the Batapa or Batapon Group (Torchine 1976). Villeneuve (1984) correlated it to the Mali Group of the Komba basin. Gazda and Kowalsky (1986) wrote that "the performed mineralogical and petrological studies indicate that the claystones and wackes of this group were formed in a shelf environment". This conclusion is supported by the marine character of the mixtites in the Kayan area. The thickness of this group is probably more than 1500 m. The upper part is composed of conglomerates and feldspathic sandstones.

Conglomerates present a graded bedding with quartz pebbles in the base, but on top of the basin, they present red argillite pebbles. All these rocks are grouped in the Youkounkoun Group. We have no information regarding the depth of this group but geophysical investigations assume a depth of 1000 to 1500 m. No complete lithostratigraphic succession has been done. Gazda and Kowalsky (1985 b) assume that "the clastic rocks of the Youkounkoun Group are diagenetic sediments of foothill alluvial cones, alluvial plains and lake beach forms." After them, "sequences of this type of sediments are characteristic of postorogenic forms". The large number of very poorly rounded frag-

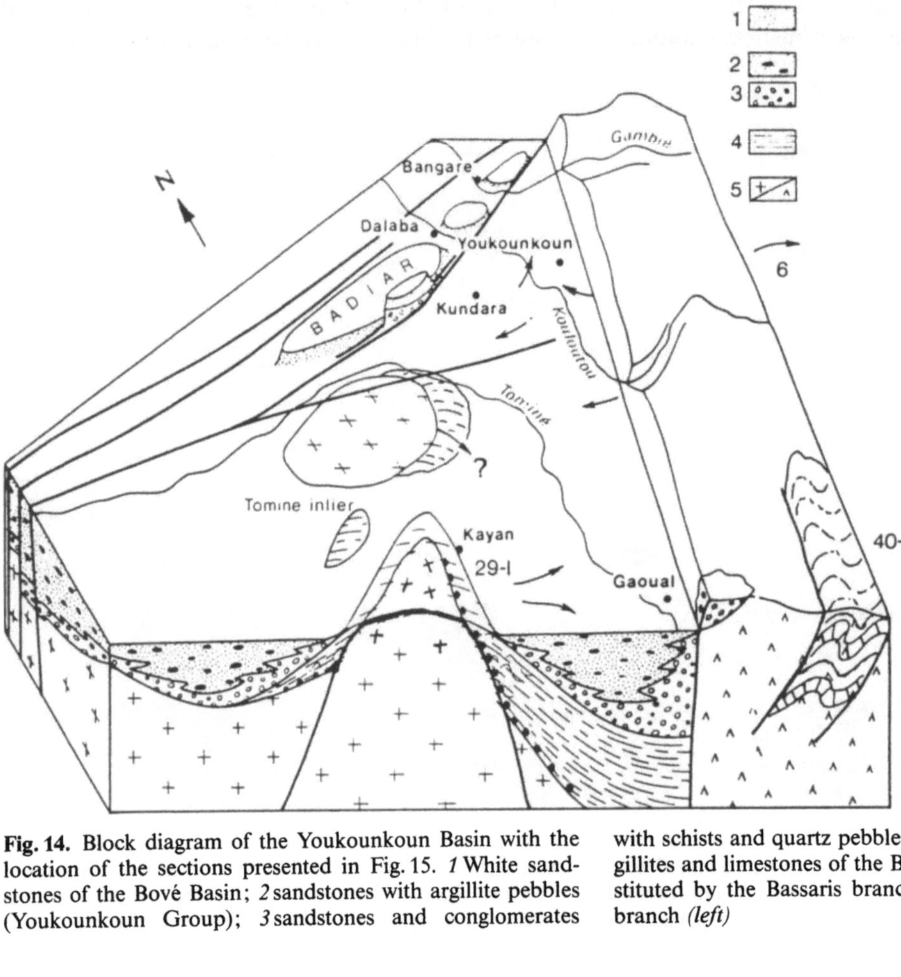

Fig. 14. Block diagram of the Youkounkoun Basin with the location of the sections presented in Fig. 15. *1* White sandstones of the Bové Basin; *2* sandstones with argillite pebbles (Youkounkoun Group); *3* sandstones and conglomerates with schists and quartz pebbles (Youkounkoun Group); *4* argillites and limestones of the Batapa Group; *5* basement constituted by the Bassaris branch *(right)* and the Koulountou branch *(left)*

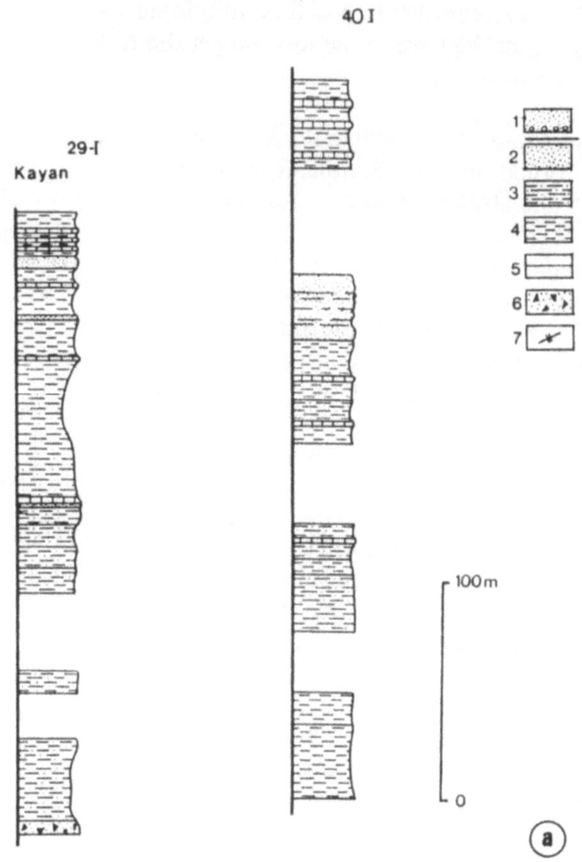

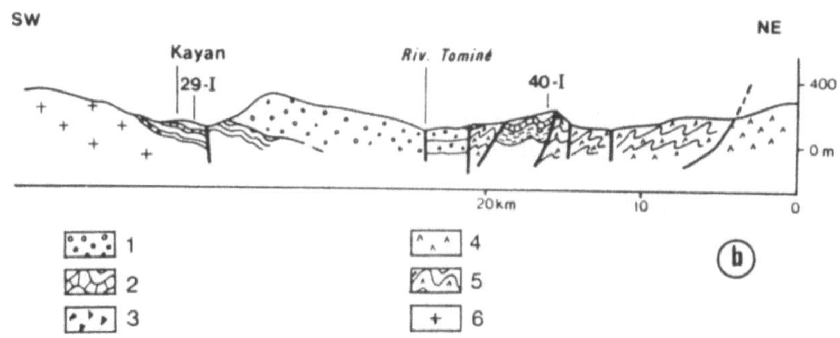

Fig. 15 **a** Examples of lithologic sequences in the Batapa Group of the Youkounkoun Basin. *1* Conglomerates; *2* sandstones; *3* siltstones; *4* argillites; *5* limestones; *6* mixtites; *7* Pan-African basement. **b** Geological section across the southern part of the Youkounkoun Basin. *1* Youkounkoun Group; *2* Batapa Group; *3* mixtites of the Batapa Group; *4* Termesse Group (Bassaris branch); *6* Guinguan Group (Bassaris branch); *7* granites of the Kouloúntou branch (location of sections in Fig. 14)

ments of schists suggests the proximity of the source area. In fact, the direction of transportation deduced from the paleocurrents shows a centripet alimentation of the differents subbasins (Fig. 14). No fossils have been found in these two groups but we consider that the basal mixtite is an extension of the glacial deposits which occur on the basis of the Mali Group and, consequently, the subsequent sediments of the Batapa Group could be considered as a westward equivalent of the Mali Group. Hence, the Youkounkoun Group, covered by the Paleozoic sediments of the Bové basin, could be Cambrian or Ordovician.

2.4 The Bové Basin

The Bové basin, which overlies all the geological units described above, covers the greatest part of western Guinea and central Bissau-Guinea (Fig. 1). Some residual hills are found in the Youkounkoun basin (see Fig. 14). The map in Fig. 1 shows the cartographic unconformity between this basin and the Youkounkoun basin, the Bassaris branch and the Komba basin. This unconformity is confirmed by an erosional unconformity observed on the southern edge of the Badiar hill. A subdivision into three groups is proposed (Roman'ko 1974; Villeneuve and Da Rocha Araujo 1984), based on lithologic associations. From the base to the top, we get the following succession:

1. The Pita Group, 250 to 600 m thick, is subdivided into two formations: the Kindia formation at the base and the Mount Gangan formation above (Fig. 16).

The Kindia Formation consists mainly of white conglomeratic sandstones which constitute the high cliffs of the Kindia area. These sandstones form a flag, thicker to the south (600 m) than to the north (250 to 400 m) of the Bové basin. No argillaceous sediment has been found, except in Guinea Bissau. In the northern part of the Basin, the research of Gazda and Kowalski (1985 a) indicates an environment where fluvial, eolian and marine shelf (beach) conditions are melting (presence of glauconite in rare places). A study of current directions indicates a general transport towards the NW (Villeneuve 1983). On the whole of the basin, the accordance of the transport way shown by an unimodal grouping on the directionnal diagram, is characteristic of a "braided fluviatile environment", according to Potter (1963) and Allen (1968). Thus, we consider that this sandstone corresponds to an alluvial plain that opened northwestwards on an epicontinental marine area borne out by scolithus in the sandstones of Mauritania. The upstream was located in the south while the downstream of the alluvial plain was situated in the north. The disappearance of brittle minerals is indeed progressive from south to north of the basin. This allows us to suppose a permanent recycling of detrital minerals in this direction.

The Mount Gangan Formation (0–80 m) was observed only in the Mount Gangan cliff, near Kindia, and it forms a lenticular body which seems to fill a paleovalley 150 to 200 m wide. It consists of sandstones with isolated, angular quartz pebbles and argillites with isolated quartz clasts looking like a diamictite. A glacial origin for these sediments was suspected (Villeneuve 1984), but not demonstrated.

No fossils occur in these rocks but the Pita group is more ancient than the first graptolite levels of the following Télimélé group, which indicates a Llandoverian age. Comparisons with the fossiliferous formations of the lower part of the "Suwanee basin" in the Florida subsurface, led us to propose an Ordovician age for the Pita group (Villeneuve 1987 b). If the Mount Gangan formation has a real glacial origin, a correlation with the upper Ordovician tillite of the Taoudeni basin (Deynoux 1980) could be envisaged.

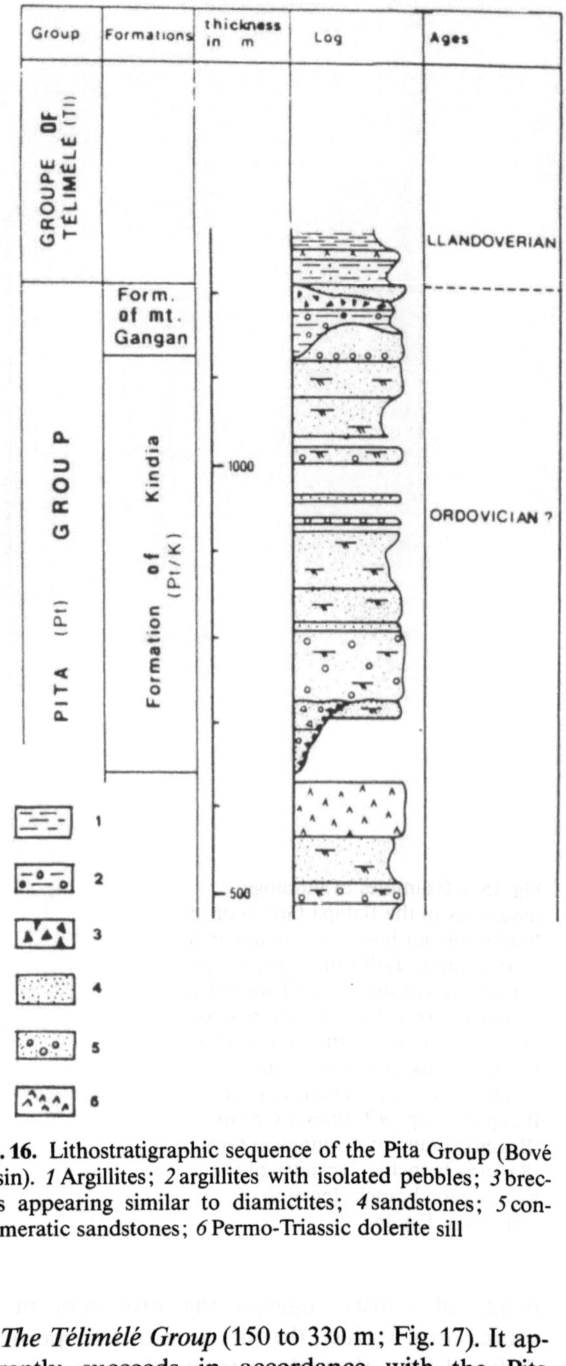

Fig. 16. Lithostratigraphic sequence of the Pita Group (Bové Basin). *1* Argillites; *2* argillites with isolated pebbles; *3* breccias appearing similar to diamictites; *4* sandstones; *5* conglomeratic sandstones; *6* Permo-Triassic dolerite sill

2. The Télimélé Group (150 to 330 m; Fig. 17). It apparently succeeds in accordance with the Pita group, but in places we observed a pediplain remnant with ferruginous conglomerates. We distinguish two formations from the lower to the upper part:

The Lower Formation. It is composed mainly of argillites and siltstones including, however, several levels of green and pyritic sandstones. Fifteen fossiliferous layers with graptolites and microfossils have been recovered. They indicate a Middle and Upper Llandoverian age, with the exception of one

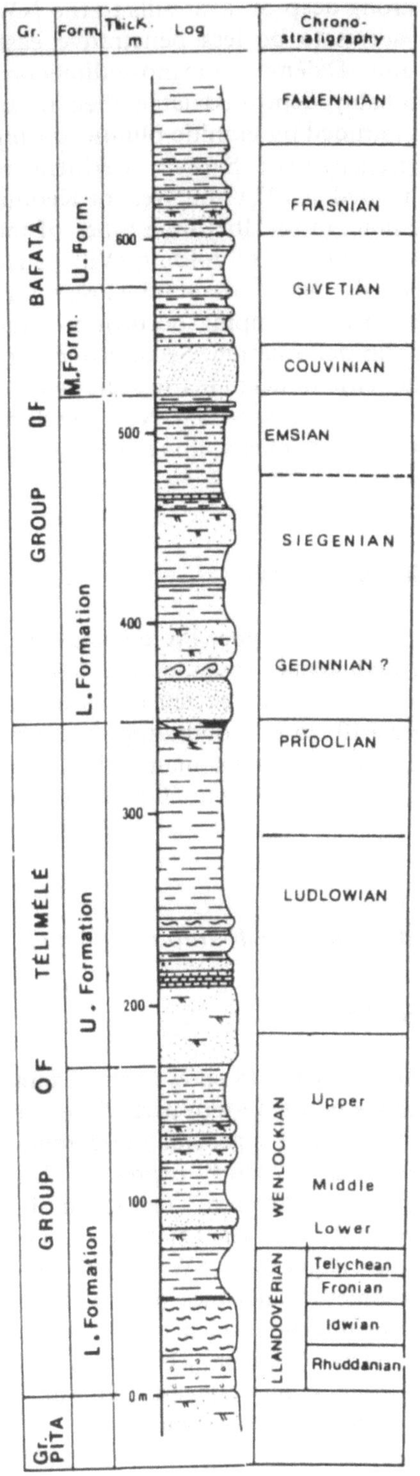

Fig. 17. Lithostratigraphic sequence of the Télimélé and Bafata Groups (Bové Basin). Same lithologic legend as in Fig. 16

which indicates a Lower Wenlockian age. A listing of these fossils is given by Villeneuve (1984). The global sequential analysis of this formation could not be done continuously. However, Villeneuve (1984), according to the current marks (wave rippling, sandstone mounds and planolites traces) assumed a shallow water environment. The marine environment is demonstrated by the presence of graptolites.

The Upper formation begins with a sandstone level 20 to 50 m thick, on top of which we found brachipods of Lower Ludlovian (Villeneuve et al. in press). The upper part of the succession is composed of black and grey shales with graptolites and microfossils (chitinozoans). These fossils indicate a Ludlovian to Pridolian age. We note, in one fossil layer, the presence of *Monograptus Yukonensis* (Jackson & Lenz), *Monograptus telleri* (Lenz) and *Monograptus thomasi* (Jaeger) associated with bivalves, acritarch and chitinozoans of Ludlovian or Early Gedinian age (Baudelot et al. 1980). These fossils occur generally in the Praguian (Gedinnian-Siegenian) of the Yukon region in Canada (Jackson et al. 1978). We do not known how to solve this stratigraphic problem. The presence of many fossil traces of worms and the association of fauna including graptolites, bivalves such as "Cardiola", and arthropod fossil traces, indicates a "shallow marine environment" (Baldwin 1977; Reading 1980).

3. *The Bafata Group* (150–430 m). It is subdivided (Fig. 17) into three formations, from the lower to the upper part:

The Lower formation is composed, in its lower parts, of two levels of sandstones separated by an argillaceous and silty level. The lower sandy level includes some brachiopods of Lower Lochkovian (Villeneuve and Rachebeuf, in press) previously considered as Pridolian (Drot et al. 1978) while the upper sandy level contains some brachiopods of Siegenian (Drot and Villeneuve 1985). In the upper part of this formation we can see grey siltstone levels with brachiopods from Upper Siegenian to Lower Emsian (Roman'ko 1974; Villeneuve 1984). In the sandstone levels we note a disturbed lamination and micro-slumps alternating with "load" or "ball and pillow" structures, which indicate a shallow marine environment. This environment is confirmed by the presence of worm fossil traces and scolithus associated with ferruginous oolitic beds and lumachelles of brachiopods. The presence in sandstone body of channels, ripple marks and cross-bedding indicates a "tidal flat area".

The middle formation begins with a thick yellow sandstone level with brachiopods, indicating a Couvinian (Eifelian) age and it is topped by pink siltstones including brachiopods of Givetian (Villeneuve 1984). The depositional environment is very similar to that of the lower formation.

The Upper formation is composed of argillites and siltstones with several levels of sandstones (2-6 m wide). Bechennec (1980) found some brachiopods that indicate Frasnian and Famennian age. The aquatic environment is proved by brachiopod and fossil traces like worm path traces (neretic zone). Again the environment seems to be shallow marine.

Gazda and Kowalski (1985 c), who studied the granulometry of part of the Bafata Group to the north of the Bové basin, supposed a shelf marine environment or a continental slope of an open marine basin. This basin probably opened on the "Theic Ocean" which separated. North America and the Gondwana during the Paleozoic.

3 Structure

The interpretative geological section of the northern Bassarides (Fig. 2) takes into account gravimetric and magnetotelluric interpretations discussed later. This section shows the spacial arrangement of the various geological units already described. The assumed thrusting of belt units on the West African Craton is possible, but this thrusting is concealed by the Komba and Faleme basins. We note, however, an unconformity between the Mali and Batapa Groups of the Komba, Feleme and Youkounkoun basins and the volcano-sedimentary groups of the Bassaris and Koulountou branches.

3.1 Foreland

3.1.1 Kedougou Inlier

Except for some data available in eastern Senegal (Bassot 1966; Bassot and Caen-Vachette 1983 b; Ndiaye et al. in press) structures of the West African Craton are not well known. Birimian formations of the Kedougou inlier have a polycyclic evolution (Bassot 1966). The western Mako Group, characterized by a dominant basic volcanism associated with calc-alkaline plutonism, was emplaced between 2250 and 2150 Ma. This group is folded and we note a strong deep of lavas sills to the NW. The schistosity seems to be less penetrative eastward. Diallé and Daléma volcano-sedimentary groups were probably deposited later, then metamorphosed and intruded by granitic plutons during the Eburnean orogeny. The Saraya synkinematic granite has been dated at 1970 Ma. West of Kedougou (Fig. 1), we note, in the Ibel limestones of the Diale Group, three phases of folding (Villeneuve 1984). First folds do not have a schistosity and could be related to the slumping occurring in the early stage of basin deformation. Second folds are cylindrical (centimetric to hectometric scale) with a vertical flow schistosity directed SW-NE. The third fold proceeds from movements along the SW-NE faults.

3.1.2 The Madina-Kouta Basin

In the foreland, the craton Upper Proterozoic cover is undeformed. Sedimentary beds of the Madina Kouta basin are generally flat except on its northern margin where a 10° to 20° south dipping can be noted. However, vertical faults mainly in SSW-NNE direction, can produce 10 to 50 m of vertical movement.

3.2 The Bassaris and Koulountou Branches

3.2.1 The Bassaris Branch

The Termesse Group. Figure 18 shows some examples of deformations in the western part of the Termesse Group. The bedding is generally discontinuous and the schistosity is poorly expressed. This schistosity (S 1) is of a "fracture" or "strain slip cleavage" type (crenulation). The axial direction of these isoclinal folds is generally SW-NE. We note decametric to kilometric anticlines and synclines with a SW-NE axial direction which are related to a second folding phase (without schistosity). The thrusts towards the east or southeast and with a west or northwest dipping are very numerous.

The Guingan Group. It suffered two deformations (Fig. 19). The first one (D 1) is ductile and responsible for the tight isoclinal folding with a 10° to 40° N dipping axial plane cleavage (slaty cleavage). A mylonitic foliation is developed frequently along narrow parallel zones. A stretching lineation appears sometimes and dips 5° to 40° to the north in

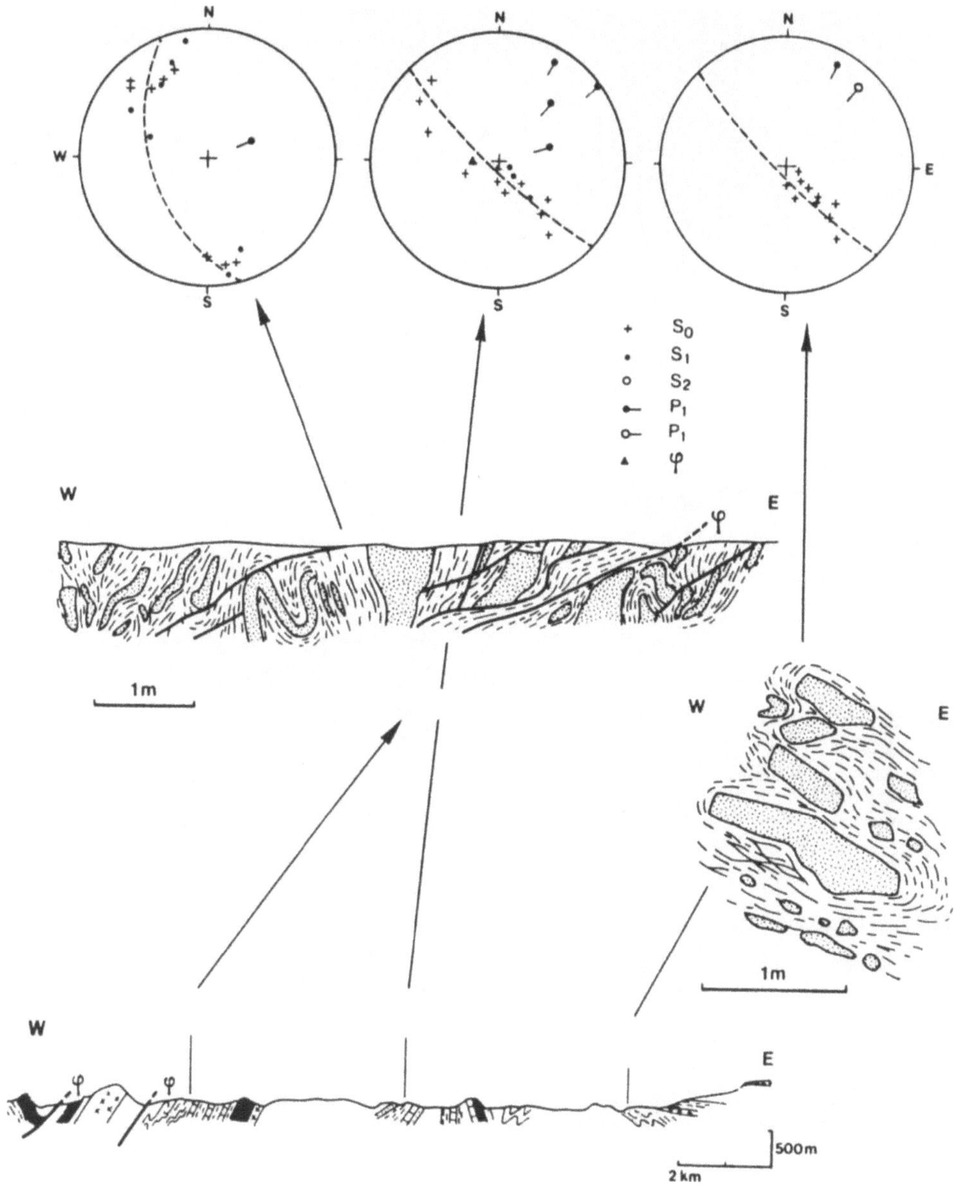

Fig. 18. Some examples of deformations in the Termesse Group. *S0* Bedding; *S1* schistosity 1; *S2* schistosity 2; *P1* axis of F1 folds; *P2* axis of F2 folds

the cleavage plane (Fig. 20). The second deformation (D 2) is characterized by an open folding (F 2) and associated crenulation, with NNE-SSW horizontal axes and subvertical axial planes. These two deformations are coaxial (see stereographic diagrams, Fig. 19). A fracture cleavage can appear locally, as well as conjugated kink bands. NNE-SSW vertical strike-slip faults and low dipping minor thrusts implicate a late fracturation phase indicating a NW-SE compressive tensional constraint (Fig. 21).

3.2.2 The Kouluntou Branch

The Niokolo-Koba Group. Depending on their origin, Niokolo-Koba Group materials present different tectonic features. The geological sections (Fig. 9) show a gently folding (NE-SW axis) for the calc-alkaline volcanics and volcano-sedimentary rocks. These open folds are hectometric to kilometric and their limbs have only a dip of 20° to 40°. The shape is often conical. Cleavage is usually of a fracture type. Meanwhile, a more penetrative foliation (strain-slip cleavage or crenulation cleavage) has been noted locally along shear zones 10 to 50 m wide. These shear zones have a SW-NE direction

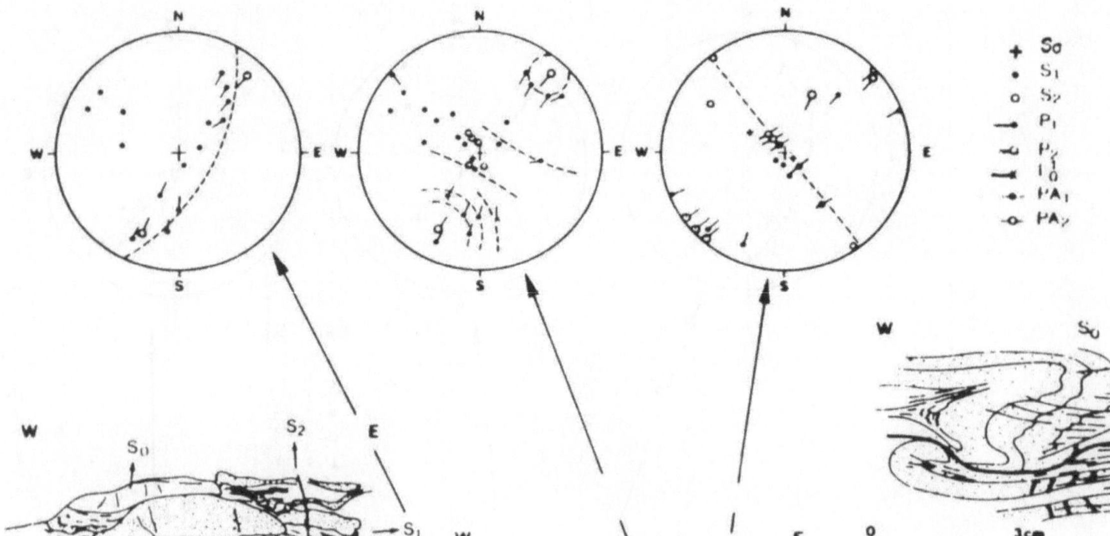

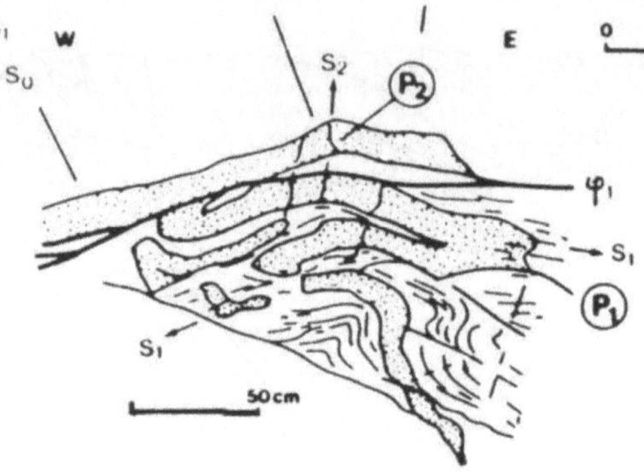

Fig. 19. Some examples of Pan-African deformations in the Guinguan group. *S0* Bedding; *S1* foliation; *S2* schistosity related to the P2 folds; *P1* axis of F1 folds; *P2* axis of F2 folds; *PA1* axial plan of F1 folds; *PA2* axial plan of F2 folds

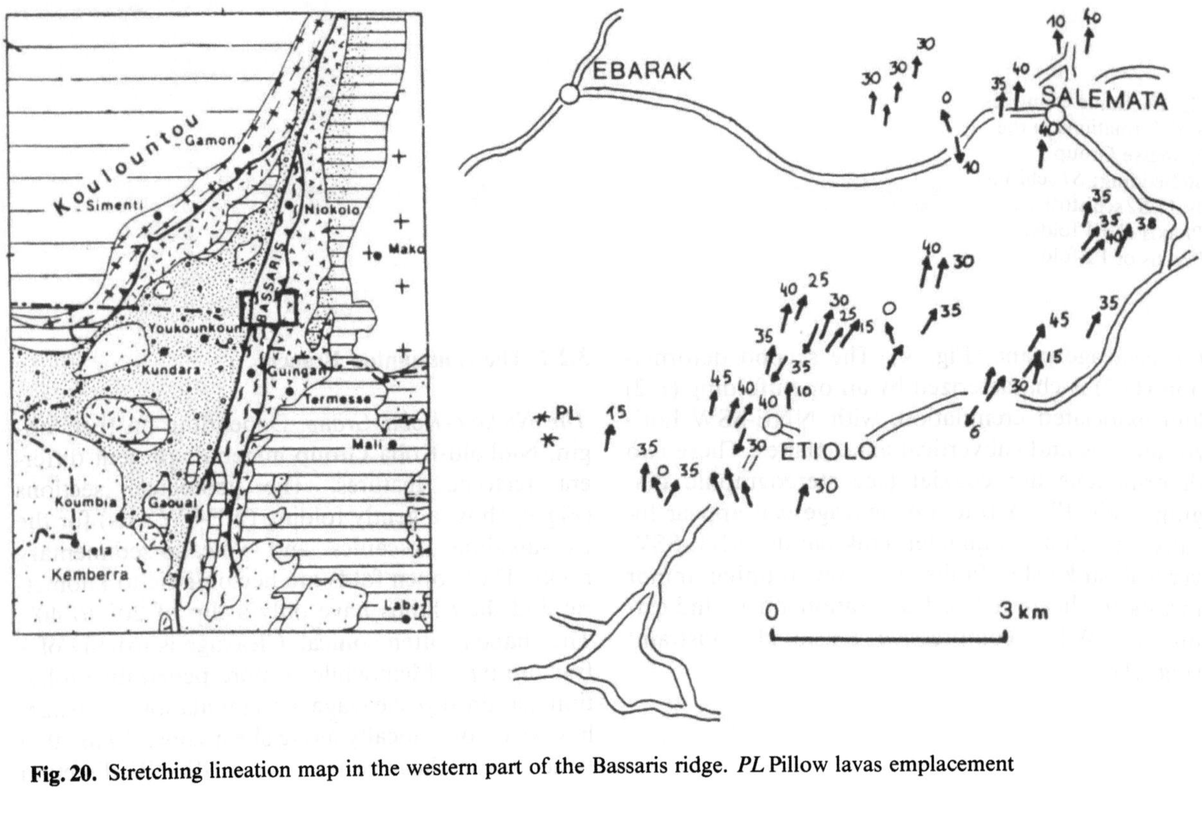

Fig. 20. Stretching lineation map in the western part of the Bassaris ridge. *PL* Pillow lavas emplacement

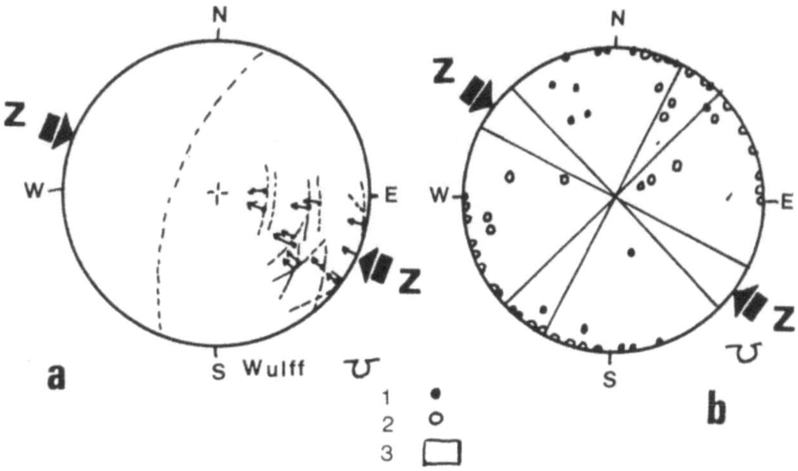

Fig. 21. Wulf diagrams showing the shortening direction **a** with measures along the reverse faults located in Fig. 5; **b** with measures on faults of the whole area. *1* Sterographic symbols for poles of dextral strike-slip faults; *2* stereographic symbols for poles of sinistral strike-slip faults

and a dip of 50° to 70° to the NW. Synkinematic muscovites recorded a 550 Ma thermal event (Dallmeyer and Villeneuve 1987). As in units of the Bassaris branch, fractures parallel to the main structural trend (NNE-SSW) are usually west-dipping, reverse faults. The Linki-Kountou granite is slightly deformed.

The Koulountou Group. The geological sections (Fig. 10) show a vertical or a south-eastward dipping. But we did not observe folds on the field. However, rocks show a penetrative cleavage mainly in the tuffs. Some rocks present two penetrative cleavages, while others present only one cleavage. In the first case, the first cleavage (S 1) is of a "flow" type with synkinematic muscovites and the second cleavage (S 2) is of a "strain-slip" type with synkinematic sericites. These two cleavages have a parallel direction (NE-SW) with a strong dipping to the NW. In this sections, we observed isoclinal and recumbent microfolds cut by the S 2 cleavage. Actually, we cannot correlate the cleavages of the Koulountou group with that of the Niokolo-Koba Group. The Damantan and Dienoundiala orthogneisses (Fig. 8) present a NE-SW mineral lineation parallel to a tight SE dipping of 35° cleavage.

3.3 The Komba, Faleme and Youkounkoun Basins

3.3.1 The Komba Basin

In the main part of the Komba basin, sediment strata are very flat except for the metric slumps along the Touba trough. However, at the western rim, close to the Bassaris branch, we observed some metric gentle open folds with a N-S axial direction. The axial cleavage of the "fracture type" is almost vertical. In some places, we observed metric horizontal fault thrusts.

3.3.2 The Faleme Basin

At the eastern rim, the sedimentary strata are horizontal on the Bandasassi platform and on the eastern edge of the Namel trough (Fig. 13). But, 15 km north of Namel, the sedimentary beds present a dip of 15° to 25° towards the NNW or SSW. No significant cleavage has been observed associated with these strata deformations. To the west along the Tiokoye river, Bassot (1966) noted a northwestward dip of 30° in the Faleme basin argillites and also strong deformations in the red sandstones occurring in the western part of the Faleme basin (Fig. 1). A large lateritic plateau did not allow us to observe the western rim of the Faleme basin.

3.3.3 The Youkounkoun Basin

Figure 22 shows some examples of the deformation occurring in the Batapa Group sediments. In the western part (Kembera inlier), the bedding of the argillites and sandstones of the Batapa Group are almost vertical and we observed a tight cleavage underlined by sericites (Villeneuve 1979). This cleavage, parallel to small thrusts, shows a 30° west dipping. More eastward, we observed open cylindrical metric folds with an axial fracture cleavage. Fold axes are directed to the NE. In the Bassaris ridge, a sheet of argillites linked to the Batapa Group shows

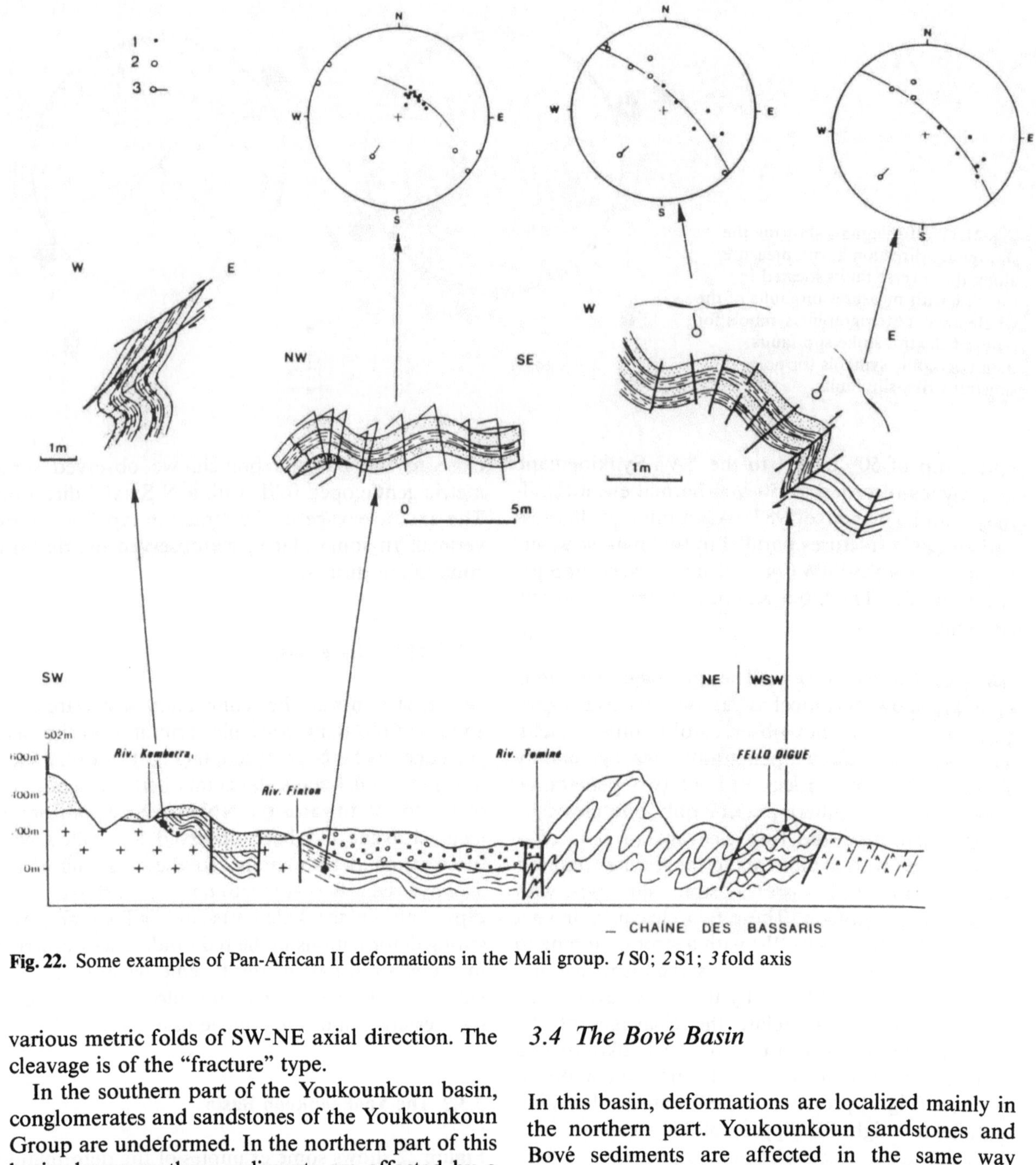

Fig. 22. Some examples of Pan-African II deformations in the Mali group. *1* S0; *2* S1; *3* fold axis

various metric folds of SW-NE axial direction. The cleavage is of the "fracture" type.

In the southern part of the Youkounkoun basin, conglomerates and sandstones of the Youkounkoun Group are undeformed. In the northern part of this basin, however, these sediments are affected by a weak deformation generating gentle folds with a SW-NE to E-W axial direction. These folds are affected by reverse faults showing a moderate north dipping. We will soon see that these deformations are linked to the Hercynian tectonic event.

3.4 The Bové Basin

In this basin, deformations are localized mainly in the northern part. Youkounkoun sandstones and Bové sediments are affected in the same way (Fig. 23) by E-W open folds with low dipping limbs. Northeastward, fold axes become progessively parallel to the NE-SW Koulountou branch. Westwards, in the Gabu area, these folds have a NW-SE axis. An axial plane fracture cleavage and reverse faults (south or southeast thrusting) are commonly noted, associated with the first deformation (D 1). A second folding occurs and could be responsible for the axial direction variations of the first folds and also of the bow of thrusts. These second folds

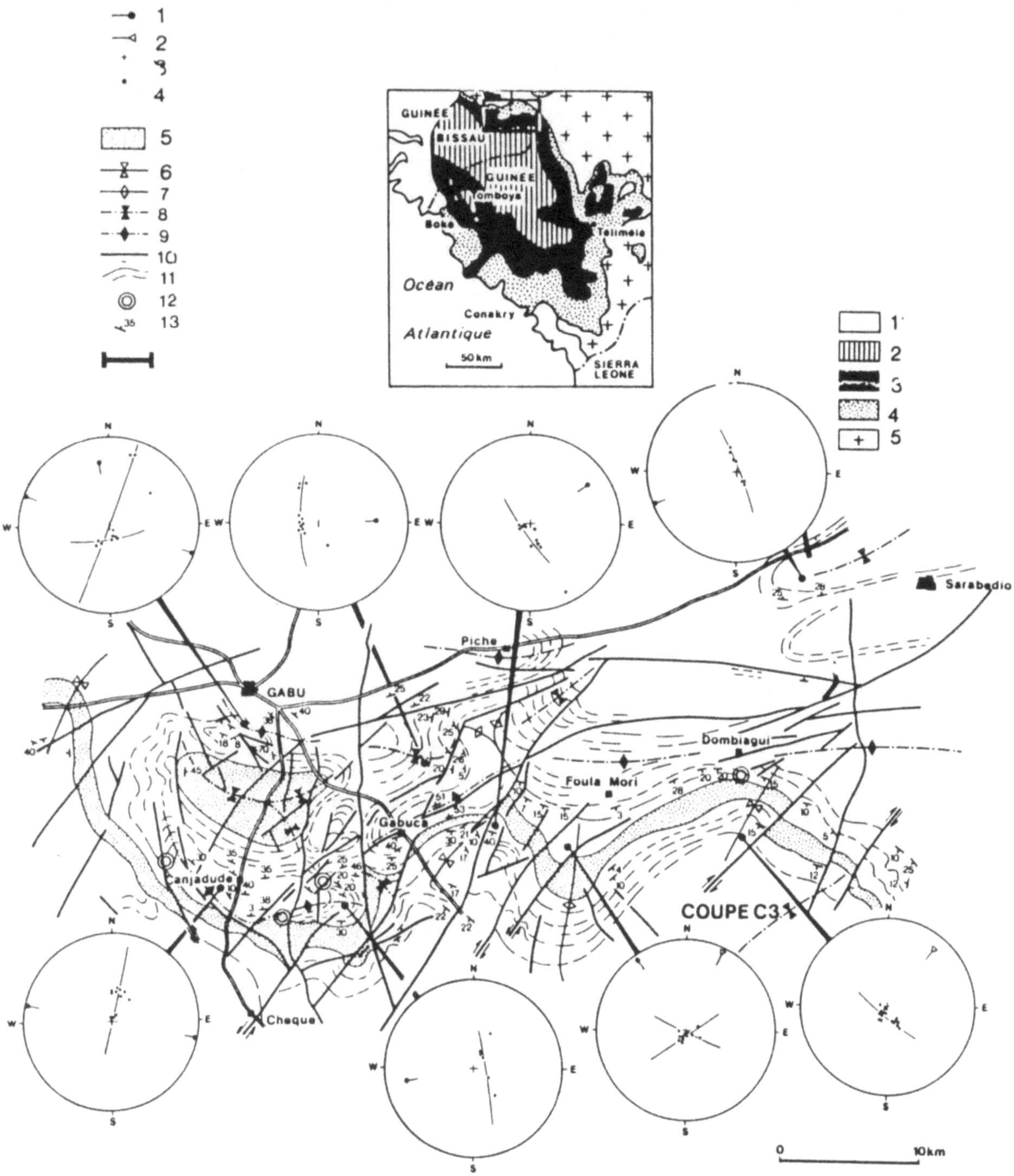

Fig. 23. Structural map of Hercynian deformations in the northern part of the Bové Basin. *Caption of location map. 1* Meso-Cenozoic basin; *2* Bafata Group; *3* Télimélé Group; *4* Pita Group; *5* basement. *Caption of the figure: 1* Axis of F1 folds; *2* axis of F2 folds; *3* stereographic symbols of bedding; *4* stereographic symbols of cleavage; *5* Télimélé Group; *6* F1 syncline; *7* F1 anticline; *8* F2 syncline; *9* F2 anticline; *10* faults; *11* structural trends; *12* local mapping unconformity; *13* direction and depth of structural planes; *14* reverse faults

have a NS to NNE-SSW axial direction and consist of plurikilometric gentle folds. A new vertical fracture cleavage appears to be associated with this deformation (D 2). The D 2 structures seem to be associated with strike-slip motions along SSW-NNE to SW-NE faults. Some of these strike-slip movements are senestrial but some of them have a dextral expression. All of these tectonic structures disappear progressively towards Guinea where the Bové basin is flat. The other part of the Bové basin is affected only by normal faults.

As all of the sedimentary groups of the Bové basin, including the Bafata Group (Devonian), are involved in this deformation, we place these two successive deformations in the Hercynian orogeny. After Villeneuve (1984, 1988 b), these deformations indicate a N-S shortening in this part of the Bové basin.

4 Metamorphism

4.1 Foreland

4.1.1 The Kedougou Inlier

Metamorphism in the Kedougou inlier Birmian groups was studied by Bassot (1966). In the Mako group, metamorphism is less intense and characterized by muscovites and sericites. In the Dialé and Dalema groups, we note the presence of micaschists with biotite.

4.1.2 The Madina-Kouta Basin

A study of metamorphism based on the degree of illite crystallization, following the Esquevin method (1969), proves that sediments of the Madina-Kouta basin do not undergo metamorphism (Villeneuve 1984). Only two samples are plotted in the anchizonal domain but we explain that as an effect of the thermal metamorphism which followed the doleritic Permo-Triasic intrusions.

4.2 The Bassaris Branch

The Termesse Group. The Esquevin method (1969) applied to the argillites of the Termesse Group shows no metamorphism. However, sericites and chlorites have been observed in some thin sections. These minerals indicate a greenschist facies.

The Guingan Group. A higher grade appears in the western part of the Guingan Group and synkinematic growth of green hornblende, epidote and garnet indicates an upper amphibolite facies. The presence of slightly or unmetamorphosed volcanic rocks (such as spilitic basalt) in the same area is problematic. Either a magmatic event occurred after the Guingan Group metamorphism, or parts of the Termesse Group were preserved in this group.

4.3 The Kouloutou Branch

The Niokolo-Koba Group. We observed the growth of sericite, chlorite, epidote, actinote and albite in the pyroclastic rocks and of epidote in the granitic rocks. These minerals indicate a greenschist facies.

The Koulountou Group. Two mineral parageneses are associated with the two cleavages observed in part of the Koulountou Group. The first paragenesis (quartz, plagioclases, epidote and muscovite) is not very different to the second paragenesis (quartz, sericite). Both indicate an upper greenschist facies. However, some metadacites and mylonites produced biotite that could indicate, locally, a lower greenschist facies.

4.4 The Komba, Faleme and Youkounkoun Basins

The Komba and Faleme Basins. The Esquevin method applied to the illites from the argillites of the Mali Group indicates an anchizonal domain of metamorphism except for the argillites sampled near the dolerite sills where we note thermal metamorphism.

The Youkounkoun Basin. In the Kemberra inlier we saw the presence of sericite along the slaty cleavage of the Batapa Group argillites. Gazda and Kowalski (1986) assume that the folding of the Batapa Group argillites, in the Bassaris area, is accompanied with chlorite (thuringite). However, in other areas of this basin, argillites of the Batapa Group are unmetamorphosed. The upper greenschist facies observed in some places could be related to the presence of important thrust faults. No metamor-

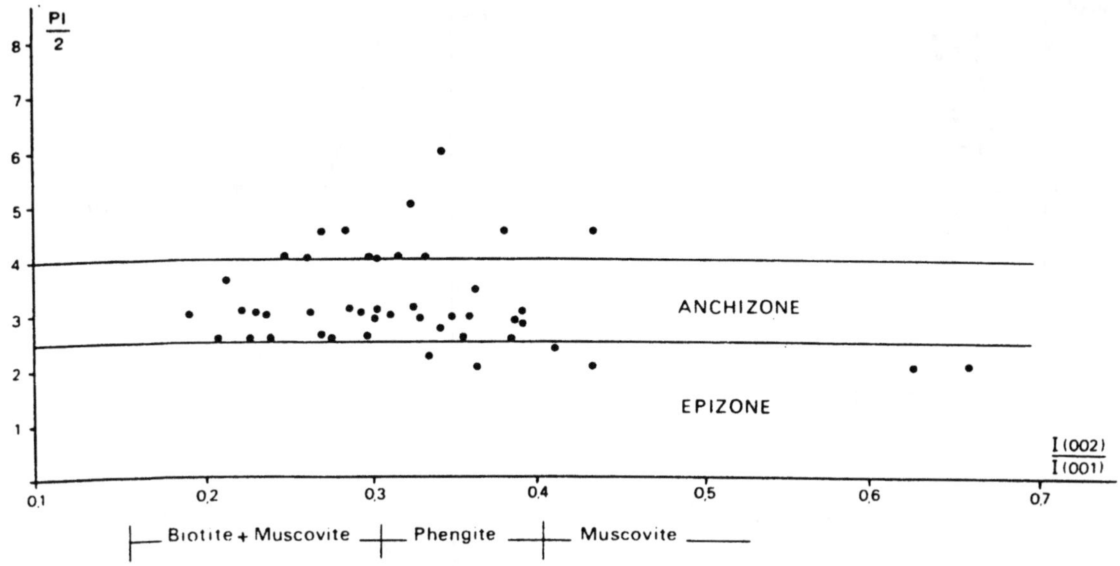

Fig. 24. Esquevin diagram (1969) for argillites of the Télimélé and Bafata Groups

phism has been recorded in the sandstones of the Youkounkoun Group.

The Bové Basin. The Esquevin method applied to illites from the Télimélé and Bafata Group argillites indicates an anchizonal domain for the most important part of these argillites (Fig. 24). This could be linked to a diagenetic evolution of these sediments.

The Permo-Triassic "Static" Metamorphism. Some of the undeformed sedimentary covers have suffered important diagenetic processes. In the Silurian and Devonian argillites of the Bové basin, for instance, epizonal recrystallizations may occur. This phenomenon, observed also in the Taoudeni basin (Girard 1985), is related to Permo-Triassic dolerite intrusion in the first stages of Atlantic rifting and produces temperatures of 250° to 450°.

5 Magmatic Activity

5.1 The Foreland

Kedougou Inlier. The last magmatic event of the Lower Proterozoic is marked by postkinematic granites emplaced around 1950 Ma (Bassot and Caen Vachette 1983 a). After a long phase of isostatic balance and erosion, the stable craton is intruded by doleritic and gabbroic dykes between 1350 and 1000 Ma (Bassot et al. 1986). Dykes have a prominent NNE direction.

Madina-Kouta Basin. We found two types of magmatic rocks in this basin (Villeneuve 1984). Rhyolites, which occur beneath the Segou Group along the Senegal/Guinea boundary and dolerites, which are interbedded in the Madina-Kouta sediments. Ponsard (1984) supposed a large amount of rhyolites concealed in the Labé area (Fig. 4). We suppose a setting of these volcanic rocks during the early stage of the basin structural formation. The second type of volcanic rocks was emplaced during the Permo-Triassic Atlantic Ocean opening stage.

5.2 The Bassaris Branch

The Termesse and Guingan Groups are characterized by abundant basalts and metabasalts with a fluidal texture and lapilli tuffs.

Basalts display a porphyric or doleritic texture with clinopyroxenes and amphiboles. Serpentine, recognized by Chiron (1964) and Bassot (1966) in the Guinguan Group, could originate from peridotite after hydrothermal processes. Oxides (Fig. 25) and rare earth discrimination diagrams (Fig. 26) give a clear affinity with MORB-type tholeiites for some lavas, confirmed by mineralogic studies (Dupont 1983). Meanwhile, the occurrence of titano-augite basalt of alkaline affinity (Dupont et al. 1984) indicates a continental influence. The rare earth discrimination diagram shows two types of lavas. The first shows a decrease in light rare earth and an increase of MgO (amphibolites of the Guinguan

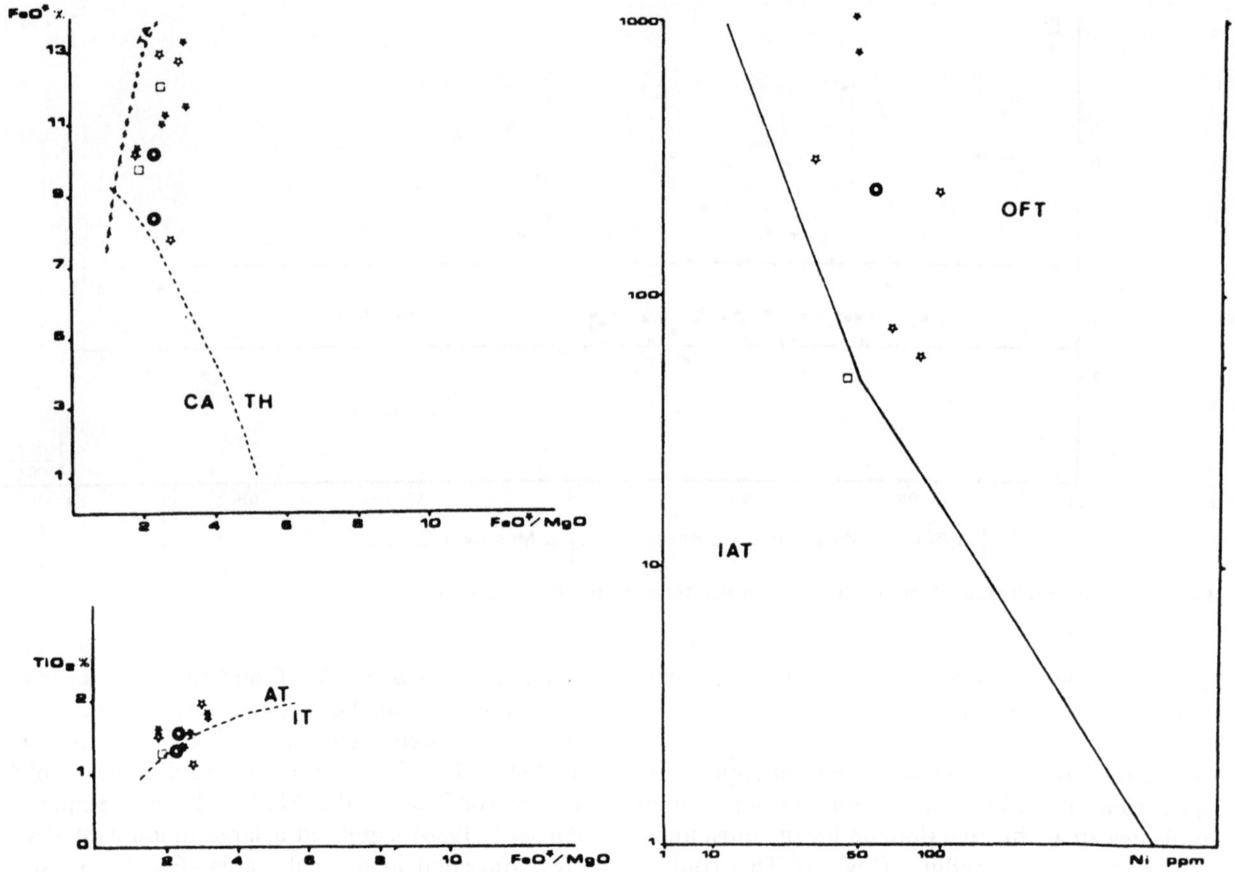

Fig. 25. FeO and TiO2 versus FeO/MgO plots for the basalts and pyroclastites of the Termesse and Guinguan groups. *AT* Antisotitaneous field; *IAT* island arc tholeiites field; *OFT* ocean floor baslat field

Group). The second shows an increase of light rare earth and of TiO2 (3,22%) and appears to be transitional (Termesse Group). The mixing of tholeiites and transitional lavas indicates a marginal or back-arc basin rather than an oceanic setting. We note the lack of harzburgites, gabbros and peridotites which are commonly associated with ophiolites.

5.3 The Koulountou Branch

Only the Niokolo-Koba volcano-plutonic complex has been studied. Volcanics rocks associated with the Linki-Kountou granite consists in a few aluminous basalts, some andesite, dacite with well-developed biotite, and abundant rhyolite and ignimbrite. Coarse-grained lapilli tuffs are associated with these lavas. Geochemical studies (Angeli 1983; Villeneuve 1984) indicate a clear calc-alkaline affinity (Fig. 27), with a decrease of FeO2 and TIO2. These characteristics are typical of orogenic series (Bebien 1980). The presence of important K2O ratios suggests a thick continental crust below. The rare earth discrimination diagram (Fig. 28) shows larger variations in light rare earth (LREE) than in heavy rare earth (HREE). This confirms the calk-alkaline affinity (Gill 1981). Granodiorites and monzogranites recognized in the Linki-Kountou granitic pluton (Bassot and Caen Vachette 1983 a) also have a calc-alkaline affinity.

5.4 Komba, Faleme and Youkounkoun Basins

No magmatic intrusion except the Permo-Triassic dolerites has been found in the Komba Basin. However, we found (Villeneuve 1984) that olivine basalts intruded in the Batapa Group argillites located in the Bassaris ridge area. These basalts could represent a magmatic activity occurring before the deposition of the Youkounkoun Group. In the Fa-

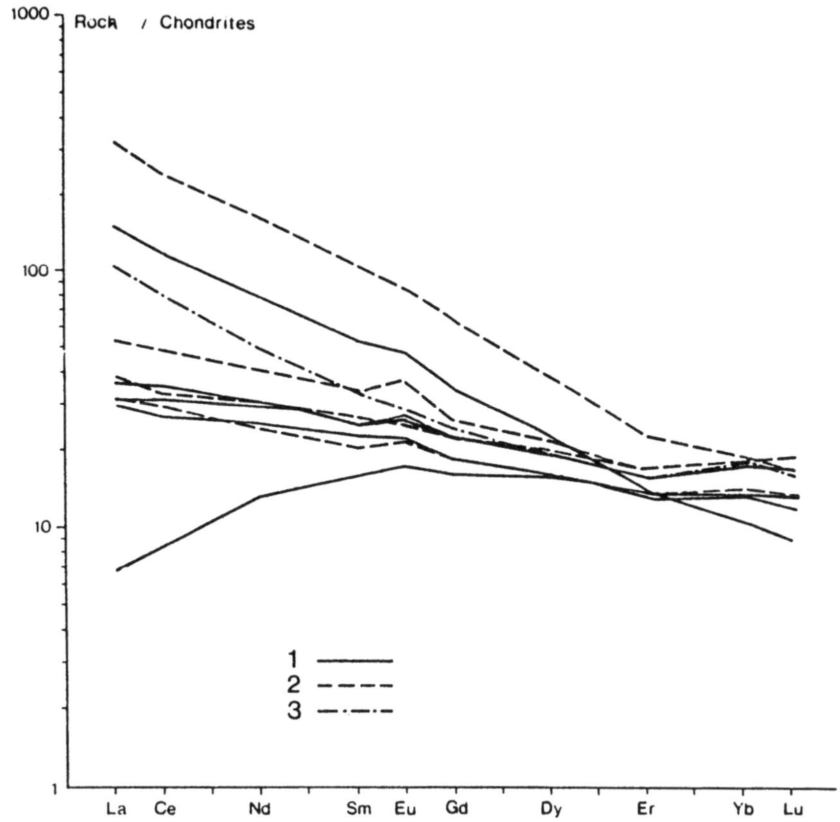

Fig. 26. Rare earth diagram of basalts and amphibolites of the Bassaris ridge (Termesse and Guinguan Groups)

leme Basin, Bassot (1966) assumed the presence of basic volcanic lavas interbedded with the argillites of the Mali Group. These intrusive rocks are very poorly documented, especially in geochemical and geochronological data.

5.5 *Permo-Triassic Magmatic Activity*

Along the West African Atlantic margin, Permo-Triassic dolerite massifs were emplaced mainly between 150 and 250 Ma. This activity started during the Carboniferous. Dykes and sills are particularly abundant in Guinea and largely controlled by SW-NE fractures (Bertrand and Villeneuve 1989) and also, in some cases, by SE-NW faults. Very thick sills (up to 1 km) can cover surfaces of several thousands of square kilometers. The petrography of these dolerites is rather homogeneous. Their texture is intergranular to gabbroic. Dominant minerals are plagioclase and augite often associated with pigeonite. Orthopyroxene may occur in some samples and olivine is only present in fine-grained dolerites. Several samples contain late crystallizing biotite and amphibolites. Geochemical analysis (Bassot et al. 1986; Bertrand and Villeneuve 1989) indicate typical continental tholeiite enriched in incompatible elements with a negative Nb anomaly. They have the same chemical composition as the oldest rift Liassic tholeiites of Morrocco. These results indicate a widespread tensional mechanism of a passive type, induced by the geodynamic condition of Pangea at the onset of the Jurassic.

5.5 *Concluding Remarks*

Except for the Birimian Groups of the Kedougou inlier, where geochemical information is lacking, we can recognize an oceanic-like tholeiitic magmatism associated with the Bassaris branch, a calc-alkaline magmatism associated with the Koulountou branch, a possible alkaline magmatism associated with the Youkounkoun and Faleme basins, and a Permo-Triassic continental tholeiitic magmatism represented by a large volume of dolerites and associated with the opening of the Atlantic Ocean. The Bassaris branch volcanism has not been dated but we expect a Pan-African age (around 700 Ma). Villeneuve (1984) and Dupont et al. (1984) suggest a

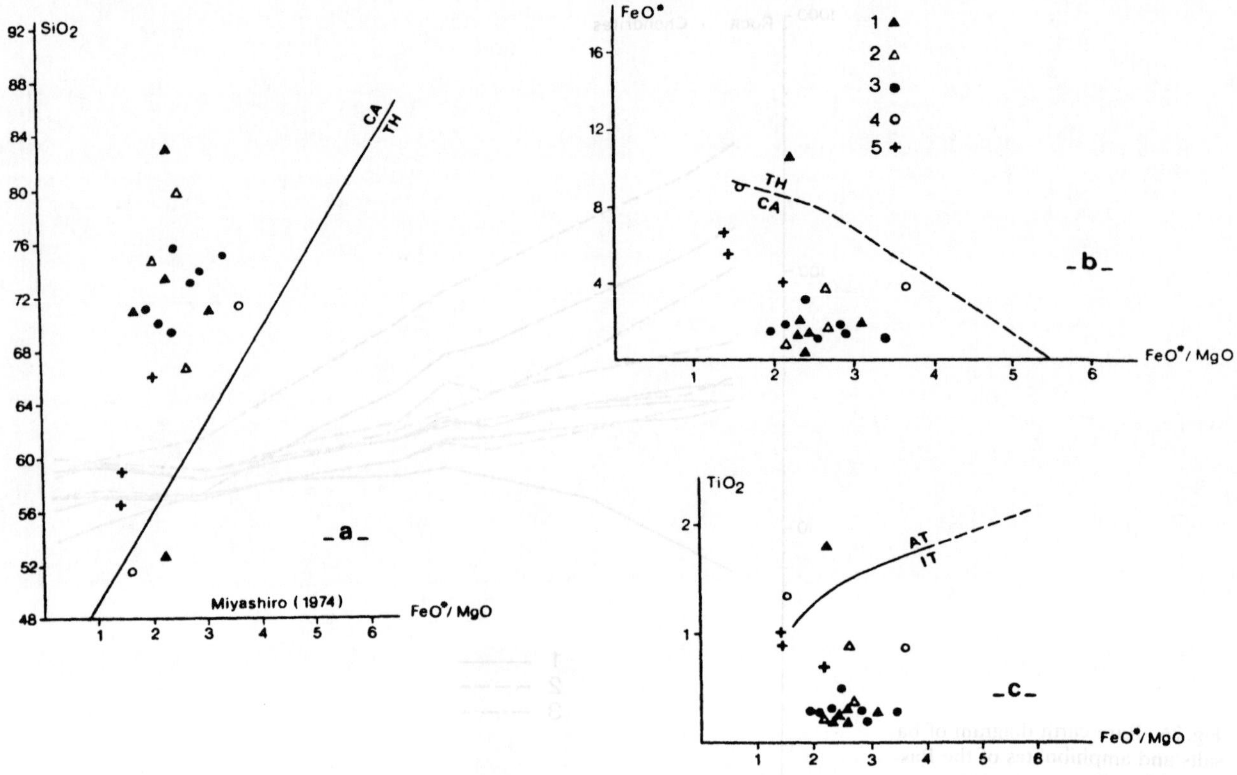

Fig. 27. Geochemical diagrams for the volcanic rocks of the Niokolo-Koba and Kouloutou Groups. Same legend as Fig. 25

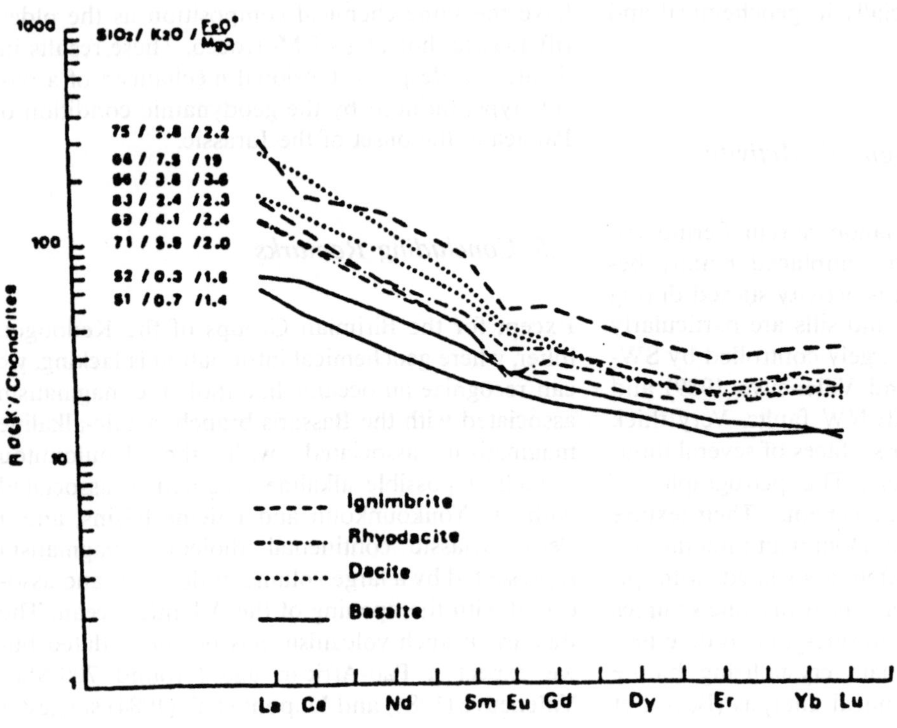

Fig. 28. Rare earth diagram of volcanic rocks of the Niokolo-Koba Group

"Red Sea-type rift" environment. The Koulountou branch is probably contemporary with the Linki-Kountou granite which has been dated at 680 Ma by the Rb/Sr method (Bassot and Caen-Vachette 1983 a). This calc-alkaline complex is linked to the active western margin activity of this Bassaride rift. Until now, the geodynamic significance of the Youkounkoun and Faleme basin magmatism has not been studied.

6 Local Geophysical Characteristics

Recent gravimetric (Ponsard 1984), seismic (Dorbath et al. 1983) and magnetotelluric (Ritz and Robineau 1986, 1988) studies have been carried out in this region.

The Bouguer gravity anomaly map presented in Fig. 29 has been compiled with a contour interval of 10 mgal from data collected by Creen and Rechenmann (1965) in southern Senegal and Gambia, Amorin Ferreira (1966) in Guinea Bissau and Akhmetjanov et al. (1976) in western Guinea. The main geological units (Villeneuve 1984) have been superimposed on the gravity map. The gravity anomaly

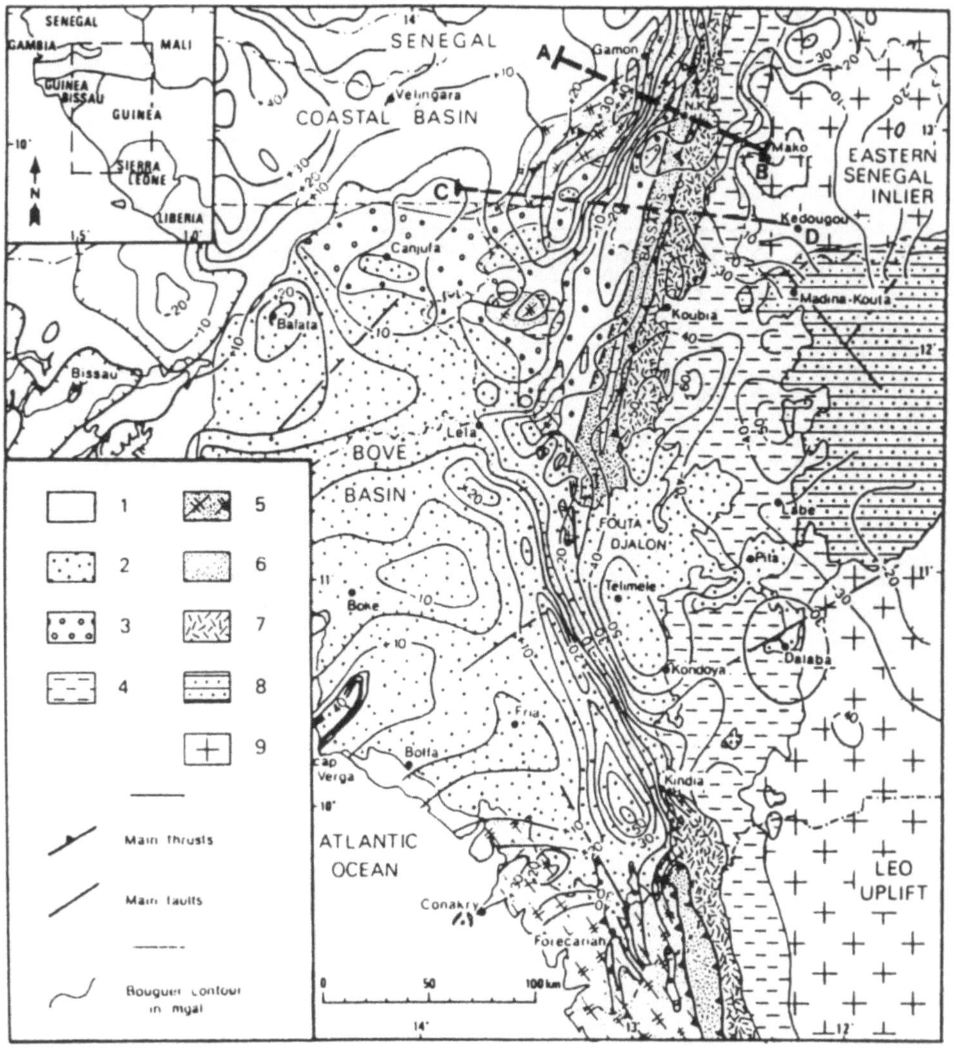

Fig. 29. Bouguer gravity anomaly map (contour interval 10 mgal) of southern Senegal and western Guinea with superposed geological data. *1* Post-Paleozoic cover; *2* Bové Basin super group (Ordovician to Devonian); *3* Youkounkoun Group (Cambrian-Ordovician); *4* Mali Group (Upper Proterozoic); *5* Upper Proterozoic reactived basement; *6* Guinguan Group; *7* Termesse Group; *8* Madina-Kouta Group; *9* Birrimian (Eburnéan) or Archean basement

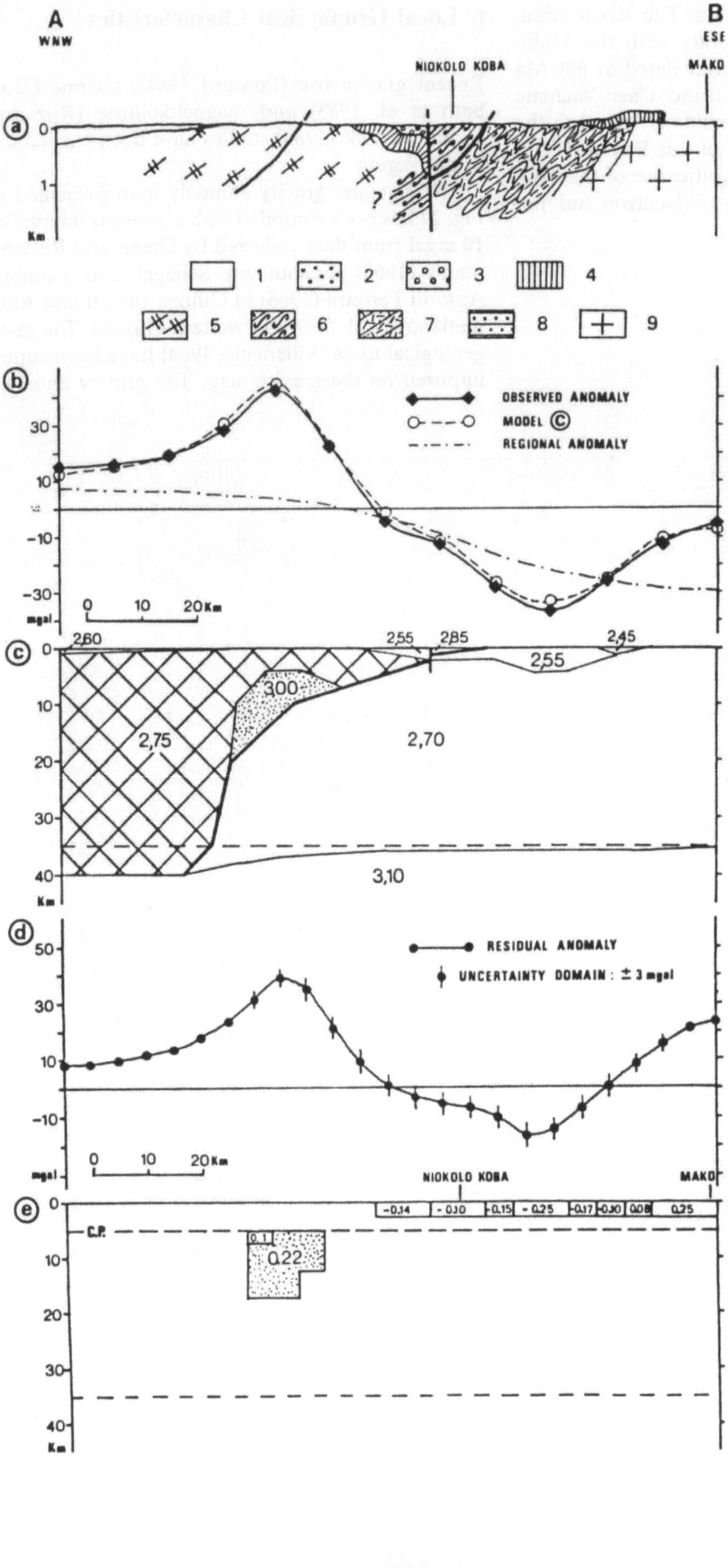

Fig. 30. Section across the Basaride orogen (Ponsard et al. 1982; Ponsard 1984). See Fig. 29 for location. **a** Geological interpretation (see legend to Fig. 29); **b** Bouguer anomaly and regional anomaly; **c** two-dimensional calculated model (densities in g/cm^3); **d** residual anomaly; **e** inverse problem approach. Ideal body under confining plane *C.P.* confining plane; density contrasts in g/cm^3

pattern reflects three principal domains: (1) an eastern domain corresponding to the margin of the cratonic Kedougou inlier and Léo uplift and marked by a broad, low tendency gravity: (2) a western domain corresponding to the region overlain by Paleozoic (Bové Basin) and post-Paleozoic (Senegalese Basin) sediments and characterized by a generally positive anomaly field; (3) a central domain represented by a gravity gradient zone which is superimposed on the Pan-African belt. The major feature in this central domain is a narrow linear belt of a highly positive anomaly with short wavelengths, elongated in a nearly N-S direction. It is flanked just to the east by a belt of linear gravity lows that extend from the north of Senegal to the south of Guinea.

The regional anomaly (Fig. 30 b), derived from the contour smoothing method along Bouguer profiles, is clearly distinguished on the eastern gravimetric plateau (−30 mgal) and a western one where the regional level is noticeably higher (+10 mgal). The gradient between both plateaus is superimposed on the short wavelengh highs and lows. Such a pattern is similar to the gravity configuration observed on Proterozoic belts in Canada (Gibb et al. 1983), in the Hoggar-Dahomey Pan-African chain on the eastern border of the West African Craton (Bayer and Lesquer 1978; Black et al. 1979) and on the margin of the Sao Francisco craton of Brasil (Lesquer et al. 1984). A model that explains such anomalies consists of two crustal blocks of contrasting mean density and thickness in isostatic equilibrium and separated by a suture formed by plate collision (Thomas 1983).

Using available density measures (Akhmetjanov et al. 1976) and taking into account geological boundaries (Fig. 30 a), seismological studies (Briden et al. 1981; Dorbath et al. 1983) and magnetotelluric interpretations (Ritz 1982; Ritz and Robineau 1988), a two-dimensional gravity model has been constructed (Fig. 30 c) across the anomaly of Gamon in southern Senegal (Koulountou branch). This model indicates that the suture separating the two crustal blocks of different density and thickness dips westward (45°) beneath the denser western, remobilized block. The density of the anomalous body that underlies the suture zone along the boundary of the reactived block has been deduced (Ponsard 1984) through an inverse problem approach (Fig. 30 e) using linear programming (Safon et al. 1977).

Eastern gravity lows are mainly explained using geological boundaries and density measures of the volcano-sedimentary rocks of the Termesse Group and the sedimentary deposits of the Mali group (Fig. 30 d, e).

A magnetotelluric traverse across the Bassarides and its foreland (Fig. 31) has led to a two-dimensional geolectric model at crustal and lithosperic levels (Ritz and Robineau 1988). This model confirms the west-dipping suture between the craton and the western margin of the Bassaride orogen. They provide new information on the depth and lateral extent of the belt units.

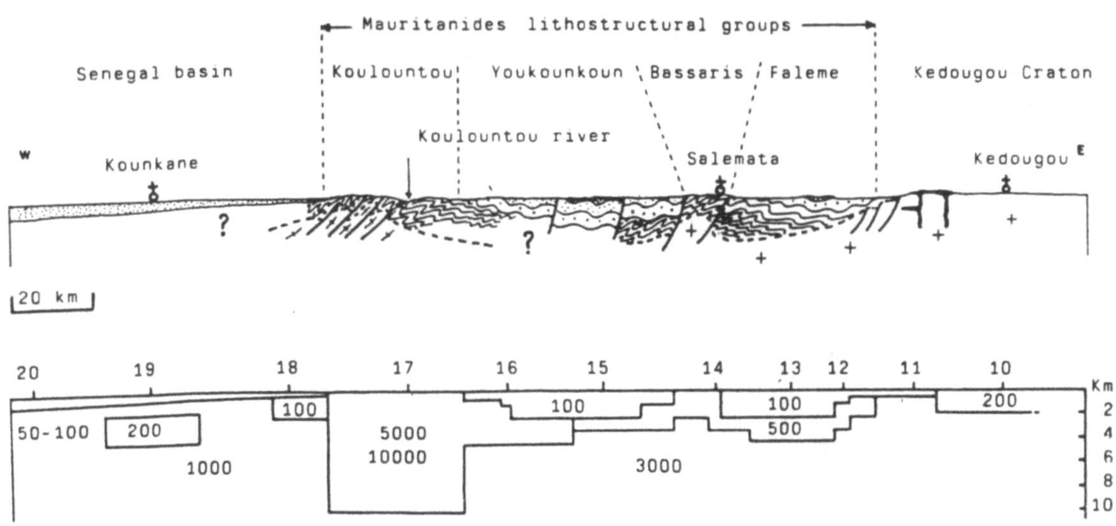

Fig. 31. Magnetotelluric section across southern Senegal (after Ritz and Robineau 1988). This section shows the difference in electric conductivity between the Komba and Youkounkoun Basins (100 ohm/m) and the Bassaris branch (Termesse and Guinguan Groups) which display 500 ohm/m

7 Geodynamic Evolution

7.1 Introduction

Finally, from the structural investigations four main tectonic events can be distinguished, thus allowing us to distinguish four distinctive fold belts. The Eburnean tectonic event corresponds to the Birrimian belt of the Kedougou inlier (Bassot 1966), the Pan-African I tectonic event corresponds to the folding of the Bassaris and Kouloutou branches (Villeneuve 1984), the Pan-African II tectonic event corresponds to the folding of part of the Komba, Faleme and Youkounkoun basin (Bassot 1966; Villeneuve 1984) and the Hercynian tectonic event corresponds to the folding of the northern part of the Bové Basin (Bassot 1966; Bechennec 1980 and Villeneuve 1984). These four belts are partly superimposed and some geodynamic models have been proposed by several authors for each of them. However, the paucity of information, especially geochronological data, does not allow us to propose a complete and sophisticated geodynamic model.

7.2 Geochronological Data

7.2.1 The Foreland

The Kedougou Inlier. We already have discussed the main information concerning the magmatic intrusions linked to the Eburnean orogeny, but we have no information on the age of metamorphism in the differents groups of the Kédougou inlier. In most of Africa this orogeny occurs around 2000 Ma.

The Madina-Kouta Basin. Only one datation is known. Bassot et al. (1963) gave an age of 1020 Ma obtained on argillites of the lower part of the Madina-Kouta basin by the Rb/Sr method. Calculation of the new Rb constant gave 1056 Ma (Dallmeyer and Villeneuve 1987). This indicates the starting of the sedimentation around 1000 Ma.

7.2.2 The Bassaris Branch

The muscovites of three samples from the Guinguan Group quartzites were analyzed by the 39 Ar/40 Ar method (Dallmeyer and Villeneuve 1987). They yielded three ages at about 660 Ma. They are interpreted as dating post-metamorphic cooling through argon retention temperatures. These ages could be linked to the tectonothermal event which corresponds to the Pan-African I tectonic event.

7.2.3 The Kouloutou Branch

The Niokolo-Koba Group. The Niokolo-Koba granite (Linki-Kountou granite) yields an age of 683 Ma by the Rb/Sr method (Bassot and Caen Vachette 1983 a) interpreted as the date of pluton emplacement. The hornblende datation in this granite displays an age of 751 to 627 Ma the 39 Ar/40 Ar method (Dallmeyer and Villeneuve 1987). But this age yields no useful geological information because of the complexities of these various age interpretations. The rhyolites of the Niokolo-Koba granite display an age of 324 Ma which may reflect a metamorphic disturbance of the initial igneous isotopic system.

The Kouloutou Group. Bassot et al. (1963) gave an age of 448 Ma for the pelites of the Simenti formation (Rb/Sr method). Dallmeyer and Villeneuve (1987) indicate an age of 538 Ma displayed by muscovite of the Simenti formation argillite and an age of 254 Ma displayed by whole rocks on the same rocks (39 Ar/40 Ar). The biotite age is given to date the initial post-metamorphic cooling through temperatures required for intracrystalline argon retention. Hornblende of the Dienoundiala orthogneiss (or Gamon granite) displays an age of 673 Ma by the 39 Ar/40 Ar method, which is difficult to interpret. But an age of 278 Ma from muscovite of the Damantan mylonite is considered to reflect Hercynian tectonic activity in this area (Dallmeyer and Villeneuve 1987).

7.2.4 The Komba and Youkounkoun Basins

Only one datation of the argillites of these basins is given by Bassot et al. (1963). It originated from the Niokolo-Koba area. The argillite fraction (<4) displays an age of 580 Ma which is consistent with the datation of the Walidiala dolomites by microfossils (Culver et al. 1988).

7.2.5 The Bové Basin

No geochronological data is available on the Paleozoic sediments of the Bové Basin. However, the involvement of Devonian deposits until the Famennian (360 Ma), in the northern Bové Basin deformation, clearly shows a post-Devonian age for the Hercynian tectonic event.

7.2.6 Concluding Remarks on Geochronological Data

We could distinguish several age groups in the Bassaride orogen:

1. Around 2000 Ma and corresponding to the magmatic and probably metamorphic events of the Eburnean orogeny.
2. 1000 to 1050 for the first deposits in the Madina Kouta basin.
3. 690 to 680 Ma with the emplacement of the Niokolo-Koba granites.
4. 660 to 655 Ma linked to a tectonometamorphic event in the Bassaris branch and recorded in plagioclases of the Niokolo-Koba granite (Pan-African I event).
5. 560 Ma corresponding to the diagenesis of the Batapa group argillites.
6. 550 to 530 Ma recorded in the muscovites of the Simenti Group schists and linked to a tectonothermal event (Pan-African II event).
7. 278 to 254 Ma recorded in the muscovites of the Damantan mylonite, phyllites of the Simenti formation (Koulountou Group) and linked to a Hercynian tectonothermal event.

7.3 Geodynamical Model

As the topic of this chapter is the Bassaride orogen, we will not discuss the geodynamic interpretations proposed for the Birimian belt of the Kedougou inlier.

A sketch of the geodynamic evolution of the Bassarides is proposed (Villeneuve 1984; Villeneuve and Dallmeyer 1987) in the light of available geological, geochemical, geophysical data and radiometric dating (Fig. 32). In this sketch, the various sedimentary and igneous formations emplaced on the western edge of the craton experienced several orogenic cycles.

Eastward, a Lower Proterozoic craton is exposed in the Kedougou inlier. It does not outcrop westward, but geophysical investigations suggest a change in the nature of the continental crust beneath the Youkounkoun Basin.

1. Before 1050 Ma
A first stretching of the crust occurred with rhyolite intrusions in the West African Craton margin and formation of Upper Proterozoic basins (Madina Kouta).

2. Between 800 to 700 Ma
Fracturation of the western edge of the West African Craton and formation of a continental rift. A reduced oceanization possibly occurred in the northern Bassarides. Termesse and Guinguan groups correspond to rift materials. Koulountou branch Groups represent the western margin of the rift, remobilized during a west-dipping subduction, and finally folded and metamorphosed.

3. 700 to 675 Ma
Closure of the rift, with a suspected continental subduction under its western margin, and emplacement of calc-alkaline complexes around 680–650 Ma in this western margin.

4. 650 to 675 Ma
Collision between the two edges of a continental rift. Termesse and Guinguan groups which represent the rift material are tightly folded, metamorphosed and partly thrusted on the West African Craton. Ignimbrites of the Niokolo-Koba and Koulountou Groups could be related to the crustal fusion of the subducted West African Craton margin.

4. 650 to 555 Ma
a) Erosion and molassic deposition of sediments in small basins as, for example, in the Kemberra inlier (Villeneuve 1979).
b) Stretching phase contemporaneous with the Late Precambrian glacial deposits in West Africa.
c) Subsidence of the Faleme, Komba and Youkounkoun Basins leading to the deposition of detrital materials.

5. 555–550 Ma
Folding and fracturing (usually Pan-African I reworked faults) of the Mali and Batapa Groups accompanied locally by a slight metamorphism dated at 550 Ma. Some of the volcano-sedimentary rocks of the Niokolo-Koba and Koulountou Groups have suffered this metamorphism.

6. 550 Ma to Cambrian/Ordovician Boundary
Erosion and deposition of molassic sediments in the Youkounkoun Basin and probably in small basins covered later by Bové Basin Paleozoic sediments.

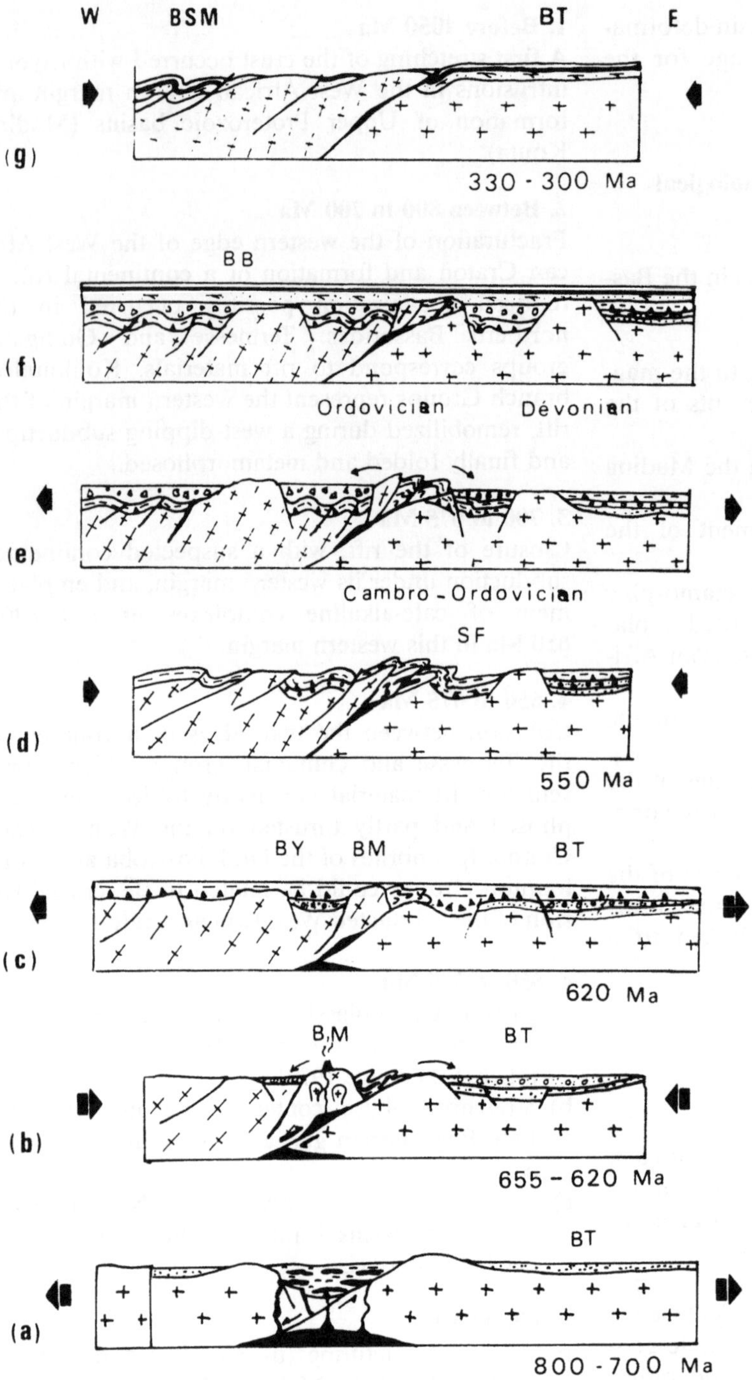

Fig. 32. The geodynamic evolution of the Bassaride orogen. **a** Rift stage; **b** closure of the rift and deposition of molasses in the Taoudeni and Youkounkoun Basins (Pan-African I tectonic event); **c** deposition of tillites and argillites of the Mali Group; **d** folding of the Mali Group (Pan-African II tectonic event); **e** deposition of sandstones and conglomerates of the Youkounkoun Group (Pan-African II molasses); **f** deposition of the Bové Basin super group; **g** Hercynian deformations; *BT* Taoudeni Basin; *BM* Bassaride fold belt; *BY* Youkounkoun Basin; *SF* faleme trough; *BB* Bové Basin; *BSM* Senegalo-Mauritanian Basin

7. Ordovician to Devonian

After the complete erosion of Pan-African belts and the molassic infilling of basins, the West African Craton was covered by Cambro-Ordovician fluviatile sandstones (Pita Group) with a NW transport direction. A glacial episode is suspected in the Late Ordovician. Then, starting in the Middle Llandoverian, several marine transgressions occurred until the Devonian (Villeneuve and Da Rocha Araujo 1984).

8. Carboniferous to Permian

By the end of the Devonian and during the Carboniferous, the final Africa - North America collision (Alleghenian orogeny) reworked Pan-African weak fracture zones and sutures. The northern part of the Bové Basin was folded and fractured at this time. Hercynian tectonics are distinguished in to a southeastward motion of the Senegalese microplate in the Kouloutou area (Venkatakrishanan and Culver 1988). Villeneuve (1988 b) argues for a compressive phase with a N-S shortening axis.

9. Post-Permian Evolution

The post-Hercynian evolution of West Africa is closely related to the Central Atlantic opening. By the beginning of Jurassic, a rifting stage leads to the emplacement of dolerite dykes along fractures and sills in Paleozoic or Proterozoic sediments. The northern part of the Bové Basin is overlain by Mesozoic-Cenozoic sediments of the Senegalo-Mauritanian basin. Despite the lack of recent orogeny, the area is not tectonically inactive. The Fouta Djalon (eastern part of the Bové Basin) has been uplifted recently, at least after the formation of the Mesozoic bauxites. Hot springs in the north of the Fouta Djalon and repeated earthquakes, the last one in 1984, prove present activity.

8 Conclusions

The *Bassaride belt,* described in Senegal and Guinea, consists of two main superimposed Pan-African orogenic events: the Pan-African I (<650 Ma) and the Pan-African II (<550 Ma). The first one agrees with the closure of a rift while the second agrees with the folding of sedimentary basins parallel to the Pan-African I belt. These two structures were later partially overlain by Lower Paleozoic sediments, then partly remobilized during the Carboniferous Hercynian orogeny.

The main difference between the Bassaride belt and its neighbouring belts lies in the good conservation of the Pan-African I structures slightly deformed by the more recent tectonic events. We can actually observe very clearly the unconformities between the Pan-African I structures and those of the more recent ones (Villeneuve 1984).

In the *Mauritanide belt* the Pan-African I event has been demonstrated indirectly by geochronological data (Dallmeyer and Lécorché 1987). The deformation produced by the Pan-African II and by the Hercynian orogenesis have completely concealed this first Pan-African fold belt.

In the *Rockelides,* where there is no evidence of Hercynian tectonic deformations, the radiometric data indicate Pan-African II metamorphism. However, the geophysical data (Bonvalot et al. 1988) show a gravimetric signature similar to the Guinean Pan-African I gravimetric suture (Ponsard et al. 1982). Thus, we believe that the Pan-African II deformations were important enough to cancel the Pan-African I structures.

The Bassaris area seems to be the only place where the first Pan-African I structures are not intensively deformed by more recent tectonic events.

References

Adeguelou F, Fall M (1979) Etude géologique du secteur de Guemedji. Rapport de fin de campagne de sondage. COGEMA-CRE, Dakar, p 82

Affaton P (1987) Le bassin des Volta (Afrique de l'Ouest): une marge passive d'âge Proterozoique supérieur tectonisée au Panafricain. Thèse Doct Etat, Univ Aix-Marseille III, p 170

Akhmetjanov BS, Loutsenko VF, Diallo M (1976) Carte des anomalies de Bouguer (1/500000) et résultats des travaux géophysiques en Guinée occidentale. Rapp Technoexport, Conakry, pp 184-205

Allen JRL (1968) Current ripple. North Holland, Amsterdam, p 433

Amorin Ferreira H (1966) Observacoes gravimetricas no territorio da Guinea Portuguese. Serv Météorol Nat, Lisboa, p 9

Angeli H (1983) Le magmatisme Proterozoique de l'ensemble Niokolo-Koba-Kouloutou. Temoins d'une marge continentale active au Panafricain. DEA, Nancy, p 30

Arnould M, Aymé JM, Guillaume R (1960) Nouvelle stratigraphie des séries primaires du Nord du Fouta Djalon (Guinée-Sénégal). Bull Soc Geol Fr 1: 631-634

Baldwin CT (1977) Trace fossils, 2, P. Crimes edit, Geol J 9 (Liverpool): 9-40

Bassot JP (1966) Etude géologique du Sénégal oriental et de ses confins guinéomaliens. Mem Bur Rech Géol Min, Paris 40: 322

Bassot JP, Caen-Vachette M (1983 a) Données géochronologiques et géochimiques nouvelles sur les granitoïdes de l'Est du Sénégal. 12e Coll, Geol Afr, Bruxelles, p 8

Bassot JP, Caen-Vachette M (1983 b) Données nouvelles sur l'âge du massif du granitoïde du Niokolo-Koba (Sénégal oriental), implications sur l'âge du stade précoce de la chaine des Mauritanides. J Afr Earth Sci 1: 159-165

Bassot JP,Bonhomme M, Roques M, Vachette M (1963) Mesures d'âges absolus sur les séries précambriennes et paléozoiques du Sénégal oriental. Bull Soc Géol Fr 7, 5: 401-405

Bassot JP,Cantagrel JM, Jamond C, De Carvalho H (1986) Nouvelles données géochronologiques sur les dolérites et les gabbros de l'Est Sénégal et de la Guinée-Bissau. Reflexions sur l'âge de mise en place et la répartition du magmatisme tholéitique continental à l'échelle de l'Afrique de l'Ouest. Garcia de Orta, Sér Géol, Lisboa 9: 1-14

Baudelot S, Diallo K, Villeneuve M (1980) Presence d'un microplancton du Ludlovien supérieur dans les argilites d'âge discuté de Tomboya (Guinée). C R somm Soc Géol Fr 6: 236-241

Bayer R, Lesquer A (1978) Les anomalies gravimétriques de la bordure orientale du craton ouest-africain: géométrie d'une suture panafricaine. Bull Soc Géol Fr 7: 863-876

Bebien J (1980) Magmatismes basiques dits "orogéniques" et teneurs en TIO2. J Vol Geoth Res 8: 337-342
Bechennec F (1980) Etude géologique du Nord Est de la Guinée Bissau. Comm Res Nat BRGM, Rapp 79, ROM 060 AF, Dakar, p 41
Bertrand H, Villeneuve M (1989) Témoins de l'ouverture de l'Atlantique central au début du Jurassique: les dolérites tholéitiques continentales de Guinée (Afrique de l'Ouest). C R Acad Sci, Paris 308: 93-98
Black R, Caby R, Moussine-Pouchkine A et al. (1979) Evidence for late Precambrian tectonics in West Africa. Nature, London 278: 223-227
Briden JC, Withcombe DN, Stuart GW, Fairhead JD, Dorbath C, Dorbath L (1981) Depth of geological contrast across the West African craton margin. Nature, London 292: 123-128
Bonvalot S, Villeneuve M, Legeley A, Albouy Y (1988) Levé gravimétrique du Sud Ouest du craton Ouest Africain. C R Acad Sci, Paris 307: 1863-1868
de Chételat E (1935) Recherches géologiques et morphologiques dans le Nord Est de la Guinée française. Rev Geogr Phys Géol, Dyn 6: 91-116
de Chételat E (1939) Sur l'existence du Dévonien inférieur en Guinée française. C R Acad Sci, Paris 209: 371-372
de Chételat E (1946) Notice explicative sur la feuille Kindia Ouest au 1/500000. Dir Min A O F, Paris, p 38
Chiron JC (1964) Etude géologique du Pays Bassaris. Rapp Bur Rech Géol Min 64: 110
Creen Y, Rechenman J (1965) Mesures gravimétriques et Magnetiques au Sénégal et en Mauritanie occidentale. Cah ORSTOM, Géophys 6: 59
Culver SJ, Pojeta J, Repetski JE, Robineau B (1988) Signification stratigraphique des microfossiles du Cambrien inférieur dans le bassin de Taoudeni, à la frontière Guinée-Sénégal. C R Acad Sci Paris 307: 651-656
Dallmeyer RD, Lécorché JP (1987) 40 Ar/39 Ar Polyorogenic mineral age record within the central Mauritanide orogen. West Africa. In: Dallmeyer RD, Lécorché JP (eds) Colloq on Tectonothermal evolution of West Africa orogens and circum Atlantic terranes linkages. Nouakchott, pp 69-71
Dallmeyer RD, Villeneuve M (1987) $^{40}Ar/^{39}Ar$. Mineral age record of a polyphase tectonothermal evolution in the southern Mauritanide orogen. Southeastern Sénégal. Bull Geol Soc Am 98: 602-611
Deynoux M (1980) Les formations glaciaires du Précambrien terminal et de la fin de l'Ordovicien en Afrique de l'Ouest. Deux exemples de glaciation d'inlandsis sur une plateforme stable. Trav Lab Sci Terre St Jérôme Marseille 17: 554
Diallo DP (1983) Contributions à l'étude géologique de la série du Dialé dans les monts Bassaris (Sénégal oriental). Thèse 3° cycle, Dakar, p 185
Dorbath C, Dorbath L, Le Page A, Gaulon R (1983) The West-African craton margin in eastern Sénégal: a seismological study. Ann Géophys, Paris 1: 25-36
Drot J, Villeneuve M (1985) Présence de Pleurothyrella Knodi en Guinée Bissau. Implications paléogéographiques. Géobios 18, 4: 525-531
Drot J, Lardeux H, Le Page A (1978) Sur la découverte de Silurien supérieur au sommet de la série de Youkounkoun au Sénégal oriental: implications paleogéographiques et structurales. Bull Soc Géol Miner Bretagne, Rennes Fr 10: 7-30
Dupont PL (1983) Les unités volcano-sédimentaires protérozoiques de Termesse et de Guinguan (Sénégal-Guinée): des reliques océaniques au sein d'une paléosuture panafricaine, D E A, Univ Nancy Fr, p 35
Dupont PL, Villeneuve M, Lapierre H (1984) Mise en évidence de reliques océaniques au sein de la chaine panafricaine des Mauritanides dans la région des Bassaris (Guinée-Sénégal). C R Acad Sci, Paris 2: 65-70
Esquevin J (1969) Influence de la composition chimique des illites sur leur cristallinité. Bull Centr Rech Pau 3, 1: 147-153
Gazda L, Kowalski WM (1985 a) Lithology and sedimentology of clastic rocks of the Télimélé and Faro series from the Gaoual administrative region (Republic of Guinea). Bull Polish Acad Sci, Warsawa 33, 1-2: 63-76
Gazda L, Kowalski WM (1985 b) Lithology and sedimentology of clastics rocks of the Pita series from the Gaoual and Kundara administrative region. Bull Polish Acad Sci, Warsawa 33, 3-4: 147-158
Gazda L, Kowalski WM (1985 c) Lithology and sedimentology of clastic rocks of the Télimélé and Faro series from the Gaoual administrative region (Republic of Guinea). Bull Polish Acad Sci, Warsawa 33, 1-2: 77-85
Gazda L, Kowalski WM (1986) Clastic rocks of the Mali Group (Vendian) from the Gaoual administrative region. Bull Polish Acad Sci, Warsawa 34, 2: 209-218
Gibb RA, Thomas MD, Lapointe PL, Mukhudopay M (1983) Geophysics of proposed Proterozoic sutures in Canada. Prec Res 19: 349-384
Gill JB (1981) Geochemistry of Viti Levu (Fidji) and its evolution as an island arc. Contr Miner Petrol 27: 179-203
Girard JP (1985) Diagénèse hydrothermale tardive des sédiments gréso-argileux du Protérozoique supérieur du bassin de Taoudeni (Afrique de l'ouest). Thèse Univ Poitiers Fr, p 250
Jackson DE, Lenz AC, Pedder AEH (1978) Late Silurian and early Devonian graptolites, brachiopod and coral faunas from northwestern and Arctic Canada. Geol Assoc Can Spec Pap 17: 159
Lesquer A, Beltrao JF, De Abreu FAM (1984) Proterozoic links between northeastern Brazil and West Africa: a plate tectonic model based on gravity data. Tectonophysics 110: 9-26
Marchand J, Bassot JP, Simon B (1987) Notice explicative de la carte photogéologique au 1/200000 de Kossanto (Mali). Dir Nat Geol Min Bamako, p 28
Ndiaye PM, Robineau B, Moreau C (in press) Déformation et métamorphisme de séries Birimiennes en relation avec la mise en place du granite de Saraya, Sénégal Oriental Bull Soc Geol Fr, Paris
Ponsard JF (1984) La marge du craton ouest africain du Sénégal à la Sierra Léone: interprétation géophysique de la chaine panafricaine et des bassins du Protérozoïque à l'actuel. Thèse Univ Aix-Marseille Fr, p 198
Ponsard JF, Lesquer A, Villeneuve M (1982) Une suture panafricaine sur la bordure occidentale du craton Ouest Africain? C R Acad Sci 295: 1161-1164
Potter PE (1963) Late Paleozoic sandstones of Illinois basin (Illinois). Geol Surv Rep 217: 92
Reading HG (1980) Sedimentary environment and facies. Blackwell Oxford, p 557
Ritz M (1982) Etude régionale magneto-tellurique des structures de la conductivité électrique sur la bordure occidentale du craton Ouest africain en république du Sénégal. Can J Earth Sci 19: 1408-1416
Ritz M, Robineau B (1986) Crustal and upper mantle electrical conductivity structures in West Africa: geodynamic implications. Tectonophysics 124: 115-132

Ritz M, Robineau B (1988) Tectonic interprétation of electrical structures beneath the West African craton edge in eastern Sénégal. Am J Sci 288: 756-776

Roman'ko YF (1974) New data on the Paleozoic stratigraphy of West Africa. Dokl Acad Nauk SSSR, Moscow 217: 420-423

Safon C, Vasseur G, Cuer M (1977) Some applications of linear programming in the inverse gravity problem. Geophysics 42: 1215-1229

Thomas MD (1983) Tectonic signifiance of paired gravity anomalies in the southern and central Appalachians. In: Hatcher RD Jr, Williams H, Zietz I (eds) Geol Soc Am Mem 158: 113-124

Torchine NS (1976) Notice explicative de la carte géologique de la République de Guinée au 1/200000. Feuille Youkounkoun, Conakry, p 165

Venkatakrishanan R, Culver SJ (1988) Plate boundaries in West Africa and their implications for Pangean continental fit. Geology 16: 322-325

Villeneuve M (1979) Etude geologique de la boutonnière de Kemberra. Trav Lab Sc Terre, Marseille Fr X 34: 17

Villeneuve M (1980) Schéma géologique du Nord de la Guinée (Afrique de l'Ouest). C R Somm Soc Géol Fr 2: 54-57

Villeneuve M (1983) Les épandages fluviatiles gréseux du Paléozoique inférieur de Guinée. 5° Congr Eur Sédimentol, Marseille Fr, pp 450-451

Villeneuve M (1984) Etude géologique de la bordure sud-ouest du craton ouest africain. La suture panafricaine et l'évolution des bassins sédimentaires protérozoiques et paléozoiques de la marge nord-ouest du continent de Gondwana. Thèse Sci Univ Aix-Marseille III Fr, p 552

Villeneuve M (1987 a) Mise en évidence de deux évènements tectoniques panafricains dans les Bassarides (Guinée-Sénégal). Le diachronisme des trois chaines de la bordure occidentale du craton ouest-africain. C R Acad Sci 304: 49-54

Villeneuve M (1987 b) Corrélations et évolution des bassins sédimentaires de la marge NW du continent de Gondwana au cours du Paléozoique inférieur. J Afr Earth Sci 7, 2: 451-461

Villeneuve M (1988 a) Evidence of late Precambrian glaciation in Guinea (West Africa). In: Deynoux M (ed) Earth glacial record. Cuiaba, Brasil, pp 24-26

Villeneuve M (1988 b) Arguments en faveur d'un raccourcissement N-S dans le NE du Bassin paléozoique Bové (Mauritanides méridionales), lors de la collision Afrique/Amérique du Nord. C R Acad Sci 306: 915-920

Villeneuve M (1988 c) Evolution géologique comparée du bassin de Taoudeni et de la chaîne des Mauritanides, en Afrique de l'Ouest. C R Acad Sci, Paris 307: 915-920

Villeneuve M, Dallmeyer RD (1987) Geodynamic evolution of the Mauritanide, Bassaride and Rokelide orogens (West Africa). Prec Res 37: 19-28

Villeneuve M, Racheboeuf P (in press) *Australostrophia senegalensis* n. sp.: first occurrence of chonostrophid brachiopod (Chonetacea) from NW Africa. N Jahrb Paleont, Tubingen, Germany

Villeneuve M, Da Rocha Araujo PR (1984) La stratigraphie du bassin paléozoique de Guinée (Afrique de l'Ouest). Bull Soc Géol Fr 26, 6: 1033-1039

Villeneuve M, Diallo MC, Keleba F, Kourouma S, Paris F, Racheboeuf P (in press) Données paléontologiques nouvelles sur le Paléozoique du bassin Bové (Guinée-Afrique de l'Ouest): conséquences stratigraphiques. C R Acad Sci, Paris

The Mauritanide Orogen and Its Northern Extensions (Western Sahara and Zemmour), West Africa

J. P. Lécorché,[1] G. Bronner,[1] R. D. Dallmeyer,[2] G. Rocci[3] and J. Roussel[1]

1 Introduction

The West African fold belt (Sougy 1962 b), has generally been presented as including the Mauritanide, Bassaride and Rokelide orogens (Fig. 1). In this paper we shall consider the Mauritanide belt sensu stricto, that is to say the "Akjoujt-Bakel bow" and its northern, poorly known, extensions in the Western Sahara and Zemmour. The orogen is bordered on the west by unconformably overlying Mesozoic and Cenozoic sedimentary sequences of the Senegal-Mauritanian and El Ayoun coastal basins. Large segments of the orogen are covered by Quaternary sand dune complexes.

Prior to 1962, the Akjoujt-Bakel bow was interpreted (Blanchot 1955; Renaud 1961) as a part of the early Proterozoic (Eburnean) orogen representing an extension of the Reguibat shield. Therefore the epicratonic "Infracambrian" to Paleozoic Taoudéni Basin was assumed to rest in major unconformity on this basement. Delpy (1959, 1960), and Bassot and Delpy (1960) suggested that at least portions of the Cambrian-Ordovician sequences exposed in the epicontinental Taoudéni Basin (Fig. 1) could be traced westward into the "Akjoujt and Bakel series", requiring a Paleozoic age for at least some of the metamorphism and folding recorded. Tessier et al. (1961), in the Akjoujt region, proposed that a major thrust fault separated unmetamorphosed, little deformed, Taoudéni-type cover sequences (then thought to be of Eocambrian age) from structurally overlying metamorphic sequences of unknown provenance. Sougy (1962 b) elaborated on this idea, suggesting that the "West African fold belt" or Mauritanide orogen, on the western rim of

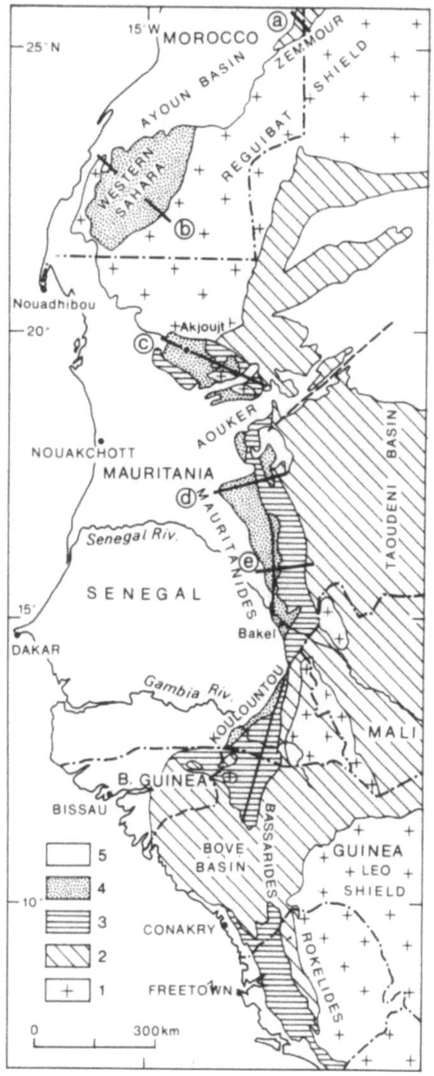

Fig. 1. Location of the Mauritanide, Bassaride and Rokelide orogens in West Africa. *1* Basement (older than 1800 Ma); *2* upper Proterozoic-Paleozoic sedimentary cover; *3* Pan-African belts; *4* late Paleozoic emplaced nappes; *5* Mesozoic to present cover. **(a)** to **(e)** location of sections of Fig. 10

[1] Laboratoire de Géophysique et Géodynamique case 422 Faculté des Sciences et Techniques de St-Jérôme, Université d'Aix-Marseille III, 13397 Marseille Cedex 13, France

[2] Department of Geology, University of Georgia, Athens, GA. 30602, USA

[3] Laboratoire de Pétrologie, Université de Nancy I – B. P. 239 Vandœuvre-lès-Nancy, Cedex, France

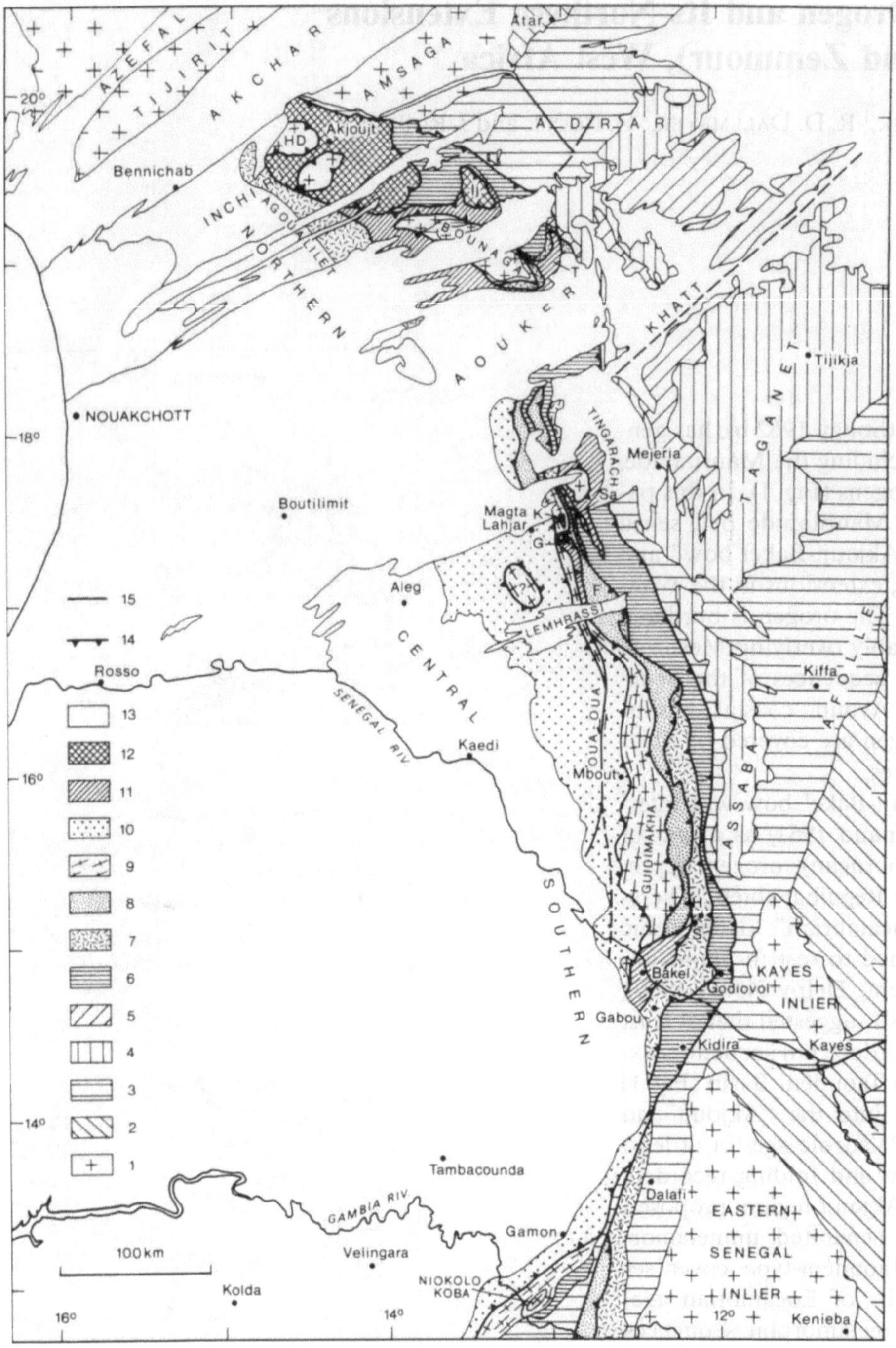

Fig. 2. Geological map of the northern, central and southern Mauritanides, after Chiron (1973), Dia (1984), Lécorché (1980), Le Page (1983). *1* Basement; *2* Supergroup 1; *3* Supergroup 2; *4* Supergroup 3; *5* Supergroup 4; *6* parautochthon; *7* imbricated ophiolitic units; *8* continental margin rift-facies association; *9* eastern calc-alkaline igneous complex; *10* western calc-alkaline igneous complex; *11* suprastructural allochthons; *12* nappes; *13* Mesozoic to recent cover; *14* main thrust; *15* main faults. *F* Farkaka, *G* Gaoua, *H* Hamdallaye, *HD* Hajjar Dekhen, *K* Kelbe, *Ij* Ijibiten, *S* Sélibabi, *Sa* Sangarafa, *T* Tamkarkart

the Precambrian shield was not comprised of a Precambrian basement but, instead, was largely of Hercynian (Late Paleozoic) age. Subsequent works (Bassot 1966, 1969; Sougy 1969; Lécorché 1969; Chiron 1973; Lécorché and Sougy 1978; Dia et al. 1979) have locally outlined the polyphased character of the Mauritanide orogen and documented at least local records of several late Proterozoic (Pan-African) and late Paleozoic events.

2 Stratigraphy-Paleontology

Compared with many mountain belts, miogeoclinal sediments are practically missing in the Mauritanides. The major features of the stratigraphy of the foreland are linked with the tectonic evolution of the belt and have been used, before the recent geochronological studies, as guides for the relative timing of the tectonic events.

In descriptions of these regions, authors did not always use correct terminology. We have chosen to use the terms "Group" for a lithostratigraphic unit, "Supergroup" for an ensemble of Groups limited at the bottom and at the top by a major unconformity, and "Unit" for a typical tectonic package of rocks. All other terms such as Series (or series) and not capitalized "group" (or "unit") are cited in the sense of their author.

2.1 The Precambrian Basement, Part of the West African Craton

The West African craton is largely concealed below the Proterozoic to Paleozoic sediments of the Tindouf and Taoudeni basins (Fig. 3). It is exposed only in the Reguibat and Leo uplifts, and in two smaller inliers east of the southern Mauritanides (Fig. 2).

2.1.1 The Reguibat shield

The West African craton crops out along a ENE-WSW axis, in the Reguibat shield or uplift (Fig. 3). Its western half consists mainly of metamorphic rocks in granulite or hornblende facies, late Archean in age. Barrère (1967) shows that these rocks consist of: (1) various types of gneiss (hypersthene, sillimanite-garnet, leptynitic-garnet), (2) pyroxene amphibolites, (3) magnetite quartzites, and (4) amphibolites and marbles, interpreted as a series of sedimentary to volcaniclastic rocks, isoclinally folded, striking north-south and verging west. Before metamorphism, this Saouda Series was intruded by basalts, gabbros and anorthosites and by conformable serpentinites.

Rocks of the Saouda Series appear to be restites within a complex of magmatitic and anatectic gran-

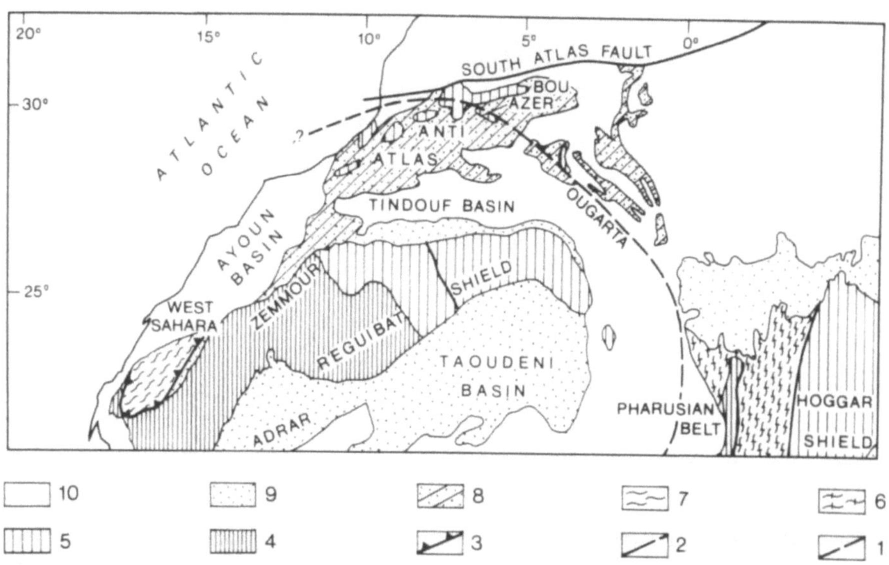

Fig. 3. Geological sketch map of northern West Africa. *1* Limit of the West-African craton; *2* main faults; *3* main thrust faults; *4* late Archean; *5* early Proterozoic; *6* Pan-African belt; *7* late Paleozoic nappes; *8* upper Proterozoic and Paleozoic folded cover; *9* upper Proterozoic and/or Paleozoic horizontal cover; *10* Mesozoic to present cover

ites known as "Rag el Abiod complex" (Barrère 1967). An 8-point Rb-Sr whole rock isochron age of 3016±135 Ma (all radiometric ages cited in this report are based on the decay constants and isotopic abundance ratios listed in Steiger and Jäger 1977) was reported by Vachette and Bronner (1975). Biotites generally record Rb-Sr and K/Ar ages between 1600 and 1870 Ma (Bessoles 1977). Clauer et al. (1988) reported $^{40}Ar/^{39}Ar$ plateau ages of 2400 Ma (hornblende) and 1850 Ma (biotite) from rocks of the Saouda Series in northern Mauritania. Therefore, at least portions of the West African shield were affected by a later thermal event, ca. 1900 Ma, responsible for recrystallization of the biotite. This event is related with the Eburnean orogeny which corresponds to the latest structuration of the craton.

2.1.2 The Kayes and Kenieba Inliers

The crystalline Precambrian basement exposed within the Kayes and Kenieba inliers (Fig. 2) is generally polydeformed, variably metamorphosed (low grade) volcaniclastic sedimentary rocks intruded by syntectonic or post tectonic granites (Bassot 1966). The granites have yielded Rb-Sr ages ca. 2050 Ma which correspond to the Eburnean event (Bassot et al. 1963). Pebbles of microgranites and microsyenites in folded conglomerate of a metasedimentary unit (Bassot 1969) could be remnants of an older (Archean?) basement. Isoclinal-type folds with cleavage parallel to the bedding are NE-SW striking and SE vergent.

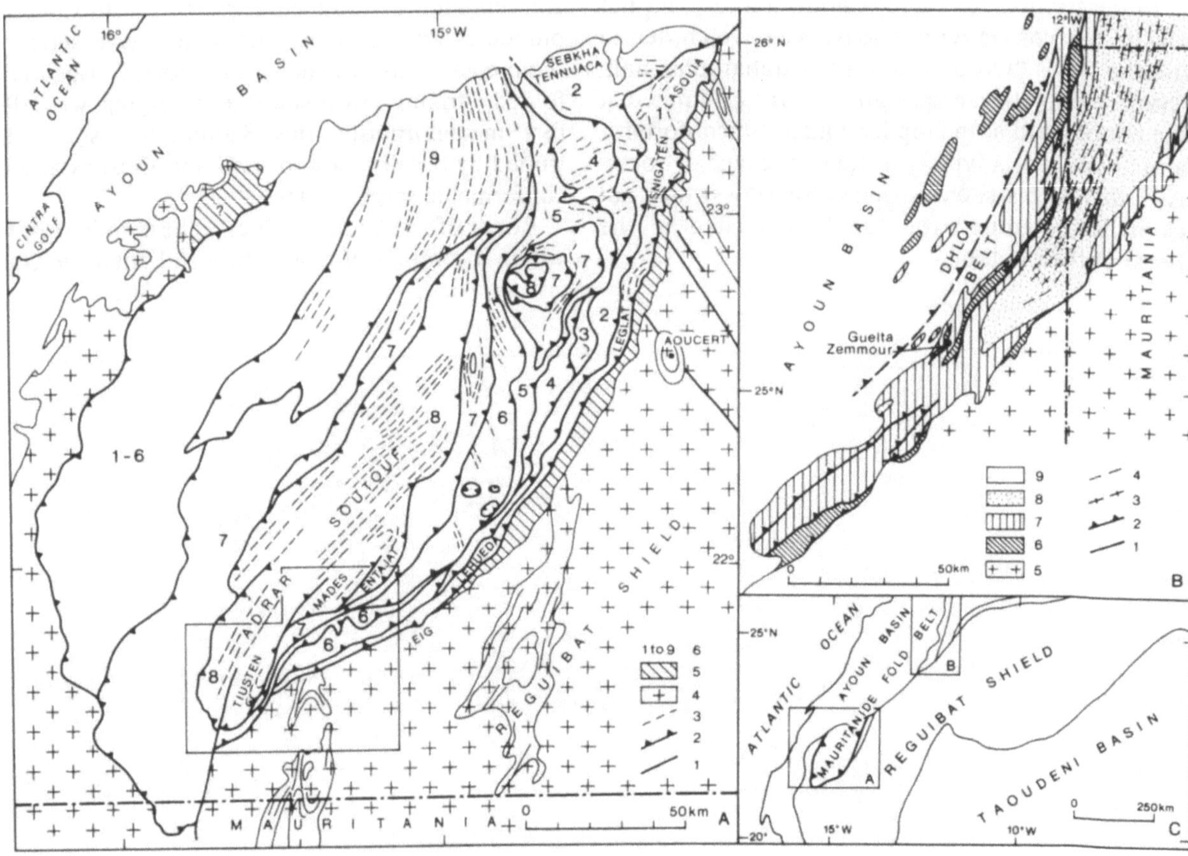

Fig. 4. The northern extension of the Mauritanide belt. **A** Structural setting of the Western Sahara nappes; **B** geological map of the Zemmour; **C** location of **A** and **B**. **A**-*1* Faults; *2* thrust faults; *3* structural trends; *4* basement; *5* Upper Ordovician to Devonian sedimentary cover; *6* nappes; *frame* location of Fig. 11. **B**-*1* Faults; *2* thrust faults; *3* anticline axis; *4* syncline axis; *5* basement; *6* Upper Proterozoic cover (Supergroup 1); *7* Cambrian to Silurian cover (Supergroup 2 and 3); *8* Devonian (Supergroup 4); *9* post-Paleozoic cover

2.2 The Proterozoic to Paleozoic Epicratonic Cover

Two major sedimentary basins are recognized on either side of the Reguibat uplift. The E-W trending Tindouf basin, lying to the north, can be traced into the Moroccan Anti-Atlas. The Taoudeni basin, in the south (Figs. 1, 3) can be traced as far as the Basaride-Rokelide orogens.

2.2.1 The Southwestern Part of the Tindouf Basin

2.2.1.1 The Zemmour Noir

The Tindouf basin (Fig. 3) swings southwestward in the Zemmour Noir region, at the boundary between the northeastern Mauritania and the southern Provinces of Morocco. In contrast with the southern rim of the basin, where Upper Ordovician tillites unconformably overlie the Reguibat basement, progressively older strata onlap with the southwestern rim of the basement (Fig. 4 B). The oldest of these units is composed of stromatolitic dolostones of the El Thlethyate Group of "Infracambrian" or upper Proterozoic age (Sougy 1964). The overlying Cambro-Ordovician Oumat el Ham Group (Fig. 5) has been studied by Destombes et al. (1969) who reported fragments of the oldest Trilobite fossils known in West Africa, *Fallotaspis* (Georgian) in the "Lower scolitic sandstones"; the Upper Tremadocian is characterized by shales containing *Acrothele* and *Anisograptidae;* the limit with the Lower Arenigian is fixed by the occurrence of *Dictyonema gr. murrayi;* the Lower Arenigian Trilobites are the youngest fauna characteristic of this group, below the "upper scolitic sandstones" which only contain undetermined Lamellibranchs.

The Upper Ordovician, Garat el Hamoueïd Group (Fig. 5), lies in pronounced erosional disconformity on the Oumat el Ham Group or on the basement. It is attributed to glaciogenic deposits (Sougy and Lécorché 1963) which, by comparison with the Anti-Atlas (Destombes 1968), would be upper Ashgillian in age.

Together, these three lithologic groups do not exceed 300 m in thickness. They are overlain by the Silurian (80 m) and the Devonian (1000 m), chronostratigraphically defined (Sougy 1964) at least up to the Frasnian (Fig. 5). Westward, these sequences become thicker; in the Dhloa region (Fig. 4 B), the pre-Silurian formations reach about 1500 m in thickness.

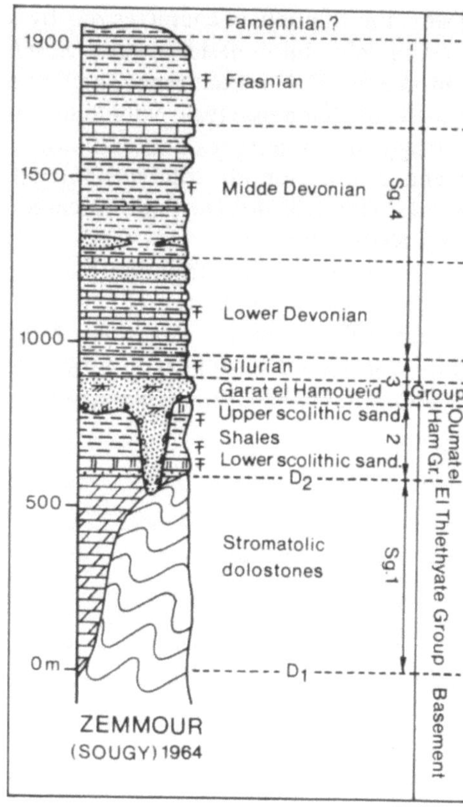

Fig. 5 Lithostratigraphic column of the Zemmour Noir

2.2.1.2 The Western Sahara

Southwest of Zemmour, these formations disappear below the Cretaceous cover of the Ayoun basin (Fig. 3), and reappear 200 km southwestward, in the Western Sahara (Fig. 4 A), as a thin NNE-SSW trending band 1 to 5 km wide and 200 km long along the metamorphic formations of the northern extension of the Mauritanide belt. They were first explained by a graben isolated in the basement (Alia Medina 1958; La Viña and Muñoz Cabezon 1958; Arribas 1960, 1968; Rod 1962). For Sougy (1962 a) and Sougy and Bronner (1969), they represent the eastern edge of a sedimentary cover almost entirely overthrust by the Western Sahara nappes and they rest in major unconformity on the Reguibat shield. A discontinuous basal formation, composed of conglomerates and argilaceous sandstones, is overlain by a continuous escarpment, 30 m high, of sandstones with local tillites (Bronner and Sougy 1969) at the base, like the Garat et Hamoueid Group of Zemmour. The basal formation (0 to 30 m) could be the infilling of the erosional Upper Ordovician erosional disconformity before deposition of the glacial sediments or an older formation (Lower Ordovician). Upon these formations, the Silurian is characterized by some shales and mainly blue limestones with *Scyphocrinus* and Orthoceras of Zemmour type. The Lower Devonian is also of Zemmour type with shales, more or less ferrugineous limestones with crinoidal debris, fragments of brachiopods, trilobites and ostracods, siltstones with *Tropidoleptus*, and calcareous siltstones with *Acrospirifer*.

The same type of formation has been observed on the aerial photographs on the western side of the metamorphic complex (Fig. 4 A). Southwestward, these formations disappear in turn, with unknown relations to the Paleozoic sequences farther south.

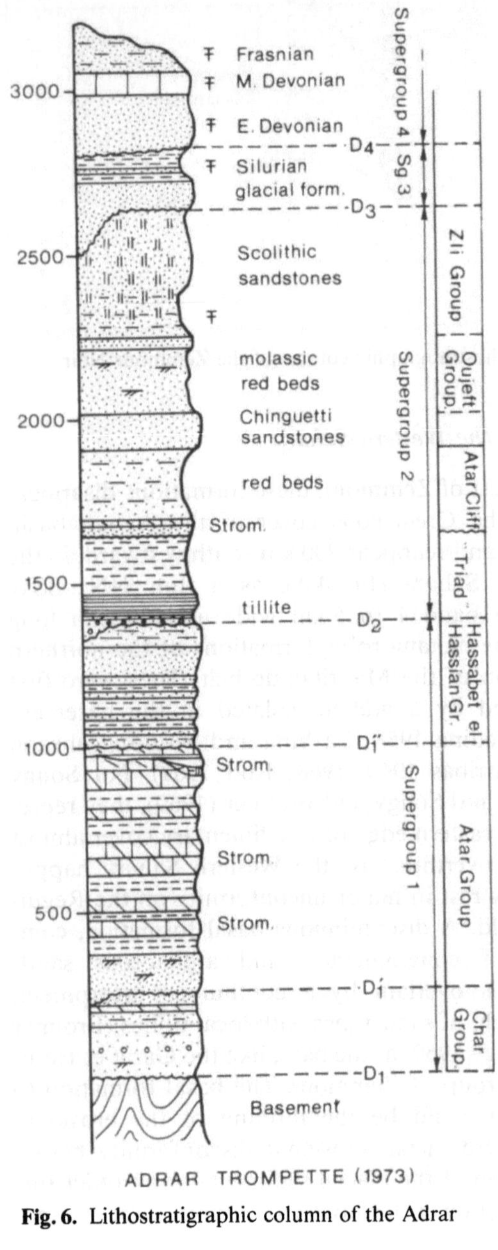

Fig. 6. Lithostratigraphic column of the Adrar

2.2.2 The Western Rim of the Taoudeni Basin: Adrar, Taganet, Assaba

2.2.2.1 Adrar

The continental to shallow marine Taoudeni basin does not exceed 5000 m in thickness, its basement acting as a rigid platform. Its stratigraphy is mainly known from field work at its edges for lack of information on the few oil exploratory drillings.

The Adrar, studied by Monod (1952) and by Trompette (1973), comprises the most complete sedimentary sequence of the basin and is generally used as a reference for correlations with the other regions (Figs. 2, 6).

Trompette (1973) has divided the Adrar succession in three lithostratigraphic "ensembles", further called "supergroups", controlled by significative events. A fourth, chronostratigraphic, ensemble corresponds to the Devonian. Carboniferous is only known in the Taoudeni region.

The first ensemble or Supergroup 1 (Fig. 6) rests with major unconformity on the Reguibat basement. It is composed of three main groups: the Char and Atar Groups made of alternating sandstones and shales with levels of stromatolitic dolostones and the Assabet el Hassian gritty-shaly Group. These groups are separated by erosional disconformities. All these rocks were deposited in continental or shallow marine conditions, clastics coming from NNE. The stromatolite reefs (mainly in the Atar Group) needed warm, calm, high-salinity, shallow water with few sediments (Trompette 1969). Bertrand-Sarfati and Raaben (1970) proposed a Vendian age for the stromatolites. Rb-Sr dating on the shaly sediments (Clauer 1976) yielded several diagenesis ages from 1000 Ma in the Char Group to 750 Ma at the top of the Assabet el Hassian Group, placing Supergroup 1 in the middle and late Proterozoic. The thickness of the sediments varies from 0 m in the neighbourhood of the Mauritanide belt to 3500 m in the Richat region. The NNW-trending Richat-Tiris trough northeast of Atar (Bronner et al. 1980) is probably related to a weakness zone of the craton. Consequently, the sedimentation rate was variable but roughly equal to a low subsidence rate of 4 m/Ma (Bronner et al. 1980).

Supergroup 2 (Fig. 6) rests unconformably on Supergroup 1 or on the basement. The slightly angular unconformity is increased by a strong erosional disconformity related to the glacial character of the basal sediments. Tillites, first demonstrated by Zimmermann (1960), are generally associated with barytic carbonates, silexites and shales. This association tillite-carbonate-silexite, Trompette's "triad", is characteristic of the basis of Supergroup 2 in the major part of the Taoudeni basin. It is succeeded in Adrar by thin conglomerates and quartzitic sandstones and by dolostones and dolomitic limestones with stromatolitic levels. SW of Atar and at the front of the Mauritanide belt, tillites and associated rocks are not observed. Along the northern front of the belt, the metamorphic formations overthrust a thin sedimentary veneer (tégument), which consists of conglomerates with reworked tillite pebbles, shales with some phosphatic levels and dolostones.

Elsewhere, the dolostone beds with stromatolites progressively pass up into red, mainly gritty, formations, molassic at the top, intercalated with the 100-m-thick, white to pink quartzitic Chinguetti sandstone that marks the top of the Atar Cliff and the structural surface of the first "tassili" (plateau) on which rests the molassic upper part of the red beds, the Oujeft Group. It is succeeded by scolithic sandstones, the Zli Group, the base of which contains some shaly micaceous levels with the inarticulate brachiopod *Westonia Chudeaui,* which could be assigned to the Cambrian-Ordovician limit (Legrand 1969). Compared with the continental red beds, these sandstones are shallow marine in origin and probably emerged during the Early (?) to Middle Ordovician. They have been strongly altered prior to deposition of Supergroup 3.

The basal, glaciogenic "triad" is derived from the NNE, while material in the red beds and upper sandstones come from the SSE. The mean thickness of Supergroup 2 is about 1250 m.

Supergroup 3 (Fig. 6) rests in strong erosional disconformity on Supergroup 2. The basal formations, more or less typically glacial, are assumed to be Late Ordovician in age. They consist essentially of sandstones of outwash type with large-scale cross bedding. The glacial manifestations are mainly the U-shaped paleotopography, very long eskers and polygonal grounds. True tillites are missing over the entire region. These formations are succeeded by sandstones with rare brachiopods and by shales with Silurian graptolites showing characteristics of a condensed sequence. The thickness of the Upper Ordovician sandstones varies from 0 to several hundreds of meters, while the Silurian does not exceed 80 m. The average thickness of the Supergroup 3 is about 200 m.

Supergroup 4 (Fig. 6) is transgressive upon the Ludlovian with coarse ferrugineous sandstones of Siegenian age and a silty Emsian. Eifelian and Givetian consist of limestones. The Frasnian is mainly composed of alternating siltstones and

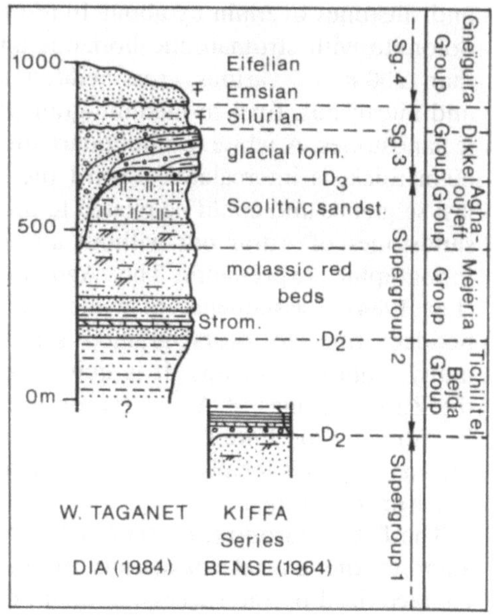

Fig. 7. Lithostratigraphic column of the Taganet

shales. Middle and Upper Devonian fossils (mainly brachiopods) show affinities with American and European faunas (Le Maître 1952). The thickness of the exposed Devonian is about 380 m (Bronner et al. 1980).

2.2.2.2 *Taganet*

Supergroup 4 is exposed along the NE-SW trending Khatt area (Fig. 2), the southern boundary of the Adrar. The Khatt is a depression invaded by the Aouker dune complex. It corresponds to the core of a major syncline of which the Taganet constitutes the southern flank and shows a sequence similar to the Adrar. But some differences must be noted (Fig. 7).

Supergroup 1 does not crop out in the Taganet but it could be preserved in a trough, similar to the Richat-Tiris trough, with a different, NE-SW trend (Bronner et al. 1980). Southeastward, in the Afollé area, it rests in major unconformity on the basement of the Kayes inlier.

Supergroup 2 (Fig. 7), roughly similar to that of the Adrar, is composed of three groups. The Tichilit el Beïda Group (Dia 1984) or Kiffa Series of Bense (1964) corresponds to the glacial sequence but tillites are never observed in the neighbourhood of the orogenic belt. The major part of the group consists of green shales and buff-coloured argillaceous siltstones with some levels of phosphatic cherts. The Méjéria Group (Dia 1984) consists of 30 m of conglomerates and mauve feldspathic sandstones and siltstones overlain by about 10 m of calcareous dolomite with stromatolitic horizons and by more than 200 m of various cross-bedded argillaceous and micaceous, fine- to medium-grained feldspathic sandstones. A white to pink quartzitic sandstone, 60 m thick, is intercalated toward the base of the last sequence and could be related to the Chinguetti sandstones of Adrar but without any comparable topographic expression. The Aghaoujeft Group (Dia 1984) is a sequence of white, plane or cross-bedded, fine- to coarse-grained sandstones with many scolithic horizons. It is quite comparable to the Zli sandstones of Adrar. Its top is truncated as in Adrar by the Late Ordovician erosional disconformity, which is locally established on a former angular unconformity.

The Dikkel Group (Fig. 7) of Dia (1984) is equivalent to the glacial formations of Supergroup 3, characterized by the succession of six main formations separated from each other by five internal erosional disconformities. Tillites have not been observed and a glacial origin is based on the paleomorphology (shape of the linked paleosurfaces, mega and micro-bars that can be interpreted as sandy proglacial deposits) rather than on the character of the sedimentation. The Gneïguira Group represents both the Silurian part of Supergroup 3 and the Devonian Supergroup 4. Dia distinguishes two formations. The lower, with Llandoverian faunas is mainly composed of argillaceous sandstones and micaceous shales. The Upper Silurian levels are unknown. The upper formation consists of microconglomeratic (sometimes oolithic) micaceous sandy shales or iron sandstones capped with massive fine- to coarse-grained sandstones. Some brachiopods indicate ages from Emsian to Eifelian.

These deposits are characteristic of a continental platform on which the sea transgressed rapidly. Sedimentation was more littoral than in Adrar, in a shallow marine context.

2.2.2.3 *Assaba*

In the Assaba (Figs. 2, 8), similar series present more continental characters. As already indicated, Supergroup 1 occurs only on the eastern edge of the Kayes inlier. Supergroup 2 occurs mainly in the Afollé plain, east of the Assaba (Kiffa Series of Bense 1964) and in the cliff of Assaba. Tillites, baritic dolostones, cherts and greenshales, graywackes and stromatolitic dolostones form the Bouly Group of Le Page (1983, 1986). Stromatolitic dolostones occur at the base of the eastern cliff of the Assaba plateau but, in the west, red beds rest directly on

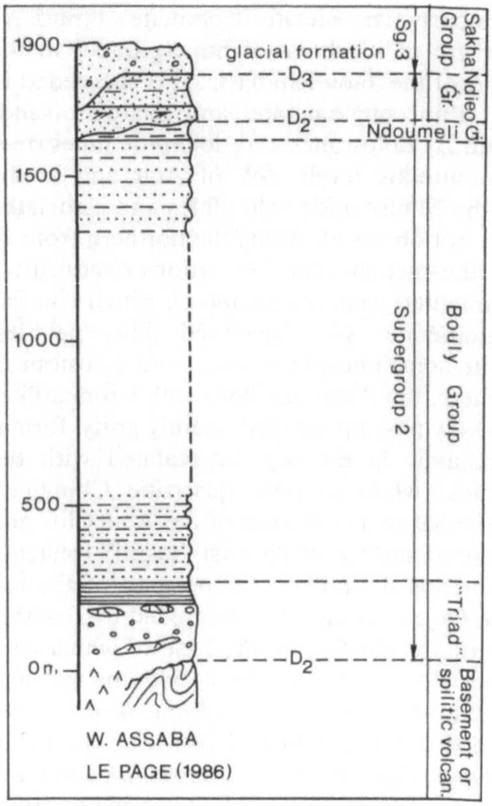

Fig. 8. Lithostratigraphic column of the Assaba

the graywackes. Their thickness (about 50 m in the south Assaba) increases gradually northward to reach 200 m in the south Taganet. They correspond to the Ndoumeli Group (Fig. 8) of Le Page (1983, 1986). In angular unconformity the Ndieo Group is mainly composed of largely cross-bedded white sandstones with some intercalations or red coarse-grained sandstones which may indicate short emersions. For Le Page (1983), the Ndieo Group (enriched northward with scolithic levels) may be compared to the Aghaoujeft Group of Dia.

Therefore the Bouly, Ndoumeli and Ndieo Groups, respectively similar to the Tichilit el Beïda, Méjéria and Aghaoujeft Groups of Dia, constitute the equivalent of Supergroup 2.

The top of the Assaba cliff is composed of sandstones that rest in strong erosional disconformity on any level of the Ndieo Group. These largely cross-bedded, coarse-grained quartzitic sandstones do not show typical glacial characters except their topography and the presence of long (several kilometers), esker-shaped, sandy bars. They constitute the Assaba Supergroup 3. Silurian and Devonian are unknown in this area. Sandstones of Silurian age reappear in residual hills north of the Bové basin

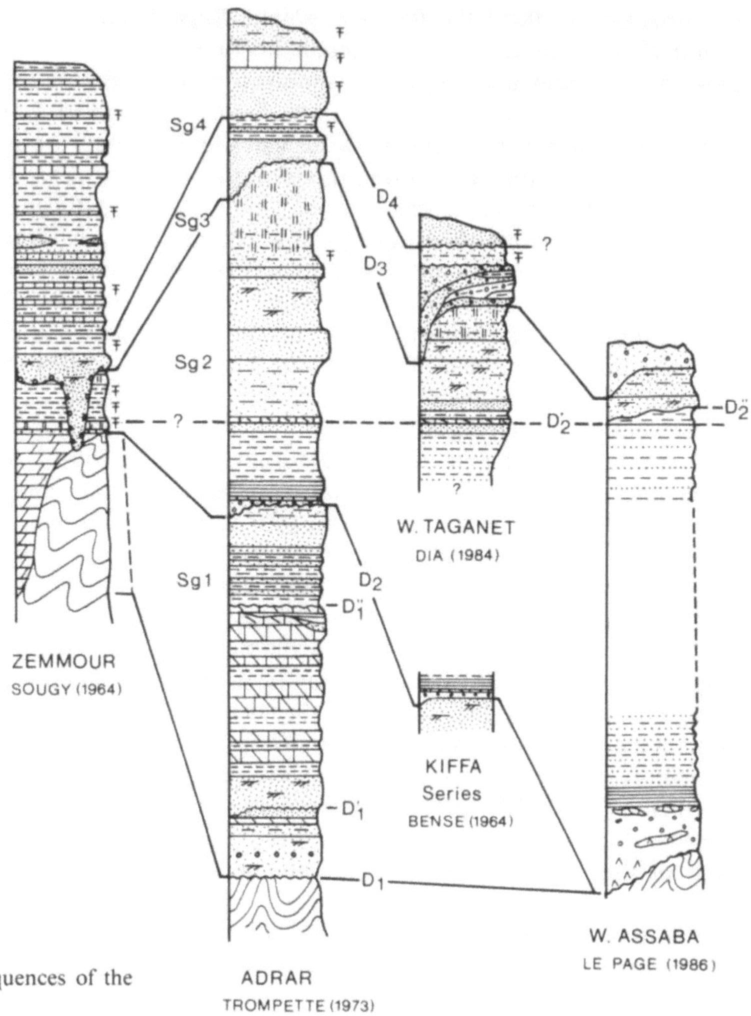

Fig. 9. Comparisons between the different sequences of the foreland

(Fig. 1). They are attributed to Pridolian (Drot et al. 1979) and particularly contain brachiopods, mainly *Diabolirhynchia Hollardi* as in western Adrar (Drot and Lécorché 1971) and in Moroccan Anti-Atlas.

2.2.3 A Common Lithostratigraphic Scale of the Foreland of the Mauritanide Belt

Correlations may be made with the southwestern flank of the Tindouf syncline (Fig. 9). The El Thlethyate stromatolite-bearing dolostones of Zemmour may be correlated with the Atar Group of Supergroup 1. But there are few analogies between Supergroup 2 and the Oumat el Ham Group which does not display early glaciogenic sediments and red beds but does expose a similar top sequence of scolithic sandstones with Lamellibranchiate fauna. If the recent record of Early Cambrian shelly microfossils at the top of glaciogenic sediments, at the base of Supergroup 2 in northern Guinea (Culver et al. 1988), applies in Mauritania, there is a good chronological correspondence with the base of the marine Oumat el Ham Group.

Correlations are very good for Supergroup 3. The basal glacial formations, characterized by true tillites and local striated floors (Sougy and Lécorché 1963) are present all along the southern and southwestern edge of the Tindouf basin, as far as the Western Sahara (Bronner and Sougy 1969). But in all these areas the glaciogenic morphology of the basal erosional disconformity is the most typical feature.

A thin, condensed, transgressive fossiliferous Silurian caps Supergroup 3. It thickens westward in Zemmour and Adrar. It appears more continental in Taganet and probably did not deposit in Assaba (Fig. 9).

North and south of the Reguibat shield, Supergroup 4 is composed only of Devonian, up to Frasnian, in Zemmour and Adrar and faunas are very similar. Lithologies (littoral facies, reefs, transgression - regression alternations) indicate that the Reguibat shield was probably emerged and partly surrounded by the Devonian sea.

Between and within these supergroups, the most important discontinuities are: D_1 at the base of Supergroup 1, on the basement; D_2 between the base of Supergroup 2 and Supergroup 1 or the basement; D_3 between the base of Supergroup 3 and Supergroup 2 or the basement; D_4 between the base of the Devonian and Supergroup 3.

D_1 is always a major, angular unconformity. D_2 is a complex unconformity: low-angle on Supergroup 1, of major-type on the basement, it is complicated by the effects of glaciogenic erosion before deposition of tillites (striated floors are common). D_3 is also complex; it seems to develop during Middle Ordovician after emersion of the scolithic sandstones and to result in the formation of a desert crust, eroded during the following glacial episod, when a new paleotopography was modeled. Locally, mainly in Taganet, Supergroup 2 was folded when emerging and D_3 is clearly an angular unconformity. But this character is limited to the western limit of the Taoudeni basin. It has not been observed in Zemmour.

Within Supergroup 1, Trompette (1973) indicates an erosional, glacial-type, disconformity (D'1) between the Char and Atar Groups, and a parallel disconformity (D''_1) between the Atar and Assabet el Hassiane Groups (Fig. 9).

Within Supergroup 2, there is a fundamental change between the basal, glacial, deposits derived from the NNE and the upper deposits derived from SSE. Generally these are in apparent conformity. Nevertheless, in the western foreland, the dolostones at the base of the red beds are more or less strongly folded whereas the red beds are, at the most, gently folded. As the contact is never seen, Lécorché (1980) interprets this difference in Adrar as a disharmony while Dia (1984) in Taganet assumes an angular unconformity D'_2 within the base of the Méjéria Group. For Le Page (1983) in Assaba, an angular unconformity D'_2 is higher, at the top of the red beds (Ndoumeli Group), before deposition of the Ndieo Group (Fig. 9).

Within Supergroup 3, Trompette (1973) finds evidence for a clear disconformity of the transgressive Silurian above the glacial sandstones. Such a disconformity seems to be present everywhere the Silurian was deposited.

2.2.4 Chronology

Before the Silurian, the series are very poor in characteristic fossils (Zemmour). In Adrar, some Rb-Sr ages on argillaceous sediments (Clauer 1976) yield ca. 1000 Ma on shales above the basal unconformity (Char Group), 900 to 800 Ma in the shales of the Atar Group, about 700 Ma in the Assabet el Hassiane Group. These ages indicate that Supergroup 1 developed from the end of the middle Proterozoic and during the upper Proterozoic.

The only Rb-Sr age on Supergroup 2, ca. 600 Ma, has been measured in the shales of the green formations of the glacial Btaat Ergil Group. Thus, Supergroup 2 tillites may have been deposited between 700 and 600 Ma but probably a little before 600 Ma.

The evolution of the foreland has long been the only link to date the tectonic events. Thus 650 Ma (Pan-African), Middle Ordovician ("Taconic"), post-Frasnian ("Hercynian") events have all been related to the main unconformities D_2 and D_3 and with the folding of the Devonian.

2.2.5 Conclusions

The foreland of the Mauritanide belt is characterized by a remarkable homogeneity of lithologies from the end of the middle Proterozoic to the Devonian. The major event reflected by this evolution is located ca. 650 Ma ago, before glaciation and deposition of Supergroup 2. The D_2 unconformity seems to be of regional extent and marks an important change between a ancient low-rate sedimentation on a rigid platform affected only by zones of weakness such as the Richat-Tiris trough and a new normal-rate cratonic sedimentation (Bronner et al. 1980).

Disconformity D'_2 marks a change between marine, rythmic green series and red continental sedimentation. It could be due to an orogenic event in the vicinity.

The D_3 disconformity marks a very important and extended climatic change but seems only locally to follow some folding of the foreland Supergroup 2. The origin of these folds is not elucidated.

The other disconformities only indicate fluctuations of the cratonic sedimentation, probably due to slight tilting of the large stable craton.

2.3 The External Zone of the Mauritanide Belt

Along the Mauritanides, metasedimentary units are generally comparable to the foreland and do not clearly indicate a miogeocline deepening.

2.3.1 The Basement-Related Formations

These are generally located in the eastern part of the belt and only south of the Reguibat uplift. The most extended is the Bou Naga window (Lécorché 1980), in the south-east of the Akjoujt area (Fig. 2), characterized by an association of high-grade metamorphic rocks of Reguibat type: gneisses, amphibolites, migmatites (Marcelin 1965, 1975). In the southeastern part of the window, these rocks are intruded by an alkaline, syenitic complex. U-Pb age determinations on surrounding gneiss yielded an Archean age of 2800 Ma and, on intruded syenites and associated trachytes, younger ages of 680 Ma (Blanc 1986). Therefore, this magmatic alkaline complex could be related to the rifting of the western part of the West African craton. The window would be itself thrust eastward during the evolution of the belt (Lécorché 1980), as suggested by its location 1000 to 1500 m above the present level of the autochthonous craton.

Two other windows, west and southwest of Méjéria, in the central Mauritanides, consist of poorly known retrogressed basement of various types. Compared with the Nouedgui-Bou Naga window it seems that they are only small slices of basement.

In the southern Mauritanides, the N-S trending belt, 120 km long, of Guidimakha granites has been interpreted (Lille 1967, 1969; Rocci 1962, 1969) as retrogressed Birrimian (Eburnean) granites 1800-2000 Ma).

In the Akjoujt area, the uppermost of the internal nappes of Lécorché (1980) has long been compared with the Archean basement of the Reguibat shield. Recent $^{40}Ar/^{39}Ar$ data (Dallmeyer and Lécorché, 1990a) show that the material is older than ca. 2000 Ma.

2.3.2 The Cover-Related Formations

The parautochthonous formations of the eastern, external part of the belt are generally folded, slightly metamorphosed, more or less schistose and very similar to the glaciogenic formations of the Supergroup 2. As the glacial characters in such terranes are not clearly established, authors (Le Page 1983; Dia 1984) generally use the term "mixtites" instead of tillites, for the polygenic conglomerates or conglomeratic schists.

In the Regs Unit of the Akjoujt area, Lécorché (1980) describes such a sequence quite comparable to Trompette's "triad" in Adrar.

In the central Mauritanides, the Djonaba Unit (Dia 1984) shows in its eastern part polygenic conglomerates, conglomeratic psammites (mixtites) and varved slates with flame structures and drop stones.

In the southern Mauritanides, the Mbalou Unit (Le Page 1983) contains "mixtites" with polygenic pebbles and dolostones below green graywackes and shales. The Hamdallaye Unit consists of microconglomeratic psammites intercalated in basic volcanics and green shales (plus some graywackes?).

The upper sequence (mainly sandstones) of the Supergroup 2 and the two other groups have not been observed in the belt, in spite of an abundance of quartzites.

Nevertheless, Crévola et al. (1974) observed Lower Devonian at Godiovol in the Mbalou Unit of the Southern Mauritanides. This led Le Page (1983) to distinguish, below the Devonian, three main lithologic sequences: the green, red and white "series". The green series is correlated with the glacial base of the Supergroup 2 up to the upper dolostones (equivalent of the dolostones with stromatolitic levels of the Adrar). The red series may be compared to the red beds below the scolithic sandstones of Adrar. The white series would be equivalent of the scolithic sandstones (upper Supergroup 2). These series are more or less present in all units of the southern belt.

2.4 The Internal Zone of the Belt

All units west of the cover-related parautochthonous sequences belong to the internal zone of the belt. Three main structural units are distinguished in view of their origin and/or of their geologic history: the infrastructural allochthons, particularly developed in the central and southern Mauritanides, the suprastructural allochthons, from the Western Sahara to Bakel, and the nappes located in the Western Sahara and in the Akjoujt area (Figs. 2, 4).

2.4.1 The Infrastructural Allochtons

They have been defined (Dallmeyer and Lécorché 1989) in the central Mauritanides and include the "ultrabasic and basic complex" (UB-B) and the "western acidic plutonic-volcanic complex" (Dia 1984). They are composed of imbricated ophiolitic units, a continental margin rift-facies association and a calc-alkaline igneous complex (Fig. 2).

2.4.1.1 The Imbricated Ophiolitic Units

Three units in the central Mauritanides were interpreted by Dia (1984), to represent members of a fragmented ophiolitic sequence. Since Rémy's work (1987) only two of them, the El Aoueïja and Oued Amoûr Units, are considered to possess possible ophiolitic affinities. The El Aoueïja Unit is largely comprised of polydeformed, metamorphosed (middle to upper greenschist facies) mafic units that include metabasalt, metagabbro, chlorite-sericite schists, calcite-chlorite (±talc) schist, minor limestone and local serpentinites. Southward, in the equivalent less metamorphosed Hamdallaye Unit of Le Page (1983), a detailed geochemical study of the volcanic components (Rémy 1987) indicates that protoliths were intraplate alkaline basalts followed by a second generation of T-type MORB composition.

The Oued Amoûr Unit is more heterogeneous and includes a variety of polydeformed, metamorphosed (middle to upper greenschist facies) lithologies: serpentinized metaperidotites, metagabbros, minor metabasalts, quartzcarbonates, chlorite-epidote schists and iron carbonate-chlorite schists.

In the southern Mauritanides, these two Units are grouped in the Mauritanian Hamdallaye Unit and the Senegalese Gabou Unit (Le Page 1983). In the northern Mauritanides, these units may not be defined in continuity. Nevertheless, such lithologies (Lécorché 1980) occur in the SE where they directly overthrust the Frasnian and probably in the SW (Agoualilet Unit). They are unknown in the Western Sahara and, of course, in Zemmour.

2.4.1.2 Continental Margin Rift-Facies Association

Initially interpreted as the deepest part of the dismembered ophiolite sequence (Dia 1984), the Farkaka Unit is comprised of a diverse group of internally imbricated, polydeformed, comparatively high-grade (middle-lower amphibolite facies) polymetamorphosed metasedimentary rocks: variably micaceous quartzites, iron-quartzites, kyanite-staurolite-garnet-mica schists and banded iron formations. It also includes metamorphosed igneous rocks such as serpentinites and associated epidote-actinolite schists, chlorite schists and variably foliated amphibolites.

A detailed geochemical investigation on the amphibolites (Rémy 1987) indicates that they are hypersthene and quartz normative and characterized by a marked negative Nb anomaly and a significant light-REE enrichment. Rémy (1987) suggests that these characteristics imply a low-Ti continental tholeiite protolith.

The Farkaka Unit may be followed in continuity from the Aouker dune complex, where it occurs in some unsanded areas with a high-grade (amphibolite to granulite) metamorphic facies to the latitude of Sélibabi with a southward decreasing intensity of metamorphism (albitic micaschists of the Diala-Bouanzé series of Lille 1967, and Chiron 1973).

These lithologies are unknown in the northern Mauritanides and in the Western Sahara.

2.4.1.3 The Calc-Alkaline Igneous Complex

Defined as western acid plutonic-volcanic complex by Dia (1984), it is divided into the Kelbé and El Hneïkat Units which form the western part of the exposed belt. Intrusive components are represented in the Kelbé Unit, which is largely comprised of variably deformed muscovite leucogranite and two-mica granites. Local textural variants range from pegmatitic to aplitic. Most exposures are penetratively deformed and display variably annealed, low-temperature mylonitic or protomylonitic textures.

Extrusive elements of the calc-alkaline suite have been included in the poorly exposed El Hneïkat Unit: metamorphosed (upper greenschist facies), polydeformed volcaniclastic schists of various types with chemical characteristics suggesting protoliths of rhyolite, dacite, andesite and basaltic tuffs (Dia 1984). Subordinate, locally intercalated metavolcanic horizons have similar chemical characteristics. These are typically quartz and hypersthene normative, relatively high in Al_2O_3 and CaO and relatively low in TiO_2 and $Mg/Mg+Fe^{2+}$. Metasedimentary rocks, mostly quartzites, are other components of the El Hneïkat Unit.

The calc-alkaline complex extends southward and is largely exposed in the southern Mauritanides, where the Guidimakha granodiorite (Toukobra Unit of Le Page 1983), and the metarhyolites, metatuffs and metaandesites of the Mbout Unit, imbricated with the metasedimentary sequence of the western Oua-Oua Mountains, show similar characteristics. Relatively undeformed calc-alkaline granite and related rhyolites (Bassot 1966; Villeneuve 1984), rhyodacite and andesites occur SSW of the Bakel area at the junction between the Mauritanides and Bassarides. With the Niokolo-Koba and Simenti Units of Villeneuve (1984), they could represent the southwestern extension of the same calc-alkaline complex.

2.4.2 The Suprastructural Allochthons

The suprastructural allochthons consist of metasedimentary rocks, mainly quartzites, that constitute more or less extended structural units overthrusting any term of the former parautochthonous or infrastructural units. They are mostly developed in the northern Mauritanides (quartzite nappe of Lécorché 1980) where they largely conceal parts of the parautochthonous and allochthonous sequences. Internally, these very mature feldspar-free quartzites display different kinds of lithologies: metaconglomerates with elongated quartz or quartzite pebbles, coarse- to fine-grained metasandstones, siltstones. Intercalated quartzitic schists are strongly cleaved.

In the southeastern Aouker, west of Méjéria (Fig. 2), the partly exposed Tingarach massif is characterized by a sequence of slightly metamorphosed quartzites at the top of which a deformed fossiliferous succession has been found. Poorly preserved brachiopodes probably indicate Early or Middle Devonian. Correlation with the foreland led Dia (1984) to suggest a Cambrian to Middle Devonian age for the Tingarach succession. From Tingarach, quartzites attributed to Upper Ordovician extend to the south in the Sangarafa massif where they structurally rest on the parautochthonous Djonaba Unit.

West of Sangarafa, the Gaoûa allochton is comprised of white, variably micaceous middle-grade mylonitic quartzites supposed to be lower in the Paleozoic succession (Dia 1984).

Such quartzites, different from the parautochthon, are found all along the belt as far as the Bakel area but their structural setting is yet to be discussed.

2.4.3 The Nappes

The nappes are located in the Akjoujt area of the northern Mauritanides (Fig. 2) and in the Western Sahara (Figs. 3, 4 A).

In the northern Mauritanides, the internal nappes (Lécorché 1980) are situated in structural setting on the basement or on a very thin sedimentary cover of Taoudeni affinity in the north along the Reguibat shield, on the parautochthon or on the suprastructural allochthons elsewhere. They consist of a pile of various units comprising: low-grade (upper to middle greenschist) metasediments, sericite schists, metasandstones, hematite quartzites, siliceous carbonates; volcanics and volcaniclastics, metadolerites and metabasalts, chlorite schists, mylonitized metarhyolites; high-grade (amphibolite to granulite facies) rocks similar to the basement of the Reguibat shield, gneisses, leptynites, amphibolites, pyroxenites, magnetite quartzites (banded iron formations) at the bottom as well as at the top of the pile. The easternmost nappes are preserved in the Ijibiten klippen (Fig. 2) ten kilometers west of the front of the belt.

In Western Sahara (Figs. 3, 4 A, 11), the main structure is a NNE-SSW trending mega-synform of nappes. These nappes rest either on the basement (in the south and southwest) or on the Upper Ordovician to Devonian foreland in the southeast. It seems that such a "tégument" re-appears locally below the northwest facing edge of the structure.

The geological and structural data have been mainly established through photogeological and remote sensing studies (Marchand et al. 1984 a, b; Bronner et al. 1983; Boeri 1985). Field data are poor, located along two or three partial cross sections (Sougy 1962 a; Sougy and Bronner 1969). Therefore, the following descriptions are preliminary results and must be used with caution.

From the middle of the nappe synform to the east, i. e. from the top to the bottom, nine main units have been recognized especially through their dominant facies, their structure and the thrust contacts. Nappe 9 appears to be made of steeply dipping beds of greenschists. Nappe 8, one of the thickest and most largely exposed of the pile, is characterized by dark exposures of hypersthene gneisses, amphibolites, gabbros and norites that constitute the Adrar Soutouf range. These high-grade (granulite facies) rocks show many similarities with the Archean basement of the Reguibat shield. Nappe 7 is composed of amphibolites, sheared granites, marbles, leptynites (amphibolite to granulite facies), also similar to Archean rocks and to the Archean-type association of the highest nappe of the Akjoujt pile. Nappe 6 is made of sheared granites trending N-S. Clear quartzites of nappe 5 occur as inselbergs contrasting with such surrounding soft material as micaceous quartzites and schists. Nappe 4 consists of amphibolites, marbles and various schists. Its photogeological signature is dark, with complex tectonic features due to piling-up of nested imbrications. The granitic nappe 3 is limited to the Guelb Dahamane and its surroundings. The nappe 2 is comprised of iron quartzites, greenstones, leptynites, amphibolites and schists; it is largely exposed in the north. Nappe 1 is composed of low-grade (middle greenschist facies) metasedimentary conglomerate, quartzites, sericite schists, ferrugineous quartzites and could be compared to the suprastructural allochthons described in the south. It may be traced almost all around the nappe synform.

Nappes 1 to 8 have been controlled in the field and in part sampled within the eastern half of the nappe synform; the western half is very poorly known from a few Spanish surveys.

2.4.4 Conclusions

From south to north, the Mauritanide belt shows a series of changes that provide valuable informations on its history.

The southern and central Mauritanides are characterized by a readily observable continuity: from E to W, the parautochthon and the infrastructural allochthons are present in all areas. The poorly extended suprastructural allochthons or related formations appear in different structural setting along the segment. Nappes of the northern type are unknown.

The northern Mauritanides display an obvious change; the parautochthon tends to disappear northward; the infrastructural allochthons are poorly exposed, scattered and very incomplete; the supracrustal allochtons and the nappes are largely extended; on the northern front, the latter rest directly on the foreland.

In the Western Sahara, the parautochthon and the infrastructural allochthons have entirely disappeared. The nappes take on a major extent and rest directly on the foreland.

3 Structure

From south to north, we present sections in the southern Mauritanides, central Mauritanides, northern Mauritanides, Western Sahara and Zemmour (Fig. 10).

3.1 Structure of the Southern Mauritanides

The southern Mauritanides have been studied mainly by Bassot (1966), south of the Senegal River, by Lille (1967) and Chiron (1973) north of the Senegal River in the Guidimakha area, by Le Page (1983) from Kidira (Senegal) to the latitude of Mbout (Mauritania), but more carefully in the Kidira-Bakel area. From east to west we cross the belt from the foreland to the edge of the overlying Meso-Cenozoic basin (Fig. 10 e).

3.1.1 The Foreland and the Parautochthon

South of the Senegal River, the foreland consists of the northern part of the Birrimian Kenieba Inlier and its basal, glaciogenic cover (Supergroup 2), which extends largely northward.

According to Le Page (1983, 1988), this autochthonous cover is progressively folded westwards, as far as the NE-SW trending Kidira thrust which marks the beginning of a parautochthonous foreland domain, the Kidira Unit. This unit consists of the same formations, but the basal mixtites are locally superposed on older volcaniclastic spilitic formations, and the "green series" (lower Supergroup 2) is particularly well developed and could reach more than 1000 m. Red beds of molassic affinity occur in the core of the synclines.

The deformation increases westward with the appearance of a slaty cleavage linked with flat shear-zones and folded by east-vergent N-S to NE-SW trending folds. Limited thrusting seems to close this evolution (Table 1). In the westernmost part of the unit, sequences are completely overturned.

North of the Senegal River (Le Page 1983, 1988; Fig. 10 e) the scheme is the same along the Birrimian Kayes Inlier. The Kidira Unit becomes the parautochthonous Mbalou Unit separated from the autochthon by the Massar thrust. The autochthon enlarges northwestwards with the development of the horizontal upper Supergroup 2 and Supergroup 3 on the Assaba plateau. In the Mbalou Unit, the folded Devonian of Godiovol (Crevola et al. 1974) rests directly on the upper sandstones of Supergroup 2; these sandstones may themselves rest directly on the green basal series, indicating that the disconformities increase largely toward the Mauritanide belt. The deformation is comparable to the Senegal section although the most internal, overturned, sequences are missing.

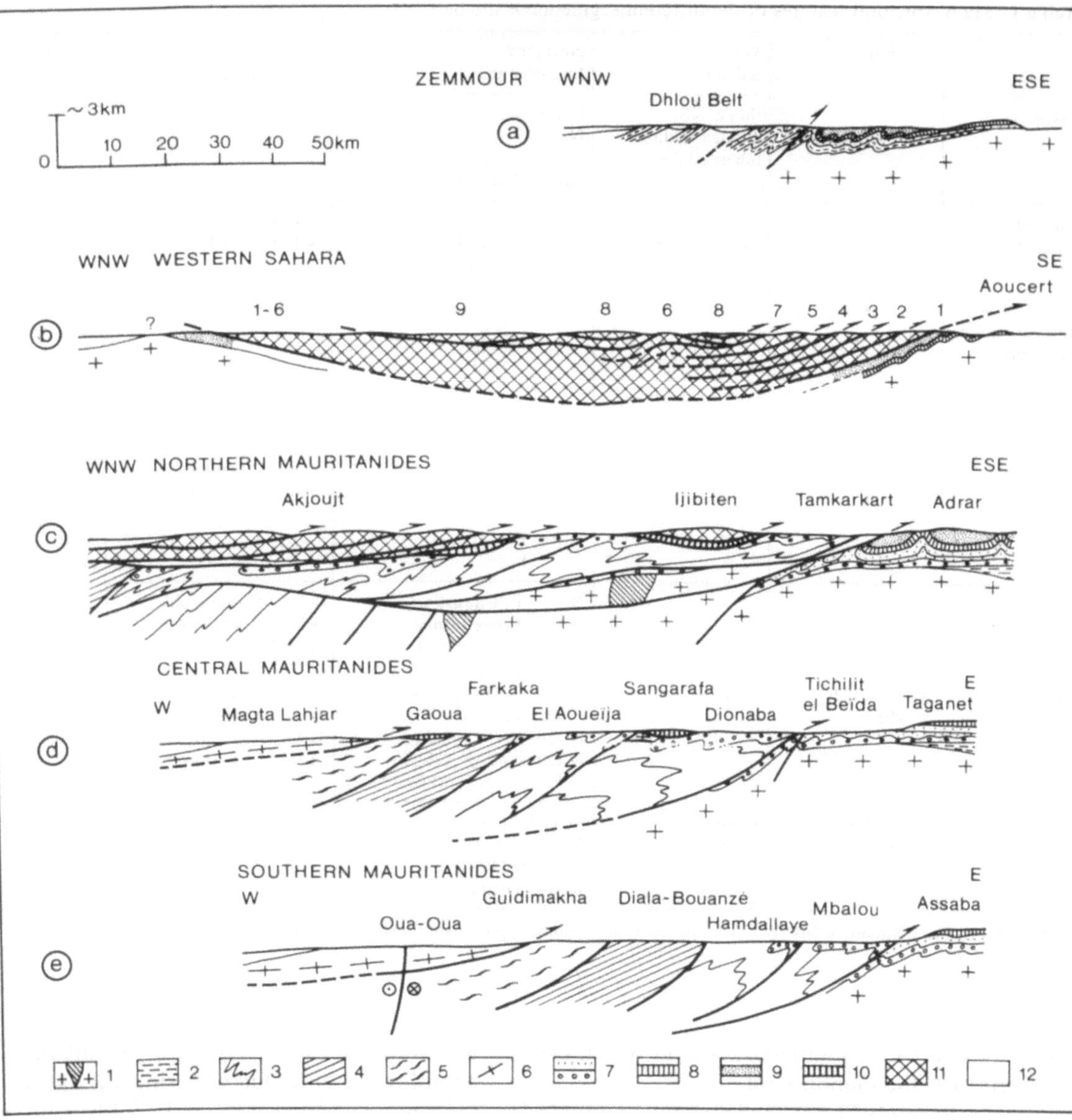

Fig. 10. Cross-sections of the different parts of the Mauritanide orogen (location in Fig. 1). *1* Basement; *2* Supergroup 1; *3* imbricated ophiolitic units; *4* continental margin rift facies association; *5* eastern calc-alkaline igneous complex; *6* western calc-alkaline igneous complex; *7* Supergroup 2; *8* Supergroup 3; *9* Supergroup 4; *10* suprastructural allochthons; *11* nappes; *12* Mesozoic to present cover

3.1.2 The Infrastructural Allochthons

West of the autochthonous and parautochthonous foreland, Le Page (1983, 1988) describes a pile of metamorphosed thrust units that he separates into external and internal units (Fig. 10 e).

The external units may in part be compared to the imbricated ophiolitic units and to the continental margin rift-facies association of central Mauritanides with which they are in continuity. The internal units correspond to the western calc-alkaline complex.

East of the external zone, the Hamdallaye Unit is composed of various low-grade lithologies with occurrence of pillow-lavas, interpreted as possible ophiolitic constituants (Rémy 1987). It extends south of the Senegal river as the Gabou Unit. Both of them overthrust the parautochthon. Serpentinites locally bound the thrust contact. The Diala-Bouenzé Unit (continental margin rift-facies association) overthrust in turn the Hamdallaye Unit, but does not extend south of the latitude of Sélibabi. According to Le Page (1983) it overthrusts the

Table 1. Main structural features of the different segments of the belt

	Zemmour	Western Sahara	Northern Mauritanides	Central Mauritanides	Southern Mauritanides
Nappes		thrust →S various folding inherited def.	thrust →E N 0°-45° F_2 →E N 30°-60° F_1 →SE inherited deform.		
Suprastruct. Allochthons		? stretch. L→S	N 0° N 150°-0° kink →E stretch. L →N 80° $S_1//S_0$	E-W+Thrust →S N 160° F_3 →E	
Infrastructural Allochthons[a]			(IOU) ? N 160° F_2 F_1	F_3 open folds thrusting N 45° F_2 knee folds thrusting E N 135° F_1 recumbent	(W-CAIC) Senestr. N 0° strike-sl. N 0° F_4 vertic. S_4 thrusting N 0° F_3 →E thrusting S_2 Late thrusting F_3 open folds thrusting N 0° F_2 recumbent N 0° F_1 isoclinal
Parautochthon	thrusting N 15° →E		N 0° F_4+thrusting N 160° F_3 overturned ? F_2 recumbent	F_3 open folds N 0° F_2 knee folds N 135° F_1 recumbent	thrusting N 0° F_2 recumbent
Foreland	N 10°-20° F_4 →ESE	N 30° F_4 →SE	N 30°-50° F_5 open folds thrusting →E N 130°-160° F_4 (Sg 2-4) N 120°-140° F_{2-3} (Sg 2) N 120° F_1? (lower Sg 2)	N 0° F_4 (Sg 2-4) N 30°-60° F_3 (upper Sg 2) N 45° F_2 (lower Sg 2) N 160° F_1 (lower Sg 2)	N 0° F_1 cylindrical
Authors	Sougy (1964) Dacheux (1967)	Bronner (unpubl.)	Marcelin (1964) Rippert (1973) Lécorché (1980) Kessler (1986)	Dia (1984)	Le Page (1983)

[a] IOU: imbricated ophiolitic Units; W-CAIC: western calc-alkaline igneous complex. F: fold axes; S: schistosity; L: lineation; Sg: supergroup

southern quartzitic Gounia-Bakel Unit which clearly overthrusts the Gabou Unit.

For Le Page (1983, 1988) deformation (Table 1) begins by N-S F_1 folds, isoclinal with S_1 slaty cleavage in the external zone, cylindrical in the parautochthon. Such folds are unknown in the internal zone. Deformation increases with curvature of F_1 folds, development of F_2 recumbent folds (S_2 composite cleavage) and reactivation of S_1 in shear zones leading to ESE thrusting, as shown by a widespread stretching lineation. This thrusting episode is the first event recorded in the internal zone (our calc-alkaline igneous complex). It is followed by overturned, east-vergent F_3, folds which are in turn strongly arcuated in the eastern part of the zone by a WNW-ESE stretching linked to a stretching lineation. In the external zone F_3 folds trending NS are open folds with rough fracture cleavage.

Last events in the parautochthon and in the external zone are respectively horizontal shear zones and thrusting. In the western calc-alkaline igneous complex the N-S Oua-Oua sinistral ductile wrench-fault is related to late N-S F_4 folds, with strong S_4 penetrative cleavage and recrystallization of phyllites along the S/c planes.

3.2 Structure of the Central Mauritanides

Work within the central Mauritanides have been carried out by Chiron (1973) and Dia (1984; see Fig. 10 d).

3.2.1 Structure of the Foreland

The Taganet foreland, which displays a large part of the Paleozoic succession, has recorded the history of the belt better than the southern Mauritanide foreland.

Dia (1984) indicates four phases of deformation (Table 1). The two first phases are recorded in the Tichilit el Beïda Group (lower Supergroup 2) immediately east of the front of the belt. They display interference pattern related to cross folding. The first F1 set, associated within shales to a thin fracture cleavage, trends WNW-ESE, decametric to kilometric. The F2 folds NE-trend SW with wave length on kilometers. The third event, not recorded in the Tichilit el Beïda Group, consists of F3 folds, N30 to 60° trending developed in the molassic Méjéria Group up to scolithic sandstones before deposition of the Upper Ordovician glacial sediments. A last event is recorded in the whole sequence up to the Devonian. It consists of F4 box folds of large wave length, trending NNW-SSE to N-S, and of limited eastward thrusting.

In our opinion, and after visiting this area with an IGCP 233 excursion (December 1987), phases 2 and 3 could be the same but disharmonic between less competent sequences (Tichilit el Beïda Group: lower Supergroup 2) and a more competent one (Méjéria and Aghaoujeft Groups: upper Supergroup 2).

3.2.2 The Parautochthon

Dia (1984) distinguished three main deformational episodes (Table 1):

1. Overturned to recombent, NE vergent folds with an associated penetrative cleavage reactivated by subsequent shearing and imbrication; stretching lineations, cigar-shaped pebbles are associated with this episode;
2. N-S trending deformation of knee fold-type with development of a fracture cleavage and boudinage; friction striae are related to limited thrusting;
3. slight deformation leading to open folds or simple corrugations:

They have mainly been observed in the low-grade (upper to middle greenschist facies) imbricated Djonaba Unit of the external part of the belt. Other wedge-shaped western units, mainly sericite quartzites, imbricated in the infrastructural allochthons are interpreted as dismembered parts of this parautochthon (Fig. 10 d).

3.2.3 The Infrastructural Allochthons

In the infrastructural allochthons (El Aoueïja, Farkaka and Magta Lahjar Units: Fig. 10 d), Dia (1984) groups two ductile, tangential, shearing episodes (D_1) and (D_2) in a progressive deformation. (D_1) is characterized by isoclinal and/or sheath F_1 folds with a penetrative schistosity, reactivated by a mylonitic foliation, and a stretching mineral lineation in a generally amphibolite facies metamorphic climate; (D_2) is characterized by east-vergent F_2 knee folds with associated fracture cleavage and stretching lineation in a retromorphic (middle greenschist) climate. This deformational sequence resulted in a general transport and imbrication of the units first northeastward (D_1) and then southeastward (D_2). In Dia's opinion, transportation is eastward-directed. Both episodes would represent a series of events conveying the stages of a progressive tectonic deformation during the same orogeny.

A third phase consists of large N-S F_3 open folds, kink bands or bending.

3.2.4 The Suprastructural Allochthons

The quartzitic Sangarafa and Gaoua Units are interpreted (Dia 1984) as klippen overlying respectively the parautochthon and the infracrustal allochthons (Fig. 10 d). Such a structural setting appears plausible on aerial photographs but has not been clearly observed on the field.

The Sangarafa Unit displays two deformations and a low-grade (lower greenschist facies) metamorphism. In the Sangarafa s. s. area, NNW-SSE trending east vergent folds are truncated by a flat basal thrust fault. Within the Aouker (Tingarach area), Cambrian (?) to fossiliferous Devonian metasediments are folded in similar fashion but appear to have been subsequently imbricated by a phase of south directed thrusting associated with secondary folds trending E-W.

The Gaoûa Unit, for Chiron (1973), overthrusts the "axial zone" but is rooted westward. For Dia (1984), the Gaoûa Unit is a klippen set up upon the infrastructural allochthons; its metamorphic grade, higher than in the Sangarafa Unit, indicates that it is more internal than Sangarafa. Nevertheless, it has been clear in the ground for many geologists of the 1987 IGCP excursion that the Gaoûa quartzites were imbricated with the associated schists. The imbrication is related to the penetrative cleavage in the schists. Closed or arcuated structures are observed on aerial photographs. They seem similar to those described by Le Page (1988) in the southern quartzites (Bakel Unit) that we consider to be rooted.

3.3 *Structure of the Northern Mauritanides*

After the first evidence for large overthrusting in the Akjoujt region (Tessier et al. 1961), the western area was studied by Giraudon (1964), Michaud (1964), Marcelin (1964, 1975), while Lécorché (1965, 1969, 1980) studied the unknown eastern area.

3.3.1 Structure of the Foreland

As the belt turns westward, its front, which runs along the westernmost part of the Adrar, comes in contact with the Reguibat shield. From W to SE, the frontal units of the belt successively overthrust (Fig. 2) the Archean basement, a thin veneer of Adrarian lower Supergroup 2 affinities, the different terms of upper Supergroup 2, Supergroup 3 and the Devonian.

The general trend of the unfolded foreland is NE-SW, dipping gently (4 to 6°) southeastward. General trends of the folded area parallel the frontal zone of the belt, i. e. N-S to NW-SE.

On the northern front, the basement seems unaffected. The autochthonous sedimentary veneer displays, only in the contact zone, asymmetric E-W to N120° microdragfolds indicating a north-northeastward motion of the overthrusting complex.

The western edge of Adrar displays various types of deformations (Table 1). To the north-west, the deformed area is narrow (1 km width) along the NW-SE trending front of the belt. NW-SE folds, reverse faults, back thrusting, curvature and rotations are observed. Southward, along the border, the folded zone widens (30 to 40 km). Large (several kilometers) box folded synclines develop to the east, separated by sharp narrow N-S anticlines (Fig. 10 c). Reverse N-S faults and limited horizontal thrusts may appear in the steeply dipping flanks of these folds.

3.3.2 Structure of the Parautochthon

Reconstitution of the lithostratigraphy and polarity criteria suggest that this lower Supergroup 2-like complex has suffered a first tectonothermal deformation interpreted as large recumbent folds, under low-grade (upper greenschist) metamorphic climate with development of an almost penetrative cleavage (Table 1). A second phase of ENE vergent overturned folds with reactivation of the former cleavage but without metamorphic recrystallization is in turn deformed by a third approximately N-S set of asymmetric to overturned east-vergent folds. Large-scale thrusting upon different terms of the foreland seems closely related to the lithology of the parautochthon, especially carbonates (underlying the red beds), which overlie the main thrust for at least 100 km.

3.3.3 The Infrastructural Allochthons

The infrastructural allochthons are very poorly exposed. They occur within the parautochthon in the southeastern area (Tamkarkart, Figs. 2, 10 c), close to and even in contact with the Frasnian folded foreland. According to Rippert (1973), these rocks have suffered at least a tectonothermal event before deposition of the tillites of Supergroup 2. Their structural position suggests that they have been transported together with the parautochthon, during the last phase of thrusting, over the Frasnian. Other rocks of this type occur in the southwestern part of the region in the Agaoualilet area. They display highly regular NW-SE trending, vertical dipping, penetrative cleavage and a middle to upper greenschist facies grade of metamorphism.

3.3.4 The Suprastructural Allochthons

The "quartzite nappe" of Lécorché (1980) is characterized by a penetrative cleavage, generally parallel to the stratification, often evolving into mylonitic foliation. Related folds have never been observed. In the conglomeratic beds, schistosity surrounds elliptic pebbles, elongated in a N 80° direction as the stretching lineation. Simple kink-bands deform the schistosity. Their axial plane is oriented N-S to N 150° with westward dips ranging from 60° to 80°, showing a relative westward motion of the upper part. Late N-S irregular, angular, asymmetric folds with broken hinges filled with quartz breccia seem to be related to late brittle deformation.

The structural setting of these quartzites has never been clearly demonstrated due to the lack of three-dimensional exposures. Lécorché (1980) interprets them as overthrusting both the parautochthon and the infrastructural allochthons, like a thin and flattened unrooted sole below the nappes (Fig. 10 c).

3.3.5 The Nappes

The nappes are mainly developed in the Akjoujt area. Some are preserved in the small klippen called Ijibiten (Ij on Fig. 2, Fig. 10 c). Mylonites are poorly developed.

Structural analysis (Marcelin 1964; Lécorché 1980; Kessler 1986) shows that only three phases of deformation are common to the entire pile. The first consists of isoclinal folds developing a penetrative general schistosity with a strong dispersion of fold axis. Kessler (1986), after demonstrating sheath folds and local mineral lineations related to this deformation, concludes that motion was directed N30°. The second phase, directly linked to thrusting, consists of metric to decametric, NE to E vergent, very asymmetric knee folds with fracture cleavage and small drag folds below the thrust plane. The stretching lineation varies from N40° in the northern Akjoujt area to 80° in Ijibiten suggesting a fan-type motion. A third phase in Ijibiten (Lécorché 1980) consists of N170° to N20° trending kink-bands, while in the Akjoujt area Kessler (1986) indicates open upright folds with an average NE-SW trend.

The frontal contact of the nappes, which may be easily seen in some exposures such as at the Guelb el Hajed klippen, north-east of Akjoujt, or in Ijibiten, does not exhibit mylonites but only a few millimeters or centimeters of crushed rocks and local large amounts of quartz breccia. The contact surface varies from a very flat, almost horizontal plane, to a succession of smoothed metric to decametric knobs with various dips, 30° westward in average. True mylonites formed in ductile zones (greenschist facies) are observed at the base of the uppermost high-grade basement nappes.

3.4 *Structure of the Western Sahara*

Structural observations in the Western Sahara combine few field studies with photo-interpretation and show that the foreland is directly overlain by the nappes. Intermediate units, parautochthon and infrastructural allochthons, are unknown and very probably missing. The suprastructural allochthons could be represented by nappe 1 (Fig. 10 b).

3.4.1 Structure of the Foreland (Field Observations)

Whereas the bottom of the Ordovician resting on the autochthonous basement does not show any deformation, the Devonian at the top is strongly deformed and changes progressively to mylonites near the main thrust plane that separates the autochthonous and the allochthonous material. Decametric to hectometric NNE-SSW trending folds, with horizontal axe, are mainly observed in the Upper Ordovician and Silurian. They are overturned eastward and may extend for several kilometers in length. A fracture cleavage is developed in some shaly formations. In Devonian sandstones, shearing appears with an eastward motion of the upper blocks over the lower ones. It may characterize a relative parautochthonous domain. The last movements were directed southeastward (Table 1).

3.4.2 The Nappes (Extension of Limited Field Observations)

The deformations observed in the nappes vary according to the involved material and its possible inherited deformations. Near the main thrust plane, folds are very similar to those in the foreland, but they are superimposed on preexisting deformation such as flow cleavage and isoclinal folds of various wave lengths. The structural map (Fig. 11) shows that many folds in nappe 8 for example existed prior to thrusting; they are cut by the main thrust plane at the base of the nappe and the secondary thrust planes within the nappe. Such hectometric to kilometric folds are very similar to those of the autochthonous Archean basement.

Many deformations affecting sedimentary markers (stretching, boudinage of beds, stretching and flattening of pebbles) may be linked to the shearing

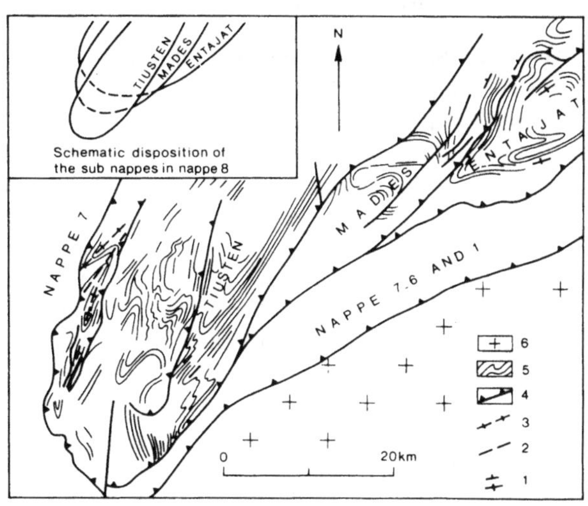

Fig. 11. Structural setting of the West Saharan nappe (location on Fig. 4 A) 8 (Adrar Soutouf). *1* Dip of foliation; *2* synform; *3* antiform; *4* thrust fault; *5* structural trend after aerial photographs; *6* autochthonous basement

of granites (nappe 3) and attributed to a southward-directed nappe displacement.

Limits between the nappes and inside them are generally characterized by mylonites, always limited in thickness. They appear on the aerial photographs as white rounded hills along the thrust traces. Late tectonic events may be recognized: an E-W trending gentle folding of plurikilometric wavelength that affects the whole area and a brittle tectonic with associated N30° dolerite dykes cutting nappes, sedimentary foreland and autochthonous basement. Such dykes may be compared with Liassic dykes in South Morocco.

3.5 Structure of the Zemmour

Although the Mauritanide belt proper is unknown in this area, deformation in the sedimentary cover displays all the characters of a belt boundary (Fig. 10 a; Table 1).

3.5.1 Structure of the Foreland

The eastern edge of the mainly Paleozoic sedimentary sequence resting on the Precambrian basement is undeformed. Deformation develops westward in a folded zone, 15 km wide, where the Devonian succession is increasingly deformed by N10° to N30° trending folds, generally asymmetric, with steep eastern lims of the anticlines.

3.5.2 The Parautochthon (?)

West of the folded zone, the Dhlou belt (Sougy 1964; Dacheux 1967), 15 to 20 km wide, is characterized by a more intense N15° folding, leading to slicing and thrusting of, for example, the Upper Proterozoic dolostones upon the Upper Ordovician. The linear control between the two zones is rarely exposed but is of reverse fault- or thrust-type with an apparent 50° to 30° west dip. The folds have suffered a southward-increasing reorientation, which could indicate an eastward final stress.

3.6 Conclusion on the Structure of the Mauritanides

Correlations between the different works confirm the thrust character of the belt. But it is not easy to correlate deformations that are interpreted in different ways in terms of chronological events (see Sect. 7.2). Table 1 is an attempt to correlate these data in terms of relative chronology.

4 Metamorphism

4.1 Introduction

A regional metamorphism has imprinted with variable strength the numerous structural units of the mobile zone of the Mauritanide orogen. Some of these metamorphic episodes have been dated by $^{40}Ar/^{39}Ar$ methods.

4.2 General Features of the Metamorphism

From Zemmour to Senegal, the foreland is generally unmetamorphosed except for the folded Tichilit et Beïda Group in the central Mauritanides, which displays anchizonal metamorphism (Dia 1984). The parautochthon displays low-grade (upper to middle greenschist facies) metamorphism, often increasing in intensity from east to west. A few parautochthonous subunits are only anchizonal.

The infrastructural allochthons display three grades of metamorphism: upper to middle greenschist facies in the imbricated ophiolitic units, northward increasing upper greenschist to granulite facies, in the continental margin rift-facies association, variable greenschist facies in all calc-alkaline complex. The supracrustal allochthons are generally anchizonal near the front of the belt. Metamorphism increases westwards up to upper greenschist grade or locally (Gaoua?) higher. The nappes display all grades of metamorphism from lower greenschist grade to amphibolite and granulite grade, in a globally reverse way.

4.3 Anchizonal Metamorphism

In the north, in the Nouedgui-Bou Naga tégument, Lécorché (1980) pointed out that the matrix of the sandstones has recrystallized in a sericite - quartz - apatite assemblage, and the micas are reoriented along an earlier slaty cleavage, sericite growing in a first gentle metamorphic phase contemporaneous with a first folding stage. Later brittle schistosity, induced by a crenulation phase, and cataclasis are the results of later deformation in cold conditions.

The other example that can be mentioned was successively studied in detail by Chiron (1973) and Dia (1984) in the Sangarafa (suprastructural allochthons) and the Djonaba (parautochthon) Units.

In the Sangarafa quartzites, from east to west, the matrix recrystallizes, sericite clusters grow, detrital minerals are incorporated in the interstitial recrystallized mosaic, the impurities disappear, and finally blastic muscovite appears. In the Djonaba tillite, the argillaceous matrix, the most sensitive part of the rock, shows a similar evolution.

4.4 Greenschist-Facies Metamorphism

Most of the structural units lie within the greenschist facies. Parageneses are linked to the tectonic phases which, occasionally, during the evolution of the orogen, are displayed on the same rocks. In this case, it is rather difficult to determine the relationship of the paragenetic sequence.

Metamorphic phenomena are listed under regional and structural setting in Table 2 (Western Sahara-northern Mauritanides) and Table 3 (central-southern Mauritanides).

Table 2. Greenschist-facies parageneses in Western Sahara and Northern Mauritanides

Western Sahara nappes	Northern Mauritanides nappes	Suprastructural allochthons	Infrastructural allochthons	Parautochthon
Schists qz - mu - (chl) Greenstones ab - chl - bi - (ser) Iron quartzites qz - (hem) - ma - mu Carbonates cal - dol - (qz)	Quartzites and schists qz - ser (mu) Chlorite schists qz - chl - ser Greenschists chl - bi - (cld) - (ser) Carbonates dol - (alb) - (qz) cal - dol - hem Spilites ab - act - (hbl) - chl ep - cal Basic metavolcanics ab - act - chl - ep - cal Acidic metavolcanics qz - pla - bi - mu - chl	Quartzites and conglomerates qz - ser - chl - (bi) Schists qz - ser - chl	Chlorite schists qz - chl - (ser) Metaandesites ser - chl - pla - act Metatrachyte ser - chl - ab - or - (qz)	Conglomerates qz - ser - chl - stp Quartzites qz - ser - ap - tur Schists qz - ser - chl Carbonates cal - dol - (qz)

Table 3. Greenschist-facies parageneses in central and southern Mauritanides

Suprastructural allochthons	Calc-alcaline igneous complex	Cont. margin rift facies (South of Lemrhassi)	Imbricated ophiotitic Units	Parautochthon
Quartzites and schists E: qz - ser W: qz - mu	Metaandesites ab - act - ep Metarhyolites ab - mu - bi Metatufs-quartzites qz - (ab) - mu - (bi) - (act) Metagranites qz - ab - mu - bi - chl	Micaschists qz - mu - bi - ab - grt Prasinites ab - bi - act - ep Serpentinites Soapstones atg - mag/chl - act - mag	Micaschists qz - mu - bi - fuschite Chlorite schists qz - chl - (ser) Serpentinites Soapstones atg - mag/chl - act - mag Metabasalts alb - chl - act - ep Jasperoid qz - hem - fuschite Carbonates cal - dol Calc schists qz - ab - chl - mu - stp - cal	Quartzites qz - ser Schists qz - ser - (chl)

4.5 Amphibolite-Facies Metamorphism

The continental margin rift-facies association underwent medium- to high-grade metamorphism. In the M'bout area, it corresponds to the Diala-Bouanzé Series of Lille (1967) or Gadel Series (Chiron 1973), relayed to the north of the 17° parallel by the Farkaka Unit (Dia 1984; Rémy 1987) as far as the Aouker dunes. No equivalent is known farther to the north.

4.5.1 Metamorphic Parageneses

4.5.1.1 Micaschists

In the south of the central Mauritanides, this facies features various rock-types (Chiron 1973):

1. Muscovite micaschists, forming the main part of Gadel Series with quartz - phengite - chloritized biotite - rare albite - tourmaline - iron oxydes;
2. Chloritoid micaschists with quartz - muscovite - chlorite - chloritoid. The chloritoid forms idiomorphic clusters, probably posterior to a first metamorphic foliation;
3. Micaschists with porphyroblastic minerals, common in the Farkaka Unit, with muscovite in large crystals - (biotite) - staurolite - kyanite - garnet in a quartzose matrix. A retrogressed overprint is related to the change of staurolite to chloritoid, kyanite to sericite, garnet to chloritoïd + sericite and biotite to chlorite. Muscovite crystals appear in thick and highly folded septa. Staurolite is poikiloblastic, garnet (up to 30%) is either phenoblastic or in stretched patches. Under this particular facies of higher grade, the micaschists are subordinate in the Farkaka area;
4. Micaschists with albite augen are widespread in the Mbout - Sélibabi region, and have been termed albite-micaschists of Diala-Bouanzé (Rocci 1962; Lille 1967, who thought these rocks must belong to a Birrimian basement). Chiron (1973, p. 49) records that these formations belong in fact to the axis of the Mauritanide orogen, and are the southern equivalent of Gadel series. Albite augen, 0.1-1 mm in size, are clearly visible. The paragenesis is mu - bi - qz - ab - grt ± tur. Albite phenoblasts are sometimes full of minute inclusions, up to 50%, and are consequently later than a first schistosity but earlier than a retrograde metamorphism.

Finally, micaschists exhibit metamorphic transformations during an upper mesozonal climax, of somewhat higher grade to the north, and several deformation phases with a change in metamorphic conditions. The older deformation occurred under high-pressure and low-temperature conditions during an isoclinal folding with stretching lineation. A retromorphic overprint is contemporaneous with the folding of the main schistosity (Dia 1984).

4.5.1.2 Quartzites

According to Rémy (1987), quartzites form up to 60% of the total volume in the north of the Farkaka Unit.

Different parageneses have been recorded: namely, (1) Qz-mu ± pla ± ep, very frequent, (2) qz-mu-ky (kyanite crystals 0.1-0.8 up to 2 mm wrapped in muscovite), (3) qz-ma + hem (50%), often very strongly folded as "itabirite", limited to the Gaoua "window", (4) qz-chl ± st ± cld (staurolite is synkinematic, in big crystals up to 3 mm, converting into chloritoid) and (5) qz-ep ± amph ± grt, very dark rock sometimes mistaken for amphibolites.

Thus these quartzites may be interpreted to be metasandstones, more or less iron-rich, deposited in a continental setting, metamorphosed in mesozonal conditions, perfectly in accordance with the micaschists.

4.5.1.3 Prasinites (Metabasites)

These rocks were mapped by Chiron (1973) in the southern Mauritanides and described as coarse-grained dark rocks with ab-(bi)-chl-ep-act-ttn. They derive from lavas or basic plutonites that underwent a metamorphism high enough to erase the previous magmatic textures. The parageneses are, in fact, typical of a medium grade retromorphic stage.

4.5.1.4 Amphibolites

In the Farkaka area, amphibolites are intercalated with the quartzites and form about 20% of the unit. They are schistose, homogeneous or banded (white and black beds). Magmatic textures are obliterated. In places the rock suggests a gabbro. Rémy (1987) points out that various textures are easily distinguished, but parageneses remain simple: amphibole - plagioclase - garnet - epidote. Proportions of minerals are highly variable: amphibole 10-65%, plagioclase 0-68%, garnet 0-25%, épidote 0-45%. Accessory minerals are qz-chl-pg-cpx-bi-cal-ttn-Fe-Ti oxydes, chiefly of retrograde origin.

Maximum intensity of metamorphism developed during D_1 and D_2 phases of deformation with

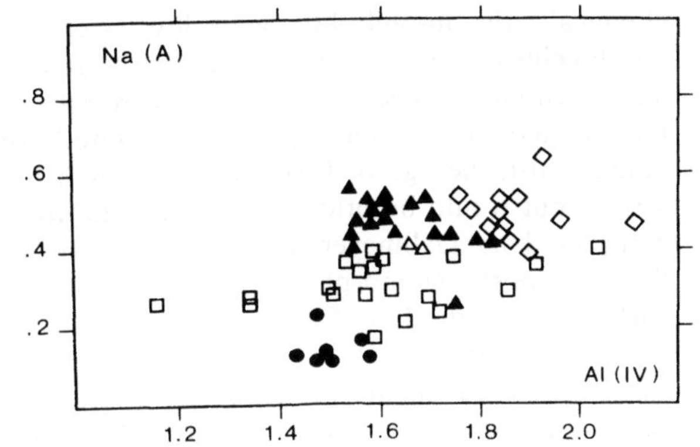

Fig. 12. Na_A versus Al^{IV} in the amphiboles from Farkaka amphibolites (after Rémy 1987). *Diamonds* O-amphiboles; *circles* G-amphiboles; *triangles* A-amphiboles; *squares* T-amphiboles

crystallization of garnet-hornblende-epidote, whereas the growing of poikilitic plagioclase is subsequent.

4.5.2 Metamorphic Conditions

Only the Farkaka amphibolites have been studied from a mineralogical point of view (Fig. 12). Microprobe analyses were performed by Rémy (1987).

Amphiboles from the southern part of the Farkaka Unit (T, Fig. 12) are variable in composition: tschermakitic - hornblende, magnesio-hornblende and ferroan equivalent. Northward (A), they are chiefly ferroan-pargasitic hornblende and ferroan-pargasite. Those from Gaoua (G) are magnesio-hornblende up to ferroan-hornblende. Finally the amphiboles from the northernmost part of the unit (O, Fig. 12) are ferroan-pargasite and pargasite to tschermakite. According to Fabriès (1963), these variations of composition are related to substitutions in the tschermakitic and edenitic components that are related to the metamorphic grade (Laird and Albee 1981; Fig. 13).

Garnets show Mg and Ca zoning. T-garnets are grossular-rich and pyrope-poor. A-garnets are almandine-grossular with a strong Ca and Fe zoning. G-garnets show also a remarkable zoning in which Fe, Mn and Ca decrease and Mg increases from core to rim. O-garnets, homogeneous and the most pyrope-rich, are of very high grade.

Epidotes show regular component variation, well documented on Fe^{3+} vs Al^{VI} diagram (Fig. 14). Oxidation degree goes down regularly from T-epidotes with high Fe^{3+} to O-epidotes with weak Fe^{3+} and high Al^{VI}.

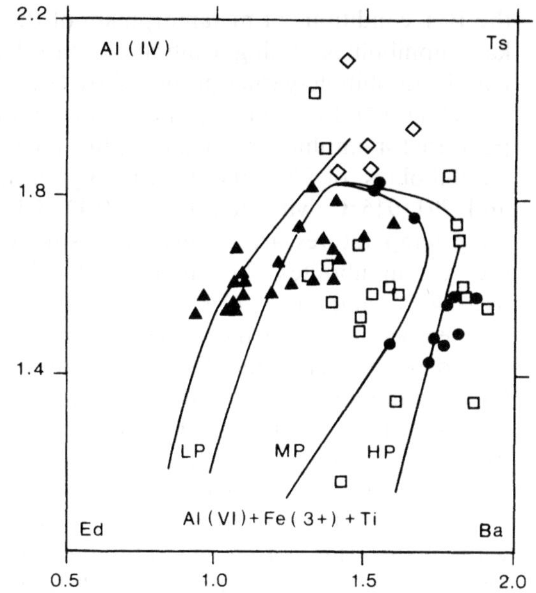

Fig. 13. $Al^{VI}+Fe^{3+}+Ti$ versus Al^{IV} in the amphiboles from Farkaka amphibolites (after Rémy 1987). *Symbols* as in Fig. 12

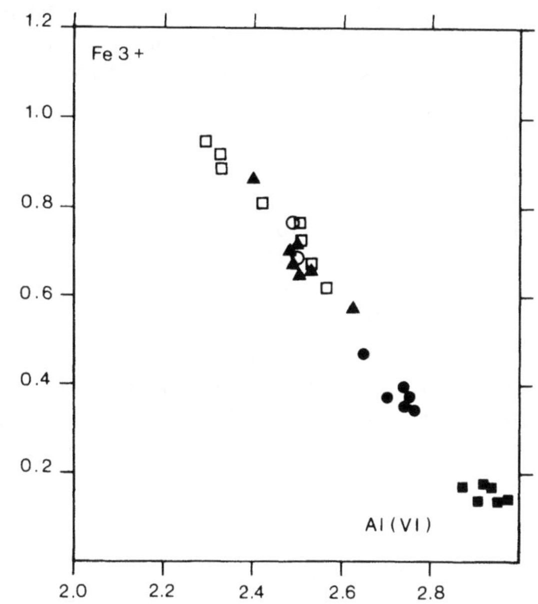

Fig. 14. Fe^{3+} versus Al^{VI} from Farkaka epidotes (after Rémy 1987). *Filled circles* rim of O-epidotes; *filled squares* core of O-epidotes; *open circles* G-epidotes; *triangles* A-epidotes; *open squares* T-epidotes

Plagioclases are everywhere albite, except in O-amphibolites where more basic undetermined and cloudy plagioclases were recorded.

Using the geothermo-barometer based on the couple hornblende-garnet and amphibole-glagioclase, it is not possible to establish with confidence the P/T conditions of metamorphism of the Farkaka amphibolites. Using calibration based on the couple amphibole-garnet proposed by Graham and Powell (1984), Rémy (1987) infers a maximum temperature for amphibolite metamorphism, namely O-amphibolites, of 961-1030 °C with X_{Ca} garnet 0.32 and 791-915 °C with X_{Ca} garnet 0.43. Obviously, these temperatures are greatly over-estimated. For the G-amphibolites, he infered 444 °C, for A, 612-640 °C and for T, 679-754 °C. Although these last temperatures are more likely, they cannot be considered to represent the actual conditions. In fact, the amphibolites from the three southern areas (G), (A), (T) show albite and not basic plagioclase. According to Spear (1980), for the concerned paragenesis, albite needs a temperature of 490° ± 20 °C. Therefore hornblende is probably no in equilibrium with albite. The paroxysmal phase of metamorphism certainly reaches the high-grade amphibolite facies but the albite-epidote-amphibolite facies was overprinted as a retrograde stage.

Finally, garnet zoning and variations in composition of hornblende and epidote show that a prograde regional metamorphism developed from the south to the north. However, the Gaoua region has undergone a somewhat different history, with either an abrupt increase in pressure in the peak metamorphism or cooling of the formations nearly isobarically.

As for the pressure, kyanite in micaschists and quartzites indicates rather high pressure, over 3.5 kb for temperatures of 450 °C (Holdaway 1972). Nevertheless, preliminary investigation on three different metamorphic assemblages on metapelites of this area lead Caby (1988) to propose $P > 12$Kb for temperature of 550° as initial metamorphic conditions for two of them. He interprets these data as the petrological signature of an Alpine-type subduction/obduction geodynamic setting.

4.5.3 Concluding Remarks

Southern Farkaka Unit equivalents (Fig. 2) underwent a metamorphism of upper greenschist grade. But northward, Farkaka Unit contrasts with low-grade contiguous units and with the low-temperature parageneses of the many lenses of the associated serpentinites. Probably these are tectonic slabs or diapiric intrusions, and their present paragenesis may be due to the retromorphic D_3 tectonic phase. However it is obvious that the Farkaka Unit does not possess the same significance as the other units of the axial zone. Rémy (1987) established that chemically the amphibolites were low-Ti continental tholeiites and were associated with quartzites and micaschists. These formations certainly have participated in the building of a thickened continental crust, the age of formation of which is not known but would be prior to 700-720 Ma, the age considered by Dallmeyer and Lécorché (1989) as the first post-metamorphic cooling following an early phase of the Late Proterozoic tectonothermal event. Serpentinite may be interpreted in this case as intrusions of mantle material and could be co-genetic with these metabasites. Finally, the juxtaposition of high and low grade terranes is interpreted by Rémy (1987, p. 220) as resulting from E-W subduction-collision and eastward thrusting of an oceanic domain situated to the east of a western microcontinent.

4.6 Granulite Facies Metamorphism

Except for its local occurrence in the northernmost part of the continental margin rift-facies association, the granulite facies grade is only known in the basement nappes. The Hajjar Dekhen nappe, at the top of the nappes of the Akjoujt area (Fig. 2, HD) displays some high grade gneisses, pyroxenites, marbles and leptynites (Marcelin 1975). Nappes 7 and 8 of the Western Sahara (Fig. 4 A), have been sampled (Sougy and Bronner 1969) and provide high-grade sequences: biotite-bearing leptynites; forsterite-bearing marbles; hypersthene, brown biotite and clinopyroxene plagioclasites; hypersthene, clinopyroxene and garnet granites. Such assemblages are overprinted by low-grade (epidote, sericite, stilpnomelane) metamorphism directly related with late shear zones. Such high-grade rocks had been interpreted as autochthonous Archean basement. Their old basement character is confirmed by $^{40}Ar/^{39}Ar$ dates on hornblendes from amphibolites sampled in the Akjoujt nappes, which yield ages in a range of 1.1 to 2.5 Ga (Lécorché and Dallmeyer 1987; Dallmeyer and Lécorché, 1990a).

4.7 Discussion and Summary

In the central part of the Mauritanides (Fig. 2), the belt units have undergone a regional prograde metamorphism from the foreland to the internal zone. But on the whole, this metamorphism is not higher than the greenschist facies. The belt is asymmetrical, since the foreland and the parautochthon, although strongly deformed throughout, are only weakly recrystallized in the anchizone and epizone and chiefly mylonitized in numerous shear zones. On the contrary, the infrastructural allochthons suffered a polyphase, polycyclic metamorphism. Nevertheless, the mineral parageneses remain in the greenschist facies, except locally. The remarkable exception is the continental margin rift-facies association (Farkaka Unit), which underwent a prograde metamorphism more and more intense from south to north, developing parageneses of high temperature and pressure.

5 Magmatism

5.1 Magmatism of Unknown Age

In the central part of Mauritanides, south of Aouker (Fig. 2), several massifs of highly deformed and often highly altered acidic rocks were mapped by Dia (1984) in his calc-alkaline igneous complex.

5.1.1 Leucogranites of the Kelbé Unit

This unit is a big slice, overthrust on the western side of the Farkaka Unit (continental margin rift complex) of the axial zone. In the great "window" of Gaoua, in the Kelbé area proper, north of Gaoua, and in several more southern locations, leucogranites appear as a basement for the Gaoua quartzites Unit.

Leucogranite is coarse-grained, heterogeneous, bearing muscovite, sericite and chlorite, quartz-rich and orthoclase-poor (generally albitized). The amount of silica is very high (>74%), the alkali content is variable and sometimes the amount of K_2O is high (>7%). Rb and Sr are low and Rb/Sr and K/Rb ratios are high (Table II.2.2, in Rémy 1987). The amounts of Zr, Nb and Y are particularly low. In a multielement diagram, Rémy (1987) compares this granite with granitoids occuring in various geodynamic settings. The more likely situation seems to be its emplacement during continent-continent collision or in an island-arc setting. The actual geochemical lineage of this highly transformed granite is difficult to establish. Chiron (1973) was convinced they are alkaline because of their albite - microcline - muscovite paragenesis, and their very low percentage in CaO. He inferred from this character that the leucogranites were post-orogenic. But Dia (1984) and Rémy (1987) attribute these mineralogical and geochemical characteristics to the strong deformation and secondary hydrothermal alterations. They conclude that the leucogranites are calc-alkaline and synorogenic.

5.2 Pan African Magmatism

5.2.1 Syenites of Bou Naga

In the southern part of the Bou Naga window (see Sect. 2.3.1; Fig. 2), south-east of the northern Mauritanides, an intrusive alkaline and partly effusive complex discovered by Malavoy (1931) was described with precision by Marcelin (1965). Rare-earth concentrations were studied by Pouit (1968) and, during the 1970s, cesium, ytrium and thorium were exploited.

According to Marcelin (1965), two separate massifs occur in the Bou Naga inlier: a western one, chiefly syenitic and an eastern one, rhyolitic with volcanic breccias, trachyte tuffs, microsyenites and trachytes. In a recent work (Blanc 1986), Blanc et al. (1986) give the precise date of emplacement of the syenite massif: 680 ± 10 Ma.

The syenite pluton is subcircular, 12 km in diameter. It is composite and formed by an undersaturated unit made of feldspathoidal syenites and phonolites with orthoclase or sanidine, albite, nepheline ± sodalite ± cancrinite ± melanite, abundant biotite, aegyrite and taramite. This pluton is intrusive into a saturated or oversaturated unit composed of syenites, quartz-syenites, perthite-granites and rhyolites, containing ferroan biotite (annite) and sometimes aegyrite. The accessory phases crystallized in two stages: an early and common group associated with clusters of biotite and amphibole, chiefly magnetite and ilmenorutile, apatite and titanite; a late group comprised of zircon widespread in the oversaturated unit and rare in the foïdites, and xenomorphic and intensely coloured fluorite. This last mineral shows that fluor-rich fluids accompanied the mineralization.

Mineralization (Pouit 1968) is concentrated in a set of veins emplaced within the alkaline consolidated complex. The most interesting minerals are xenotime, thorite, bastnaesite, generally associated with fluorite in great amount. Thorianite and uranothorite have been noted.

Geochemical data confirm the alkalinity of the complex. But the evolution of the two units is different. On the whole, the complex is highly differ-

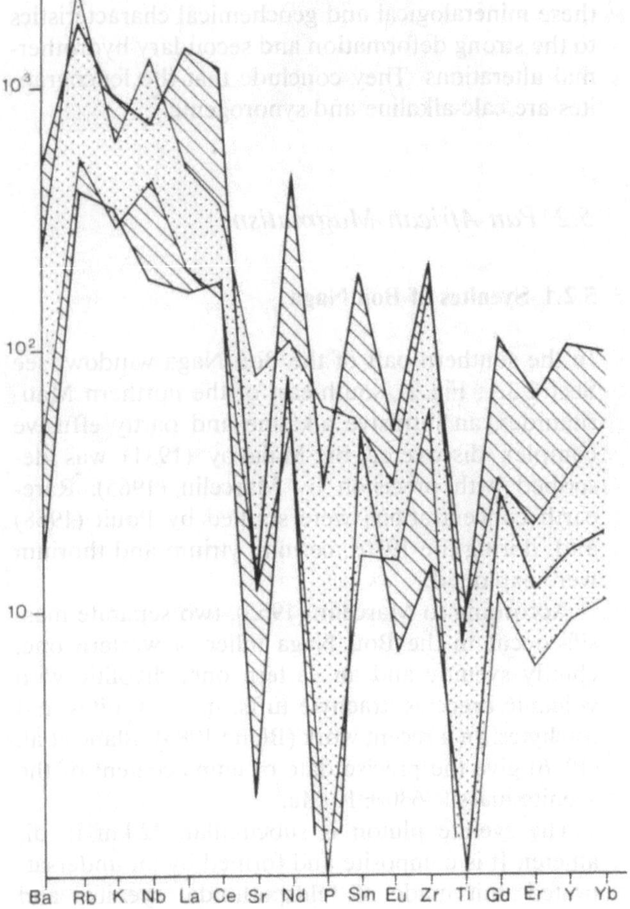

Fig. 15. Chondrite-normalized multielemental patterns (after Blanc 1986). *Dots* undersaturated complex. *Hatching* saturated and silicic complexes

entiated and basic entities are absent. The multielement diagram (Fig. 15) shows enrichment in highly incompatible elements (Rb, K, Nb, Ce), negative anomalies in Sr, P and Ti, positive anomalies in Rb, Nd, Zr, Gd, Y and Yb. The diagram (Fig. 15) shows that these behaviours are distinct in the two units for some elements such as Nb or Sm.

The geochemistry of the Bou Naga massif is similar to other alkaline massifs of the West African craton and of the mobile belts on the eastern side. It is typical of an end-evolution of an alkaline trend. Moreover, the magmatic evolution of the saturated unit of Bou Naga is very similar to the evolution of the anorogenic complexes of Niger-Nigeria (Bowden and Turner 1974).

In conclusion, the Upper Proterozoic intrusions of Bou-Naga correspond to an anorogenic magmatism, which can be related to a rifting stage prior to the emplacement of a great ultramafic/mafic unit in the Mauritanide orogen, the evidence of an ocean opening.

5.2.2 Niokolo-Koba Granite

In Eastern Senegal (Fig. 2), Bassot (1966) describes in the Kouloutou Series (southern calc-alkaline complex) the Niokolo-Koba granite, elongated along the strike of the folds. The rocks are dioritic to calc-alkaline granite in composition. Their relations with the acid volcanics and sediments of the series are in part tectonic (Villeneuve 1984). Bassot and Caen-Vachette (1983) report for these granites a Rb/Sr whole rock isochron age of 683 ± 18 Ma, which could be their emplacement age (Dallmeyer and Villeneuve 1987).

5.2.3 Ultramafic/Mafic Magmatism

In the infrastructural allochthons, serpentinite bodies are scattered all along the belt. They are often associated with chlorite schists and talc schists enclosed in a soapstone package. Furthermore, there are orthoamphibolites, metagabbros with preserved magmatic pyroxenes, basalts either in massive flows or in pillow-lavas followed by breccias, more or less coarse tuffs and volcaniclastic sediments.

The above rock association led Dia (1984) to confirm the previous hypotheses of Lille (1967) and Chiron (1973) about the ophiolitic character of the ultramafic/mafic assemblage. He defended the model of a dismembered and reversed ophiolitic suite in the core of the Mauritanide orogen. From a very detailed study on the amphibolites (see Sect. 4.5) and on the basic lavas, Rémy (1987) concluded that the ophiolitic assemblage model was not sufficiently documented.

In the Mauritanides, from Akjoujt to Bakel (Fig. 2), ultramafic/mafic formations occur in narrow structural units, parallel to the orogen. Chiron (1973), located them in his Gadel and El Aouidja groups, whereas Dia (1984) subdivided the Gadel group in the Farkaka Unit to the west, and the Oued Amoûr Unit to the east, keeping the external group as the "El Aoueïja" Unit.

5.2.3.1 *Ultramafites*

Ultramafites occur as lenticular bodies, more or less widespread, in small black hills often capped with iron-hats when they are serpentinites, and as soapstones in the depressions or at the base of the hills.

The massive **serpentinites** display a classical brownish alteration coat, with a cavernous structure, a dense network of ferrugineous veinlets and some remnants of chromite more or less transformed to magnetite (Kane 1986). The primary textures also are obliterated. Only some cumulate textures remain. The mineral association is antigorite - chrysotile - chlorite - talc-chromite. The chemical composition (SiO_2 37-39%, MgO 35.5-38.5%, total iron 7-12%) is harzburgitic (Kane 1986; Rémy 1987). Their structural position seems to correspond, for the most part, to thrust contacts. According to Rémy (1987) the serpentinites constitute about 20% of the whole Farkaka Unit. They also occur in the Oued Amoûr Unit and farther south in the Sélibabi region (Fig. 2).

Rocci (1962) discovered the first massive serpentinites, described later on by Lille (1967), between Bakel and Sélibabi. They were associated with NNW or NW fractures. But in most outcrops, a schistose crumpled serpentinite is scattered along tectonic contacts inside any sedimentary or magmatic horizons. In the "External zone" of Le Page (1983, 1988), i. e. the eastern infrastructural allochthon, the contacts between the different units are commonly marked by serpentinite. No olivine remnant has been observed in these rocks but sometimes pyroxenes are seen.

Intrusions or lenses of diabases, gabbros and acidic rocks occur in some serpentinite bodies. They are metamorphosed but their primitive magmatic features remain obvious. The most remarkable are rodingitized diabase dikelets. Moreover, very blastomylonitic leptynites (ortholeucogneisses) have been reported by Chiron (1973) and analysed by Dia (1984), who interprets them as metamorphosed and deformed plagiogranites.

The presence in the Farkaka Unit of serpentinite bodies chiefly of harzburgitic but partly lherzolitic composition, and also of rodingites and plagiogranites of oceanic affinity, supports Dia's hypothesis of the existence of fragments of an ophiolitic assemblage in this unit.

Pyroxenites and amphibolo-pyroxenites. North of Magta Lahjar (Fig. 2) are exposed ensanded widespread garnet amphibolites associated with garnet pyroxenites and amphibole-pyroxenites, with local concentrations of kyanite and corundum. They are interpreted by Dia (1984) as eclogites (see Sect. 4.5.3).

Soapstones are widespread in the Farkaka and Oued Amoûr Units where they are predominant in the serpentinites, and also in the El Aoueïja Unit and in the Hamdallaye Unit (Le Page 1983). They are chiefly chlorite schists, chloritites with martite cubes, tremolitites and talcschists.

Chromite. Kane (1986, 1987) lists two types of chromite in the serpentinites. The first type has a mineralogical composition ranging from chromite to ferrochromite and magnetite depending on the degree of alteration. The second type is represented by chromitite lenses in primitive serpentinized dunites. Chemically, discriminated chromites and chromitites are very similar to podiform chromites of ophiolitic type, aluminium-rich and titanium-poor.

In conclusion, the serpentinites are the most characteristic rock unit assemblage of the ultramafic group. Schistose and dike-emplaced serpentinites are probably of secondary origin. The origin of the massive serpentinites is more difficult to determine. In the Gabou-Hamdallaye Unit and its northern extension, the El Aoueïja Unit, the association with gabbros and tholeitic basalt, may represent the lower part of an ophiolite of upper Proterozoic age.

5.2.3.2 Basic Magmatism

Gabbros

The gabbros are far less common than the ultramafites, but they are very interesting because their pyroxenes are often fresh. They are exposed chiefly in the El Aoueïja and Hamdallaye Units where they are well preserved. Outcrops of gabbro are also present in the Oued Amour Unit.

Metagabbros occur (Rocci 1962) in the Anietir region, east of Mbout (Fig. 2). The rock is a coarse-grained gabbro with saussuritized plagioclase and pyroxenes transformed into amphibole. Rémy (1985) described some of them in the Hamdallaye area (southern Mauritanides) with wellpreserved pyroxenes. In a cross-section (Fig. 16), Rémy (1985) describes gabbros immediately overlain by basaltic pillow-lavas, with large fresh cpx crystals (4-5 mm), moulding albitized plagioclase laths (2-5 mm) in ophitic texture. Some ferro-titanium oxides ($<5\%$) have crystallised later on. The cpx crystals are sector-zoned and present a diopsidic core rich in MgO and CaO, and iron-rich augite rims. Chemical analyses of gabbros and their pyroxenes are consistent with abyssal tholeites.

Hypabyssal Rocks and Lavas

These basic rocks are related to gabbros in dykes, pillows and massive flows and are often interbedded with volcaniclastics, comprising large massifs as in the Sélibabi area (Fig. 16).

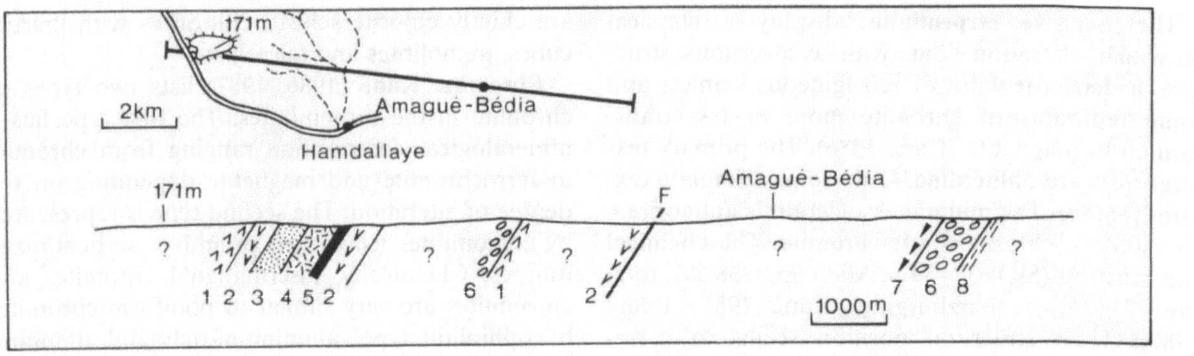

Fig. 16. Schematic cross-section in El Aoueïja (Hamdallaye) unit (adapted from Rémy 1987). *1* Dolerite; *2* basalt; *3* tuffite; *4* coarse-grain tuff; *5* quartzite; *6* pillow-lavas; *7* gabbro; *8* umbers

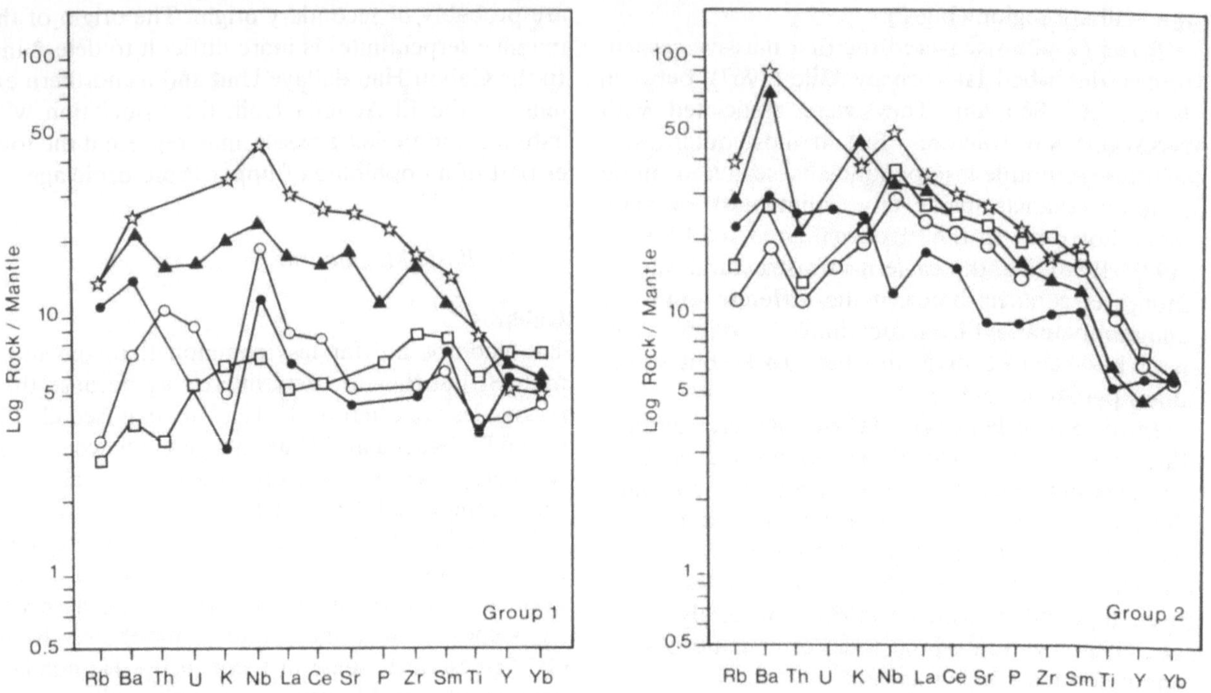

Fig. 17. Multielement diagrams normalized to mantle (Wood et al. 1979) of El Aoueïja basalts of groups 1 and 2, compared to different basalts (after Rémy 1987). Group 1: *filled circles* El Aoueïja basalt; *squares* N-MORB, average (Holm 1985); *open circles* E-MORB, Philippines basins (Wood et al. 1980); *triangles* tholeiitic basalt of the Ethiopian rift (Piccirillo et al. 1979); *stars* P-MORB, Indian ridge (Le Roex et al. 1985). Group 2: *stars* El Aoueïja basalt; *open circles* Hawaii OIB (Gladney and Goode 1981); *squares* Hawaii OIB (basaltic volume); *filled circles* continental tholeiites, Antarctic (Kyle 1980); *triangles* alkali basalt of the Rio Grande rift (basaltic volume)

Basalt pillows are rather small (<0.5 m) with a chloritic schistose matrix, albite phenocrysts (1–2 cm), chlorite-calcite-epidote glomerules. In the mesostasis small cpx crystals (0.2 mm) and chrome-spinel are visible in a paragenesis of calcite-chlorite-epidote.

Massive basalts are porphyritic. They contain vesicular blebs of calcite, microphenocrysts of augite, 1 mm long, and albitized plagioclase. Primitive olivine crystals are replaced by epidote clusters. Major elements, traces and REE are characteristic of two types of basalts (Fig. 17). The first group is enriched MORB-type, the second (group 2) is of alkaline tendency and may represent a continental rift-facies associated to limited ocean opening.

Dolerites occur in dykes and sills and exhibit an intersertal texture with minute albitized crystals (up to 60%), chlorite and epidote. Remnants of small clinopyroxenes (10–15%) show that it is a pyroxene diabase.

Magmatic breccias with fragments of spilite and keratophyre occur in the southern Mauritanides, Sélibabi area (Rocci 1962; Lille 1967).

Metadiabases and metavolcanics. In the nappes of the Akjoujt area, a volcano-sedimentary unit, in the vicinity of the copper mine, may be subdivided (Michaud 1964) into lower volcanics, middle schisto-quartzitic formations and upper volcanics. The copper ore-body is localized at the base of the upper volcanics. Metabasites are massive or sometimes schistose, interbedded with schists. Magmatic paragenesis is locally preserved: basic plagioclase + ferroan horblende or ferroan tschermakite. Metatuffites transformed into amphibole-chlorite schists or albite-chlorite schists, and metasiliceous cinerites are also exposed. From geochemical data, Ba Gatta (1982) and Kessler (1986) infer that the lower volvanics are typical basalts and metabasalts ($45.4 < SiO_2 < 50.6$), the upper volcanics beeing much more siliceous. Trace elements data allow them to assign to this volcano-sedimentary nappe material an island-arc environment.

In the central Mauritanides, basalts and metabasalts are mainly associated with the El Aoueïja Unit, but are also inferred in the Oued el Amoûr Unit where they are schistose and metamorphosed to chlorite schists and prasinites (Dia 1984).

In the southern Mauritanides, Le Page (1983, 1986, 1988) describes in the "External Zone" (green series), mafic volcanics with turbidite sequences. In the two main lithostratigraphical sequences: Hamdallaye and Gabou Units, the green series represent an important submarine basaltic volcanism episode sometimes associated with iron-rich quartzites. The rocks are strongly deformed and metamorphosed to low greenschist facies.

Orthoamphibolites

Although much more metamorphic than the former rocks, the amphibolites of the Farkaka Unit (continental margin rift-facies association) are lavas and cumulates of basaltic composition (Rémy 1987). They occur as schistose or, more seldom, massive beds, in all the sequences, and are often interbedded with quartzites. Their mineralogy is simple: amphibole, plagioclase, garnet, epidote, but the proportions of each mineral is highly variable. Two groups of rocks have been differentiated. The most important certainly represents magmatic liquids of basaltic composition with low amounts of TiO_2, P_2O_5, Zr, Nb and Y. The second represents primitive cumulates. The geochemical characteristics of the Farkaka amphibolites - $TiO_2/Zr = 100$, $Zr/Nb = 15$, $Zr/Y = 5$ and enrichment in LREE - are in accordance with low-Ti continental tholeiites.

5.3 Conclusions

The ultramafic/mafic formations of the central Mauritanides are variable in petrology and chemistry: some elements as chromitites, serpentinites, gabbros and some basalts present oceanic tholeiite affinities, others such as amphibolites and some basalts have continental tholeiite features or alkaline tendencies. Dia's (1984) model (obducted ophiolitic reversed and dismembered assemblage) is certainly too simple and we need to date the main protoliths. Although the tholeiitic basalts and alkaline basalts of the El Aoueïja-Hamdallaye Unit are contemporaneous, it is not demonstrated that they have the same age as the metabasalts of the Farkaka Unit. Therefore, we can only consider the highly metamorphic Farkaka Unit, with its particular continental magmatism, as a crustal fragment of a thick continental slab, whereas the magmatic components of the Oued Amoûr and El Aoueïja Units may correspond to paleo-ophiolites as those of the Pan-African pericratonic mobile zone.

5.3.1 Orogenic Volcanism of the Calc-Alkaline Complex

All along the Magta Lahjar Unit (Dia 1984) and Mbout Series (Lille 1967; Chiron 1973), the detrital and argillaceous metasediments are intercalated with schistose lavas with well preserved magmatic structures. In the north, these lavas are poorly individualized because of the overlying nappes. In the central Mauritanides they are preserved in the Magta Lahjar and El Hneïkat Units, in the southern Mauritanides in the Ouechkech and Rhabra groups (Chiron 1973) of the Mbout Unit (Le Page 1983) and farther southwest in the Niokolo-Koba and Koulountou Units (Villeneuve 1984). According to Chiron, in the Mbout region, from the eastern limit of the Senegal-Mauritanian Basin to the Oua-Oua Mountains, a volcanosedimentary complex is poorly exposed with meta-andesites and prasinites forming the Rhabra group. It is followed to the east by porphyritic metarhyolites of the Ouechkech group. A second stripe of meta-andesites and prasinites is limited to the east by a complex of tuffs and conglomerates that overthrust the Farkaka Unit (or its equivalent Gadel group). Dia (1984) showed that these highly folded volcanic formations, metamorphosed into greenschist facies, have a geochemistry in conformity with calc-alkaline orogenic suites.

To the south of the southern Mauritanides, orogenic volcanism constitutes the main part of the

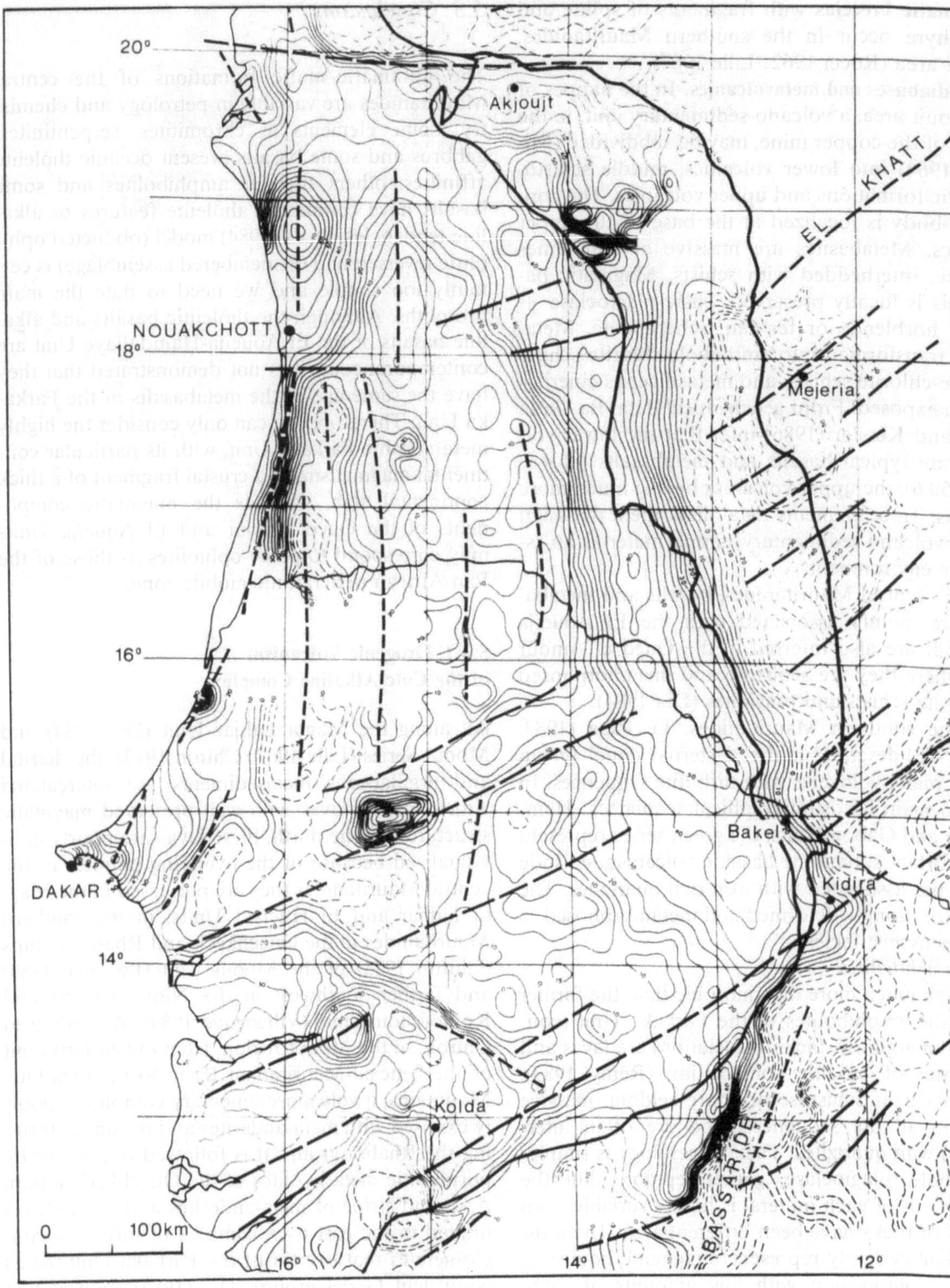

Fig. 18. Bouguer gravity anomaly map associated with the main Mauritanide orogen. Contour interval: 5 mgal. Gravity contours below 0 mgal are *dashed*. Gravity trends and transverse discontinuities are shown

southwestern branch of the orogen (Villeneuve 1984) in the Niokolo-Koba and Kolountou Units. According to Angeli (1984) a volcanic complex of basalts, andesites, dacites and rhyolites alternates with pyroclastites, breccias, jaspers and a few sedimentary horizons of sandstones and argillites. The lavas are chiefly porphyritic with a residual magmatic paragenesis of diopsidic clinopyroxene, amphibole, biotite, in a metamorphic assemblage of albite - chlorite - actinote ± calcite, typical of the greenschist facies. The geochemical composition - major elements, traces and REE - is in accordance with a highly potassic calc-alkaline suite.

In conclusion, this typical orogenic volcanism is attributed to the evolution of an active continental margin. However, if it is not yet clearly shown that these formations are synchronous all along the orogen, several stages of orogenic volcanism probably took place; the older in the south could be Pan-African in age (680 Ma according to Bassot and Caen-Vachette 1983), the younger in the central part, Hercynian?

6 Local Geophysical Characteristics

6.1 *Introduction*

Geophysical data concerning the Mauritanide belt, its cratonic foreland and its internal margin masked beneath the Mesozoic and younger sediments of the coastal plain, consist essentially of gravity data obtained by ORSTOM's surveys (Crenn and Rechenmann 1965). These data have contributed to demonstrate gross crustal structures, main structural trends and province boundaries in the absence of seismic profiling. Despite the well known non-uniqueness of gravity modeling, they also impose, to a certain extent, some bounds on geodynamical models.

6.2 *The Major Gravity Trends*

Figure 18 presents the Bouguer gravity anomaly map compiled with a contour interval of 5 mgal. Positive contours (gravity values over 0 mgal) are drawn in thin full lines, negative contours in thin dashed lines. The boundaries of the exposed orogen (eastern limit of the coastal plain and eastward extension of the orogenic front) are outlined on the map by heavy lines. Heavy dashed lines mark observed geophysical trends.

The prominent gravimetric feature is the so-called Mauritanides anomaly (Guétat 1981), a long-wavelength positive anomaly that belongs to the circum-West African belt of positive anomalies (Lesquer et al. 1984). It extends approximately 700 km from south of the Reguibat uplift, near Akjoujt, to eastern Senegal. The NNW-SSE axis of the anomaly parallels the main exposed segment of the Mauritanides orogen slightly west of the axis of the outcrop. The Mauritanides anomaly, which probably marks a major crustal discontinuity along the West African craton margin, delimits two gravimetrically well-defined crustal provinces (Lécorché et al. 1983; Roussel et al. 1984).

East of the Mauritanides anomaly is a broad regional negative anomaly over the western border of the West African craton and its sedimentary cover (Taoudeni basin). This gravity low area is also characterized by NE-SW gravity trends and elongated NE-SW anomalies that Guétat (1981) and Lécorché et al. (1983) interpretet to reflect largely the structure of the Precambrian basement and to have controlled sedimentation in the Upper Precambrian and Paleozoic. These distinctive NE-SW gravity signatures are obviously prolonged westward beneath the easternmost exposed portions of the Mauritanide orogen clearly revealing its allochthonous nature (Lécorché et al. 1983).

West of the Mauritanides anomaly is a generally positive Bouguer anomaly field in the whole area between the exposed orogenic belt and the Atlantic coast, contrasting with the gravity lows observed over the West African craton and the Taoudeni basin. This wide positive area corresponds with the hidden internal margin of the Mauritanides belt and with the Mesozoic and Tertiary Mauritanian-Senegal coastal basin. In fact there is no direct correlation between the sedimentary sequences within the basin and the positive anomaly, which is therefore related to deeper sources. Furthermore, gravimetry reveals in this positive area two principal groups of anomalies characterized by NE-SW and N-S major trends.

Gravity anomalies elongated in a north-south direction predominate to the north of about 15° N latitude. Some of them appear to be offset by transverse gravity discontinuities. N-S gravity anomalies are associated with magnetic lineations (Liger 1979; Roussel and Liger 1983). They probably reflect the existence of tensional features related to the Triassic or Early Mesozoic rifting and subsequent opening of the northern and central Atlantic Ocean. Unpublished refraction data from petroleum companies support this interpretation (see for example Chanut and Micholet 1988).

Gravity trends with a NE-SW orientation, as in the cratonic zone to the east of the Mauritanides anomaly, are particularly well developed in the southern part of the Mauritanian-Senegal basin (south of 15° N). They are locally associated with extensive dense intrusives, probably of basalt or dolerite, that may result from the reactivation of ancient transverse fracture zones during the break-up of Atlantic Ocean (Roussel and Liger 1983).

6.3 Interpretation

The fact that the western domain is locally marked by gravity trends structurally analogous to that of the West African craton led Lécorché et al. (1983) to propose that a coastal crustal block similar in character to the West African craton exists beneath the coastal plain west of the Mauritanide orogen. The contrasting positive anomaly field over this western block suggests, however, a strongly reworked crust thinner or denser than the cratonic basement beneath the Mauritanides foreland (Lesquer et al. 1984; Roussel et al. 1984). This interpretation is consistent with the models proposed for the Bassaride orogen (Ponsard et al. 1982) and is also supported by the existence between the two blocks of a steep discontinuity in seismic velocities (Briden et al. 1981). This led Lesquer et al. (1984) to suggest that the structure of the western crustal block is essentially the result of the Pan-African orogeny.

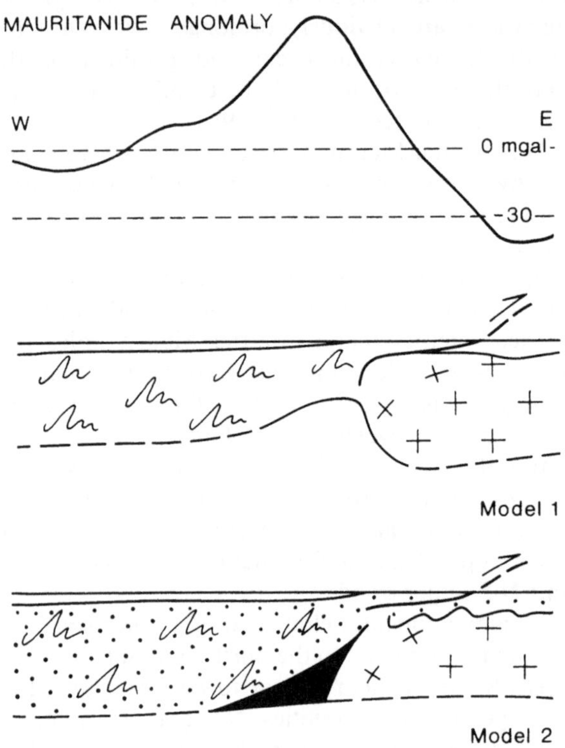

Fig. 19. Alternative models for crustal suture across the Mauritanides anomaly (after Roussel and Lesquer, this Vol.). *Model 1* Crustal thinning beneath the western coastal block (Lécorché et al. 1983); *model 2* thicker and denser western coastal block (Lesquer et al. 1984)

Thus, interpretative models of the regional anomaly compiled from profiles across the West African craton margin, from the coastal plain (positive background ~+10 mgal) through the belt and onto the craton (negative background ~−30 mgal), consists of two crustal blocks of contrasting mean density and thickness. The blocks are therefore separated by the positive Mauritanides anomaly which likely reflects a remnant west-dipping crustal suture.

In contrast with the short-wavelength gravity highs reflecting unrooted dense bodies associated with the Pan-African suture in the Pharusian belt (Black et al. 1979) or in the Bassaride orogen (Ponsard et al. 1982), the Mauritanides anomaly is a huge long-wavelength positive anomaly and its main source must be a deep-rooted dense mass, as suggested by interpretation using the inverse approach and linear programming. Finally, the anomaly has been interpreted as an asymmetric mantle-rooted massif, oversteepened eastward, whose crest is at a depth of ca. 15 km (Lécorché et al. 1983).

Figure 19 shows two possible models explaining the positive gravity background of the western block and the long-wavelength characteristics of the Mauritanides anomaly. Lécorché et al. (1983) interpreted the positive regional anomaly as reflecting thinner crust beneath the western coastal block. Crustal resistivity structures derived from magnetotelluric soundings are in agreement with this interpretation (Ritz and Robineau 1987). According to Lesquer et al. (1984), the reworked western block is thicker and denser than the cratonic margin, reflecting a crustal structure analogous to that proposed for the Bassarides orogen (Ponsard et al. 1982).

The continuity of the Mauritanides anomaly is only disturbed at the latitude of Nouakchott to the west of the El Khatt fracture zone where a marked deflection of anomalies seems to indicate a E-W dextral displacement of about 50 km. The approximate co-linearity of the gravity discontinuity and the major faults in the Khatt area, which cut Devonian rocks, suggests that the displacement was post-Devonian (Lécorché et al. 1983).

It is interesting to note that the two transverse gravity discontinuities that mark the northern (to the north of Akjoujt along latitude 20° N) and southern (Kidira-Kolda lineament in eastern Senegal and Casamance) boundaries of the main deep-seated Mauritanides anomaly coincide with the northern and southern limits of the major segment of the exposed Mauritanide belt. These major structural boundaries probably played an important role in the evolution of the belt, as emphasized by recent magnetotelluric soundings (Ritz and Robineau 1986, 1987).

6.4 Conclusion

Recent geochronological data (Dallmeyer and Lécorché 1989, 1990a, 1990b) indicate that a late Paleozoic tectonothermal event is superimposed on the Pan-African orogen exactly in this exposed segment of the belt. Thus, we suggest that the specific characteristics and geometry of the Mauritanides anomaly, which contrast strongly with those of the other circum-West African craton gravity highs, resulted from compressional stresses or from repeated tensional and compressional stresses associated with a renewed significant eastward translation of the western block during collision of Gondwana and eastern North America, the 20° N (Akjoujt) and Kidira - Kolda lineaments probably both acting as transform boundaries. In a recent paper (Venkatakrishnan and Culver 1987), the northern and southern boundaries of the Senegal block are suggested to be NW-SE to preserve the integrity of the post-Paleozoic fit of Gondwana-Laurasia. The existence of this western coastal block or Senegal microplate is consistent with paleomagnetic data (Lefort and Van der Voo 1981), indicating that a crustal block was probably located between the Mauritanide and Appalachian orogens during the late Paleozoic amalgamation of Pangea.

7 Geodynamic Evolution

7.1 Introduction

The various authors who have worked on the Mauritanide belt have proposed personal geodynamical syntheses that differ considerably, first in terms of tectonics between autochthonists and allochthonists, then in terms of chronology, when most people were convinced that the Mauritanides were a thrust belt. The main reason is probably that nowhere in the belt is there clear evidence to date the principal phases of its evolution.

7.2 The Interpretations

Chiron's (1973) interpretation of the central and southern Mauritanides considered an internal rifting of the West African craton without complete oceanization but accompanied by horst-and-graben tectonics with abundant magmatism (infrastructural allochthons). Closing leads to the uplift of reworked parts of the craton (Guidimakha granodiorites, serpentine belts) along the "axial zone" striking along the former rifting zone. From geochronological data (Rb/Sr mineral ages of ca. 620 Ma) on the Guidimakha granodiorites (Lille 1967), Chiron (1973) inferred a Pan-African age. "Eocambrian" (Supergroup 2) and Paleozoic covers (our parautochthon and suprastructural allochthons), deposited E and W of the former belt, have suffered a Late Devonian tectonothermal event ca. 350 Ma, as indicated by the younger Rb/Sr ages yielded by the Guidimakha granodiorites (Lille 1967). This event was linked with local limited thrusting. Chiron (1973) admitted that this scheme could not be extended in the northern Mauritanides where general overthrusting occurred.

Compared with Chiron's area, the "axial zone" in the northern Mauritanides is almost entirely missing, but the basis of the cover (Lécorché's external nappe: parautochthon) is affected by a tectonothermal event prior to the Late or post-Devonian emplacement of the nappes. Lécorché (1980) pointed out that the only large gap of sedimentation linked with local folding took place in the Middle Ordovician and could be linked to the closing of a western trough in which were deposited the present parautochthon and orogenic volcanics. The Upper Ordovician (and later?) cover would then act as a sole during the piling up of the nappes coming from a more western oceanic (?) area closed during the collision of Gondwana - Laurentia. Large-scale late-stage subhorizontal shearing led to the final emplacement of both the parautochthon and the nappes upon the western edge of the craton. Nevertheless, Dia et al. (1979) and Lécorché (1983) do not reject the possibility of Pan-African events south of the Aouker, prior to the "Caledonian" phase.

Le Page's (1983, 1988) interpretation in the southern Mauritanides was built on the recognition in all his structural units of the "green", "red" and "white" Series, with progressive folding from the older (green) to the younger (white). Then Le Page (1983) proposes a triple rifting, during Pan-African times, of the western edge of the craton individualizing three microplates which progressively converged from west to east from Cambrian to Upper Devonian. The final collision of Laurentia led to N-S sinistral strike-slip faulting along the westernmost colliding microplate, i. e. in the western part of the belt.

In the northern part of the central Mauritanides, Dia (1984) observed an ultramafic to mafic complex (part of Chiron's axial zone) that he interpreted as a dismembered ophiolitic complex, overthrust along its western edge by calc-alkaline formations of post collisional type. The basic sequences developed during rifting and oceanization in the western edge of the West African craton. Westward subduction and closing of this ocean, together with deposition of tillites and flysch-like sediments (lower Supergroup 2) on the eastern margin of the craton, led to obduction of the oceanic crust, overthrusting by the calc-alkaline complex, and imbrication with the metasedimentary cover during a last stage of shortening. The participation of this Uppermost Proterozoic cover in the orogenic events implied progressive tectonics of Pan-African age ca. 650-600 Ma. Post-orogenic deposits and the Paleozoic sequence (up to Middle Devonian) have been folded during Late Paleozoic times. Cambrian (?) to Devonian supracrustal allochthons overthrusting the former belt would be remnants of this tangential event.

7.3 Geochronological Data

A set of Rb-Sr measurements on minerals (Bassot et al. 1963) from rocks of Eastern Senegal have long been the only reference for the whole belt. A second set of Rb-Sr and K/Ar measurements on minerals of the Guidimakha granites and granodiorites (Lille 1967, 1969) was more of local interest.

The first recent Rb-Sr isochron concerns the Niokolo-Koba calc-alkaline granite of Eastern Senegal (Bassot and Caen-Vachette 1983). It yielded an age of 683 ± 18 Ma while intrusive rhyolites within this granite had given another isochron at 324 ± 5 Ma, interpreted as a complete rejuvenation of older rhyolites.

U/Pb dates on zircons in alkaline syenites of Bou Naga (Blanc et al. 1986) in the northern Mauritanides have yielded a mean age of 680 ± 10 Ma, which has already been discussed (see Sect. 5.2.1).

K/Ar age determinations on < 2 µm fraction of rocks close to the contact of the basal nappe at Guelb el Hadej (Lécorché and Clauer 1984) in the northern Mauritanides, have yielded a mean age of 307 Ma, interpreted as the age of the thrust.

But since 1985, $^{40}Ar/^{39}Ar$ incremental release mineral ages, from different parts of the belt suggest a more complex tectonothermal evolution involving several late Precambrian-Early Cambrian events and a variable late Paleozoic overprint (Dallmeyer and Lécorché 1989, 1990a, 1990b; Lécorché and Dallmeyer 1987; Lécorché et al. 1989).

The average ages obtained in the different units of the orogen, mainly in the central Mauritanides are given in Table 4. These first ages suggest a complex, polyphase tectonothermal history for the Mauritanide belt, especially during Pan-African events.

The earliest history is recorded within the infrastructural allochthons, in amphibolites from the continental margin rift-facies association (Farkaka Unit), 700-720 Ma and 670-683 Ma cooling ages, have been recorded by a primary type of metamorphic hornblende and paragonitic white micas. The age difference could reflect diachronous cooling through contrasting argon closure temperatures related to compositional types but following the same tectonothermal event.

Study of a second type of hornblende from the same rocks yields younger cooling ages of 644-647 Ma and 611-624 Ma. These younger ages are interpreted to date cooling following a second distinct late Proterozoic tectonothermal event. Muscovites sampled in the same unit yield cooling ages between 585 and 595 Ma. Continuous diffusional loss of radiogenic ^{40}Ar from muscovite during prolonged maintenance above closure temperatures between the estimated age of crystallization of ca. 650 Ma and the cooling age of ca. 595 Ma likely resulted in development of an intergrain vapor phase, from which were probably derived in part the extraneous argon components that contaminate the hornblende grains.

In the eastern calc-alkaline complex (Kelbé Unit, eastern El Hneïkat Unit) and in the western suprastructural allochthons (Gaoûa quartzites), muscovite records cooling ages of 570-580 Ma suggesting that these structural units suffered the effects of the second late Proterozoic tectonothermal event and were also maintained for a long time above closure temperatures. The Gaoûa quartzites would therefore be older than ca. 650 Ma, which precludes

Table 4: Synthetic geochronological data in the different units of the belt[a]

Nappes (structural setting)	Suprastructural allochthons		Infrastructural Allochthons				Parautocht.	Foreland	Time scale
			calc-alkaline		Continent. margin rift complex	Ophioliticunits			
	W	E	W	E					
h 2,000 (1,100)									
△ m310 △	m300 (ovp)		m300 (rec.)				m 300 325 (ovp)		300
△ — — △									
h 2,480 m 2,260								Sg 4 D4 Sg 3	400
△ △								D3	
△ — — △ m 320 (rec.)							m 550–540	Sg 2	500
△ 307* △	m 580–570			m 580–570	m 595–585	m 595			
Late Prot.\Paleoz.					h 627–611 h 647–644			D2	600
b 1,800					h 689–670				700
h 2,400					h 720–700			Sg1	

* K/Ar age; [a] Mineral ages on h: hornblende; b: biotite; m: muscovite. Ovp: overprint; rec.: recrystallization; Sg: Supergroup; D: disconformity

their correlation with the Cambrian (?) - Devonian suprastructural allochthons as suggested by Dia (1984) and allows us to consider them as metasedimentary equivalents of the Supergroup 1, till now unknown in the belt.

Within parautochthonous sequences, 540-550 Ma post-metamorphic cooling ages are recorded by paragonitic white mica along the mylonitic foliation close to a basal tectonic contact with a retrogressed basement. Elsewhere in the parautochthon, muscovite displays discordant spectra in which apparent ages systematically increase through low-temperature portions of the analyses to reach 470 to 540 Ma in the high-temperature portion. These spectra suggest a polymetamorphic history involving initial cooling through argon closure temperature prior to 470 Ma and a variable but significant thermal overprint circa 300-325 Ma.

Affects of this 300-325 Ma thermal overprint are recorded throughout the belt. Its most extensive effects may be seen in the western calc-alkaline complex (El Heneïkat and Mbout Units) where muscovite displays 300 Ma plateau cooling ages, meaning a complete recrystallization. Eastward, extensive effects of the overprint are only recorded near the major thrust faults, including thrusts within the calc-alkaline sequence and the bordering thrusts of the parautochthon slabs. But these thrusts have only been partially rejuvenated and are therefore considered to be much older structures.

Effects of the 300-325 Ma tectonothermal overprint are spatially related to the thrust at the base of the western suprastructural allochthon (Gaoûa and Bakel quartzites). Within structurally higher levels of Gaoua quartzites the overprint is minimal. It is maximal close to the contact, within the quartzites and within the imbricate zone developed immediately beneath.

Eastward, the most extensive 300-325 Ma overprint is recorded within mylonitic quartzites near the basal thrust of the eastern suprastructural allochthon (Sangarafa unit).

In the northern Mauritanides, such an overprint is recorded within the parautochthon close to the basal contact of the supracrustal allochthons (quartzite nappe). As already related, K/Ar age of 307 Ma has been measured at the contact of the nappes upon the foreland. Similar $^{40}Ar/^{39}Ar$ plateau-ages have been obtained on muscovites within ductile shear zones at the base of the upper basement nappe. A slightly older, 320 Ma plateau age, interpreted as a primary age of metamorphism, has been recorded within sericite schists of a metasedimentary nappe. On the contrary, several high-grade basement nappes display 2-2.4 Ga ages without overprint.

For political reasons, it has not been possible to sample the Western Sahara area, which would be of major interest for comparisons with the northern Mauritanides.

7.4 An Attempt at Synthetic Interpretation in the Light of Geochronological Data

In the northern central Mauritanides, an early phase of metamorphism prior to 720 Ma led to diachronous cooling of paragonite and hornblende of the continental margin rift-facies association, between 720 and 670 Ma. This episode is not recorded southward.

A second tectonothermal episode affected the continental margin rift association, the ophiolitic complex, eastern components of the calc-alkaline igneous complex and the Gaoua allochthons (and Bakel). This event partially rejuvenated paragonitic white mica from the first episode at ca. 600 Ma (cooling age). It would be older than 647 Ma. It results in a very slow cooling mainly for the muscovite.

A third, very low-grade, event appears to have occured ca. 540–550 Ma and first affected the parautochthonous sequences. This was likely followed by uplift and erosion with resultant molasse deposition in the foreland (Dia 1984) although characters of this sequence do not suggest a clear relation with the erosion of the belt.

Local deformation of the foreland sequences in the Early-Middle Ordovician, underlined by the angular unconformity of the Supergroup 3, led Lécorché and Sougy (1978), Lécorché (1980, 1983) and Lécorché et al. (1983) to see there a foreland record of a major event. Hitherto, thermal evidence for such an event has not been noticed in the thermal history.

Effects of post-Frasnian folding, observed in the foreland, are also widespread throughout the western metamorphic units, thus involving emplacement of supracrustal allochthons and nappes of various material into their present positions in the Late Carboniferous (ca. 300–325 Ma), and local reactivation of older thrust faults. Recent $^{40}Ar/^{39}Ar$ plateau ages on the volcanic calc-alkaline complex east of Mbout (Dallmeyer and Lécorché, 1990b) indicate that the volcanics are either Pan-African in age and entirely recrystallized or are related to a late Carboniferous ensialic arc and are thrust over the internal zones of the earlier orogenic belt. Such an interpretation was already inferred from the muscovite plateau ages recorded in the western calc-alkaline igneous complex (Magta Lahjar Unit).

7.5 Geodynamic Implications and Conclusions

Field relationships, regional variations in radiometric ages, igneous geochemical affinities and geophysical characteristics suggest a polyphase tectonothermal evolution for the Mauritanide (and Bassaride) orogen. The following tectonic development, after Lécorché et al. (1989) and Dallmeyer and Lécorché (1989, 1990a, 1990b), is proposed (Fig. 20):

1. Middle-late Proterozoic (ca. 1100–700 Ma; Fig. 20A): deposition of continental sedimentary sequences (Supergroup 1) on crystalline basement of the West African shield.

2. Late Proterozoic (ca. 700–680 Ma; Fig. 20B): intracontinental rifting leads to eventual formation of oceanic lithosphere of uncertain but generally limited extent. Initial phases of extension were marked by emplacement of dykes of low-Ti continental tholeiite (protolith of Farkaka amphibolite) into continental margin sedimentary sequences. Blanc (1986) reported 680 ± 10 Ma U-Pb zircon crystallization ages for various phases of the syenitic suite of the Bou Naga window (see Sects. 2.3.1, 5.2.1), considered to have originated during Late Proterozoic rifting. Attenuation of continental crust eventually led to mantle depressurization resulting in partial melting and formation of alkalic basalts (Rémy 1987). These evolved to T-type MORB basalts and formation of oceanic lithosphere represented by imbricated ophiolitic units. A slightly older (700–720 Ma) compressional tectonothermal event is locally recorded within continental margin rift-facies association in the central Mauritanides. This likely developed while the western crustal block was separated from the West African craton and probably resulted from imbrication of sedimentary sequences along the margin of the western block during initial phases of west-directed subduction.

3. Late Proterozoic (ca. 680–640 Ma; Fig. 20C, D): compression and resultant imbrication of oceanic and continental crust during Pan-African I orogenesis. This led to partial melting at deep structural levels and formation of a western ensialic volcanic arc. Bassot and Caen-Vachette (1983) and Dallmey-

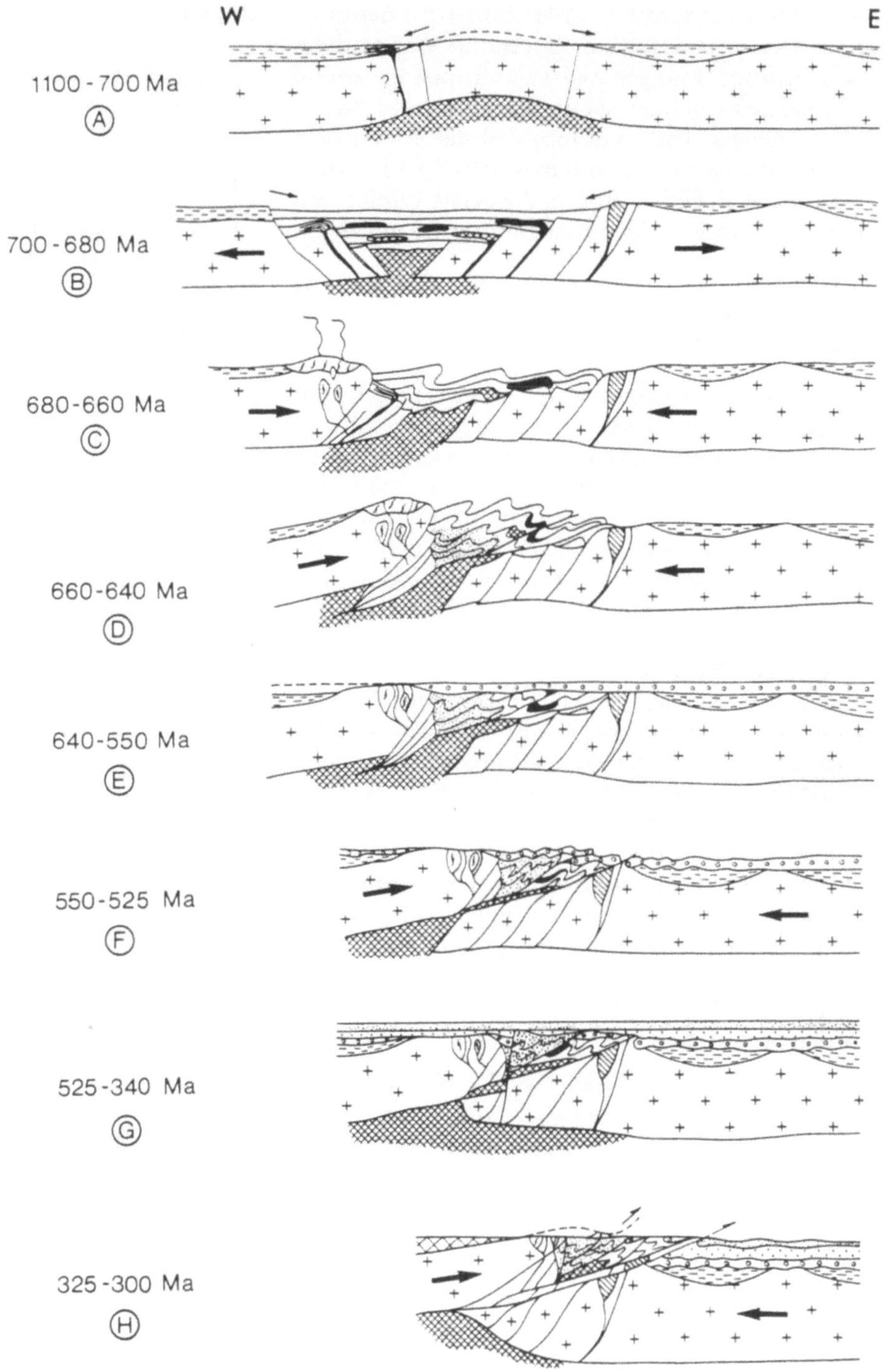

Fig. 20. Geodynamic evolution of the Mauritanide orogen. Refer to text for discussion

er and Villeneuve (1987) reported Rb-Sr whole-rock isochron and $^{40}Ar/^{39}Ar$ hornblende ages suggesting ca. 680–685 Ma crystallization ages for calc-alkaline granites of this suite (Niokolo-Koba, southern Mauritanides). Perhaps the isothermic pressure increase suggested to have developed between the two late Proterozoic metamorphic events recorded in the Farkaka amphibolites of central Mauritanides resulted from thickening of continental crust during the continental collision.

4. Late Proterozoic-Early Cambrian (ca. 640–550 Ma; Fig. 20E): uplift and cooling of the Pan-African I orogen. Deposition of foreland flysch sequences with basal tillites (e.g. Tichilit el Beïda and Bouly Groups; Mali Group in the Bassarides).

5. Cambrian (ca. 550-525 Ma; Fig. 20F): deformation and metamorphism in central Mauritanides (Pan-African II orogenesis) with an apparent southward increase in intensity (in Bassarides and mainly in Rokelides). This may record distal effects of a continent-continent collision which led to formation of the Rokelide orogen. Associated uplift and erosion resulted in deposition of foreland molasse.

6. Cambrian-Late Devonian (ca. 525-360 MA; Fig. 20G): deposition of clastic successions in the Taoudeni and Bové basins.

7. Early Carboniferous (ca. 325-300 Ma; Fig. 20H): collision of Gondwana and Laurentia results in eastward relative movement of the western continental block. This produces: 1) variable metamorphism and internal imbrication of predeformed western sequences (nappes); 2) ductile imbrication of Late Proterozoic rift sequences; 3) folding and thrusting of intracontinental foreland sequences (foreland fold and thrust belt and parautochthon); 4) emplacement of suprastructural allochthons, and extensive reworking of deeper crustal levels along the western margins of the orogen. The nappes and the western calc-alkaline complex may be directly related to this phase. Lefort (1984) has advanced an hypothesis of indentation of the Reguibat shield within the Laurentian border and Vauchez et al. (1987) have further supported it. The nappes of northern Mauritania and Western Sahara could be directly related to this process and would represent imbricated volcanic-arc material, metasediments and old basement thrust upon the West African craton to relatively high structural position.

8. Early Permian (ca. 300-275 Ma): ductile deformation along faults bordering coastal block during terminal stages of the eastward translation, especially in the southern Mauritanides.

Even if the Pan-African history of the Mauritanide orogen is not completely elucidated, it is clear that this belt was at first a pure product of Gondwanaland tectonics. Nevertheless it looks incomplete, for Florida chose an American nationality after the Hercynian-Alleghanian war, and the Pan-African link disappeared (if it ever existed) between the northern Mauritanides and the Anti-Atlas inliers, around the western rim of the Reguibat shield. The answer is not in Africa. Perhaps it is in America, around the Chesapeake Bay as proposed by Lefort (1984) and argued by Vauchez et al. (1987), as part of an Avalon terrane annexed by the expansionist Laurentia.

References

Alia Medina M (1958) Esquema geológico del Sahara español. Instituto de Estudios africanos, Madrid

Alia Medina M (1960) La tectonica del Sahara español. Rep 21th Int Geol Congr, Norden, Copenhagen, Denmark, part XVIII, pp 193-202

Angeli H (1984) Le magmatisme protérozoïque supérieur de l'ensemble Niokolo-Koba - Koulountou (Sénégal-Guinée). Témoin d'une marge continentale active au panafricain. DEA, Univ Nancy I, France, 35 pp (unpubl)

Arribas A (1960) Las formaciones metamorficas del Sahara español y sus relaciones con el Precambrio de otras regiones africanas. Rep 21th Int Geol Congr, Norden, Copenhagen, Denmark, part IX, Sect 9: 51-58

Arribas A (1968) El Precambrio del Sahara español y sus relaciones con las series sedimentarias mas modernas. Boll Geol Mineral, Madrid 79-5: 445-480

Ba Gatta AA (1982) Contribution à l'étude géologique et minéralogique du gisement d'Akjoujt (Mauritanie). 3rd cycle thesis, Univ Orleans, Fr, 162 pp (unpubl)

Barrère J (1967) Le groupe précambrien de l'Amsaga entre Atar et Akjoujt (Mauritanie). Etude d'un métamorphisme profond et de ses relations avec la migmatisation. Mém Bur Rech Géol Min, Paris 42: 215

Bassot JP (1966) Etude géologique du Sénégal oriental et de ses confins guinéo-maliens. Thesis 1963, Univ Clermont-Ferrand, Fr. Mém Bur Rech Géol Min, Paris 40: 322

Bassot JP (1969) Aperçu sur les formations précambriennes et paléozoïques du Sénégal oriental. Bull Soc Géol Fr XI (2): 160-169

Bassot JP, Caen-Vachette M (1983) Données nouvelles sur l'âge du massif de granitoïde du Niokolo-Koba (Sénégal oriental); implications sur l'âge du stade précoce de la chaîne des Mauritanides. J Afr Earth Sci 1 (2): 159-165

Bassot JP, Delpy J (1960) Sur l'âge cambro-ordovicien de la série d'Akjoujt (Mauritanie). CR Acad Sci, Paris 250: 3206-3208

Bassot JP, Bonhomme M, Roques M, Vachette M (1963) Mesures d'âges absolus sur les séries précambriennes et paléozoïques du Sénégal oriental. Bull Soc Géol Fr 3: 401-405

Bense C (1964) Les formations sédimentaires de la Mauritanie méridionale et du Mali nord-occidental (Afrique de l'Ouest). Thesis 1961, Univ Nancy, Fr. Mém Bur Rech Géol Min, Paris 26: 270

Bertrand-Sarfati J, Raaben ME (1970) Comparaison des ensembles stromatolitiques du Précambrien supérieur de Sahara occidental et de L'Oural. Bull Soc Géol Fr XII (3-4): 364-371

Bessoles B (1977) Géologie de l'Afrique: le craton ouest-africain. Mém Bur Rech Géol Min, Paris 88: 402

Black R, Caby R, Moussine-Pouchkine A et al. (1979) Evidence for Precambrian plate tectonics in West Africa. Nature (Lond) 278: 223-227

Blanc A (1986) Le magmatisme du complexe alcalin minéralisé en terres rares, Yttrium et Thorium de Bou Naga (Mauritanie); environnement géologique et géochimique, radiochronologie et signification géodynamique. Thesis, Univ Nice, Fr, 247 pp (unpubl)

Blanc A, Caruba C, Caruba R, Dars R, Ohnenstetter D, Peucat JJ (1986) Age archéen du socle de la fenêtre de Bou Naga (Mauritanie), âge panafricain des massifs intrusifs alcalins. 11ème Réun Annu Sci Terre, Clermont-Ferrand, Fr (abstr), p 19

Blanchot A (1955) Le Précambrien de Mauritanie occidentale

(esquisse géologique). Thesis 1953, Univ Nancy, Fr. Bull Direction Min Afr Occident Française, Dakar, Sénégal 17: 208
Boeri F (1985) Traitement numérique d'images. Application à la téléanalyse d'un secteur test du tronçon septentrional de la chaîne des Mauritanides au Sahara occidental. DEA, Univ Aix-Marseille III, Fr, 62 pp (unpubl)
Bowden P, Turner DC (1974) Peralkaline and associated ring-complexes in the Nigeria-Niger Province, West Africa. In: Sorensen (ed) The alkaline rocks. Wiley, NY, pp 330-351
Briden JC, Whiteombe DN, Stuart GW, Fairhead JD, Dorbath C, Dorbath L (1981) Depth of geological contrast across the West African craton margin. Nature (Lond) 292: 123-128
Bronner G, Sougy J (1969) Extension de la glaciation fini-ordovicienne à la région d'Aoussert (Sahara espagnol méridional). Ann Fac Sci Clermont-Ferrand, Fr, 41, Géol Minéral 19: 79-80
Bronner G, Roussel J, Trompette R, Clauer N (1980) Genesis and geodynamic evolution of the Taoudeni cratonic basin (Upper Precambrian and Paleozoic), West Africa. In: Bally AW, Bender PL, McGetchin TR, Walcott RI (eds) Dynamics of plate interiors. Geodyn Ser, Am Geophys Union, Washington 1: 81-90
Bronner G, Marchand J, Sougy J (1983) Structure en synclinal de nappes des Mauritanides septentrionales (Adrar Souttouf, Sahara occidental). 12th Colloq Afr Geol, Bruxelles, Tervuren (abstr), p 15
Caby R (1988) High-pressure metamorphic assemblages from Central Mauritanides, West Africa. Geol Soc Am Annu Meet Abstr Progr (abstr) 25599
Chanut C, Micholet J (1988) Le stade rift dans le bassin sédimentaire du Sénégal. J Afr Earth Sci 7 (2): 397-400
Chiron JC (1973) Etude géologique de la chaîne des Mauritanides entre le parallèle de Moudjéria et le fleuve Sénégal (Mauritanie). Thèse de Doctorat d'Etat, Univ Lyon, Fr. Mém Bur Rech Géol Min 84: 284
Clauer N (1976) Géochimie isotopique du Strontium des milieux sédimentaires. Application à la géochronologie de la couverture du craton ouest-africain. Sci Géol, Strasbourg, Fr, Mém 45: 256
Clauer N, Dallmeyer RD, Lécorché JP (1988) The chronology and extent of Late Paleozoic tectonothermal activity in the North Mauritanide foreland. Geol Soc Am Bull (in press)
Crenn Y, Rechenmann J (1965) Mesures gravimétriques et magnétiques au Sénégal et en Mauritanie occidentale. Cah ORSTOM, Sér Géophys, Paris 6: 59
Crévola G, Dars R, Le Page A, Quin JP (1974) Découverte de Dévonien fossilifière plissé à Godiovol (Guidimakha - République Islamique de Mauritanie). CR Somm Soc Géol Fr 2: 33-35
Culver SJ, Pojeta J, Repetski JE (1988) First record of Early Cambrian shelly microfossils from West Africa. Geology 16: 596-599
Dacheux A (1967) Etude photogéologique de la chaîne de Dhlou (Zemmour-Mauritanie septentrionale). Rapp Lab Géol Fac Sci Univ Dakar 22: 45
Dallmeyer RD, Lécorché JP (1989) $^{40}Ar/^{39}Ar$ polyorogenic mineral age record within the central Mauritanide orogen, West Africa. Geol Soc Am Bull 101: 55-70
Dallmeyer RD, Lécorché JP, 1990a, $^{40}Ar/^{39}Ar$ polyorogenic mineral age record in the northern Mauritanide orogen, West Africa: Tectonophysics 177: 81-107
Dallmeyer RD, Lécorché JP, 1990b, $^{40}Ar/^{39}Ar$ polyorogenic mineral age record within the southern Mauritanide orogen (M'Bout-Bakel Region), West Africa: American Journal of Science 290: 1136-1168
Dallmeyer RD, Villeneuve M (1987) $^{40}Ar/^{39}Ar$ mineral age record of polyphase tectonothermal evolution in the southern Mauritanide orogen, Southeastern Senegal. Geol Soc Am Bull 98: 602-611
Delpy J (1959) Attribution au "Primaire" de formations métamorphiques estimées précambriennes en Mauritanie centrale. CR Acad Sci, Paris 249: 1374-1376
Delpy J (1960) Attribution au Primaire de la Série de Bakel (Mauritanie, Sénégal) considérée auparavant comme précambrienne. CR Acad Sci, Paris 250: 3035-3037
Destombes J (1968) Sur la nature glaciaire des sédiments du groupe du 2ème Bani, Ashgill supérieur de l'Anti-Atlas. Maroc. CR Acad Sci, Paris, D 267: 684-686
Destombes J, Sougy J, Willefert S (1969) Revision et découvertes paléontologiques (Brachiopodes, Trilobites et Graptolites) dans le Cambro-Ordovicien du Zemmour (Mauritanie septentrionale). Bull Soc Géol Fr XI (2): 185-206
Dia O (1984) La chaîne panafricaine et hercynienne des Mauritanides face au bassin protérozoïque supérieur à dévonien de Taoudeni dans le secteur-clé de Méjéria (Taganet, Sud RIM); Lithostratigraphie et tectonique. Un exemple de tectoniques tangentielles superposées. Doctorat d'Etat thesis, Univ Aix-Marseille III, Fr, 516 pp (unpubl)
Dia O, Lécorché JP, Le Page A (1979) Trois événements orogéniques dans les Mauritanides d'Afrique occidentale. Rev Géogr Phys Géol Dyn, Paris 31 (5): 403-409
Drot J, Lécorché JP (1971) Sur la présence de *Diabolirhynchia Hollardi* (Brachiopodes, *Rhynchonellida*) dans un niveau calcaire de l'Adrar mauritanien et ses conséquences sur la stratigraphie du Siluro-Dévonien de cette région. Ann Univ Provence - Sci, Marseille, Fr XLVI: 181-188
Drot J, Lardeux J, Le Page A (1979) Découverte de Silurien fossilifère dans le Parc National du Niokolo-Koba (Sénégal). Conséquences paléogéographiques et structurales. Bull Soc Géol Minéral Bretagne, C/X (1): 7-30
Fabriès J (1963) Les formations cristallines et métamorphiques au NE de la province de Séville (Espagne). Doctorat d'Etat thesis, Univ Nancy, Fr, 276 pp (unpubl)
Giraudon R (1964) Etude du granite des Hajar Dekhen et des schistes cristallins environnants (Région d'Akjoujt, Mauritanie occidentale). Rep Bur Rech Géol Min, Dakar, Sénégal, DAK 64-A13: 153 (unpubl)
Giraudon R, Sougy J (1963) Position anormale du socle granitisé des Hajar Dekhen sur la série d'Akjoujt et participation de ce socle à l'édification des Mauritanides hercyniennes (Mauritanie occidentale). CR Acad Sci, Paris 257: 937-940
Gladney ES, Goode WE (1981) Elemental concentrations in eight new United States Geological Survey rock standards: a review. Geostandards News 5: 31-64
Graham CM, Powell R (1984) A garnet-hornblende geothermometer: calibration testing and application to the Palona schists, southern California. J Metamorphic Geol 2: 13-33
Guétat Z (1981) Etude gravimétrique de la bordure occidentale du craton ouest-africain. 3rd cycle thesis, Univ Sci Tech Languedoc, Montpellier, Fr, 185 pp (unpubl)
Holdaway MJ (1972) Thermal stability of Al-Fe epidote as a function of O_2 and Fe content. Contrib Mineral Petrol 37: 307-340
Holm PE (1985) The geochemical fingerprints of different tectonomagmatic environments using hygromagmatophile element abundances of tholeiitic basalts and basaltic andesites. Chem Geol 51: 303-323
Kane M (1986) Sur les caractères ophiolitiques de complexe

ultra-basique - basique du Protérozoïque supérieur des Mauritanides centrales. Metallogenèse et géochimie des indices de métaux de base associés et guides de prospection (Pour une ophiolitisation au Précambrien). 3rd cycle thesis, Univ Nancy I, Fr, 310 pp (unpubl)
Kane M (1987) Evidence for an ultrabasic-basic complex with ophiolitic characteristics in the central part of the panafrican Mauritanides. Tectonothermal evolution of the West African orogens. IGCP 233 Annu Meet Abstr Progr, Nouakchott, Mauritania, pp 113-115
Kessler SF (1986) Etude structurale et pétrologique sur les nappes internes des Mauritanides dans la région d'Akjoujt (RI de Mauritanie). Thesis, Univ Aix-Marseille III, Fr, 2 vol, 108 and 91 pp (unpubl)
Kretz R (1983) Symbols for rock-forming minerals. Am Mineral 68: 277-279
Kyle PR (1980) Development of heterogeneities in the subcontinental mantle: evidence from the Ferrar group, Antarctica. Contrib Mineral Petrol 73: 89-104
Laird J, Albee AL (1981) Pressure temperature and time indication in mafic schist; their application to reconstructing the polymetamorphic history of Vermont. Am J Sci 281: 127-175
La Viña DJ de, Muñoz Cabezon DC (1958) Mapa geologico del Sahara español y zonas limotrofes. Inst Geol Mineral España, Madrid, scale 1:500000
Lécorché JP (1965) Premières observations sur la région d'Ijibiten (Est d'Akjoujt, République Islamique de Mauritanie). Rapp Lab Géol Fac Sci Univ Dakar 10: 29
Lécorché JP (1969) Structure de la région d'Ijibiten (feuille El Gleïtat, E d'Akjoujt, Mauritanie occidentale). Bull Soc Géol Fr XI: 222-232
Lécorché JP (1980) Les Mauritanides face au craton ouest-africain. Structure d'un secteur-clé: la région d'Ijibiten (Est d'Akjoujt, RI de Mauritanie). Thèse de Doctorat d'Etat, Univ Aix-Marseille III, Fr. In: Trav Lab Sci Terre Saint-Jérôme, Marseille, Fr, B 26: 446
Lécorché JP (1983) Structure of the Mauritanides. In: Schenk PE (ed) Regional trends in the geology of the Appalachian-Caledonian-Hercynian-Mauritanide orogen. Reidel, Dordrecht, pp 347-353
Lécorché JP, Clauer N (1984) First radiometric date (K/Ar) on the front of the Mauritanides in the Akjoujt region (Mauritania). In: Le Maroc et l'orogénie paléozoïque. IGCP Proj 27, Ministére de l'Energie et des Mines. Annu Meet. Rabat, Morocco (abstr), pp 23
Lécorché JP, Dallmeyer RD (1987) Hercynian-Panafrican relationships in the northern part of the central Mauritanides. In: Tectonothermal evolution of the West African orogens and circum-Atlantic terrane linkages. IGCP Proj 233, University of Georgia. Annu Meet, Nouakchott, Mauritania, Abstr Progr, pp 131-133
Lécorché JP, Sougy J (1969) Relations des formations d'Akjoujt (Mauritanides) avec le Paléozoïque de l'Adrar (Bassin de Taoudeni) dans la région d'Iriji (E d'Akjoujt, Mauritanie occidentale). Bull Soc Géol Fr XI: 233-250
Lécorché JP, Sougy J (1978) Les Mauritanides, Afrique occidentale. Essai de synthèse. In: Tozer ET, Schenk PE (eds) Caledonian-Appalachian orogen of the North Atlantic region. Geol Surv Can, Pap 78-13: 231-239
Lécorché JP, Roussel J, Sougy J, Guétat Z (1983) An interpretation of the geology of the Mauritanides orogenic belt (West Africa) in the light of geophysical data. Geol Soc Am Mem 158: 131-147
Lécorché JP, Dallmeyer RD, Villeneuve M (1989) Definition of tectonostratigraphic terranes in the Mauritanide, Bassaride, and Rokelide orogens, West Africa. In: Dallmeyer RD (ed) Terranes in the Circum-Atlantic Paleozoic orogens. Geol Soc Am Spec Pap 230: 131-144
Lefort JP (1984) Mise en évidence d'une virgation carbonifère induite par la dorsale Réguibat (Mauritanie) dans les Appalaches du Sud (USA). Arguments géophysiques. Bull Soc Géol Fr XXVI (6): 1293-1303
Lefort JP, Van der Voo R (1981) A kinematic model for the collision and complete suturing between Gondwanaland and Laurasia in the Carboniferous. J Geol 89: 537-550
Legrand P (1969) Description de *Westonia chudeaui nov. sp.*, Brachiopode inarticulé de l'Adrar mauritanien (Sahara occidental). Bull Soc Géol Fr XI (2): 251-256
Le Maître D (1952) Contribution à l'étude des faunes paléozoïques de l'Adrar mauritanien (Sahara occidental). Bull Direction Min HAOF Dakar, Sénégal 15 (II): 298-383
Le Page A (1983) Les grandes unités des Mauritanides aux confins du Sénégal et de la Mauritanie. L'évolution structurale de la chaîne, du Précambrien supérieur au Dévonien. Doctorat d'Etat thesis, Univ of Aix-Marseille III, Fr, 518 pp (unpubl)
Le Page A (1986) La lithostratigraphie des grandes zones structurales des Mauritanides, entre le 14e et le 16e parallèle nord (Sénégal oriental et Rép Isl de Mauritanie). J Afr Earth Sci 5: 119-134
Le Page A (1988) Rock deformation associated with the displacement of allochthonous units in the central segment of the Caledono-Hercynian Mauritanide belt (Islamic Republic of Mauritania and eastern Senegal). J Afr Earth Sci 7: 265-283
Le Roex AP, Dick JB, Erlank AJ, Reid AM, Frey FA, Hart SR (1983) Geochemistry, mineralogy and petrogenesis of lavas erupted along the southwest Indian ridge between the Bouvet Triple Junction and II degrees East. J Petrol 24: 267-318
Le Roex AP, Dick HJB, Reid AM, Frey FA, Erlank AJ, Hart SR (1985) Petrology and geochimistry of basalts from the American-Antarctic ridge, Southern ocean: implications for the westward influence of the Bouvet mantle plume. Contrib Mineral Petrol 90: 367-380
Lesquer R, Beltrao JF, De Abreu FAM (1984) Proterozoic links between North eastern Brazil and West Africa. A plate tectonic model based on gravity data. Tectonophysics 110: 9-26
Liger JL (1979) Structure profonde du bassin côtier sénégalo-mauritanien. Interprétation de données gravimétriques et magnétiques. 3rd cycle thesis, Trav Lab Sci Terre Saint Jérôme, Marseille, Fr B 16: 158
Lille R (1967) Etude géologique du Guidimaka (Mauritanie). Essai de résolution structurale d'une série épimétamorphique. Thesis Univ Grenoble, Fr, Mém Bur Rech Géol Min 55: 397
Lille R (1969) Précambrien et Cambro-Ordovicien du Guidimaka (Mauritanie orientale). Bull Soc Géol Fr XI (2): 257-267
Malavoy J (1931) Géologie de la Mauritanie du Nord. CR Acad Sci, Paris 193: 184-186
Marcelin J (1964) Plissements superposés, nappes et métamorphisme dans la région d'Akjoujt (Mauritanie). CR Acad Sci, Paris 259: 3291-3293
Marcelin J (1965) Recherches géologiques dans l'Amsaga et dans l'Aftout de Faye (Mauritanie). Secteur Aftout de Faye. Rep Bur Rech Géol Min, Dakar, Sénégal, DAK 65-A17: 135 (unpubl)
Marcelin J (1975) La chaîne des Mauritanides. Partie nord (Région entre l'Akchar et l'Aouker). In: Notice explicative de la carte géologiqe à 1/1000000 de la Mauritanie. Bur Rech Géol Min, Paris, pp 117-120

Marchand J, Bronner G, Sougy J (1984a) Notice explicative de la feuille Chalwa. Notes Mém Serv Géol Maroc 344: 7

Marchand J, Bronner G, Sougy J (1984b) Notice explicative de la feuille Bir Amzarane. Notes Mém Géol Maroc 342: 18

Michaud JG (1964) Contribution à l'étude géologique et métallogénique des environs d'Akjoujt (Mauritanie). 3rd cycle thesis, Univ Paris, 142 pp (unpubl)

Monod T (1952) L'Adrar mauritanien (Sahara occidental). Esquisse géologique. Bull Direction Min Afr Occident Française, Dakar, Sénégal 15 (I): 285

Piccirillo EM, Justin-Visentin E, Zanetin B, Joron JL, Treuil M (1979) Geodynamic evolution from plateau to rift: major and trace element geochemistry of the central eastern Ethiopian plateau volcanics. Neues Jahrb Geol Paleontol Abh 158: 139-179

Ponsard JF, Lesquer A, Villeneuve M (1982) Une suture panafricaine sur la bordure occidentale du craton ouest-africain? CR Acad Sci, Paris 295 (2): 1161-1164

Pouit G (1968) Le gisement de Terres Rares (Y) de Bou Naga (RIM). Rep Direction Min Géol Mauritanie, 20 pp (unpubl)

Rémy P (1985) Caractères tholéiitiques et alcalins dans le matériel ophiolitique du groupe d'El Aoueïja (Mauritanides centrales). Ophioliti 10: 391-410

Rémy P (1987) Le magmatisme basique des Mauritanides centrales: une ouverture océanique limitée d'âge protérozoïque supérieur en Afrique de l'Ouest. Thesis, Univ Nancy I, Fr, 281 pp (unpubl)

Renaud L (1961) Le Précambrien du Sud de la Mauritanie et du Sénégal oriental. Thesis 1958, Univ Clermont-Ferrand, Fr, Mém Bur Rech Géol Min 5: 141

Rippert JC (1973) Le Tamkarkart: la chaîne des Mauritanides contre la bordure du bassin de Taoudeni (République Islamique de Mauritanie). Etude structurale d'un bord de craton. 3rd cycle thesis, Univ Aix-Marseille I, Marseille, Fr, 190 pp (unpubl)

Ritz M, Robineau B (1986) Crustal and upper mantle electrical conductivity structures in West Africa: geodynamic implications. Tectonophysics 124: 115-132

Ritz M, Robineau B (1987) Lithospheric structures of the West African craton derived from magnetotelluric soundings. In: Tectonothermal evolution of the West African orogens and circum-Atlantic terrane linkages. IGCP Proj 233 Annu Meet, Nouakchott, Mauritania Abstr Progr, pp 167-169

Rocci G (1962) Réinterprétation du Birrimien du Guidimaka (Mauritanie méridionale). CR Acad Sci, Paris 255: 3200-3202

Rocci G (1969) Rôle des granites dans les édifices ayant subi un rajeunissement radiométrique. Geol Assoc Can, Spec Pap 5: 33-45

Rod E (1962) Fault pattern, northwest corner of Sahara shield. Bull Am Assoc Petrol Geol 46 (4): 529-534

Roussel J, Liger JL (1964) A review of deep structure and ocean-continent transition in Senegal basin (West Africa). Tectonophysics 91: 183-211

Roussel J, Liger JL (1983) A review of deep structure and ocean-continent transition in the Senegal basin (West Africa). Tectonophysics 91: 183-211

Roussel J, Lécorché JP, Ponsard JF, Sougy J, Villeneuve M (1984) Panafrican to Hercynian deformations in the Mauritanides and tectonic significance of gravity anomalies. Tectonophysics 109: 41-59

Spear FS (1980) NaSi-CaAl exchange equilibrium between plagioclase and amphibole. Contrib Mineral Petrol 72: 33-41

Sougy J (1962a) Contribution à l'étude géologique des guelbs Bou Leriah (Région d'Aoucert, Sahara espagnol). Bull Soc Géol Fr IV: 436-445

Sougy J (1962b) West African fold belt. Geol Soc Am Bull 73: 871-876

Sougy J (1964) Les formations paléozoïques de Zemmour noir (Mauritanie septentrionale). Etude stratigraphique, pétrographique et paléontologique. Thèse de Doctorat d'Etat 1961, Univ Nancy, Fr, Ann Fac Sci Univ Dakar, Sénégal 15: 679

Sougy J (1969) Grandes lignes structurales de la chaîne des Mauritanides et de son avant-pays (socle précambrien et sa couverture infracambrienne et paléozoïque), Afrique de l'Ouest. Bull Soc Géol Fr XI (2): 133-149

Sougy J, Bronner G (1969) Nappes hercyniennes au Sahara espagnol méridional (tronçon nord des Mauritanides). 5th Int Colloq Afr Geol. Ann Fac Sci Clermont-Ferrand Fr 41, Géol Minéral 19: 75-76

Sougy J, Lécorché JP (1963) Sur la nature glaciaire de la base de la série de Garat el Hamoueïd (Zemmour, Mauritanie septentrionale). CR Acad Sci, Paris 256: 4471-4474

Steiger RH, Jäger E (1977) Subcommission on geochronology; convention on the use of decay constants in geo- and cosmochronology. Earth Planet Sci 36: 359-362

Tessier F, Dars R, Sougy J (1961) Mise en évidence de charriages dans la série d'Akjoujt (République Islamique de Mauritanie). CR Acad Sci, Paris 252: 1186-1188

Trompette R (1969) Les Stromatolites du "Précambrien supérieur" de l'Adrar de Mauritanie (Sahara occidental). Sedimentology 13: 123-154

Trompette R (1973) Le Précambrien supérieur et le Paléozoïque inférieur de l'Adrar de Mauritanie (bordure occidentale du bassin de Taoudeni, Afrique de l'Ouest). Un exemple de sédimentation de craton. Etude stratigraphique et sédimentologique. Thèse de Doctorat d'Etat, Univ Aix-Marseille III. Trav Lab Sci Terre Saint-Jérôme, Marseille, Fr B 7: 702 (3 vol)

Vachette M, Bronner G (1975) Ages radiométriques Rb/Sr, de 2900 et 2700 Ma des séries précambriennes de l'Amsaga et du Tiris, dorsale Reguibat (Mauritanie). 7th Int Colloq Afr Geol, Firenze, Italia. Trav Lab Sci Terre St-Jérôme, Marseille, Fr B 11: 142-146 (french-english)

Vauchez A, Kessler SF, Lécorché JP, Villeneuve M (1987) Southward extrusion tectonics during the Carboniferous Africa-North America collision. Tectonophysics 142: 317-322

Venkatakrishnan R, Culver SJ (1987) The northern and southern margins of the Senegal microplate. In: Tectonothermal evolution of the West African orogens and circum-Atlantic terrane linkages. IGCP Annu Meet, Nouakchott, Mauritania, Abstr Progr, pp 203-204

Villeneuve M (1984) Etude géologique sur la bordure sud-ouest du craton ouest-africain: la suture panafricaine et l'évolution des bassins sédimentaires protérozoïques et paléozoïques de la marge NW du continent de Gondwana. Doctorat d'Etat thesis, Univ Aix-Marseille III, Fr, 552 pp (unpubl)

Wood DA, Joron J, Treuil M (1979) A re-appraisal for the use of trace elements to classify and discriminate between magma series erupted in different tectonic settings. Earth Planet Sci Lett 45: 326-336

Wood DA, Joron JL, Marsh NG, Tarney J, Trevil M (1980) Major ridge. Initial Rep DSDP 58: 873-894

Zimmermann M (1960) Nouvelle subdivision des séries antegothlandiennes de l'Afrique occidentale (Mauritanie, Soudan, Sénégal). Rep 21th Int Geol Congr, Norden, Copenhagen, Denmark VIII: 26-36

The Moroccan Hercynides

A. Pique,[1] J.-J. Cornee,[2] J. Muller,[2] J. Roussel[3]

1 Introduction

1.1 Development of Geological Research in Morocco

Geological exploration of Morocco began at the end of the last century and developed rapidly. The first reliable synthesis of much of the country was made by Gentil and illustrated by a map at a scale of 1/500000 (Gentil 1920). Several regional descriptions were carried out by the Geological Survey, progressing roughly from north to south. Choubert summarized the geology with a synoptic map at 1/500000 (Choubert 1952), followed in 1985 by the current 1/1000000 map, which includes the southern provinces. More recent work has often involved collaboration between the Geological Survey and other organizations such as the rapidly growing Moroccan universities. Present research is devoted to thematic studies that are largely structural, and detailed mapping at scales of 1/100000 and 1/50000.

1.2 Structural Domains of Morocco

Choubert and Marçais (1956) defined several structural domains in Morocco which, from south to north, form the basis of the following division (Fig. 1 a):

1.2.1 The Saharian Domain

This composes the West African shield (Dorsale Réguibat in the Saharian domain) (Fig. 1 b) and its Paleozoic and Meso-Cenozoic platform (Tindouf basin) and has remained stable since the Eburnean orogeny, at ca. 2000 Ma.

1.2.2 The Ouled Dhlim (Western Sahara) Domain

This includes Precambrian and Lower Paleozoic rocks that were thrust eastward upon the West African craton during the Mauritanide orogeny. The western part of this domain underlies the Atlantic coastal plain.

1.2.3 The Anti-Atlas and Zemmour Domain

This forms the peripheric part of the West African craton and was deformed during the Pan-African orogeny, mildly folded during the Variscan orogeny and, at most, slightly tilted during Cenozoic times.

1.2.4 The Western and Eastern Meseta

The Paleozoic rocks of the Meseta were deformed during the Variscan orogeny and are unconformably overlain by undeformed Mesozoic and Cenozoic rocks.

1.2.5 The Atlas Domain

This comprises a relatively thick cover of deformed Paleozoic rocks that were mildly folded and strongly uplifted during Mesozoic and Cenozoic times.

[1] Institut de Géologie, 1, rue Blessig, 67084 Strasbourg Cedex, France. *Present address:* Département de Géologie, Université de Bretagne occidentale, 29287 Brest Cedex, France
[2] URA 1208, Sédimentologie - Paléontologie, Université de Provence 13331 Marseille Cedex 3, France
[3] Laboratoire de Géophysique, et Geodynamique, Faculté ST Jérome 13397 Marseille Cedex 13, France

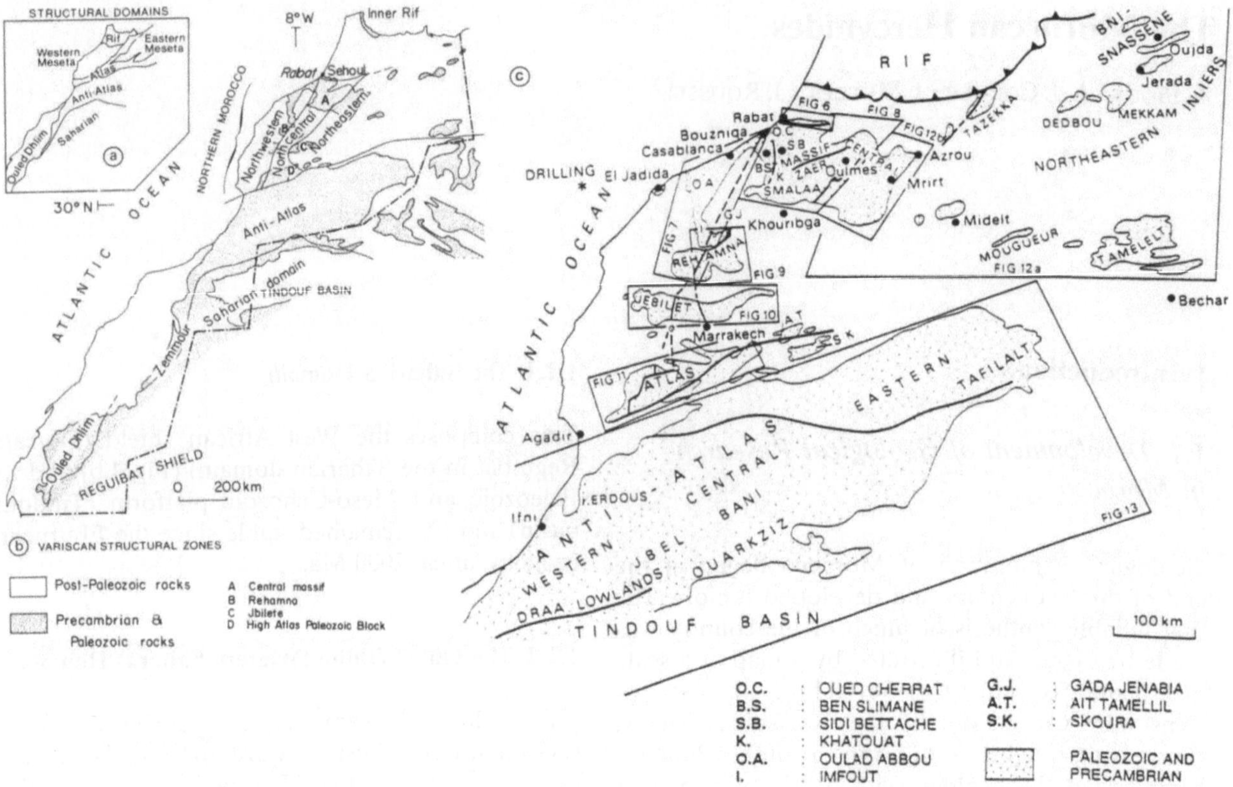

Fig. 1. a Structural domains and b Variscan structural zones of Morocco. c Index map with location of Figs. 6–13

1.2.6 The Rif Domain

Here, the Mesozoic and Cenozoic cover rocks have been folded and thrust southward upon the rest of Morocco as a part of the western Mediterranean Alpine system.

Paleozoic rocks are represented in each of these structural domains and, almost everywhere, they were deformed prior to the deposition of the Mesozoic cover. Hence, the Variscan belt extends throughout Morocco, except in the African domain. The intensity of the Variscan deformation is highly variable, ranging from essentially undeformed and unmetamorphosed to highly sheared rocks. In addition, Variscan deformation has been shown to have occurred at different times throughout Morocco, a fact which led Piqué (1988) to define different Variscan structural zones. The main characteristics of these zones are briefly described below. They will be extensively described in the chapter devoted to the Variscan structural evolution (Sect. 3).

1.3 The Variscan structural zones (Fig. 1 b)

1.3.1 Northern Morocco Zones

This is the principal domain of the Variscan belt in Morocco. The internal Variscan belt (Piqué 1983) is located in northeastern Morocco; there, Middle Paleozoic turbiditic sequences were deformed at the end of Devonian times. North-central Morocco is characterized by relatively thick Late Devonian and Early Carboniferous sequences that were more or less strongly deformed during Middle Carboniferous times. Northwestern Morocco was weakly deformed, except along its eastern margin.

1.3.2 Sehoul and Inner Rif

These two zones are in tectonic contact with the Variscan zones of northern Morocco. The Sehoul zone is composed of Lower Paleozoic rocks that were thrust upon the northern Morocco zones during Paleozoic times. The Inner Rif is part of the Alpine Rif allochthon, and shows similarities with northeastern Morocco.

2 Stratigraphy and Paleontology

Paleozoic rocks crop out extensively in the various domains of Morocco (location map, Fig. 1c). The continuity of these outcrops, the frequent absence of metamorphism or pervasive deformation, and the scarcity of vegetation has made Morocco a classic area for Paleozoic stratigraphy. This is particularly true for the Lower Paleozoic sequences of the Anti-Atlas (Destombes et al. 1985). Despite differences in the stratigraphic column and Paleozoic environments between the Anti-Atlas and northern parts of the country, they are less important than the similarities, indicating that the southern and northern parts of Morocco were close and connected throughout the Paleozoic.

2.1 Latest Proterozoic

Deposition of the latest Proterozoic succession followed the main phase of the Pan-African orogeny and rocks of this age outcrop in the Anti-Atlas and in the cores of some anticlinoria in northern Morocco. Above the main post-Pan-African unconformity, the succession constitutes three sequences (Choubert 1963):

2.1.1 Siroua - Saghro Sequence (P II-III)

This sequence is mainly represented in the northeastern Anti-Atlas. It is composed of red beds which lie unconformable upon older strata and felsic volcanic rocks. The detrital sediments were deposited in intramontane basins fed by the erosion of Pan-African highlands. In the western Anti-Atlas, Hassenforder (1987) describes the Anezi Group that represents a 2000-m-thick continental sequence, showing evidence of the Late Precambrian glaciation. The Mesti granite has been dated at 681±2 Ma and 688±16 Ma (Rb-Sr, whole rock). Field evidence shows that the emplacement of this granite postdates the oldest rocks of the Anezi Group, thereby providing a minimum age for the base of the Siroua-Anezi sequence. These rocks are generally undeformed, except in narrow zones where they have been folded and faulted during an episode of "Pan-African II" deformation.

2.1.2 Ouarzazate Sequence (P III)

In the central Anti-Atlas, this is a fluviatile and lacustrine sequence, with massive flows of andesites, dacites and calc-alkaline basalts. In the southeastern Anti-Atlas, the sequence is characterized by ignimbrites; isotopic dating indicates an age between 618±22 Ma and 573±21 Ma (whole rock, Rb-Sr technique, Jeannette et al. 1981).

2.1.3 Adoudounian

This sequence transitionally follows the Ouarzazate sequence with detrital basal strata being transgressive toward the south-east. These are followed by the "Lower Limestones", 1000 m thick. The sea responsible for the deposition of these limestones did not cover all of the Anti-Atlas, but was restricted to a gulf that extended over northern Anti-Atlas. The overlying "Série lie-de-vin" is 300-400 m thick and marked a slight regressive event. It comprises maroon and purple shales with detrital continental rocks (Tikirt sandstones) to the east. In the central Anti-Atlas, the deposition of these strata was contemporaneous with the alkaline volcanic flows of Jbel Boho (Alougoum) which have been dated at 534±10 Ma (U/Pb, Ducrot and Lancelot 1977). The "Upper Limestones" of the Adoudounian are of Early Cambrian age.

2.2 Lower Paleozoic Rocks

2.2.1 Cambrian

In the Anti-Atlas, the Cambrian succession is well documented and comprises several lithostratigraphic groups dated by abundant Trilobites and Archeocyathidae faunas (Fig. 2):

1. *Upper limestones formation of the Adoudounian Group.* These Early Cambrian stromatolitic limestones correspond to a quiet, shallow marine environment.

2. *Limestones, shales, tuffs and sandstones group.* These Early Cambrian shales and limestones are dated by Archeocyathidae (*Robustocyathus, Erismacoscinus* etc.), and are overlain by the regressive "terminal sandstones". From west to east, the thickness of this group decreases and the proportion of clastic sediments increases, suggesting the presence

of terrestrial uplands in the eastern regions of the Anti-Atlas.

3. *Inner Feïjas Group*. This group comprises argillites and siltstones that are readily weathered and occupy a topographic depression (Feija) between more resistant ridges composed of Terminal sandstone and the overlying Tabanit Group. The Inner Feijas Group which begins with the transgressive *Micmacca* breccia, comprises grey and green siltstones of the "Schistes à *Paradoxides*", which are intercalated with Goulimine quartzites in the western Anti-Atlas. The thickness of the group reaches 1300 m in the southwestern Anti-Atlas which includes and decreases toward the East where the facies are coarser. The fauna, *Ellipsocephalus, Paradoxides,* etc. indicates a Middle Cambrian age.

4. *Tabanit Group*. This group comprises 300 m of quartzites, dated by *Bailella* as Middle Cambrian in age. The Tabanit sandstones are considered a regressive facies. Despite the discovery of the endemic brachiopod *Saccogonum saccatum* (Havlicek 1971), it was generally accepted that the entire Anti-Atlas was emergent during Late Cambrian times. Recently, however, Late Cambrian deposits have been identified above the Tabanite sandstones where they are dated by trilobites *(Olentella africana, Seletella lagitena)* and brachiopods *(Billingsella destombesi)* from the same level as *Saccogonum* (Destombes and Feist 1987).

In northern Morocco Cambrian rocks are comparable to those of the Anti-Atlas (Corsini et al. 1988 a). Their facies are similar, and both the main transgressive episodes that deposited shales and siltstones, and the regressive ones indicated by coarser sandstones, are present (Destombes et al. 1985). However, two differences should be noted:

1. In northwestern Morocco (Fig. 1), the deposits are thicker than those of the Anti-Atlas and other parts of northern Morocco and accumulated within a north-northeast trending intracontinental graben (Fig. 3) from Early Cambrian times (Mayol 1987) until the end of the Middle Cambrian (Bernardin et al. 1988). Thicknesses reaches about 13 000 m in the western Paleozoic High Atlas (Cornée, in prep.).

2. Volcanic activity was locally important along graben margins in northwestern Morocco. Volcanic rocks include rhyolitic, andesitic and basaltic flows with tuffs and breccias, especially in the Western High Atlas (De Koning 1957; Schaer 1964 a). Their chemical affinity is presently being examined. Late Cambrian rocks, previously thought to be absent in

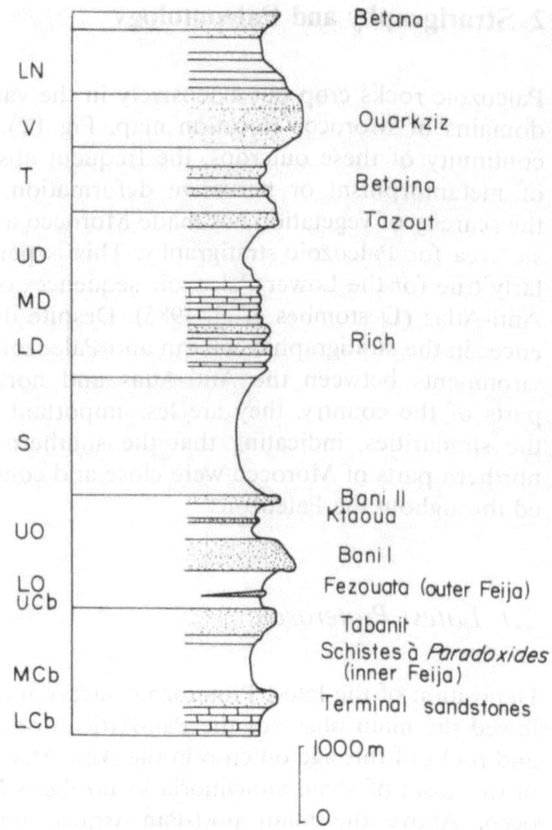

Fig. 2. Schematic lithostratigraphic column of the Paleozoic rocks of the western Anti-Atlas. LN: Lower Namurian – V: Visean; T: Tournaisian – UD: Upper Devonian; MD: Middle Devonian; LD: Lower Devonian; S: Silurian; UO: Upper Ordovician; O: Lower Ordovician; UCb: Upper Cambrian; MCb: Middle Cambrian; LCb: Lower Cambrian

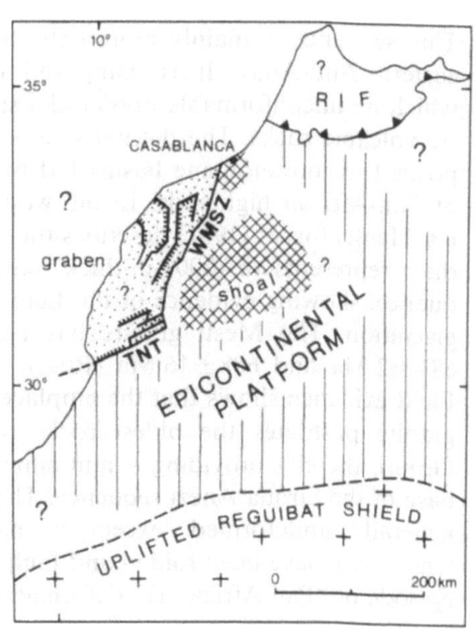

Fig. 3. Middle Cambrian graben in northwestern Morocco (Bernardin et al., 1988). TNT: Tizi-n' Test fault; WMSZ: Western Meseta Shear Zone

northern Morocco, have recently been discovered in the Casablanca area (André et al. 1987) and in the western High Atlas (Cornée et al. 1987 b). As in the Anti-Atlas, Late Cambrian sequences (clays and sandstones) are thin ($\leq$200 m) and were sometimes eroded prior to the Ordovician transgression. The Upper part of the Tabanit sandstones of the Casablanca region (= El Hank quartzites) is yet to be considered Late Cambrian in age (André et al. 1987). These Late Cambrian deposits fossilize the previous Cambrian graben.

In the Sehoul zone, Cambrian rocks are represented. The lower part of the exposed sequence comprises alternating beds of greywackes and carbonates with exactly the same facies as the basal part of the Middle Cambrian sequence in the rest of Morocco.

2.2.2 Ordovician

In the Anti-Atlas, above the Tabanit ridge, Ordovician rocks constitute several sequences comprising alternatively argillites and sandstones (Fig. 2):

1. *Outer Feijas Group*. These rocks underlie a depression bordered by ridges of the Tabanit and overlying Lower Bani groups (Jbel Bani). They are transgressive and often rest disconformably upon various Cambrian horizons. They comprise four formations that represent the Early and Middle Ordovician (Tremadoc, Arenig, Llanvirn). The rocks are mainly argillites and siltstones (Fezouata and Tachilla formations) separated by the quartzites and sandstones of the Zini formation. The Fezouata formation is dated by the graptolites *Dictyonema flabelliforme, Clonograptus* and *Didymograptus,* and by numerous Trilobites such as *Asaphellus, Symphysurus, Ampyx.* The Tachilla formation has yielded the graptolites *Didymograptus hirundo* and *D. murchinsoni,* the trilobites *Colpocoryphe* and *Placoparia,* as well as echinoderms and brachiopods. The Outer Feijas Group reaches a maximum thickness of 1300 m within a trough that trends east-west in the central Anti-Atlas and northeast-southwest to north-south in the Zemmour. Elsewhere, for instance in northeastern Anti-Atlas, the group is considerably thinner (300 m).

2. *First Bani Group*. The Outer Feijas Group grades progressively upward into the First Bani Group with an increase of coarse detrital material and oolitic ironstones. The First Bani Group mainly comprises regressive sequences of sandstones (sometimes ferruginous), and clean quartzitic beds. Its thickness reaches 400 m in the central Anti-Atlas but decreases toward the northeast. Trilobites (e. g. *Dalmanitina, Colpocoryphe* and *Selenopeltis*), Brachiopods (e. g. *Orthambonites*) and echinoderms (e. g. *Synocystis, Aristocystites*) indicate a Llandeilo age perhaps extending to the base of the Caradoc.

3. *Ktaoua Group*. The Ktaoua Group represents another transgressive unit, comprising argillaceous beds intercalated with sandstones (Rouid - Aissa sandstones) in the western Anti-Atlas which, in the eastern Anti-Atlas (Tafilalt), contain conglomerates. In the easternmost Anti-Atlas, the sandstones grade into Bryozoan biostrome. The Ktaoua Group has yielded numerous trilobites (e. g. *Dalmanitina, Calymenella, Kloucekia* and *Brongniartella*), Brachiopods (e. g. *Heterorthis, Drabovia, Howellites* and *Eochonetes*) and Echinoderms (e. g. *Destombesia, Codiacystis* and *Maghrebocystis*) which indicate a Caradoc to Ashgill age.

4. *Second Bani Group*. This group underlies the last topographic ridge of the Anti-Atlas, before the Draa lowlands. It is composed of two formations. The older, which rests conformably upon the Ktaoua Group, is quartzitic and represents the Hirnantian (Late Ashgill). The second formation disconformably overlies both the Ktaoua and First Bani Groups. The sedimentary facies of the Second Bani Group, which include "argiles microconglomératiques" with small pebbles in a clayey matrix and conglomerates containing exotic pebbles with striated surfaces, indicate a glacial depositional environment. A reduced fauna (e. g. *Hirnantia sagittifera*) allows this formation to be dated as Late Ashgill and links its facies to the well-known Late Ordovician glaciation of Sahara (Deynoux et al. 1985).

In northern Morocco, Ordovician marine sedimentation began during the Arenig. The Ordovician facies are similar to those of the Anti-Atlas with argillites, sandstones and quartzites, minor occurrences of oolitic ironstones and very few carbonates. In the western and central parts of northern Morocco, the sediments become progressively coarser during the Late Ordovician, and the Caradocian is represented by thick beds of clean quartzites. In the eastern parts of northern Morocco, however, the Upper Ordovician beds are finer grained and more pelitic, and are distributed in turbiditic sequences described in the eastern Jebilet by Hamoumi (1988).

The northernmost part of northcentral Morocco (Rabat-Tiflet zone) is characterized by ordovician magmatic activity during which effusive and intrusive mafic rocks were emplaced. In the Sehoul zone, the Ordovician is represented by detrital and turbiditic cycles of shales and greywackes that overlie the sequences attributed to the Cambrian.

Sedimentological studies (Hamoumi 1988) show that during the Ordovician, the Anti-Atlas and northern Morocco were parts of the same shallow marine platform which was itself connected to the West African craton. Clastic sediments that originated to the south were deposited on the platform and reworked as storm deposits. The Sehoul zone constituted a slope to the north of the platform. At the beginning of the Late Ordovician, the western part of the country from the Anti-Atlas to northwestern Morocco, was invaded by coarse detrital material that originated in uplifting areas to the west of present Morocco. Meanwhile, turbidites were deposited in deeper, subsiding areas in eastern Morocco. The end of the Ordovician was marked all over Morocco by the development of glacial facies.

2.2.3 Silurian

In the Anti-Atlas, Silurian beds underlie the Draa lowlands, south of the Jbel Bani ridge. In the central Anti-Atlas, the Early Llandovery is not represented. The first Silurian deposits are finegrained sandstones that are overlain by a thick (1100 m) pile of dark blue argillites with abundant graptolites, trilobites and molluscs of Wenlock age. Toward the top of the sequence, carbonate concretions and beds are intercalated with the argillites. The associated fauna, with the crinoïd *Scyphocrinites* and the bivalve *Cardiola interrupta,* indicates a Ludlow age. The thickness of this Silurian sequence decreases in the western Anti-Atlas by a reduction of the argillites, and toward the east, where the sandstones disappear.

In northern Morocco, study of the Silurian rocks is difficult because they are exposed very restricted areas and ordinarly do not constitute continuous sections. In northwestern Morocco the Silurian is represented by dark blue shales with some carbonates (120 to 20 m). From the Casablanca area to Western Paleozoic High Atlas, Early Silurian deposits are reduced to a few meters only, and Late Wenlock to Pridoli rocks are transgressive (Destombes et al. 1985; Cornée et al. 1987 a) with a rich graptolite fauna and *Scyphocrinites* beds in limestones. Basaltic flows have also been recognized southwest of Casablanca (Cornée et al. 1985). In eastern Morocco the facies are similar except in the lack of sandstones and in the presence of siliceous shales which are almost true phtanites. Here, the Silurian sequence locally begins with the basal Llandovery (Destombes et al. 1985). On the regional scale, Silurian deposits are transgressive and related to melting of the Late Ordovician glaciation. The presence of basal Llandovery faunas in eastern Morocco suggests that these zones were not emergent during this period. The uniform facies of the Silurian rocks throughout Morocco, with dark shales and limestones in the upper part of the sequence, indicates that the country was covered by a shallow, mainly euxinic epicontinental sea, that was fed by very little clastic material. Significant sandstones and quartzide beds occur only in western Morocco, again suggesting the presence of tectonic uplands west of present Morocco.

2.2.4 Early and Middle Devonian

In western Anti-Atlas (Draa lowlands), the Early Devonian is represented by sandy argillites which rest conformably on Silurian strata. They are dated as Gedinnian and Siegenian by the trilobites *Acastella heberti* and *A. tiro levis* and the brachiopods *Howellella mercurii* and *H. inchoans,* as well as some by the "Rich sequences". Each Rich is composed of some 200–300 m of limestones, argillites and sandstones. Four Rich are described in the western Anti-Atlas (Hollard 1967), and represent the Emsian-Middle Devonian. Limestones are black and pyritous. They are characterized by the absence of any detrital material and are dated by the ammonoids *Sellanarcestes wenkenbachi, Anarcestes leteseptatus, Agoniatites fulguralis* and *A. obliquus* and the brachiopod *Uncites gryphus.*

In eastern Anti-Atlas, the Gedinnian and Siegenian are represented by black limestones and argillites with a Bohemian fauna (e. g. *Paranowakia bohemica*). Volcanic tuffs occur locally at Hammar Laghdad. After a local regression, the Emsian and Middle Devonian are represented by the pelagic and locally reefal Tafilalt limestones, dated by the trilobites *Cheirurus gibbus, Odontochile* and *Phacops* (lower limestones), the ammonoids *Sellanarcestes wenkenbachi* and trilobite *Cornuproetus africanus* (argillites) and the ammonoids *Maenioceras terebratum, Pharciceras* and *Tornoceras simplex* (upper limestones).

The different Lower and Middle Devonian sedimentary rocks clearly indicate different domains within the Anti-Atlas. The western part was a subsiding domain during Early Devonian in which clastic sediments of southern (and western?) origin were deposited. The east (Tafilalt, Maider) was a more stable platform that was not reached by detrital sediments. Here, the Lower and Middle Devonian deposits were relatively thin cephalopod limestones.

In most parts of Northern Morocco, Early and Middle Devonian facies are characteristic of shallow water conditions but, as in the Anti-Atlas, subsidence rates were highly variable. In northwestern Morocco, for instance, the Ouled Abdou area shows a thick (more than 1000 m) Lower and Middle Devonian sequence composed mainly argillites and marls, alternating with dolomitic limestones and reefal limestones of Late Emsian - Eifelian and Givetian ages. These reefs were distributed along a northeast trending ridge (the Imfout axis) east of which was a graben controlled by north-north-eastern trending faults. The tilted blocks were covered by Early Devonian terrestrial redbeds, conglomerates and very coarse clastic material (El Kamel 1987). This graben system, which is broad in the Rehamna massif, narrows progressively southwestward in the western High Atlas. In northeastern Morocco, Early and Middle Devonian facies are different. Bensaïd (1979) described a composite stratigraphic sequence comprising, in decreasing age, dark argillites with *Acastella tiro* and *Nowakia acuaria;* green argillites and thin sandy limestones with plants remnants and *Anetoceras advolveris* (Lower Emsian) quartzites, marls and limestones with *Anarcestes lateseptatus* (Eifelian); and chert limestones with Stromatopores and Corals (Givetian). Farther east, coeval strata are turbiditic sequences with slumped beds and sedimentary breccias (Hoepffner 1987; Bouabdelli, work in progress). These detrital facies are known all along the northeastern Morocco Zone (Fig. 1 b). They contain spores and Acritarches that have been dated from Middle Devonian in Debdou (Marhoumi and Rauscher 1984).

In Morocco, whether in the Anti-Atlas or the northern part of the country, Early and Middle Devonian times mark the end of the clastic sedimentation that had characterized Morocco during the Early Paleozoic (Jeannette and Piqué 1981). By this time, an important part of the West African shield was covered by the sea and clastic material no longer reached Morocco. Consequently, the Moroccan shelf developed carbonates and, owing to its subtropical location (?), reef building developed. As with earlier periods, however, western parts of the country, such as the Anti-Atlas and northern Morocco, were more depressed and received detrital sediments. Eastern Morocco acted as one or several interconnected basins in which pelagic and turbiditic sediments were deposited.

2.3 Upper Paleozoic Rocks

2.3.1 Upper Devonian and Tournaisian

By the beginning of Late Devonian times, the sedimentary pattern that had prevailed throughout the Early Paleozoic changed fundamentally. Detrital supplies henceforth came from uplifting areas within Morocco itself (Fig. 4) as is recorded in the Anti-Atlas, since Frasnian times.

In the western Anti-Atlas, Frasnian rocks are black limestones similar to those deposited during the Givetian, but containing thin beds of sandstones. These sands came from the north, at the location of the present axis of the Anti-Atlas, which began to uplift at this time. This movement continued, during the Famennian with the deposition of 2000-2500 m of fine clastic sediments argillites and shales, together with sandstones and shelly limestones. This was followed during the Tournaisian by the Tazout sequences comprising three successive sandstone horizons separated by argillaceous intercalations.

In eastern Anti-Atlas, a system of uplifting platforms and downwarping basins was established during the Late Devonian and demonstrated from sedimentological observations (Wendt 1985). The platforms are characterized by disconformities and by the deposition of condensed Cephalopod limestones that constitute the upper part of the "Tafilalt limestones". In the basins, marls and green argillites with *Clymenia* and goniatites were deposited. During Tournaisian times, margins of the uprising areas were progressively covered by a transgressive sea. Brachiopod sandstones and *Gattendorfia* argillites were deposited. In the northernmost part of the Anti-Atlas (Tineghir area), Upper Devonian conglomerates disconformably overlie Ordovician shales and siltstones.

In northern Morocco, coeval paleogeographic changes (the "Late Devonian crisis", of Piqué and Michard 1988) were more important. In northeastern Morocco, the Midelt-Oujda turbidites were deformed by an early Variscan event (see below). This regional compression was contemporaneous with the development of a sedimentary basin in north central Morocco. This Sidi Bettache basin (Piqué 1984) opened inside the former Paleozoic shelf in

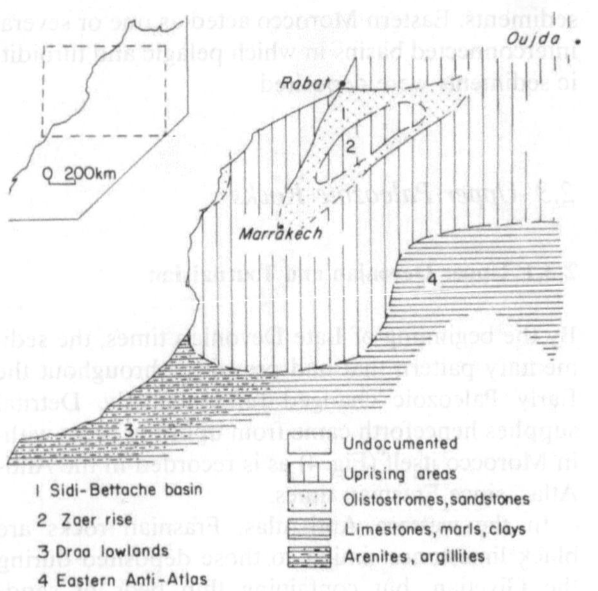

Fig. 4. Distribution of sedimentary facies during Late Devonian/Tournaisian times

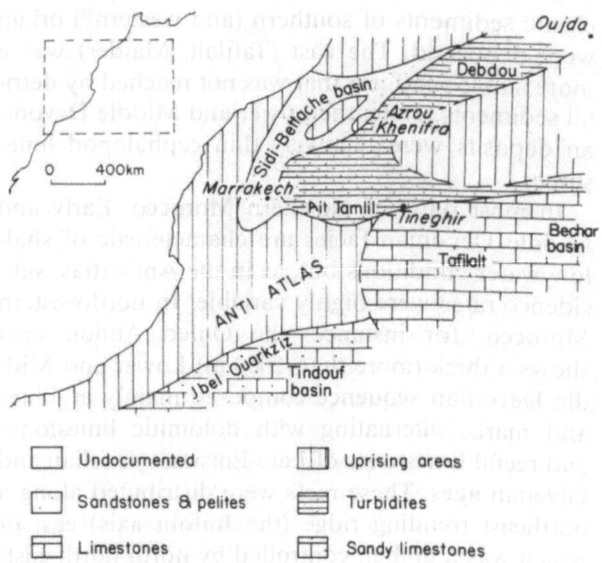

Fig. 5. Distribution of sedimentary facies during Late Visean times. Location on Fig. 1 c

north central Morocco. Several sections around the basin (Piqué 1979) show comparable successions: marls and limestones of Early and Middle Devonian age are overlain by Famennian conglomerates and breccias that are the first deposits at the basin margins. Outside the basin, for instance in the coastal block of northwestern Morocco, fine-grained sands, argillites and quartzites were deposited during the Famennian, before a regression that probably occurred during the Tournaisian. The margins of the basin were outlined by the breccias mentioned above, by olistostromes and mudflows (Padgett et al. 1977) and by proximal turbidites dated by brachiopods such as *Syringothyris, Spirifer tornacensis* and *Schellwiennella*. A central part of the basin acted as a sedimentary ridge: the Zaer ridge (Piqué 1984; Piqué et al. 1985). Bimodal tholeiitic to alkaline volcanic flows and tuffs (Piqué and Kharbouch 1983) are known at different horizons in these strata, when accessible to observation, more central parts of the basin show more distal facies without volcanic intercalations. The Sidi-Bettache basin occupied a V-shaped area over 400 km in length from Rabat to the High Atlas Paleozoic block. Its development may have been progressively younger toward the south. This basin could be either a pull-apart basin bordered by dextral north-northeast and east-southeast trending faults, or an extensional basin.

2.3.2 Visean to Lower Westphalian (Fig. 5)

Anti-Atlas. Lower Visean rocks underlie the Betaina, a topographic depression located between the Tazout and the Ouarkziz ridges. Here they comprise 700 to 1100 m of goniatite argillites that are transgressive upon the Tazout sandstones. In the eastern Anti-Atlas (Tafilalt) reefs developed. In northernmost Anti-Atlas, the Tineghir faulted basin was filled by turbidites and olistostromes that originated south. The Ouarkziz sequence is a thick paralic unit comprising marine limestones with crinoids, brachiopods and corals, alternating with continental sandstones. It is dated as Late Visean to Early Namurian. The overlying Betana sequence is 1500 m thick and composed of continental sandstones and argillites. The eastern Anti-Atlas became continental during the Namurian, but the sea remained, until Early Westphalian times, in the Bechar Basin where paralic coal sequences were deposited.

A turbiditic basin developed along the western part of *northeastern Morocco* (Beauchamp and Izart 1987). It extended from the Aït Tamellil inlier (High-Atlas) to the eastern Jebilet, farther to the northeast, to the Azrou region (central Morocco) and the Dedbou - inlier. In this northeastern area, the turbiditic sequence was deposited on Middle - Upper Visean transgressive limestones. In the Jebilet, the turbiditic sequence (the "Kharrouba formation", Huvelin 1977; Graham 1982) is followed by regressive Late-Visean reefal limestones (André

1986). These strata were then mildy folded and eroded before the deposition of an olistostrome that is probably Namurian in age. This olistostrome and the whole turbiditic basin is finally covered by allochthonous units. From Azrou to the eastern Jebilet, these previously mildy folded nappes were derived from the southeast and the east (Allary et al. 1972; Huvelin 1977; Oukeïmeni 1987). In the Aït Tamellil inlier, on the other hand they were derived from the north-northeast (Jenny and Le Marrec 1980). In eastern Morocco, this turbiditic sedimentation was accompanied by an andesitic calc-alkaline volcanism (Kharbouch et al. 1985; Hoepffner 1987) and, from Aït Tamlil to Debdou, developed within a foreland basin along the early Variscan belt. The basin moved progressively westwards during Carboniferous in the Oujda area and its latest marine strata are Early Westphalian in age. A similar evolution is recorded in the Inner Rif.

Meanwhile, in *northcentral Morocco* the Sidi Bettache basin was progressively filled by the Oued Korifla pelitic strata dated as Early Visean, and then by the late Visean Oued Mechra sandstones and the Late Visean - Early Namurian Bou Rzim greywackes (Piqué 1984). The cumulative thickness of these detrital strata reaches 4000 to 5000 m. They have been dated by brachiopods, goniatites and, overall, by *Archeodiscidae*. In the Rehamna and Jebilet massifs, magmatic activity lasted troughout Visean times; in the central Jebilet bimodal effusive and intrusive rocks were emplaced into Visean sedimentary rocks of the Sarhlef Formation (Huvelin 1977; Michard 1982; Bordonaro 1983; El Kamel 1987; Aarab and Beauchamp 1987). The latest marine beds were deposited in a restricted trough, during the Early Westphalian (Bensaïd et al. 1980).

2.3.3 Late Paleozoic

The first unconformable, terrestrial strata are the Upper Westphalian Sidi-Kacem conglomerate and lacustrine limestones. Other redbeds associated with felsic volcanic rocks are known in various parts of the Variscan belt. They were deposited in faulted intramontane basins. Most of them are still poorly dated, and represent either the Stephanian or Permian and were locally tilted before Triassic times. In the western High Atlas, however, well documented flora indicate Stephanian deposits with coal. These rocks rest unconformably upon Variscan structures (Neltner 1938; De Koning 1957). The upper part of the sequence has recently been dated as Lower Permian (Broutin et al. 1989).

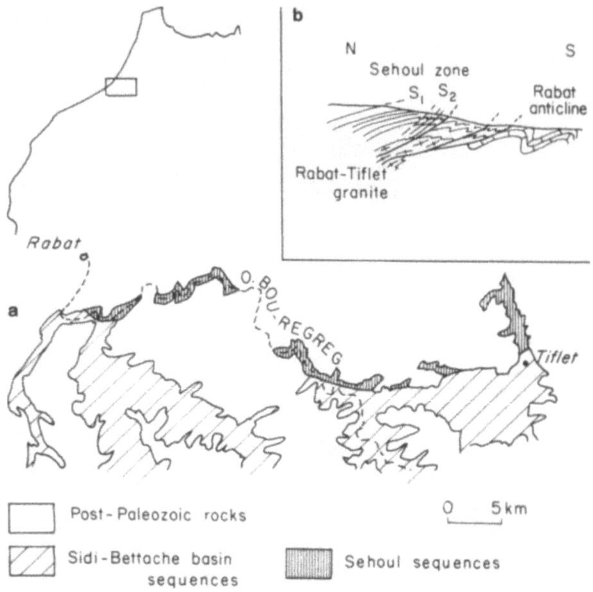

Fig. 6. The Sehoul zone. **a** map; **b** cross-section. Location indicated in Fig. 1 c

3 Structure and Metamorphism

3.1 The Sehoul Zone

This zone is located north of the Northcentral Morocco zone with which it is in tectonic contact (Fig. 6). To the north, it is covered by recent deposits and its eastern and western ends are concealed. The zone is underlain by a thick Cambrian to Early Ordovician sequence comprising, in its upper part, deep-sea facies. In spite of its limited exposure the zone is important in that it shows Early Paleozoic deformation that is not represented elsewhere in Morocco.

This Early Paleozoic (D_1) deformation developed east-west isoclinal folds contemporaneous with a low-grade metamorphism. Folding is recumbent and south verging and was followed by an episode of southward vergent thrusting. This D_1 deformation affected the Middle Cambrian sequence, and it occurred before the Early and Middle Devonian, which are represented by unconformable and transgressive deposits. The Rabat-Tiflet granite, which has been dated at 430±2 Ma (Charlot et al. 1973, Rb-Sr method, recalculated), postdates the main metamorphic event. Hence, D_1 was a Late Cambrian to Late Ordovician event. These Ordovician structures were not subsequently deformed except at the southern part of the Sehoul zone. There, a

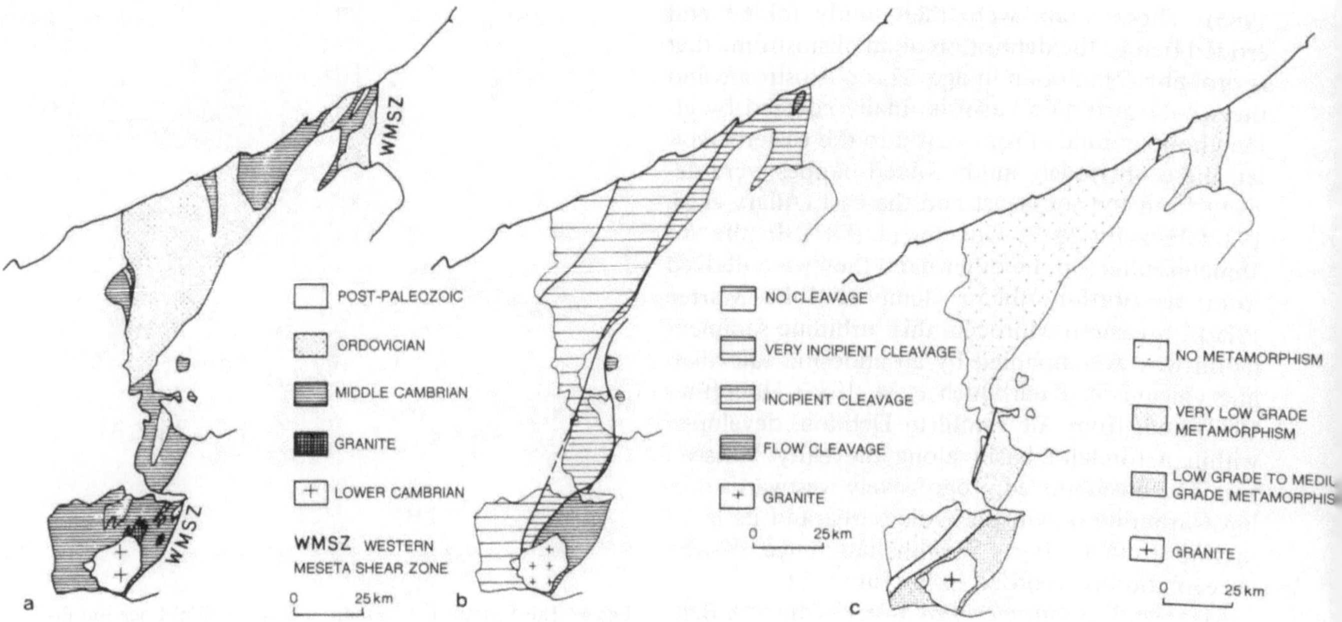

Fig. 7. The Coastal Block (including western end of Rehamna massif). Location indicated in Fig. 1 c. **a** Schematic geological map; **b** structural map; **c** metamorphic map

second deformation (D_2), developed an east-trending S_2 crenulation cleavage, axial planar to east-west chevron folds. This D_2 deformation is related to the Variscan orogeny since it affects the Devonian strata of the Rabat anticlinorium (see below). The Rabat-Tiflet granite was sheared during south directed thrusting at the end of the (D_2) deformation. This Variscan episode predates the deposition of the Visean Jbel Bakach conglomerate (El Hassani, work in progress) and corresponds to the early Variscan thrusting of the whole Sehoul zone upon northcentral Morocco.

3.2 The Coastal Block (Northwestern Morocco)

The northwestern Morocco structural zone is usually called the Coastal Block (Michard 1967; Piqué 1979) to emphasize its relative stability during the Variscan orogeny. It includes the western parts of the Central Morocco, Rehamna, Jebilet and High-Atlas Paleozoic massifs (Fig. 1). Its eastern limit is a complex, discontinuously faulted zone that trends south-southwest from Bouzniqa to the High Atlas Paleozoic massif. Its western limit is concealed offshore on the Atlantic margin of Morocco. The underlying rocks of the Coastal Block are Late Proterozoic felsic lavas and Paleozoic rocks that are mainly sedimentary and range in age from Cambrian to Upper Devonian. During Late Devonian and Carboniferous times, the Coastal Block underlay the western margin of the Sidi-Bettache basin. Imprecisely dated Variscan structures developed during the Carboniferous and probably predate the Stephanian. These are broad, north-south to north-northeast trending anticlines and synclines (Fig. 7 a). Their axial plans are upright with a slight tendency toward westward vergence. The deformation was controlled both by the thickness of the Cambrian sequences and by the fault-system inherited from the Cambrian rifting. The intensity of the deformation decreases when the thickness of the sediments increases. In the Jebilet, for instance, the Bou Gader Unit comprises an unfossiliferous carbonaceous rocks attributed to the Early Cambrian (Huvelin 1977; Mayol et Muller 1985; Mayol 1987). This unit belongs to the eastern margin of the former Cambrian rift and, in its northern part, is thrusted westward slightly on the reverse limb of the adjacent recumbent syncline. The thrust plane is nearly horizontal (Sougy 1976), but becomes progressively steeper southward where it corresponds to the main reactivated faulted margin of the former rift (Mayol 1987).

At the scale of the outcrop, however, it is generally impossible to see any structure since folding developed above the cleavage front and the rocks show no evidence of metamorphism.

Toward the east, however, the tectonothermal gradient steepens. An axial plane cleavage appears and becomes progressively more penetrative (Fig. 7 b). At the same time, metamorphism increases eastward, from very low to low-grade and medium-grade (Fig. 7 c; Piqué 1979; Piqué and Wybrecht 1987). At its eastern limit, the Coastal Block comprises highly deformed slates that form part of the Western Meseta Shear Zone (WMSZ: Piqué et al. 1980) described below.

At the western limit of the Coastal Block, Kreuzer et al. (1984) and Ruellan (1985) describe samples drilled on the El Jadida escarpment, about 80 km west of the shore. These are granodioritic rocks that crystallized around 515 Ma and were subjected to an important mylonitic event at 455 Ma and to successive thermal reactivations at 360, 336 and 315 Ma. This history does not correspond to that of the Coastal Block and it is probable that the El Jadida escarpment represents the block's western limit.

3.3 North Central Morocco

3.3.1 Western Central Massif

As described by Termier (1936), the following structures occur from west to east within the massif (Fig. 8):

1. *Ben Slimane syncline.* Here a Late Devonian - Early Carboniferous sequence (Fadli, work in progress) is affected by west-northwest (N110-120E) trending folds, whose axial planes are often upright.

2. *Cherrat anticline.* This structure is developed on a former synsedimentary ridge at the western part of the Sidi Bettache basin (Piqué et al. 1985). The rocks that crop out in the center of the anticline are Silurian argillites and Lower to Middle Devonian argillites and limestones. They are affected by polyphase deformation. The first episode (D_1), is contemporaneous with very low- to low-grade metamorphism and developed north-trending folds. Their axial planes are outlined by a slaty cleavage but their axes show widely differing orientations within the axial plane. This D_1 deformation was followed by successive episodes of reverse and strike-slip faulting (Chalouan 1981). The broad extent of Silurian pelitic rocks within the Cherrat anticline and the fact that Ordovician rocks are never visible in the core of the structure suggest a décollement exits in the Paleozoic sequence at the level of the Silurian shales and argillites.

3. *Rommani synclinorium.* Here rocks of the Sidi Bettache basin are Dinantian and Namurian in age, except in the small Ain-Hallouf anticline where Famennian series crop out. Fold trends within the synclinorium are highly variable. In the westernmost part of the structure (Oued Yquem valley), they are northwest-trending, like those of the Ben-Slimane syncline; in the eastern part, they are east-northeast (N70E), and in the center of the synclinorium (e. g. Oued Korifla valley), fold orientations vary within single outcrops (Piqué 1987). In the Khatouat area, which forms the southwestern prolongation of the Rommani synclinorium, F_1 folds

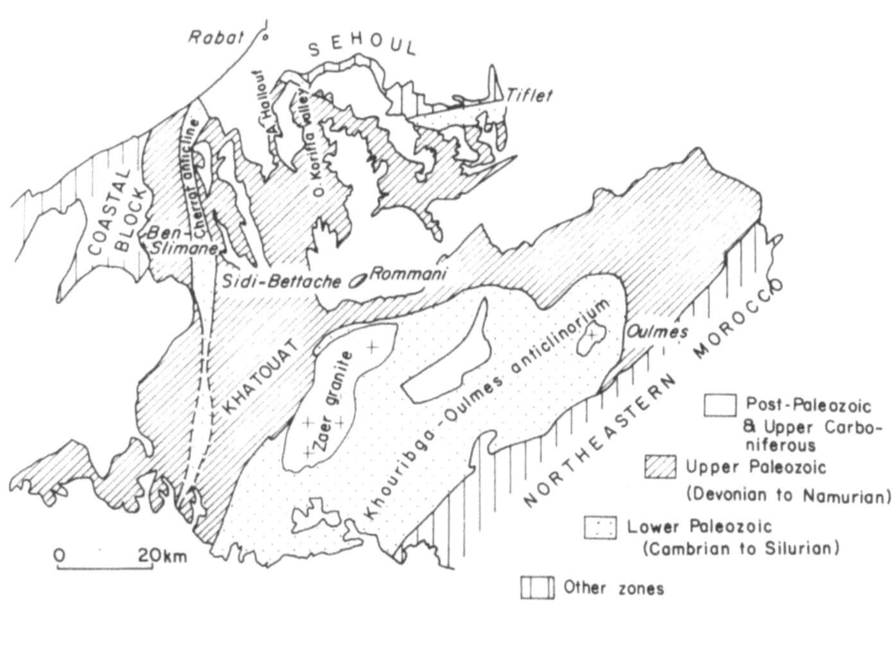

Fig. 8. The northern part of northcentral Morocco. See Fig. 1 c for location

and the associated S_1 cleavage turn progressively from the NW-SE direction of the Sidi Bettache area, to the N70E direction of the Rommani area. Piqué et al. (1985) have shown that this curvature does not reflect a second folding episode, but rather, is original and marks the local response of D_1 stresses to preexistant ridges like those Zaer and Cherrat. Metamorphic conditions reach low grade at the margins of the synclinorium but progressively decrease toward the center of the structure (Piqué 1982). The age of the metamorphism and associated deformation is certainly post-Namurian in the center of the synclinorium. On the western margin of the structure, however, a preliminary K-Ar date on fine and very fine fractions (illites) indicate a 320 Ma age (Huon et al. 1988).

4. *Khouribga - Oulmes anticlinorium*. This structure was divided by Termier (1936) and Cailleux (work in progress) into several anticlines (e. g. Ouled Khalouk, Ouled Aïssa and Mserser), in which Cambrian and Lower Ordovician rocks crop out, and synclines, which contain Upper Ordovician and Silurian rocks. Like that of Cherrat, the Khouribga - Oulmes anticlinorium was developed on a former Late Devonian to Early Carboniferous structure, the Zaer rise. The general trend of the structure is northeast but local curvature, again probably original, has produced some east-west directions. Most of the synmetamorphic folds are upright or inclined deversed to the southeast. The degree of metamorphism is greater than that of the Rommani synclinorium and often reaches low-grade. Syn- to late-kinematic granites (e. g. Oulmes, Zaer, Moulay - Bou Azza) were emplaced during Stephanian times. The eastern limit of the Khouribga - Oulmes anticlinorium is the Oulmes-Smaala fault. This was a sinistral ductile fault (Aït Omar 1985; Diot et al. 1987; Tahiri, work in progress) during Stephanian times. Post-metamorphic movement of probable Permian age along the southwestern part of this fault was right-lateral (Cailleux 1978).

5. *Rabat-Tiflet anticlinorium*. It extends roughly east-west between the Sehoul zone and the Rommani synclinorium. The cores of individual anticlines are occupied by Ordovician and Silurian rocks, whereas rocks as young as Late Visean are present in some synclines. The folds are open and upright, although a southern vergence is sometimes conspicuous, especially in the western half of the structure where the Rabat anticlinorium is overthrust by the Sehoul zone. Metamorphism is generally diagenetic to very low-grade. To the south, the anticlinorium is reservated from the Rommani syncline by several east-southeast (N110° E) to east-northeast (N70° E) trending dextral faults that are probably Permian in age (Piqué 1975; Cailleux et al. 1984).

3.3.2 Rehamna Massif

East of the Coastal block (Fig. 7), several structural elements are distinguished in the central and eastern portions of the massif (Fig. 9):

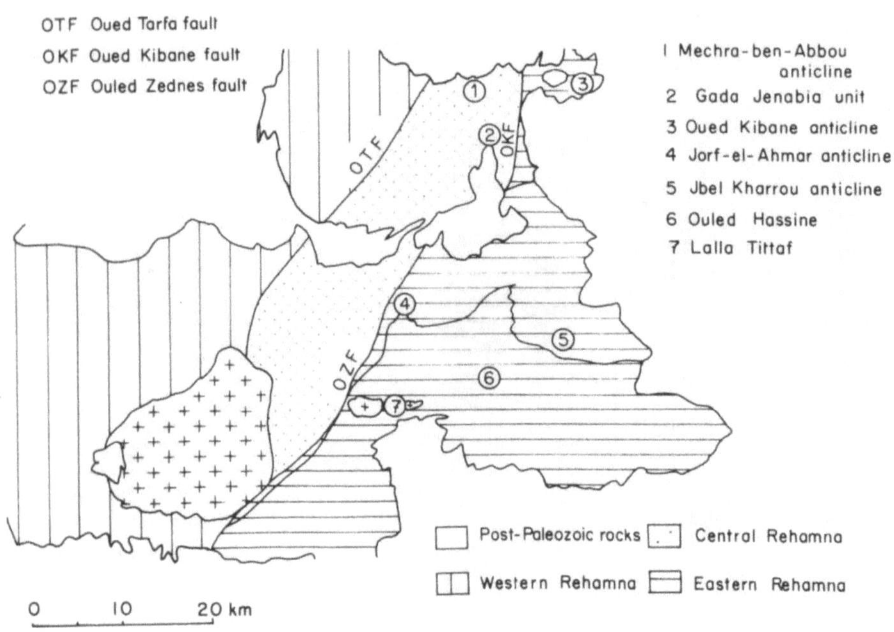

Fig. 9. The Rehamna massif, see Fig. 1 c for location

1. Parautochthonous units which lie on the easternmost border of the Coastal Block. The folds and associated cleavage in these units are truncated by flat-lying thrust planes (Rais-Assa et al. 1983). This parautochthonous zone is bound to the east by the Oued Tarfa fault.

2. The Oued Tarfa fault (OTF, Fig. 9), trends north-northeast and dips steeply eastward. During Early Devonian times it was a normal fault and controlled sedimentation. During the main (mid-Carboniferous) deformation it acted as a reverse fault. It later acted as a sinistral strike-slip fault during deposition of the continental Stephano-Permian Mechra-Ben-Abbou redbeds. Finally, it became a normal fault during the deposition of the uppermost redbeds which are associated with volcanic rocks (El Kamel 1987).

3. The central Rehamna zone lies between the Oued Tarfa and Oued Kibane fault zones (OKF). In *northern Rehamna,* this complex zone comprises several structural units that are dislocated and dismembered by early glide nappes and post-cleavage thrusts (El Kamel et al. 1985), by strike-slip faults (Michard et al. 1978; El Kamel 1987) and by shear faults related to the main folding event. These units are also affected by numerous normal faults related to Stephano-Permian extension. The principal units are north-northeast to north-south anticlines, which contain Ordovician to Early Devonian strata, and the north-northeast trending Gada Jenabia syncline. This last unit is an early glide nappe composed of Early Devonian and Upper Visean sedimentary strata and Late Visean intrusive rocks. Farther north, in the Khatouat, this sequence would have its equivalent in the southwestern prolongation of the Rommani synclinorium. The western margin of the Oued Kibane fault zone is characterized by en-echelon folds related to dextral strike-slip movement during the main deformation event (El Kamel 1987). To the south, the central Rehamna zone comprises slates and schists, quartzites and metaconglomerates, respectively attributed to the Cambrian, Ordovician and Early - Middle Devonian.

4. The Eastern Rehamna. In the metamorphic part of the massif, this zone is separated from Central Rehamna by the north-northeast trending Ouled Zednes fault zone, a branch of which is connected with the Oued Kibane fault. This Ouled Zednes zone (OZF, Fig. 9) is underlain by highly deformed and sheared limestones of probable Devonian age, and quartzites that are attributed to the Ordovician. The quartzitic beds belong to the extremely disrupted western limb of the Jorf-el-Ahmar anticline. Toward the east, its trend switches to an east-west orientation and, in the Jbel Kharrou, to a northwest-southeast direction. Here, its inverted southern limb is thrust upon the Eastern Rehamna schists. These schists, which are probably Devonian (Ouled Hassine formation) and Visean (Lalle Tittaf formation) occupy a synformal position (Cornée 1982; Hoepffner et al. 1982; Piqué et al. 1982). At the southeastern extremity of the Rehamna massif, an allochthonous unit (Cornée 1982) contains unmetamorphosed Visean rocks.

From a structural point of view, four major events are evident in the metamorphic core of the massif (Piqué et al. 1982). The first deformational phase (D_1) developed northeast trending folds (F_1) which verge northwest and are contemporaneous with a low-grade metamorphism. The second (D_2) phase occurred during the culmination of the regional metamorphism. In the eastern Rehamna, the S_1 foliation was refolded and F_2 isoclinal folds developed, with a composite S_{1-2} foliation. Most of these are sheath folds and indicate westerly directed ductile thrusting. The Sidi-Abdallah conglomerates were flattened during this D_2 episode (Piqué 1975). In the central Rehamna, the obliquity between D_1 and D_2 structures reduces and, during the D_2 deformation, the S_1 foliation was reactivated by ductile shear rather than being refolded. This foliation is flat-lying in the most metamorphic parts of the massif. The stretching lineation, which is well displayed in the Kef-el-Mouneb deformed conglomerates (Piqué 1973), trends north-northeast parallel to the fold axes. In the central Rehamna, the D_2 phase was contemporaneous with dextral ductile shear (Piqué et al. 1982; Lagarde and Michard 1986; Corsini et al. 1988 b). The metamorphic grade was significant and allowed the development of staurolite and kyanite (Hoepffner et al. 1982). The third (D_3) post-metamorphic phase is represented by thrusts, especially in the eastern Rehamna (Hoepffner et al. 1982; Cornée 1982). Compared to the structure of the central Massif which they extend southward, the Rehamna structures show a width reduction of about 50%. The Rommani syncline, for instance, is several tens of kilometers wide at its center but has a width of 15 km at its southwestern end in the Khatouat. In northern Rehamna, between the Mechra-ben-Abbou anticline and the Oued Kibane fault, the syncline is 3 km wide and eventually disappears in the Ouled Zednes region of the metamorphic Rehamna. It is probable that part of this narrowing was original and that the Sidi-Bettache

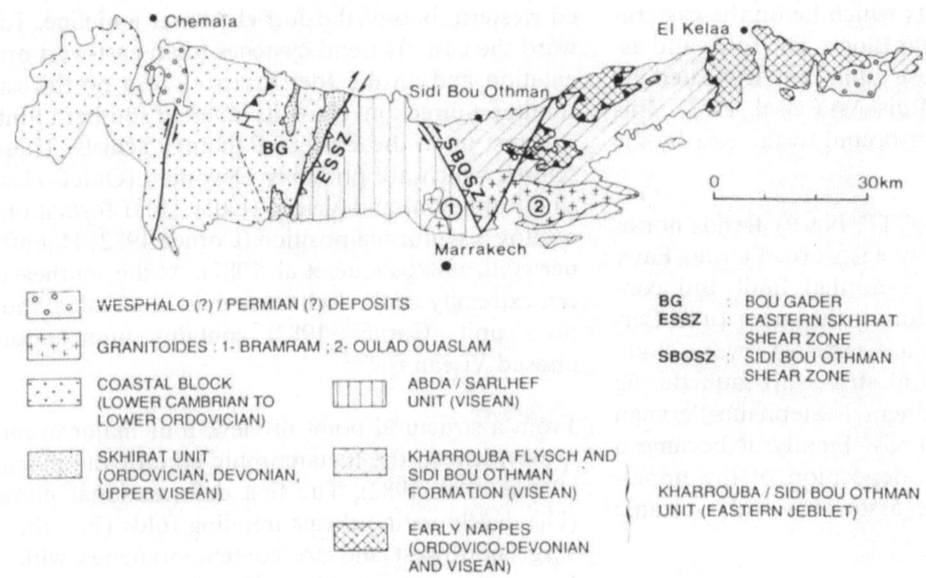

Fig. 10. The Jebilet massif. Location indicated in Fig. 1 c.

basin effectively narrowed toward the southwest (Piqué 1984; Piqué et al. 1985). However, part of this narrowing is due to deformation, either as a result of the ductile phases or the post metamorphic thrusts, since flattening becames progressively more important toward the south. A further problem concerns the significance of the Barrovian metamorphism, since the pressure required for kyanite development in the Devonian metaconglomerates far exceeds that provided by a normal sedimentary load. It is therefore necessary to invoke tectonic overloading that is probably related to the ductile thrusting of the eastern Rehamna onto the central part of the massif (Corsini et al. 1988 b). At deeper structural levels, this thrust could correspond to the décollement that is suspected in the Oued Cherrat valley.

3.3.3 The Jebilet Massif (Fig. 10)

This massif is the less metamorphic southern equivalent of the Rehamna massif.

3.3.3.1 Structural Domains

Eastward from the western Jebilet, which forms part of the Coastal Block (Fig. 7), several structural units are distinguished:

1. *The Skhirat unit* is composed of brecciated centimeter to hectometer sized blocks of Ordovician, Devonian and late Visean age. The larger ones exhibit some elements of a stratigraphic succession comparable with that of the Rehamna north central Zone. These blocks are disrupted and the most competent are clearly embedded in a non-stratified muddy and sandy matrix of unknown age. This matrix directly overlies the Lower Cambrian rocks of the Bou Gader unit. Eastward, the brecciated Skhirat unit is stratigraphically overlain by the quartzites and quartzophyllites of the Jbel Rhira formation which could be Upper Visean in age (Mayol 1987). The "Skhirat breccias" have been interpreted both as the sole of the Abda-Sarlhef nappe (Sougy 1976; Poutchkovsky 1978) and as a sedimentary formation directly transgressive on the Bou Gader unit (Bordonaro et al. 1979). However, a recent detailed study by Mayol (1987) allows the unit to be interpreted as an olistostrome originating in the Coastal Block during the opening of a Carboniferous intracontinental basin that involves the central and eastern Jebilet (Mayol and Muller 1985; Cornée and Ferrandini 1985) and, farther to the north, the Sidi-Bettache basin (Piqué 1984).

2. *The Abda-Sarlehf* unit, which is thrust on the Skhirat unit (Bordonaro 1983), comprises basal conglomerates and breccias deposited on the tilted blocks of the western margin of the Carboniferous intracontinental basin. Tholeitic intrusives are located along the main faults which were later reactivated as reverse faults. These basal deposits are overlain by Late Visean limestones and muddy iron-rich sulfurous sediments corresponding to a distal shelf environment (Aarab and Beauchamp 1987).

3. *The Sidi Bou Othman – Kharrouba unit* (SBOK). This unit, which is thrust upon the Sarhlef unit (El

Hassani and Zahraoui 1982), is composed of a distal flysch sequence (Kharrouba formation), Upper Visean in age (Huvelin 1977), stratigraphically overlain by conglomerates, sandstones, carbonaceous limestones and siltstones (Sidi-Bou-Othman formation). The latter formation contains olistoliths, disrupted beds and slumps suggesting a slope environment. The Kharrouba flysch occupies the Marrakech-Azrou trough. At the top of the sequence there are slumps with Early Carboniferous and pre-Carboniferous olistoliths. Some of the youngest slumps contain Early Devonian rocks typical of the northeastern Morocco trough and indicating an eastern source for the materials (Oukeïmini 1987). The flysch sequence is overlain by regressive Late Visean reefal limestones (Graham 1982; André 1986), which indicate that the basin was filled at this time. The relative positions of these units roughly fit the Late Visean Jebilet basin proposed by Bordonaro et al. (1979).

3.3.3.2 Variscan Deformation

Variscan deformation postdated Late Visean times and predated the deposition of the Stephano-Permian continental redbeds. Hence, it is commonly referred to the "Namuro-Westphalian" deformation.

The first Variscan deformation occured in the Kharrouba - Sidi Bou Othman unit, where gentle kilometer-scale upright and northeast-to-east trending folds developed. This folding was followed by an erosional period that peneplained the topography. Huge olistostromes were then emplaced, followed by nappes that moved westward. These nappes are composed of previously folded, Ordovician to Devonian sequences and are overlain by klippes with Late Visean rocks. These early folds have also been recognized under the Gada Jenabia nappe of northern Rehamna (El Kamel 1987). In the eastern Jebilet, the olistostrome and allochthonous slabs were intruded and metamorphosed by the Oulad Ouaslam granitoid which has been dated as Late Visean to Lower Namurian (Mrini 1985). Folding, erosion and olistostrome and nappes emplacement are therefore likely to be early Namurian in age.

The main Variscan deformation (D_1) is marked by the development of north-trending folds with an S_1 axial planar cleavage. This fan-shaped cleavage is mostly upright. A crenulation cleavage (S_2) occurs only in the most deformed areas. In the Skhirat and Western Sidi Bou Othman - Kharrouba units, strong tectonothermal activity is concentrated in ductile shear-zones:

a) To the east, the Sidi Bou Othman Shear zone (SBOSZ, Fig. 10) is, north-northwest-trending (El Hassani et Zahraoui 1982) and associated with the syntectonic emplacement of the Bramram granitoid (Lagarde 1985; Le Corre and Saquaque 1987). This shear zone presently separates the Sarhlef and Sidi Bou Othman - Kharrouba units and shows east-west stretching lineations and a sigmoïdal S_1 trajectory (Muller, unpubl. observ.). It acted both as a sinistral strike-slip zone (Lagarde and Choukroune 1982) during late granitoïd intrusion and as a westward thrust.

b) To the west, the eastern Skhirat Shear Zone (ESSZ, Fig. 10) (Mayol 1987) seems to be a branch of the Western Meseta Shear Zone (WMSZ). As in the Rehamna inlier, movements are dextral (Mayol 1987; Le Corre and Bouloton 1987) and narrow, late metamorphic mylonitic dextral zones have been recognized. This shear zone, which forms the eastern limit of the Coastal Block, lies precisely on a step in the Precambrian basement (Bernardin 1987).

The end of Variscan deformation was marked by thrusting with west-northwest to northwest vergence. Some of the thrusts are old synsedimentary faults reactivated as reverse-faults. Displacement on these thrusts is not great and does not exceed 5 km, as shown by the Bou Gader thrust (Mayol and Muller 1985; Bouloton and Le Corre 1985; Mayol 1987).

3.3.4 The High Atlas Paleozoic Block

3.3.4.1 General Presentation

The High Atlas Paleozoic Block (Fig. 1) extends into the southern part of northern Morocco where, despite previous work (Roch 1930; Neltner 1938; Schaer 1962, 1964 a, b, 1981; De Koning 1957, Proust 1973), its Paleozoic stratigraphic and structural evolution is still poorly known. This is partly due to important post-Variscan orogenesis with several uplifts during Permian (Ferrandini et al. 1987) and Cenozoic times (Schaer 1967, 1982; Proust 1973).

3.3.4.2 General Structure

Structurally, the Paleozoic High-Atlas is divided into two domains (Schaer 1982): the Ouzerllah massif and the western Paleozoic High-Atlas (Fig. 11 a).

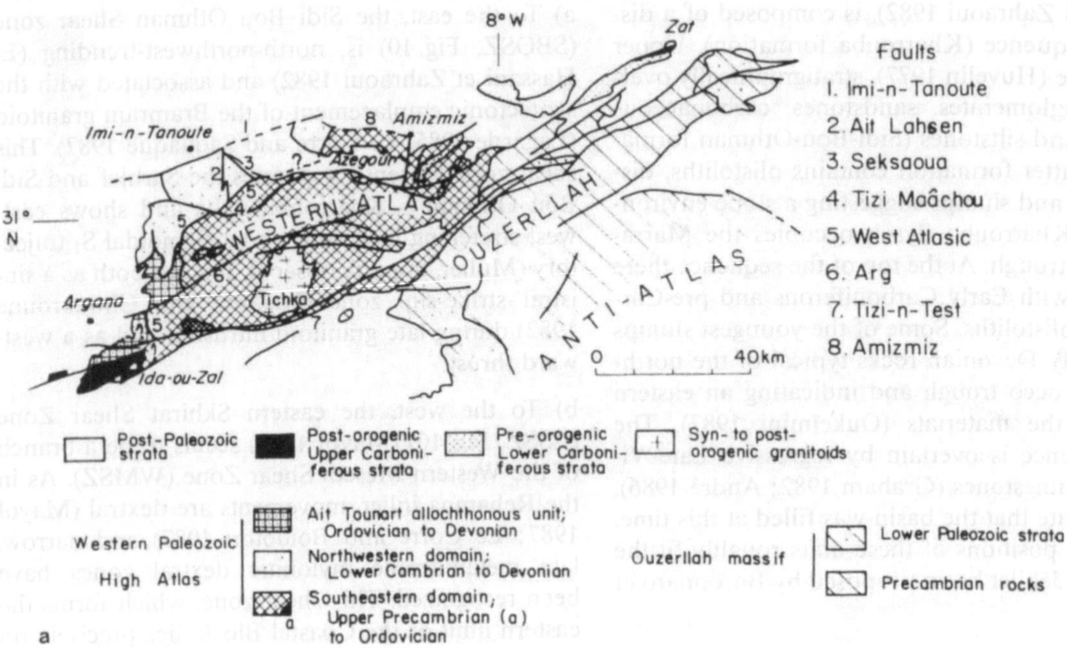

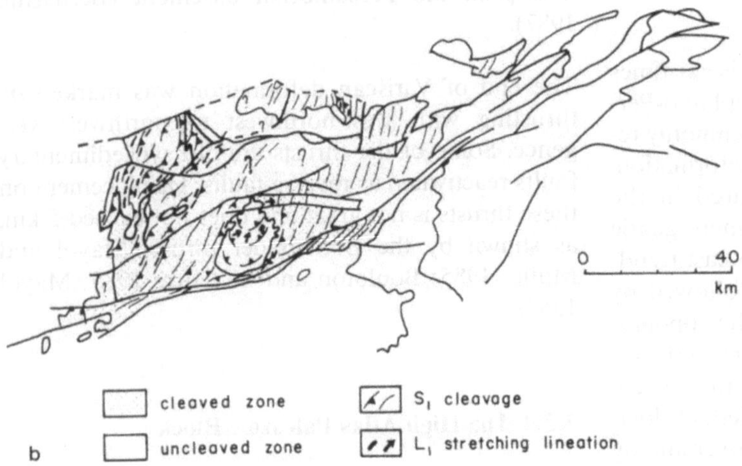

Fig. 11. The High Atlas paleozoic block **a** Schematic structural map; **b** schematic S 1 and stretching lineation trajectories

The Ouzerllah massif is mainly composed of Precambrian granites and migmatites, covered by thick uppermost Precambrian lavas. Paleozoic strata were eroded during Mesozoic times and are preserved only on the margins of the massif. These strata indicate that the Ouzerllah massif was a shoal during Early Paleozoic (Schaer 1964 a; Destombes et al. 1985; Binot et al. 1986) and Early Carboniferous (Beauchamp and Izart 1987) times. No clearly Variscan deformation is recorded in these rocks and the Ouzerllah massif is consequently considered as a part of the Anti-Atlas.

The western Paleozoic High Atlas, in contrast, is underlain by thick Lower Paleozoic sedimentary rocks of mainly Cambrian age, covered in its northeastern part by thin Lower Carboniferous deposits (Beauchamp and Petit 1981). The effects of the Variscan deformation and metamorphism are obvious. Granitoid intrusions were emplaced during Carboniferous and Permian times (Termier et al. 1972; Mrini 1985), and several fault zones, often inherited from Lower Paleozoic times, were active during the Variscan orogeny and during Permian and Cenozoic times as well. The Tizi-n'Test fault zone trends east-northeast and separates the western Anti-Atlas domain from northern Morocco. Eastward, however, it crosscuts the Ouzerllah massif and probably extends into the Aït Tamellil inlier (Jenny 1983 a). Rather than being a major transcontinental structure (Mattauer et al. 1972), it was a Variscan dextral shear zone (Petit 1976; Lagarde and Roddaz 1983) west of the Ouzerllah massif where displacements are limited to some kilometers (Petit 1976; Proust et al. 1977; Schaer 1982; Binot et al. 1986; Cornée et al. 1987 c). North of the massif, the Imi-n-Tanoute and Amizmiz fault zones, which were strongly active during the Alpine orogeny, were probably already active during the Variscan deformation. The west Atlasic fault zone (WAFZ; Cornée et al. 1987 c) has various north and east-northeast trending segments (Fig. 11 a) and separates a mildly folded northwestern area from a mildly to strongly deformed southeastern area. The former area shows affinities with northwestern Morocco (Cornée et al. 1987 a, b), while the latter seems to be an extension of north central Morocco.

3.3.4.3 *Variscan Evolution*

Variscan evolution is divided by Cornée (1989) into several successive stages. The age of deformations are not directly known, but from field data and from relations between deformation and granitoid emplacement, the Variscan orogeny is dated from Early Namurian to Westphalian times:

1. The main Variscan deformation, (D_1) developed north-northeast-trending F_1 folds that are upright or overturned to the west-northwest (Fig. 11 b). To the northwest deformation remained weak and cleavage is at most incipient except adjacent to the WAFZ which acted as a ductile shear zone. To the southeast deformation was intense, especially around the granitoid massifs. Here the associated stretching lineation, which is ordinarly north-northeast trending and subhorizontal, shows steep plunges (Lagarde and Roddaz 1983; Aït Ayad 1987). In strongly deformed zones, D_1 is locally marked by synmetamorphic slip along S_1 planes and by ductile boudinage and sheath folds. Synmetamorphic motions occured along ductile shear zones that includes the sinistral Azegour fault (Aït Ayad 1987), the dextral Tizi-n-Test fault (Lagarde and Roddaz 1983) and the dextral WAFZ (Cornée et al. 1987 b).

A regional metamorphism developed contemporaneously with the D_1 deformation. In the northwestern area metamorphism decreases northeastward from low-grade to very low grade and even diagenesis. The southeastern area, on the other hand, is characterized by generally low-grade metamorphism but is locally medium-grade around granitoïd intrusions where synkinematic biotite, garnet, cordierite, andalusite and even sillimanite have been recorded (Termier and Termier 1971).

As a whole, D_1 kinematics reflect the north-northeastward stretching and extrusion of the Western High-Atlas in response to dextral movement on northeast to east-northeast shear zones. The regional shortening direction was west-northwest-east-southeast (N100-110° E; Cornée et al. 1987 b).

2. Transtensional motion along fault zones in the western High-Atlas, was responsible for the reactivation of the WAFZ and the Aït-Lahsen fault as normal faults. This extension on north trending faults (Cornée and Ferrandini 1985) resulted from left-lateral motion (?) along the Imi-n-Tanoute and Amizmiz fault zones (Cornée, Tayebi, work in progress).

3. Several slices of Lower and Middle Paleozoic rocks (Aït-Tounart unit) previously folded during D_1, were thrust southeastward to form klippes resting on northwestern and southeastern areas (Cornée et al. 1990). The displacement, which opposes the regional vergence of the F_1 folds, is at least 20 km. It is thought to represent the reactivation, as a thrust, of northwest-dipping faults inherited from the previous transtensional episode and presently concealed beneath the Mesozoic strata of the Argana valley.

4. A subsequent deformation (D_2) affected both autochthonous and allochthonous units. This episode is characterized by the reactivation of north-south faults (e. g. WAFZ, Aït Lahsen) into thrusts, and northeast and east-northeast trending faults (e. g. Tizi-n'Test) into dextral wrench faults. At the same time, upright and recumbent F_2 folds developed with an incipient S_2 cleavage which generally trends north-northeast but is locally northwest-southeast trending (conjugate zone). Regional shortening was west-northwest to east-southeast (N100-120° E).

5. The Tizi Maâchou and Arg faults (Fig. 11 a) are east-west to west-southwest-east-northeast-trending brittle dextral faults, indicating a northwest-southeast (N120-140° E) shortening direction. Their movement affected the Tichka granitoid and marked the end of the main Variscan orogeny.

6. Stephano-Autunian continental strata were deposited unconformably upon the Variscan structures, in fault-controlled intramontane basins (De Koning 1957; Broutin et al. 1989).

7. During upper Permian and prior to upper Triassic times, the fault pattern was reactivated. East-northeast trending faults (e. g. Imi-n-Tanoute and Arg) acted as sinistral reverse faults, whereas east-west to east-southeast-trending faults (e. g. Tizi

Maâchou) became reverse faults. These movements and associated folds indicate a NNE-SSW (N20° E) shortening direction. This Permian activity has been studied only in the westernmost part of the Paleozoic High-Atlas (Ferrandini et al., 1987) and its possible extension eastward is not yet documented. In the western Paleozoic High-Atlas it resulted in uplift of the region. It could be contemporaneous with the late leucogranitic intrusion of the Azegour region (Mrini 1985) and with the last heating episode of the Tichka granitoid.

3.3.5 Conclusions

3.3.5.1 The Western Meseta Shear Zone

There are important differences in the tectonothermal evolution of the western Coastal Block which is weakly deformed and weakly metamorphosed except at its eastern part, and north-central Morocco, which is characterized by pervasive deformation and metamorphism. The boundary between these two domains trends south-southwest from Bouzniqa toward the High-Atlas Paleozoic block and existed during the sedimentary stage. During the Late Devonian and Early Carboniferous, for example, it represented the western limit of the Sidi Bettache basin and its southern extension (Piqué 1979; Piqué and Michard 1988). There are also indications (Piqué et al. 1980; Lagarde and Michard 1986; Corsini et al. 1988 b) that the boundary was a dextral shear zone during the Variscan orogeny. To the north, where the metamorphic grade is relatively low, the boundary is marked by wide zones with ESE-trending en-échelon folds, (e. g. in the Ben Slimane syncline), and narrower zones where the folds are steeply plunging and axial planes trend northerly (e. g. in the Oued Cherrat anticline). In the south, where the metamorphic grade is higher, the boundary is a ductile shear zone. Various structures, (e. g. rotated garnets) indicate dextral movement and high values ($\gamma > 6$) of shear strain. Hence the western limit of the sedimentary basin was reactivated during Early Carboniferous times in a transpressional regime, locally with low-angle dextral strike-slip (Corsini et al. 1988 b).

Southward, in the Jebilet massif, the WMSZ is more difficult to identify as it probably corresponds to several faulted zones in the central part of the massif. To the west, it probably coincides with the eastern Skhirat/Abda zone at the margin of the Early Carboniferous basin. Here, after early thrusting (Mayol and Muller 1985; Cornée and Ferrandini 1985), limited syn- to late metamorphic dextral movements are recorded (Mayol and Muller 1985; Mayol 1987; Le Corre and Bouloton 1987). To the East, sinistral westward displacement (Muller, pers. observ.) occurred toward the end of granitoïds emplacement along the southeast-trending Sidi Bou Othman shear zone. In the western High-Atlas the WMSZ is a wide, strongly deformed NNE-trending zone with dextral movements along shear zones and strong stretching in the southeastern domain. This deformational pattern ends against the Tizi-n'Test fault zone, whose presence is only clearly demonstrated west of the Ouzerllah massif.

3.3.5.2 The Role of the Old Structures

The preceding description attests to the long-lived evolution of the boundary between the Coastal Block and northcentral Morocco. Within northcentral Morocco itself, many examples also suggest a strong control of preexisting structures upon the Variscan deformation. The Khouribga-Oulmes anticlinorium, for example, was established on a former Devonian rise. Here, as in the Rommani synclinorium, arguments can be made (Piqué et al. 1985) to suggest that the structural vergence and width of the structures were, at least in part, controlled by Devonian paleogeography. A consequence of these arguments is that the shear zones which developed along ancient sedimentary fault zones are deep-seated structures. That does not exclude, however, the possibility for synmetamorphic thrust stacking of the metamorphic units, nor does it exclude the possible existence of deep crustal décollements that seismic profiles could document. El Kamel (1987) has also shown that block faulting was an important feature in the Rehamna Massif during Devonian times. In fact, major Variscan faults in the western High-Atlas Paleozoic blocks, such as the WAFZ and the Tizi-n'Test fault, were active during Cambrian times (Binot et al. 1986; Cornée et al. 1987 b, c).

3.4 Northeastern Morocco

In contrast to western Morocco, the Paleozoic terrains of northeastern Morocco do not form a broad continuous belt. Instead, due to higher levels of erosion, the Paleozoic massifs are smaller. Moreover, some of them have been incorporated in the Atlas chain. However, the Atlasic shortening is nowhere sufficient to impede correlation from one Paleozoic massif to another. These massifs form four subzones in Northeastern Morocco as described below.

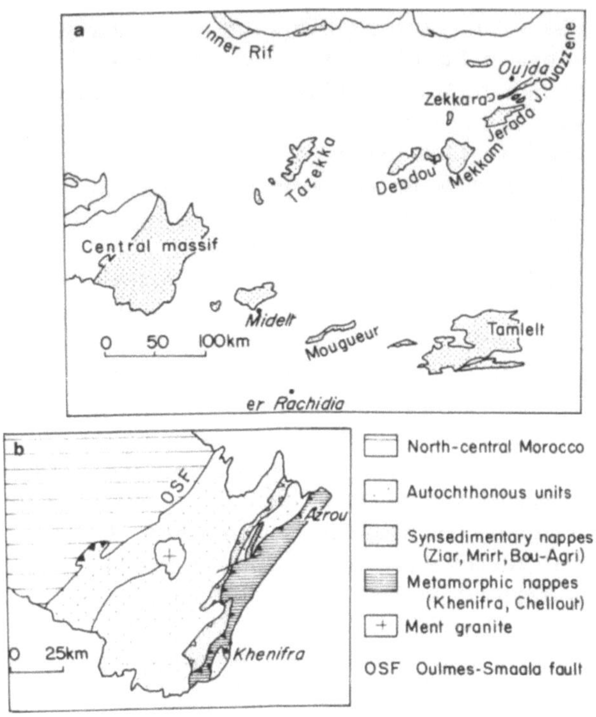

Fig. 12. The northern part of northeastern Morocco. Location indicated in Fig. 1c. **a** the Paleozoic massif; **b** eastern portion of the central massif

3.4.1 Midelt and Northeastern Inliers

These massifs are located east of the Middle Atlas and north of the High Atlas, inside the eastern (Oran) Meseta (Fig. 12). The most important are the Midelt, Dedbou, Mekkam and Jerada massifs. They show polyphase Variscan deformations, the first phase of which is represented by generally, isoclinal and recumbent F_1 folds that trend northwest-southeast (N130E-N110E) in the Midelt area. Their axes are often widely dispersed in the flat-lying axial planes, but the mineral lineation has a well-defined, west-southwest (N70E) orientation, concordant with the general westward vergence of the folds. F_2 folds, which developed during the same metamorphic episode, represent a heterogeneous deformation marked by ductile shear and the development of kinematic indicators (e.g. S/C planes) of westward vergence. During this shearing, the F_2 folds developed a sheath geometry. The metamorphism reached low-grade with white micas + biotite + garnet and has been dated at 366 ± 7 Ma at Midelt (Sb-Sr, Clauer et al. 1980) and 372 ± 3 and 368 ± 8 Ma at Dedbou and Mekkam respectively (K-Ar on very fine fractions; Huon et al. 1988). Their data allow an accurate determination of the age of synmetamorphic deformation. The regional, syntectonic metamorphism was followed by the emplacement of granitoids dated between 330 and 300 Ma (Mrini 1985).

In the northeasternmost inliers of Morocco (e.g. Zekkara and Jorf Ouezzene), Hoepffner (1987) described imbricated thrusts which stacked several slices of Lower Paleozoic strata. He assigned these movements to the D_1 and D_2 deformations of Debdou-Mekkam and Midelt, developed here at a structurally higher level, or to younger synsedimentary glides in the Visean basin.

Following this first, Late Devonian deformation, Early Carboniferous sediments were deposited unconformably. The Carboniferous sequence is well exposed in the Jerada inlier and extends to the Early Westphalian. The folds have a wavelength of several kilometers, are east-northeast (N70E) trending, and show a conspicuous northern vergence. The thermal evolution remained weak. The oldest granitoid mentioned above were followed, during Late Carboniferous times, by younger granites (see below, Chap. 4).

A similar and perhaps contemporary evolution is recorded in the High-Atlas inliers (Mouguer massif and northern part of the Tamlelt massif). In these areas, Hoepffner (1987) described a polyphase, synmetamorphic deformation in slates attributed to the Cambrian. Isoclinal and recumbent F_1 fold are followed by F_2 shear folds. The F_2 folds trend northwest-southeast (Mougueur) to east-west (Tamlelt) and are thought to correspond to the Late Devonian, Eo-Variscan phase. However, the east-west folds with their steeply dipping axial planes, could be related to the Late Carboniferous phase.

3.4.2 The Tazekka Massif

On the basis of their structural style, Hoepffner (1987) distinguished three subzones in the Tazekka massif. The zones are underlain by Cambro-Ordovician rocks, unconformably covered by Carboniferous strata:

1. The eastern subzone is underlain by the Ordovician Tazekka Slates. When they are not affected by subsequent deformation, the first folds (F_1), are north-northeast (N20°E) trending, west-vergent isoclinal structures. Their axial planes are outlined by a slaty cleavage that developed contemporaneously with a low-grade metamorphism. Most of the S_1 foliation planes were reactivated during later episodes, either as shear

planes or during F_2 folding. The D_2 episode corresponds to a continuation, still under metamorphic conditions, of the D_1 deformation.
2. The central (Bab-Azhar) zone shows, in its eastern half, F_1 folds similar to those of the Tazekka slates. These folds have been subsequently affected by the D_2 episode during which they were sheared and the fold hinges were separated from their thinned limbs and rotated inside the S_1 parallel shear planes. The western half of the Bab-Azhar zone shows a decrease in metamorphic grade. The only visible structures are small-scale folds which developed in the incompetent layers during the stacking of more competent beds.
3. The western zone is unmetamorphosed. However, the stratigraphic column is vertically repeated several times by a process attributed by Hoepffner (1987) to westward oriented duplex structures.

The Tazekka massif therefore comprises a metamorphosed and polydeformed eastern domain and a non metamorphic western zone, whose structural style recalls foreland-type formation. The central zone represents a 5-km-wide shear zone that corresponds to the ductile thrust of the eastern Tazekka upon the western part. The age of the synmetamorphic D_1 and D_2 episodes is not well bracketed stratigraphically. It postdated the Ordovician and it predated the Upper Visean. However, preliminary isotopic dating (Huon et al. 1988), has yielded a 329–330 Ma age (K-Ar on mica). Later on, the unconformably overlying Late Visean to Namurian and perhaps Early Westphalian cover, was deformed by north-northeast (N30° E) trending folds associated with a incipient subvertical cleavage developped under very weak diagenetic conditions. In the previously deformed rocks, structures related to D_1 and D_2 were refolded in small chevron folds, often recumbent to the southeast. Shear planes were steepened and sometimes acted as right lateral strike-slip faults. This deformation has been dated at around 300 Ma (Huon et al. 1988).

3.4.3 Eastern Central Massif and Jebilet

Altrough several areas of the eastern Central Massif have been previously studied (Allary et al. 1976), the entire region from Azrou to Khenifra (Fig. 12 b) is presently under reinvestigation (Bouabdelli, 1989). It is thus hoped that the interpretation presented here, which is based upon published work, will be improved upon in the new future.

The first regional deformation (D_1) is well displayed from Azrou to Khenifra. It affected rocks as young as Middle-Upper Devonian. It was marked by tight to isoclinal folds most of which are recumbent to the northwest, and was contemporaneous with a very low to low-grade metamorphism. A flat-lying S_1 cleavage is a general characteristic of this zone, although some areas, in the particular Carboniferous allochthons of Bou Agri, Ziar and Mrirt (see below), were not affected by this first deformation. A westward vergence is typical of the D_1 structures except in the Zaian area, near Kenifra, where Cambrian quartzites apparently verse eastwards. The age of the synmetamorphic D_1 deformation has been determined in the Azrou region at 334 ± 8 Ma and 330 ± 6 Ma for Ordovician and Visean slates respectively (Huon et al. 1988, K-Ar method on fine fractions). Hence, deformation which was coeval with the $D_1 + D_2$ episodes of the Tazekka massif, corresponds to the "Pre-(Late) Visean Phase" of Allary et al. (1972).

The D_1 deformation was followed rapidely by erosion and the subsequent deposition of an unconformable Late visean cover, itself overlain by allochthonous units. Allary et al. (1972) distinguished two different nappes. The Kenifra nappe and its correlative counterpart in Azrou region (the Chellout slices) is a compressive nappe that was thrust northwestward penecontemporaneous with cleavage development. It was followed by the emplacement of the Ziar-Mrirt nappe and the correlative Bou Agri unit in Azrou region. This second allochthon is thought to be a synsedimentary gravity nappe. Allary et al. (1972) propose the emplacement of the two sets of allochthons to have occurred separately and under compressive and distensive regimes respectively. It is possible, however, that the two units belong to a same nappe, the distentional nose of which corresponds to the Ziar, Mrirt and Bou Agri units, while its compressive rear corresponds to the Khenifra and Chellout units. Following their emplacement into the Late Visean and Namurian basin, both the nappes and autochthonous rocks were folded by broad, open north-northeast (N30° E)-trending folds. Later, east-west strike-slip faults affected the whole region.

The allochthonous units of eastern Jebilet correspond to the Ziar-Mrirt and Bou Agri nappes of the eastern Central Massif. Doubtlessly, a zone of uplift existed during the Late Visean at the eastern limit of the Late Visean-Namurian sedimentary trough which extended from eastern Jebilet to the eastern Central Massif. Exotic units slid into the basin, in response to an initial compression which was relayed by gravitational sliding.

3.4.4 The Skoura and Ait Tamlil High-Atlas Inliers

These massifs (Fig. 5) have been respectively described by Laville (1980), Jenny and Le Marrec (1980) and Jenny (1983 b). Both are underlain by Late Precambrian and Paleozoic rocks as young as Visean. The Skoura inlier and the southern part of the Ait Tamlil massif are weakly deformed by broad open east-west folds which belong to the Anti-Atlas structural domain. In contrast, the central and northern part of the Ait Tamlil inlier shows a more complicated evolution. Here, the first phase of folding created east-southeast (N110° E-130° E) trending tight to isoclinal folds, overturned toward the south. The age of the folding is Early Carboniferous. It was followed by the emplacement of a gravity nappe, originating in the north and comparable to the eastern Central Massif and Jebilet allochthons. The second episode of folding was east-west directed and deformed both the nappe and its substratum. The limit between the flat-lying Skoura and southern Ait Tamlil, and the deformed central and northern Ait Tamlil is an extension of the Tizi n'Test fault (Jenny 1983 a), on the southern margin of the Early Carboniferous trough.

3.5 The Anti-Atlas

In the Anti-Atlas, the Variscan orogeny is most intense in the western part of the massif (Hassenforder 1987). In the Jbel Bani (Fig. 13), Lower Paleozoic rocks were affected by northeast-southwest folds which are locally deflected near Precambrian inliers ("Boutonnières") that pierce the Paleozoic cover. This deformation remained weak, with an estimated shortening of about 15%. Toward the west, around the Kerdous or Bas Draa inliers (Fig. 13), however, Variscan deformation was stronger. Here, north-south to northeast-southwest folds developed during low-grade metamorphism, accompanied by a strong axial-plane cleavage with 55% shortening. Variscan folds and late metamorphic thrusts are east-vergent. Locally, the fold axes are deflected along east-west right-lateral faults (e. g. Tata fault, Hassenforder 1987) that were active during Late Carboniferous and Permian times. The Variscan structures of Western Anti-Atlas continue to the Zemmour region in the south, where Sougy (1964) described north-northeast-trending folds and thrusts, with the same eastward vergence. The age of the deformation in the western Anti-Atlas is not accurately dated. The Jbel Bani Ordovician rocks and Devonian Rich are affected by the folding altough the Lower Carboniferous sequence of the Ouarkziz area is undeformed. However its does not represent the unconformable cover of the Anti-Atlas folded strata since the Ouarkziz sequence belongs to the Tindouf platform of the West African shield, located outside the Variscan orogen. Based on Rb-Sr and K-Ar methods on clay-size fractions, Bonhomme and Hassenforder (1985) determined two thermal episodes at around 370 and 300 Ma.

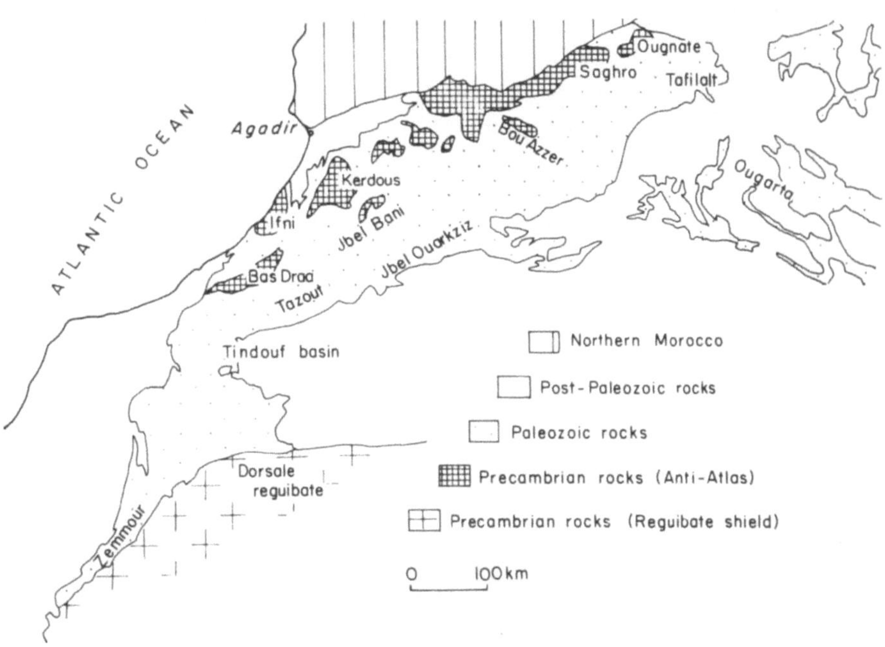

Fig. 13. Geological sketch map of the Anti-Atlas. Location indicated in Fig. 1c.

In central and southeastern Anti-Atlas Variscan deformation was very weak and is marked only by broad folds which trend east-west in central Anti-Atlas and northwest-southeast in the eastern Anti-Atlas where they extend into the Algerian Ougarta. In the northeastern Anti-Atlas, Michard et al. (1982) described east-west folds and southeastward-directed thrusts that affect Visean rocks. They correspond to similar structures described farther east (Ball et al. 1975) that effect Namurian strata and are covered unconformably by Bashkirian conglomerates. In the Atlasic Tamlelt inlier (Fig. 12) Cornée (pers. observ.) assigned these structures to a major southward thrusting event where they are marked by numerous repetitions in Ordovician-Silurian rocks.

The structural style of the Anti-Atlas differs strongly from the style of northern Morocco. This suggests that the zone presently occupied by the High-Atlas belt represented an important boundary, during the Hercynian times (Atlas Paleozoic Transform fault of Michard et al. 1983).

3.6 The Internal Rif

This part of the Variscan belt is dominated by the overprint of the Alpine orogeny. Not only was the entire internal Rif moved westward and thrust upon the rest of Morocco during Cenozoic time, but it was also divided into several Alpine allochthonous units; namely, the metamorphic Sebtides (Kornprobst 1974; Saddiqi 1988), and the Ghomarides, where relatively low-grade metamorphism allows the recognition of Paleozoic rocks (Fig. 14). The Ghomarides are broken into several Alpine nappes (Kornprobst 1974; Chalouan 1986) that show the same Variscan deformation.

The first episode of deformation affected the Paleozoic sequence up to the Upper Devonian. Upper Visean strata (and Tournaisian strata in related areas of the Spanish Malaguides and Algerian Kabylia) unconformably overlie the folded older Paleozoic rocks. This first, Eo-Variscan, deformation therefore represents a Late Devonian orogenic phase, corresponding to the first episode recorded in northeastern Morocco. During this phase, two successive deformations occurred which probably reflect a single, continuous process. Isoclinal, north-northeast (N30° E)-trending F_1 folds are recumbent to the northwest. They were contemporaneous with low-grade metamorphism that continued during the development of the ENE (N70E)-trending F_2 folds. The associated stretching lineation is oblique to the fold axes and trends to the northeast (N50° E). This D_2 deformation was heterogeneous and developed in a shear regime. Toward the base of the sequence, the folds are more flattened, the cleavage is more penetrative, and shearing is more important. This suggests a ductile décollement at the lowest levels of the Paleozoic pile, along the interface between the Paleozoic cover and a suspected Precambrian basement. The Late Carboniferous phase developed north-northwest-trending open folds contemporaneous with a very low grade metamorphism.

In terms of its Variscan structural evolution and its stratigraphic and sedimentological history, the

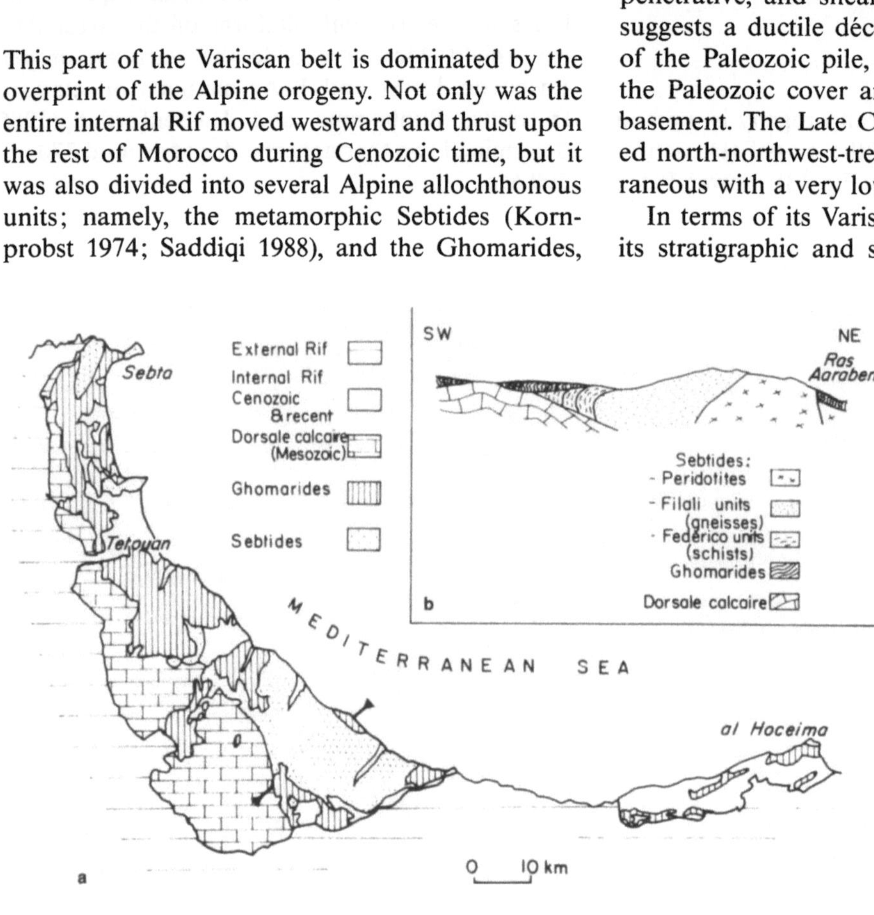

Fig. 14. Geological sketch map of the Internal Rif **a** and **b** schematic cross-section (Chalouan 1986), located on the map. Location indicated in Fig. 1 b.

Inner Rif is similar to northeastern Morocco. It is therefore suggested that these areas, together with Malaguides and Kabylia, were located along a same tectonic zone during Paleozoic times.

4 Magmatic Activity

From the late Proterozoic to late Paleozoic, the sedimentary column all over Morocco is frequently interrupted by traces of a volcanic activity (e. g. flows and pyroclastic rocks). These volcanic are sometimes widespread and conspicuous (e. g. during the latest Proterozoic - earliest Paleozoic, the late Devonian - Dinantian and the Permian) and sometimes minor but still present (e. g. the early Paleozoic). Most of these rocks are well-defined petrographically, but few have been studied in terms of their geochemical affinities. However, this is not the case for granitoid bodies that have been isotopically dated and chemically studied.

4.1 Pre-Orogenic and Orogenic Volcanic Activity

4.1.1 Latest Proterozoic - Eaerliest Paleozoic

After its main episode of deformation, the Pan-African orogeny was followed by volcanic and plutonic activity. The volcanic rocks were mostly felsic flows and pyroclastic rocks and as ignimbrites. Those of the Ouarzazate sequence are calc-alkaline and related to the last stages of Pan-African collision (Boyer et al. 1978). Some volcanic rocks, however, indicate crustal distention. These include the flows and mafic dykes of the Amane Issougri microrift, (Boyer and Leblanc 1977), the Ouarzazate sequence, the younger Jbel Boho (Alougoum) flows, in Adoudounian sequence, the latter being alkaline (Choubert 1952; Leblanc 1975). These latest Proterozoic volcanic rocks are best known in the Anti-Atlas, where they crop out extensively. However, their equivalents occur in restricted outcrops in northern Morocco; at El Jadida (Gigout 1951), Goaîda and Jbel Hadid in northeastern Morocco (Cailleux, Bouabdelli, work in progress) and have been recently discovered in the Rehamna massif at Sidi Bahilil (Corsini et al. 1988 a).

Lower Cambrian deposits of the southeastern domain of the western paleozoic High-Atlas show strong evidence of volcanic activity with lavas on the western margin of the Ouzerllah massif and volcano-sedimentary rocks westward as far as the northern flank of the Anti-Atlas (Schaer 1964 a). These lavas are rhyolitic, andesitic and basaltic in type.

The greywackes and siltstones that form the Middle Cambrian of Morocco frequently show faint evidence of coeval volcanism in that they usually contain pyroclastic beds and volcanic-derived materials dispersed throughout the matrix of the greywackes. In some places, volcanic rocks themselves are present and include trachytic, andesitic or basaltic flows, and tuffs and breccias beds, indicating proximity to volcanic centers. The chemical trend of these Cambrian rocks is presently being investigated.

4.1.2 Early Paleozoic

Volcanic activity had ceased at the end of the Middle Cambrian, and from the Ordovician to Middle Devonian, evidence of volcanism is restricted to a few minor occurrences.

1. Rabat-Tiflet area, where Ordovician mafic flows are several meters thick and comprise spilites emplaced in marine waters. Their chemical trend needs more detailed analysis to be defined and is presently being examined.
2. Ouled-Abbou (Cornée et al. 1985) described thin basaltic flows in the Coastal Block. However these deeply weathered rocks have not been otherwise characterized.
3. Inner Rif, where relatively important magmatic activity is represented by mafic, spilitic flows, trachytes and dolerites which are often associated with radiolarites. Volcanism was more or less continuous from at least the Late Ordovician to Early Devonian times. However, detailed study of these rocks has not yet been untertaken.
4. Anti-Atlas, where the earliest Devonian Hammar Laghdad basalts rocks are known in the Tafilalt, but have no known equivalent elsewhere in the Anti-Atlas.

4.1.3 Late Devonian - Early Carboniferous

During this period volcanic activity was again especially important in northern Morocco. Detailed studies (Piqué and Kharbouch 1983; Kharbouch et al. 1985; Kharbouch, work in progress, etc.) have led to the recognition of two magmatic provinces in northern Morocco during the late Devonian and Dinantian.

Western and Central Morocco
Here volcanism is restricted to the margins of the Sidi-Bettache basin (Piqué and Kharbouch 1983). In the vicinity of Rabat it is represented by spilites, keratophyres and dolerites. Primary parageneses (andesine-labrador, clinopyroxene and opaque minerals) are preserved in the dolerites and a secondary association (albite, chlorite, quartz, calcite, epidote, sphene) is developed in the lavas. The oldest, Late Devonian and Tournaisian lavas are alkaline, while the youngest Visean lavas are transitional toward true tholeiites. These rocks are similar to those of extensional tectonic environments, and their presence there confirms the extensional regime of northcentral Morocco, as indicated by the sedimentary facies of the Sidi-Bettache basin. Farther to the south, in the Jebilet massif, Visean bimodal volcanic rocks and gabbros represent the same tectonic setting.

Northeastern Morocco
A thick volcanic sequence is intercalated within the Late Visean and Namurian rocks that unconformably overlie the folded rocks of the Eo-variscan belt. Some of these rocks are lavas (andesites, dacites and rhyolites) but most of them are pyroclastic (tuffs, breccias and cinerites). The original mineralogy is often preserved and comprises K-feldspars, intermediate plagioclases and quartz, with pyroxenes, amphiboles and/or biotite, ilmenite and magnetite. The major and rare earth elements indicate a calc-alkaline trend, comparable to the volcanics of present active margins.

4.2 The Granitoids

4.2.1 Latest Proterozoic Granitoids

In the Anti-Atlas, the main stage of the Pan-African orogeny was followed by the emplacement of successive granitic bodies that are comagmatic with the latest Proterozoic volcanic series. Many of these provides valuable information on the age of the sedimentary sequences. For instance, in the Ifni inlier (Fig. 13) the Mesti and Ifni granites, respectively dated by Rb-Sr method at 621 ± 2 and 660 ± 22 Ma, are intrusive in the Siroua (PII–III) sequence and they are covered by the Ouarzazate (PIII) sequence. The Taoulecht granite, dated by Rb-Sr method at 595 ± 15 Ma, crosscuts the Ouarzazate sequence and is covered by the Adoudounian strata (Jeannette et al. 1981).

4.2.2 Pre-Variscan Paleozoic Granitoids

1. The mylonitic El Jadida granodiorite was drilled in the course of oceanological programs about 80 km west of El Jadida (Fig. 17), at the western limit of the Moroccan Atlantic shelf. Its emplacement has been dated at around 515 Ma and its mylonitization occurred at about 455 Ma (Kreuzer et al. 1984).
2. The Rabat-Tiflet granite (Fig. 6) was emplaced in the Sehoul zone at 430 ± 2 Ma (Charlot et al. 1973, recalculated) and was mylonitized during an early stage of the Variscan orogeny.

4.2.3 The Variscan granitoids

These are located exclusively in northern Morocco, the most important of which (Fig. 15) are as follows:

1. Zaer, Oulmes, Ment (Central Massif);
2. Sebt - Brikyine (Rehamna);
3. Bramram, O. Ouaslam (Jbilete);
4. Tichka, Azegour (High-Atlas Paleozoic massif);
5. Bou Mia, Aouli (Midelt);
6. Alaouene, Soulouina etc. (northeastern inliers).

Their ages, petrography and geochemical trend are highly variable, both from one massif to another and within a single massif. These data are summarized for the main granitoids on Table 1.

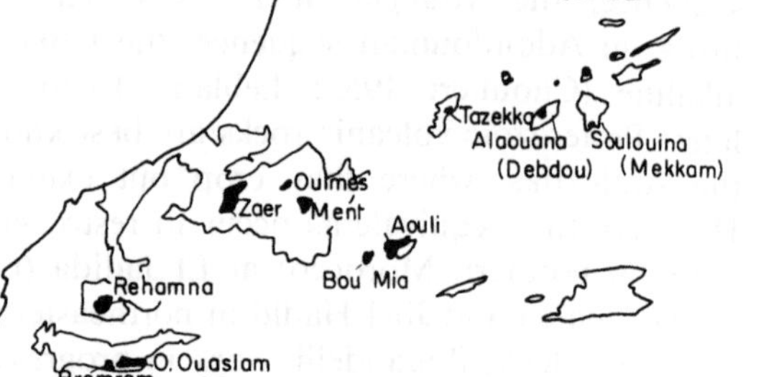

Fig. 15. Variscan granitoid massifs in northern Morocco.

Table 1. Variscan granitoids. Data from Mrini (1985) and Rosé (1987)

Granitoid	Type[a]	Mantel material (%)	Age
Zaer	γD-CA	50-60	300
	L	15-25	280
Oulmes	L	0-10	265
Ment	CA	20-50	280
	L	5-12	265
Rehamna	L		280
	M		265
Jbilete	γD-CA	40-80	300-330
	L	15-30	
Tichka	γD-CA	75-100	310-320
	L		295
Azegour	M	40-60	270
Bou-Mia	M		290
Aouli	γD-CA	35-60	310-330
	M		290
	L?		270?
Alaouana	CA		290
	L		
Soulouina	CA		290

[a] *γD* Granodiorite; *CA* calc-alkaline; *L* leucogranite; *M* monzonite.

Two sets of granitoids are distinguished. The granodiorites are calc-alkaline and have low $^{87}Sr/^{86}Sr$ initial ratios, while the leucogranites have higher $^{87}Sr/^{86}Sr$ initial ratios, suggesting a higher component of crustal origin. The oldest granodiorites (Tichka and Aouli) are located in southeastern and eastern Morocco. They were cut by younger, leucogranitic magmas that are roughly contemporaneous with the granodiorites of Central Morocco (e. g. Zaër, Ment etc.). Hence, the evolution from granodioritic to leucogranitic magmas varies both in time (from Carboniferous to Permian) and space (from southeastern to central Morocco). This dual pattern also extends to models for the tectonic evolution of Morocco. Boushaba et al. (1987) proposed that the generation of the granodioritic magma resulted from Late Devonian and Dinantian crustal thinning. Alternatively, Rosé (1987) has suggested it was due to continental subduction that occurred between 320 and 290 Ma in southeastern Morocco. These models will be discussed below (Sect. 6).

4.3 Late Orogenic Volcanism

In the Permian intramontane basins, volcanic rocks are present at various levels within the terrestrial sedimentary column. They are andesite, rhyolites and ignimbrites, with important occurrences of tuffs and cinerites. The thickness of the volcanic intercalation is locally significant (e. g. 300 m at Tiddas, 250 m at Khenifra). These rocks are calc-alkaline and comagmatic with the youngest Variscan granitoids (Boushaba et al. 1987).

Subsequent Triassic volcanism with its tholeiitic dolerites and basalts, has no genetic relations with the Permian volcanics. Instead, it belong to the extensional tectonic cycle that created the Atlantic ocean.

5 Local Geophysics

5.1 Existing Data

A Bouguer gravity anomaly map of Morocco compiled from petroleum and regional academic surveys has been published by Van den Bosch (1971). This gravity map has been recently interpreted in great detail (Bernardin 1987). Regional anomalies suggest the existence of deep structural features in the crust together with changes in the depth of the Moho discontinuity. But gravity data alone cannot provide reliable information on crustal thickness. Residual gravity anomalies have also been computed in order to interpret shorter wavelength anomaly features according to geological and structural data.

In addition to petroleum seismic surveys, which are restricted in any case to certain basin areas and the results of which are not totally published, deep seismic studies have been performed with mobile seismic stations (Thiessen 1979; Makris et al. 1985). Some confidential profiles have provided very reliable information on crustal thicknesses in particular beyond the High-Atlas, the Anti-Atlas and the Moroccan Meseta.

5.2 Outline of the Crustal structure of Morocco

The main outline of the deep crustal structure of Morocco may be derived from seismic soundings and from the combined analysis of gravity and seismic data.

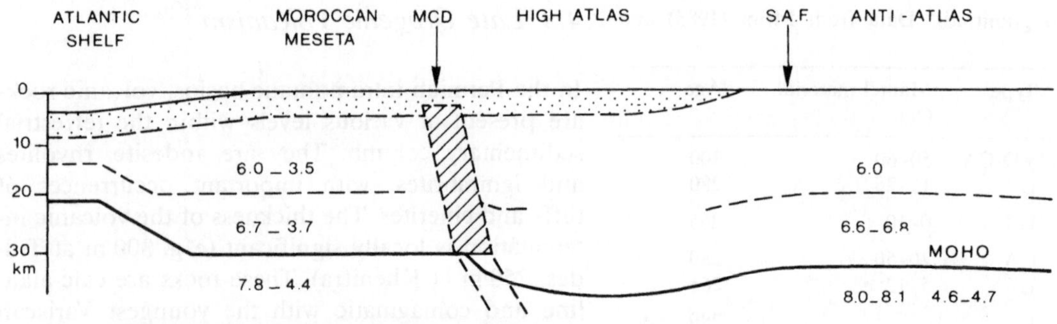

Fig. 16. Schematic crustal section from the Atlantic shelf, through the Moroccan Meseta, High Atlas and Anti-Atlas. P and S wave velocities in the crust and upper mantle are indicated respectively. *SAF* "South Atlas Fault". *MCD* Major Crustal Dislocation (After Makris et al. 1985). Location indicated in Fig. 1 b.

Models compiled from the deep seismic data reveal (Fig. 16) some interesting features (Makris et al. 1985): 1) The crust is rather homogeneous, with an average thickness of 30 km beneath the Moroccan Meseta; 2) The crustal thickness beneath the western High-Atlas does not differ significantly from that of the Meseta; it remains at about 30 km and shows no crustal root; 3) A westernly crustal thinning occurs toward the Atlantic margin where the Moho rises at about 20 km; and 4) Beneath the Anti-Atlas the Moho depth is similar to that of the Meseta (around 30 km). However, the crystalline basement portion of the crust must be thicker in the Anti-Atlas, since the Meseta crust includes post-Paleozoic cover while the Anti-Atlas has none.

Two-dimensional gravity models, constrained by the results of seismic soundings, have been compiled to fit both the Bouguer anomalies (Makris et al. 1985) and the regional anomalies produced by the smoothing method (Bernardin 1987). The resulting crustal models are comparable and the main features are as follows:

1. The High-Atlas crust is asymmetric with the greatest depth to the Moho occurring between 34 and 38 km according to the constraints of the alternative models. Hence the High-Atlas appears to be shallow-rooted, considering its altitude of over 3000 m. This is consistent with the intracontinental origin proposed for the chain, which has suffered only moderate shortening (Mattauer et al. 1977).
2. The steep gravity gradient which occurs in the vicinity of the northwest flank of the High-Atlas is interpreted as a major crustal discontinuity (MCD, Fig. 16) or boundary which separates the western Meseta block from the High-Atlas domain (Fig. 1).
3. In contrast, the gently dipping root zone beneath the southeast flank of the High-Atlas chain thins gradually toward the Anti-Atlas. Interestingly, the South Atlas fault (SAF: Fig. 16) has neither a gravimetric not a seismic expression, as no velocity variation occurs across the fault zone. This does not support the idea suggested by Mattauer et al. (1972) that important horizontal shear movements occurred along this fault.
4. The Anti-Atlas and the Moroccan Meseta which exhibit Pan-African and Hercynian deformations respectively, appear to be in isostatic equilibrium.

5.3 Deep Structural Features of the Moroccan Meseta

Gravity pattern analysis of the Bouguer anomaly map allows three gravimetric provinces to be distinguished.

A western gravimetric province, corresponding to northwestern Morocco (Fig. 1), is characterized by a generally positive gravity field. NNE-SSW trending axes are recognized both in the Western part of the Paleozoic inliers (Jebilet, Rehamna, central Morocco) and under the Mesozoic cover of the Atlantic Coastal basins. Such trends are probably Variscan in age. The eastern limit of this western province is the West Meseta Shear Zone. East of the WMSZ, a *central gravimetric province* shows a generally NE-SW-trending gravity gradient. This province corresponds to northcentral Morocco and to the western edge of northeastern Morocco. There, ovoid or circular gravity lows reflect numerous granitic intrusions. The eastern edge of the province is outlined by the strong gravity gradient which is likely to correspond to the major crustal boundary as mentioned above. *In the eastern gravimetric province* a

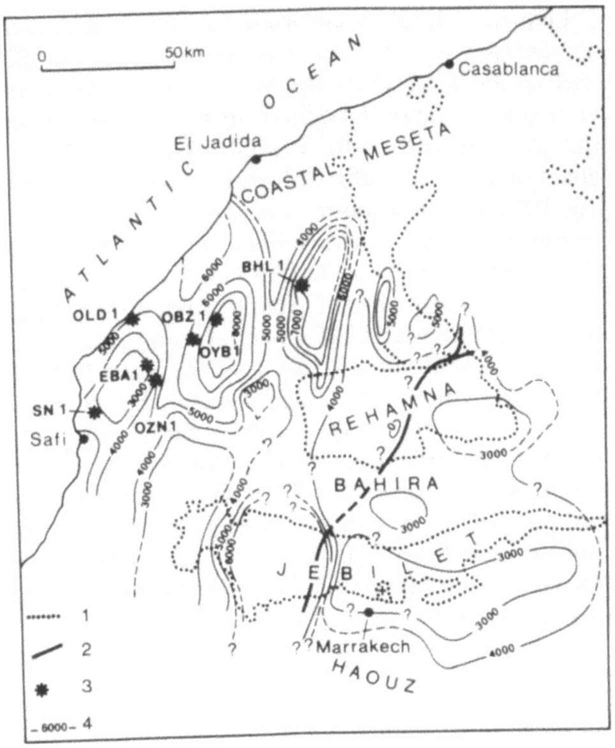

Fig. 17. Isopach map for the Paleozoic in the western Moroccan Meseta (after Bernardin et al. 1988). *1* limit of Paleozoic inliers; *2* WMSZ; *3* deep borehole; *4* thickness in meters

regional low gravity field is associated with the alpine High-Atlas and its root zone.

In the Hercynian inliers, depth estimates for the granitic intrusions have been made using spectral analysis of numerous residual anomaly profiles. Two-dimensional models shows that these granites correspond to superficial shallow rooted batholiths (3 to 10 km thick). The deepest rooted massifs (10 km) appear to be located in the central Morocco inlier.

Interpretation of the Bouguer and residual anomalies of the Moroccan Meseta suggest that the basement is divided into a series of horsts and grabens controlled by a N-S trending fault system. The grabens are filled with sedimentary deposits which seem in places to be very thick.

In northwestern Morocco (Barbu 1977; Leridon 1986), boreholes data and seismic data exist to constrain gravity modelling and more elaborate quantitative interpretations have been established. Here an isobath map of the Hercynian unconformity was first drawn and the gravimetric effect of the thick Permo-Triassic deposits was then computed and substracted from the Bouguer anomaly. Next, the stripped anomaly related to deeper sources was interpreted by direct calculation along a series of two-dimensional profiles. Density values used for the models are from Tcherviakov and Ivanov (1980) for the Meseta rocks and from Leridon (1986) for the coastal basin series. The results of this interpretation consists of an isopach map for the Paleozoic series in the western Meseta (Bernardin 1987). This reveals a series of elongate north-south horsts and grabens in the Coastal Block (Fig. 17). The grabens are filled with sedimentary deposits between 2000 m and 9000 m in thickness. Sedimentary thickness do not exceed 3000–4000 m eastward in the WMSZ where the Precambian basement is generally less deep (northcentral Morocco). The final step in the interpretation has involved estimating the thickness of the Cambrian sequence alone in the western Meseta (Bernardin et al. 1988). The Cambrian graben in the Coastal Block (Fig. 3) whose sediment thickness may in places reach more than 7000 m, are thought to reflect a late stage of evolution of a Late-Precambrian-Lower Cambrian intracontinental rift that is likely to be related to the opening of the Iapetus ocean.

6 Geodynamic Evolution

6.1 Paleozoic Platforms and Basins

6.1.1 Latest Proterozoic - Early Paleozoic

The effects of the Pan-African orogeny are clearly visible in the Anti-Atlas. Various structures developed, during low-grade metamorphism, in response to the closure of an oceanic domain. The main deformation was followed, during latest Proterozoic times, by the deposition of subaerial detrital sedimentary sequences and the emplacement of effusive and intrusive magmas. Similar sediments and magmatic rocks occur sparcely in northern Morocco, suggesting that the Pan-African belt extended north of the Anti-Atlas and that the basement of all of Morocco beneath the Paleozoic cover was sialic.

The petrography and sedimentology of the latest Proterozoic sedimentary rocks suggest they are "molassic" sequences of the Pan-African orogeny, deposited in intramontane basins. Coeval calc-alkaline magmatic rocks would then represent subsequent, late orogenic magmatic activity. However, the presence of rift-type rocks and alkaline magmatism indicates that an extensional regime was already active at the very end of Late Proterozoic times.

The Early Cambrian was a very quiet period, marked by the deposition of archeocyathidae and trilobite limestones in a transgressive sea. The presence of similar deposits in various parts of Morocco indicates that the Early Cambrian shelf was widespread, extending from the Anti-Atlas toward northern Morocco and probably including the Sehoul zone. However, a major trough began to develop in the western High-Atlas and Jbilete inliers that was controlled in the south by the Tizi-n'Test fault (Mayol 1987; Bernardin et al. 1988).

6.1.2 Middle Cambrian

During the Middle Cambrian, most of Morocco was covered by a shallow sea in which detrital sedimentary rocks, such as greywackes and quartzites, were deposited. In many places the intervention of volcanic activity is marked, especially in some areas of western Morocco where volcanic centers were located. The mid-Cambrian shelf was subjected to crustal extension at its periphery: in western Morocco a north-northeast-trending rift developed on the Lower Cambrian one. In northernmost Morocco a continental margin environment, indicated by the deep-sea Sehoul facies, succeeded the shelf. This extension was contemporanous with crustal thinning in the Avalon zone of the eastern Appalachians and indicates a tendancy fo the peripheral and marginal parts of Morocco to separate from the West African shelf. However, this rifting was rapidly aborted, at the end of the Middle Cambrian in western Morocco and at the end of Ordovician times in northernmost Morocco.

6.3 Ordovician to Middle Devonian

From the Ordovician to Middle Devonian, most of Morocco was covered by an epicontinental sea. Broadly speaking, the sediments that were deposited were more or less coarse sands during the Ordovician, fine clays during the Silurian, and limestones during the Early and Middle Devonian. This "positive megasequence" (Jeannette and Piqué 1981) outlined the progressive erosion of the West African shield, from where, on the basis of thickness changes and sedimentary structures, the detrital sediments were derived. During the Devonian, the shield which was flat and covered by the epicontinental sea, and no longer sent any clastic supply to the northern platform which was therefore a carbonate shelf.

Like the Middle Cambrian, however, there was an increase in subsidence from the center of the shelf to the west, where thicker and relatively coarser sequences were deposited, in part derived from the west. To the north, the Ordovician shelf was characterized by mafic volcanic rocks. In the Rabat-Tiflet area and north, the shelf grades into the continental slope of the Sehoul zone. At that time, the peripheral parts of the West African shield constituted the northwestern Gondwanian miogeocline. At its northern margin (the Sehoul zone), crustal extension ceased and was replaced by Late Ordovician deformation that created a mobile belt north of present Morocco. In the western part of the miogeocline, an Ordovician deformation is recorded in the mylonitization of the El Jadida granitoid. It is not marked in the Coastal Block, except in the input of west-derived, Late Ordovician clastic sediments. In the Anti-Atlas, high sedimentation rates and coarse facies persisted in the western part of the miogeocline, contrasting with the thin shelf deposits of the central and eastern Anti-Atlas.

During Lower Devonian times, a new tectonic basin began to develop in northcentral Morocco, just east of the former Cambrian rift. This new north-northeast-trending basin is wider in its northern part (Rehamna, Sidi Bettache) than in its southern part. Another subsiding zone, the so-called Midelt-Oujda trough, appeared within the shelf itself from the Eastern Jebilet to Azrou-Khenifra, and farther northeast, to the Debou and Mekkam inliers (Piqué 1983). Here, Devonian pelagic limestones and turbidites differ from the marls and the reef limestones of the shelf. This trough or, more probably, this succession of basins, could indicate the existence of either a transtensional or an extensional regime during the Early and Middle Devonian, and perhaps since the Late Ordovician (Hamoumi 1988). The Midelt-Oujda basin ist the first manifestation of a permanent zone of weakness in the Moroccan crust during the Paleozoic. It can be extended into the northwestern Algerian inliers and, possibly, into the Kabylia-Inner Rif – Malaguides blocks which were located at the northern part of African Gondwana during the Paleozoic (Bourrouilh and Gorsline 1979). We do not know whether crustal extension was important enough to appreciably thin the continental crust beneath the Midelt-Oujda basins. In present Morocco, it is unlikely that true oceanic crust was created, but there is evidence that suggests crustal thinning was progessively more important toward the northeast along the Middelt-Oujda axis. Early Paleozoic volcanic activity which is very weak in northeastern Morocco, is quite noticeable in the Inner Rif while Silurian and Lower Devonian rocks are deep-sea facies. This indicates a considerably thinned sialic crust at the northern continental rise of Africa.

6.1.2 The Late Devonian Crisis

6.1.2.1 Northeastern Morocco

At least in the Midelt and Debdou-Mekkam, and probably in other parts of northeastern Morocco, the main Variscan orogenic episode occurred at the end of Devonian. At this time, Lower Paleozoic rocks, including the Lower and Middle Devonian turbidites of the eastern part of the former Midelt-Oujda basin, were deformed. Dated isotopically at around 360 Ma, this deformation is commonly referred to as the Eovariscan episode. The Eovariscan folds and thrusts resulted from a roughly east-west (N70E) regional compression, and their structural vergence is westward. The early age of this deformation compared with other parts of the Variscan belt of Morocco, the generally low-grade metamorphism, and the general vergence toward the other structural zones, suggest that the internal zones of the Moroccan Variscan belt were located in northeastern Morocco (Piqué 1983).

This orogenic episode probably occurred with a lateral component between two regions, namely the main part of Morocco to the northwest and Africa to the southeast, which had been incompletely separated each from the other during the Early Paleozoic. The closure of the thinned crust between these two regions was realized by continental subduction along a zone trending roughly ENE-WNW but whose dip is still debated. The "suture" may be located east for present Morocco, under the Meso- and Cenozoic cover of the Algerian Hauts-Plateaux. However, there is little direct evidence for the collisional nature of the Eovariscan orogeny. The main evidence is the calc-alkaline nature of the earliest granitoids (e. g. Tichka, Midelt), which is indicative of a subduction process and an active margin setting. The production of these intrusive (Rosé 1987) and extrusive (Kharbouch et al. 1985) magmas, would then be related to the development of continental subduction.

6.1.2.2 Northcentral Morocco

The opening of the Sidi Bettache basin is precisely dated as Famennian, and hence coïncides with the Eovariscan deformation of northeastern Morocco. Several arguments have been presented, suggesting it was a pull-apart basin, opened by the motion of N30° E-trending dextral strike-slips faults like the future western Meseta Shear Zone (Piqué 1979; Piqué et al. 1980). The dextral motion along these faults is compatible with the stress field that was active in northeastern Morocco, where the shortening direction trends N70° E. It is thus thought that the Sidi Bettache basin opened in a transtensional regime controlled by these dextral faults. Its development can be considered as the consequence, in the Variscan external zone of Morocco, of the same regional compression that was the cause of the Eovariscan folding in the internal zones. However, the general kinematic pattern for the opening of Late Devonian basins at the scale of Morocco remains unknown.

6.1.2.3 Anti-Atlas

In the Anti-Atlas, Variscan deformation has not yet been satisfactorily dated by isotopic methods, so it is not possible to determine the timing of Variscan events precisely. However, it is apparent that the central part of the Anti-Atlas began to rise during the Late Devonian, since this region began to act as a sediment source for adjacent areas. During the Late Devonian, the eastern Anti-Atlas also underwent a block-faulting episode that caused the disruption of the former shelf. Whether or not this episode was related to a purely tensional setting (Wendt 1985) or a transtensional regime as in the Sidi Bettache basin, extension was moderate. No deep, turbiditic facies like those of northern Morocco were generated in the Anti-Atlas basins and no magmatic rocks were emplaced along the basin margins. This indicates that disintegration of the former shelf was less important than in northern Morocco.

6.1.3 Carboniferous and Permian

6.1.3.1 Northeastern Morocco

After the closure of the deep Midelt-Oujda basin during the Late Devonian, the Eovariscan deformation continued into Dinantian times with the stacking of metamorphic units, indicated by the flat-lying foliations and thrusts. This resulted in the thickening of the crust along the orogenic belt, coeval with continental subduction and the creation of calc-alkaline magmas (see above). The visco-elastic response of the lithosphere induced by the crustal thickness led to the development of a foreland basin that bordered the orogenic belt to the northwest. This Azrou foreland basin extended from eastern Jbilet to at least Khenifra and Azrou, and its development recalls the evolution of the western margin of the Appalachians during Ordovician times. The occasionally reefal Middle-Late Visean limestones of the shelf deepened and turbidites and olistostromes were deposited from Aît Tamellil to the Azrou-Khenifra region. At the same time, eastern parts of the foreland basin were subjected to regional deformation. The Chellout unit of the Azrou region, for instance, was cleaved and metamorphosed to low-grade at 330 Ma (Late Visean) and thrust westward. The shortening direction was N120-130°. Insufficient isotopic data exists to decide and whether or not this late Visean deformation was a distinct and clearly separate event from that of the Midelt Eovariscan deformation or, as is more probable, if it was a stage in a continuous tectonic evolution. In any case, the basin evolution followed the usual development of foreland basins. While its southeastern limit was deformed, the shelf that bordered it to the northwest began to deepen. As more and more of its southeastern regions were involved in the deformation, the basin itself moved northwestward.

6.1.3.2 Northcentral Morocco

During the evolution of the eastern foreland basin, the Sidi Bettache basins continued to fill. The initial stage of its development, which was marked by tectonic subsidence, was followed about 30 Ma later, by a thermal subsidence stage during which subsidence was no longer controlled by faults but by the thermal contraction of the lithosphere (McKenzie 1978). As a result, the surface of the basin widened and the former bounding faults were sealed by Upper Visean sandstones. This period also marks the end of magmatic activity. In the southern extension of the Sidi Bettache basin, however, each stage of the basin development occurred later. There, it was still fault-bounded during Late Visean times in the High-Atlas Paleozoic block, and bimodal tholeiitic magmas (gabbros and lavas) were emplaced in the Jbilete. This is precisely the time that basin opening began to be replaced by a transtensional regime. There are indications in the northern Sidi Bettache basin and in the Jbilete that Variscan deformation began to affect the margins of the basin at the end of Visean times (320 Ma). It is probably at this time that the western limit of the Sidi Bettache basin (the Western Meseta Shear Zone) started acting as a ductile transcurrent fault system in combination with westward thrusting, and the development of possible décollements within the Silurian argillites. Special mention should also be made of the northern margin of the Sidi Bettache basin, that is the Sehoul zone. Early in the course of the Variscan orogeny and certainly prior to the Visean, the Sehoul zone was thrust southward upon the northern part of the Sidi Bettache basin (i. e. the Rabat-Tiflet anticlinorium). During Namurian and Early Westphalian times, the marine basin narrowed progressively and rocks that were deposited in the residual through were deformed during the Early-Middle Westphalian. Post-metamorphic folds and thrusts also affected previously deformed rocks. The Permian basins were induced by faults whose trend and motion indicate roughly north-south shortening.

6.1.3.3 Anti-Atlas

In the central and eastern Anti-Atlas, Variscan deformation remained weak but was responsible for the creation of a very large anticlinorium, which trends east-northeast in central Anti-Atlas, and which broadened during Carboniferous times. This anticlinorium confined the marine domain to a narrow faulted basin in the Tineghir and Ben Zireg regions, and to the Bechar and Tindouf basins which emerged progressively from west to east. In the western Anti-Atlas, Variscan deformation developed with an eastward vergence, during Carboniferous, comparable to the external zones of the Mauritanides.

6.2 Kinematics of the Variscan Deformation and Interpretative Model (Fig. 18)

In northern Morocco, Variscan deformation migrated westwards from uppermost Devonian times in northeastern Morocco until Late Visean-Early Namurian and finally Mid-Westphalian times in northcentral Morocco. There, complete closure of the Paleozoic basin was achieved during Stephanian and even Permian times. The Eovariscan deformation episode is marked by westward vergence, and a roughly east-west stretching lineation that is oblique onto regional arcuate folds trends. The Upper Visean to Lower Namurian Variscan deformation, on the other hand, is characterized by a stretching lineation which is often parallel to fold axes that trend regionally north-northeast to northeast both along the eastern margins of the Coastal

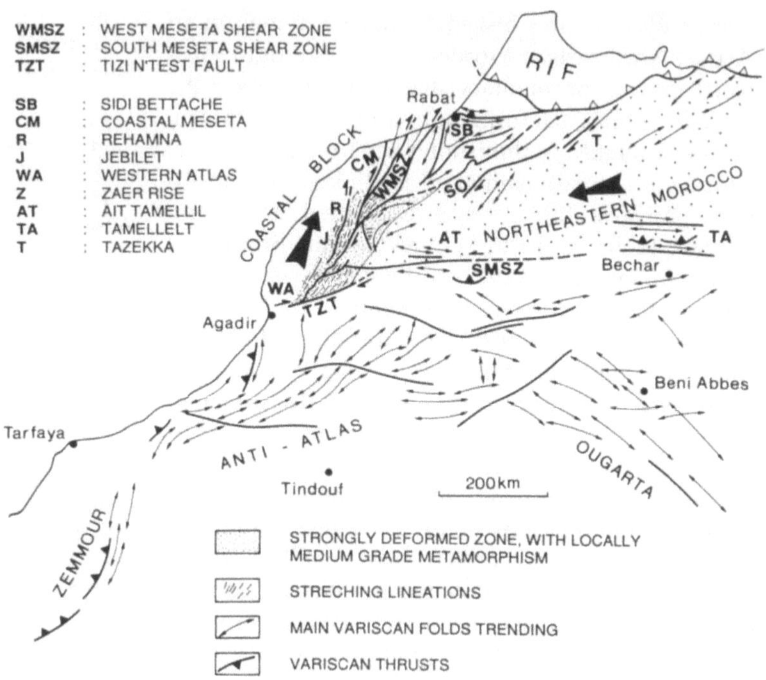

Fig. 18. Kinematics of the Variscan (Namuro-Westphalian) deformations: a corner effect along the northern margin of the African shield.

Block and northcentral Morocco. Stretching oblique to fold axes is rare and limited to synkinematic granitoïd intrusions or NW-SE shear zones. Such a pattern contrasts with the mild east-vergent deformation, in the Anti-Atlas.

We propose here a collisional model for the Variscan belt of Morocco in which intracontinental deformation was controlled by crustal blocks inherited from the previous Paleozoic basin development. As northeastern Morocco suffered maximum crustal thinning, it was a preexisting weakness zone and *Eovariscan deformations* were concentrated there first, resulting from an intracontinental subduction-collison zone farther southeast.

At the same time, the Sidi Bettache, northern Rehamna and eastern Anti-Atlas basins were created. As the crust was progressively thickened, deformation migrated westwards, towards the previously disrupted zone of both northwestern and northcentral Morocco. To the south, on the other hand, the Anti-Atlas domain acted as the homogenous northern margin of the West African shield. *During Upper Visean/Lower Namurian times,* northeastern Morocco, which had been folded and metamorphosed during the uppermost Devonian, acted as a rigid block with a thin early Carboniferous cover (2500 m maximum). At the same time, northcentral and northwestern Morocco suffered more intense deformation, which was permitted by a thick sedimentary cover of Cambrian age in northwestern Morocco and Late Devonian-Early Carboniferous age in northcentral Morocco. Deformation was concentrated along the margins rather than the center of these regions, since the former were located on old zones of crustal weakness that were active during Lower Paleozoic times.

The margins of northeastern Morocco were strongly active during Upper Visean/Lower Namurian times, as follows:

1. To the north, the margin is marked by important dextral strike-slip faults that trend northeast (Allary et al. 1976), perhaps as far as the Smaala-Oulmes fault zone.
2. To the south, we define the "South Meseta Shear Zone" (SMSZ), which trends N70–80° E as the sheared northern margin of the West African shield. In the Aït Tamellil inlier, a structural study from Lesage (in Jenny 1983 b) indicates that the SMSZ is marked by sinistral shear-zones associated with WNW-ESE trending en échelon folds. Similar observations can be deduced eastward from Hoepffner (1987) (Mougueur and Tamelelt inliers).
3. To the west, northcentral Morocco and the eastern margin of northwest Morocco are strongly deformed and metamorphosed from the northern Rehamna massif to the western High Atlas in association with the WMSZ.

Consequently, we interpret the Late Visean/Early Namurian Variscan deformational pattern *as a corner effect* at the disrupted northern margin of Africa. Northeastern Morocco acted as a rigid westward moving promontary that impinged upon the Coastal Block. Deformation and metamorphism are most intense in the frontal zone of northcentral Morocco. Laterally, inherited north-south to northeast-southwest faults were reactivated as dextral wrench faults, as in the entire Coastal Block and along the margins of the Sidi Bettache basin. These movements, which were accompanied by generally north-northeast to north-east trending stretching lineations, indicate that the Coastal Block and northcentral Morocco zone were laterally stretched and extruded northeastward, in response to the oblique westward movement of northeastern Morocco. These intracontinental block tectonics were controlled by the margins of the Devono-Carboniferous basins. Such a pattern may explain, on a regional scale, the heterogenous distribution of both deformation and metamorphism, as well as the location of widespread granitoid emplacement along the margins of the indentation. This corner effect result from a crustal structure that was active during the whole Variscan orogeny. It explains the early thrusts and olistostromes in preorogenic Carboniferous basins, with the northwestward vergence from Azrou to eastern Jebilet and southwestward vergence in the Aît Tamellil inlier (Jenny et Le Marrec 1980). Following synmetamorphic deformation, late thrusts developped. These are northwest-vergent in both the Rehamna and Jebilet massifs, east-west to northwest-southeast-vergent in the western Atlas, and southward vergent along SMSZ. The end of Variscan orogeny is more poorly known. Main faulted zones were reactivated *during Stephanian and Permian times*. Presently available data suggest that the shortening direction suffered regional rotation, but the kinematic pattern for this period is still under study.

References

Aarab El, Beauchamp J (1987) Le magmatisme pré-orogénique des Jebilet Centrales (Maroc). Précision pétrographiques et sédimentologiques; implications géodynamiques. CR Acad Sci Paris, Ser II, 304: 169-174

Aït Ayad N (1987) Etude des relations entre déformation et intrusion granitique. L'exemple du granite hercynien d'Azegour (Haut-Atlas de Marrakech). Thèse 3ème cycle, Univ Marrakech, Maroc, 153 pp

Aït Omar S (1985) Le pluton granitique hercynien d'Oulmès (Maroc central): schémas de déformation finie et cinématique de mise en place. CR Acad Sci Paris, Ser II, 301: 1299-1302

Allary A, Andrieux J, Lavenu A, Ribeyrolles M (1972) Les nappes hercyniennes de la Meseta sud-orientale (Maroc central). CR Acad Sci Paris 274: 2284-2287

Allary A, Lavenu A, Ribeyrolles M (1976) La chaîne hercynienne du Maroc central sud-est. Notes Mém Serv Géol Maroc 261: 170

André JP (1986) Bioconstructions carbonatées et faciès associés au Koudiat Lahmara, Jebilet orientales. Eclogae Geol Helv 79: 295-317

André JP, Boissin JP, Corsini M, Renard JP (1987) Sur le Cambrien de la région de Casablanca (Maroc): la série de Dar Bou Azza. Bull Soc Géol Fr III: 1161-1170

Ball E, Fabre J, Guellel S, Megard M, Moussine-Pouchkine A (1975) Sur la présence de cisaillements plats d'âge hercynien dans le Carbonifère de Bechar (Algérie). CR Acad Sci Paris 280: 2721-2724

Barbu A (1977) Le concept de zone pétrolière potentielle dans l'exploration du bassin des Doukkala (Maroc central). Mines Géol Energ Rabat 42: 49-57

Beauchamp D, Izart A (1987) Early Carboniferous basins of the Atlas-Meseta domain (Morocco): Sedimentary model and geodynamic evolution. Geology 15: 797-800

Beauchamp J, Petit JP (1981) Sédimentation et déformations synsédimentaires au Carbonifère inférieur dans le Haut-Atlas de Marrakech: Généralités des glissements intraviséens dans le domaine atlaso-mésétien. C R Acad Sci Paris, Ser II, 293: 707-710

Bensaïd M (1979) L'Ordovicien supérieur, le Silurien et le Dévonien dans l'Est du Maroc central. Mines Géol Energ Rabat 46: 83-86

Bensaïd M, Janvier P, Termier M, Termier G (1980) Présence du Westphalien inférieur dans la série de Fourhal (Maroc central). CR Acad Sci Paris 290: 1329-1332

Bernardin C (1987) Interprétation gravimétrique et structure profonde de la Meseta marocaine et de sa marge atlantique. Trav Lab Sci Terre St-Jérôme, Marseille, Fr, Ser B, 29: 162

Bernardin C, Cornée JJ, Corsini M, Mayol S, Muller J, Tayebi M (1988) Variations d'épaisseur du Cambrien moyen en Meseta marocaine occidentale: Signification géodynamique des données de surface et de subsurface. Can J Earth Sci 25: 2104-2117

Binot F, Dresen G, Stets J, Wurster P (1987) Die Tizi-n'Test Verwerfungszone im Hohen Atlas (Morokko). Geol Rundsch 75: 647-664

Bonhomme M, Hassenforder B (1985) Le métamorphisme hercynien dans les formations tardi- et post-panafricaines de l'Anti-Atlas occidental (Maroc). Données isotopiques Rb-Sr et K-Ar des fractions fines. Sci Géol Bull 38: 175-183

Bordonaro M (1983) Tectonique et pétrographie du district à pyrrhotine de Kettara (Paléozoïque des Jebilet centrales). Thèse 3ème cycle, Strasbourg 132 pp

Bordonaro M, Gaillet L, Michard A (1979) Le géosynclinal carbonifère sudmésétien dans les Jebilet (Maroc). Une corrélation avec la province pyriteuse du Sud de l'Espagne. C R Acad Sci Paris, Ser D, 233: 707-710

Bouloton J, Le Corre C (1985) Le problème de la tectonique tangentielle dans les Jebilet (Maroc hercynien). Hercynica 1: 121-129

Bouabdelli M (1989) Tectonique et sédimentation dans un bassin orogénique: le Aillon viséen d'Azrou-Khenifra (Est du massif hercynien central du Maroc). Thèse Doctorat Sci., Univ. Pasteur, Strasbourg, Fr.

Bourrouilh R, Gorsline D (1979) Petriassic fit and alpine tectonics in the western Mediterranean. Geol Soc Am Bull 90: 1074-1083

Boushaba A, Cailleux Y, El Wartiti M (1987) Les granitoïdes hercyniens et le volcanisme permien du domaine mesetien marocain. Bull Sci Terre Fac Sci Rabat 3: 103-138

Boyer C, Leblanc M (1977) Les appareils émissifs de la formation volcanique infracambrienne d'Ouarzazate, Anti-Atlas (Maroc). C R Acad Sci Paris 285: 641-644

Boyer C, Chikhaoui M, Dupuy C, Leblanc M (1978) Le volcanisme calcoalcalin précambrien terminal de l'Anti-Atlas (Maroc) et ses altérations. Interprétation géodynamique. C R Acad Sci Paris 287: 427-430

Broutin J, Ferrandini J, Saber H (1989) Implications stratigraphiques et paléogéographiques de la de'couverte d'une flore permienne euraméricaine dans le Haut Atlas occidental (Maroc). C R Acad Sci Paris, Ser II, 308: 1509-1515

Cailleux Y (1978) Géologie de la région des Smaala (Massif central marocain). Stratigraphie du Primaire et tectonique hercynienne. Notes Mém Serv Géol Maroc 40 (275): 7-106

Cailleux Y, Deloche C, Gonord M, Rolin P (1984) Mise en évidence de deux couloirs de cisaillement dans la zone paléozoïque de Rabat-Tiflet (Maroc septentrional). C R Acad Sci Paris, Ser II, 299: 569-572

Chalouan A (1981) Stratigraphie et structure du Paléozoïque de l'Oued Cherrat: un segment du couloir de cisaillement de Meseta occidentale (Maroc). Notes Serv Géol Maroc 42 (308): 33-100

Chalouan A (1986) Les nappes ghomarides (Rif septentrional, Maroc). Un terrain varisque dans la chaîne alpine. Thèse Doctorat Sci, Univ Pasteur Strasbourg

Charlot R, Rhalib M, Tisserant D (1973) Etude géochronologique préliminaire des granites de la région de Rabat-Tiflet (Maroc occidental). Note Serv Géol Maroc 33 (249): 55-58

Clauer N, Jeannette D, Tisserant D (1980) Datation isotopique des cristallisations successive d'un socle cristallin et cristallophyllien (Haute-Moulouya, Moyen-Maroc). Geol Rundsch 169: 63-83

Choubert G (coord) (1952) Carte géologique du Maroc au 1/500000

Choubert G (1952) Le volcan géorgien de la région d'Alougoum (Anti-Atlas). C R Acad Sci Paris 234: 350-352

Choubert G (1963) Histoire géologique du Précambrien de l'Anti-Atlas, Tome I. Notes Mém Serv Géol Maroc 162: 352

Choubert G, Marçais J (1956) Les grands traits de la géologie du Maroc. Introduction géologique au lexique stratigraphique du Maroc. Notes Mém Serv Géol Maroc 134: 5-38

Cornée JJ (1982) Etude lithostratigraphique et tectonométamorphique des Rehamna sud-orientales. Plissements et nappes. Contribution à la connaissance de la chaîne hercynienne en Meseta marocaine. Trav Lab Sci Terre St-Jérôme, Marseille, Ser B, 20: 175

Cornée JJ (1989) Le Haut Atlas occidental paléozoïque: un reflet de l'histoire hercynienne du Maroc occidental. Stratigraphie, sédimentation et tectonique - Thèse Doctorat Sci., Univ. Aix-Marseille III, Fr., 2 tomes, 901 pp

Cornée JJ, Ferrandini J (1985) Sur la tectonique précoce hercynienne et le rôle des accidents anté-schisteux dans le Haut Atlas occidental paléozoîque au Sudd'Imi-n-Tanoute (Maroc). Relations avec les Jebilet occidentales. CR Acad Sci Paris, Ser II, 301: 1157-1162

Cornée JJ, Costagliola C, Léglise H, Willefert S, Destombes J (1985) Précisions stratigraphiques sur l'Ordovicien supérieur et le Silurien du synclinal d'Oulad Abbou (Meseta marocaine occidentale). Manifestations volcaniques au Silurien. Ann Soc Géol Nord 104: 141-146

Cornée JJ, Destombes J, Willefert S (1987 a) Stratigraphie du Paléozoïque de l'extrémité nord-ouest du Haut-Atlas occidental (Maroc hercynien); interprétation du cadre sédimentaire du Maroc occidental. Bull Soc Géol Fr III: 327-335

Cornée JJ, Tayebi M, Havlicek V (1987 b) Découverte de Saccogonum cf saccatum, brachiopode du Cambrien (supérieur?), dans le Haut-Atlas occidental (Maroc). Conséquences stratigraphiques et structurales. Geobios, Lyon, Fr 20: 517-527

Cornée JJ, Muller J et Tayebi M (1990) Mite en évidence d'un chevauchement post-schisteuse tardi-hercynien à caracteré suppressif dans le Haut Atlas occidental paléozoïque (Maroc) - CR Acad Sci Paris 310: 1521-1526

Cornée JJ, Ferrandini J, Muller J, Simon B (1987 c) Le Haut-Atlas paléozoïque: un graben cambrien moyen entre deux décrochements dextres N60° E hercyniens (Maroc). CR Acad Sci Paris, Ser II, 305: 499-503

Corsini M, Muller J, Cornée JJ, Diot M (1988 a) Découverte de la série basale du Cambrien et de son substratum dans les Rehamna centraux, haut-fond au Cambrien (Meseta marocaine). Prémices de l'orogenèse hercynienne. CR Acad Sci Paris, Ser II, 306: 63-68

Corsini M, Cornée JJ, Muller J, Vauchez A (1988 b) Déplacements tangentiels synmétamorphes vers le SW dans les Rehamna centraux (Meseta marocaine occidentale, Maroc hercynien). C R Acad Sci Paris, Ser II, 306: 751-755

de Koning G (1957) Géologie des Ida ou Zal (Maroc). Stratigraphie, pétrographie et tectonique de la partie SW du bloc occidental du massif ancien du Haut-Atlas. Leidse Geol Meded, Hollande, 23: 209

Destombes J, Feist R (1987) Découverte du Cambrien supérieur en Afrique (Anti-Atlas central, Maroc). C R Acad Sci Paris 304: 719-724

Destombes J, Hollard H, Willefert S (1985) Lower Paleozoic rocks of Morocco. In: Holland C (ed) Lower Paleozoic rocks of northwestern and west-central Africa. Wiley, NY, pp 91-336

Deynoux M, Sougy J, Trompette R (1985) Lower Paleozoic rocks of West Africa and the western part of Central Africa. In: Holland C (ed) Lower Paleozoic rocks of northwestern and west-central Africa. Wiley, NY, pp 337-495

Diot H, Bouchez JL, Boutaleb M, Macaudière J (1987) Le granite d'Oulmès (Maroc central): structure de l'état solide et modèle de mise en place. Bull Soc Géol Fr 8 (3): 157-168

Ducrot J, Lancelot J (1977) Problème de la limite Précambrien-Cambrien. Etude radiochronologique par la méthode U-Pb sur zircons du volcan du Jbel Boho (Anti-Atlas marocain). Can J Earth Sci 14: 2771-2777

El Hassani A, Zahraoui M (1982) Evolution tectonométamorphique des unités structurales du segment hercynien des Jebilet Centrales. Bull Inst Sci Rabat 6: 45-51

El Kamel F (1987) Géologie du Paléozoïque des Rehamna nord-orientaux (Maroc). Thèse Sci, Univ Aix-Marseille III, Fr, 311 pp

El Kamel F, Muller J, Sougy J (1985) Dispositif sédimentaire et évolution tectonique hercynienne du Paléozïque de Mechra-ben-Abbou, Rehamna (Maroc). C R Acad Sci Paris, 301: 1361-1364

Ferrandini J, Cornée JJ, Saber H (1987) Mise en évidence d'une compression subméridienne d'âge permien probable dans le massif ancien du Haut-Atlas occidental (Maroc). Evolution de l'ellipsoïde de déformation depuis d'orogenèse hercynienne jusqu'à la distension triasique. CR Acad Sci Paris, Ser II, 304: 1243-1248

Gentil L (1920) Carte géologique du Maroc à l'échelle du 1/500000. Ministère des Mines, Maroc

Gigout M (1951) Etudes géologiques sur la Meseta maro-

caine occidentale (arrière-pays de Casablanca, Mazagan et Safi). Trav Inst Sci Chérifien 3, Notes Mém Serv Géol Maroc 86: 507
Graham JR (1982) Transition form basin-plain to shelf deposits in the Carboniferous flysch of Southern Morocco. Sediment Geol 33: 173-194
Hamoumi N (1988) Les milieux de sédimentation dans l'Ordovicien du Maroc. Thèse Doctorat Sci, Univ Pasteur, Strasbourg, Fr
Hassenforder B (1987) La tectonique panafricaine et varisque de l'Anti-Atlas dans le massif du Kerdous (Maroc). Thèse Doctorat Sci, Univ Pasteur, Strasbourg, Fr
Havlicek V (1971) Brachiopodes de l'Ordovicien du Maroc. Notes Mém Serv Géol Maroc 230: 135
Hoepffner C (1987) La tectonique hercynienne dans l'Est du Maroc. Thèse Doctorat Sci, Univ Pasteur, Strasbourg, Fr
Hoepffner C, Jenny P, Piqué A, Michard A (1982) Le métamorphisme hercynien dans le massif des Rehamna. Notes Mém Serv Géol Maroc Rabat 303: 130-150
Hollard H (1967) Le Dévonien du Maroc et du Sahara nord-occidental. Int Symp Devonian Syst Calgary 1967 (1), Alberta Soc Petrol Geol, pp 203-244
Huon S, Piqué A, Clauer N (1987) Etude de l'orogenèse hercynienne au Maroc par la datation K-Ar de l'évolution métamorphique des schistes ardoisiers. Sci Géol Bull 40: 273-284
Huvelin P (1977) Etude géologique et gîtologique du massif hercynien des Jebilet (Maroc occidental). Notes Mém Serv Géol Maroc, pp 232-308
Jeannette D, Piqué A (1981) Le Maroc hercynien: plate-forme disloquée du craton ouest-africain. CR Acad Sci Paris 293: 79-82
Jeannette D, Benziane F, Yazidi A (1981) Lithostratigraphie et datation du Protérozoïque de la boutonnière d'Ifni (Anti-Atlas, Maroc). Precambrian Res 14: 363-378
Jenny J (1983 a) Les décrochements de l'Atlas de Demnate (Haut-Atlas central, Maroc): prolongation orientale de la zone de décrochement du Tizi-n'Test et clef de la compréhension de la tectonique atlasique. Eclogae Geol Helv 76: 243-251
Jenny J (1983 b) Livret-guide de l'excursion B 6: l'orogénie calédonienne et hercynienne au Maroc, IGCP 27 (unpubl)
Jenny J, Le Marrec A (1980) Mise en évidence d'une nappe à la limite méridionale du domaine hercynien dans la boutonnière d'Aït Tamlil (Haut-Atlas central, Maroc). Eclogae Geol Helv 73: 681-696
Kharbouch F, Juteau T, Treuil M, Joron JL, Piqué A, Hoepffner C (1985) Le volcanisme dinantien de la Meseta marocaine nord-occidentale et orientale. Caractères pétrographiques et géochimiques et implications géodynamiques. Sci Géol Bull 38: 155-163
Kornprobst J (1974) Contribution à l'étude pétrographique et structurale de la zone interne du Rif (Maroc septentrional). Notes Mém Serv Géol Maroc 251: 256
Kreuzer H, Musser P, Reinecke T, Wissmann G (1984) Petrography and K-Ar dating of the mazagan granodiorite DSDP Leg 79, holes 544 A and 547 B. In: Hinz K, Winterer E et al. (eds) Initial Rep DSDP 79, Washington (US Government Printing Office), pp 543-549
Lagarde JL (1985) Cisaillements ductiles et plutons granitiques contemporains de la déformation hercynienne post-viséenne de la Meseta marocaine. Hercynica 1: 29-37
Lagarde JL, Choukroune P (1982) Cisaillement ductile et granitoïdes syntectoniques: l'exemple du massif hercynien des Jebilet. Bull Soc Géol Fr XXIV: 299-307
Lagarde JL, Michard A (1986) Stretching normal to the regional thrust displacement in a thrust-wrench shear zone, Rehamna massif, Morocco. J Struct Geol 8: 483-492
Lagarde JL, Roddaz B (1983) Le massif plutonique du Tichka (Haut-Atlas occidental, Maroc): un diapir syntectonique. Bull Soc Géol Fr XXV: 389-395
Laville E (1980) Tectonique et microtectonique d'une partie du versant sud du Haut-Atlas marocain (Boutonnière de Skoura, Nappe de Toundout). Notes Serv Géol Maroc 41 (2854): 81-183
Le Corre C, Bouloton J (1987) Un modèle de "structure en fleur" associant décrochement et convergence: les Jebilet centro-occidentales (Maroc hercynien). CR Acad Sci Paris, Ser II, 304: 751-755
Le Corre C, Saquaque A (1987) Comportement d'un système pluton-encaissant dans un champ de déformation régional: le granite du Bramram (Jebilet, Maroc hercynien). Bull Soc Géol Fr 8 (3): 665-673
Leblanc M (1975) Ophiolites précambriennes et gîtes arséniés de cobalt (Bou Azzer, Maroc). Thèse Doctorat Sci, Univ Paris 6
Leridon J (1986) Combined interpretation of gravity and seismic data from the coastal basins of Morocco. First Break 4: 15-19
Makris J, Demnati A, Klubman J (1985) Deep seismic soundings in Morocco and a crust and upper mantle model deduced from seismic and gravity data. Ann Geophys 3: 369-380
Marhoumi R, Rauscher R (1984) Un plancton dévonien de la Meseta orientale au Maroc. Rev Palaeobot Palynol 43: 237-253
Mattauer M, Proust F, Tapponnier P (1972) Major strike-slip fault of late Hercynian age in Morocco. Nature (Lond) 237: 160-162
Mattauer M, Tapponnier P, Proust F (1977) Sur les mécanismes de formation des chaînes intracontinentales. L'exemple des chaînes atlasiques du Maroc. Bull Soc Géol Fr 7 (19): 521-526
Mayol S (1987) Géologie de la partie occidentale de la boutonnière paléozoïque des Jebilet, Maroc. Trav Lab Sci Terre St-Jérôme, Marseille, Ser B, 27: 241
Mayol S, Muller J (1985) Mise en évidence d'une unité allochtone hercynienne précoce (antéschisteuse) dans les Jebilet occidentales (Maroc). Etude de la structuration de la zone de contact. C R Acad Sci Paris, Ser II, 300: 369-372
McKenzie D (1978) Some remarks on the development of sedimentary basins. Earth Planet Sci Lett 40: 25-32
Michard A (1967) Plissement concentrique, plissement synschisteux et granitisation dans le Cambro-Ordovicien des Rehamna occidentaux (Maroc). C R Somm Séances Soc Géol Fr pp 347-349
Michard A (coord) (1982) Le massif paléozoïque des Rehamna (Maroc). Notes Mém Serv Géol Maroc Rabat 303: 180
Michard A, Hoepffner C, Jenny P (1978) Le couloir de cisaillement calédono-hercynien de la Meseta occidentale sur la transversale de Mechra Ben Abbou (Rehamna, Maroc). Bull Soc Géol Fr 20: 889-894
Michard A, Yazidi A, Benziane F, Hollard H, Willefert S (1982) Foreland thrusts and olistostromes on the pre-Saharan margin of the variscan orogen, Morocco. Geology 10: 253-256
Michard A, Cailleux Y, Hoepffner C (1983) La structure de las Meseta marocaine, une revue. In: Le Maroc et l'orogénie paléozoïque: The Caledonide Orogen. Rabat Symp 1983 (abstr)
Mrini Z (1985) Age et origine des granitoîdes hercyniens du Maroc: Apport de la géochronologie et de la géochimie isotopique (Sr, Nd, Pb). Thèse Univ Clermont-Ferrand, Fr

Neltner L (1938) Etudes géologiques dans le Sud marocain. Notes Mém Serv Géol Maroc 42: 296
Oukeïmeni D (1987) Les nappes hercyniennes des Jebilet orientales et leur substratum carbonifère. Thèse 3ème cycle, Univ Aix-Marseille III, Fr 145 pp
Padgett C, Ehrlich R, Moody M (1977) Submarine debris flow deposits in an extensional setting, Upper Devonian of Western Morocco. J Sediment Petrol 47: 811-818
Petit JP, (1976) La zone de décrochement du Tizi-n'Test et son fonctionnement depuis le Carbonifère. Thèse 3ème cycle, Univ Sci Techn Languedoc, Montpellier, Fr, 116 pp
Piqué A (1983) La déformation des conglomérats dans les zones métamorphiques des Rehamna (Meseta marocaine). I: Passage de la fracturation au flux plastique. Sci Géol Bull 26: 219-244
Piqué A (1975) La déformation des conglomérats dans les zones métamorphiques des Rehamna (Meseta marocaine). II: Passage du plissement au cisaillement. Sci Géol Bull 26: 205-213
Piqué A (1979) Evolution structurale d'un segment de la chaîne hercynienne: La Meseta marocaine nord-occidentale. Sci Géol Mém 56: 243
Piqué A (1982) Relations between stages of diagenetic and metamorphic evolution and the development of a primary cleavage in the northwestern Moroccan Meseta. J Struct Geol 4: 491-500
Piqué A (1983) Structural domains of the Hercynian belt in Morocco. In: Schenk P (ed) Regional trends in the geology of the Appalachian - Caledonian - Hercynian - Mauritanide orogen. Reidel, Dordrecht, pp 339-345
Piqué A (1984) Faciès sédimentaire et évolution d'un bassin: le bassin dévono-dinantien de Sidi-Bettache (Maroc nord-occidental). Bull Soc Géol Fr 26: 1019-1023
Piqué A (1987) Un élément majeur de la Meseta marocaine nord-occidentale: le bassin dévono-dinantien de Sidi-Bettache. Notes Mém Serv Géol Maroc 43 (321): 41-64
Piqué A (1988) Variscan terranes in Morocco. In: Dallmeyer RD (ed) Tectonothermal evolution of the West African orogens and circum Atlantic terrane linkage. IGCP 233, Nouakchott (abstr) pp 161-164
Piqué A, Kharbouch F (1983) Distension intracontinentale et volcanisme associé. La Meseta marocaine nord-occidentale au Dévono-Dinantien. Bull Cent Rech Exploration-Production Elf-Aquitaine 7: 377-387
Piqué A, Michard A (1989) Moroccan Hercynides: a synopsis: The Paleozoic sedimentary and tectonic evolution at the northern margin of West Africa. Am J Sci 289: 286-330
Piqué A, Wybrecht E (1987) Origine des chlorites de l'épizone. Héritage et cristallisation synschisteuse. Exemple des grauwackes cambriennes du Maroc occidental. Bull Minéral 110: 665-682
Piqué A, Jeannette D, Michard A (1980) The western Meseta shear zone, a major and permanent feature of the Hercynian belt in Morocco. J Struct Geol 2: 55-61
Piqué A, Hoepffner C, Jenny J, Guezou JC, Jeannette D, Michard A (1982) Tectonique du massif des Rehamna (Maroc). Evolution de la déformation dans les zones métamorphiques hercyniennes. Notes Mém Serv Géol Maroc 303: 86-129
Piqué A, Chalouan A, Fadli D (1985) Des rides paléogéographiques aux anticlinaux. Permanence des directions structurales dans la Meseta marocaine occidentale au cours du Paléozoïque. Sci Géol Bull 38: 147-153
Poutchkovsky A (1978) Etude géologique et structurale des formations paléozoïques de la région de Bou Gader. Thèse 3ème cycle, Univ d'Aix-Marseille III, Fr, 119 pp
Proust F (1973) Etude stratigraphique, pétrographique et structurale du bloc oriental du massif ancien du Haut-Atlas (Maroc). Notes Mém Serv Géol Maroc 254: 541-551
Proust F, Petit JP, Tapponnier P (1977) L'accident du Tizi-n'Test et le rôle des décrochements dans la tectonique du Haut-Atlas occidental (Maroc). Bull Soc Géol Fr 19: 541-543
Rais-Assa R, Cornée JJ, Muller J, Sougy J (1983) Les unités allochtones du massif des Rehamna, Maroc hercynien. C R Acad Sci Paris, Ser II, 296: 783-786
Roch G (1930) Etudes géologiques dans la région méridionale du maroc occidental. Notes Mém Serv Mines Carte Géol Maroc 9: 542
Rosé F (1987) Les types granitiques du Maroc hercynien. Thèse Univ Curie, Paris
Ruellan E (1985) Géologie des marges continentales passives. Evolution de la marge atlantique du Maroc (Mazagan); étude par submersible, Seabeam et sismique réflexion. Comparaison avec la marge N.O. africaine et la marge homologue E. américaine. Thèse Univ Bretagne Occident, Brest, Fr, 297 pp
Saddiqi O (1988) Tectonique de la remontée du manteau: les peridotites des Beni Bousera et leur enveloppe métamorphique, Rif interne, Maroc. Thèse Univ Pasteur, Strasbourg, Fr, 171 p
Schaer P (1962) Problèmes de granitisation dans la partie occidentale du massif ancien du Haut-Atlas. C R Acad Sci Paris pp 2270-2272
Schaer JP (1964 a) Volcanisme cambrien dans le massif ancien du Haut-Atlas occidental. CR Acad Sci Paris 258: 2115-2117
Schaer JP (1964 b) Aspects de la tectonique dans le massif ancien du Haut-Atlas. CR Acad Sci Paris 258: 2353-2356
Schaer JP (1967) Interférences entre les structures du socle et celles de la couverture dans le Haut-Atlas marocain. In: A la Baconnière Publ, Etages tectoniques, pp 297-306
Schaer JP (1981) Carte géologique Imi-n-Tanoute 1/100000, partie paléozoïque. Notes Mém Serv Géol Maroc 203
Schaer JP (1982) Le Haut-Atlas marocain. In: Colloque de Zurich (in press)
Sougy J (1964) Les formations paléozoïques du Zemmour noir (Mauritanie occidentale). Etude pétrographique, stratigraphique et paléontologique. Ann Fac Sci Univ Dakar 12: 695
Sougy J (1976) Existence d'une importante tectonique tangentielle, avec nappes repliées, dans les Jebilet occidentales et centrales (Meseta marocaine occidentale). Trav Lab Sci Terre St-Jérôme, Marseille, Ser B, 12: 41
Tcherviakov IB, Ivanov TI (1980) Propriétés densimétriques des roches et des terrains du domaine atlasique. Bull Mines Géol Energ Rabat 48: 51-55
Termier H (1936) Etudes géologiques sur le Maroc central et le Moyen Atlas septentrional. Notes Mém Serv Géol Maroc 33: 1566
Termier H, Termier G (1971) Le massif granodioritique du Tichka (Haut-Atlas occidental, Maroc). Notes Mém Serv Géol Maroc 216
Termier H, Termier G, Leutwein F, Sonet J (1972) Sur l'âge varisque des plutonites du Tichka (Maroc). Bull Soc Fr Minéral Cristallogr 94: 124-127
Thiessen J (1979) Seismische und gravimetrische Untersuchungen der Krustenstruktur am marokkanischen Kontinentalrand. Thesis, Univ Hamburg, FRG
Van den Bosch JWH (1971) Carte gravimétrique du maroc au 1/500000. Notes Mém Serv Géol Maroc 234, 6 feuilles
Wendt J (1985) Disintegration of the continental margin of north-western Gondwana: Late Devonian of the eastern Anti-Atlas (Morocco). Geology 13: 815-818

Chapter 6

Circum-Atlantic Correlatives

Terranes within the Iberian Massif: Correlations with West African Sequences

C. Quesada,[1] F. Bellido,[1] R. D. Dallmeyer,[2] I. Gil-Ibarguchi,[3] J. T. Oliveira,[4] A. Perez-Estaun,[5] A. Ribeiro,[6] M. Robardet,[7] and J. B. Silva[6]

1 Introduction

The Iberian Peninsula consists of two very different parts: (1) the eastern half which constitutes the westernmost extension of the Alpine orogen; and (2) the western half (Iberian Massif) representing the southwesternmost and largest piece of the so-called European Hercynian Foldbelt. Although Precambrian and Paleozoic inliers are scattered within the former, they were generally so strongly reactivated during the Alpine orogeny that any attempt of correlation with the Iberian Massif has as yet failed. The latter, on the other hand, has been extensively surveyed in recent decades, and a summary of the present state of the art constitutes the main objective of this contribution. A tectonostratigraphic terrane analysis perspective is preferred for this purpose. This will allow a more accurate unravelling of each individual terrane evolutionary history, which will in turn greatly facilitate correlation.

In general terms, the Iberian Massif consists of several terranes, with varied palinspastic origins (see following section), accreted at various times during the Paleozoic as a consequence of a convergent tectonic process which led to the closure of a Proto-Atlantic ocean. At least two of those terranes include variably reactivated Precambrian basement remnants and one of them (so-called Iberian Terrane) shows strong evidence for a late Precambrian orogenic event, probably correlative of the Pan-African, Cadomian and Avalonian orogenies of other Circum-Atlantic borderlands.

2 Tectonostratigraphic Terranes Within the Iberian Paleozoic Orogen

The application of the concepts of tectonostratigraphic terrane analysis (Coney et al. 1980) to the geology of the Iberian Massif has allowed the distinction of several major units, the characteristics of which make them likely candidates to be regarded as suspect exotic terranes with respect to another unit (so-called Iberian Terrane) which constitutes most of the Iberian Massif area (Ribeiro et al. 1987; Quesada and Ribeiro 1989). Figure 1 shows the spatial distribution of tectonostratigraphic terranes within the Iberian Massif as well as the currently used zonal subdivision (Julivert et al. 1974; modifying a previous proposal by Lotze 1945), from which it differs markedly.

Obvious exotic units are those of an oceanic origin (ophiolites s. l.). They occur in the northwest of the Massif, where they constitute the lowermost thrust sheets of the Galicia-Tras-os-Montes crystalline nappe complexes, which structurally overlie the imbricated northern part of the Iberian Terrane; and in the southwest (Pulo do Lobo unit, which includes the Beja-Acebuches ophiolite: Munhá et al. 1986). At this preliminary stage of terrane analysis and for the aims of this review they have been grouped together, although much more research is required in order to determine whether or not they represent remnants of the same ocean, the same or different obducted ophiolite sheets, or even have further subdivisions corresponding to distinct oceanic environments.

Those continental units separated from the local parautochthon (Iberian Terrane) by intervening oceanic units are being regarded as likely suspect

1 Instituto Tecnologico GeoMinero de España: Rios Rosas 23, 28003 Madrid, Spain
2 Department of Geology, University of Georgia, Athens, Georgia 30602 USA
3 Departamento de Mineralogia-Petrologia, Universidad del Pais Vasco, 48080 Bilbao, Spain
4 Serviços Geológicos de Portugal, 1200 Lisboa, Portugal
5 Departamento de Geología, Universidad de Oviedo, 33005 Oviedo, Spain
6 Departamento de Geologia, Faculdade de Ciencias, 1294 Lisboa, Portugal
7 Institut de Géologie, Université de Rennes, Campus de Beaulieu, 35042 Rennes, France

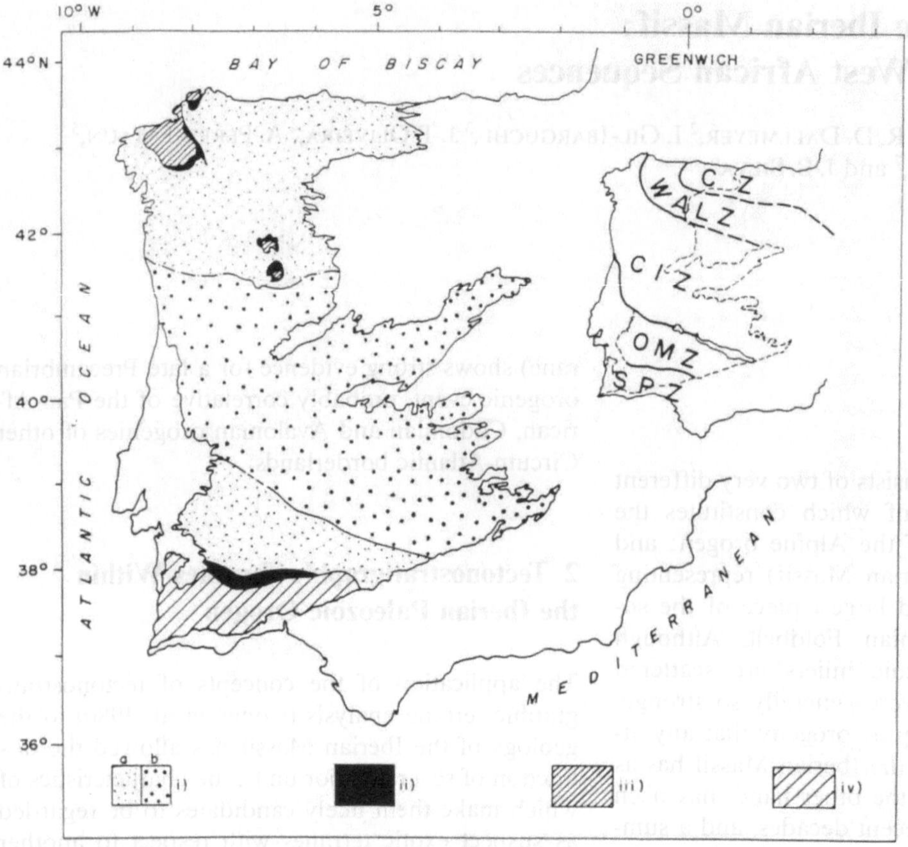

Fig. 1. Terrane map of the Iberian Massif. *1* Iberian Terrane: *a* imbricated margins, *b* parautochthonous core; *2* Oceanic Exotic Terranes; *3* continental suspect terranes in NW Iberia; *4* South Portuguese Terrane. *Inset* Zonal subdivision (Julivert et al. 1974). *CZ* Cantabrian Zone; *WALZ* West Asturian-Leonese Zone; *CIZ* Central Iberian Zone; *OMZ* Ossa-Morena Zone; *SPZ* South Portuguese Zone

exotic terranes, and no known evidence contradicts this assumption as yet. However, at the present level of knowledge nothing can be said about the degree of exotism or possible correlatives of these units. Even the correlation of "exotic" continental units within the Massif itself is extremely difficult, due to the lack of exposed common parts of the geologic record in the two areas where they occur. For this reason the crystalline continental units constituting the uppermost allochthons of the stacked Galicia-Tras-os-Montes complexes, which consist mostly of Precambrian basement and lower Paleozoic cover imbricates, will be dealt with separately from the South Portuguese terrane in which only upper Paleozoic cover sequences are exposed. The possibility that both were correlatives cannot be, however, definitely disregarded.

The Iberian Terrane in its turn, which constitutes more than 80% of the Iberian Massif surface and is here considered as the reference autochthonous unit, could well be itself an exotic element with respect to one of the Paleozoic macrocontinents (Gondwana?, Armorica?) or even be a part of them.

3 Precambrian Evolution of Iberian Terranes

With the obvious exception of the Paleozoic oceanic terranes, Precambrian rocks constitute the basement of the rest of the Paleozoic Iberian terranes. However, they are only exposed at the surface in the Iberian Terrane and the Galicia-Tras-os-Montes upper allochthonous terrane. The existence of a continental basement underneath the exposed cover of the South Portuguese terrane is out of the question according to the evidence supplied by the widespread igneous rocks of this unit (Munhá 1979) and the platformal nature of various sedimentary sequences (Oliveira 1983).

3.1 *Iberian Terrane*

Upper Proterozoic rocks showing evidence of a late Precambrian orogenic event occur in profusion throughout this unit, having been variably (generally strongly) reactivated during the Hercynian orogeny. This fact together with the present dispersion of the original relationships make even more difficult an analysis of its Precambrian evolution. The vast majority of the Precambrian occurrences display rather monotonous sedimentary sequences with, presumably, the same origin. A greater variability and complexity exists, however, in the southern part (Ossa-Morena zone, see Fig. 1) in which the presence of ocean-derived and volcanic arc units along with platformal sedimentary sequences makes it a likely candidate to be a Precambrian suspect composite terrane.

3.1.1 Stratigraphy

Two types of Precambrian sequences may be considered according to their relationships to the aforementioned late Precambrian orogenic event: (1) pre-orogenic, and (2) syn-orogenic (Quesada et al. 1990, Quesada, 1990a–b).

Pre-orogenic Sequences
Most of the pre-orogenic occurrences are located in southern parts of the Iberian Massif (Ossa-Morena zone) with only a few scattered exposures in southern and northwestern parts of the Central Iberian zone. They belong to three sequences with probably different palinspastic origins whose mutual original relationships are unknown yet:

1. Valencia de las Torres-Cerro Muriano Supergroup (Quesada et al. 1987). This is by far the most exposed (>90% area) pre-orogenic sequence. It consists (Fig. 2) of an upper, mainly metasedimentary, succession (Serie Negra group, Alia 1963: pelites, greywackes, black cherts and mafic metavolcanics) which includes acritarch assemblages of middle-to-upper Riphéan age (Chacón 1979; Chacón et al. 1984), underlain by a lower, mainly metaigneous, succession (Azuaga Gneiss Group, Delgado 1971: gneisses, amphibolites and some metasediments) whose age is as yet unknown.

Detailed synthetic overviews can be found in Quesada et al. (1990) and Quesada (1990a, b). Relevant data supporting the interpretation proposed on Fig. 2 are the bimodal nature of the igneous activity represented by the Azuaga Gneiss Group, and the evolution in source area and sedimentary environment from the lower part of the Serie Negra (shelf environment with a mature continental source) to its upper part (coarsening upwards turbidite sequence with a calc-alkaline volcanic source area).

2. Serpentinite sequences which appear in generally small, scattered occurrences within the Ossa-Morena zone. Although most of them are fault-bounded and difficult to interpret, some others define thrust sheets separating Valencia de las Torres-Cerro Mu-

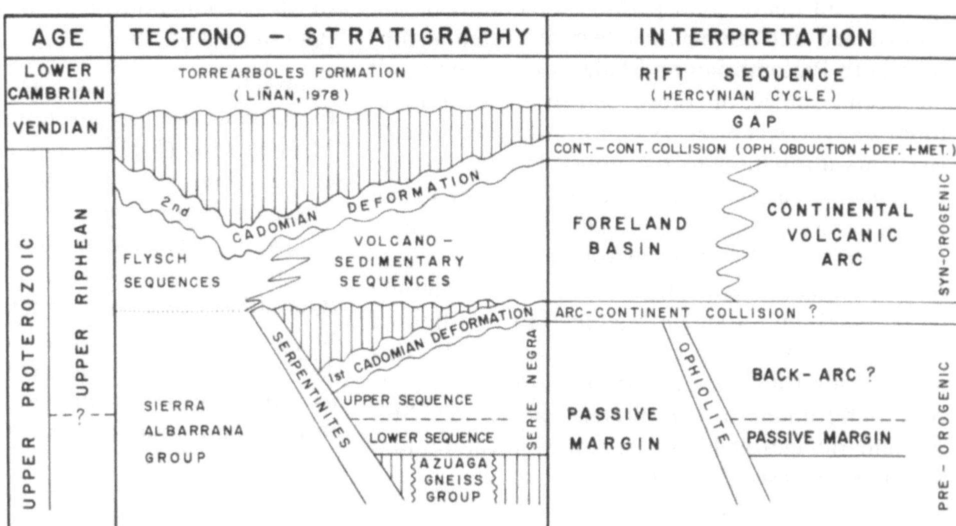

Fig. 2. Tectono-stratigraphic cartoon and interpretation of Precambrian sequences comprising the Iberian Terrane

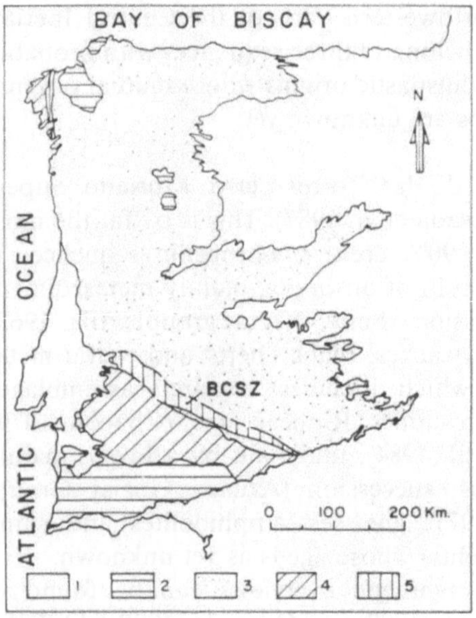

Fig. 3. Suspect Precambrian Terranes in the basement of the Paleozoic Iberian Terrane. *1* post-Paleozoic sequences; *2* undifferentiated Paleozoic suspect terranes; *3, 4, 5* Iberian composite terrane; *3* Sierra Albarrana terrane; *4* Valencia de las Torres-Cerro Muriano arc terrane; *5* imbrication zone ~cryptic suture ~Badajoz-Córdoba Shear Zone (BCSZ)

riano Supergroup rocks from underlying Sierra Albarrana Supergroup rocks (Apalategui et al. 1983), or appear as olistholiths within upper Proterozoic syn-orogenic sequences (Apalategui and Quesada 1987). They have been interpreted collectively as dismembered remnants of a once continuous ophiolite sheet (Quesada et al. 1987).

3. Sierra Albarrana Supergroup (Quesada et al. 1987). This is a metaterrigenous shelf sequence (pelites and quartzofeldspathic sandstones) of unknown age and ambiguous tectonic relationships with both Precambrian and Paleozoic sequences. It has been a subject of controversy in the local literature (see Quesada et al. 1990, for a detailed discussion). Its attribution to the upper Proterozoic has been mostly based on the existence of lower Paleozoic intrusives (Lancelot and Allegret 1982; García Casquero et al. 1985) postdating a regional metamorphic event which affects these rocks (Quesada et al. 1987), and a single reference to non-figured acritarchs of Riphéan age by Chacón (1979). This interpretation has received strong support in a recent publication of late Proterozoic (c. 600 Ma) radiometric ages yielded by granites intruding into this sequence (Schäfer et al. 1989).

These three pre-orogenic Precambrian sequences have been tentatively interpreted as representing different tectonostratigraphic terranes, amalgamated during the Cadomian orogeny (Quesada 1989). Their areal distribution is shown on Fig. 3 in which the cryptic suture zone, as deduced by serpentinite occurrences and HP metamorphic rocks within the overlying unit (Valencia de las Torres-Cerro Muriano Supergroup), roughly coincides with the so-called Badajoz-Córdoba Shear Zone, a complex structure that has been subsequently reactivated several times in Paleozoic times.

Syn-orogenic Sequences

These are separated from the underlying pre-orogenic sequences by an important deformational event and their origin is intimately bound to the orogenic processes that affected the region in the late Precambrian (Upper Riphéan-Vendian). Two types of syn-orogenic sequences may be recognized (Fig. 2): (1) volcanosedimentary successions, and (2) flysch successions.

1. Volcanosedimentary successions occur exclusively in the Ossa-Morena Zone of the Iberian Terrane (Fig. 1). They consist of lava, pyroclastic and epiclastic rocks ranging in composition from basalt to rhyolite but with a remarkable predominance of intermediate rocks (andesite, dacite). Both clastic and carbonate sediments are interbedded in the volcanic pile. Scarce findings of poorly preserved fossil assemblages allow its attribution to the Upper Riphéan-Lower Vendian period (Liñán and Schmitt 1980; Quesada et al., 1990). The orogenic calc-alkaline geochemical signature of the volcanic rocks (Sánchez Carretero et al. 1989) supports their interpretation as a continental volcanic arc developed during subduction of an upper Proterozoic ocean (Quesada et al. 1987).

2. Flysch successions are widely distributed throughout the Iberian Terrane north of the Ossa-Morena zone (Fig. 1). They have been roughly dated as upper Riphéan-Vendian (Palacios 1986). These terrigenous, mostly turbiditic, deposits are arranged in several depositional sequences, the depocenters of which seems to migrate to the north (Quesada, 1990a). Carbonates and seismic-related megaturbidites are locally important (Moreno 1977). These flysch sequences have been interpreted as a foreland basin fill postdating and closely following the evolution of a collisional event (Quesada, 1990b).

3.1.2 Structure

The severe overprint suffered by the Precambrian basement of the Iberian Terrane during the Hercynian orogeny renders quite uncertain the recognition and characterization of the structures developed during the late Proterozoic orogenic event. In most cases it is only by means of indirect pieces of evidence that the existence of a Precambrian polyphase deformation can be deducted [deformed clasts in overlying dated formations (Fig. 4) and intrusive relationships of dated plutons]. A few areas exist, however, in which the local characterization of the Precambrian structure is possible, due to the non-penetrative nature of the Hercynian overprint as demonstrated by mapping (Fernández et al. 1983; Eguiluz 1987) and dating (Cueto et al. 1983). This is particularly well known in the central part of the Monesterio Antiform (Fig. 5) in which a polyphase structuration of the Precambrian rocks can be demonstrated to predate deposition of lower Cambrian weakly or uncleaved rocks. This structuration involved the formation of a first generation of isoclinal recumbent folds, noticeable at both the map (Fig. 5) and micro and mesoscopic scales (Fig. 4), with associated schistosity, and only affecting the Valencia de las Torres-Cerro Muriano pre-orogenic sequence. This was followed by a second

Fig. 4. Black chert clast in lowermost Cambrian conglomerate; note superposition of two folding phases

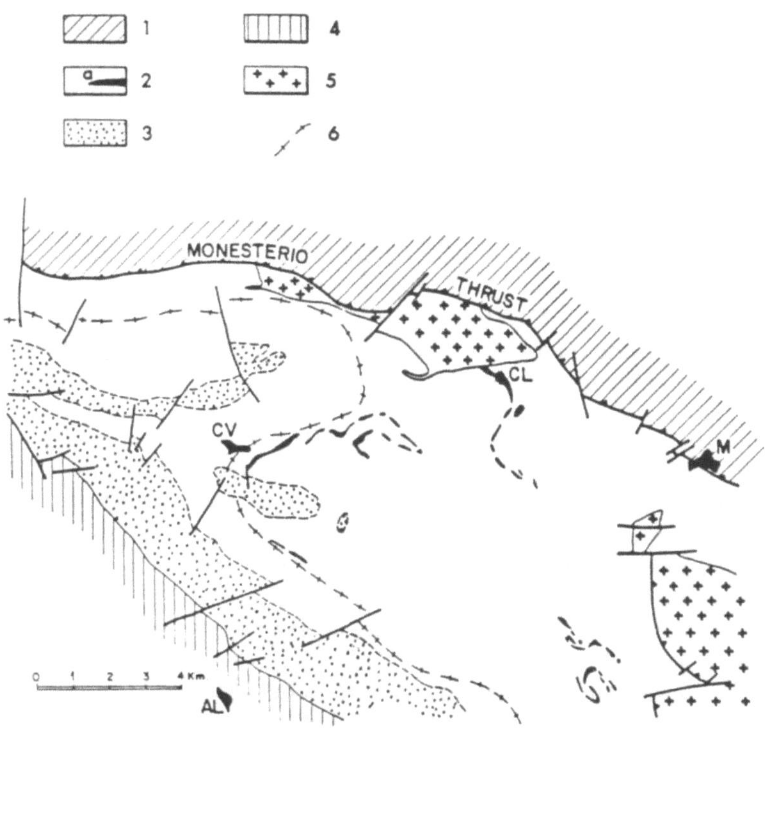

Fig. 5. Cadomian structural elements in the Arroyomolinos unit (Ossa-Morena Zone). *1* Monesterio unit; *2, 3, 4* Arroyomolinos unit; *2* Serie Negra Group (*a* black cherts); *3* synorogenic volcanosedimentary sequences; *4* Paleozoic rocks; *5* late Proterozoic and/or Paleozoic plutons; *6* biotite isograd

generation of more open SW-verging folds with associated crenulation cleavage which affected both the pre and syn-orogenic sequences. This structuration, which is the only one affecting penetratively the Precambrian rocks in this unit except for variably wide Hercynian strain bands, was synchronous to a regional metamorphic event, reaching locally high-grade conditions. The peak of metamorphism can be estimated to lie close to 552 ± 16 Ma (Rb-Sr whole rock isochron; Quesada, unpubl. results) age of the Monesterio Granodiorite (Quesada 1975), an *in situ* anatectic granitoid, coring one of the metamorphic domes developed at the latest stages of the Precambrian orogenic event.

The non-reactivated nature of this unit is also supported by the post-kinematic intrusion of undeformed plutonic rocks of late Precambrian or lower Paleozoic age (Pallares pluton: 572 ± 70 Ma, Cueto et al. 1983; Barcarrota pluton: 505 ± 10 Ma, Galindo et al. 1988; Tablada pluton: 494 ± 18 Ma, Quesada, unpubl.).

In central autochthonous parts of the Iberian Terrane (Fig. 1), in which the only Precambrian exposures correspond to syn-orogenic flysch sequences, macro and mesoscopic scale folds are known, generally without associated cleavage and asymmetric geometry. Nevertheless, the Precambrian structure is also very poorly known, due to the lack of detailed studies along with the absence of continuous marker horizons.

3.1.3 Magmatism

As for the stratigraphy, pre-orogenic and syn-orogenic stages may be distinguished when dealing with the Precambrian igneous activity of the Iberian Continental Margin. Pre-orogenic magmatism, briefly referred to above, is represented by alkaline to tholeiitic (Pascual 1981; García Casquero, in press) bimodal suites which characterize the Azuaga Gneiss Group, and by tholeiite basalts (Eguiluz 1987) interbedded in the lower sequence of the Serie Negra Group (Fig. 2). Both igneous suites have been collectively interpreted as witnessing for a continental rifting event which eventually led to oceanic crust formation, represented by the sparse serpentinite occurrences of the Ossa-Morena zone (Ribeiro et al., 1990). Some authors however, have argued that the Azuaga Gneiss Group could represent an older Precambrian basement (Pérez Lorente 1979; Pascual 1981).

Syn-orogenic magmatism, on the other hand, may be divided into subduction-related and collision-related types. The first indication of subduction-related magmatism appears in the upper part of the Serie Negra Group in which mainly andesitic, calc-alkaline rock clasts are increasingly fed into the basin. Then, following the first deformational event, an important subduction-related (Sánchez Carretero et al. 1989) volcanic activity spread throughout the Ossa-Morena zone which became a continental volcanic arc or an Andean-type margin. Numerous cogenetic tonalite-to-granite plutons are scattered across this zone, intruding into the pre-orogenic sequences or even the lower part of the volcanic edifice. A southward increase in K_2O within these calc-alkaline igneous sequences has been reported by Sánchez Carretero et al. (1989). This could be indicative of southward directed subduction.

Collision-related magmatism, not fully characterized, due to the lack of extensive radiometric dating, is mostly represented by peraluminous granitoids the formation of which is interpreted in close connection with decompression (uplift) processes in the latest stages of the late Precambrian orogeny, subsequent to the main episodes of crustal thickening.

3.1.4 Tectonothermal Evolution

The metamorphic pattern shown by the Precambrian rocks constituting the basement of the Iberian Terrane allows the adding of further precisions to its evolution during the Cadomian orogeny. Apart from local thermal domes developed during the latest orogenic stages, mainly concentrated in southern portions of the Iberian Terrane (Ossa-Morena Zone), the only exception to the generally very low to low metamorphic grade affecting the Precambrian sequences occur within the so-called Badajoz-Córdoba Shear Zone (Fig. 3) in which a complex metamorphic evolution, involving HP stages (eclogite and granulite formation), has been described (Mata and Munhá 1986). This is restricted to the Valencia de las Torres-Cerro Muriano pre-orogenic sequence in the same belt in which the serpentinite sequence also occurs. A recently published age of 617 ± 6 Ma (U-Pb zircon in granulite; Schäfer et al. 1989) provides an approximative estimate for the age of the granulite part of the P-T path.

3.1.5 Concluding Remarks on the Precambrian Evolution of the Iberian Terrane

Any attempt to undertake analysis of the various Precambrian sequences which occur within the Paleozoic Iberian terrane under the perspective of the tectonostratigraphic terrane theory (Coney et al. 1980) must face the numerous problems and limitations imposed by the severe tectonothermal reactivation that such rocks suffered as a consequence of subsequent tectonic processes in the Paleozoic, which resulted in many cases in dispersion of the original relationships, penetrative deformation, and reset of the metamorphic and isotopic systems, among others. Nevertheless, the amount of new data made available in the last decade, as shown in previous paragraphs, allow a preliminary approach to be made, although the results of such an analysis must be forcefully schematic.

A first tentative approach to the terrane analysis of Precambrian sequences in the Iberian terrane was made by Quesada et al. (1987). The main point of their argument lies in the existence in the northern Ossa-Morena Zone of a structural superposition of three Precambrian sequences with very different stratigraphy (the intermediate one being probably ophiolitic in nature) and very different tectonothermal history (the uppermost one showing evidence of a high-pressure metamorphic event). The serpentine unit was interpreted as a remnant of a dismembered ophiolite, therefore representing a cryptic suture. A straightforward conclusion of this interpretation is the consideration of the two units separated by the ophiolite as belonging to two different continental blocks, i. e. two different tectonostratigraphic terranes. The serpentinite sheet itself is obviously an exotic terrane with respect to any of the other two. Quesada et al. (1987) labelled such terranes by the name of the corresponding tectonosedimentary miogeoclinal sequences, i. e. Valencia de las Torres-Cerro Muriano terrane and Sierra Albarrana terrane, leaving the oceanic terrane unnamed. The Cadomian orogeny is to be held responsible for the amalgamation of these terranes in a complex process involving subduction and subsequent continent-continent collision.

The likely areal distribution of the three Precambrian terranes constituting the basement of the Paleozoic Iberian terrane has been shown on Fig. 3, although a great deal of uncertainty exists in NW areas due to the strongly penetrative nature of the Hercynian overprint and abundance of Late Paleozoic intrusives there. A summary of the relevant features of the three terranes is presented below to-

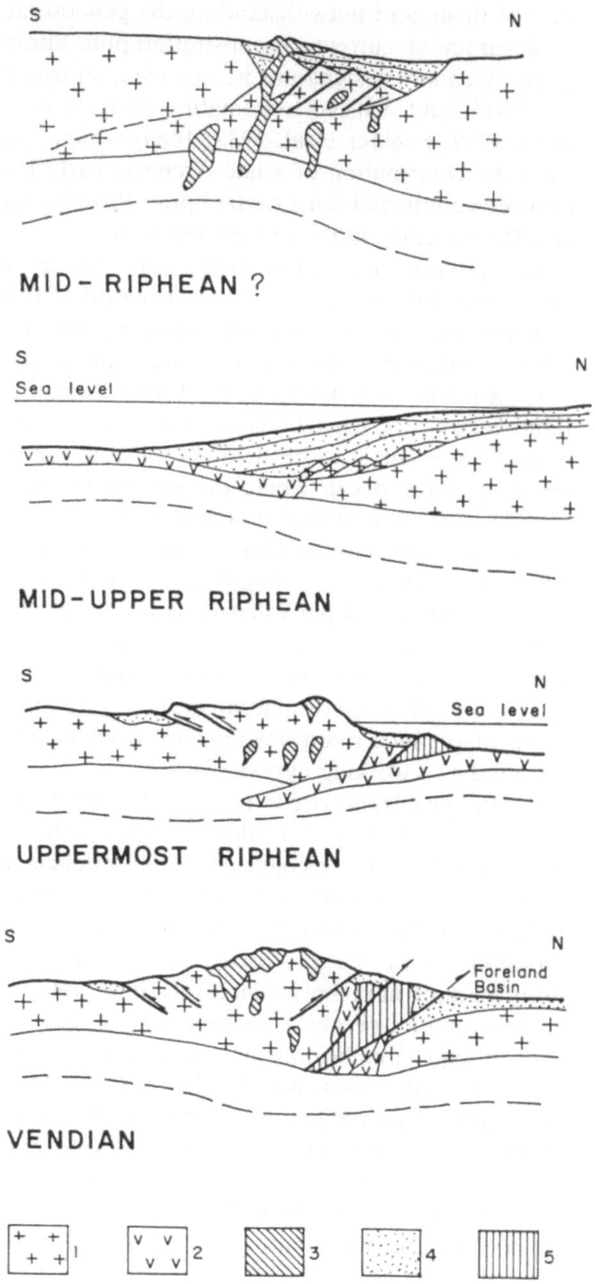

Fig. 6. The presumed evolution of the Iberian terrane during the upper Proterozoic

gether with the constraints that they impose on the characteristics of the amalgamation process (a schematic evolutionary model is shown in Fig. 6).

The continental nature of both the Valencia de las Torres-Cerro Muriano and Sierra Albarrana terranes is beyond any doubt, according to their respective sedimentary and/or igneous pre-orogenic records, despite the fact that no direct evidence of old basement occurrence has ever been found in either of them, and notwithstanding the general lack of accuracy of current chronostratigraphic knowledge. A common signature is, however, shown by the U-Pb and Sm-Nd systematics (Nägler et al. 1988, 1989; Schäfer et al. 1988, 1989) which indicates the contribution of a late Archean/early Proterozoic continental crust to the upper Precambrian and Paleozoic evolution of both terranes.

The pre-orogenic upper Proterozoic history of both continental terranes includes a period of miogeocline growth (sea-floor spreading?) postdating what has been interpreted as a rifting sequence, only recorded in the Valencia de las Torres-Cerro Muriano terrane (Azuaga Gneisses Group). Whether this rifting event is to be held responsible for the opening of the ocean where the serpentines were formed is as yet a matter of speculation. The same is true with regard to the consideration of the Sierra Albarrana terrane as a hypothetical counterpart, across that ocean, of the Valencia de las Torres-Cerro Muriano terrane. In the absence of any opposing evidence, and bearing in mind the possibility that large-scale strike-slip displacements could well have taken place, such interpretations are from here on accepted as a working hypothesis.

Several problems arise in trying to interpret what has been called first Cadomian deformational event, which is only recognized in the Valencia de las Torres-Cerro Muriano terrane, as well as the latest part of the terrane pre-orogenic sedimentary evolution (Serie Negra Group upper sequence). Both features have been tentatively interpreted as witnessing for a period of magmatic arc emergence and subsequent collision with a continental margin, presumably the Valencia de las Torres-Cerro Muriano terrane margin (Quesada, 1990b). Unfortunately, no traces of such an arc have been as yet found, except for indirect pieces of evidence such as calc-alkaline igneous clasts within Serie Negra upper sequence conglomerates and greywackes. The possibility that this first Cadomian deformation could represent an early phase of the continent-continent collision that is neatly marked by the second Cadomian deformation is hardly envisagable since, postdating the first deformation, an extensive phase of magmatic arc development, as shown by the syn-orogenic volcanosedimentary sequences and related plutonic rocks, clearly suggests a period of active subduction beneath the Valencia de las Torres-Cerro Muriano terrane.

The Cadomian orogenic evolution of the Paleozoic Iberian terrane, as has been previously advanced, involved an initial phase of subduction and a subsequent phase of continent-continent collision after consumption of the intervening ocean. The subduction phase was characterized by the building up of an arc onto the overriding plate (Valencia de las Torres-Cerro Muriano terrane). An Andean-type active margin is regarded as the most likely environment during this phase. A southward dip of the subduction zone is suggested by the areal distribution of syn-orogenic sequences as well as K_2O content variations across the zone.

The second Cadomian phase most likely represented the collision between both continental blocks and resulted in the underthrusting of the Sierra Albarrana terrane beneath the Valencia de las Torres-Cerro Muriano terrane. This interpretation is supported by: (1) preserved geometrical relationship; (2) existence of high pressure metamorphic relics in the neighbourhood of the suture; (3) evidence of ophiolite obduction; (4) development of a foreland basin onto the overridden block, the successive depocenters of which migrated towards the north presumably as a consequence of the migration of deformation in the same direction; (5) evolution of the geometrical characteristics of the structures developed during this second Cadomian phase, which are antithetic with the subduction zone in the upper block and synthetic in the overridden block. In the latter, they envolve from flat-lying near the suture, through north-verging to upright in northern positions.

Every tectonic activity associated to the Cadomian orogeny seems to have ceased by the lowermost Cambrian, as it is demonstrated by the unconformable relationships with basal Cambrian clastic formations (partly red beds) marking the initiation of a new rifting event. The axis of maximum extension during this new event was located south and west of the Ossa-Morena Zone and resulted in the break-up of the continent with newly accreted terranes that had been formed during the Cadomian collision. The Paleozoic Iberian terrane was formed in this process, including pieces of the various amalgamated Precambrian terranes, among which the suture was reactivated several times in the Paleozoic, as normal fault with a strike-slip component during the lower Paleozoic rift process and as wrench and thrust fault during the Hercynian orogenesis (Ribeiro et al., 1990). Collectively, all these tectonic processes (Precambrian and Paleozoic) have resulted in the formation of a very complex belt which is currently known in the literature as the Badajoz-Córdoba Shear Zone (see Fig. 3).

3.2 *Suspect Exotic Terranes in NW Iberia*

The allochthonous units which occur in the northwest of the Central Iberian Zone (Fig. 1) constitute a nappe pile, largely removed from its respective original setting, which was thrust, with a roughly eastward sense of displacement, onto parautochthonous sequences belonging to the Iberian Terrane (Ribeiro et al., 1990). A lower allochthonous sheet of ophiolites and an upper continental nappe complex are distinguished in this region, both interpreted as remnants of exotic terranes (Ribeiro et al. 1987a, b, c). The latter, which occurs at the Cabo Ortegal, Ordenes, Bragança and Morais massifs, consist may of both Precambrian basement and lower Paleozoic cover sequences.

A telescoped section across a continental lithospheric block has been interpreted to occur in the Morais Massif (Ribeiro 1983; Ribeiro et al. 1987b) in which a basal garnet lherzolite sheet, interpreted as a subcontinental mantle slice, is overlain by a mafic granulite thrust sheet (lower crust) in turn succeeded by a granitic augen-gneiss (upper crust) covered unconformably by a micaschist succession including Upper Cambrian to Lower Ordovician fossils (Timofeev, in Chacón 1979), which is interpreted as a lower Paleozoic cover sequence.

All this sequence underwent a severe tectonothermal overprint by the Variscan orogeny, during which its obduction onto the Iberian miogeocline took place. This overprint renders rather difficult the recognition of Precambrian events in these sequences. The main lines of evidence suggesting that their interpretation as a basement may be correct are: (1) existence of a foliation in the augengneisses that predates emplacement of mafic dykes and sills that also intrude into the overlying lower Paleozoic micaschists. These mafic dikes were subsequently affected by the Variscan deformation and transformed into amphibolites; (2) the mafic granulite and metaperidotite sheets show at least two phases of penetrative deformation, accompanied by granulite facies metamorphism, prior to the formation of dilation veins in which hornblende crystals grew with their long axis parallel to the stretching lineation in the Variscan structures.

An alternative interpretation of the mafic granulites and metaperidotites has been recently presented by Peucat et al. (1990) and Gil-Ibarguchi et al. (in press) on the basis of geochemical and geochronological results from similar rocks in the Cabo Ortegal complex. They suggested their formation in an arc environment in the Lower Ordovician having been metamorphosed under eclogitic/granulitic conditions soon after formation (in the 490–470 Ma interval) in a subduction tectonic setting. The apparent contradiction between these data and the former interpretation (Ribeiro et al. 1987a, b) clearly shows the need for further research. Two aspects with likely important connotations concerning the palinspastic origin of these exotic terranes must be emphasized: (1) the lack of Pan-African ages and (2) the frequent reference to ca. 2–2.5 Ga zircon ages (Kuijper 1979; Kuijper et al. 1982; Peucat et al., 1990) interpreted as the age of the continental source for some of the rocks in these terranes, which is similar to the source age of many late Precambrian and Paleozoic rocks within the Iberian Terrane (Schäfer et al. 1988; Nägler et al. 1988; Schäfer et al. 1989; Nägler et al. 1989).

4 Paleozoic Evolution of Iberian Terranes

The recognition of the various tectonostratigraphic terranes which constitute the Iberian Massif, as outlined above (see under Sect. 2), is mainly based on their respective Paleozoic histories, a summary of which is presented in the following sections. As for the Precambrian evolution, the exotic terranes which occur in the peninsular northwest are also subjects involving great discrepancies and questions opened to further research. The Paleozoic evolution of the other suspect terrane (South Portuguese Terrane, see Fig. 1) is, however, very well known and provides excellent constraints on the characteristics and relative timing of the convergence process (Hercynian orogeny) which resulted in the present-day terrane arrangement.

4.1 *Iberian Terrane*

The large surface of exposure of this terrane, together with the complete record of both pre-orogenic (pre-Hercynian) and syn-orogenic sequences and igneous and metamorphic events allows a better knowledge of its evolution, as compared to the more fragmentarily exposed exotic terranes.

4.1.1 Stratigraphy

Three main phases may be considered when confronting the Paleozoic evolution of the Iberian continental margin terrane from a tectono-stratigraphic point of view; e.g. (1) rift phase (Cambrian-Lowermost Ordovician); (2) passive margin phase (Ordovician-Devonian); (3) syn-orogenic phase (Middle Devonian-Lower Permian).

Rift phase

Starting in the lowermost Cambrian, and probably uppermost Vendian locally (Torreárboles Fm., Liñán 1978) with shallow marine and terrestrial red beds, this extensional phase is characterized by an initial transgression leading to the establishment of shelf sedimentary conditions over a wide area in which both (mostly) clastic and carbonate shelf sequences were deposited. The most significant features of the sedimentary sequences characterizing this phase are: (1) remarkable facies and thickness changes, which allow recognition of highly subsiding belts (i. e. the West-Asturian Leonese Zone, Fig. 1) with respect to others more stable ones, and (2) existence of bimodal volcanic and subvolcanic activity mainly concentrated in the most external, western parts of the margin. This volcanism evolved both in space (outwards) and time from alkaline bimodal sequences to tholeiite basalt sequences, which is typical of continental rifting processes leading to oceanic crust formation.

The transition to the next, passive margin phase, which is locally marked by a break-up unconformity (Sardic unconformity), is markedly diachronous, ranging between basal Ordovician in the southwesternmost areas and lower (?) Silurian in the northwest (Centro-Transmontano domain, Ribeiro 1974). This diachronism strongly suggests the sequential opening of the bordering ocean basin from south to north.

Passive margin phase

The characteristic sedimentary record on the Iberian continental margin terrane during this more stable phase consists of relatively thick (2–3 km) clastic shelf sequences, covering the entire terrane, with local intercalations of carbonates. The environmental picture that all these sequences collectively define is that of a very wide continental shelf (minimum figure of about 1000 km may be realistic if structural restoration is added to the present distribution area), very gently sloping to the west where more distal, outer shelf and slope deposits predominate, as compared to the more proximal eastern part.

Noteworthy sedimentary features, relevant for their paleographical significance, are: (1) persistence of certain facies over very large areas and within a very narrow timespan (namely the Armorican Quartzite formation and other formations), which may be indicative of a wide tectonically very stable shallow shelf in which facies variations were mainly controlled by eustatic sea level changes, and (2) sporadic occurrence of glacio-marine deposits, mostly of upper Ordovician age, which may be correlative of time-equivalent deposits bordering the African ice sheet, and may provide a close connection between the Iberian Massif and Gondwanaland during the lower Paleozoic (Robardet and Doré 1988). The presence of storm-dominated sequences (i. e. the middle Ordovician Monte da Sombadeira Formation, Brenchley et al. 1986) may also have important paleogeographical connotations.

From a biostratigraphic point of view, there are noticeable differences between the proximal areas (mainly the Central Iberian regions) and the more external southwestern and western fringe (particularly the southwestern Ossa-Morena Zone). This has been established for the Ordovician and Silurian periods (Robardet and Gutiérrez-Marco, 1990). During the Lower Devonian the proximal areas are characterized by "rhenish" faunal assemblages, while the more external fringe has yielded "hercynian" faunas (Racheboeuf and Robardet 1986), with an intermediate mixed area in the Central Ossa-Morena Zone (Valle Syncline area, Robardet et al. 1986). This particular arrangement may be interpreted as representing the transition from proximal (benthic) to distal (pelagic) environments, and provides little support to previous interpretations regarding the Badajoz-Córdoba shear zone as a Hercynian suture (Matte 1983, 1986; Brun and Burg 1982; Burg et al. 1981; Chacón et al. 1983; etc.) separating two different continental blocks (Robardet 1976; Paris and Robardet 1977) accreted during the Hercynian orogeny.

Syn-orogenic phase

Passive margin conditions prevailed up to the upper Paleozoic, when the Iberian continental margin was progressively affected by a deformation wave, moving inward, as a consequence of collision with another continental block. The transition is generally marked in the sedimentary record by sudden deepening and inversion in the sediment supply source (now the growing orogen). Obviously, as deformation progressed diachronously towards the core of the terrane so did the appearance of syn-orogenic deposits which are younger in the same direction, being commonly separated from the underlying sequences by a sedimentary gap and locally an unconformity. In general terms syn-orogenic sequences in this terrane record the flysch/molasse

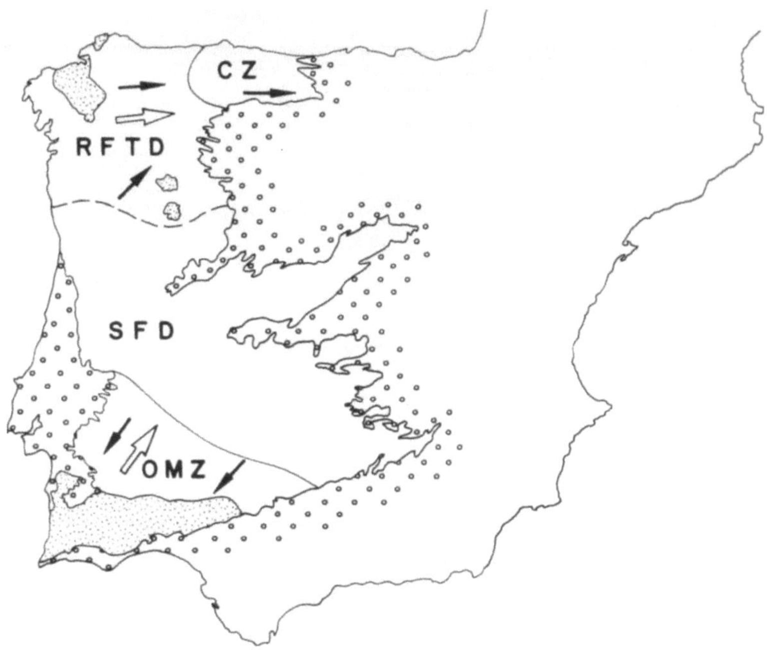

Fig. 7. Major structural domains within the Iberian Terrane. *CZ* Cantabrian Zone; *RFTD* Recumbent fold and thrust domain; *SFD* Subvertical Fold Domain; *OMZ* Ossa-Morena Zone; *Dotted areas* = Indifferentiated Suspect Terranes; *small arrows* = vergence of structures; *large arrows* = direction of deformation propagation

infill of foreland (namely the Cantabrian -N- and Pedroches -S- basins and intermontane basins, migrating from SW and W to NE and E. In most cases there is strong evidence of strike-slip activity along their bounding faults all along their evolution (Gabaldón et al. 1985; Gabaldón and Quesada 1986), which is strongly indicative of deformation taking place under a neat transpressional regime (Quesada et al. 1987; Apalategui and Quesada 1987).

4.1.2 Structure

From a structural point of view the Iberian Terrane may be considered as consisting of four major domains (Fig. 7).

1. Cantabrian zone, which constitutes the foreland thrust and fold belt of the northern half of the orogen;
2. Recumbent fold and thrust domain (Diez et al., 1990);
3. Subvertical fold domain (Diez et al., 1990);
4. Ossa-Morena zone, which constitutes the southern imbricated margin of the Iberian terrane.

A brief description of the internal structure of each of these domains, as well as their mutual relationships, may be summarized as follows:

1. Cantabrian Zone
The Cantabrian Zone displays a thin-skinned type geometry developed under non-metamorphic to anchymetamorphic conditions, being penetrative cleavage generally absent. The rotation of nappe emplacement direction with time, from a northerly direction initially, through easterly to southerly displacement finally, is responsible for the complex arcuate shape of the structure which generally followed a forward and downward propagation pattern, locally complicated by out of sequence or back-thrusting. Figure 8 shows a schematic cross-section of the northern part of the Iberian terrane in which the eastern third corresponds to the Cantabrian zone, where it is worthwhile noting the important development of syn-orogenic clastic wedges, intimately tied to the progressive emplacement of nappes.

2. Recumbent fold and thrust domain
A recumbent fold and thrust domain has been recently defined by Diez et al. (1990) and includes the West Asturian-Leonese zone (Lotze 1945), a part of the Central Iberian zone (Julivert et al. 1974) and the Galicia-Tras-os-Montes Schistose Domain (Farias et al. 1987). The eastern boundary of this domain coincides with the Narcea Antiform (an antiformal stack cored by Precambrian rocks), which separates it from the adjacent Cantabrian zone (Fig. 7). The southern boundary is difficult to trace, owing to the profusion of granitic intrusives in its vicinity. This recumbent fold and thrust domain is structurally overlain by the exotic, oceanic and continental terranes dealt with elsewhere in this paper. It is worthnoting that the structure in this domain,

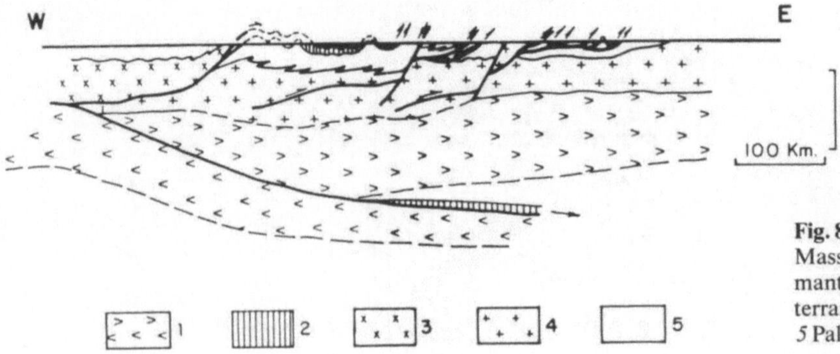

Fig. 8. Cross-section across the NW Iberian Massif (after Iglesias et al. 1983). *1* Upper mantle; *2* oceanic crust; *3* continental exotic terrane crust; *4* Iberian terrane crust; *5* Paleozoic sediments

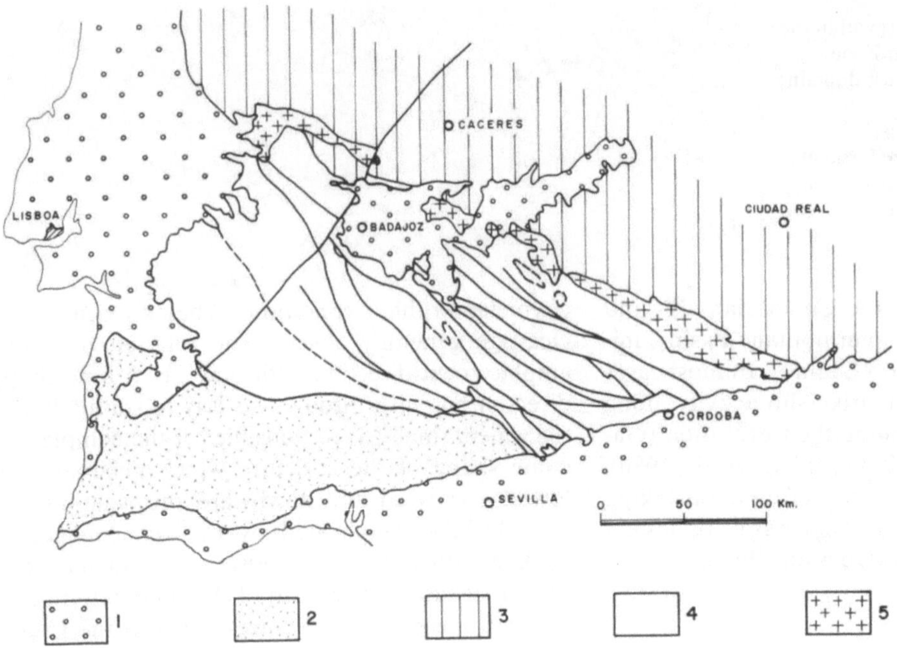

Fig. 9. Map of the Ossa-Morena Zone showing the most important wrench faults splitting it. *1* post-Paleozoic cover; *2* indifferentiated exotic terranes; *3* subvertical fold and thrust domain; *4* Ossa-Morena Zone; *5* Pedroches Batholith

which is characterized by thrusts and recumbent folds, sometimes with more than 20 km of reverse limb or displacement, verges towards the core of the arc and, presumably, deformation propagated in the same direction.

The geometry of this domain is the result of superposition of three or more successive types of structures (see Fig. 8, western and central parts): (1) east-verging recumbent folds; (2) east-verging thrust faults; (3) large open folds with steep axial surfaces; and (4) some normal faults. The amount of shortening in a E-W section is much greater than in a N-S section. These structures were developed under generalized regional metamorphic conditions (locally reaching the amphibolite grade: e.g. core of the Mondoñedo recumbent fold; Martínez Catalán 1985) with development of several cleavages.

3. Subvertical Fold Domain

The subvertical fold domain corresponds to most of the Central Iberian zone (Julivert et al. 1974) and is characterized by wide anticlinal areas separated by narrow synclines. Perhaps the most salient characteristic of the structure within this domain is its non-coaxial nature, as evidenced by generalized transection of folds by cleavages and subsequent bending of the previous structures, along which crenulation cleavages were formed locally. The superposition of these structures sometimes gave rise to basin and range interference patterns. Metamorphism in this domain is of a generalized low greenschist grade.

4. Ossa-Morena Zone

The Ossa-Morena Zone is separated from the adjacent subvertical fold domain by the Badajoz-Córdoba Shear Zone (Fig. 7), a complex structure with a long history as thrust, wrench and perhaps normal fault (Ribeiro et al., 1990). The internal structure geometry is characterized by a profusion of wedge-shaped blocks separated from each other by anastomosing subvertical shear zones and strike-slip faults (Fig. 9). The main structure of individual blocks is made up of the superposition of westerly-verging recumbent to asymmetric folds and thrusts refolded by subvertical to southwest-verging axial plane folds (Fig. 10). The latter shows a progressive decreasing in interlimb angle to the SW, as well as an increasing in cleavage development, which changes from slaty cleavage to schistosity. Apart from local thermal domes, metamorphic grade also increases in the same direction. As opposed to the northern domains, the propagation sequence of structures in the Ossa-Morena Zone is inward, as shown by their relationship with the syn-orogenic deposits (Quesada et al., 1990).

Timing of deformation

The Cantabrian Zone, as a typical foreland thrust and fold belt, was deformed later in the history than the other domains described. The climax was reached during Westphalian times (Julivert et al. 1974). In the other, more internal domains, the age of deformation is older, ranging from middle/upper Devonian to lower Carboniferous (Ribeiro et al., 1990). Nevertheless, it is worthwhile emphasizing that the vergence of structures, as stated above, seems to be opposite in the northern domains with respect to the southern one, despite their similar propagation sequence towards the core of the Iberian Terrane.

4.1.3 Magmatism

Paleozoic igneous rocks within the Iberian Terrane include, in broad terms, two very different groups with rather distinctive characteristics: (1) pre-oro-

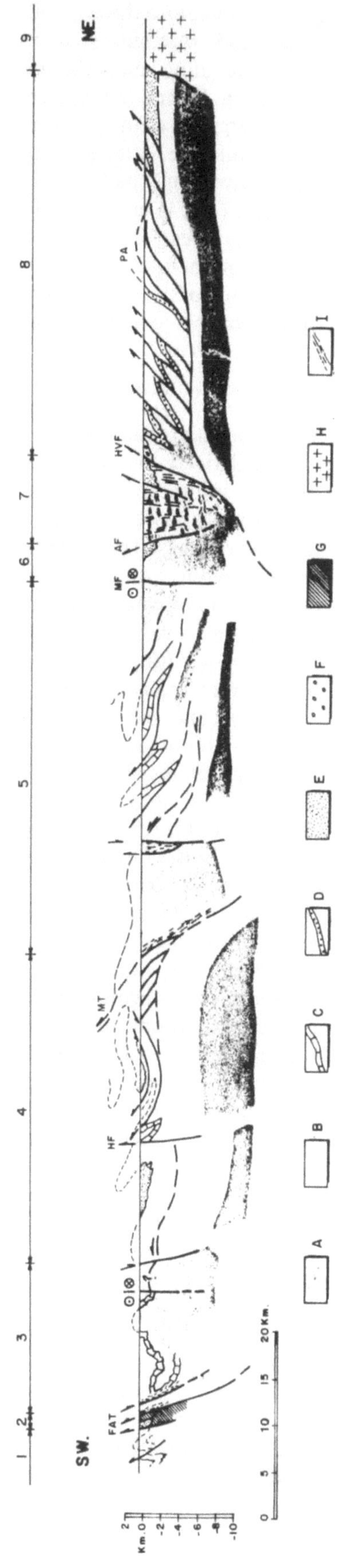

Fig. 10. Cross-section of the Ossa-Morena. South-Portuguese Zone: *1* South-Portuguese Terrane; *2* Pulo do Lobo Terrane. Ossa-Morena Zone: *3* Southern Crystalline Massifs; *4* south central belt; *5* north central belt; *6* Sierra Albarrana Domain; *7* Badajoz-Córdoba shear belt; *8* Obejo-Valsequillo-Puebla de la Reina Domain; *9* Pedroches Batholith. *A* Precambrian basement; *B* Paleozoic cover; *C* Cambrian limestones; *D* Ordovician Armorican Quartzite; *E* Mid-Devonian to Lower Carboniferous flysch; *F* Upper Carboniferous molasse; *G* Beja-Acebuches ophiolite; *H* Pedroches batholith; *I* major shear bands; *FAT* Ficalho-Almonaster Thrust; *HF* Hinojales fault; *MT* Monesterio Thrust; *MF* Malcocinado Fault; *AF* Azuaga Fault; *HVG* Hornachos-Villaharta Fault; *PA* Peraleda Antiform

genic peralkaline through alkaline to tholeiitic volcanic and plutonic suites; and (2) syn-orogenic calc-alkaline metaaluminous to peraluminous suites. The former are concentrated towards the terrane margin, i. e. in the presently imbricated parts of it, while the latter spread across a much wider area, involving parautochthonous foreland sequences (Fig. 1) as well as the already accreted exotic terranes. The following description will be centred on the pre-orogenic magmatism exclusively, leaving that of the syn-orogenic period, not exclusive of this terrane, to a later paragraph (see Sect. 4.5).

4.1.3.1 *Pre-orogenic Magmatism*

This is mostly characterized by frequently bimodal mantle-derived or mantle-induced volcanic and plutonic suites ranging in age from lower Cambrian to Silurian, and developed under neatly extensional tectonic regimes. The largest volume of this kind of rocks is concentrated in the Ossa-Morena Zone (southern imbricated margin) where they characterize, together with the sedimentary and structural evolution, a rifting event covering the lower Cambrian-lower Ordovician time span.

The second volumetrically important area in terms of pre-orogenic magmatism consists of largely allochthonous sheets in the northwest (Malpica-Tuy and Galicia-Tras-os-Montes units), presumably derived from further west positions, in which a rifting event in the Upper Ordovician? and Silurian is clearly shown (Ribeiro et al. 1987a). Whether both rift-related igneous suites were associated with the same rifting process, sequentially opening an oceanic basin from north to south as also suggested by the sedimentary evolution (see Sect. 4.1.1), or to two distinct extensional events, is still open to further research. In any case, both consist of bimodal (basalt-rhyolite) suites, mostly deposited in submarine environments and interbedded with mainly clastic sedimentary sequences. Genetically related subvolcanic and plutonic rocks intruding into the local basement are common in both areas.

From a geochemical point of view they evolved from alkaline with peralkaline end members, to tholeiitic compositions (Ribeiro 1986; Quesada et al. 1987), which is typical of rift-related volcanism leading to ocean crust formation. This evolution of the geochemical signature with time is also noticeable in space, in the Ossa-Morena zone during the Cambrian, where the rocks are progressively younger and more evolved in a southwesterly direction (Quesada et al. 1987), which is consistent with the spreading axis being located in that direction. Moreover, this interpretation is strongly supported by the occurrence of the Pulo do Lobo oceanic terrane (including the Beja-Acebuches ophiolite; Munhá et al. 1986) immediately south of the Ossa-Morena zone. Presumably, this could be the situation as well in the northern units, in which the vertical transition in geochemical character is neatly shown (Ribeiro 1986) and, furthermore, they are tectonically overlain by ophiolitic complexes (Ribeiro et al. 1987a).

Cogenetic plutonic rocks are represented by variably deformed and metamorphosed alkaline gabbros and alkaline and peralkaline granites yielding radiometric ages in the range 505–430 Ma (Priem et al. 1970; Galindo et al. 1988; Lancelot et al. 1985).

4.2 *Oceanic Exotic Terranes*

Paleozoic ocean-derived sequences have been recognized (Fig. 1) both in northern (Cabo Ortegal, Ordenes, Bragança and Morais thrust complexes) and southern (Pulo do Lobo unit) parts of the Iberian Massif, having been interpreted as ophiolitic terranes, exotic with respect to the reference Iberian Terrane which they overlie (Ribeiro et al. 1987a). Despite the fact that neither their ages nor their respective palinspastic situation prior to the Hercynian collision are known, it has been suggested that they could represent various parts of the same ocean; perhaps the final product of the rifting process recorded in the underlying Iberian Terrane in the lower Paleozoic (Iglesias et al. 1983; Ribeiro 1987; Ribeiro et al., 1990).

4.2.1 Northwestern Ophiolites

They frequently appear as dismembered imbricates of a once presumably continuous sheet, being most completely represented in the Ordenes and Morais complexes. There, a stratigraphy comprising harzburgites (serpentinites), flaser and banded gabbros, sheeted dykes and basalts can be reconstructed despite the generally amphibolite grade metamorphism affecting them. The geochemical characteristics of the mafic and ultramafic rocks (N- to T-MORB) are consistent with an oceanic origin. These northwestern Iberia ophiolites show the effects of a single prograde metamorphic event, dated at ca. 380–390 Ma, and interpreted in close connection with the process of their obduction onto the Iberian Terrane (Peucat et al., 1990; Dallmeyer and Gil-Ibarguchi, 1990).

4.2.2 Southern Ophiolitic Terranes

They have been grouped together in the preliminary terrane analysis of the Iberian Massif undertaken by Ribeiro et al. (1987 c) under the name of Pulo do Lobo oceanic terrane. This included the formerly defined Beja-Acebuches ophiolite (Munhá et al. 1986) as well as the Pulo do Lobo and Ribeira de Limas metasedimentary formations which occur in the Pulo do Lobo antiform (Fig. 11). All together they define a continuous band of oceanic rocks separating the Iberian Terrane from the South Portuguese Terrane to the south, that has been interpreted as a major suture within the Iberian Hercynian Foldbelt (Munhá et al. 1986; Ribeiro et al. 1987 c).

The Beja-Acebuches ophiolite comprises from bottom to top: serpentinites, flaser gabbros, massive and banded gabbros, sheeted dyke complex and locally pillowed basalts, recovered by pelites of the Pulo do Lobo formation with local chert interbeds (Munhá et al. 1986). This stratigraphy along with the T- to N-MORB geochemical signature of these rocks (Munhá et al. 1989) provides strong support to the interpretation of the Beja-Acebuches ophiolite as a piece of ocean floor. Its obduction onto the Iberian Terrane is suggested by field relations as well as by kinematic indicators (namely, S to N stretching lineations; Fonseca, pers. commun. 1986).

The Pulo do Lobo and Ribeira de Limas formations, on the other hand, correspond to extremely sheared phyllites and metagreywackes with minor intercalations of felsic volcanics and N-MORB tholeiites towards the base (Munhá 1983 b). These sequences have been recently interpreted as belonging to an accretionary wedge developed at the margin of the Iberian Terrane as a consequence of north-directed subduction (Silva et al., in press).

The precise age of both the sea-floor rocks and overlying sediments is unknown; they must have been obducted, however, prior to the upper Devonian, since upper parts of unconformable overstepping sequences (namely, the Horta da Torre Formation) have yielded spores and acritarchs of lower to middle Famennian age (Oliveira et al. 1987).

The Beja-Acebuches ophiolite underwent a low-pressure metamorphic event reaching locally the granulite grade (Castro et al. 1987), the cooling of which has been dated at ca. 350 Ma (Dallmeyer and Quesada, unpubl.). The Pulo do Lobo terrane in its turn shows widespread greenschist grade metamorphic conditions.

4.3 Northwest Iberia Exotic Continental Terranes

These constitute the uppermost allochthonous sheets in the nappe complexes of Cabo Ortegal, Ordenes, Bragança and Morais (Fig. 1) overlying the aforementioned ophiolitic nappe. As was stated when dealing with the Precambrian evolution of these units, they are the subject of a strong controversy mostly due to their structural and metamorphic complexity. The only proven Paleozoic rocks in these complexes correspond to metasedimentary (clastic) sequences, which occur in the Morais and Ordenes Massifs, overlying metagranitic rocks interpreted as part of the Precambrian basement on the basis of field and structural evidence (Ribeiro et al. 1987 a). These metasediments, locally affected by amphibolite grade metamorphism, have yielded in the Morais complex acritarch assemblages of upper Cambrian to lower Ordovician age (Mitrofanov and Timofeev, in Chacón 1979). The presence in this sequence of mafic dikes and sills, which also cut across the underlying granitic gneisses, has been interpreted as indicating a lower Paleozoic rifting event in this unit (Ribeiro et al. 1987 a).

Catazonal parts of these terranes, which include partly retrograded eclogitic and granulitic assemblages, are the main subject of discussion. An alternative interpretation to their consideration as basement slivers has been put forward recently by Peucat et al. (1990), and Gil-Ibarguchi et al. (in press) on the basis of geochemical and geochronological data. The absence of Precambrian ages other than late Archean-early Proterozoic upper intercepts in the U-Pb zircon systematics (suggesting the contribution of continental crust of that age to the formation of some of these rocks), together with the profusion of ages indicating two separate HP metamorphic events at ca. 490 Ma and ca. 420 Ma respectively, and the geochemical signature of the mafic protoliths, have led these authors to propose a lower Paleozoic active margin environment, removed from the Iberian Terrane, in which the complex evolution of these units, which could represent various palinspastic settings, would have taken place sequentially as a consequence of subduction of the intervening ocean between this hypothetical margin and the Iberian Terrane margin. A retrogressive metamorphic event dated at ca. 390 Ma is considered in connection with the obduction of the catazonal rocks onto the ocean floor, in turn subsequently obducted onto the Iberian Terrane, presumably as a consequence of the complete closure of the intervening ocean. (Peucat et al. 1990; Dallmeyer and Ibarguchi, 1990).

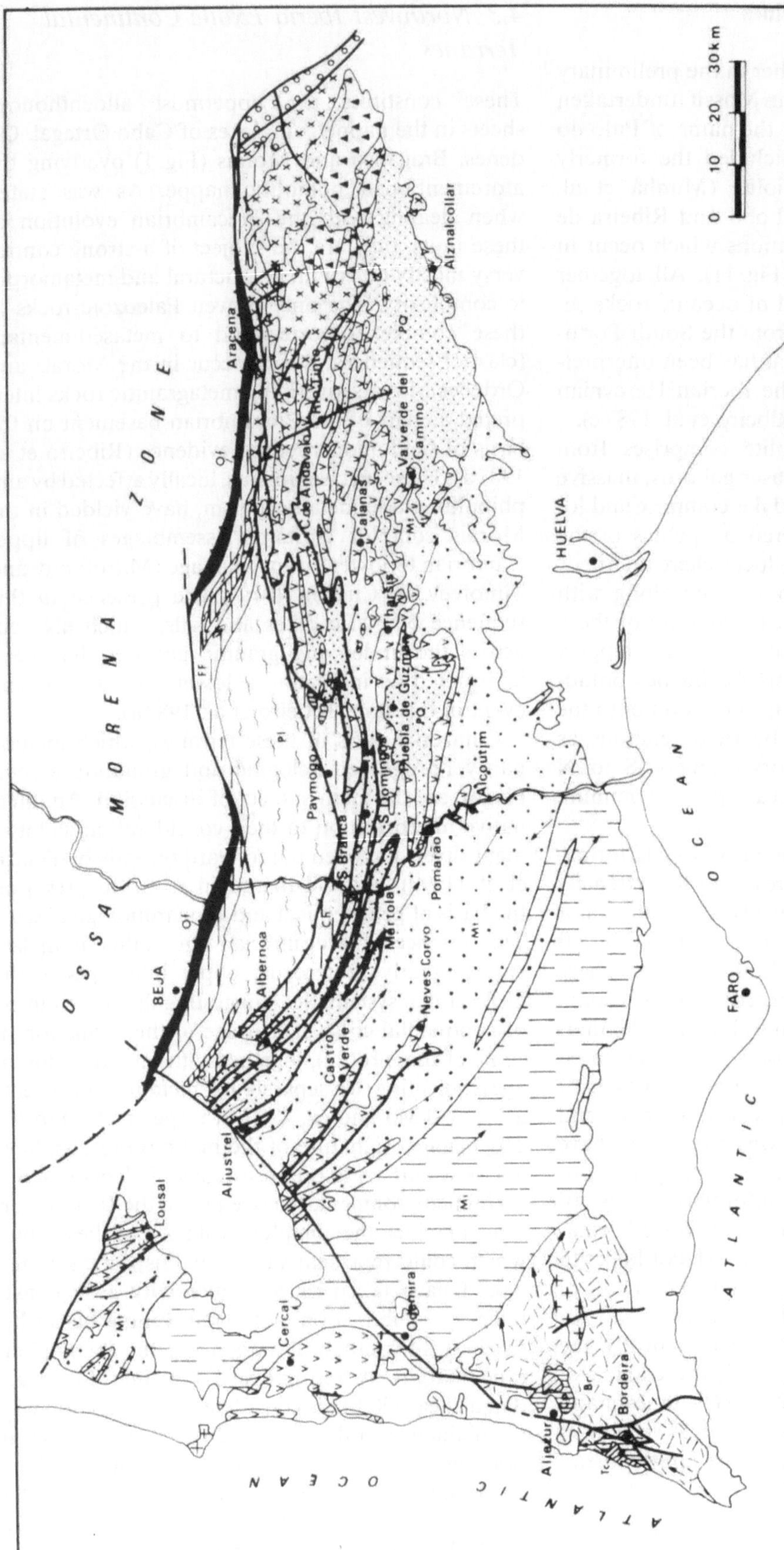
OSSA MORENA ZONE
BEJA
HUELVA
FARO
ATLANTIC OCEAN
Aracena
Alznalcollar
Valverde del Camino
Calañas
Tharsis
Puebla de Guzmán
S. Domingos
Alcoutim
Paymogo
S. Branca
Mértola
Pomarão
Neves Corvo
Castro Verde
Albernoa
Aljustrel
Lousal
Cercal
Odemira
Aljezur
Bordeira
0 10 20 30 km

This latter interpretation is in conflict with some field relationships, in particular the meaning and environment of formation of the granitic augengneisses, deformed prior to the intrusion of mafic dikes and the deposition of the lower Paleozoic metaterrigenous sequences, as well as the meaning of both mafic rocks and sediments, which are, at least in part, coeval to the older episode of HP metamorphism recorded in the catazonal rocks by Peucat et al. (1990).

4.4 South Portuguese Terrane

Opposite the NW suspect exotic terranes dealt with in the previous section, the South Portuguese Terrane, which occupies the southern edge of the Iberian Massif, shows a much simpler picture. The lack of exposed rocks older than the Upper Devonian does not preclude its characterization as a continental unit (see below), which became a suspect exotic terrane after the first recognition of the Pulo do Lobo Terrane as an oceanic unit by Munhá et al. (1986).

4.4.1 Stratigraphy

In the South Portuguese Terrane three main tectonostratigraphic domains can be recognized from north to south (Fig. 11): Pyrite Belt, Baixo Alentejo Flysch Group, and southwest Portugal (Oliveira 1983).

The stratigraphy of the *Pyrite Belt* consists of a detritic basement represented by the Phyllite-Quartzite Formation, and the overlying Volcano Sedimentary Complex (Schermerhorn 1971). The complete lithostratigraphic pile can only be observed in its south branch (Fig. 12b, c, d, e, f, g). The Phyllite-Quartzite Formation is composed of phyllites, siltstones, quartzwackes and quartzites and, in many places, dark shales with interbedded lenses of detritic limestones at the top. The limestones provided cephalopods, trilobites and conodonts of middle to upper Famennian age (Boogaard 1963; Boogaard and Schermerhorn 1980, 1981). The Phyllite-Quartzite Formation is conformably overlain by the VS Complex which forms an assemblage of bimodal volcanics (mostly felsic pyroclastics, basalts and minor andesites) and interfingering dark and red shales, siliceous shales, siltstones, scattered limestones and interbedded lenses of cherts (exhalites), sideritic and phosphatic nodules. The red shales constitute stratigraphic markers of regional significance. Intrusions of diabases are interpreted as feeders of the basic volcanics. In the central area (Puebla de Guzmán-Pomarão Anticline and parts of the Valverde Anticline, Fig. 11) three episodes of acidic volcanism are identified. However, outside this area there are important lateral facies changes which make it impossible to correlate units of the VS Complex from one region to another. The thickness of the VS Complex also changes dramatically from up to 500 m near the volcanic centers, to few meters away from these centers. Faunas of conodonts and few cephalopods

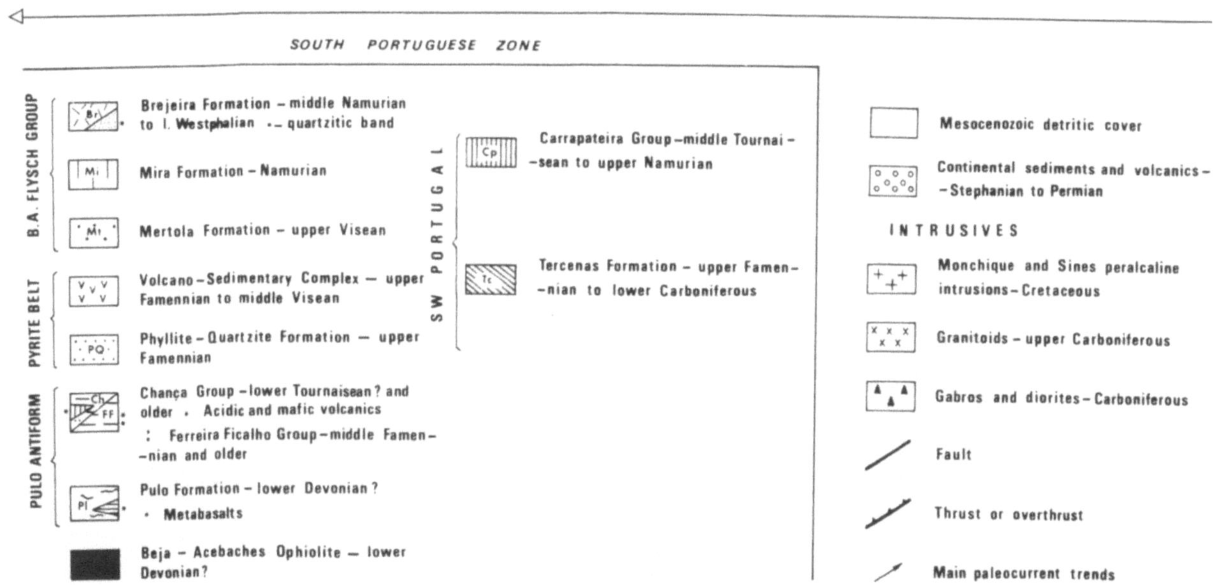

Fig. 11. Geological map of the South Portuguese Zone (including the South Portuguese and Pulo do Lobo Terranes).

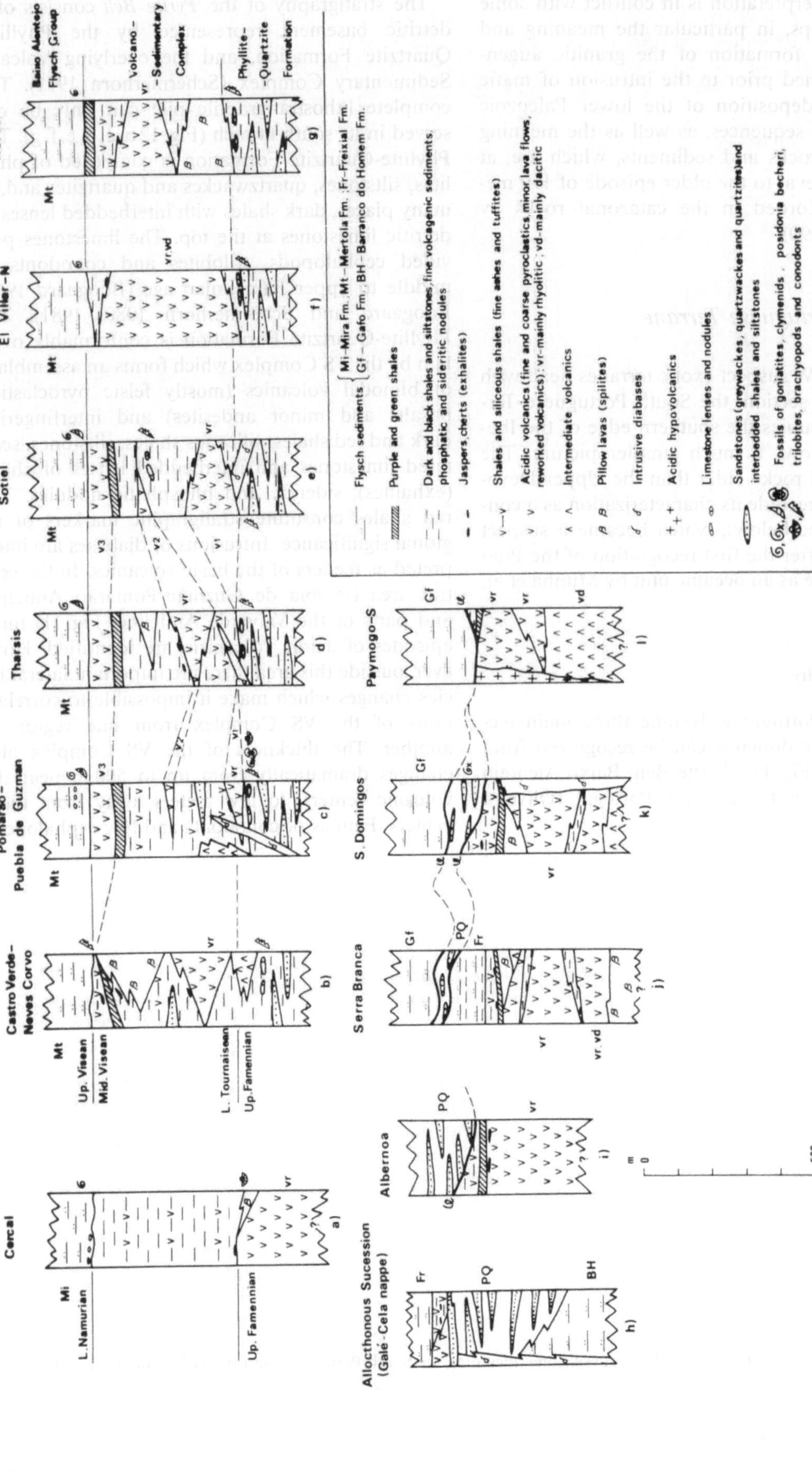

Fig. 12. Selected stratigraphic sections from representative localities within the South Portuguese Terrane

indicate an age range from the upper Famennian to the lower upper Viséan (Boogaard 1963; Oliveira 1983). The age of the Cercal volcanics is doubtful, due to the lack of well-preserved faunas. The VS Complex is conformably overlain by the turbidites of the Mertola Formation (see below).

In the north branch of the Pyrite Belt the VS Complex is also composed of bimodal volcanics and interfingering volcanogenic sediments, but the acidic volcanics, including lavas, are more common. The stratigraphic sequence is truncated by overthrusts which bring varied allochthonous units over the VS Complex (Fig. 12h, i, j, k, l).

The *Baixo Alentejo Flysch Group* comprises mostly turbidites that form a continuous detritic cover more than 5 km thick, which prograded southward from the Upper Viséan to the lower Westphalian. Three basin wide formations are recognized (Oliveira et al. 1979): Mértola Formation, with a sandstone/shale ratio >1, and the sandstones displaying sedimentary features typical of proximal turbidites; Mira Formation composed mostly on thin bedded turbidites and a sandstone/shale ratio <1; Brejeira Formation with two different sedimentological belts (Fig. 11): one to the north, 5 to 10 km wide, composed of immature quartzites and shales; the other, to the south, made up of greywackes and shales. Paleocurrents flowed predominantly to SSE but eastward and northward flow directions also occur in southwestern Portugal.

In *southwest Portugal* the stratigraphic sequence (Fig. 13) begins with the Tercenas Formation, a detrital basal unit of late Famennian age, composed of shallow water sandstones and interbedded shales. This detritic basement is followed by the units of the Carrapateira Group. The lithological succession of this group comprised the shales and siltstones of the Bordalete Formation, which grade upward to dolomitic limestones, marls and dark shales of the Murraçao Formation, and these in turn are overlain by the black shales, dolomitic limestones and phosphatic nodules of the Quebradas Formation. Faunas of goniatites, trilobites and corals indicate a Middle Tournaisian to Late Namurian age for the mud-carbonate shelf sediments of the Carrapateira Group (Oliveira et al. 1985).

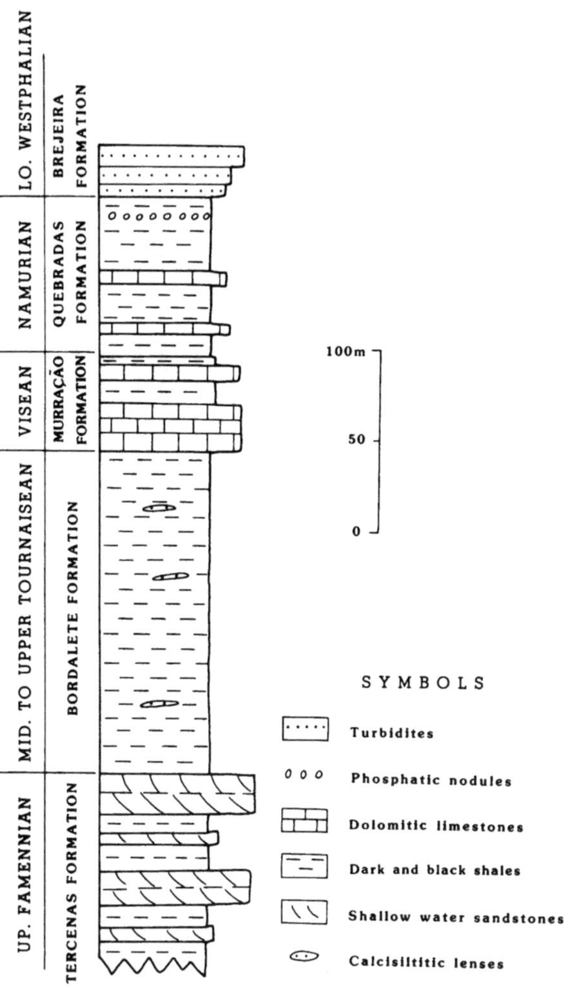

Fig. 13. Representative stratigraphic section of the Southwest Portugal Domain, showing the persistence of platformal facies (passive margin) until the Lower Devonian.

4.4.1.1 Depositional Interpretation

The Pyrite Belt and SW Portugal domains share the same type of detritic basement that formed a large epicontinental sea during the upper Devonian. During the lower Carboniferous this detritic platform was affected by a transtensive tectonic regime (see below), which gave rise to grabens and half grabens that became the locus of the bimodal volcanics and sediments of the Pyrite Belt (Oliveira, 1990). In the upper Viséan the transtensive regime turned to a transpressive one (tectonic inversion) leading to the development of the syn-orogenic flysch sediments of the Baixo Alentejo Flysch Group. The southwestward progression of the orogenic wave and related flysch sediments reached the mud-carbonate shelf of SW Portugal only in the Lower Westphalian.

4.4.2 Structure

As a whole, the arcuate thrust belt of the South Portuguese Terrane shows tectonic deformation of a thin-skinned type. This results from the oblique clo-

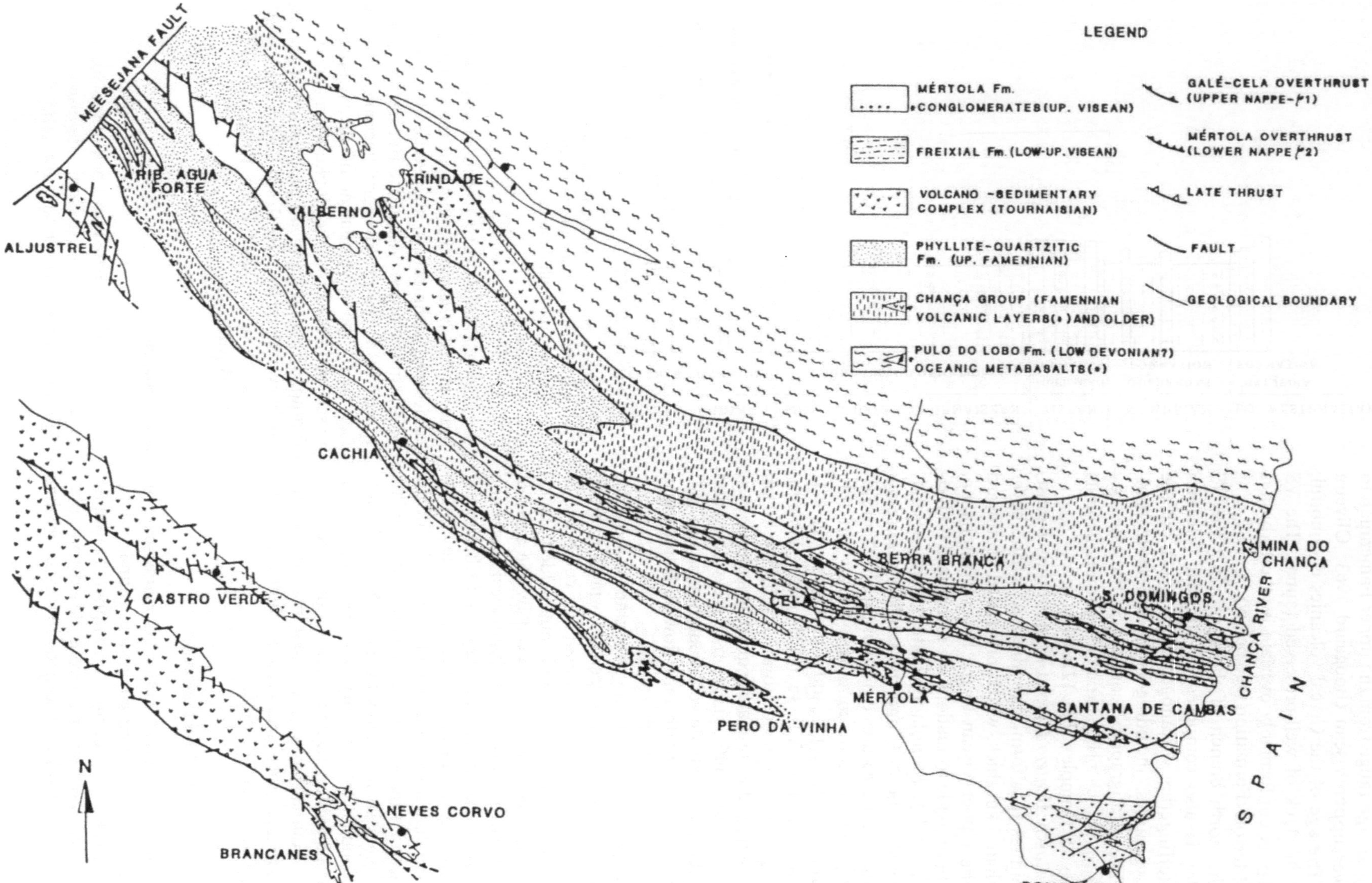

Fig. 14. Structural map of part of the Portuguese Pyrite belt (after Silva et al., 1990)

sure of a pre-Famennian oceanic basin, which separated this terrane from the now adjacent Iberian Terrane, in the interval upper Famennian-upper Tournaisian, followed by transpressive deformation from the upper Viséan to the Lower Westphalian (Silva et al., 1990). From a geotectonic point of view this region has been variably interpreted as: (1) intracontinental orogen (Schermerhorn 1975; Rothier et al. 1980); (2) accretionary prism related to a north dipping subduction located southwestward of the South Portuguese Terrane (Bard 1971; Carvalho 1972; Bard et al. 1973; Vegas and Muñoz 1976; Monteiro and Carvalho 1986); (3) back-arc basin (Munhá 1983; Ribeiro and Silva 1983); and (4) oblique-slip belt (Badham 1982; Andrews 1983).

The same subdivisions considered in the South Portuguese Terrane in the former section, are useful in order to better describe the structure; i. e. Pyrite Belt, Baixo Alentejo Flysch Group and Southwest Portugal. In general terms, metamorphism grades from the prehnite-pumpellyite facies in the Pyrite Belt, through a very low anchymetamorphic regime in Southwest Portugal (Schermerhorn 1975; Munhá 1976, 1983a). A tectonometamorphic and paleogeographic polarity is displayed by this terrane. In fact, thrusting and folding exhibit a southwesterly vergence, and the age of deformation and flysch deposition becomes younger towards the southwest.

1. Pyrite Belt:
Structures in thrust footwall and hanging-wall positions can be identified in this zone (Fig. 14).

Footwall Structures (Parautochthonous). Those situated in the hinterland zones, consist of anticlines like Serra Branca and Albernoa (Fig. 14), underlying the hanging-wall structures.

In foreland zones of the Pyrite Belt, only the footwall structures are exposed, and they consist of anticline structures as Pomarão and Neves Corvo (Fig. 14).

Hanging-Wall Structures (Allochthonous). The Galé-Cela nappe, involving mostly lithologies of the Barranco do Homem and PQ Formations (Fig. 14), overlies the Mértola nappe including mostly lithologies of the VS Complex. These allochthonous units are superposed to the Upper Viséan turbidites of the Mértola Formation.

The D1 phase, coincident with D2 in the adjacent Pulo do Lobo Terrane, in the South Portuguese Terrane is linked to ductile shear zones in the hinterland (Fig. 15C, between the Galé Overthrust and the Cerro do Vau Thrust), and is related to overthrust emplacement in the Pyrite Belt. The generation of this thrust (Galé-Cela nappe) is associated with recumbent folding and S1a cleavage, and overlies a less deformed footwall (Mértola nappe). Induced by this emplacement, and in the uppermost structural level, synsedimentary thrusts developed in the foreland zones (Mértola nappe), during upper Viséan flysch deposition, and associated to a wild flysch conglomerate. With a maximum length of 20 km before folding (Ribeiro and Silva 1983), the overthrust emplacement complies to a forward sequence.

During the D2 tectonic phase, which produced crenulation cleavage in the Pulo do Lobo Terrane, the thrust surfaces in the South Portuguese Terrane were refolded with the development of regional prevasive S1b cleavage. Cogenetic with the folding phases D1 and D2, and related to the generation of the Iberian Variscan Arc, a sinistral wrench component was added to the main thrust component, producing transection of folds in a non coaxial deformation regime (Silva 1981, 1985).

2. Baixo Alentejo Flysch Group:
The three main lithostratigraphic units: Mértola, Mira and Brejeira formations (Oliveira 1983), were deformed by D2 regional folding, vergent to the southwest. Cleavage and thrusting become more gentle and less intense southward (Ribeiro 1981). When S1b cleavage is more flat, postmetamorphic crenulation cleavage is developed.

3. Southwest Portugal:
D1 is the most important episode of deformation in this area (Ribeiro 1983; Ribeiro and Silva 1983). The associated NW-SE-trending cleavage changes from less intense and steeply dipping (northeastward) in the higher structural levels, to more intense and flat near the Carrapateira thrust plane. This, as commonly happens in thin-skinned belts, was a ductile shear in the north (lower structural level) and a more brittle fault in the south (high structural level). It dies out in the core of a recumbent fold, beyond which chevron folds with subhorizontal axial planes were also developed. The D1 structures were also refolded in this zone by two less penetrative deformational episodes (Ribeiro 1983).

4.4.3 Magmatism

Apart from scarce postcollisional subvolcanic and shallow plutonic rocks, concentrated in the eastern part of the South Portuguese Terrane (Simancas

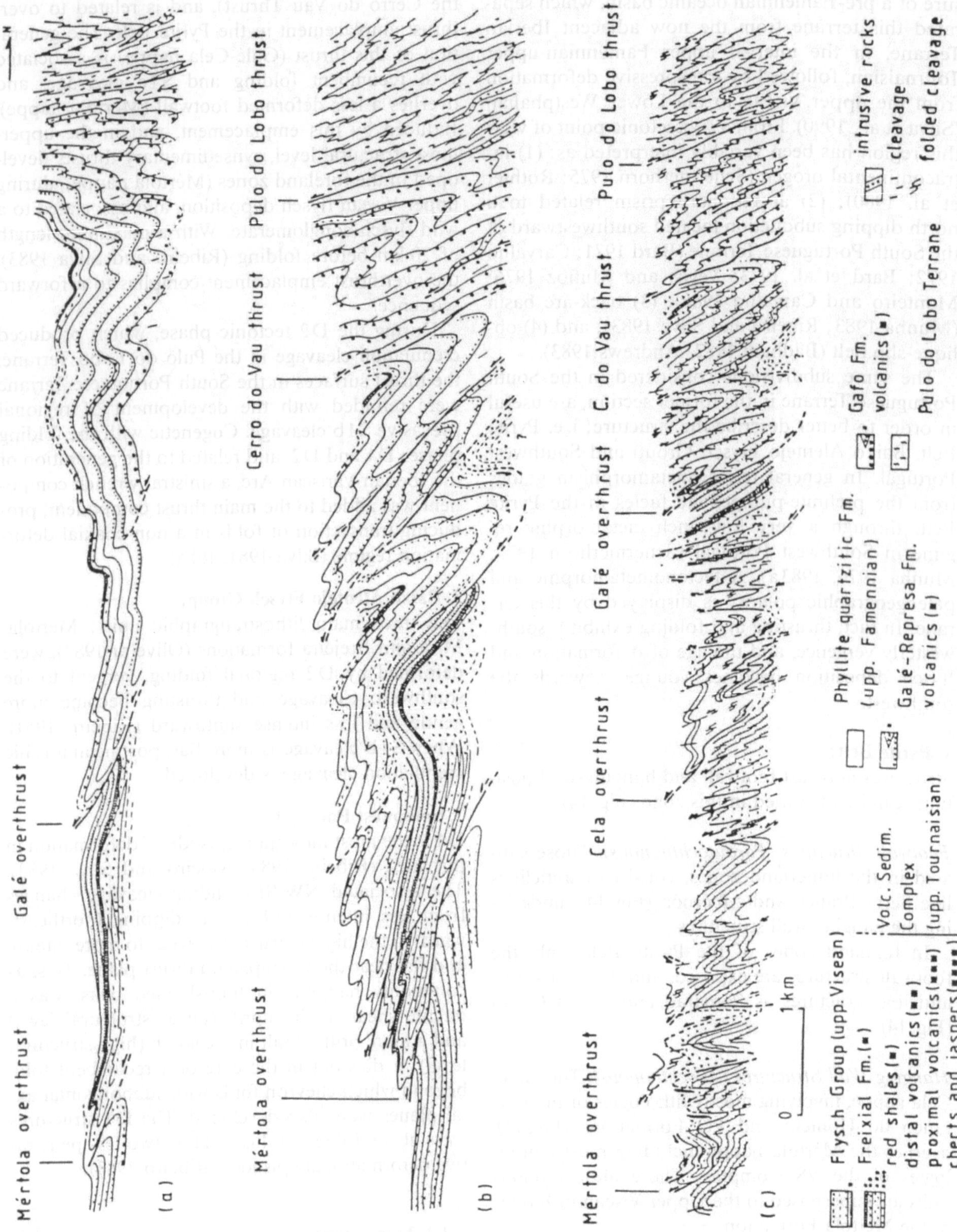

Fig. 15. Schematic representation of the main units south from the Pulo do Lobo Terrane, including overstep sequences of Chança Group and Pyrite Belt (After Silva, in press). **a** After D1 deformation episode, and beginning of D2 phase; **b** Overthrust emplacement during D2 phase, and S1a cleavage generation; **c** Refolding of overthrusts and generation of S1b regional cleavage

1984), igneous activity in this unit is mainly represented by the lower Carboniferous bimodal volcanic suites which constitute the so-called VS Group (see above). This consists of lavas, pyroclastics and epiclastics with basaltic, andesitic (minor) and predominantly rhyolitic compositions. Petrographic and geochemical data indicates that these rocks types are not linked by fractional crystallization (Munhá 1983b). Despite their close association both in space and time, the felsic and mafic (tholeiitic to alkalic basalts; Munhá 1983b) volcanics originated and evolved separately: no lithological transitions occur and the volcanic centers are distinct. The source for the mafic magmas must be sought in the upper mantle, whereas the felsic volcanics derive from magma chambers developed by melting in the crust, possibly by heat supplied by underplating of the mafic magmas (Munhá 1983b).

Combining the geochemical and geological data of the VS group volcanics, suggesting an extensional regime, and the fact that the age of volcanism appears to decrease southwestward, Silva et al. (1990) have suggested the migration of pull-apart basins in a transtensional regime, across the previously continuous South Portuguese Terrane shelf, as the most likely environment for the formation of the VS group.

4.4.4 Tectonothermal Evolution

The South Portuguese Terrane shows a complete record of its evolution from a passive margin stage, characterized by the platformal sequences of the PQ group which persisted until the lower Westphalian in the southwest Portugal area, through a period characterized by the sequential southwestward collapse of that platform and formation of the VS group under neat transtensional conditions, to the final transpressive thin-skinned deformation and related flysch deposition which formed an orogenic wave migrating to the southwest. This evolution has been graphically schematized in Fig. 16 (Silva et al., 1990) and records the sequential approximation of the South Portuguese continental Terrane to the Iberian Terrane as a consequence of the oblique closure of the intervening ocean (source of the oceanic Pulo do Lobo and Beja-Acebuches exotic terranes). The oblique nature of the convergent regime is supported by non-coaxial nature of the deformation as well as by the coeval existence of bimodal volcanism in the Pyrite Belt with subduction-related volcanism in the Ossa-Morena Zone of the Iberian Terrane (namely, the Upper Tournaisian Sao Cristovão calc-alkaline volcanics, see Fig. 16 B). The latter, together with the northward obduction-related fabrics in the Beja-Acebuches ophiolite referred to in previous sections, suggest that subduction (and obduction) proceeded from south to north; i. e. below (and above) the Iberian Terrane, defining a flake geometry as suggested by Iglesias et al. (1983), Ribeiro et al. (1987b) and Ribeiro et al. (1990). Post-collisional reactivation of the former subduction zone, acting then as a kind of type A subduction (Fig. 16C), resulted in a major detachment beneath the South Portuguese Terrane (8-10 km depth) and the thin-skinned type deformation (Ribeiro and Silva 1983).

4.5 *Accretion Process: The Hercynian Orogeny*

Although many relevant features of the convergent process that led to the accretion of the suspect exotic terranes to the reference Iberian Terrane have been previously advanced, it is worthwhile emphasizing those aspects that are crucial for the understanding of the accretionary process, which is currently known by the name of Hercynian Orogeny. In general terms, this involved subduction (and eventual obduction) of oceanic crust and subsequent continent-continent collision under a neatly oblique regime, in the interval Lower-Middle Devonian through Lower Permian.

Despite the strong asymmetry existing between the northern and southern imbricated parts of the Iberian Terrane, concerning vergence, structural style and amount of shortening, both areas share other common features such as: (1) propagation of the deformation and related syn-orogenic wedges towards the core of the terrane; (2) rough synchroneity of the first deformation in the internal zones and the first syn-orogenic deposits; (3) apparent synchroneity of ophiolite obduction. This coincidence, together with the consistently non-coaxial strain regime, suggest a single collisional event between a promontory (the NW edge of the Iberian Terrane) and a northwestern continental block (Brun and Burg 1982), now represented by the continental exotic terranes, along which the Cantabrian arc could well have been developed (Matte and Ribeiro 1975; Ries and Shackleton 1976). Sinistral strike-slip along the southwestern margin of the arc could account for the sequential closure of the oceanic branch south of the promontory, coeval to large-scale overthrusting in the latter, as well as for

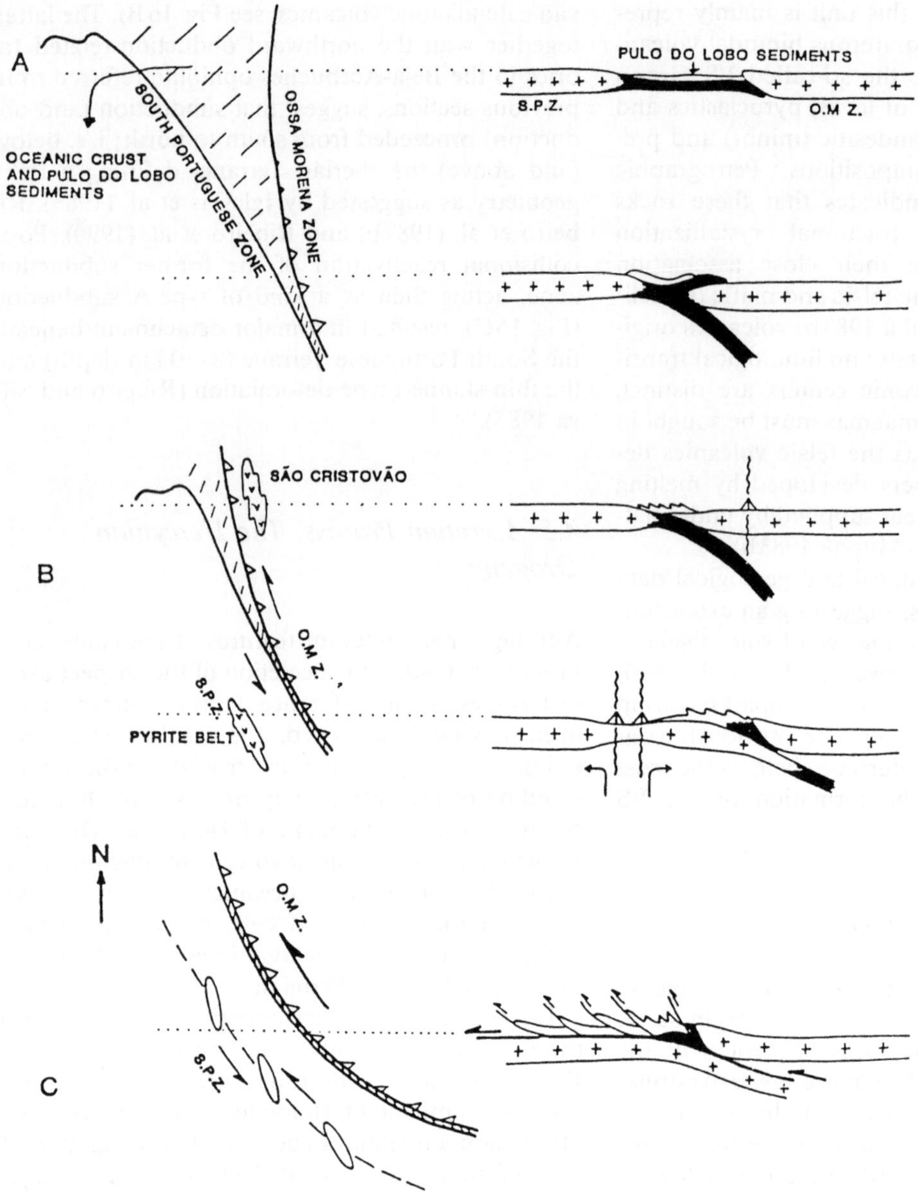

Fig. 16. Geodynamic evolution of the South Portuguese Zone: (After Silva et al., 1990). **A** During the upper Devonian (pre-Famennian): oblique collision and partial obduction; **B** during the upper Famennian to lower Carboniferous: subduction and collision (São Cristovão) and beginning of the intracontinental distension in the foreland (Pyrite Belt); **C** Middle Carboniferous (Upper Viséan to Middle Westphalian): southwestward transpressive deformation induced by subduction reactivation (type A subduction) and generation of the arcuate belt

the reactivation as left-lateral wrench faults of old features within the Iberian Terrane: namely, the presumed Cadomian cryptic suture (Badajoz-Córdoba Shear Zone).

The significant crustal thickening shown by the northern imbricated margin of the Iberian Terrane and overlying exotic terranes led, by decompression associated to the collapse of the nappe pile and erosion, to widespread partial melting of infracrustal levels. This gave rise to the large volume of rather monotonous peraluminous granite plutons which characterize the central and northwestern parts of the Iberian Massif.

The formation of intermontane basins (partly pull-aparts) in Upper Carboniferous to Lower Permian times shows the latest, waning effects of the Hercynian orogeny in this part of the Circum-Atlantic Paleozoic orogenic realm (Gabaldón and Quesada 1986).

Acknowledgements

The ideas expressed in this contribution have been largely improved through many constructive discussions with J. Munhá, O. Apalategui and L. A. Cueto. The professional skills of J. Vallejo (artwork) are also greatly appreciated.

References

Alia M (1963) Rasgos estructurales de la Baja Extremadura. Bol R Soc Esp Hist Nat (Secc Geol) 70: 247-262

Andrews JR (1983) The Iberian Pyrite Belt, an oblique-slip mobile zone. Notes Mem Serv Géol Maroc Rabat 335, 1986

Apalategui O, Quesada C (1987) Transversal Geológica Zona Ossa-Morena. Excursion Field Guide, pp 1-99

Apalategui O, Higueras P, Cueto LA, Quesada C (1983) Mapa Geológico de España, E 1:50000, 2ª serie (Magna). Sheet No 855: Usagre. IGME

Badham JPN (1982) Strike-slip orogens - an explanation for the Hercynides. J Geol Soc Lond 139: 493-504

Bard JP (1971) Sur l'alternance des zones métamorphiques dans le segment hercynien sud-ibérique; comparaison de la variabilité des caractères géotectoniques de ces zones avec les orogènes 'orthotectoniques'. Bol Geol Min 82 (3-4): 324-345

Bard JP, Capdevilla R, Matte P, Ribeiro A (1973) Geotectonic model for the Iberian Variscan Orogen. Nature (Lond) 241: 50-52

Boogaard MV den (1963) Conodonts of upper Devonian and lower Carboniferous age from Southern Portugal. Geol Mijnbouw 42 (8): 248-259

Boogaard MV den, Schermerhorn LJG (1980) Conodont faunas from Portugal and southwestern Spain. Part 4 - A Famennian conodont fauna near Nerva (Rio Tinto). Scr Geol 56: 1-4

Boogaard MV den, Schermerhorn LJG (1981) Conodont faunas from Portugal and southwestern Spain. Part 6 - A lower Famennian conodont fauna at Monte Forno da Cal (South Portugal). Scr Geol 63: 1-16

Brenchley PJ, Romano M, Gutierrez Marco JC (1986) Proximal and distal hummocky cross-stratified facies on a wide Ordovician shelf in Iberia. Can Ass Petrol Geol Mem 2: 241-255

Brun JP, Burg JP (1982) Combined thrusting and wrenching in the Ibero-American Arc: a corner effect during continental collision. Earth Planet Sci Lett 61: 319-332

Burg JP, Iglesias M, Laurent P, Matte P, Ribeiro A (1981) Variscan intracontinental deformation: The Coimbra-Córdoba shear zone (SW Iberian Peninsula). Tectonophysics 78: 161-177

Carvalho D (1972) The metallogenetic consequences of plate tectonics and the upper Paleozoic evolution of southern Portugal. Estud Notas Trabalhos Serv Fomento Mineiro (Porto) 20 (3/4): 297-320

Castro P, Quesada C, Munhá J (1987) Metamorphic regime in the Beja-Acebuches Variscan Ophiolite (SW Iberia). Extended abstract in intern conference: Tectonothermal evolution of the West African orogens and Circum-Atlantic terrane linkages. IGCP Project 233, Nouakchott, Mauritania (abstr), pp 57-60

Chacón J (1979) Estudio geológico del Sector Central del Anticlinorio Portalegre - Badajoz - Córdoba. PhD Thesis. Univ Granada, pp 1-728

Chacón J, Oliveira V, Ribeiro A, Oliveira JT (1983) La estructura de la zona de Ossa-Morena. In: Geologia de España. Libro Jubilar, IGME 1: 490-503

Chacón J, Fernandez J, Mitrofanov F, Timofeev BV (1984) Primeras dataciones microfitopaleontológicas en el sector de Valverde de Burguillos-Jerez de los Caballeros (Anticlinorio de Olivenza-Monasterio). Cuad Lab Xeol Laxe 8: 211-220

Coney P, Jones DL, Monger JWH (1980) Cordilleran suspect terranes. Nature (Lond) 288: 329-333

Cueto LA, Eguiluz L, Llamas JA, Quesada C (1983) La granodiorita de Pallares, un intrusivo precámbrico en la alineación Olivenza-Monesterio (Zona de Ossa-Morena). Comm Serv Geol Portugal 69 (2): 219-226

Dallmeyer RD, Gil-Ibarguchi I (1990) Age of amphibolitic metamorphism in the ophiolitic unit of the Morais Allochthon (Portugal): implications for early Hercynian orogenesis in the Iberian Massif. J Geol Soc Lond 147: 873-878

Delgado M (1971) Esquema geológico de la hoja número 878 de Azuaga (Badajoz). Bol Geol Min 82: 277-286

Diez MA, Vegas R, Gonzales-Lodeiro F (1990) Structure of the autochthon of the Central Iberian Zone. In: Dallmeyer RD, Martínez E (eds) Pre-Mesozoic geology of Iberia. Springer, Berlin Heidelberg New York Tokyo

Eguiluz L (1987) Petrogénesis de rocas ígneas y metamórficas en el antiforme Burguillos-Monesterio. Macizo Ibérico Meridional. PhD Thesis Univ Pais Vasco, pp 1-694

Farias P, Gallastegui G, Gonzales Lodeiro F et al. (1987) Aportaciones al conocimiento de la litoestratigrafía y estructura de Galicia Central. Mémorias Fac Cienc, Univ Porto 1: 401-430

Fernández J, Coullaut JL, Eguiluz L, Garrote A (1983) Mapa Geológico de España, 1:50000. IGME, Monesterio, Ser 2a No 897

Gabaldon V, Quesada C (1986) Exemples de bassins houillers limniques du sud-ouest de la Péninsule Ibérique: évolution sédimentaire et contrôle structural. Mem Soc Géol Fr NS 149: 29-38

Gabaldon V, Garrote A, Quesada C (1985) El Carbonífero inferior del norte de la Zona de Ossa-Morena (SW de España). CR X Int Carboniferous Congr 3: 173-186

Galindo C, Casquet C, Portugal Ferreira M, Macedo CAR (1988) Geocronología del Complejo Plutónico Táliga-Barcarrota (CPTB). (Badajoz-España). In: Bea F, Carnicero A, Gonzalo JC, López-Plaza M, Rodríguez Alonso MD (eds) Geología de los granitoides y rocas asociadas del Macizo Hespérico. Ed Rueda, Madrid, pp 385-392

García Casquero JL (in press) Continental Troudhjemites in displaced terranes of the Oporto-Badajoz-Córdoba Belt (Hesperian Massif). Geol Rundsch

García Casquero JL, Boelrijk NAIM, Chacón J, Priem HNA (1985) Rb-Sr evidence for the presence of Ordovician granites in the deformed basement of the Badajoz-Córdoba Belt, SW Spain. Geol Rundsch 74 (2): 379-384

Gil-Ibarguchi JI, Mendia M, Girardeau J, Peucat JJ (in press) Eclogites and granulites in the Cabo Ortegal Complex (northwestern Spain). Lithos

Iglesias M, Ribeiro ML, Ribeiro A (1983) La interpretación

aloctonista de la estructura del noroeste peninsular. In: Geología de España. Libro Jubilar, IGME 1: 459-466
Julivert M, Fontbote JM, Ribeiro A, Conde L (1974) Mapa Téctónico de la Península Ibérica y Baleares. Escala 1: 1000000. IGME
Kuijper RP (1979) U-Pb systematics and the petrogenetic evolution of infracrustal rocks in the Paleozoic basement of western Galicia, NW Spain. Vehr ZWO Lab Isotopen Geol, Amst 5: 1-101
Kuijper RP, Priem HNA, den Tex E (1982) Late Archean-Early Proterozoic source areas of zircons in rocks from the Paleozoic orogen in western Galicia. Precambrian Res 19 (1): 1-29
Lancelot JR, Allegret A (1982) Radiochronologie U/Pb de l'ortogneiss alcalin de Pedroso (Alto Alentejo, Portugal) et évolution anté-hercynienne de l'Europe occidentale. Neues Jahrb Min Abh Mh 9: 385-394
Lancelot JR, Allegret A, Iglesias M (1985) Outline of Upper Precambrian and Lower Paleozoic evolution of the Iberian Peninsula according to U-Pb dating of zircons. Earth Planet Sci Lett 74: 325-337
Liñán E (1978) Bioestratigrafía de la Sierra de Córdoba. PhD Thesis Univ Granada 191: 1-212
Liñán E, Schmitt M (1980) Microfósiles de las Calizas precámbricas de Córdoba (España). Temas Geol Min 4: 171-196
Lotze F (1945) Zur Gliederung der Varisciden der iberischen Meseta. Geotech Forsch 6: 78-92
Martínez Catalán JR (1985) Estratigrafía y estructura del Domo de Lugo (Sector Oeste de la Zona Asturoccidental-Leonesa). Corpus Geol Gallaeciae, 2nd Ser 2: 1-29
Mata J, Munhá J (1986) Geodynamic significance of high-grade metamorphic rocks from Degolados-Campo Maior (Tomar-Badajoz-Córdoba Shear Zone). Maleo 2 (13): 28
Matte P (1983) Two Geotraverses across the Ibero-Armorican Variscan Arc of Western Europe. In: Rast N, Delany M (eds) Profiles of orogenic belts. AGU Geodyn Ser 10: 53-81
Matte P (1986) Tectonics and plate tectonics models for the Variscan Belt of Europe. Tectonophysics 126: 329-374
Matte P, Ribeiro A (1975) Forme et orientation de l'ellipsoïde de déformation dans la virgation hercynienne de Galice. Rélations avec le plissement et hypothèses sur la genèse de l'arc ibéro-armoricain. CR Acad Sci Paris 280: 2825-2828
Monteiro JH, Carvhalho D (1986) Sea floor volcanism and polymetallic sulfide deposits in ancient active margins: the case of the Iberian Pyrite Belt. In: Teleki DG, Robson MR, Moore JR, von Stockelberg V (eds) Marine minerals advances in research and resource assessment. Reidel, Dordrecht
Moreno F (1977) Estudio geológico de los Montes de Toledo Occidentales. PhD Thesis Univ Complut Madrid, pp 1-526
Munhá JM (1976) Nota preliminar sobre o metamorfismo na Faixa Piritosa portuguesa. Comun Serv Geol Portugal 60: 151-161
Munhá JM (1979) Blue amphiboles, metamorphic regime and plate tectonic modelling of the Iberian Pyrite Belt. Contrib Miner Petrol 69: 279-289
Munhá JM (1983a) Low-grade regional metamorphism in the Iberian Pyrite Belt. Comun Serv Geol Portugal 69 (1): 27-41
Munhá JM (1983b) Hercynian magmatism in the Iberian Pyrite Belt. In: Sousa MJL, Oliveira JT (eds) The Carboniferous of Portugal. Mem Serv Geol Portugal 29: 39-81
Munhá JM, Oliveira JT, Ribeiro A, Oliveira V, Quesada C, Kerrich R (1986) Beja-Acebuches Ophiolite: characterization and geodynamic significance. Maleo 2 (13): 31
Munhá JM, Oliveira JT, Ribeiro A, Quesada C, Fonseca P, Castro P (1989) Accreted terranes in southern Iberia: The Beja-Acebuches ophiolite and related oceanic sequences. 28th Int Geol Congr Washington (abstr) 2: 481
Nägler TF, Gebauer D, Schäfer HJ, von Quadt A (1988) Sm-Nd and Pb-isotope geochemistry of Paleozoic sediments of the Almadén and Guadalmez synclines (Central Iberian Zone, Spain). In: Cinturones Orogénicos, II Congr Geol España Symp Vol, pp 45-50
Nägler TF, Gebauer D, Schäfer HJ (1989) Nd-, Sr- and Pb-isotope geochemistry of two pre-Permian sedimentary profiles from the Spanish Meseta: evidence of both crustal recycling and crustal growth. Terra Abstr 1 (1): 336
Oliveira JT (1983) The marine Carboniferous of South Portugal: a stratigraphic and sedimentological approach. In: Sousa MJL, Oliveira JT (eds) The Carboniferous of Portugal. Mem Serv Geol Portugal 29: 3-37
Oliveira JT (1990) Stratigraphy and syn-sedimentary tectonism in the South Portuguese Zone. In: Dallmeyer RD, Martínez E (eds) Pre-Mesozoic geology of Iberia. Springer, Berlin Heidelberg New York Tokyo
Oliveira JT, Horn M, Paproth E (1979) Preliminary note on the stratigraphy of the Baixo Alentejo Flysch Group, Carboniferous of Portugal, and on the palaeogeographic development compared to corresponding units in northwest Germany. Comun Serv Geol Portugal 65: 151-168
Oliveira JT, Horn M, Kullman J, Paproth E (1985) Stratigraphy of the upper Devonian and Carboniferous sediments of southwestern Portugal. X Congr Int Stratigr Géol Carbonif CR Madrid 1983 (1): 107-120
Oliveira JT, Cunha T, Streel M, Vanguestaine M (1987) Dating the Horta da Torre Formation, a new lithostratigraphic unit of the Ferreira-Ficalho Group, South Portuguese Zone: Geological consequences. Comun Serv Geol Portugal 72: 26-34
Palacios T (1986) Acritarcos Proterozoico Superior - Vendiense en el Complejo Esquisto-Granváquico de Extremadura. PhD Thesis Univ Extremadura, pp 1-530
Paris F, Robardet M (1977) Paléogéographie et rélations ibéroarmoricaines au Paléozoïque anté-carbonifère. Bull Soc Géol Fr Ser 7, 19 (5): 1121-1126
Pascual E (1981) Investigaciones geológicas en el sector Córdoba-Villaviciosa de Córdoba (Sector Central de Sierra Morena). PhD Thesis Univ Granada, pp 1-521
Pérez Lorente F (1979) Geología de la zona Ossa-Morena al Norte de Córdoba (Pozoblanco-Belmez-Villaviciosa de Córdoba). PhD Thesis Univ Granada 281: 1-340
Peucat JJ, Bernard-Griffiths J, Gil-Ibarguchi I et al. (1990) Geochemical and geochronological cross section of the deep Variscan crust: the Cabo Ortegal high-pressure nappe (NW Spain). Tectonophysics 177: 263-292
Priem HNA, Boelrijk NAIM, Vershure RH, Hebeda EH, Verdurmen EAT (1970) Dating events of acid plutonism through the Paleozoic of the western Iberian Peninsula. Eclogae Geol Helv 63 (1): 255-274
Quesada C (1975) Geología de un sector de la parte central del anticlinorio Olivenza-Monesterio. Alrededores de Monesterio (Badajoz). Thesis Licenc Univ Granada, pp 1-128
Quesada C (1989) Late-Proterozoic Terranes within the basement of the Paleozoic Iberian Terrane. Extended Abstr Int Conf Tectonothermal Expression of Terrane Accretion within the Appalachian Orogen, IGCP Proj 233, Athens, Georgia, USA (abstr), pp 83-86

Quesada C (1990a) Precambrian succession in the Iberian Hercynian Foldbelt. In: Strachan RA, Taylor GK (eds) Avalonian and Cadomian Geology of the North Atlantic. Blackie, Glasgow, 109-133

Quesada C (1990b) Precambrian successions in SW Iberia: their relationship to Cadomian orogenic events. In: The Cadomian Drogeny, D'Lemos RS, Strachan RA, Topley CG (eds), Geol. Soc. Spec Publ 51: 353-362

Quesada C, Ribeiro A (1989) Terrane map of the Iberian Peninsula. In: Keppie JD, Dallmeyer RD (eds) Tectonic map of Pre-Mesozoic terranes in Circum-Atlantic Phanerozoic orogens. IGCP Proj 233. Terranes in Circum-Atlantic Paleozoic Orogens

Quesada C, Apalategui O, Eguiluz L, Liñan E, Palacios T (1990) Precambrian stratigraphy of the Ossa-Morena Zone. In: Dallmeyer RD, Martínez E (eds) Pre-Mesozoic geology of Iberia. Springer, Berlin Heidelberg New York Tokyo 252-266

Quesada C, Larrea FJ, Florido P et al. (1987) Mapa Geológico Minero de Extremadura. Dir Gen Ind Energ Min Junta Extremadura, pp 1-131

Racheboeuf PR, Robardet M (1986) Le Pridoli et le Dévonien inférieur de la zone d'Ossa-Morena (sud-ouest de la Péninsule Ibérique). Etude des Brachiopodes. Geol Paleontol 20: 11-37

Ribeiro A (1974) Contribution à l'étude tectonique de Tràsos-Montes Occidental. Mem Serv Geol Portugal, pp 1-168

Ribeiro A (1981) A geotraverse through the Variscan Fold Belt in Portugal. In: Zwart HJ, Dornsiepen UF (eds) The Variscan orogen in Europe. Geol Mijnbouw 60 (1): 41-44

Ribeiro A (1983) Structure of the Carrapateira Nappe in the Bordeira area, SW Portugal. In: Sousa MJL, Oliveira JT (eds) The Carboniferous of Portugal. Mem Serv Geol Portugal 29: 91-97

Ribeiro A, Silva JB (1983) Structure of the South Portuguese Zone. In: Sousa MJL, Oliveira JT (eds) The Carboniferous of Portugal. Mem Serv Geol Portugal 29: 83-89

Ribeiro A, Dias R, Pereira E et al. (1987a) Guide-book for the Miranda do Douro-Porto Excursion. Conf Deformation Plate Tectonics, Gijón (Spain), pp 1-50

Ribeiro A, Pereira E, Iglesias M (1987b) Flake tectonics in the NW Iberian Variscides. Conf Deformation Plate Tectonics, Gijón (Spain) (abstr), 59 pp

Ribeiro A, Quesada C, Dallmeyer RD (1987c) Tectonostratigraphic terranes and the geodynamic evolution of the Iberian Variscan Fold Belt. Conf Deformation Plate Tectonics, Gijón (Spain) (abstr), pp 60-61

Ribeiro A, Quesada C, Dallmeyer RD (1990) Geodynamic evolution of the Iberian Massif. In: Dallmeyer RD, Martínez E (eds) Pre-Mesozoic geology of Iberia. Springer, Berlin Heidelberg New York Tokyo 399-410

Ribeiro ML (1986) Geologia e petrologia da região a SW de Macedo de Cavaleiros (Tràs-os-Montes oriental). PhD Thesis Univ Lisbon, pp 1-202

Ribeiro ML (1987) Petrogenesis of early Paleozoic peralkaline rhyolites from the Macedo de Cavaleiros region (NE Portugal). Geol Rundsch 76 (1): 147-168

Ries AC, Shackleton RM (1976) Pattern of strain variation in arcuate fold belts. Phil Trans R Soc Lond A283: 281-288

Robardet M (1976) L'originalité du segment hercynien sud-ibérique au Paléozoïque inférieur: Ordovicien, Silurien et Dévonien dans le Nord de la province de Seville (Espagne). CR Acad Sci Paris D 283: 999-1002

Robardet M, Doré F (1988) The late Ordovician diamictic formations from southwestern Europe: north-Gondwana glaciomarine deposits. Palaeogeogr Palaeoclimatol Palaeoecol 66: 19-31

Robardet M, Gutiérrez-Marco JC (1990) Sedimentary and faunal domains in the Iberian Peninsula during the Lower Paleozoic. In: Dallmeyer RD, Martínez E (eds) Pre-Mesozoic geology of Iberia. Springer, Berlin Heidelberg New York Tokyo

Robardet M, Weyant M, Laveine JP, Racheboeuf P (1986) Le Carbonifère inférieur du synclinal du Cerrón del Hornillo (Province de Seville, Espagne). Rev Paléobiol 5 (1): 71-90

Routhier P, Aye F, Boyer C (1980) La ceinture sud-ibérique à amas sulfurés dans sa partie espagnole. Tableau géologique et métallogénique. Synthèse sur le type amas sulfurés volcano-sédimentaires. Mém Bur Rech Géol Minières 94 (26 Congr Int Géol Paris, 1980)

Sánchez Carretero R, Carracedo M, Eguiluz L, Garrote A, Apalategui O (1989) El magmatismo calcoalcalino del Precámbrico terminal en la Zona de Ossa-Morena (Macizo Ibérico). Rev Soc Geol España 2 (1-2): 7-21

Schäfer HJ, Gebauer D, Nägler TF, von Quadt A (1988) U-Pb zircon and Sm-Nd studies of various rock types of the Ossa-Morena Zone (southwest Spain). In: Cinturones Orogénicos, II Congr Geol España, Granada, pp 51-57

Schäfer HJ, Gebauer D, Nägler TF (1989) Pan-African and Caledonian ages in the Ossa-Morena Zone (southwest Spain): a U-Pb and Sm-Nd study. Terra Abstr 1 (1): 350-351

Schermerhorn LJG (1971) An outline stratigraphy of the Iberian Pyrite Belt. Bol Geol Min 82 (3/4): 239-268

Schermerhorn LJG (1975) Spilites, regional metamorphism and subduction in the Iberian Pyrite Belt: some comments. Geol Mijnbouw (Den Haag) 54 (1): 23-35

Silva JB (1981) Estudio estructural dum sector da Faixa Piritosa (Região de Mértola). Comun ao 2 Encontro de Geociências, Coimbra, 1981, pp 18-19

Silva JB (1985) Structure of the Pyrite Belt: the Mértola Traverse. Papers on the Carboniferous of the Iberian Pyrite Belt. An Fac Ciênc Porto (Suppl) 64 (1983): 41-47

Silva JB, Oliveira JT, Ribeiro A (1990) Structural outline of the South Portuguese Zone. In: Dallmeyer RD, Martínez E (eds) Pre-Mesozoic geology of Iberia. Springer, Berlin Heidelberg New York Tokyo

Simancas JF (1984) Geología de la extremidad oriental de la Zona Sud-Portuguesa. PhD Thesis Univ Granada, pp 1-447

Vegas R, Muñoz M (1976) El Contacto entre las zonas Surportuguesa y Ossa-Morena en el SW de España. Una nueva interpretation. Comun Serv Geol Portugal 60: 31-51

Central-Western Europe: Major Stages of Development During Precambrian and Paleozoic Times

C. PIN[1]

1 Introduction

The pre-Mesozoic basement of central-western (CW) Europe is exposed in several scattered massifs (Fig. 1). This is because of covering by post-hercynian sedimentary basins, superimposition of the Alpine fold belt, and disruption by the opening of the Bay of Biscay and the Mediterranean system.

In spite of numerous detailed studies, conducted for more than a century, the knowledge of this wide domain remains quite incomplete, and no clear, comprehensive picture is available as yet. Abundant geochronological work has, however, been performed during the last 20 years. Though sometimes equivocal, these isotopic data highlight the major stages of the evolution of the lithosphere in this segment. Combined with this geochronological framework, the results of tectonic, geophysical, and geochemical studies enable geodynamic inferences to be drawn. Therefore, this chapter puts special emphasis on the geochronological data, and the recognition of possible markers of plate tectonic processes, to outline a tentative evolutionary scheme.

[1] Département de Géologie et U. A. 10 C. N. R. S., Université Blaise Pascal, 5 Rue Kessler, 63038 Clermont-Ferrand Cedex, France

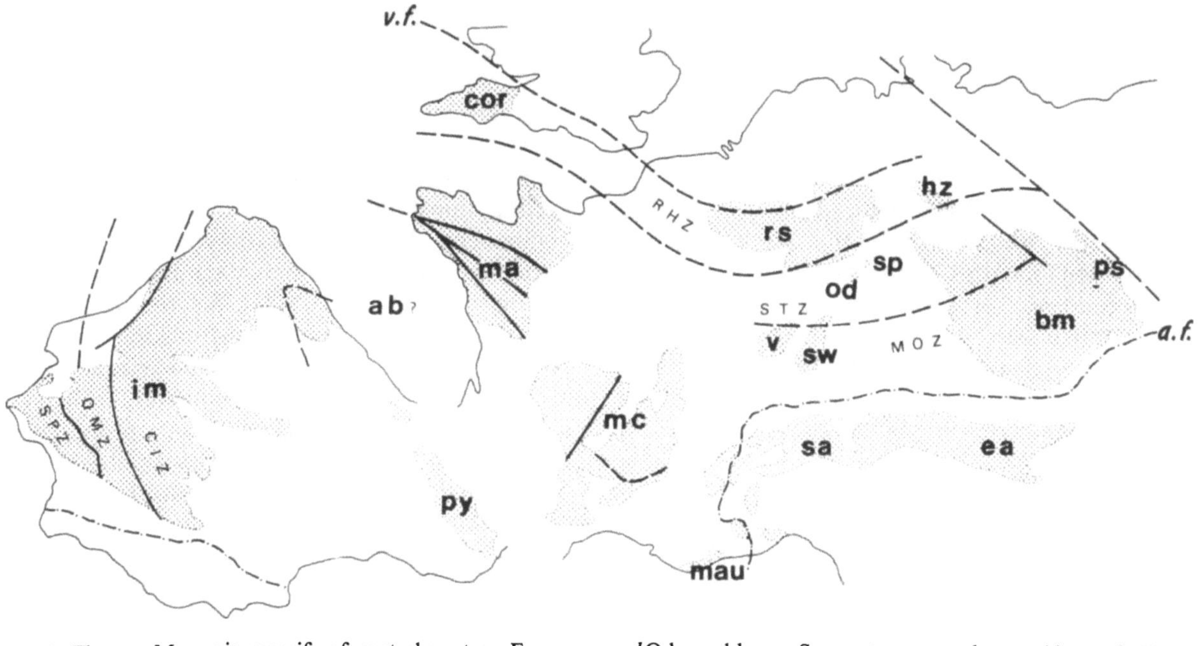

Fig. 1. The pre-Mesozoic massifs of central-western Europe *(lower case symbols)*. Major tectonic zones *(upper case symbols)* are also delineated. Pre-Mesozoic Massifs (from west to east): *ab* Aquitan Block (inferred); *ma* Massif Armoricain; *cor* Cornwall; *py* Pyrénées; *mc* Massif Central; *mau* Maures; *rs* Rheinische Schiefergebirge; *v* Vosges; *sw* Schwarzwald; *od* Odenwald; *sp* Spessart; *sa* southern Alps; *hz* Harz; *ea* eastern Alps; *bm* Bohemian Massif; *ps* Polish Sudetes. Tectonic zones: *RHZ* Rheno-Hercynian Zone; *STZ* Saxo-Thuringian Zone; *MOZ* Moldanubian Zone; *MSZ* Moravo-Silesian Zone; *VF* Variscan Front (northern boundary of the Hercynian belt); *AF* Alpine Front

2 Old Precambrian Relics

Twenty years ago, the central-western European domain was commonly thought to consist predominantly of an old Precambrian basement, more or less extensively reworked during Late Precambrian and Paleozoic orogenies. More specifically, the high-grade, internal zones of the Hercynian Belt (so-called Moldanubian Zone) were ascribed to Middle Proterozoic or even older events (e.g. Zoubek 1974). This view, however, was not confirmed by isotopic dating. Most radiometric data point unambiguously to Paleozoic ages for both the igneous protoliths and metamorphic overprints (e.g. Bernard-Griffiths et al. 1977; Gebauer and Grünenfelder 1979; Pin and Lancelot 1982; van Breemen et al. 1982; Gebauer et al. 1981). Scarce micropaleontological findings in associated metasediments further corroborate this fact (e.g. Konzalova 1980; Pacltova 1980). Various old U-Pb upper intersection ages have been obtained on zircons extracted from several mafic rocks which underwent eclogite-facies metamorphism during the Paleozoic. At first, these upper intersection data were considered as evidence for Precambrian ages of the magmatic protoliths (Gebauer and Grünenfelder 1982; Peucat et al. 1982; Paquette et al. 1985). However, the regognition in these rocks of crustal contamination processes lead to a reassessment of the upper intersection ages. With a few exceptions (Gebauer et al. 1987), these are now interpreted as merely reflecting the occurrence of an inherited crustal component (Paquette 1987). Indeed, zircon xenocrysts have been documented in mafic rocks (Compston et al. 1986; Wright and Wyld 1986).

Therefore, it is most likely that the highly metamorphosed outcrops exposed throughout median Europe do not represent a cystalline basement of Precambrian age. Rather, they consist of Uppermost Proterozoic to Lower Paleozoic series which suffered high-grade metamorphism during the Hercynian mega-cycle, in Middle and Late Paleozoic times.

Well-documented relics of old Precambrian crust occur, however, in western Europe. They have been identified for many years in N. Brittany and W. Normandy (Pentevrian), both on geological grounds and isotopically (e.g. Adams 1976; Leutwein et al. 1973). They consist of medium-grade, metasedimentary and metavolcanic rocks, associated with intrusive, meta-igneous gneisses (Auvray 1979). U-Pb zircon upper intersection ages of the orthogneisses cluster around 2 Ga, but ages close to 1.8 Ga have also been reported (Calvez and Vidal 1978; Auvray et al. 1980; Vidal et al. 1981).

Granulite facies rocks have been dredged from the continental margins of the ocean crust-floored Bay of Biscay (Capdevila et al. 1974; Didier et al. 1977). Early Rb-Sr work (Capdevila and Vidal 1975) suggested an exceedingly ancient age for these rocks. This has been confirmed by U-Pb and Sm-Nd work which point to Early Proterozoic (1.9 Ga), or even Late Archean (2.7 Ga) ages (Guerrot et al. 1987; Peucat et al. 1988a). Interestingly, and in contrast with the lower grade Pentevrian domains, these granulites do not display noticeable Hercynian overprinting, as indicated by the ages (ca. 1400–1600 Ma) obtained on the weakly resistant biotite Rb-Sr system (Capdevila and Vidal 1975). It seems likely, therefore, that these high-grade relics were not involved in the Hercynian orogenic events. They might belong to a terrane which remained outside the Hercynian realm and might tentatively be ascribed to the so-called Aquitan block (Autran and Cogné 1980), thought to be responsible for the pronounced bending of the Hercynian Belt along the Ibero-Armorican arc. The bottom parts of this micro-craton might have been exposed during the continental break-up episode which initiated the anti-clockwise rotation of the Iberian plate, in the Mesozoic.

Direct isotopic evidence for other old Precambrian igneous or metamorphic outcrops elsewhere in CW Europe is lacking. Nd isotope characteristics of lower crustal xenoliths and the widespread hercynian granitoids provide useful information for assessing the possible presence of any similar, unexposed ancient crust at depth. The data clearly point to recycling (most likely sedimentary) of old crustal components. However, and in contrast with an interpretation of Pb isotope data from K-feldspar (Michard-Vitrac et al. 1981), the results generally preclude a *massive* derivation from Early Proterozoic or older basement sources (Peucat et al. 1988b; Liew and Hofmann 1988; Duthou and Pin 1988). A similar conclusion had already been proposed on the basis of Sr isotope geochemistry (Jäger 1977, 1983; Vidal 1977). Recent ion probe studies of zircons in Uppermost Precambrian and Lower Paleozoic (meta)sediments reveal the occurrence of detrital grains with inherited cores as old as 3.8 Ga (Gebauer et al. 1989). In contrast with the interpretation of Gebauer et al., it is considered here that these ancient zircons do not provide compelling evidence for an extremely ancient, Early Archean development of the *local continental crust*. This is because of the likelihood of detrital zircons having been involved in multiple sedimentary recycling, and having travelled very large distances during every erosion-sedimentation cycle.

In summary, comparatively small, old Precambrian cratonic blocks occur in NW France and at the margins of the Bay of Biscay. They were not significantly remobilized during the major Hercynian orogeny. Although old detrital components do occur in the metasedimentary reservoir, circumstantial isotopic evidence does not favour the occurrence of reworked, Early Proterozoic or Archean *crystalline basement* in the Hercynian domain itself.

3 The Middle Proterozoic Hiatus

There is a conspicuous absence of radiometric dates obtained on meta-igneous rocks between 1800 Ma and 700 Ma. A possible exception are the plagioclase amphibolites and banded amphibolites from the Berisal Complex, Pennine Alps (Stille and Tatsumoto 1985). These rocks are interpreted in terms of mixed tholeiitic basalts and dacites, and they give Sm-Nd whole-rock isochron ages of ca. 1 Ga. Closely associated hornblendites yield a much younger Sm-Nd age (475+/−81 Ma), as do nine samples from the same rock suite by the Rb-Sr method (500+/−80 Ma). The geological interpretation of these data is not entirely clear. Stille and Tatsumoto favoured the hypothesis that the ca. 1 Ga ages refer to the time of the initial volcanic event, whereas the ca. 500 Ma ages would be linked to a remelting episode. Alternatively, the possibility that the ca. 1 Ga Sm-Nd linear arrays result from crustal contamination of younger, ca. 500 Ma mantle magmas cannot be totally dismissed.

Ironically, the almost complete lack of unequivocal evidence for magmatic events in the 1800-700 Ma range contrasts markedly with the abundance of Nd model ages falling in this time interval. These have been obtained from exposed (meta)sediments of Late Proterozoic or Paleozoic ages, Paleozoic orthogneisses, Hercynian granitoids, and samples from the lower crust (Ben Othman et al. 1984; Stosch and Lugmair 1984; Michard et al. 1985; Downes and Leyreloup 1986; Pin and Sills 1986; Voshage et al. 1987; Downes and Duthou 1988; Liew and Hofmann 1988; Peucat et al. 1988b; Duthou and Pin 1988). As is usual for model ages, their geological interpretation is not straightforward. In an idealized and most likely simplistic hypothesis, they could suggest that a major crust-forming episode occurred about 1500 Ma ago. However, the absence of meta-igneous relics of this age does not support this hypothesis (Peucat et al. 1988b). In fact, the common occurrence among the sedimentary zircon populations of detrital grains displaying ages of 2 Ga or more (e.g. Grauert et al. 1973; Gebauer and Grünenfelder 1977; Vidal et al. 1980; Kröner et al. 1988 among many others) provides evidence for recycled components significantly older than the Nd model ages. This may suggest a more realistic scenario: the Middle Proterozoic model ages are not linked to any particular geological event; rather, they represent average values resulting from broad sedimentary and magmatic mixing processes involving sedimentary components (very prone to multiple recycling, as shown by Veizer and Jansen 1979), and younger (Late Proterozoic and Paleozoic) juvenile end-members. This interpretation is supported by ion probe dating of detrital zircons (Gebauer et al. 1989) which point to the absence of newly formed grains for a period of ca. 500 Ma at around 1.5 Ga.

4 Late Precambrian Crustal Growth

While the oldest Early Proterozoic crystalline basement relics consist only of volumetrically minor inliers, there is widespread evidence for a major episode of crustal growth of Late Proterozoic age in CW Europe. Upper Precambrian series are best exposed in those areas which were not strongly affected by Hercynian tectonic processes, i.e. SW Britain and N Armorican Massif in Western Europe, the Barrandian domain of Central Bohemia and the Brno block (Moravia) in Central Europe. In all these areas, the major common feature is the abundance of "juvenile" tholeiitic and calc-alkaline volcanic and intrusive rocks (e.g. Thorpe 1979; Auvray et al. 1980; Dissler et al. 1988; Chantraine et al. 1988; Waldhausrova 1984; Dudek 1980). These have been dated between ca. 700 Ma and 580 Ma in SW Britain and NW France (see reviews in Thorpe et al. 1984; Guerrot et al. 1989). Associated sediments consist of volcaniclastics, greywackes, graphitic shales and black cherts, and younger turbidites. These "Brioverian" (Armorican Massif) and "Algonkian" (Bohemian Massif) systems were folded during the Cadomian orogeny. At the same time, they suffered metamorphism, generally of low grade, but reaching anatexis in the northern Armorican Massif (Peucat 1986a), and possibly the Eastern Alps (Söllner and Hansen 1987). Abundant granitoids were emplaced in these terrains between ca. 580 Ma and 540 Ma (e.g. Graviou et al. 1988; van Breemen et al. 1982). Moreover, exposed ca. 550 Ma old (e.g. Vitrac-Michard and Allègre 1975; Bernard-Griffiths et al. 1977; Pin and Lancelot 1978) calc-alkaline orthogneisses are documented in zones which underwent strong overprinting during the Hercynian events. Likewise, sparse data from the lower crust of the Hercynian Belt suggest that igneous rocks of similar age may occur at depth (Supply 1981; Pin and Sills 1986). Moreover, Gebauer et al. (1989) found a peak of ages around 600 Ma in the detrital zircon populations they studied with the ion probe. These provide further evidence for the widespread occurrence of igneous rocks of Late Precambrian age in CW Europe.

In summary, although sedimentary recycling of older components played an important role, it appears that significant continental growth characterized the Latest Proterozoic. This occurred by opening and closure of marginal basins and accretion of island arcs and sedimentary wedges, with incorporation of subordinate micro-cratonic blocks (e.g. Vidal et al. 1981; Thorpe et al. 1984). This scenario bears broad analogy with the Pan-African crustal evolution documented in a huge domain extending from the Arabo-Nubian Shield, through NW Africa (Morocco) to the Avalon terrane of Newfoundland and New Brunswick. It is inferred that this evolution resulted in the formation of new continental crust over most of the mid-European domain. This probably occurred at the "northern" margin of the Gondwana supercontinent.

5 The Late Cambrian-Early Ordovician Continental Break-up

The Late Cambrian-Early Ordovician period (ca. 500–480 Ma) is characterized by major bimodal or mafic magmatism documented throughout CW Europe by stratigraphic or radiometric evidence (e.g. Ricci and Sabatini 1978; Gebauer et al. 1981; Ducrot et al. 1982; Pin and Lancelot 1982; Paquette et al. 1984). In some cases, the bimodal igneous activity was preceded by felsic magmatism of alkaline affinity, reminiscent of anorogenic settings (e.g. Ducrot et al. 1979). Although tholeiitic basalts and crustal rhyolites usually predominate, alkali metabasalts and meta-rhyolites are also documented (e.g. Seyler 1986; Narebski et al. 1986). These metaigneous rocks provide compelling evidence for ensialic rifting processes over the whole of CW Europe (e.g. Weber 1984). Other bimodal suites of similar age (so-called leptyno-amphibolitic groups) are common in the high-grade, internal zones of the Hercynian Belt. Their geodynamic significance is controversial. On the basis of the mild calc-alkaline tendencies of the mafic rocks, most authors favour a back-arc model (e.g. Giraud et al. 1985; Piboule and Briand 1985; Bodinier et al. 1988; Briand et al. 1988). However, isotopic evidence suggests that crustal contamination processes could account for the geochemical characteristics (Pin and Marini, in prep.). A passive margin setting, created by ensialic rifting unrelated to subduction, could better satisfy the metamorphic evolution (Pin and Vielzeuf 1988). Scarce mafic-ultramafic complexes of similar (ca. 500 Ma) age occur (Gebauer and Grünenfelder 1979; El Ageed et al. 1980; Bodinier et al. 1981; Dubuisson et al. 1988; Müller-Sohnius et al. 1987; Pin and Carme 1987; Ménot et al. 1988). In a general way, these display distinct geochemical and isotopic affinities with oceanic basalts. Taking into-account the limited preservation potential of ophiolitic remnants in ancient domains, these mafic-ultramafic associations suggest that the Early Paleozoic extensional episode reached a seafloor spreading stage. While in marked contrast with classical concepts of the Hercynian Belt (e.g. Krebs and Wachendorf 1973), this view is corroborated by paleomagnetic data (e.g. Perroud et al. 1984) which require the existence of significant oceanic areas in the Paleozoic.

6 The Mid-Paleozoic HP-Metamorphic Event

6.1 *Northern and Central Hercynides*

A large number of radiometric ages in the 500–380 Ma timespan have been reported from the high-grade, axial zone of the Hercynian Belt (see review by Dornsiepen 1979). These data include Rb-Sr whole-rock isochrons obtained not only on meta-igneous gneisses, but also on metasediments of various grade, including anatexites. U-Pb zircon lower intersection ages on metasediments are also common.

The geological interpretation of these dates in terms of the ages of metamorphic events (e.g. Gorochov et al. 1983; Gebauer et al. 1989) raises several methodological problems. It is claimed here that neither whole-rock Rb-Sr linear arrays nor U-Pb zircon lower intersections on metasediments provide entirely reliable estimates of the age of a metamorphic episode for the following reasons. As Dornsiepen (1979) pointed out, all the Rb-Sr data were obtained from polymetamorphic terrains which suffered, in many cases, a strong HT Hercynian overprinting, as documented by a wealth of Carboniferous mineral ages. Partial opening of the Rb-Sr system might have occurred leading to underestimates of the true ages. This is more especially possible in meta-igneous rocks with high Rb/Sr ratios (e.g. Bell and Blenkinsop 1978) and such a behaviour may be sometimes demonstrated in CW Europe (e.g. Paquette et al. 1984). Another important objection is centered on the problem of large-scale homogenization of Sr isotopes during metamorphic events. Unless special assumptions are made concerning the relative scale of the initial variations of the Rb/Sr ratio and the Sr isotope equilibration (Roddick and Compston 1977), the interpretation in terms of metamorphic ages of whole-rock isochrons based on a regional sampling requires that large scale isotope homogenization occurred during metamorphism or anatexis. This is not supported by diffusion parameters of Sr in silicate melts (Magaritz and Hofmann 1978), even in hydrous magmas (Watson 1981). In fact, large-scale homogenization seems to be more likely during sedimentation, dewatering or diagenesis. The whole-rock Rb-Sr systematics could then survive metamorphism and even partial melting. This hypothesis deserves serious consideration as suggested, for example, in the Bohemian Massif by 470 Ma "isochrons" in granitoids of well documented Carboniferous (ca. 320 Ma) emplacement ages (Holl et al. 1989).

Likewise, U-Pb data on detrital zircons are potentially misleading. Allègre et al. (1974) demonstrated that the ca. 450 Ma lower intersection ages obtained on metasediments do not necessarily require a Caledonian event, since multiepisodic models involving lead losses at ca. 600 and 300 Ma could easily fit the data. In addition, the mixing of different zircon populations, each with discordant U-Pb ages, is likely to occur in detrital sediments. In this case, mixing lines are produced where both upper and lower intersections with the Concordia curve are devoid of geological significance.

Taking into account the difficulties inherent in the interpretation of radiometric dates on metasediments, or even S-type granitoids, it seems safer to rely more on isotopic information obtained from meta-igneous, preferably mafic rocks (although these are not problem-free!). On geological grounds, it appears that the eclogites commonly found in the high-grade zones of the Hercynian Belt record the oldest metamorphic episode. Interestingly, U-Pb zircon measurements on most of these rocks are consistent with a major HP-HT metamorphism around 430–380 Ma (Gebauer and Grünenfelder 1979; Peucat et al. 1982; Pin and Lancelot 1982; Ducrot et al. 1983; Müller-Sohnius et al. 1987; Paquette et al. 1987; Quadt and Gebauer 1988). This is in agreement with the 420 Ma age of the HP-LT metamorphism recorded by blueschists at Ile de Groix, S Armorican Massif (Peucat 1986b). Subduction-related processes are suggested by the very high pressures reached (12–18 kb; Pin and Vielzeuf 1983, 1988). They are also supported by scarce arc volcanics of similar age (Carpenter et al. 1982; André et al. 1986; Narebski et al. 1986; Thiéblemont and Cabanis 1986; Thiéblemont et al. 1987; Quadt and Gebauer 1988).

All these isotopic data have often been taken as evidence for a mid-European branch of the Caledonides (e.g. Kvale 1976). However, geological evidence (e.g. molasse-type deposits) for a mid-Paleozoic *orogeny* in CW Europe is lacking and the interpretation of the isotopic data remains equivocal. The absence of coarse clastic rocks in the Silurian–Early Devonian sedimentary record precludes a model of continent-continent collision and associated crustal thickening. This occurred only in Carboniferous times. Rather, several lines if evidence favour a scenario involving arc-continent collisional processes: the subduction of oceanic crust and attached passive margins beneath mostly oceanic arcs could account for the HP (both low- and high-temperature) metamorphism without requiring any crustal root. The probable occurrence of remnant back-arc oceanic zones would also explain why suturing through continent-continent collision did not occur immediately after this early HP phase of the Hercynian cycle.

6.2 Eastern and Southern Alps

A wealth of radiometric ages around 450 Ma has been supplied from the pre-Mesozoic basement of the Austro-Alpine domain and the southern Alps (review by Sassi et al. 1985). These were obtained by the Rb-Sr method on metasediments and metagranitoids, and interpreted as an evidence for an intra-Ordovician event whose geological significance has been strongly debated (e.g. Bögel et al. 1979). Geodynamic interpretations proposed for this "Caledonian" episode include: (1) a thermal event in an extensional context (e.g. Heinisch and

Schmidt 1976; Schmidt and Söllner 1983); (2) a compressional tectono-metamorphic episode (Sassi et al. 1987 for a review); (3) magmatism and deformation in a subduction setting (Loeschke 1977).

Bearing in mind the methodological problems outlined in the previous section, special emphasis is placed upon radiometric results obtained on meta-igneous rocks. As elsewhere in Europe, a ca. 500 Ma magmatic episode is documented by U-Pb and Rb-Sr dating of metagranites (Neubauer et al. 1987) and several mafic-ultramafic associations, sometimes of ophiolitic affinities (Frank et al. 1976; Quadt 1985; Söllner and Gebauer in Söllner and Hansen 1987). Some of these rocks were transformed into eclogites. Geological observations indicate that a metamorphic and deformation episode occurred prior to the Caradocian (Ebner et al. 1987). This is in close agreement with U-Pb data on pre- and post-tectonic granitoids which bracket the first isoclinal folding between ca. 440 and 430 Ma (Cliff 1980). The emplacement of numerous granite plutons and widespread ignimbrites (up to 1500 m thick) accompanied this event (Pecerillo et al. 1979; Heinisch 1981).

In summary, the geological record of the Eastern Alps displays a sequence of events which compares well with what is inferred for the remainder of CW Europe. It appears, however, that the first compressive phase occurred in the Late Ordovician rather than in the Silurian. A tentative plate tectonic scheme involving an island arc and/or active continental margin setting is favoured by the available geological and geochemical evidence (Loeschke 1977; Frisch et al. 1987).

7 The Devonian Period

In contrast with the plate convergence inferred from the previous HP-metamorphic event, an extensional regime prevailed during the Middle-Late Devonian. This is evidenced by the initiation of new sedimentary basins, facies relationships (Sawkins and Burke 1980), and more specifically, by abundant bimodal or mafic magmatism (e.g. Perekalina 1981). Scarce occurrences of metabasalts with MORB affinities, a few ophiolitic remnants such as the Lizard complex (Davies 1984), and the mafic-ultramafic massifs from the Polish Sudetes (Pin et al. 1988) show that new oceanic crust was created during this Devonian rifting episode. This mainly occurred in the northern part of central-western Europe (so-called Rheno-Hercynian and Saxo-Thuringian Zones), but similar extension in southern Europe cannot be excluded (Heinisch et al. 1987).

Magmatic rocks with subduction zone affinities are also documented from the same epoch in the N Massif Central (e.g. Sider and Ohnenstetter 1986), Vosges (Rasamimanana and Bébien 1985; Rizki and Baroz 1988), and Moravia (Patocka 1987). Further, circumstantial evidence for subduction processes is provided by a very low geothermal gradient (ca. 10°/km) in the southern Rheno-Hercynian Zone (Massonne and Schreyer 1983) and the occurrence in Cornwall of sediments analogous to those from recent active margins (Floyd and Leveridge 1987). It may be inferred that remnant oceanic crust, preserved during the Mid-Paleozoic compression, was involved in these Late Devonian subduction zones. Therefore, at least some of the newly-formed oceanic domains might be related to back-arc spreading.

8 The Carboniferous Continental Collision

The major compressional deformation experienced by the pre-Alpine CW Europe occurred in Tournaisian-Visean times (ca. 360–330 Ma) as a result of the final suturing of several Early Paleozoic and Devonian oceanic zones. The continent-continent collision process is documented by the tectonic style, dominated by large-scale intra-crustal thrusting (e.g. Matte and Burg 1981; Behr et al. 1982; Matte 1986) and widespread synorogenic sedimentation (Franke and Engel 1986). In the northern flank of the belt (Rheno-Hercynian and Saxo-Thuringian Zones), northward-verging folds and thrusts are documented by field and seismic profiling data (e.g. Kenolty et al. 1981; Meissner et al. 1981; Shackleton et al. 1982). At the southern margin of this domain, near the boundary with the Moldanubian Zone, slices of ca. 400 Ma old high-grade metamorphic basement (e.g. the Münchberg Massif) were involved in this nappe tectonics (Behr et al. 1984). In the south (Southern Armorican Massif, Massif Central, Southern Bohemian Massif), an opposite, broadly southward vergency is documented (e.g. Carme 1975; Matte and Burg 1981; Rajlich et al. 1986). Again, a clear orogenic polarity may be defined. It grades from deep-seated thrust-nappes, associated with a ca. 340 Ma old (e.g. Pin and Lancelot 1982; van Breemen et al. 1982; Aftalion et al. 1989) locally inverted (e.g. Burg et al. 1984) amphibolite- to granulite-facies metamorphism, to somewhat younger high-level recumbent folds, gravitational nappes and Late Visean-Namurian flysch basins. These are well exemplified in the classical section of Montagne Noire, S. Massif Central (Franke and Engel 1986). Thus, the Hercynian collision tectonics was characterized by the development of two roughly symmetric "crustal accretionary wedges" (Mattauer 1986). This was accompanied by both prograde metamorphism (in the external zones) and retrogression of the ca. 400 Ma high-grade basement in the inner zones. Copious crustal melting occurred, as indicated by abundant granitoids. Such pervasive melting might have resulted from underthrusting of fluid-rich, heat-producing element-rich, sedimentary rocks.

9 The Late Hercynian Period

The late-stage evolution of the Hercynian Belt (ca. 320–280 Ma) was dominated by pervasive transcurrent shearing (Arthaud and Matte 1977) coeval with uplift and erosion, and filling of coal-bearing sedimentary troughs and intra-montane basins. At the same time, a major thermal event occurred characterized by medium- to low-pressure granulite facies metamorphism of the lower crust (Pin and Vielzeuf 1983). This is well documented by both xenoliths scavenged by Neogene volcanoes and slices of the pre-Alpine lower crust uplifted in Mesozoic-Cenozoic times (e.g. Ivrea Zone). Abundant ca. 280 Ma old deep-seated mafic intrusions (e.g. Rivalenti et al. 1981; Pin 1986) and higher level granitoids provide additional evidence for this late-stage thermal event. Intra-crustal differentiation into a cumulative and restitic lower crust, and a granite-rich upper crust occurred at that time. It is believed that this major process marked the end of the transformation of a broad, heterogeneous mobile zone into a segment of stable, mature continental crust.

10 Conclusion and Comparison with West African Orogens

Notwithstanding Early Archean detrital components (Gebauer et al. 1989), the geological record of CW Europe goes back as far as the Early Proterozoic and the Late Archean. However, preserved relics of such early pieces of continental crust are scarce, and volumetrically minor. Upper Precambrian series are more widely documented. They correspond to a significant episode of continental growth, analogous to the Pan-African evolution recorded all around the northern margin of the African craton. The Paleozoic history was dominated by a complex succession of extensional and compressive regimes within a wide mobile zone. Rifting, continental break-up and ocean opening occurred in the Late Cambrian–Early Ordovician. A major, probably diachronic, eclogite- and blueschist-forming event is characteristic of the 440–380 Ma (Late Ordovician to Early Devonian) period. This metamorphic episode was not associated with significant mountain building and it might be interpreted in terms of arc-continent collisional processes. The Mid–Late Devonian was characterized by renewed extension and subduction of oceanic lithosphere. The major deformation occurred in the Early Carboniferous. This involved large-scale nappe tectonics with symmetrical outward vergencies, related to continent-continent collision along at least two oceanic suture zones. Abundant synorogenic sedimentation, and copious crustal melting documented by very abundant granites are the most typical characteristics of this Hercynian evolution. The abundance of granites is tentatively linked to the widespread occurrence of "fertile" (Clemens and Vielzeuf 1988) sedimentary and hydrated volcanic protoliths. A pervasive late-stage thermal event is indicated by coeval (ca. 300–280 Ma) lower crustal granulites and high-level granitoids. Together with the previous collision tectonics, this event was a major agent of the intra-crustal differentiation and consolidation of the CW European domain.

Attempting large scale correlations with the Western African orogens is beyond the scope of this chapter, as it would require consideration of the intervening areas of the Iberian Massif and Morocco. However, comparative geodynamics may be a fruitful approach and it is of interest to highlight the main differences and similarities of orogenic belts such as the European Hercynides and the Mauritanides *s. l.*

These developped roughly contemporaneously, during the Late Proterozoic and the Paleozoic, in broadly similar large scale contexts, respectively at the western and northern edges of the West African Craton. High-grade crystalline basement with ages of ca. 2.7 and 2.0 Ga is documented in both areas. While these form only minor inliers (Bay of Biscay and Pentevrian relics) in CW Europe, Liberian and Eburnean crystalline domains constitute the foreland of the West African orogens. In both belts, Late Proterozoic volcanic and sedimentary sequences are important components.

The timing of pre-Paleozoic tectonic and metamorphic events is also similar. Polyphase Pan-African activity (Dallmeyer and Villeneuve 1987) might correlate, albeit diachronic, with the two-stage (ca. 580 and 540 Ma) Cadomian cycle recorded in the Armorican Massif (Graviou et al. 1989).

In contrast with the similarities of the Precambrian histories, the overall Early and Middle Paleozoic evolutions differ markedly. The major continental break-up recorded near the Cambro-Ordovician boundary throughout CW Europe has no counterpart in Western Africa. No true Paleozoic ophiolites are exposed, although Late Precambrian oceanic remnants are inferred from geophysical data in the westward extension of the Mauritanides. Traces of a glaciation (Deynoux 1978; Franke and Engel 1986, p. 33) in the Late Ordovician suggest high latitudes for both areas.

The Mid-Paleozoic compression documented in CW Europe by a high-pressure metamorphism might correlate with the intra-Ordovician nappe tectonics recognized by Lécorché (1980) in the northern segment of the Mauritanides. However, no high-grade rocks (eclogites, granulites, blueschists) are connected with the Paleozoic evolution of the West African orogens.

The Late Paleozoic compressional events also differ both in time and style. In the European Hercynides, a continent-continent collision occurred in the Early Carboniferous after subduction of oceanic crust in the Devonian. This initiated pervasive intracrustal deformation (first nappe tectonics coeval with high-grade metamorphism, then strike-slip faulting) until the end of the Carboniferous. Instead, the major collision tectonics in West Africa occurred in the Permian (270-240 Ma) and is mainly recorded by thrusting within foreland sequences (Villeneuve and Dallmeyer 1987). Presumably, more internal domains, including inferred ophiolites, occur westward beneath the Mesozoic cover and the Atlantic margin.

The absence of granite in the Mauritanides contrasts remarkably with their unusual abundance in the Hercynian Belt. This might be explained by the contrasting relative roles of high-grade crystalline rocks vs sedimentary wedges in both belts, assuming that cratonic basements and their cover sequences have a limited potential for granite production during a subsequent thermal event.

In summary, although both chains are characterized by the same general tectonic style characterized by thrust and nappes, and clearly formed as a result of collisional processes, they differ significantly. In fact, because of the opening of the Atlantic ocean, only a comparatively small segment (ca. 200 km wide) is exposed in the Mauritanides. This corresponds to the external zones of the eastern flank of a much broader two-sided collision orogen. Clearly, any geodynamic comparison with the European Hercynides (ca. 1000 km wide) should consider the entire Mauritanides-Appalachian system.

Acknowledgements
I thank J. P. Bassot for an introduction to the geology of West Africa, and D. Gebauer and F. Neubauer for critical reviews. Corrections of the English text by A. Woodland are gratefully acknowledged.

References

Adams CJ (1976) Geochronology of the Channel Islands and adjacent french mainland. J Geol Soc Lond 132: 233-250

Aftalion M, Bowes DR, Vrana S (1989) Early Carboniferous U-Pb zircon age for garnetiferous, perpotassic granulites, Blansky Les Massif, Czechoslovakia. N Jahrb Mineral Monatsheft H 4: 145-152

Allègre CJ, Albarède F, Grünenfelder M, Köppel V (1974) 238U/206Pb-235U/207Pb-232Th/208Pb zircon geochronology in Alpine and non-Alpine environment. Contr Mineral Petrol 43: 163-194

André L, Hertogen J, Deutsch S (1986) Ordovician-Silurian magmatic provinces in Belgium and the Caledonian orogeny in middle Europe. Geology 14: 879-882

Arthaud F, Matte P (1977) Late-Paleozoic strike-slip faulting in southern Europe and northern Africa: result of a right-lateral shear zone between the Appalachians and the Urals. Geol Soc Bull 88: 1305-1320

Autran A, Cogné J (1980) La zone interne de l'orogène varisque dans l'Ouest de la France et sa place dans le développement de la chaîne hercynienne. 26th Int Geol Congr, Colloq C6 Geol Eur, Mém Bur Rech Géol Min 108: 90-111

Auvray B (1979) Genèse et évolution de la croûte continentale dans le Nord du Massif Armoricain. Dr Sci Thesis, Univ Rennes, 681 pp

Auvray B, Charlot R, Vidal P (1980) Données nouvelles sur le protérozoique inférieur du domaine nord-armoricain (France): âge et signification. Can J Earth Sci 17: 532-538

Behr HJ, Engel W, Franke W (1982) Variscan wildflysch and nappe tectonics in the Saxo-Thuringian zone (northeast Bavaria, West Germany). Am J Sci 282: 1438-1470

Behr HJ, Engel W, Franke W, Giese P, Weber K (1984) The variscan belt in central Europe: main structures, geodynamic implications, open questions. Tectonophysics 109: 15-40

Bell K, Blenkinsop J (1978) Reset Rb/Sr whole-rock systems and chemical control. Nature (Lond) 273: 533-534

Ben Othman D, Fourcade S, Allègre CJ (1984) Recycling processes in granite-granodiorite complex genesis: the Querigut case studied by Nd-Sr isotope systematics. Earth Planet Sci Lett 69: 290-300

Bernard-Griffiths J, Cantagrel JM, Duthou JL (1977) Radiometric evidence for an Acadian tectonometamorphic event in Western Massif Central français. Contrib Mineral Petrol 61: 199-212

Bodinier JL, Dupuy C, Dostal J, Carme F (1981) Geochemistry of the ophiolite from the Chamrousse complex (Belledonne Massif, Alps). Contrib Mineral Petrol 78: 379-388
Bodinier JL, Burg JP, Leyreloup A, Vidal H (1988) Reliques d'un bassin d'arrière-arc subducté, puis obducté dans la région de Marvejols (Massif Central). Bull Soc Géol Fr (8) IV: 21-33
Bögel H, Morteani G, Sassi FP, Satir M, Schmidt K (1979) The Hercynian and pre-Hercynian development of the eastern Alps. Report on a meeting. N Jahrb Geol Paläontol Abh 159: 1, 87-112
Briand B, Piboule M, Bouchardon JL (1988) Diversité des métabasites des groupes leptyno-amphiboliques du Rouergue et de Marvejols (Massif Central). Origine et implications. Bull Soc Géol Fr (8) IV: 489-498
Burg JP, Leyreloup A, Marchand J, Matte P (1984) Inverted metamorphic zonation and large-scale thrusting in the Variscan Belt: an example in the French Massif Central. In: Hutton DHW, Sanderson DJ (eds) Variscan tectonic of the North Atlantic Region. Geol Soc Spec Publ 14: 47-61
Calvez JY, Vidal P (1978) Two billion years old relicts in the Hercynian Belt of western Europe. Contrib Mineral Petrol 65: 395-399
Capdevila R, Vidal P (1975) Données géochimiques et radiométriques sur les granulites et charnockites de la marge continentale nord-espagnole (Golfe de Gascogne). 3ème RAST, Montpellier, Soc Géol Fr, p 89
Capdevila R, Lamboy M, Le Pretre JP (1974) Découverte de granulites, de charnockites et de syénites néphéliniques dans la partie occidentale de la marge continentale nord-espagnole. C R Acad Sci Paris 278: 17-20
Carme F (1975) Idées nouvelles sur le style et l'ampleur de la tectonique hercynienne à la marge Sud de la zone Arverne (Massif Central). C R Acad Sci Paris 278: 2613-2616
Carpenter MSN, Peucat JJ, Pivette B (1982) Geochemical and chronological characteristics of Palaeozoic volcanism in the Saint-Georges sur Loire synclinorium (S. Armorican Massif). Evidence for Pre-Hercynian tectonic evolution. Bull Bur Rech Géol Min I (1-2): 63-79
Chantraine J, Chauvel JJ, Balé P, Denis E, Rabu D (1988) Le Briovérien (Protérozoïque supérieur à terminal) et l'orogenèse cadomienne en Bretagne (France). Bull Soc Géol Fr (8) IV: 815-829
Clemens JC, Vielzeuf D (1988) Constraints on melting and magma production in the crust. Earth Planet Sci Lett 86: 287-306
Cliff RA (1980) Uranium-lead isotopic evidence from zircons for Lower Palaeozoic tectonic activity in the Austroalpine Nappe, the Eastern Alps. Contrib Mineral Petrol 71: 283-288
Compston W, Williams IS, Campbell IH, Gresham JJ (1986) Zircon xenocrysts from the Kambalda volcanics: age constraints and direct evidence for older continental crust below the Kambalda-Norseman greenstones. Earth Planet Sci Lett 76: 299-311
Dallmeyer RD, Villeneuve M (1987) 40Ar/39Ar mineral age record of polyphase tectonothermal evolution in the southern Mauritanide orogen, southeastern Senegal. Geol Soc Am Bull 98: 602-611
Davies GR (1984) Isotopic evolution of the Lizard complex. J Geol Soc Lond 141: 3-14
Deynoux M (1978) Les formations glaciaires du Précambrien terminal et de la fin de l'Ordovicien en Afrique de l'Ouest. Deux exemples de glaciation d'islandsis sur une plateforme stable. Dr Sci Thesis, Univ Marseille
Didier J, Guennoc P, Pautot G (1977) Granodiorites, granulites et charnockites de l'éperon de Goban (marge armoricaine) au contact du domaine océanique. C R Acad Sci Paris 284: 713-716
Dissler E, Doré F, Dupret L, Gresselin F, Le Gall J (1988) L'évolution géodynamique cadomienne du Nord-Est du Massif Armoricain. Bull Soc Géol Fr (8) IV: 801-814
Dornsiepen UF (1979) Rb/Sr whole rock ages within the European Hercynian. A review. Krystalinikum 14: 33-49
Downes H, Duthou JL (1988) Isotopic and trace-element arguments for the lower-crustal origin of hercynian granitoids and pre-hercynian orthogneisses, Massif Central (France). Chem Geol 68: 1-18
Downes H, Leyreloup A (1986) Granulitic xenoliths from the French Massif Central-petrology Sr and Nd isotope systematics and age estimates. In: Dawson JB, Carswell DA, Hall J, Wedepohl KH (eds) The nature of the lower continental crust. Geol Soc Spec Publ 24: 319-330
Dubuisson G, Hirn A, Girardeau J, Mercier JCC, Veinante (1988) Multiple Variscan nappes in Limousin, western Massif Central, France: geophysical constraints to the geological model and geodynamic implications. Tectonophysics 147: 19-31
Ducrot J, Lancelot JR, Reille JL (1979) Datation en Montagne Noire d'un témoin d'une phase majeure d'amincissement crustal caractéristique de l'Europe prévarisque. Bull Soc Géol Fr (7) XXI: 501-505
Ducrot J, Allégret A, Lancelot JR (1982) Prevariscan evolution of the Western Europe at the light of U/Pb zircon dating of orthogneisses and eclogite. 5th Int Conf Geochronol Cosmochronol Isotope Geol, Nikko (JPN), (abstr) p 83
Ducrot J, Lancelot JR, Marchand J (1983) Datations U-Pb sur zircons de l'éclogite de la Borie (Haut-Allier, France) et conséquences sur l'évolution anté-hercynienne de l'Europe Occidentale. Earth Planet Sci Lett 62: 385-394
Dudek A (1980) The crystalline basement block of the outer Carpathians in Moravia: Bruno-Vistulicum. Rozpr Cesk Akad Ved 90: 1-85
Duthou JL, Pin C (1988) Sr, Nd isotope geochemistry of Hercynian granitoids from Western Massif Central (France). Rend Soc Ital Mineral Petrol 43: 525-530
Ebner F, Neubauer F, Statteger K (1987) The Caledonian event in the Eastern Alps: a review. Pre-Variscan and Variscan events. Flügel HW, Sassi FP, Grecula P (eds) Pre-Variscan and Variscan events in the Alpine-Mediterranean mountain belts. Alfa, Bratislava, pp 169-182
El Ageed A, Saager R, Stumpfl EF (1980) Pre-alpine ultramafic rocks in the eastern central Alps, Styria Austria. In: Panayioton A (ed) Ophiolites. Proc Int Ophiolite Symp, Cyprus, 1979, Cyprus Geological Survey Dept, pp 600-606
Floyd PA, Leveridge BE (1987) Tectonic environment of the Devonian Gramscatho basin, south Cornwall: framework mode and geochemical evidence from turbiditic sandstones. J Geol Soc Lond 144: 531-542
Frank W, Klein P, Nowy W, Scharbert S (1976) Die Datierung geologischer Ereignisse im Altkristall in der Gleinalpe (Steiermark) mit der Rb/Sr-Methode. Tschermarks Mineral Petrograph Mitt 23: 191-203
Franke W, Engel W (1986) Synorogenic sedimentation in the Variscan Belt of Europe. Bull Soc Géol Fr (8) II: 25-33
Frisch W, Neubauer F, Bröcker M, Brückmann W, Haiss N (1987) Interpretation of geochemical data from the Caledonian basement within the Austroalpine basement complex. In: Flügel HW, Sassi FP, Grecula P (eds) Pre-Variscan and Variscan events in the Alpine-Mediterranean mountain belts. Alfa, Bratislava, pp 209-226
Gebauer D, Grünenfelder M (1977) U-Pb systematics of

detrital zircons from some unmetamorphosed to slightly metamorphosed sediments of Central Europe. Contrib Mineral Petrol 65: 29-37

Gebauer D, Grünenfelder M (1979) U-Pb zircon and Rb-Sr mineral dating of eclogites and their country rocks. Example: Münchberg gneiss massif, northeast Bavaria. Earth Planet Sci Lett 42: 35-44

Gebauer D, Grünenfelder M (1982) Geological development of the Hercynian Belt of Europe based on age and origin of high-grade and high-pressure mafic and ultramafic rocks. 5th Int Conf Geochronol Cosmochronol Isotope Geol, Nikko (JPN), (abstr) pp 111-112

Gebauer D, Bernard-Griffiths J, Grünenfelder M (1981) U-Pb zircon and monazite dating of a mafic-ultramafic complex and its country-rocks. Contrib Mineral Petrol 76: 292-300

Gebauer D, Quadt A, Williams IS, Compston W (1987) Archean zircons in a retrograted eclogite of the Gotthard Massif. Rundgespräch „Geodynamik des Europäischen Variszikums". Kaledonisch-Variszische Strukt Alpen, Fribourg/Schweiz, (abstr) p 7

Gebauer D, Williams IS, Compston W, Grünenfelder M (1989) The development of the Central European continental crust since the early Archaean based on conventional and ion-microprobe dating of up to 3.84 b.y. old detrital zircons. Tectonophysics 157: 81-96

Giraud A, Marchand J, Dupuy C, Dostal J (1985) Geochemistry of leptyno-amphibolite complex from Haut-Allier (French Massif Central). Lithos 17: 203-214

Gorochov IM, Melnikov NN, Varsavskaja ES, Kutjavin EP (1983) Rb-Sr dating of magmatic and metamorphic events in the eastern part of the Bohemian Massif. Cas Mineral Geol 28: 349-361

Grauert B, Hänny R, Soptrajanova G (1973) Age and origin of detrital zircons from the pre-Permian basements of the Bohemian Massif and the Alps. Contrib Mineral Petrol 40: 105-130

Graviou P, Peucat JJ, Auvray B, Vidal P (1989) The cadomian orogeny in the Northern Armorican Massif: petrological and geochronological constraints on a geodynamic model. Hercynica IV: 1-13 (in press)

Guerrot C, Peucat JJ, Capdevila R (1987) The oldest granulitic crust involved in the Hercynian belt: preliminary U-Pb and Sm-Nd isotopic data. Terra Cognita, (abstr) 7: 159

Guerrot C, Peucat JJ, Dupret L (1989) Données nouvelles sur l'âge du système briovérien (Protérozoïque supérieur) dans le Nord du Massif armoricain. C R Acad Sci Paris 308: 89-92

Heinisch H (1981) Zum ordovizischen „Porphyroid"-Vulkanismus der Ost- und Südalpen, Stratigraphie, Petrographie, Geochemie. Jahrb Geol Bundesanst 124 (1): 1-109

Heinisch H, Schmidt K (1976) Zur kaledonischen Orogenese in den Ostalpen. Geol Rundsch 65: 459-482

Heinisch H, Sprenger W, Weddige K (1987) Devonian basaltic volcanism and turbiditic sediments from the East Alpine Paleozoic, Northern Graywacke Zone (Tyrol/Salzburg, Austria): implications for Paleozoic plate tectonics. Terra Cognita (abstr) 7 (2-3): 179

Holl PK, v Drach V, Müller-Sohnius D, Köhler H (1989) Caledonian ages in Variscan rocks: Rb-Sr and Sm-Nd isotopic variations in dioritic intrusives from the northwestern Bohemian Massif, West Germany. Tectonophysics 157: 179-194

Jäger E (1977) The evolution of the Central and West European continent. In: La chaîne varisque d'Europe moyenne et occidentale. Colloq Int CNRS (ed), Rennes, 243: 227-239

Jäger E (1983) The age of the continental crust of Central, Southern and Western Europe - arguments from geochemistry and isotope geology. Schweiz Mineral Petrograph Mitt 63: 339-346

Kenolty N, Chadwick RA, Blundell DJ, Bacon M (1981) Deep seismic reflection survey across the Variscan Front of southern England. Nature (Lond) 293: 451-453

Konzalova M (1980) Some late precambrian microfossils from the Bohemian massif and their correlation. Precambrian Res 15: 43-62

Krebs W, Wachendorf H (1973) Proterozoic-paleozoic geosynclinal and orogenic evolution of central Europe. Geol Soc Am Bull 84: 2611-2630

Kröner A, Wendt I, Liew TC et al. (1988) U-Pb zircon and Sm-Nd model ages of high-grade Moldanubian metasediments, Bohemian Massif, Czechoslovakia. Contrib Mineral Petrol 99: 257-266

Kvale A (1976) Major features of the European Caledonodes and their development. In: Ager DV, Brooks M (eds) Europe from crust to core. Wiley, Lond, pp 81-115

Lécorché JP (1980) Les Mauritanides face au craton ouest-africain. Structure d'un secteur-clef: la région d'Ijibiten (est d'Akjoujt, Rép. Islam. de Mauritanie). Dr Sci Thesis, Univ Marseille

Leutwein F, Power G, Roach P, Sonet J (1973) Quelques résultats géochronologiques obtenus sur des roches d'âge précambrien du Cotentin. C R Acad Sci Paris 276: 2121-2124

Liew TC, Hofmann AW (1988) Precambrian crustal components, plutonic associations, plate environment of the Hercynian Fold Belt of Central Europe: Indications from a Nd and Sr isotopic study. Contrib Mineral Petrol 98: 129-138

Loeschke J (1977) Kaledonischer eugeosynklinaler Vulkanismus Norwegens und der Ostalpen im Vergleich mit rezentem Vulkanismus unterschiedlicher geotektonischer Positionen: eine Arbeitshypothese. Z Dtsch Geol Ges 128: 185-207

Massonne HJ, Schreyer W (1983) A new experimental phengite barometer and its application to a variscan subduction zone at the southern margin of the Rhenohercynium. Terra Cognita (abstr) 3: 187

Mattauer M (1986) Intracontinental subduction crust-mantle decollement and crustal stacking wedge in the Himalayas and other collision belts. Geol Soc Spec Publ 19: 37-50

Matte P (1986) Tectonics and plate tectonics model for the variscan belt of Europe. Tectonophysics 126: 329-374

Matte P, Burg JP (1981) Sutures, thrusts and nappes in the Variscan arc of Western Europe: plate tectonic implication. Geol Soc Spec Publ 8: 353-357

Meissner R, Bartelsen H, Murawski H (1981) Thin-skinned tectonics in the northern Rhenish Massif, Germany. Nature (Lond) 290: 399-401

Ménot RP, Peucat JJ, Scarenzi D, Piboule M (1988) 496 My age of plagiogranites in the Chamrousse ophiolite complex (external crystalline massifs in the French Alps): evidence of a Lower Paleozoic oceanization. Earth Planet Sci Lett 88: 82-92

Michard-Vitrac A, Albarède F, Allègre CJ (1981) Lead isotopic composition of hercynian granitic K-feldspars constraints continental genesis. Nature (Lond) 291: 460-464

Müller-Sohnius D, Drach V, Horn P, Köhler (1987) Altersbestimmungen an der Münchberger Gneismasse, Nordost-Bayern. N Jahrb Mineral Abh 156: 175-206

Narebski W, Dostal J, Dupuy C (1986) Geochemical characteristics of lower Paleozoic spilite-keratophyre series in the Western Sudetes (Poland): petrogenetic and tectonic implications. N Jahrb Mineral Abh 155: 243-258

Neubauer F, Frisch W, Hansen BT (1987) Time relations between eo-Alpine metamorphism and thrusting: evidence from the crystalline basement of the eastern Grauwacken Zone. In: Flügel HW, Faupl P (eds) Geodynamics of the Eastern Alps. Deuticke, Vienna, pp 263-271

Pacltova B (1980) Further micropaleontological data for the Paleozoic age of Moldanubian carbonate rocks. Cas Mineral Geol 25: 275-279

Paquette JL (1987) Comportement des systèmes isotopiques U-Pb et Sm-Nd dans le métamorphisme éclogitique. Chaîne hercynienne et chaîne alpine. Dr Thesis Univ Rennes I, 222 pp

Paquette JL, Marchand J, Peucat JJ (1984) Absence de tectonique cadomienne dans le complexe de Champtoceaux (Bretagne méridionale)? Comparaison des systèmes Rb-Sr et U-Pb d'un métagranite. Bull Soc Géol Fr (7) XXVI: 907-912

Paquette JL, Peucat JJ, Bernard-Griffiths J, Marchand J (1985) Evidence for old Precambrian relics shown by U-Pb zircon dating of eclogites and associated rocks in the Hercynian belt of South Brittany, France. Chem Geol 52: 203-216

Paquette JL, Balé P, Ballèvre M, Georget Y (1987) Géochronologie et géochimie des éclogites du Léon: nouvelles contraintes sur l'évolution géodynamique du Nord-Ouest du Massif Armoricain. Bull Minéral 110: 683-696

Patocka F (1987) The geochemistry of mafic metavolcanics: implications for the origin of the Devonian massive sulfide deposits at Zlaté Hory, Czechoslovakia. Mineral Dep 22: 144-150

Pecerillo A, Poli G, Sassi FP, Mezzacasa G (1979) New data on the upper Ordovician acid plutonism in the Eastern Alps. N Jahrb Mineral Abh 137: 162-183

Perekalina TV (1981) Variscan volcanism of central and western Europe. Geol Mijnbouw 60: 17-21

Perroud H, van der Voo R, Bonhommet N (1984) Paleozoic evolution of the Armorica plate on the basis of paleomagnetic data. Geology 12: 579-582

Peucat JJ (1986 a) Behaviour of Rb-Sr whole rock and U-Pb zircon systems during partial melting as shown in migmatitic gneisses from the St Malo Massif, NE Brittany, France. J Geol Soc Lond 143: 875-885

Peucat JJ (1986 b) Rb-Sr and U-Pb dating of the blueschists of the Ile de Groix. Geol Soc Am Mem 164: 229-238

Peucat JJ, Vidal P, Godard G, Postaire B (1982) Precambrian U-Pb zircon ages of eclogites and garnet pyroxenites from S. Brittany (France): an old oceanic crust in the W. European Hercynian Belt. Earth Planet Sci Lett 60: 70-78

Peucat JJ, Guerrot C, Bernard-Griffiths J, Ibarguchi JIG, Capdevila R, Iglesias M (1988 a) Occurrences of lower Proterozoic and Paleozoic high pressure rocks in N. W. Spain. Int Conf Terranes Variscan Belt Fr West Eur, Montpellier, (abstr)

Peucat JJ, Jegouzo P, Vidal P, Bernard-Griffiths J (1988 b) Continental crust formation seen through the Sr and Nd isotope systematics of S-type granites in the Hercynian belt of Western Europe. Earth Planet Sci Lett 88: 60-68

Piboule M, Briand B (1985) Geochemistry of eclogites and associated rocks of the southeastern area of the French Massif Central: origin of the protoliths. Chem Geol 50: 189-199

Pin C (1986) Datation U-Pb sur zircons à 285 Ma complexe gabbro-dioritique du Val Sesia-Val Mastallone et âge tardi-hercynien du métamorphisme granulitique de la zone Ivrea-Verbano (Italie). C R Acad Sci Paris 303: 827-830

Pin C, Carme F (1987) A Sm-Nd isotopic study of 500 Ma old oceanic crust in the Variscan belt of Western Europe: the Chamrousse ophiolite complex, Western Alps (France). Contrib Mineral Petrol 96: 406-413

Pin C, Lancelot JR (1978) Un exemple de magmatisme cambrien dans le Massif Central: les métadiorites quartziques intrusives dans la série du Lot. Bull Soc Géol Fr XX: 203-208

Pin C, Lancelot J (1982) U-Pb dating of an early Paleozoic bimodal magmatism in the French Massif Central and of its further metamorphic evolution. Contrib Mineral Petrol 79: 1-12

Pin C, Sills J (1986) Petrogenesis of layered gabbros and ultramafic rocks from Val Sesia, the Ivrea zone, NW Italy: trace element and isotope geochemistry. Geol Soc Spec Publ 24: 231-249

Pin C, Vielzeuf D (1983) Granulites and related rocks in Variscan median Europe: a dualistic interpretation. Tectonophysics 93: 47-74

Pin C, Vielzeuf D (1988) Les granulites de haute-pression d'Europe moyenne, témoins d'une subduction éo-hercynienne. Implications sur l'origine des groupes leptyno-amphiboliques. Bull Soc Géol Fr (8) IV: 13-20

Pin C, Majerowicz A, Wojciechowska I (1988) Upper Paleozoic oceanic crust in the Polish Sudetes: Nd-Sr isotope and trace element evidence. Lithos 21: 195-209

Quadt A (1985) Isotope data from the scheelite deposit Felbertal (Eastern Alps). Terra Cognita 5 (SL 11): 2-3

Quadt A, Gebauer D (1988) Sm/Nd, U/Pb and Rb/Sr dating of high-pressure ultramafic to felsic rocks from the Moldanubian area of NE Bavaria (FRG) and the Saxonian granulite massif (GDR). Conf Bohemian Massif Prague (abstr) p 71

Rajlich P, Synek J, Sarbach M, Schulmann K (1986) Hercynian-thrust related zones and deformation of the varied group on the contact of granulites (southern Moldanubian, Bohemian Massif). Geol Rundsch 75: 665-683

Rasamimanana G, Bébien J (1985) Le "trapp de Raon l'Etape" (Vosges septentrionales): un témoin important des manifestations volcaniques ayant accompagné le développement de l'orogenèse hercynienne. C R Acad Sci Paris 300: 859-862

Ricci CA, Sabatini G (1978) Petrogenetic affinity and geodynamic significance of metabasic rocks from Sardinia, Corsica and Provence. N Jahrb Mineral Monatsh 1: 28-38

Rivalenti G, Garutti G, Rossi A, Siena F, Sinigoi S (1981) Existence of different peridotite types and of a layered igneous complex in the Ivrea Zone of the Western Alps. J Petrol 22: 127-153

Rizki A, Baroz F (1988) Le volcanisme tholéïtique du massif de Schirmeck (Vosges septentrionales, France), témoins d'une zone de convergence de plaques au Paléozoïque supérieur. C R Acad Sci Paris 307: 511-516

Roddick JC, Compston W (1977) Strontium isotopic equilibration: a solution to a paradox. Earth Planet Sci Lett 34: 238-246

Sassi F, Cavazzini G, Visona D, Del Moro A (1985) Radiometric geochronology in the Eastern Alps: results and problems. Rend Soc Ital Mineral Petrol 40: 187-224

Sassi F, Zanferrari G, Zirpoli G (1987) The Caledonian event in the Eastern Alps a review. In: Flügel HW, Sassi FP, Grecula P (eds) Pre-Variscan and Variscan events in the

Alpine-Mediterranean mountain belts. Alfa, Bratislava, pp 431-434

Sawkins FJ, Burke K (1980) Extensional tectonics and mid-paleozoic massive sulfide occurrences in Europe. Geol Rundsch 69 (2): 349-360

Schmidt K, Söllner F (1983) Towards a geodynamic concept of the "Caledonian event" in Central and SW Europe. Verh Geol Bundesanst 3: 251-268

Shackleton RM, Ries AC, Coward MP (1982) An interpretation of the Variscan structures in S. W. England. J Geol Soc Lond 139: 533-541

Seyler M (1986) Petrology and genesis of Hercynian alkaline orthogneisses from Provence, France. J Petrol 27: 1229-1251

Sider H, Ohnenstetter M (1986) Field and petrological evidences for the development of an ensialic marginal basin related to the Hercynian orogeny in the north-eastern Massif Central, France. Geol Rundsch 75: 421-443

Söllner F, Hansen BT (1987) "Pan-afrikanisches" und "kaledonisches" Ereignis im Otztal-Kristall in den Ostalpen: Rb-Sr- und U-Pb-Altersbestimmungen an Migmatiten und Metamorphiten. Jahrb Geol Bundesanst 130: 529-569

Stosch HG, Lugmair GW (1984) Evolution of the lower continental crust: granulite facies xenoliths from the Eifel, West Germany. Nature (Lond) 311: 368-370

Stille P, Tatsumoto M (1985) Precambrian tholeiitic-dacitic rocks-suites and Cambrian ultramafic rocks in the Pennine nappe system of the Alps: Evidence from Sm-Nd isotopes and rare earth elements. Contrib Mineral Petrol 89: 184-192

Supply JP (1981) Géochronologie U-Pb et pétrologie des enclaves granulitiques de Bournac (Massif Central). Mém DEA, Univ Montpellier, 21 pp (unpubl)

Thiéblemont D, Cabanis B (1986) Découverte d'une association de volcanites d'arc et de basaltes de type "MORB" dans la formation paléo-volcanique silurienne de la Meilleraie, Vendée, France. C R Acad Sci Paris 302: 641-646

Thiéblemont D, Cabanis B, Wyns R, Treuil M (1987) Etude géochimique (majeurs et traces) de la formation amphibolitique de St Martin des Noyers (Vendée). Mise en évidence d'un paléo-arc insulaire dans la partie interne de l'orogène varisque. Bull Soc Géol Fr (8) III: 371-378

Thorpe RS (1979) Late Precambrian igneous activity in Southern Britain. Geol Soc Spec Publ 8: 579-584

Thorpe RS, Beckinsale RD, Patchett PJ, Piper JDA, Davies GR, Evans JA (1984) Crustal growth and late Precambrian-early Paleozoic plate tectonic evolution of England and Wales. J Geol Soc Lond 141: 521-536

van Breemen O, Aftalion M, Bowes DR et al. (1982) Geochronological studies of the Bohemian Massif, Czechoslovakia, and their significance in the evolution of Central Europe. Trans R Soc Edinb Earth Sci 73: 89-108

Veizer J, Jansen SL (1979) Basement and sedimentary recycling and continental evolution. J Geol 87: 341-370

Vidal P (1977) Limitations isotopiques à l'âge et à l'évolution de la croûte continentale en Europe moyenne et occidentale. In: La chaîne varisque d'Europe moyenne et occidentale. Colloq Int CNRS (ed) 243: 129-141

Vidal P, Peucat JJ, Lasnier B (1980) Dating of granulites involved in the Hercynian fold belt of Europe: an example taken from the granulite facies orthogneisses at La Picherais, Southern Armorican Massif. Contrib Mineral Petrol 72: 283-289

Vidal P, Auvray B, Charlot R, Cogné J (1981) Precambrian relicts in the Armorican Massif. Their ages and roles in the evolution of the Western and Central European Cadomian-Hercynian belt. Precambrian Res 14: 1-20

Villeneuve M, Dallmeyer RD (1987) Geodynamic evolution of the Mauritanide, Bassaride, and Rokelide orogens (West Africa). Precambrian Res 37: 19-28

Vitrac-Michard A, Allègre CJ (1975) 238U-206Pb, 235U-207Pb systematics on Pyrenean basement. Contrib Mineral Petrol 51: 205-212

Voshage H, Hunziker JC, Hofmann AW, Zingg A (1987) A Nd and Sr isotopic study of the Ivrea zone, Southern Italy. Contrib Mineral Petrol 97: 31-42

Waldhausrova J (1984) Proterozoic volcanites and intrusive rocks of the Jilove zone, central Bohemia. Krystalinikum 17: 77-97

Weber K (1984) Variscan events: early Paleozoic continental rift metamorphism and late Paleozoic crustal shortening. In: Hutton DHW, Sanderson DJ (eds) Variscan tectonics of the North Atlantic region. Geol Soc Spec Publ 14: 3-22

Wright JE, Wyld SJ (1986) Significance of xenocrystic Precambrian zircon contained within the southern continuation of the Josephine ophiolite: Devils Elbow ophiolite remnant, Klamath Mountains, northern California. Geology 14: 671-674

Zoubek V (1974) Quelques problèmes généraux du Précambrien de la zone varisque d'Europe. In: La chaîne varisque d'Europe moyenne et occidentale. Colloq Int CNRS (ed) 243: 121-127

Late Proterozoic and Early Paleozoic Tectonothermal Evolution of the Eastern Alps

F. NEUBAUER[1]

1 Introduction

The nature and age of the Alpine basement has been a controversial topic of the Alpine geology (e. g. von Raumer 1988). The fossil record begins in the Late Ordovician (for a review, see Schönlaub 1979) except for scarce sporomorphs and acritarchs in metamorphic areas (Giorgi et al. 1979; Reitz and Höll 1988; Sassi et al. 1984) which are not related to continuous Late Ordovician to Carboniferous sections. These data suggest the existence of a distinct late Proterozoic to Early Ordovician sedimentary cycle separated from the later one. Geochronological data from crystalline complexes structurally removed from low-grade sediments favour the presence of a basement metamorphosed and cooled during Early Paleozoic times (for a recent review, see Ebner et al. 1987; Sassi et al. 1987). The relationship between weakly metamorphosed fossiliferous Late Ordovician to Carboniferous strata and crystalline complexes is reviewed in this chapter.

Basement of the Eastern Alps is diversified in lithology, age and tectonic setting. The basement was affected by events of the Variscan and Alpine orogenies which obliterated older structures. Recently, models have been proposed to explain the successive consolidation of the Alpine basement in terms of terranes (e. g. Frisch and Neubauer 1989).

From the fossiliferous early Paleozoic strata, the existence of a Precambrian basement is suggested by the following relationships:

1. An only locally preserved unconformity occurs between crystalline basement rocks and low-grade Ordovician metasedimentary units (Neubauer 1985).
2. Siliciclastic detritus of Late Ordovician to Early Devonian sandstones, heavy minerals, and magmatic and metamorphic components of agglomerates, ortho- and paraconglomerates indicate the presence of a continental hinterland of a pre-Late Ordovician age (Ebner et al. 1987).
3. Upper intercept U/Pb ages of detrital zircons within Paleozoic metasediments are ca. 2–2.3 Ga and suggest the existence of an early Proterozoic hinterland (Grauert et al. 1973; Söllner and Hansen 1987).
4. Inherited lead isotopic components (3.2–0.8 Ga) of zircons in Paleozoic and Tertiary magmatic rocks show evidence for old crust in the region considered for magma generation and emplacement (Cliff 1980; Söllner and Hansen 1987; Neubauer et al. 1989 a).
5. Lead isotopic data from stratiform ore deposits in Devonian strata indicate the existence of a 2–3 Ga old lead component derived from U-depleted, Th-rich lower crust (Koeppel 1984).

2 Precambrian and Early Paleozoic Lithostratigraphic Units in the Eastern Alps

Late Precambrian and early Paleozoic crystalline complexes occur in five distinct settings within the Eastern Alps (Fig. 1): (1) Basement structural slices ("Kaintaleck" slices) along the basal thrust plane of fossiliferous Paleozoic strata in the Greywacke Zone (Fig. 2); (2) gneiss-amphibolite association; (3) the Speik complex and related units; (4) the Koriden and the Plankogel complex; and (5) the Stubach Complex and Habach Group of the Penninic basement in the Tauern Window. In terms of Alpine tectonics, the units (1) and (4) belong to the Austroalpine domain. The Penninic unit is a continuation of the European crust which was overriden by the Austroalpine units during Cenozoic plate collision. The Austroalpine unit is the northern margin of the Adriatic microplate (or "Apulia"), which is thought to be a promontory of Africa during the Alpine cycle.

[1] Institute of Geology and Paleontology, Karl-Franzens-University, Heinrichstraße 26, A-8010 Graz, Austria

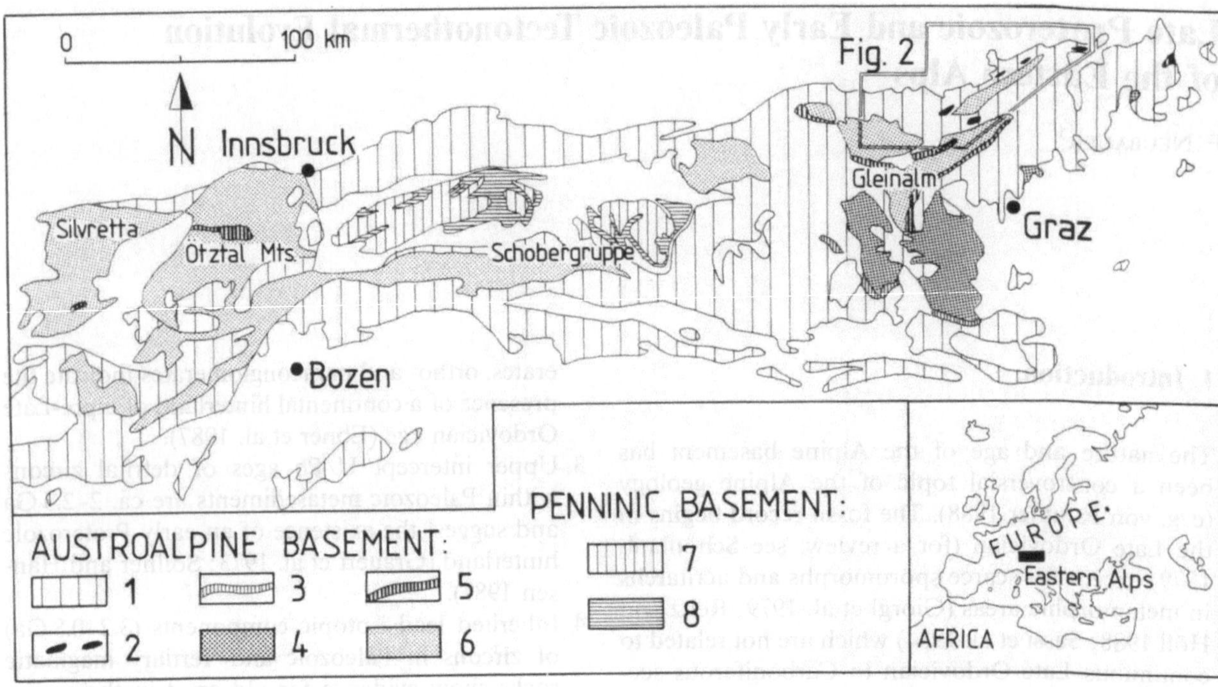

Fig. 1. Precambrian and early Paleozoic tectonostratigraphic units within the basement of the Eastern Alps. *1–6* Austroalpine basement, *1* Late Ordovician – Carboniferous tectonostratigraphic units and units with unknown age; *2* Kaintaleck basement slices; *3* Plankogel complex; *4* Koriden complex; *5* Speik complex and related units; *6* amphibolite-gneiss association; *7–8* Penninic basement; *7* late Paleozoic granitoids; *8* Stubach complex, Habach group and related units.

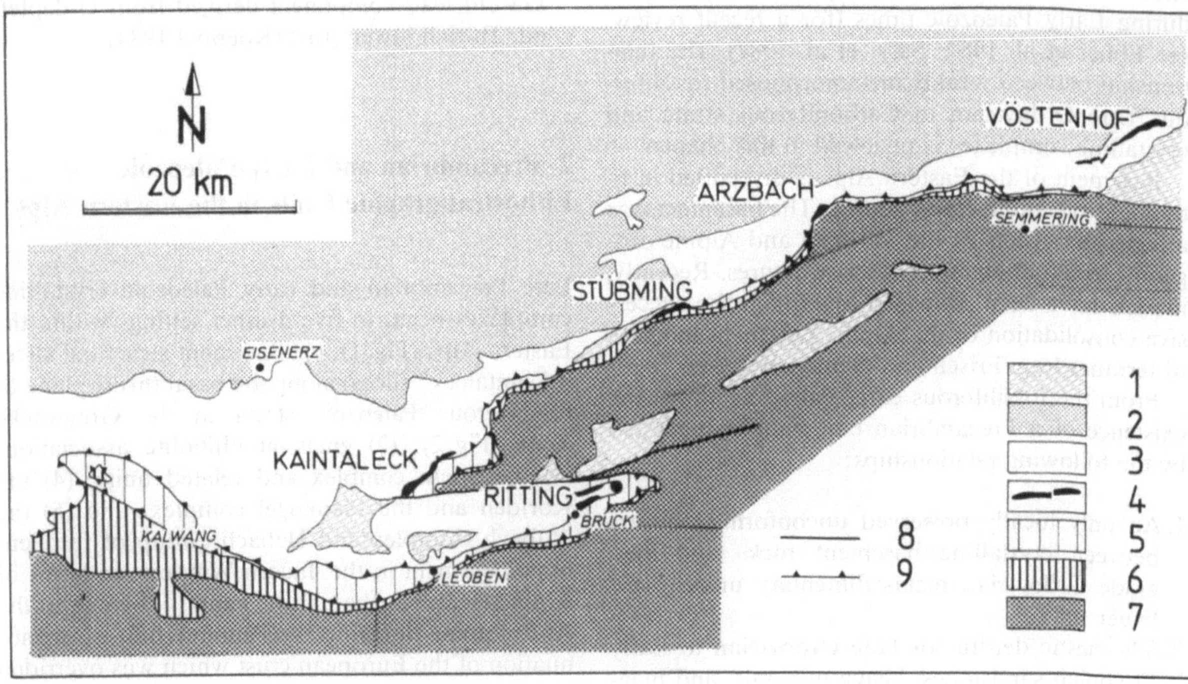

Fig. 2. Simplified geological map of major tectonostratigraphic units of the northeastern Alps. Simplified and modified after Schönlaub (1979). *1* Cenozoic sediments; *2* Permomesozoic sediments of the northern Calcareous Alps; *3–5* Graywacke Zone: *3* Late Ordovician – Visean metasediments; *4* Kaintaleck slices; *5* Visean to Early Permian metasediments; *6* Permotriassic metasediments at the top of lower tectonic units; *7* Variscan basement of lower tectonic units including some Permotriassic metasediments; *8* fault; *9* major Alpidic thrust surfaces.

2.1 Kaintaleck Slices

The Kaintaleck slices (Fig. 2) occur as thin basement structural slices in different lithologies along the basal thrust surface of a major Alpine thrust sheet. The most widespread lithologic association (type Ritting) is represented by garnet-zoisite amphibolites, serpentinites, paragneisses, marbles and quartzites. The chemistry of mafic rocks is similar to N-MOR basalts (Neubauer et al. 1989b). A disseminated Cu-Fe mineralization and the composition of Cr-spinels in serpentinites favour an ophiolite-like origin of this complex (Neubauer and Thalhammer 1989).

Another slice (Frauenberg slice east of locality Bruck of Fig. 2) is composed predominantly of plagioclase-amphibolites and intermediate orthogneisses. The chemistry of these rocks favours an alkaline differentiation trend. After the U/Pb zircon pattern of intermediate orthogneisses, Neubauer et al. (1989a) postulated a late Archean age of protolith crystallization (2.53 Ga). The lower intercept (516 Ma) suggests an early Paleozoic thermal overprint. In both lithologies, Neubauer and Thalhammer (1989) recognized three stages of metamorphism: An early high-pressure metamorphic stage is reflected by pyrope-rich garnets and phengitic muscovites. The garnets decomposed to symplectitic zoisite + amphibole during a second metamorphic stage. Estimation of P-T conditions for the second stage yielded ca. 5 kb and 550–600 °C. The ca. 516 Ma lower intercept age of zircons is interpreted as the age of high-grade metamorphism. The third stage is reflected by an Alpine thermal overprint during early Cretaceous times (Neubauer et al. 1987).

These crystalline complexes are unconformably overlain by a metaconglomerate which comprises the stratigraphic base of fossiliferous sequences (Fig. 3; Neubauer 1985). Metaconglomerates and boulder- and pebble-bearing metapsammitic phyllites are composed of fragments of the underlying basement. In addition, trondhjemitic orthogneisses are very common components (Daurer and Schönlaub 1978; Neubauer 1985). Two boulders of the trondhjemitic orthogneiss from two different localities yielded U-Pb zircon upper intercept ages of approximately 500 Ma (Neubauer et al. 1987). The upper intercept ages are interpreted as the age of protolith formation. Chemical patterns of orthogneisses argue for an origin from volcanic arc granitoids.

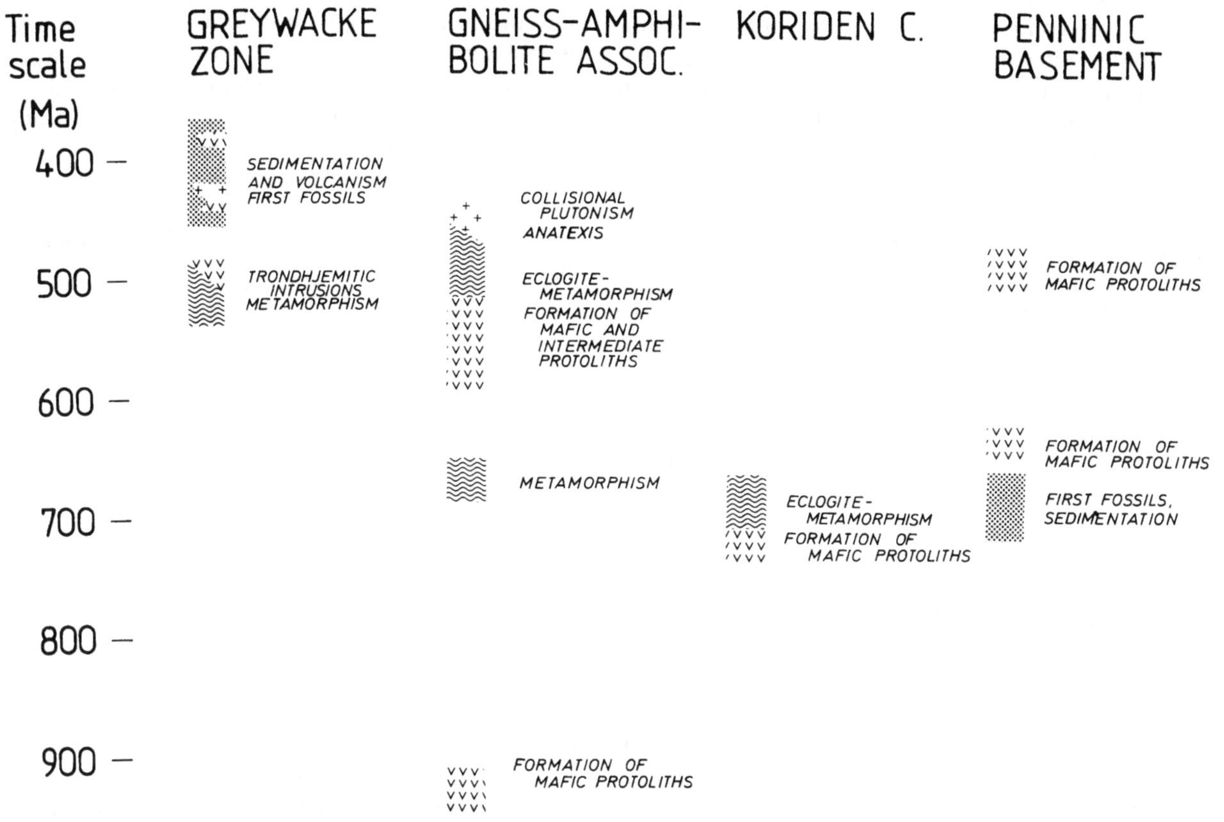

Fig. 3. Chronological order of late Proterozoic and early Paleozoic events of basement complexes in the Eastern Alps.

The age of the ophiolite-like assemblage is not known precisely; however, they must be older than Late Ordovician of overstepping sediments. The ophiolite-like assemblage and arc-related magmatic rocks preserved in the conglomerates argue for an overall arc-related environment during formation.

The basal formations of fossiliferous sediments contain horizons of pebbles, greenschists and acidic metatuffaceous material. Schäffer and Tarkian (1986) suggested that these have been formed in a rift-environment. Neubauer and Frisch (1988) argued for the existence of a back-arc basin postdating the orogeny. The magmatism evolved from calc-alkaline basic (Giese 1988) to immediately following acidic, probably collisional type (Heinisch 1981). Neubauer and Frisch (1988) compared the geodynamic setting with the intraorogenic back-arc basins like the Egean Sea within the east-Mediterranean Alpine mountain belts.

2.2 *Gneiss-Amphibolite Association*

The gneiss-amphibolite association consists of extensive areas of paragneisses (biotite-plagioclase gneisses), sheetlike orthogneisses (granodioritic to granitic in composition) and large bodies of amphibolites intercalated within gneisses. Originally, Frank et al. (1976) argued for a cogenetic evolution of orthogneisses and amphibolites. The orthogneisses yielded a Rb/Sr isochron with a low initial Sr isotopic ratio (0.7041 ± 0.0012). The amphibolites of the Gleinalm (Fig. 1) are derived from calcalkaline and tholeiitic basalts (Frisch et al. 1987; Neubauer 1988). On the other hand, Maggetti et al. (1987) and Maggetti and Galetti (1988) presented evidence for an island arc tholeiite (N-MORB-like) origin for eclogites in western exposures of the amphibolite-gneiss association. Preliminary Rb/Sr data of intermediate and mafic rocks of the Silvretta region suggest evidence of protolith formation between ca. 950–900 Ma (Flisch, pers. commun.).

The tectonothermal history of the gneiss-amphibolite association is very complex (Fig. 3). The age of protolithic sediments is Precambrian, at least in the western region (Ötztal Mts.) because Söllner and Hansen (1987) reported evidence for metamorphism in paragneisses ca. 670 Ma (see geochronological data in detail, Table 1). On the other hand, the metamorphic path in some mafic rocks starts with an eclogite metamorphism dated at ca. 500 Ma (U-Pb zircon method; Söllner and Gebauer, cited in Söllner and Hansen 1987) which immediately followed protolith formation (Söllner and Gebauer, pers. comm.). Subsequent decompression and uplift was accompanied by a high geothermal gradient which resulted in partial melts in metapelite rocks. The age of this anatectic event is well-supported by U/Pb zircon and Rb/Sr data. However, widespread plutonism of Late Ordovician age is of collisional type (Thöni 1986). This plutonism was probably related to Late Ordovician acidic volcanism preserved within basal fossiliferous formations.

2.3 *Speik Complex and Related units*

In the eastern Austroalpine area, a strongly metamorphosed and dismembered ophiolite complex (Speik complex) occurs as a nappe complex in the hangingwall of the amphibolite-gneiss association. A similar complex is incorporated within the amphibolite-gneiss complex in the western Austroalpine region (Fig. 1). The strongly metamorphosed Speik complex contains all members of an ophiolite from refractory ultramafic rocks (El Ageed et al. 1980), ultramafic and mafic cumulates to amphibolites similar to N-MOR basaltic liquids (Frisch et al. 1987; Neubauer 1988; Neubauer et al. 1989b). The age constraints are poor. Similar to an ophiolite-like assemblage within the Kaintaleck slices, the Speik complex is overlain by a metamorphosed and fossil-free section similar in lithology and lithostratigraphy to the fossiliferous Late Ordovician to Carboniferous of other Austroalpine areas. Although geochronological data are lacking, the age of the Speik complex should be older than Late Ordovician.

2.4 *Koriden Complex*

The Koriden complex (Fig. 1) is composed of monotonous paragneisses which contain lense-shaped eclogites and eclogitic amphibolites (Miller et al. 1988). The age of magmatic protoliths and the age of metamorphic events are matters of speculation. The eclogites were derived, in part, from metagabbro protoliths which are preserved in some regions (Heritsch 1973). Miller et al. (1988) reported low-Ti eclogites and eclogites of a N-MORB affinity later confirmed by Manby and Thiedig (1988). Preliminary Sm/Nd data (Manby and Thiedig 1988) indicate a late Proterozoic age for the eclogite protoliths (716 ± 30 Ma). On the other hand, a Sm/Nd garnet

Table 1. Geochronological data of late Proterozoic and early Paleozoic tectonostratigraphic units of Eastern Alps[a]

Rock type	Method[b]	Age (Ma)	Reference
1. Kaintaleck slices:			
Hornblende-garnet gneiss	U/Pb zircon u. i.	ca. 2530	Neubauer et al. (1989)[a]
	U/Pb zircon l. i.	ca. 516	
Trondhjemitic gneiss boulders	U/Pb zircon u. i.	$502 \pm {}^{6}_{5}$	Neubauer et al. (1987)
	U/Pb zircon l. i.	500 ± 20	Neubauer et al. (1987)
2. Gneiss-amphibolite association:			
Migmatitic gneisses	U/Pb zircon, diepisodic model	ca. 670	Söllner and Hansen (1987)
Plagioclase gneiss	Rb/Sr WR	518 ± 46	Frank et al. (1976)
Eclogite	U/Pb zircon	ca. 497	Söllner and Gebauer (cited in Söllner and Hansen 1987)
Quartzitic biotite gneiss	U/Pb zircon l. i.	490 ± 14	Söllner and Hansen (1987)
Schollen migmatite	Rb/Sr WR	455 ± 27	Söllner and Hansen (1987)
Monzonitic gneiss	U/Pb zircon l. i.	$455 \pm {}^{2}_{4}$	Söllner and Hansen (1987)
Augen gneiss	Rb/Sr WR	455 ± 13	Troll et al. (1976)
Augen gneiss	Rb/Sr WR	451 ± 18	Satir (1976)
Augen gneiss	Rb/Sr WR	449 ± 4	Borsi et al. (1973)
Augen gneiss	Rb/Sr WR	445 ± 24	Hammerschmidt (1981)
Granite gneiss	U/Pb zircon l. i.	444 ± 4	Söllner and Hansen (1987)
Tonalitic gneiss	U/Pb zircon l. i.	$443 \pm {}^{13}_{16}$	Cliff (1980)
Granite gneiss	Rb/Sr WR	442 ± 9	Grauert (1969)
Orthogneiss	Rb/Sr WR	432 ± 16	Scharbert (1981)
Orthogneiss	U/Pb zircon l. i.	$427 \pm {}^{10}_{11}$	Cliff (1980)
Augen gneiss	Rb/Sr WR	424 ± 9	Harre et al. (1968)
3. Koriden			
Eclogite	Sm/Nd WR	716 ± 30	Manby and Thiedig (1988)
	Sm/Nd WR-garnet	693 ± 39	Manby and Thiedig (1988)
4. Penninic basement (Tauern window):			
"Basal amphibolite"	U/Pb zircon u. i.	657 ± 14	Quadt (1989)
	Sm/Nd zircon	640 ± 7	Quadt (1989)
Amphibolite	U/Pb zircon u. i.	498 ± 4	Quadt (1984)
Garnet amphibolite	Sm/Nd m. i.	498 ± 12	Quadt (1989)
	U/Pb zircon	486 ± 5	Quadt (1989)
Hornblendite	U/Pb zircon	492 ± 2	Quadt (1984)

[a] Only data older than 420 Ma are included in the table. Note that there is a late Paleozoic and Alpidic metamorphic overprint in these units also recorded in geochronological data.
[b] *l. i.* Lower intercept age; *m. i.* mineral isochron; *u. i.* upper intercept age; *WR* whole rock isochron.

- whole rock isochron defines an age of 693 ± 39 Ma, which is close to the suggested protolith age.

2.5 Penninic Domain

The basement of the Penninic domain is exposed in the Tauern window within the Eastern Alps (Fig. 1). It contains two tectonostratigraphic units, possibly related and later intruded by Variscan granitoids. The Stubach complex is regarded either as a backarc ophiolite or as the base of a magmatic arc (Frisch and Raab 1987; Neubauer et al. 1989b). The Habach group contains metavolcanic and metasedimentary rocks. There is an agreement between all workers that the volcanic sequence is an expression of an island arc. The chemical composition of primary magmatic rocks ranges from boninite and basalt to andesite (Frisch and Raab 1987; Steyrer 1982; Steyrer and Höck 1985; Thalhammer 1987). Hornblendite and amphibolite yielded two groups of U-Pb zircon and Sm/Nd ages (Table 1), of ca. 650 and 500–480 Ma (Quadt 1984, 1989) which were interpreted to reflect protolith ages (Fig. 3). These ages contrast, in part, with fossils within the metasedimentary rocks which argue for a late Proterozoic age (Reitz and Höll 1988). This apparent discrepancy could reflect that: (1) The relationship between metavolcanic and metasedimentary rocks is not a primary one or (2) the dated mafic rocks are of plutonic origin and were intruded during the early Ordovician into the preexisting metasedimentary sequence.

3 Discussion

The data reviewed above argue for the existence of arc-related environments in the Eastern Alps during the late Proterozoic and early Paleozoic. Basement complexes (like the Koriden and the Plankogel complex which is related to the Koriden complex) may be interpreted in terms of a melange-like subduction complex which underwent subsequent HP-metamorphism. Other complexes (like large parts of the amphibolite-gneiss association) are magmatic arcs which likely reflect large masses of magmatic rocks created above a subduction zone.

The subduction process was probably an episodic one, because geochronological data favour different magmatic stages between 950 and 500 Ma. The timing of eclogite metamorphism also indicates two stages at ca. 700 and 500 Ma.

The relationship of Austroalpine basement complexes to surrounding continental lithosphere is not known. Paleomagnetic data of pre-Silurian rocks are lacking. The missing limestones in pre-Ashgill Ordovician sedimentary sequences suggest that sedimentation took place at high paleolatitudes (Schönlaub 1979). Environmental data of Late Ordovician rocks favour a relationship to the Armorica-Sardinia block and to the glacial event in northern Africa (Schönlaub 1979).

However, the subduction and subsequent orogeny due to collision indicate the relation to large-scale plate movements. Similar, sparse geochronologic data relating to metamorphic or deformation events are known from the entire Alpine belt extending from Turkey to the Eastern Alps (Kräutner 1987; Rudakow 1987; Satir and Friedrichsen 1986; Senghör et al. 1984). In the Western Alps, geochronologic data indicate formation of oceanic crust at ca. 500 Ma (Pin and Carme 1987; Menot et al. 1988). Similar island arc assemblages and ophiolites were reported from the Balkan region situated between the Alps and Turkey (Haydoutov 1987).

In the Eastern Alps, geochronologic data suggest an orogenic event similar in age to the event Cadomian I (appox. 670 Ma) and the event Cadomian II in northern Armorica of western Europe (Cogne and Wright 1980; Chantraine et al. 1988; Dissler et al. 1988; Peucat 1986; Pin, this Vol.). The Cadomian II event is also known as the local Sar-

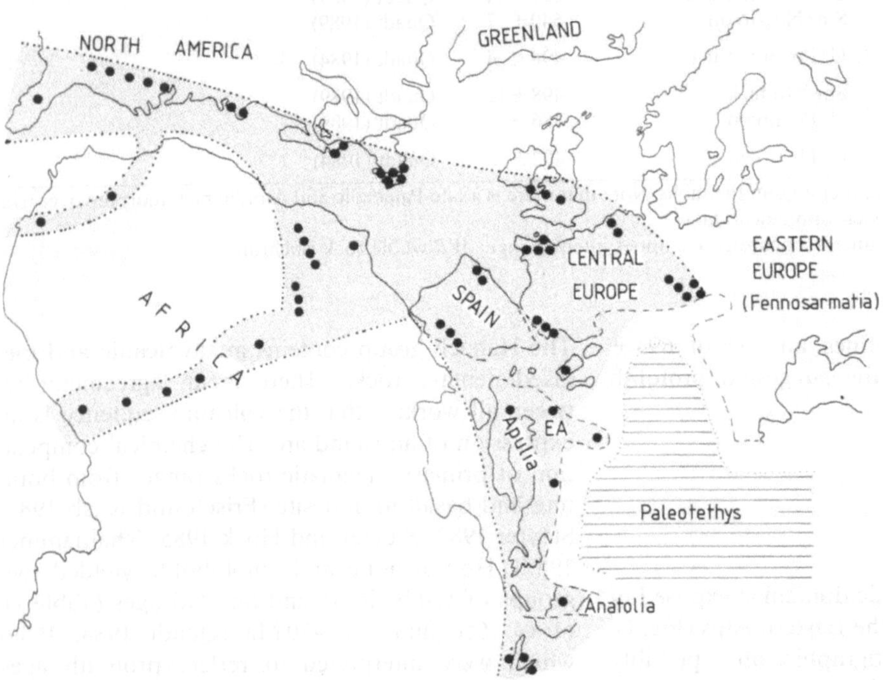

Fig. 4. Distribution of late Proterozoic and early Paleozoic orogenic events in southern Europe and northern Africa. Late Permian reconstruction, simplified and completed after Neugebauer (1988). *Dotted area* Continental lithosphere suffered some compressional plate movements between 670 and 500 Ma; *filled circles* evidence for orogenic events between 670 and 500 Ma by geochronological (and biostratigraphic) methods; *EA* Eastern Alps

dic phase in southern Europe. Orogenic events straddling the Precambrian/Cambrian boundary (Cadomian II) are known in the Bohemian massif and along its eastern margin (e.g. van Breemen et al. 1982; Scharbert and Batik 1980). Both events coincide closely with Pan-African orogenic events of Africa (e.g. Dallmeyer and Lecorche 1989, and this Vol.), and respectively with the Grampian and Penobscotian of the Caledonian-Appalachian belt. Thus, there is increasing evidence for compressional plate movements, whether in subduction zones or in orogenic belts which surround northern Africa (Fig. 4) in a broad belt.

The Apulian microplate which acted as an indenter during Alpine orogeny is related to Africa during the Alpine cycle. In the Alpidic start position of the Eastern Alps in the Permian shown in the cartoon of Fig. 4, the Austroalpine region belong to the future Apulian microplate and, therefore, to Africa. Thus, it is striking to relate the late Proterozoic/early Paleozoic evolution of the Eastern Alps to the Pan-African belt of Africa (neglecting the effects of Variscan plate movements for simplicity: Fig. 4).

Acknowledgement

The author would like to thank R. D. Dallmeyer for discussion to clear up some inconsistencies, for constructive criticism, and for linguistic improvement of an earlier version of this paper.

References

Borsi S, del Moro A, Sassi FP, Zirpoli G (1973) Metamorphic evolution of the Austridic rocks to the South of the Tauern Window (Eastern Alps). Radiometric and geopetrologic data. Memor Sci Geol Ital 12: 549-571

Chantraine J, Chauvel JJ, Bale P, Denis E, Rabu D (1988) Le Brioverien (Proterozoique superieur a terminal) et l'orogenese cadomienne en Bretagne (France). Bull Soc Geol Fr 1988 (8) IV: 815-829

Cliff RA (1980) U-Pb isotopic evidence from zircons for lower Paleozoic tectonic activity in the Austroalpine nappe, the Eastern Alps. Contrib Mineral Petrol 71: 283-288

Cogne J, Wright JB (1980) L'orogene cadomien. In: Cogne J, Slansky M (eds) Geologie of Europe. BRGM, Orleans, pp 29-55

Dallmeyer RD, Lecorche JP (1989) $^{40}Ar/^{39}Ar$ polyorogenic mineral record within the central Mauretanide orogen, West Africa. Geol Soc Am Bull 101: 55-70

Daurer A, Schönlaub HP (1978) Anmerkungen zur Basis der Nördlichen Grauwackenzone. Mitt Österr Geol Ges 69 (1976): 77-88

Dissler E, Dore F, Dupret L, Gresselin F, Le Gall J (1988) L'evolution geodynamique cadomienne du Nord-Est du Massif armorcain. Bull Soc Geol Fr 1988 (8)/IV/5: 801-814

Ebner F, Neubauer F, Stattegger K (1987) The Caledonian event in the Eastern Alps: A review. In: Flügel HW, Sassi FP, Grecula P (eds) Pre-Variscan and Variscan events in the Alpine-Mediterranean mountain belts. Alfa, Bratislava, pp 169-182

El Ageed A, Saager R, Stumpfl EF (1980) Pre-Alpine ultramafic rocks in the eastern Central Alps, Styria, Austria. Proc Int Ophiolite Symp, Cyprus, A 601-606

Frank W, Klein P, Nowy W, Scharbert S (1976) Die Datierung geologischer Ereignisse im Altkristallin der Gleinalpe (Steiermark) mit der Rb/Sr-Methode. Tschermaks Mineral Petrograph Mitt 23: 191-203

Frisch W, Neubauer F (1989) Pre-Alpine terranes and tectonic zoning in the Eastern Alps. Geol Soc Am Spec Publ 230: 91-100

Frisch W, Raab D (1987) Early Paleozoic Back-Arc and Island-Arc Settings in Greenstone Sequences of the Central Tauern Window (Eastern Alps). Jahrb Geol Bundesanst 129: 545-566

Frisch W, Neubauer F, Bröcker M, Brückmann W, Haiss W (1987) Interpretation of geochemical data from the Caledonian basement within the Austroalpine basement complex. In: Flügel HW, Sassi FP, Grecula P (eds) Pre-Variscan and Variscan events in the Alpine-Mediterranean mountain belts. Alfa, Bratislava, pp 209-226

Giese U (1988) Lower Paleozoic volcanic evolution at the northwestern border of the Gurktal nappe, Upper Austroalpine, Eastern Alps. Schweizer Mineral Petrograph Mitt 68: 381-396

Giorgi L, Giraud P, Vachard D (1979) Sur la presence de microorganismes d'age cambrien dans les schistes cristallins du versant occidental du massif cristallin externe des Grandes Rousses (Alpes occidentales). C R Acad Sci Paris 288 D: 1079-1082

Grauert B (1969) Die Entwicklungsgeschichte des Silvretta-Kristallins auf Grund radiometrischer Altersbestimmungen. Ph thesis Univ Berne, 166 pp

Grauert B, Arnold A (1968) Deutung diskordanter Zirkonalter der Silvrettadecke und des Gotthardmassivs. Contrib Mineral Petrol 20: 34-55

Grauert B, Hänny R, Soptrajanova G (1973) Age and origin of detrital zircons from the pre-Permian basement of the Bohemian Massif and the Alps. Contrib Mineral Petrol 40: 105-130

Hammerschmidt K (1981) Isotopengeologische Untersuchungen am Augengneis vom Typ Campo Tures bei Rain in Taufers, Südtirol. Mem Sci Geol XXXIV: 273-300

Harre W, Kreuzer H, Lenz H, Müller P, Wendt I, Schmidt K (1968) Rb/Sr and K/Ar-Altersbestimmungen an Gesteinen des Ötztalkristallins. Geol Jahrb 86: 797-826

Haydoutov I (1987) Ophiolites and island arc igneous rocks in the Caledonian basement of South Carpathian region. In: Flügel HW, Sassi FP, Grecula P (eds) Pre-Variscan and Variscan events in the Alpine-Mediterranean mountain belts, Mineralia slovaca - Monography. Alfa, Bratislava, pp 279-292

Heinisch H (1981) Zum ordovizischen "Porphyroid"-Vulkanismus der Ost- und Südalpen, Stratigraphie, Petrographie und Geochemie. Jahrb Geol Bundesanst 124: 1-109

Heritsch H (1973) Die Bildungsbedingungen von alpinotypem Eklogitamphibolit und Metagabbro, erläutert an Gesteinen der Koralpe, Steiermark. Tschermaks Mineral Petrogr Mitt 19: 213-271

Koeppel V (1984) Lead isotopes as tracers of the origin of metals in ore deposits and of the evolution of continental crust: examples from western and central Europe. Proc 27th Int Geol Congr, VNU Sci Press, Amsterdam, 12: 53-82

Kräutner HG (1987) The metamorphic Paleozoic of the Romanian Carpathians. In: Flügel HW, Sassi FP, Grecula P (eds) Pre-Variscan and Variscan events in the Alpine-Mediterranean mountain belts. Alfa, Bratislava, pp 329-350

Maggetti M, Galetti G (1988) Evolution of the Silvretta eclogites: metamorphic and magmatic events. Schweizer Mineral Petrograph Mitt 68: 467-484

Maggetti M, Galetti G, Stosch HG (1987) Eclogites from the Silvretta Nappe (Switzerland): geochemical constraints on the nature and geotectonic setting of their protoliths. Chem Geol 64: 319-334

Manby GM, Thiedig F (1988) Petrology of eclogites from the Saualpe, Austria. Schweizer Mineral Petrograph Mitt 68: 441-466

Menot R-P, Peucat JJ, Scarenzi D, Piboule M (1988) 496 My age of plagiogranites in the Chamrousse ophiolite complex (external crystalline massifs in the French Alps): evidence of a Lower Paleozoic oceanization. Earth Planet Sci Lett 88: 82-92

Miller Ch, Stosch HG, Hoernes S (1988) Geochemistry and origin of eclogites from the type locality Koralpe and Saualpe, Eastern Alps, Austria. Chem Geol 67: 103-118

Neubauer F (1985) Eine präoberordovizische Transgression in der Grauwackenzone (Ostalpen) und ihre Bedeutung für paläozoische Geodynamik und alpidischen Deckenbau. N Jahrb Geol Paläontol Monatsh 1985/1: 46-64

Neubauer F (1988) Bau und Entwicklungsgeschichte des Rennfeld-, Mugel- und des Gleinalmkristallins (Ostalpen). Abh Geol Bundesanst 42: 1-137

Neubauer F, Frisch W (1988) Ordovician-Silurian geodynamic evolution of the Alps - the orogeny back-arc basin model. Schweizer Mineral Petrograph Mitt 68: 351-357

Neubauer F, Thalhammer O (1989) Mineralization, deformation and metamorphism of the "Panafrican" basement of the Eastern Greywacke zone, Eastern Alps. Terra abstracts 1: 45

Neubauer F, Frisch W, Hansen BT (1987) Time relations between eoalpine metamorphism and thrusting: evidence from the crystalline basement of the Eastern Greywacke zone. In: Flügel HW, Faupl P (eds) Geodynamics of the Eastern Alps. Deuticke, Vienna, pp 263-271

Neubauer F, Frisch W, Hansen BT (1989a) A Late Archean rock in the Eastern Alps? Terra (abstr) 1/1: 5

Neubauer F, Frisch W, Schmerold R, Schlöser H (1989b) Metamorphosed and dismembered ophiolite suites in the basement units of the Eastern Alps. Tectonophysics 165: 49-62

Neugebauer J (1988) The Variscan plate tectonic evolution: an improved "Iapetus model". Schweizer Mineral Petrograph Mitt 68: 313-333

Peucat JJ (1986) Behaviour of Rb-Sr whole rock and U-Pb zircon systems during partial melting as shown in migmatitic gneisses from the St Malo Massif, NE Brittany, France. J Geol Soc Lond 143: 875-885

Pin C, Carme F (1987) A Sm-Nd isotopic study of 500 Ma old oceanic crust in the Variscan belt of Western Europe: the Chamrousse ophiolite complex, Western Alps (France). Contrib Mineral Petrol 96: 406-413

Quadt A (1984) Geochemische und U-Pb-Untersuchungen der Unteren Schieferhülle (Ostalpen). Fortschr Mineral Beih 62/1: 191-193

Quadt A (1989) Sm-Nd and U-Pb systematics of zircon fractions, garnet and whole-rock amphibolitic rocks of the Tauern window (Eastern Alps). Terra abstracts 1: 349

Reitz E, Höll R (1988) Jungproterozoische Mikrofossilien aus der Habachformation in den mittleren Hohen Tauern und dem nordostbayerischen Grundgebirge. Jahrb Geol Bundesanst 131: 329-340

Rudakov SG (1987) Classification of pre-Mesozoic zones of the Carpathian-Balkan region according to the type of evolution during the Late Proterozoic and Paleozoic. In: Flügel HW, Sassi FP, Grecula P (eds) Pre-Variscan and Variscan events in the Alpine-Mediterranean mountain belts. Alfa, Bratislava, pp 415-426

Sassi FP, Kalvacheva R, Zanferrari A (1984) New data on the age of deposition of the Southalpine phyllitic basement in the Eastern Alps. N Jahrb Geol Paläontol Monatsh 1984: 741-751

Sassi FP, Zanferrari G, Zirpoli G (1987) The Caledonian event in the Eastern Alps: a review. In: Flügel HW, Sassi FP, Grecula P (eds) Pre-Variscan and Variscan events in the Alpine-Mediterranean mountain belts. Alfa, Bratislava, pp 431-434

Satir M (1976) Rb-Sr- und K-Ar-Altersbestimmungen an Gesteinen und Mineralien des südlichen Ötztalkristallins und der westlichen Hohen Tauern. Geol Rundsch 65: 394-410

Satir M, Friedrichsen H (1986) The origin and evolution of the Menderes Massif, W-Turkey: A rubidium/strontium and oxygen isotope study. Geol Rundsch 75: 703-715

Schäffer U, Tarkian M (1986) Die Genese der stratiformen Sulfidlagerstätte Kalwang (Steiermark), der Grünsteinserie und einer assoziierten silikatreichen Eisenformation. Tschermaks Mineral Petrograph Mitt 33: 169-186

Scharbert S (1981) Untersuchungen zum Alter des Seckauer Kristallins. Mitt Ges Geol Bergbaustudenten Österr 27: 173-188

Scharbert S, Batik P (1980) The Age of the Thaya (Dyje) Pluton. Verh Geol Bundesanst 1980/3: 325-331

Schönlaub HP (1979) Das Paläozoikum. Österr Abh Geol Bundesanst 33: 1-124

Senghör AMC, Satir M, Akkök R (1984) Timing of tectonic events in the Menderes Massif, Western Turkey: Implications of tectonic evolution and evidence for Pan-African basement in Turkey. Tectonics 3: 693-707

Söllner F, Hansen BT (1987) "Pan-afrikanisches" und "kaledonisches" Ereignis im Ötztal-Kristallin der Ostalpen: Rb-Sr- und U-Pb-Altersbestimmungen an Migmatiten und Metamorphiten. Jahrb Geol Bundesanst 130: 529-569

Steyrer HP (1982) Geochemie, Petrographie und Geologie der Habach-Formation im Originalgebiet zwischen äußerem Habachtal und Untersulzbachtal (Pinzgau, Salzburg). Ph D Thesis Univ Salzburg (unpubl)

Steyrer HP, Höck V (1985) Geochemistry of metabasalts in the Habach formation (Salzburg, Hohe Tauern, Austria). A preliminary report. Ofioliti 10: 441-456

Thalhammer O (1987) Boninites as source rocks of tungsten mineralization at Mittersill, Austria? Mineral Petrol 39: 221-242

Thöni M (1986) The Rb-Sr thin slab isochron method - an unreliable geochronologic method for dating geologic events in polymetamorphic terrains? Mem Sci Geol 38: 283-352

Troll G, Forst R, Söllner F (1976) Über Alter und Metamorphose des Altkristallins der Schobergruppe, Osttirol. Geol Rundsch 65: 483-511

van Breemen O, Aftalion M, Bowes DR, Dudek A, Misar Z, Povondra, Vrana S (1982) Geochronological studies of the Bohemian massif, Czechoslovakia and their significance in the evolution of Central Europe. Trans R Soc Edinb Earth Sci 73: 89-108

von Raumer JF (1988) Caledonian-Variscan structures in the Alps. Schweizer Mineral Petrograph Mitt 68: 291-299

Northern Appalachians: Avalon and Meguma Terranes

J. D. Keppie[1], R. D. Nance[2], J. B. Murphy[3], and J. Dostal[4]

1 Introduction

Possible correlatives of the West African Orogens in the northern Appalachians occur along the eastern seaboard of North America, where they comprise the Avalon Composite Terrane and the Meguma Terrane. The Avalon Composite Terrane is defined by the presence of a lithostratigraphically correlative, subaerial-shallow marine, Cambro-Ordovician overstep sequence containing an Acado-Baltic fauna (Keppie 1985). In general, these Cambro-Ordovician rocks are only preserved in outliers, so overstep relationships are inferred from their close lithostratigraphic similarity. Where the Cambro-Ordovician rocks have been removed by erosion, it is possible to use the presence of the lithostratigraphically correlative, Silurian-Gedinnian overstep sequence containing the distinctive Rhenish-Bohemian fauna (Boucot 1975). By this definition, the Avalon Zone includes New England southeast of the Hope Valley-Bloody Bluff fault zone; Maine and New Brunswick south of the Turtle Head and Honeydale faults; most, if not all of northern Nova Scotia, and Newfoundland east of the Dover-Hermitage fault (Fig. 1). Terranes with Late Precambrian rocks northwest of these faults, such as the Hope Valley and Nashoba Terranes (O'Hara and Gromet 1985), have not been included in the Avalon Composite Terrane because the distinctive, Avalonian, Early Paleozoic rocks are absent. Should future work show that the Early Paleozoic overstep sequence was present before erosion, these terranes would need to be added to the Avalon Composite Terrane. In the Avalon Composite Terrane, the Early Paleozoic rocks rest unconformably - disconformably - conformably upon a variety of Late Precambrian igneous rocks and turbidites overlying Middle-Late Proterozoic or older metasediments and gneisses.

Contrasts in the late Proterozoic stratigraphy led to the subdivision of the Avalon Zone into several late Proterozoic terranes (Keppie 1985, 1988). The main terranes were: (1) Cape Breton Terrane: a cratonic magmatic arc; (2) Antigonish Highlands and Cheticamp Terranes: inter-arc basins floored by oceanic lithosphere; and (3) Cobequid Highlands Terrane: a cratonic magmatic arc. However, new data have allowed lithostratigraphic correlations to be made across northern Nova Scotia which suggest the presence of just one terrane consisting of alternating basins floored by thinned continental lithosphere (rather than oceanic lithosphere) and horsts (Keppie and Dallmeyer 1989b). These are interpreted as a magmatic arc-intra-arc rift complex that lacks oceanic components.

Similarly, Barr and Raeside (1986) subdivided Cape Breton Island into four, apparently contrasting, tectonostratigraphic zones or terranes, designated from southeast to northwest: Southeastern, Bras d'Or, Highlands and Northwestern Highlands. They defined these terranes as follows: (1) southeastern Terrane: Late Precambrian volcanism and plutonism followed by Cambro-Ordovician rift basin sedimentation and minor volcanism: (2) Bras d'Or Terrane: gneissic basement overlain by platform sedimentary rocks (George River Group) intruded by mainly Late Precambrian and Ordovician (?) granitoid rocks: (3) Highlands Terrane: gneissic core flanked by low grade sedimentary and volcanic rocks of probable Precambrian age intruded by diverse Precambrian-Carboniferous plutons: and (4) Northwestern Highlands Terrane: gneissic basement intruded by varied plutonic rocks including anorthosite, syenite and gabbro. Barr and Raeside (1986) inferred the Southeastern and Bras d'Or Terranes to be part of the Avalon Terrane, although Barr et al. (1987) correlated the Bras d'Or Terrane

[1] Department of Mines and Energy, P. O. Box 1087, Halifax, Nova Scotia, B3J 2X1, Canada
[2] Department of Geological Sciences, Ohio University, Athens, Ohio 45701, USA
[3] Department of Geology, St. Francis Xavier University, Antigonish, Nova Scotia, B2G 1CO, Canada
[4] Department of Geology, St. Mary's University, Halifax, Nova Scotia, B3H 3C3, Canada

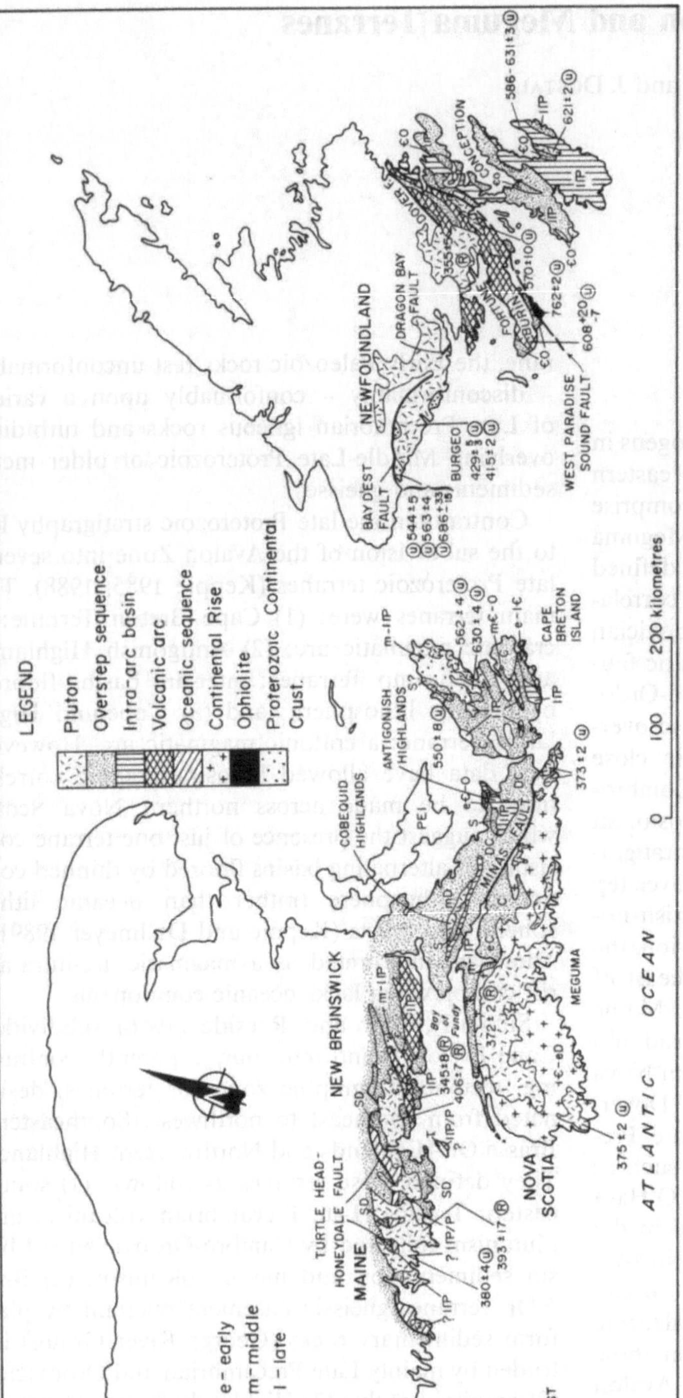

Fig. 1. Tectonic map of the Meguma Terrane and Avalon Composite Terrane in the northern Appalachians. The boundary between these terranes is the Minas Fault. The northern boundary of the Avalon Composite Terrane is the Hope Valley - Bloody Bluff - Turtle Head Honeydale - Dover Fault system

with the Gander Terrane. The Highlands Terrane was correlated with the Gander Terrane or the outboard edge of the Piedmont Terrane in Newfoundland, and the Northwestern Highlands Terrane was inferred to be an outboard part of autochthonous, Grenvillian North America (Piedmont Terrane of Williams and Hatcher 1983; North American Terrane of Keppie 1988). However, other data suggest that Cape Breton Island, like mainland northern Nova Scotia, represents a single Avalonian terrane. Keppie (1979, 1982) and Currie (1987) have shown that the felsic and mafic gneisses in the Northwestern Highlands "Terrane" may be traced through both the Highlands and Bras d'Or "Terranes". In addition, the Precambrian platformal unit characterized by the George River Group (Keppie 1979) in the Bras d'Or "Terrane" may be traced into the Highlands "Terrane" as the McMillan Flowage Formation (of Barr and Raeside 1986) and Cape North Group (of Macdonald and Smith 1980). Also, Late Proterozoic volcanic rocks and the Cambro-Ordovician overstep sequence are present in both Southeastern and Bras d'Or "Terranes", while pebbles containing Cambro-Ordovician and Silurian fauna typical of the Avalon Composite Terrane have recently been found in a Lower Carboniferous conglomerate derived from the Mabou Highlands lying within the Highlands "Terrane". Furthermore, published geochronological data show that late Proterozoic-Cambrian and Paleozoic plutons are present across most of Cape Breton Island. The basement beneath the southeastern "Terrane" is not exposed. However, the geochemistry of the late Proterozoic volcanic rocks indicates the presence of continental crust (Keppie et al. 1979). These data suggest that the rocks of Cape Breton Island, rather than representing distinct terranes in an entire cross-section of the Appalachian Orogen as proposed by Barr and Raeside (1986), expose an oblique section of the Avalon Composite Terrane with progressively deeper levels of the late Proterozoic magmatic arc complex cropping out northwestward. That the gneisses in northern Cape Breton Island are not contiguous with the outer edge of cratonic North American basement is indicated on seismic reflection profiles between Newfoundland and Cape Breton Island which show that the top of the Grenvillian basement is at a depth of >10 seconds two-way travel time in the Cabot Strait off northern Cape Breton Island (Marillier et al. 1989).

Ophiolitic rocks are only known in the Burin Terrane within the Avalon Composite Terrane of Newfoundland. The ophiolitic nature of the Burin Terrane contrasts with the continental nature of the neighbouring Fortune and Conception Terranes. However, the present geographical separation between the Avalon Composite Terrane in Newfoundland, Nova Scotia-New Brunswick-Maine, and New England, and uncertainties about their correlation, suggest that it is prudent at present to distinguish them as separate terranes.

South of the Avalon Composite Terrane in Nova Scotia lies the Meguma Terrane which consists mainly of a lithostratigraphically distinct Cambro-Ordovician sequence of turbiditic wacke and slate. These rocks are overlain by shallow marine-continental Siluro-Devonian rocks containing a Rhenish-Bohemian fauna (Boucot 1960) which links it with the Avalon Composite Terrane.

1.1 Terrane Boundaries

In simple terms, the boundary between the Avalon Composite Terrane and the Meguma Terrane is defined by the E-W Minas fault zone which swings into a southeasterly dipping listric fault in the Bay of Fundy (Keppie and Dallmeyer 1987). The Minas fault is overstepped by the Lower Carboniferous Horton Group. In Devono-Carboniferous times these two segments of the fault were dextral transcurrent and W-vergent thrust structures respectively. However, in the Mesozoic these movements were reversed in response to the birth of the Atlantic Ocean. Subsidiary listric faults exposed along the southern coast of New Brunswick display a similar movement history of Carboniferous thrusting followed by Mesozoic normal faulting (Plint and Van de Poll 1982, 1984; Nance and Warner 1986). The southeastern margin of the Meguma Terrane borders on the present Atlantic Ocean.

In general terms, the nature of the northwestern boundary of the Avalon Composite Terrane changes along strike from a vertical fault zone of Late Silurian age in Newfoundland, through a westerly-vergent thrust of Early-Middle Devonian age in southern New Brunswick, to a dextral shear zone with a component of E-vergent thrusting of Late Devonian-Early Permian age in New England (Keppie 1988). In Newfoundland, the Dover-Hermitage fault zone appears on seismic reflection profiles to be a vertical fault which extends through the entire crust (Keen et al. 1986). There is a marked contrast in intensity of deformation across the fault from intense polyphase deformation accompanied by metamorphism which reaches the upper amphibolite facies in the Gander Terrane, to milder deformation accompanied by greenschist metamorphism in the neighbouring Avalon Composite Terrane (Dallmeyer et al. 1981a). Hanmer (1981) has shown that the amphibolite-upper greenschist facies ductile strain in the Gander Terrane resulted from sinistral transpression, while Caron and Williams (1988a) inferred that early sinistral transpression in the western Avalon Composite Terrane was followed, adjacent to the Dover fault, by lower greenschist to brittle fabrics resulting from a complex movement history involving dextral shear followed by normal, reverse and dextral displacements (Caron and Williams 1988b). The development of some of the ductile fabrics is constrained between the intrusion of the Cape Freels pluton (Rb-Sr whole-rock isochron age of 414±5 Ma; Bell et al. 1977) and a $^{40}Ar/^{39}Ar$ plateau age of 397±5 Ma on hornblende from gneisses adjacent to the Dover Fault (Dallmeyer et al. 1981a). $^{40}Ar/^{39}Ar$ plateau ages on whole-rock phyllite from the western Avalon Composite Terrane range from 400±10 to 353±10 Ma and are interpreted to closely post-date the metamorphism associated with the development of the cleavage (Dallmeyer et al. 1983). An upper limit on the brittle movements on the Dover-Hermitage fault is provided by the age of the Ackley City Batholith, a stitching pluton (Rb-Sr whole-rock isochron age of 355±5 Ma; Bell et al. 1977; $^{40}Ar/^{39}Ar$ plateau age on hornblende = 356±10 Ma, Dallmeyer et al. 1981a). The northwestern boundary of the Avalon Composite Terrane in southern New Brunswick and Maine is defined by the Honeydale-Turtle Head fault, which has been interpreted as a NW-vergent thrust by Ruitenberg and Ludman (1978), Ludman (1981) and Osberg et al. (1985), although Stewart et al. (1986) infer a SE-vergence from the reflection seismic data. This boundary displaces units as young as Late Silurian and is stitched by several plutons, the oldest of which is dated at 380±4 Ma (Lucerne pluton, Loiselle et al. 1983). In southeastern New England, the northwestern boundary of the Avalon Composite Terrane (Bloody Bluff-Hope Valley shear zone) is a dextral shear zone with a component of easterly-vergent thrusting (Skehan 1983; Rodgers 1985; O'Hara and Gromet 1985). Some of the movements on this boundary are constrained between the 370±7 Ma Rb-Sr whole-rock isochron age of the Scituate pluton (Hermes and Zartman 1985) and the 273±2 Ma U-Pb monazite age and lower intercept, U-Pb zircon age of 276±7 Ma of the Narragansett Pier stitching pluton (Zartman and Hermes 1987).

This review of the Avalon Composite Terrane will focus on its Precambrian-Cambro-Ordovician geological record under the following headings: stratigraphy and paleontology, structure and metamorphism, igneous activity, geophysics, and tectonic interpretation. Other recent reviews include O'Brien et al. (1983), Rast and Skehan (1983), Keppie et al. (1985) and Nance (1986), to which the reader is referred for a more comprehensive treatment, and references therein.

2 Stratigraphy and Paleontology

The stratigraphy of the Avalon Composite Terrane and the Meguma Terrane is summarized in Figs. 2 and 3. The oldest rocks in the Avalon Composite Terrane are represented by ortho- and para-gneiss and amphibolite sheets: the Pleasant Bay and Kellys Mountain Complexes and equivalents in Cape Breton Island, the Great Village River Gneiss and Mount Thom Complex in the Cobequid Highlands of Nova Scotia, and the Brookville Gneiss in southern New Brunswick. A possible lithologically correlative unit of unknown age in southeastern Massachusetts may be the gneiss and schist exposed southeast of the Carboniferous Narragansett Basin.

At present the age of these units is poorly constrained. In New England, the only relevent data come from the Narragansett Pier Granite, which gives an Archean source age (2678±78 Ma) based upon the upper intercept of the chord on the U-Pb concordia diagram (Zartman and Hermes 1987).

U-Pb zircon analyses from a sample of the Brookville Gneiss do not plot on a single chord, which is indicative of several unresolved Pb loss events (Olszewski and Gaudette 1982). However, one analysis of a single, euhedral, clear zircon from the same sample was essentially concordant at ca. 814 Ma. A quartz diorite gneiss believed to be the product of anatexis during the metamorphism yielded U-Pb upper and lower intercept ages of 827±40 and 333±40 Ma, but is based on only two analyses (Olszewski and Gaudette 1982). A Rb-Sr whole-rock isochron for both gneiss lithologies yielded an age of 771±55 Ma. Collectively, these data suggest that the high grade metamorphism of the Brookville Gneiss took place at least 800 Ma ago.*

The Mount Thom Complex has given a poorly defined Rb-Sr whole-rock isochron of 934±82 Ma

* see footnotes on p. 329.

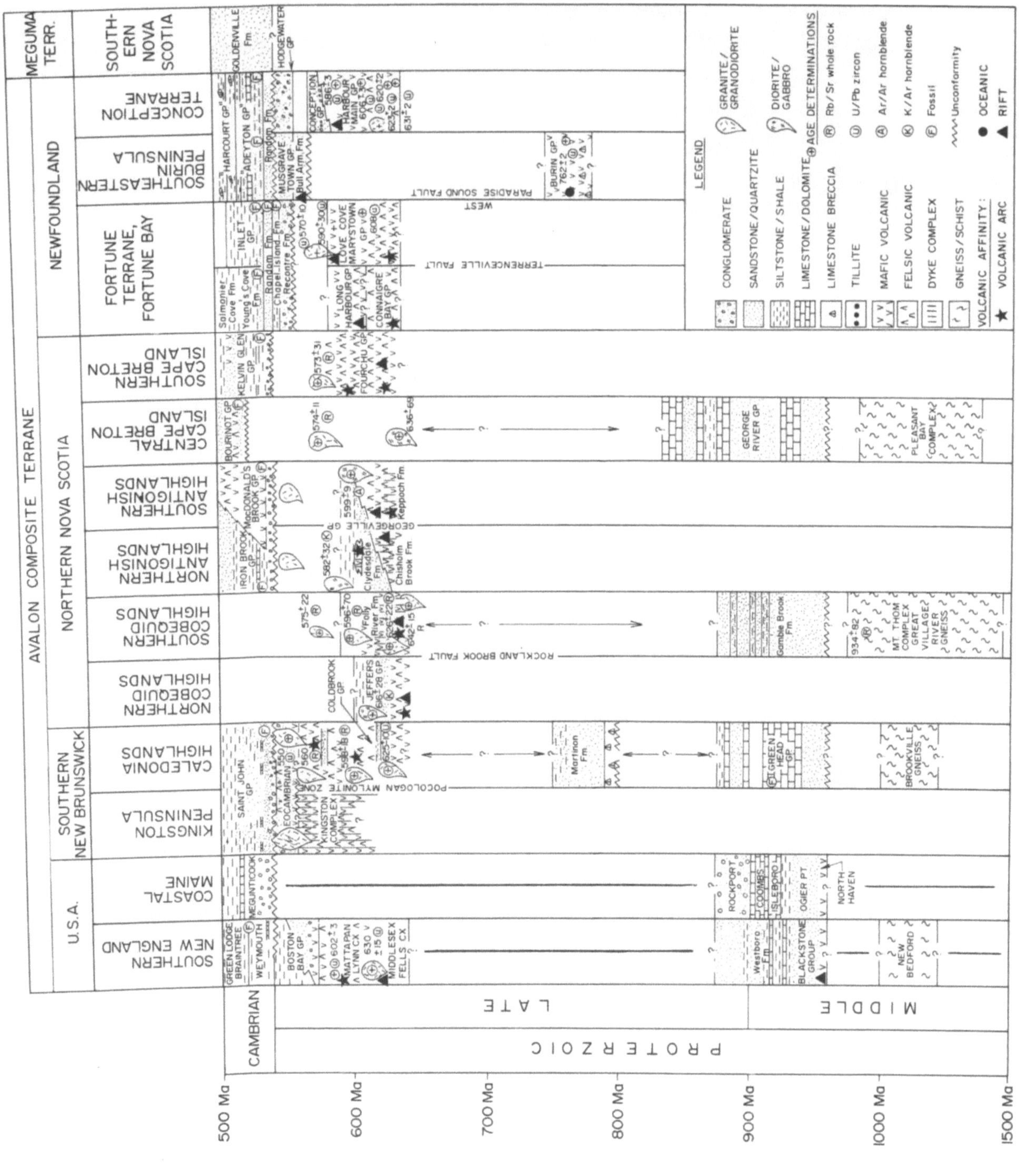

Fig. 2. Precambrian-Cambrian lithostratigraphy of the Meguma Terrane and Avalon Composite Terrane in the northern Appalachians

which may provide a minimum age for the metamorphism (Gaudette et al. 1983, 1984).

The gneisses of the Kellys Mountain Complex have yielded a Rb-Sr whole-rock error chron of 701 ± 66 Ma, which was inferred to provide a minimum age for the metamorphism by Olszewski et al. (1981), however $^{40}Ar/^{39}Ar$ plateau ages on hornblende and mica have given ca. 495 Ma ages for the Kellys Mountain Plutonic Complex and its contact aureole (Keppie and Dallmeyer 1989a). The Pleasant Bay Complex underlies large areas in the western Cape Breton Highlands and is made up of biotite gneiss, amphibolitic gneiss, gneissic granodiorite, marble, quartzite and schist cut by lit-par-lit gneiss (Currie 1987). A syenite intrusive into gneisses of the Pollets Cove Brook Group (= Pleas-

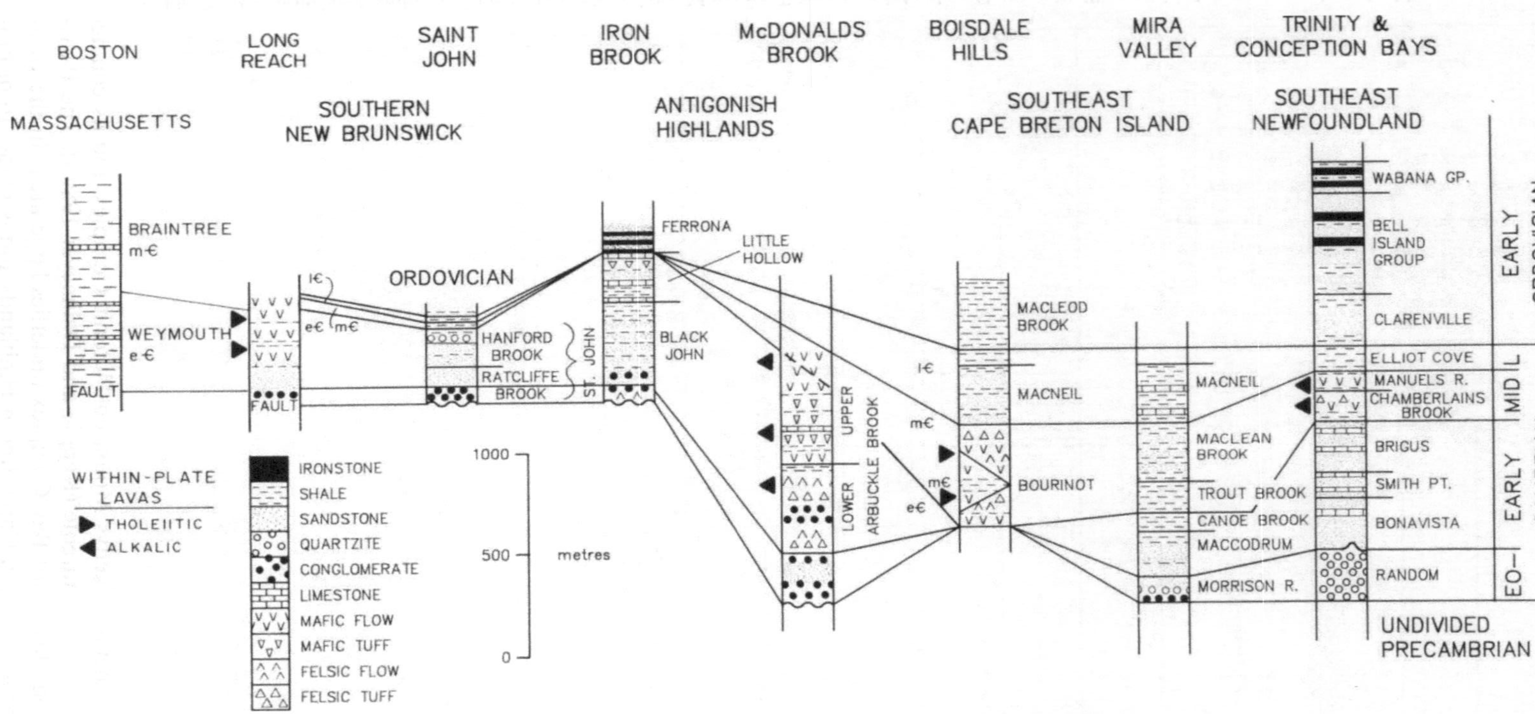

Fig. 3. Cambrian-Ordovician lithostratigraphy of the Avalon Composite Terrane in the northern Appalachians

ant Bay Complex) in northern Cape Breton Island has yielded an U-Pb chord with lower intercept age of 918 + 78 − 179 Ma and an upper intercept age of 1172 + 135 − 73 Ma (Barr et al. 1987). Using these data, Barr et al. (1987) estimated the age of amphibolite metamorphism to be 1040 + 40 − 10 Ma, although it is unclear how this estimate was derived. An alternative interpretation would have the upper intercept as the time of intrusion and the lower intercept as the time of metamorphism. The abraded suite of zircons lies towards the top end of the chord, suggesting that it is projecting toward the time of intrusion, while the euhedral shape of core of the zircons is consistent with an igneous origin. The lower intercept would then represent a minimum age for the high grade metamorphism and deformation.

While more reliable geochronological data are essential, existing data in the Avalon Composite Terrane tentatively suggest that a metamorphic event took place around the Middle-Late Proterozoic boundary.

The next lithostratigraphic unit (Fig. 2) is represented by strongly deformed, miogeoclinal metasedimentary rocks: quartzite, carbonates, greywacke and slate (± metavolcanic rocks) of the McMillan Flowage Formation and George River and Cape North Groups (Cape Breton Island), Gamble Brook Formation (Cobequid Highlands), Green Head (New Brunswick), parts of North Haven - Ogier Point-Isleboro - Coombs - Rockport - Ellesworth Formations (southern Maine) and the Blackstone Group and Westboro Formation (southeastern New England). In most places the contact between the gneisses and the metasediments is tectonic. Stomatolites occur in the Green Head Group and were assigned a Middle Riphean age by Hofmann (1974), which is poorly dated but would tentatively correspond to ca. 1000 ± 200 Ma. The close lithological similarities between the George River Group, the McMillan Flowage Formation and the Cape North Group suggest that they are direct correlatives (Keppie 1979). On the other hand, Barr and Raeside (1986), based upon correlations with Newfoundland, argued that the George River Group was Precambrian, while the McMillan Flowage and Cape North Group were Paleozoic. However, the McMillan Flowage Formation has been intruded by several plutons which have yielded 550-570 Ma U-Pb ages (Barr et al. 1988), thereby strengthening the lithostratigraphic correlation of the McMillan Flowage Formation with the George River Group. If this correlation is correct, then these Precambrian metasediments may be traced across the terrane boundary between the Bras d'Or and Highlands Terranes as defined by Barr and Raeside (1986), which are therefore acquitted of being separate terranes. In southern New Brunswick, the contact between the Green Head Group and the Martinon Formation is not exposed, but it is inferred to be an unconformity because slump breccia with fragments of the Green Head Group occur in the Martinon Formation. This breccia is overlain by turbiditic sandstone, siltstone and black chert (Leavitt 1963).

The rest of the Late Proterozoic rocks of the Avalon Composite Terrane are characterized by volcanic rocks and clastic metasediments (Figs. 2, 3). The age of these units is best defined in Newfoundland; elsewhere precise geochronology is generally lacking.

The oldest volcanic rocks occur in the Burin Group of Newfoundland where they are dominated by ophiolitic, mafic lavas and pyroclastic rocks with minor limestone, turbidites, slate, black chert and olistostromal breccia intruded by cogenetic dykes and a sill dated at 763 ± 2 Ma (U-Pb on zircon, Krogh et al. 1988). The Burin Group is a fault-bounded slice which pinches out northwards in the Paradise Sound fault. This fault has suffered Paleozoic movements which complicate the determination of the time of accretion. However, an upper limit is provided by the Rock Harbour Group which unconformably overlies the Burin Group and is correlated with the overstepping, latest Proterozoic Musgravetown Group (Hiscott 1981; Krogh et al. 1988).

Unfortunately, rocks of the same age as the Burin Group are not exposed elsewhere in the Avalon Composite Terrane. However, the geochemistry of the younger Late Proterozoic igneous rocks indicates that they were developed on continental crust (Krogh et al. 1988). This contrasts with the ophiolitic nature of the Burin Group and forms the basis for separating eastern Newfoundland into three Late Proterozoic terranes. Where the Burin Group pinches out, the Paradise Sound fault becomes either a cryptic suture or is cut out by later faulting, which appears to separate a magmatic arc terrane on the west from a cratonic, rifted basin on the east. Correlation of the latest Proterozoic alluvial rocks (Musgravetown - Signal Hill - Rock Harbour Groups) and overlying Cambro-Ordovician platformal sequences across the Paradise Sound fault implies that they represent an overstep sequence across the three terranes in eastern Newfoundland.

West of the Paradise Sound fault, the Fortune Terrane consists of the Marystown-Love Cove

Groups in the east and the Connaigre Bay and Long Harbour Groups in the west. The Marystown-Love Cove Groups are composed of subaerial volcanic rocks which evolved from bimodal and calc-alkaline or tholeiitic flood basalt to predominantly felsic rocks (Krogh et al. 1988). The Marystown Group has been dated at $608+20-7$ Ma using U-Pb on zircon from an ash-flow tuff (Krogh et al. 1988). The Connaigre Bay and Long Harbour Groups are lithologically similar to the lower and upper parts of the Marystown Group respectively. The Long Harbour Group is overlain by peralkaline rhyolite correlated with those of the Musgravetown Group and followed by platform Cambro-Ordovician rocks (Krogh et al. 1988).

East of the Paradise Sound fault, the Conception Terrane is characterized by the following succession: (1) bimodal, metaluminous flows and pyroclastic rocks with alkaline-tholeiitic, within-plate affinities of the Harbour Main Group with ages in the range 584–635 Ma (Nixon and Papezik 1979; Cameron and Papezik 1983; U-Pb zircon; Krogh et al. 1988); (2) thick northerly-derived turbidites of the Conception and Connecting Point Groups, containing an interbedded tillite near the base of the unit and Upper Riphean-Vendian metazoan fossils near the top; and (3) prodeltaic sandstone and shale of the St. Johns Group. These are overlain, locally with unconformity, by (4) alluvial deposits of the Musgravetown and Signal Hill Groups; and by (5) a Cambro-Ordovician platform sequence (O'Brien et al. 1983).

Late Proterozoic rocks (Fourchu Group and equivalents) are exposed in southern Cape Breton Island, where they are primarily composed of volcanic arc flows and pyroclastic rocks, which are interbedded with subaerial-submarine sediments (Keppie et al. 1979; Dostal et al., 1990). Farther north in Cape Breton Island, volcanic equivalents of the Fourchu Group are only preserved as outliers. However, correlatives exist in the calc-alkaline plutonic roots to the volcanic edifice (Barr et al. 1988). In southern Cape Breton Island, the Late Proterozoic rocks are overlain by Cambro-Ordovician platform rocks. Elsewhere in Cape Breton Island, Cambro-Ordovician rocks are absent. However, Precambrian rocks in the Cape Breton Highlands are overlain by Ordovician-Silurian bimodal, volcanic and sedimentary rocks. Unfortunately, fossils have not been recovered from these rocks. However, pebbles in the Lower Carboniferous Horton Group surrounding the Mabou Highlands contain Cambro-Ordovician and Siluro-Devonian fauna typical of the Avalon Composite Terrane (Norman 1935; Keppie, unpubl. data). Current directions and other pebble lithologies indicate derivation from the Mabou Highlands, suggesting the former presence of the overstep sequence across the Avalon Composite Terrane.

Late Proterozoic rocks in the Antigonish and Cobequid Highlands (Georgeville and Jeffers Groups, Folly River and Warwick Mountain Formations) consist of bimodal, calc-alkaline and tholeiitic rocks overlain by a thick succession of proximal to distal turbidites (Murphy and Keppie 1987; Pe-Piper and Piper 1987; Murphy et al. 1990). In the Antigonish Highlands, the Cambro-Ordovician platformal units rest unconformably upon deformed Precambrian rocks and are, in turn, unconformably overlain by a Siluro-Devonian succession, which are typical of the Avalonian overstep sequence. In the Cobequid Highlands, only the Siluro-Devonian overstep sequence is preserved.

In southern New Brunswick, the Late Proterozoic Coldbrook Group comprises volcanic arc lava, pyroclastic rocks, volcanogenic conglomerate and sandstone intruded in the north by the 10-km-wide and 200-km-long Kingston dyke complex of calc-alkaline to tholeiitic affinity (Giles and Ruitenberg 1977; Currie 1984; Dickson 1985). Equivalent rocks in southern Maine may be present in the Ellsworth and Columbia Falls Formations (Osberg et al. 1985). The Coldbrook Group is overlain by a platformal Eocambrian-Cambrian-Ordovician overstep sequence; lithostratigraphically similar rocks (Megunticook Formation) in southern Maine are unfossiliferous (Osberg et al. 1985). Isolated Precambrian rocks are exposed north of the Belleisle-Lubec fault, and Siluro-Devonian rocks similar to those overstepping the Avalon Composite Terrane elsewhere suggest that Avalonian rocks extend as far north as the Honeydale-Turtle Head fault (Keppie 1985, 1988).

The latest Proterozoic-Early Paleozoic succession in southeastern New England consists of: (1) bimodal, transitional alkalic volcanic rocks of the Middlesex Fells Complex (Durfee Cordoza et al. 1990); overlain by (2) intermediate-felsic, calc-alkaline flows and pyroclastic rocks of the Lynn and Mattapan complexes which have yielded an U-Pb zircon age of 602 ± 3 Ma (Kaye and Zartman 1980; Smith and Hon 1984; Hermes and Murray 1990). These are faulted against (3) the Boston Bay Group, a thick unit of glaciomarine, turbiditic conglomerate, sandstone and argillite containing Vendian acritarchs (Lenk et al. 1982; Socci and Smith 1987); which is followed by (4) Lower-Middle Cambrian shale, siltstone and minor limestone containing

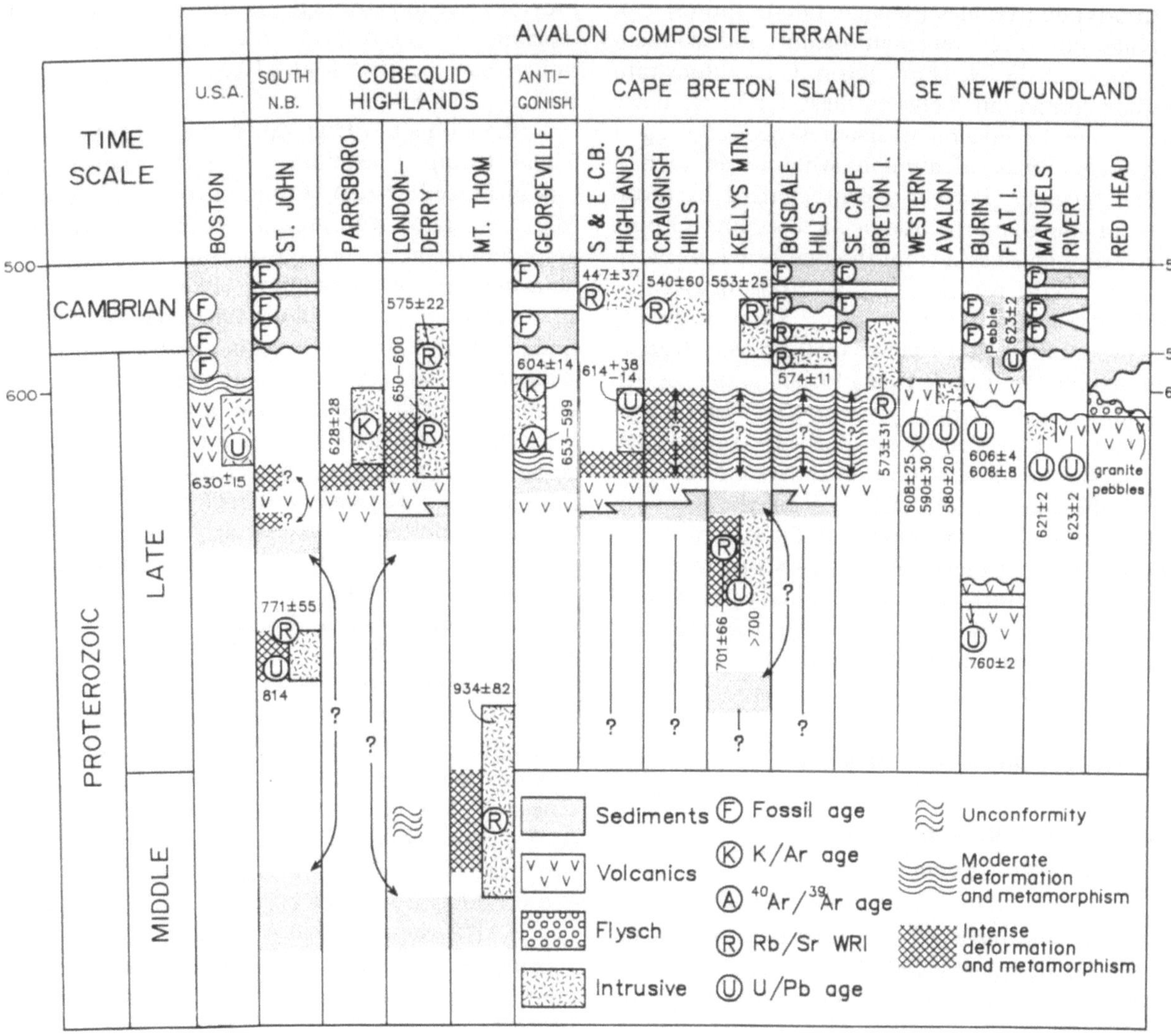

Fig. 4. Constraints on the times of deformation and metamorphism in the Avalon Composite Terrane in the northern Appalachians. *Boston* Volcanic clasts with a predepositional mild deformation occur in the Boston Bay Group; *St. John* deformed clasts of Coldbrook and Green Head Groups occur in the Early Cambrian conglomerates; *Parrsboro* polyphase deformation at greenschist facies metamorphism in Jeffers Formation is cut by Jeffers Brook pluton dated at 628±28 and 616±22 Ma (K-Ar on hornblende; Wanless et al. 1973; dates recalculated by Keppie and Smith 1978), and dated at 605-607 Ma by $^{40}Ar/^{39}Ar$ plateau ages on hornblende (Keppie et al. 1989); *Londonderry* polyphase deformation at greenschist facies metamorphism in Folly River Formation is cut by McCallum Settlement pluton (575±22 Ma, Rb-Sr whole-rock isochron; Gaudette et al. 1983, 1984), Debert River pluton (596±70 Ma, Rb-Sr whole-rock isochron; Donohoe et al. 1986), and Frog Lake pluton (622±3 Ma, $^{40}Ar/^{39}Ar$ plateau age on hornblende; Keppie et al. 1989); *Mt. Thom* polyphase deformation at amphibolite facies metamorphism dated at 934±82 Ma by Rb-Sr whole-rock isochron (Gaudette et al. 1984); *Georgeville* polyphase deformation at greenschist facies metamorphism in Georgeville Group cut by Greendale pluton dated at 604±14 (K-Ar on muscovite, R. K. Wanless, pers. commun.) and 611±3 and 620±5 Ma ($^{40}Ar/^{39}Ar$ plateau ages on hornblende; Keppie et al. 1989), unconformably overlain by Early Cambrian rocks; *Boisdale Hills* polyphase deformation at greenschist facies metamorphism cut by Shunacadie pluton (574±11 Ma Rb-Sr whole-rock isochron; Poole 1980) and Boisdale Hills pluton (563±31 Ma Rb-Sr whole-rock isochron; Cormier 1972; date recalculated by Keppie and Smith 1978; ca. 527-535 Ma $^{40}Ar/^{39}Ar$ plateau ages on hornblende; Keppie et al. 1989); *Southeastern Cape Breton Island* single phase of deformation at greenschist facies metamorphism in Fourchu Group cut by plutons dated between 548±18 Ma and 573±130 Ma (Rb-Sr whole-rock isochrons, Cormier 1972; dates recalculated by Keppie and Smith 1978; Cormier 1980), and between ca. 608-630 Ma ($^{40}Ar/^{39}Ar$ plateau ages on hornblende; Keppie et al. 1989), overlain by Cambrian rocks containing deformed pebbles; *Southeastern Newfoundland* Lower Cambrian rocks rest disconformably to unconformably and nonconformably upon various Precambrian units, while other local unconformities occur at several stratigraphic levels

Acado-Baltic trilobites (Skehan 1983); and (5) Upper Silurian-Lower Devonian volcanic and sedimentary rocks of the Newbury Basin. Close lithostratigraphic correlations between these Early Paleozoic rocks and the Avalon overstep sequence suggest they were formerly continuous with one another.

The Meguma Terrane is characterized by turbidites of the Cambro-Ordovician Meguma Group, which distinguishes it from the platformal overstep sequence in the Avalon Composite Terrane (Keppie 1985, 1988). Constraints on its age are provided by the youngest detrital titanite (560 ± 5 Ma; Keppie and Krogh 1987) and Tremadocian graptolites in the upper unit (Keppie 1985, 1988). This unit is overlain by Late Ordovician-Early Devonian shallow marine to subaerial sedimentary and volcanic rocks, the youngest of which contain the Rhenish-Bohemian fauna also common to the Avalon Composite Terrane (Boucot 1960). This sequence is unconformably overlain by predominantly terrestrial Carboniferous rocks, which overstep the boundary with the Avalon Composite Terrane.

3 Structure and Metamorphism

3.1 Middle-Late Proterozoic Tectonothermal Event

The times and intensity of deformation and metamorphism are summarized in Fig. 4. The oldest tectonothermal event is recorded in the oldest gneissic units in the Avalon Composite Terrane. In southern Cape Breton Island, the Kellys Mountain Complex displays folds in the gneissic banding with a parallel foliation (Jamieson 1984; Keppie et al. 1985). A 701 ± 66 Ma Rb-Sr whole-rock isochron from these gneisses was interpreted as the time of metamorphism by Olszewski et al. (1981). However, a low-pressure, upper amphibolite facies metamorphism including anatexis produced in the contact aureole of the Kellys Mountain Plutonic Complex post-dates the gneissic foliation. Hornblende and mica from both the plutonic complex and its aureole gave ca. 493-498 Ma $^{40}Ar/^{39}Ar$ plateau ages (Keppie and Dallmeyer 1989a). In northwestern Cape Breton Island, the gneiss of the Pleasant Bay Complex displays polyphase deformation and several episodes of metamorphism: the early metamorphism is largely obscured by a high-pressure and high-temperature upper amphibolite (-granulite) facies metamorphism overprinted by late retrograde metamorphism (Currie 1987). As discussed in the previous section, the age of the high-grade metamorphism is inferred to be ca. $918 + 78 - 179$ Ma, rather than the $1045 + 40 - 10$ Ma age estimated by Barr et al. (1987).

In the Cobequid Highlands, gneiss of the Mount Thom Complex exhibits polyphase deformation that produced an early foliation folded by isoclinal folds associated with an axial planar foliation and amphibolite facies metamorphism. This, in turn, is deformed by a crenulation cleavage (Donohoe 1976; Donohoe and Cullen 1983). The poorly defined 934 ± 82 Ma Rb-Sr whole-rock isochron age (Gaudette et al. 1984) may date the high-grade metamorphic event, although isotopic mixing and/or resetting by the last event cannot be ruled out and could account for the large error. Farther west in the Cobequid Highlands, the Great Village River Gneiss is reported to possess early fabrics, developed under amphibolite facies metamorphism, which are truncated by a shear zone forming the boundary of the Gamble Brook Formation (Cullen 1984; Murphy et al. 1988).

In southern New Brunswick, the Brookville Gneiss has been interpreted as either the highly metamorphosed equivalent of the Green Head Group (Wardle 1978), or a basement to the Green Head Gneiss (Currie et al. 1981). Currie et al. (1981) record a more complex structural style and at least two more phases of mafic dyke emplacement in the Brookville Gneiss than occur in the Green Head Group. This suggests that a period of deformation accompanied by upper amphibolite facies metamorphism and anatexis may have occurred in the Brookville Gneiss before deposition of the Green Head Group. As presented in the previous section, geochronological data suggest that this metamorphism took place at least 800 Ma ago.* The polyphase deformation in the Brookville Gneiss occurs as complex small interference patterns formed by roughly coplanar axial planes but variably oriented fold axes in narrow linear zones within more homogeneous gneiss with planar or gently curved to pinch-and-swell gneissosity (Currie et al. 1981).

3.2 Late Proterozoic-Cambrian Tectonothermal Event

Identification of Late Precambrian structures in the Avalon Composite Terrane is still fraught with uncertainties in many areas. This is largely due to a

---

* see footnotes on p. 329.

combination of factors including a dearth of critical contacts and isotopic data, and the problem of distinguishing Paleozoic from Precambrian-Cambrian structures, especially where Paleozoic deformation is relatively intense, as in northern Cape Breton Island.

The nature of the Late Precambrian tectonothermal event varies considerably. Mild and non-penetrative warping and block faulting, disconformities and unconformities and sub-greenschist facies metamorphism accompany the event in Newfoundland, in the southern parts of the Antigonish Highlands and southern New Brunswick, and in southeastern New England. In contrast, moderate single-phase to intense polyphase deformation and greenschist to amphibolite facies metamorphism accompany the event in Cape Breton Island, in the northern Antigonish Highlands and in the Cobequid Highlands. In southern New Brunswick, Ruitenberg et al. (1979) and McCutcheon (1981) reported deformed clasts of both Green Head and Coldbrook Groups in Lower Cambrian conglomerates. Such inhomogeneous deformation appears to be related to the presence or absence of both steeply and gently dipping shear zones. Few kinematic analyses of these shear zones are presently available, and they give various senses and times of movement.

1. Keppie and Murphy (1988) have inferred that the Cambro-Ordovician rocks in the Antigonish Highlands were deposited in a dextral pull-apart basin which closed in the Ordovician by sinistral movements;
2. The Precambrian structures in the northern Antigonish Highlands are inferred to have resulted from dextral transpression superimposed on early transcurrent movements of unknown sense (Murphy et al. in press);
3. Nance and Murphy (1990) interpret the structures in the Folly River Formation in the Cobequid Highlands in terms of sinistral transtension at ca. 630 Ma followed by dextral transpression at ca. 600 Ma;
4. The oblique north to northeasterly trend of the dykes relative to the axis of the Kingston complex in southern New Brunswick suggests their emplacement accompanied sinistral shear (Leger and Williams 1986): dyke emplacement is considered to have coincided with that of the Eocambrian plutonic suite at 565 ± 8 Ma (U-Pb on zircon) on the basis of their similar Th-rich composition (Currie 1988);**
5. Dextral Late Proterozoic movements on the Weston and Bloody Bluff faults in southeastern New England have been postulated by Nelson (1976) and Barosh (1984).

Clearly, more kinematic studies combined with geochronology are required before any regional conclusions can be drawn.

4 Igneous Activity

The geochemistry of the volcanic and plutonic rocks of the Avalon Composite Terrane is summarized in Figs. 1 and 2. The oldest volcanic rocks occur only in the fault-bounded Burin Group of Newfoundland dated at 763 ± 2 Ma (U-Pb on zircon; Krogh et al. 1988). Strong et al. (1978) and Strong and Dostal (1980) have shown that the submarine basalt of the Burin Group evolved from alkaline to oceanic tholeiites.

The oldest rocks exposed in the Conception Terrane occur in the Harbour Main Group, which is cut into three lithologically distinct blocks by faults (O'Brien and King 1982). From west to east the fault blocks consist of: (1) ash-flow tuff, basalt and volcaniclastic sediments with an ignimbrite dated at 606 ± 4 Ma and a rhyolite dated at 623 ± 2 Ma (U-Pb on zircon; Krogh et al. 1988); (2) felsic and mafic pyroclastic rocks intruded by comagmatic Holyrood granite dated at 621 ± 2 Ma and a rhyolitic porphyry dated at 631 ± 2 Ma (U-Pb on zircon; Krogh et al. 1988); and (3) pillow lava and aquagene tuff intruded by a porphyritic rhyolite dyke dated at 586 ± 3 Ma (Krogh et al. 1988). Geochemically most of these rocks are alkaline with metaluminous affinities, although the Holyrood pluton displays calc-alkaline characteristics (Nixon and Papezik 1979; Cameron and Papezik 1983).

The oldest rocks in the Fortune Terrane occur in the lower parts of the Love Cove (dated at 590 ± 30 Ma U-Pb on zircon; Dallmeyer et al. 1981 b) and Marystown Groups and in the Connaigre Bay Group. Generally, they consist of subaerial flows and pyroclastic rocks of mafic, intermediate and felsic composition showing calc-alkaline - tholeiitic affinities (Strong et al. 1978; O'Brien et al. 1983; O'Brien et al. 1984). An ignimbrite from the Marystown Group gave an age of 608 + 20 − 7 Ma (Krogh et al. 1988). These rocks are intruded by comagmatic calc-alkaline plutons, one of which has yielded a U-Pb age of 557 + 29 − 16 Ma (recalculated by Krogh et al. 1988 from Dallmeyer et al. 1981 b), $^{40}Ar/^{39}Ar$ plateau ages on hornblende of 560 ± 15

** see footnotes on p. 329.

and 566±15 Ma, and a Rb-Sr whole-rock isochron of 548±11 Ma (Dallmeyer et al. 1981b).

These rocks are overstepped by the Musgravetown Group and its correlatives, which contain some volcanic rocks. In the Bull Arm Formation, the volcanic rocks consist mainly of bimodal lavas and pyroclastic rocks with alkaline affinities (Hughes and Malpas 1971). In the Long Harbour Group and the upper part of the Marystown Group, the volcanic rocks are flood basalt and peralkaline rhyolite (O'Brien 1987). Volcanism in the overlying Cambrian rocks is bimodal, within-plate and alkalic (Greenough and Papezik 1985). Some peralkaline plutons in the Avalon Composite Terrane may be subvolcanic equivalents of these latter suites of volcanic rocks (Tuach 1984; O'Brien 1987). Thus, the igneous rocks of the Fortune and Conception Terranes evolved from calc-alkaline and metaluminous-alkaline in the Late Proterozoic (625-560 Ma) to alkaline in the latest Proterozoic and Cambrian. The geochemistry of these rocks may be interpreted in terms of a Late Proterozoic rifted magmatic arc followed by latest Proterozoic-Cambrian within-plate rifting.

Late Proterozoic volcanic rocks in southern Cape Breton Island occur in several fault blocks. In most of the blocks they are predominantly subaerial-submarine, mafic-intermediate-felsic lavas and pyroclastic rocks with volcanic arc affinities (Keppie et al. 1979; Zhai et al. 1987). One of the blocks contains a massive sulphide deposit that is associated with rhyolite showing a distinctive, nearly flat rare earth pattern and a negative Eu anomaly. These volcanics are associated with hornblende-biotite, I-type, epizonal, calc-alkaline diorite, tonalite and granite (Barr et al. 1982) giving ages in the range of ca. 615-515 Ma (Rb-Sr whole-rock isochrons; Keppie and Smith 1978) and ca. 630-600 Ma ($^{40}Ar/^{39}Ar$ plateau ages on hornblende; Keppie et al. 1989). Farther north in Cape Breton Island the volcanic rocks have been eroded away; however, the calc-alkaline plutons are preserved (Barr et al. 1982, 1988). Cambrian volcanic rocks are typically tholeiitic, bimodal and within-plate (Murphy et al. 1985). Thus, igneous activity changed from magmatic arc in the Late Proterozoic to within-plate, rift-related in the Cambrian.

In the Antigonish and Cobequid Highlands of Nova Scotia, the Late Proterozoic volcanic rocks are made up of two suites which, in places, are interbedded: (1) bimodal, porphyritic and calc-alkaline; and (2) bimodal, aphyric, continental tholeiites (Murphy et al. 1990, in press). These volcanic rocks are interpreted to have been extruded in an intra-arc rift basin. Associated plutonism is mainly gabbroic-dioritic and has a calc-alkaline geochemistry, and has yielded $^{40}Ar/^{39}Ar$ plateau ages on hornblende of ca. 630-600 Ma (Keppie et al. 1989). Early Cambrian volcanic rocks are bimodal, subaerial to submarine, and have alkalic, within-plate, geochemical characteristics (Murphy et al. 1985). Thus, these areas show a change from mixed, calc-alkaline and tholeiitic, intra-arc rifting in the late Proterozoic to alkalic and within-plate, rift-related in the Cambrian.

In southern New Brunswick, late Proterozoic volcanic rocks have been assigned to the Coldbrook Group, which comprises a thick sequence of mafic-intermediate-felsic lava, felsic tuff, lahar, volcanic breccia and volcanogenic sediments (Giles and Ruitenberg 1977). The mafic lava has a volcanic arc geochemistry (our unpublished data). Associated plutonism shows a diverse range of composition from gabbro, through hornblende diorite, tonalite, granodiorite, quartz monzonite to granite (Ruitenberg et al. 1979; Currie et al. 1981). Age dates are limited to 625±15 Ma (U-Pb on zircon from the Cape Spencer granite; Watters 1987), 598±18 Ma (Rb-Sr whole-rock isochron from the Fortyfive River granodiorite; Barr 1987), 530±16 Ma (Rb-Sr whole-rock isochron on the Bonnell Brook pluton; McLeod 1986) and 598±17 Ma (K-Ar on hornblende from the Alma diorite; Barr 1987). Eo-Middle Cambrian volcanic rocks are bimodal, subaerial, tholeiitic and related to within-plate rifting (Greenough et al. 1985). Thus, Late Proterozoic magmatic arc igneous activity gave way to tholeiitic, within-plate, rift-related magmatism in the Cambrian.

In southeastern New England, Late Proterozoic volcanic rocks occur in the Middlesex Fells, Lynn and Mattapan Complexes. The older Middlesex Fells Complex contains bimodal volcanic rocks, the mafic members of which are transitional alkali olivine basalt (Durfee Cordoza et al. 1990). The Lynn-Mattapan Complexes consist of felsic pyroclastic rocks, lahar, volcanic breccia and andesite (Smith and Hon 1984; Hermes and Murray 1990; Thompson and Hermes 1990), which have been dated at 602±3 Ma (Kaye and Zartman 1980). These latter complexes are slightly younger than the neighbouring plutons: (1) the calc-alkaline Dedham diorite-tonalite-granodiorite-granite has yielded an U-Pb, upper intercept age of 630±15 Ma (Zartman and Naylor 1984) and a Rb-Sr whole-rock isochron age of 595±16 Ma (Kovach et al. 1977); and (2) the calc-alkaline, Esmond, gabbro-granite pluton that gave an U-Pb, upper intercept age of 621±8 Ma

(Hermes and Zartman 1985). Thus, the Late Proterozoic igneous rocks of New England evolved from alkaline, rift-related to calc-alkaline, magmatic arc.

5 Possible Correlatives

In southern Newfoundland, south of the Bay d'Est-Dragon Bay Fault zone (Fig. 1), a belt of rocks (Burgeo Terrane) occurs that has affinities with the Avalon Composite Terrane (O'Brien et al. 1986, 1989; Keppie and Dallmeyer 1989a). This coastal belt consists of paragneiss, orthogneiss (686 + 33/ − 15 Ma, U-Pb zircon age; O'Brien et al. 1989), migmatite and amphibolitic gneiss (Grey River and Cinq Cerf Gneiss) that were (1) polydeformed and metamorphosed under amphibolite facies conditions at 579 ± 10 Ma (U-Pb zircon age; O'Brien et al. 1989); (2) intruded by the Roti granite at 563 ± 4 Ma (Dunning et al. 1988); (3) unconformably (?) overlain by low grade metasedimentary and metavolcanic rocks yielding magmatic zircon of 544 ± 5 Ma age and detrital zircon of ca. 593 Ma age (O'Brien et al. 1989); (4) unconformably overlain by a thick volcanosedimentary sequence (La Poile Group) that yielded U-Pb ages between 420 + 8/ − 2 and 428 ± 5 Ma that may be extrusive equivalents of the Burgeo pluton dated at 429 + 5/ − 3 Ma (foliated granite) and 415 ± 2 Ma (weakly foliated granite), and the strongly foliated to mylonitic Gaultois pluton dated at 422 ± 2 Ma (U-Pb zircon age, Dunning et al. 1988; O'Brien et al. 1986); and (5) post-tectonically intruded by the Chetwynd pluton (377 ± 20 Ma, U-Pb zircon age and 372 ± 5 Ma, $^{40}Ar/^{39}Ar$ biotite age; Dallmeyer 1980; Chorlton and Dallmeyer 1986) and the undated Francois pluton. Although the Burgeo Terrane lacks the Avalonian, Cambro-Ordovician overstep sequence, the La Poile Group may be equivalent to the Ordovician-Silurian volcanosedimentary sequence occurring in the Cape Breton Highlands of Nova Scotia. The older units, Late Proterozoic volcanic and sedimentary rocks and older gneisses, are comparable in both the Burgeo Terrane and the Cape Breton Highlands.

Deformation of the La Poile Group took place under greenschist facies metamorphic conditions and varies from a weakly developed fracture cleavage associated with open folds in the east to polyphase deformation involving coaxial folds associated with a slaty cleavage overprinted by a crenulation cleavage followed by cross folds (Chorlton 1980; O'Brien 1988). The two dated phases of the Burgeo pluton bracket the earliest foliation in the La Poile Group, while the Chetwynd pluton provides an upper age limit on the later phases; i.e. Middle Silurian to Middle Devonian. This deformation is inferred to be associated with docking and accretion of the Burgeo Terrane to the Newfoundland Gander Terrane, and the Burgeo pluton stitches the Bay d'Est-Dragon Bay Fault boundary between the two terranes.

6 Geophysics

6.1 *Magnetic, Gravity and Seismic Data*

The Avalon Composite Terrane is generally characterized by positive magnetic anomalies (up to 700 gammas) and positive free air gravity anomalies (up to 70 mGals) which contrast with the negative anomalies over the Devono-Carboniferous basins and the Meguma Terrane (Williams and Haworth 1984a, b). Within the Avalon Composite Terrane of Newfoundland, arcuate zones of positive magnetic and gravity anomalies may be correlated with outcrop belts of volcanic and plutonic igneous rocks, whereas the negative anomalies coincide with exposed sedimentary rocks. These and similar anomalies may be traced in the offshore area to the edge of the continental shelf and as far east as the Flemish Cap. South of Newfoundland, however, they merge with or are truncated by the Collector Anomaly interpreted as the southern edge of the Avalon Composite Terrane (Haworth and Lefort 1979). Similar geophysical features are more obscure over and around the Avalon Composite Terrane of Maritime Canada and New England, because they are partly masked by the effects of Paleozoic rocks. Reflection seismic data acquired north of Newfoundland indicate that the deep structure of the Avalon Composite Terrane is distinct and characterized by discontinuous, arcuate reflectors at ca. 5 and 9-12 seconds two-way travel time separated by transparent regions (Keen et al. 1986). The lower reflective zone probably represents the Moho, but the significance of the upper reflectors is unknown. Refraction seismic data suggest that compressional wave velocities beneath the Avalon Composite Terrane are 6 km/s on top of 7.4 km/s, which contrasts with the calculated velocites beneath the Meguma Terrane of 5.4 km/s above 6.25 km/s (Dainty et al. 1966).

6.2 Paleomagnetic Data

Paleomagnetic data for Late Proterozoic-Early Paleozoic stratified rocks in the Avalon Composite Terrane have yielded the following intermediate paleolatitudes:

1. 34–36° S (±7°) for the Late Proterozoic Marystown Group in Newfoundland, interpreted as primary magnetization (Irving and Strong 1985);
2. 45° S (±6°) for the post-folding magnetization of the Late Proterozoic Fourchu Group in Cape Breton Island, inferred to have been produced by oxidation during erosion following the latest Proterozoic folding (Johnson and Van der Voo 1986);
3. 47–49° S (±11°) for the dual-polarity, pre-folding magnetization of the Middle Cambrian Bourinot Group in Cape Breton Island, deduced to be primary magnetization (Johnson and Van der Voo 1985);
4. 42° S (±5°) for the pre-folding magnetization of the Late Ordovician-Early Silurian Dunn Point Formation in Nova Scotia, inferred as primary magnetization (Van der Voo and Johnson 1985).

Declinations associated with these paleolatitudes are highly variable, an observation which may be related to rotations about a vertical axis. Siluro-Devonian paleolatitudes for the Avalon Composite Terrane and cratonic North America are similar (Miller and Kent 1988). However, Cambro-Ordovician paleolatitudes are different: 45°–50° for the Avalon Composite Terrane and 10°–20° S for eastern North America (Van der Voo 1988). Late Proterozoic-Cambrian paleolatitudes derived for the Avalon Composite Terrane are consistent with a location adjacent to Gondwana or Armorica (Johnson and Van der Voo 1986).

7 Tectonic Interpretation

The stratigraphic and petrological data for the Late Proterozoic rocks of the Avalon Composite Terrane indicate that they were formed in a rifted magmatic arc environment in which active magmatic arc areas alternated with rifted basins. Available data indicate that the rifted basins were floored by continental crust except for the oceanic lithosphere preserved in the Burin Terrane of Newfoundland. The original width of the oceanic Burin Terrane is unknown. Its presence may be interpreted in terms of either (1) the original existence of two arcs or (2) the original presence of one arc subsequently offset by transcurrent faulting to either side of the oceanic terrane. It is also possible that the Late Proterozoic rocks from New England to Cape Breton Island formed in one contiguous magmatic arc. However, gaps in exposure allow the possibility that more than one arc is present. Structural data suggest that opening and closing of the intra-arc rifts took place by transcurrent motions, presumably related to oblique convergence. The polarity of subduction is presently uncertain. The Late Proterozoic magmatic arc rocks were replaced by latest Proterozoic-Cambrian tholeiitic-alkaline, rift-related volcanic rocks. The Cambrian rocks appear to have been deposited in pull-apart basins, suggesting a continuation of the transpressional strike-slip regime operative in the Late Proterozoic.

The geochemistry of the Late Proterozoic igneous rocks indicates that the magmatic arc was built upon continental crust. This crust is only exposed in southern New Brunswick and Nova Scotia, where it consists of a Middle-Late Proterozoic miogeocline resting unconformably upon a gneissic basement. Both of these latter units were deformed and metamorphosed in Middle-Late Proterozoic times before the development of the Late Proterozoic igneous rocks.

Faunal provinciality and paleomagnetic data indicate that the Avalon Composite Terrane had affinities with Gondwana and Armorica during Late Proterozoic-Cambrian times. In a Permian Pangaea A2 reconstruction (Van der Voo et al. 1984), the Avalon Composite Terrane and Meguma Terrane in the northern Appalachians lie adjacent to northwest Africa. Consequently, most correlations have drawn attention to the similarities between the Avalon Composite Terrane and the Pan African orogens of Morocco and the Trans-Saharan belt (e.g. O'Brien et al. 1983), while Schenk (1981) has inferred a source for the Meguma sediments in Morocco. The Bou-Azzer ophiolite in Morocco is the same age as the Burin ophiolitic sequence in Newfoundland, and the Late Proterozoic and Early Paleozoic stratigraphy are comparable. However, in northwest Africa, the Late Proterozoic orogens are built upon Archean and Early Proterozoic rocks, whereas preliminary data suggest that the Avalon Composite Terrane is underlain by Middle-Late Proterozoic basement. Only in South America are Late Proterozoic orogens built upon a Middle-Late Proterozoic basement (Keppie and Dallmeyer 1989b). This continent may therefore represent the provenance of the Avalon Composite Terrane. The synchroneity of Late Proterozoic-Cambrian subduction beneath the Avalon Composite Terrane and the opening of

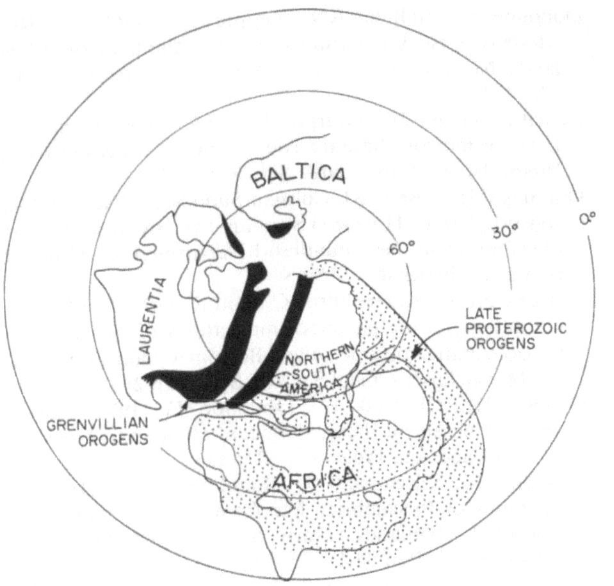

Fig. 5. Latest Precambrian-earliest Cambrian palinspastic reconstruction of the Circum-Atlantic region (modified from Bond et al. 1984) showing the locations of Pan-African (which would include the Avalon Composite Terrane) and Grenvillian orogens

Iapetus implies that the Avalon was not within the Iapetus Ocean. A possible Late Proterozoic-Cambrian reconstruction is shown in Fig. 5. The preservation of the Avalonian magmatic arc sequences contrasts with the general absence of magmatic arc rocks in continent-continent collision orogens, where they were presumably eroded away because they lie on the upper plate. This suggests that the termination of subduction beneath the Avalon Composite Terrane was not due to continental collision, but may have resulted from the relocation of subduction zones into Iapetus as it passed from a passive to an active margin ocean in the Late Cambrian.

Acknowledgement
We would like to thank Drs. P. F. Williams and O. D. Hermes for their reviews of the manuscripts, and D. MacDonald for editing and indexing. Funds for some of the research were provided by the 1984-1989 Canada-Nova Scotia Mineral Development Agreement. Permission to publish this contribution was given by the Director of the Mineral Resources Division of the Nova Scotia Department of Mines and Energy.

References

Barosh PJ (1984) The Bloody Bluff fault system. In: Hanson LS (ed) Geology of the Coastal Lowlands, Boston to Kennebunk, Maine. 46th N Engl Intercoll Geol Conf, Salem State College, Salem, Massachusetts, pp 310-324

Barr SM (1987) Field relations, petrology and age of plutonic and associated metavolcanic and metasedimentary rocks, Fundy National Park area, New Brunswick. Geol Surv Can Pap 87-1 A: 263-280

Barr SM, Raeside RP (1986) Pre-Carboniferous tectonostratigraphic subdivisions of Cape Breton Island, Nova Scotia. Marit Sediments Atl Geol 22: 252-263

Barr SM, O'Reilly GA, O'Beirne AM (1982) Geology and geochemistry of selected granitoid plutons of Cape Breton Island. Nova Scotia Dep Mines Energy 82-1: 176

Barr SM, Raeside RP, van Breemen O (1987) Grenvillian basement in the northern Cape Breton Highlands, Nova Scotia. Can J Earth Sci 24: 992-998

Barr SM, Raeside RP, Dunning GR, Jamieson RA (1988) New U-Pb ages from the Cape Breton Highlands and correlations with southern Newfoundland: a lithoprobe east contribution. Geol Assoc Can, Progr Abstr 13: A5

Bell K, Blenkinsop J, Strong DF (1977) The geochronology of some granite bodies from eastern Newfoundland and its bearing on Appalachian evolution. Can J Earth Sci 14: 456-476

Bond GC, Nickeson PA, Kominz MA (1984) Breakup of a supercontinent between 625 Ma and 555 Ma: new evidence and implications for continental histories. Earth Planet Sci Lett 70: 325-345

Boucot AJ (1960) Implications of Rhenish Lower Devonian brachiopods from Nova Scotia. 21st Int Geol Congr Part 12, pp 129-137

Boucot AJ (1975) Evolution and extinction rate controls. Dev Paleontol Stratigr 1. Elsevier, NY

Cameron KJ, Papezik VS (1983) Late Precambrian basaltic rocks of the Avalon Zone in Newfoundland: stratigraphy,

* Recently published geochronology for the Brookville Gneiss indicates: 1) that it contains detrital zircons with minimum ages of 943 ± 3 and 641 ± 3 Ma; 2) that it was intruded by quartz diorite and tonalite c. 600-630 Ma; and 3) was metamorphosed to upper amphibolite facies under low pressures at c. 570-540 Ma. Thus, rather than representing the basement to the Avalon Composite Terrane, the protolith has been inferred to be a temporal equivalent of the late Proterozoic arc sequence metamorphosed at deeper structural levels (Bevier ML, White CE, Barr SM, in press, late Precambrian U-Pb ages for the Brookville Gneiss, southern New Brunswick, and implications for stratigraphy of the Avalon Terrane, Atlantic Geology; Dallmeyer RD, Doig R, Nance RD, Murphy JB, in press, $^{40}Ar/^{39}Ar$ and U-Pb mineral ages from the Brookville Gneiss and Green Head Group: implications for terrane analysis and evolution of Avalonian "basement" in southern New Brunswick. Atlantic Geology).

** Recent U-Pb geochronology reported for the Kingston dyke complex indicates that they were intruded c. 435.5 ± 1.5 Ma and therefore record the sinistral accretion of the Avalon Composite Terrane with rest of the Appalachians (Doig R, Nance RD, Murphy JB, Casseday RP, in press, Evidence for Silurian sinistral accretion of the Avalon composite terrane in Canada. Journal of the Geological Society of London 147).

metamorphism, petrochemistry and significance. Geol Soc Am, Progr Abstr 15: 172
Caron A, Williams PE (1988a) The kinematic indicators of the Love Cove Group in northeastern Newfoundland. Geol Assoc Can, Progr Abstr 13: A17
Caron A, Williams PF (1988b) The multistage development of the Dover Fault in northeastern Newfoundland: the late stages. Geol Assoc Can, Progr Abstr 13: A17
Chorlton LB (1980) Geology of the La Poile River area (110/16), Newfoundland. Newfoundland Department of Mines and Energy, Mineral Development Division, Rep 80-3: 86
Chorlton LB, Dallmeyer RD (1986) Geochronology of Early to Middle Paleozoic tectonic development in the southwest Newfoundland Gander Zone. J Geol 94: 67-89
Cormier RF (1972) Radiometric ages of granitic rocks, Cape Breton Island, Nova Scotia. Can J Earth Sci 9: 1074-1086
Cormier RF (1980) New Rubidium/Strontium ages in Nova Scotia. Nova Scotia Dep Mines Energy Rep 80-1: 223-234
Cullen MP (1984) Geology of the Bass River Complex, Cobequid Highlands, Nova Scotia. MSc thesis, Dalhousie Univ, Halifax, Nova Scotia, 183 pp (unpubl)
Currie KL (1984) A reconsideration of some geological relations near Saint John, New Brunswick. Geol Surv Can Pap 84-1A: 193-201
Currie KL (1987) Relations between metamorphism and magmatism near Cheticamp, Cape Breton Island, Nova Scotia. Geol Surv Can Pap 85-23: 66
Currie KL (1988) The western end of the Avalon zone in southern New Brunswick. Marit Sediments Atl Geol 24: 339-352
Currie KL, Nance RD, Pajari GE, Pickerill RK (1981) Some aspects of the pre-Carboniferous geology of Saint John, New Brunswick. Geol Surv Can Pap 81-1A: 23-30
Dainty AM, Keen CE, Keen MJ, Blanchard JE (1966) Review of geophysical evidence on crust and upper mantle structure of the eastern seaboard of Canada. Am Geophys Union Monogr 10: 349-369
Dallmeyer RD (1980) Geochronology report. Current research. Newfoundland Department of Mines and Energy, Mineral Development Division Report 80-1
Dallmeyer RD, Blackwood RF, Odom AL (1981a) Age and origin of the Dover Fault: tectonic boundary between the Gander and Avalon Zones of the northeastern Newfoundland Appalachians. Can J Earth Sci 18: 1431-1442
Dallmeyer RD, Odom AL, O'Driscoll CF, Hussey EM (1981b) Geochronology of the Swift Current granite and host rocks of the Love Cove Group, southwestern Avalon Zone, Newfoundland: evidence of a late Proterozoic volcanic-subvolcanic association. Can J Earth Sci 18: 699-707
Dallmeyer RD, Hussey EM, O'Brien SJ, O'Driscoll CF (1983) Chronology of tectonothermal activity in the western Avalon Zone of the Newfoundland Appalachians. Can J Earth Sci 20: 355-363
Dickson WL (1985) Geology, geochemistry and petrology of the Precambrian and Carboniferous rocks between Saint John and Beaver Harbour, southern New Brunswick. PhD thesis, Univ New Brunswick, Fredericton, New Brunswick, Can, 409 pp (unpubl)
Donohoe HV (1976) The Cobequid Mountains Project. Nova Scotia Dep Mines Rep 76-2: 113-124
Donohoe HV, Cullen M (1983) Deformation, age and regional correlation of the Mt Thom and Bass River Complexes, Cobequid Highlands, Nova Scotia. Geol Soc Am, Progr Abstr 15 (3): 190
Donohoe HV, Halliday AN, Keppie JD (1986) Two Rb-Sr whole rock isochrons from plutons in the Cobequid Highlands, Nova Scotia, Canada. Marit Sediments Atl Geol 22: 148-154
Dostal J, Keppie JD, Murphy JB (1990) Geochemistry of Late Proterozoic basaltic rocks from southeastern Cape Breton Island. Can J Earth Sci 27: 619-631
Dunning GR, O'Brien SJ, Colman-Sadd SP, O'Neill P, Dickson WL, Krogh TE (1988) Silurian orogeny in southcentral Newfoundland. Newfoundland Department of Mines Report of Activities for 1988 48-49
Durfee Cordoza K, Hepburn JC, Hon R (1990) Geochemical constraints on the paleotectonic settings of two Late Proterozoic mafic volcanic suites, Boston-Avalon Zones, eastern Massachusetts. Geol Soc Am Spec Pap 245: 113-131
Gaudette HE, Olszewski WJ, Donohoe HV (1983) Age and origin of basement rocks, Cobequid Highlands, Nova Scotia. Geol Soc Am, Progr Abstr 15: 136
Gaudette HE, Olszewski WJ, Donohoe HV (1984) Rb/Sr isochrons of Precambrian age from plutonic rocks in the Cobequid Highlands, Nova Scotia. Nova Scotia Dep Mines Energy Rep 84-1: 285-292
Giles PS, Ruitenberg AA (1977) Stratigraphy, paleogeography and tectonic setting of the Coldbrook Group in the Caledonian Highlands of southern New Brunswick. Can J Earth Sci 14: 1263-1275
Greenough JD, Papezik VS (1985) Petrology and geochemistry of Cambrian volcanic rocks from the Avalon Peninsula, Newfoundland. Can J Earth Sci 22: 1594-1601
Greenough JD, McCutcheon SR, Papezik VS (1985) Petrology and geochemistry of Cambrian volcanic rocks from the Avalon Zone in New Brunswick. Can J Earth Sci 22: 881-892
Hanmer S (1981) Tectonic significance of the Gander Zone, Newfoundland: an Acadian ductile shear zone. Can J Earth Sci 18: 121-135
Haworth RT, Lefort JP (1979) Geophysical evidence for the extent of the Avalon zone in Atlantic Canada. Can J Earth Sci 16: 552-567
Hermes OD, Murray DP (1990) Geochemical distinctions of late Proterozoic and Paleozoic volcanism in the Avalon zone of New England. Geol Soc Am Spec Pap 245: 97-112
Hermes OD, Zartman RE (1985) Late Proterozoic and Devonian plutonic terrane within the Avalon Zone of Rhode Island. Geol Soc Am Bull 96: 272-282
Hiscott RN (1981) Stratigraphy and sedimentology of the Late Proterozoic Rock Harbour Group, Flat Islands, Placentia Bay, Newfoundland Avalon Zone. Can J Earth Sci 18: 495-508
Hofman H (1974) The stromatolite Archaezoon acadiense from the Proterozoic Green Head Group of St John, New Brunswick. Can J Earth Sci 11: 1098-1115
Hughes CJ, Malpas JG (1971) Metasomatism in the Late Precambrian Bull Arm Formation in southwestern Newfoundland: recognition and implications. Geol Assoc Can Proc 24: 85-93
Irving E, Strong DF (1985) Paleomagnetism of rocks from the Burin Peninsula, Newfoundland: hypothesis of the late Paleozoic displacement of Acadia criticized. J Geophys Res 90: 1949-1963
Jamieson RA (1984) Low pressure cordierite-bearing migmatites from Kellys Mountain, Nova Scotia. Contrib Mineral Petrol 86: 309-320
Johnson RJE, Van der Voo R (1985) Middle Cambrian paleomagnetism of the Avalon Terrane in Cape Breton Island, Nova Scotia. Tectonics 4: 629-651

Johnson RJE, Van der Voo R (1986) Paleomagnetism of the Late Precambrian Fourchu Group, Cape Breton Island, Nova Scotia. Can J Earth Sci 23: 1673-1685

Kaye CA, Zartman RE (1980) A late Proterozoic Z to Cambrian age for the stratified rocks of the Boston Basin, Massachusetts, USA. In: Wones DR (ed) The Caledonides of the USA. Virginia Polytechn Inst State Univ Mem 2: 257-261

Keen CE, Keen MJ, Nichols B et al. (1986) Deep seismic reflection profile across the northern Appalachians. Geology 14: 141-145

Keppie JD (1979) Geological, structural and metamorphic maps of Nova Scotia. Nova Scotia Dep Mines Energy, scales 1:500000 and 1:1000000

Keppie JD (1982) The Minas Geofracture. Geol Assoc Can Spec Pap 24: 263-280

Keppie JD (1985) The Appalachian collage. In: Gee DG, Sturt BA (eds) The Caledonide Orogen - Scandinavia and Related Areas, Part 2. Wiley, New York, pp 1217-1226

Keppie JD (1988) Northern Appalachian terranes and their accretionary history. Geol Soc Am Spec Pap 230: 159-192

Keppie JD, Dallmeyer RD (1987) Dating transcurrent terrane accretion: an example from the Meguma and Avalon Composite Terranes in the northern Appalachians. Tectonics 6: 831-847

Keppie JD, Dallmeyer RD (1989a) $^{40}Ar/^{39}Ar$ mineral ages from Kellys Mountain, Cape Breton Island, Nova Scotia: implications for the tectonothermal evolution of the Avalon Composite Terrane. Can J Earth Sci 26: 1509-1516

Keppie JD, Dallmeyer RD (1989b) Tectonic map of Pre-Mesozoic terranes in circum-Atlantic Phanerozoic orogens. Scale 1:5000000. IGCP Proj no 233, Halifax, Nova Scotia, Canada

Keppie JD, Krogh TE (1987) Age of detrital zircons and sphene in the Meguma Group, southern Nova Scotia: clues for the origin of the Meguma Terrane. In Tectonothermal Evolution of the West Africa Orogens and Circum-Atlantic Terrane Linkages. IGCP Proj 233 Conf Abstr Progr, Nouakchott, Mauritania, p 121

Keppie JD, Murphy JB (1988) Anatomy of a pull-apart basin: an example from the Cambro-Ordovician rocks in the Antigonish Highlands of Nova Scotia. Marit Sediments Atl Geol 24: 123-138

Keppie JD, Smith PK (1978) Compilation of isotopic age data of Nova Scotia. Nova Scotia Dep Mines Rep 78-4: 54

Keppie JD, Currie KL, Murphy JB, Pickerill RK, Fyffe LR, St Julien P (1985) Appalachian geotraverse (Canadian Mainland). Geol Assoc Can Excurs 1: 181

Keppie JD, Dostal J, Murphy JB (1979) Petrology of the Late Precambrian Fourchu Group in the Louisburg area, Cape Breton Island. Nova Scotia Dep Mines Pap 79-1: 18

Keppie JD, Dallmeyer RD, Murphy JB (1989) Proterozoic-Cambrian plutonism in the Avalon Terrane in Nova Scotia, Canada: constraints from $^{40}Ar/^{39}Ar$ hornblende ages. Geol Soc Am Progr Abstr, 21: 26

Kovach A, Hurley PM, Fairbairn HW (1977) Rb-Sr whole-rock age determinations of the Dedham Granodiorite, eastern Massachusetts. Am J Sci 227: 905-912

Krogh TE, Strong DF, O'Brien SJ, Papezik VS (1988) Precise U-Pb zircon dates from the Avalon Terrane in Newfoundland. Can J Earth Sci 25: 442-453

Leavitt EM (1963) The geology of the Precambrian Green Head Group in the Saint John, New Brunswick area. MSc thesis, Univ New Brunswick, Fredericton, New Brunswick, Can, 170 p (unpubl)

Leger A, Williams PF (1986) Transcurrent faulting history of southern New Brunswick. Geol Surv Can Pap 86-1B: 111-120

Lenk C, Stother PK, Kaye CA, Barghoorn ES (1982) Precambrian age of the Boston Basin: new evidence from microfossils. Science 216: 619-620

Loiselle MC, Eriksson S, Wones DR, Sinha AK (1983) Timing and emplacement of post-Acadian plutons in central and eastern Maine. Geol Soc Am Progr Abstr 15: 187

Ludman A (1981) Significance of transcurrent faulting in eastern Maine and location of the suture between Avalonia and North America. Am J Sci 4: 463-482

Macdonald AS, Smith PK (1980) Geology of Cape North area, northern Cape Breton Island, Nova Scotia. Nova Scotia Dep Mines Energy Pap 80-1: 60

Marillier F, Keen CE, Stockmal GS et al. (1989) Crustal structure and surface zonation of the Canadian Appalachians: implications of deep seismic reflection data. Can J Earth Sci 26: 305-321

McCutcheon SR (1981) Revised stratigraphy of the Long Reach area, southern New Brunswick: evidence for major, northwestward-directed Acadian thrusting. Can J Earth Sci 18: 646-656

McLeod MJ (1986) Contrasting geology across the Cradle Brook thrust zone: subaerial vs submarine Precambrian environments, Caledonia Highlands, New Brunswick. Marit Sediments Atl Geol 22: 296-307

Miller JD, Kent DV (1988) Paleomagnetism of the Siluro-Devonian Andreas redbeds: evidence for an Early Devonian supercontinent. Geology 16: 195-198

Murphy JB, Keppie JD (1987) The stratigraphy and depositional environment of the Late Precambrian Georgeville Group, Antigonish Highlands, Nova Scotia. Marit Sediments Atl Geol 23: 49-61

Murphy JB, Cameron K, Dostal J, Keppie JD, Hynes AJ (1985) Cambrian volcanism in Nova Scotia, Canada. Can J Earth Sci 22: 599-606

Murphy JB, Pe-Piper G, Nance RD, Turner D (1988) A preliminary report on geology of the eastern Cobequid Highlands, Nova Scotia. Geol Surv Can Pap 88-1B: 99-107

Murphy JB, Keppie JD, Hynes AJ (in press) Geology of the Antigonish Highlands. Geol Surv Can Pap 89-11

Murphy JB, Keppie JD, Dostal J, Hynes AJ (1990) The geochemistry and petrology of the Late Precambrian Georgeville Group: a volcanic arc - rift succession in the Avalon terrane of Nova Scotia. Geol Soc Lond Spec Publ No 51: 383-393. Geol Soc Lond Spec Publ

Nance RD (1986) Precambrian evolution of the Avalon Terrane in the northern Appalachians. Marit Sediments Atl Geol 22: 214-238

Nance RD, Murphy JB (1990) Kinematic analysis of the Bass River complex, Nova Scotia: Cadomian basement/cover relations in the Appalachian Avalon terrane. Geol Soc Lond Spec Publ No 51: 395-406. Geol Soc Lond Spec Publ

Nance RD, Warner JB (1986) Variscan tectonostratigraphy of the Mispec Group, southern New Brunswick; structural geometry and deformational history. Geol Surv Can Pap 86-1A: 351-358

Nelson AE (1976) Structural elements and deformational history of rocks in eastern Massachusetts. Geol Soc Am Bull 87: 1377-1783

Nixon GT, Papezik VS (1979) Late Precambrian ash-flow tuffs and associated rocks of the Harbour Main Group near Colliers, eastern Newfoundland: chemistry and magmatic affinities. Can J Earth Sci 16: 167-181

Norman GWH (1935) Lake Ainslie Map-area, NS. Geol Surv Can Mem 177: 103

O'Brien SJ (1987) Geology of the Eastport (west half) map area, Bonavista Bay, Newfoundland. Newfoundland Dep Mines Energy, Miner Dev Div Rep 87-1: 257-270

O'Brien BH (1988) Relationships of phyllite, schist and gneiss in the La Poile Bay-Roti Bay areas (parts of 110/9 and 110/16), southwestern Newfoundland. Newfoundland Department of Mines and Energy, Mineral Development Division, Report 88-1: 109-125

O'Brien SJ, King AF (1982) The Avalon Zone in Newfoundland. In: King AF (ed) Guidebook for Avalon and Meguma Zones. IGCP Project 27 - The Caledonide Orogen. Memorial Univ Newfoundland Rep 9: 1-24

O'Brien SJ, Dickson WL, Blackwood RF (1986) Geology of the central portion of the Hermitage Flexure area, Newfoundland. Newfoundland Department of Mines and Energy, Mineral Development Division, Report 86-1: 189-208

O'Brien SJ, O'Brien BH, Dunning GR (1989) Silurian and Precambrian events along the southeast margin of the Newfoundland Central Mobile Belt. Geol Assoc Can Prog Abstr 14: A10-A11

O'Brien SJ, Wardle RJ, King AF (1983) The Avalon Zone: a Pan-African Terrane in the Appalachian Orogen of Canada. Geol J 18: 195-222

O'Brien SJ, Nunn GAG, Dickson WL, Tuach J (1984) Geology of the Terrenceville (1M/10) and Gisbourne Lake (1M/15) map areas, southeastern Newfoundland. Newfoundland Dep Mines Energy, Miner Dev Div Rep 84-4: 54 p

O'Hara KD, Gromet LP (1985) Two distinct late Precambrian (Avalonian) Terranes in southeastern New England and their late Paleozoic juxtaposition. Am J Sci 285: 673-709

Olszewski WJ, Gaudette HE (1982) Age of the Brookville gneiss and associated rocks, southeastern New Brunswick. Can J Earth Sci 19: 2158-2166

Olszewski WJ Jr, Gaudette HE, Keppie JD, Donohoe HV (1981) Rb-Sr whole-rock age of the Kelly Mountain basement complex, Cape Breton Island. Progr Abstr 13 (3): 169

Osberg PH, Hussey AM, Boone GM (1985) Bedrock geological map of Maine, 1:500000. Maine Geol Surv

Pe-Piper G, Piper DJW (1987) The Pre-Carboniferous rocks of the western Cobequid Hills, Avalon Zone, Nova Scotia. Marit Sediments Atl Geol 23: 41-48

Plint AG, van de Poll HW (1982) Alluvial fan and piedmont sedimentation in the Tynemouth Creek Formation (Lower Pennsylvanian) of southern New Brunswick. Marit Sediments Atl Geol 18: 104-128

Plint AG, van de Poll HW (1984) Structural and sedimentary history of the Quaco Head area, southern New Brunswick. Can J Earth Sci 21: 753-761

Poole WH (1980) Rb-Sr age of Shunacadie pluton, central Cape Breton Island, Nova Scotia. Geol Surv Can Pap 80-1C: 165-169

Rast N, Skehan JS (1983) The evolution of the Avalonian plate. Tectonophysics 100: 257-286

Rodgers J (1985) Bedrock geological map of Connecticut. Conn Geol Nat Hist Surv, 1:125000

Ruitenberg AA, Ludman A (1978) Stratigraphy and tectonic setting of early Paleozoic sedimentary rocks of the Wirral-Big Lake area, southwestern New Brunswick and southeastern Maine. Can J Earth Sci 15: 22-32

Ruitenberg AA, Giles PS, Venugopal DV, Buttimer SM, McCutcheon SR, Chandra J (1979) Geology and mineral deposits Caledonia area. New Brunswick Miner Res Div Mem 1: 213

Schenk PE (1981) The Meguma zone of Nova Scotia: a remnant of western Europe, South America, or Africa? Can Soc Petrol Geol Mem 7: 119-148

Skehan JW (1983) Geological profiles through the Avalonian terrain of southeastern Massachusetts, Rhode Island and eastern Connecticut, USA. Am Geophys Union, Geodyn Ser 10: 275-300

Smith CJ, Hon R (1984) Geology, petrology and origin of the Precambrian igneous rocks located in the area north of Boston. In: Hanson LH (ed) Geology of the Coastal Lowlands, Boston to Kennebunk, Maine. 76th N Engl Intercoll Geol Conf, Salem State College, Salem, Massachusetts, pp 292-309

Socci AD, Smith GW (1987) Evolution of the Boston Basin: a sedimentological perspective. Can Soc Petrol Geol Mem 12: 87-99

Stewart DB, Unger JD, Phillips JD et al. (1986) The Quebec-western Maine seismic reflection profile: setting and first year results. Reflection Seismology: The Continental Crust. Geodyn Ser 14: 189-199

Strong DF, Dostal J (1980) Dynamic melting of Proterozoic upper mantle: evidence from rare earth elements in oceanic rocks of eastern Newfoundland. Contrib Mineral Petrol 72: 165-173

Strong DF, O'Brien SJ, Taylor SW, Strong PG, Wilton DH (1978) Aborted Proterozoic rifting in eastern Newfoundland. Can J Earth Sci 15: 117-131

Thompson MD, Hermes OD (1990) Ash-flow stratigraphy in the Mattapan Volcanic Complex, Greater Boston, Massachusetts. Geol Soc Am Spec Pap 245: 83-95

Tuach J (1984) Metallogenic studies of granite associated mineralization in the Ackley Granite and the Cros Hills Plutonic Complex, Fortune Bay, Newfoundland. Newfoundland Dep Mines Energy Miner Dev Div Rep 84-1: 245-253

Van der Voo R (1988) Paleozoic paleogeography of North America, Gondwana and intervening displaced terranes: comparisons of paleomagnetism with paleoclimatological and biogeographical patterns. Geol Soc Am Bull 100: 311-324

Van der Voo R, Johnson RL (1985) Paleomagnetism of the Dunn Point Formation (Nova Scotia): high paleolatitudes for the Avalon terrane in the Late Ordovician. Geophys Res Lett 12: 337-340

Van der Voo R, Peinado J, Scotese CR (1984) A paleomagnetic reevaluation of Pangea reconstructions. In: Van der Voo R et al. (eds) Plate reconstructions from Paleozoic Paleomagnetism. Am Geophys Union, Geodyn Ser 12: 11-26

Wanless RK, Stevens RD, Lachance GR, Delabio RN (1973) Age determinations and geological studies, K-Ar isotopic ages. Rep 11, Geol Surv Can Pap 79-2: 67

Wardle RJ (1978) The stratigraphy and tectonics of the Greenhead Group; its relationship to Hadrynian and Paleozoic rocks, southern New Brunswick. PhD thesis, Univ New Brunswick, Fredericton, New Brunswick (unpubl)

Watters SE (1987) Gold-bearing rocks - Bay of Fundy coastal zone. New Brunswick Dep Nat Res, Miner Res Div, Inf Circular 87-2: 41-44

Williams H, Hatcher RD Jr (1983) Appalachian suspect terranes. Geol Soc Am Mem 158: 33-53

Williams H, Haworth RT (1984a) Magnetic Anomaly Map of Atlantic Canada. Mem Univ Newfoundland, Map No 5, Scale 1:2000000

Williams H, Haworth RT (1984b) Bouguer gravity anomaly

map of Atlantic Canada. Mem Univ Newfoundland, Map No 6, Scale 1:2000000
Zartman RE, Hermes OD (1987) Archean inheritance in zircon from late Paleozoic granites from the Avalon Zone of southeastern New England: an African connection. Earth Planet Sci Lett 82: 305-315
Zartman RE, Naylor RS (1984) Structural implications of some radiometric ages of igneous rocks in southern New England. Geol Soc Am Bull 95: 522-539
Zhai M, Keppie JD, Dostal J (1987) Rare earth elements in the Stirling Belt, Cape Breton Island, as exploration guides for massive sulphides in Late Precambrian rhyolite. Nova Scotia Dep Mines Energy Rep 87-5: 143

Exotic Terranes in the Central-Southern Appalachian Orogen and Correlations With West Africa

R. D. Dallmeyer[1]

1 Introduction

Central and southern segments of the Appalachian orogen have long been subdivided into several northeast-trending lithotectonic "belts" (Fig. 1) on the basis of contrasts in lithologic character, metamorphic grade, and/or structural style (e.g. King 1955; Hatcher 1972, 1978b, 1987; Rankin 1975; Glover et al. 1983). The nomenclature and regional configuration of these belts as well as the character of intervening boundaries has changed markedly over the last 30 years (e.g. King 1955; Overstreet and Bell 1965a, b; Hatcher 1972, 1978b, 1987; Butler 1980). Such subdivision of the orogen has become increasingly more difficult to reconcile with results of recent detailed geologic mapping where, for example, protolith sequences within a low-grade belt may be traced directly into an adjacent high-grade belt (e.g. Overstreet and Bell 1965b; Griffin 1972; Secor et al. 1986a). In addition, contrasting times of regional metamorphism have been documented within different areas of the same belt (e.g. Dallmeyer et al. 1986; Noel et al. 1988). These and other similarly contradictory relationships have led to the suggestion by some workers (e.g. Zen 1981; Williams and Hatcher 1982, 1983; Secor et al. 1986b; Higgins et al. 1986, 1988; Noel et al. 1988; Dallmeyer 1986, 1989a; Horton et al. 1989, 1991) that the central-southern Appalachian orogen is comprised of several markedly different, fault-bounded structural units (Fig. 2) with tectonic significance similar to that of tectonostratigraphic terranes defined within Mesozoic-Cenozoic orogens of the Circum-Pacific realm (e.g. Coney et al. 1980; Jones et al. 1983a, b; Howell and Jones 1984).

Results of seismic reflection studies (e.g. Cook et al. 1979; Harris et al. 1981) together with regional gravity and magnetic characteristics (e.g. Hatcher and Zietz 1978, 1980; Cook and Oliver 1981; Dainty and Frazier 1984) indicate that the various central-southern Appalachian terranes are likely underlain by autochthonous/parautochthonous Laurentian basement, and largely constitute highly allochthonous crystalline structural units. Lithologic and/or faunal characteristics suggest that at least some of the allochthonous units exposed east of the Hayesville thrust (Figs. 1, 2) originated in late Precambrian-early Paleozoic palinspastic settings removed from Laurentia and therefore represent "exotic" terranes (e.g. Secor et al. 1983, 1986b; Dallmeyer 1986, 1989a; Horton et al. 1989). Structural units of internally imbricated metasedimentary rocks exposed west of the Hayesville thrust (western Blue Ridge and Talladega nappe complexes) appear to have originated within the lower Paleozoic Laurentian miogeocline, and have been termed "native" terranes by Horton et al. (1989). Because fossil-bearing overstep sequences are not present in the central southern Appalachian orogen, the detailed chronology of outboard terrane amalgamation and final accretion to Laurentia have been uncertain and controversial (e.g. Williams and Hatcher 1982; Dallmeyer 1986, 1988, 1989a; Horton et al. 1989). Paleomagnetic characteristics of several plutons within the southern Appalachian Piedmont are not consistent with significant post-Devonian latitudinal movement (e.g. Ellwood 1982; Barton and Brown 1983; Dooley 1983), suggesting that most terrane elements were at least proximal to Laurentia prior to the Carboniferous. This together with sedimentological expression of outboard (eastern) tectonic instability reflected in Ordovician-Devonian units within the Laurentian miogeocline (e.g. Chowns 1989) suggests that initial terrane accretion had occurred prior to late Paleozoic thrust transport into present relative structural positions on the Laurentian margin.

A series of basement massifs largely underlain by middle-late Proterozoic (Grenville) crystalline rocks and overlying metasedimentary cover sequences occur locally within internal, metamorphic portions of the central-southern Appalachian orogen. Horton

[1] Department of Geology University of Georgia Athens, Georgia 30602

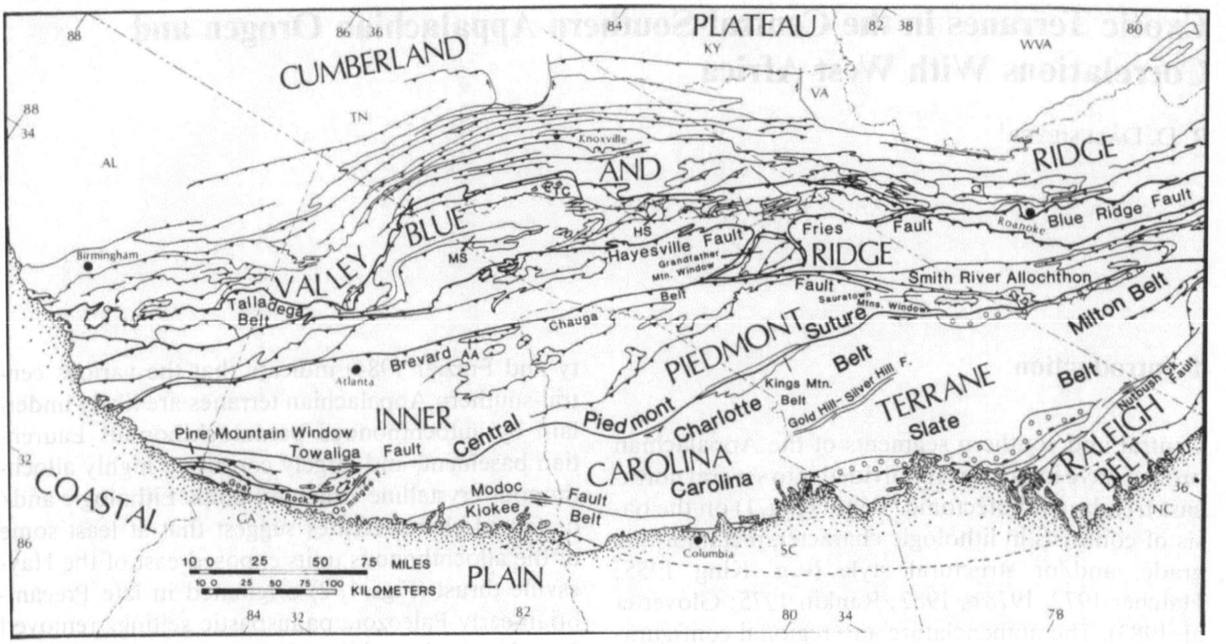

Fig. 1. Tectonostratigraphic subdivisions and major faults within central and southern portions of the Appalachian orogen (Hatcher 1987): *AA* Alto allochthon; *SMT* Six-Mile thrust; *GMW* Grandfather Mountain Window; *MCM* Mountain City window

et al. (1989) have termed these the Baltimore, Sauratown and Pine Mountain terranes (Fig. 2). These may represent horsts of Laurentian basement which were structurally elevated during initial, late Proterozoic rifting of Laurentia. Alternatively, they may represent fragments of continental crust (either rifted from Laurentia and/or of other palinspastic derivation) which accreted to Laurentia during Paleozoic orogenesis.

Horton et al. (1989, 1991) separated exotic terranes exposed in the central-southern Appalachian orogen and their extensions beneath the eastern Coastal Plains into four general categories, including (Fig. 2): (1) disrupted terranes; (2) variably fragmented remnants of oceanic lithosphere; (3) various volcanic arc terranes; and (4) metamorphic terranes with uncertain origins. This chapter presents an overview of these exotic terranes together with a description of basement units beneath the Gulf and southeastern Atlantic Coastal Plains. It also explores potential tectonostratigraphic correlations with structural elements within the various West African orogens.

2 Central-Southern Appalachians and Eastern Coastal Plain Basement

2.1 Disrupted Terranes

Howell et al. (1985) used the term "disrupted terranes" in reference to structural units characterized by "... blocks of heterogeneous lithology and age, usually set in a matrix of foliated shale, sandstone, or serpentinite ...". Some of the central-southern Appalachian tectonic elements classified as disrupted by Horton et al. (1989, 1991) do display characteristics typical of classically-defined melange complexes. These include the Falls Lake and Juliette terranes (Fig. 2) which are comprised largely of mafic and ultramafic blocks which are structurally interlayered with variably metamorphosed pelitic units (mudstone and/or graywacke). Horton et al. (1989, 1991) suggested that these terranes likely represent melange complexes which developed outboard of Laurentia within lower Paleozoic (Cambrian-Ordovician?) accretionary complexes. They also described a geophysically distinct area beneath the Virginia-Maryland Coastal Plain (their Sussex terrane) as likely composed primarily of foliated metavolcanic rocks, ultramafic units, gabbro, diorite and low-grade metapelitic units. Horton et al. (1989, 1991) included the Sussex terrane in their category of disrupted terranes. The other central-southern Appalachian tectonic units classified as disrupted terranes by Horton et al. (1989, 1991) are more regional, composite structural features comprised of internally coherent terrane elements (of volcanic, ophiolitic, and/or continental nature). These include the Jefferson, Inner Piedmont, Smith River and Potomoc composite terranes.

2.1.1 Potomac Composite Terrane

The Potomac composite terrane is comprised of internally imbricated structural units which contain a variety of protoliths, including abyssal turbidites, volcanic-arc sequences, fragmented ophiolite components and distinct melange horizons. The structural units were assembled prior to emplacement of the Ocoquan Granite which records a 494 ± 14 Ma[1] Rb-Sr whole-rock isochron crystallization age (Mose and Nagel 1982).

2.1.2 Smith River Composite Terrane

The Smith River composite terrane is as a fault-bounded structural unit within the more extensive Jefferson composite terrane. Lower structural intervals contain discontinuous, chaotic horizons of ultramafic and mafic melange within amphibolite derived from basalt. Upper structural levels are comprised of schist and gneiss. The Smith River composite terrane records a polydeformational and polymetamorphic evolution predating emplacement of the Leatherwood granite which records a U-Pb zircon age of 450 Ma (Rankin 1975) and a Rb-Sr whole-rock isochron age of 455 ± 20 Ma (Odom and Russell 1975).

2.1.3 Jefferson Composite Terrane

The Jefferson composite terrane is an extensive, internally imbricated tectonic element that constitutes eastern sectors of the Blue Ridge tectonic province. It is largely composed of metasedimentary sequences (metasandstone and metapelitic rocks) together with metabasaltic amphibolite and subordinate ultramafic rocks which likely represent fragmented ophiolitic material. Metasedimentary units display marine characteristics, and the entire tectonic unit may, at least in part, represent an accretionary wedge that formed an uncertain distance outboard of Laurentia in the early Paleozoic (Cambrian-lower Ordovician?). It was metamorphosed, further imbricated, and initially thrust westward onto the Laurential margin in the Middle Ordovician.

Nelson et al. (1987b) divided sectors of the Jefferson terrane exposed in north Georgia and southwestern South Carolina into three distinct composite allochthonous units. In ascending tectonostratigraphic order these include the Helen, Tallulah Falls, and Richard Russell nappes (Fig. 3).

Helen Nappe

The Helen nappe is predominantly characterized by metagraywacke, quartzite, variably graphitic schist, metasiltstone, phyllite, locally conglomeratic feldspathic metasandstone, manganiferous metavolcanic rocks and several chemical metasedimentary units represented by banded iron formation, manganiferous schist, and quartzite. These constitute the Helen Group of Nelson and Gillon (1985). These sequences are polydeformed and display upper greenschist to middle amphibolite facies regional metamorphic assemblages. Variably deformed granitic to granodioritic plutons (mylonitic orthogneisses) occur locally. Pillows have locally been described from intercalated amphibolite, and geochemical characteristics suggest protoliths were "near-are" basalts (Hopson 1988). The widespread association of variably graphitic schist with amphibolite indicates that volcanism occurred near shore and was frequently interrupted by clastic sedimentation.

Tallulah Falls Nappe

The Tallulah Falls nappe structurally overlies the Helen nappe along the Dahlonega fault. The Tallulah Falls nappe is largely represented by polydeformed, generally high-grade metasedimentary rocks which constitute the Tallulah Falls Formation (Hatcher 1971). This includes: (1) a lower succession of metagraywacke, schist and amphibolite; (2) intermediate levels of aluminous schist, schist and metagraywacke; and (3) an upper unit of quartzite and schist (Hatcher 1976). These metasedimentary sequences are interpreted to unconformably overlie older, predeformed gneisses exposed in the Toxoway and Tallulah Falls domes (Hatcher 1976). Odom et al. (1976) and Fullagar et al. (1979) reported Grenville radiometric ages (ca. 1-1.1 Ga) for several of these basement units (Wiley and Toxaway gneisses). Variably deformed intrusive rocks

[1] All isotopic ages discussed in this report are based upon the decay constant and isotopic abundance ratios listed in Steiger and Jäger (1977).

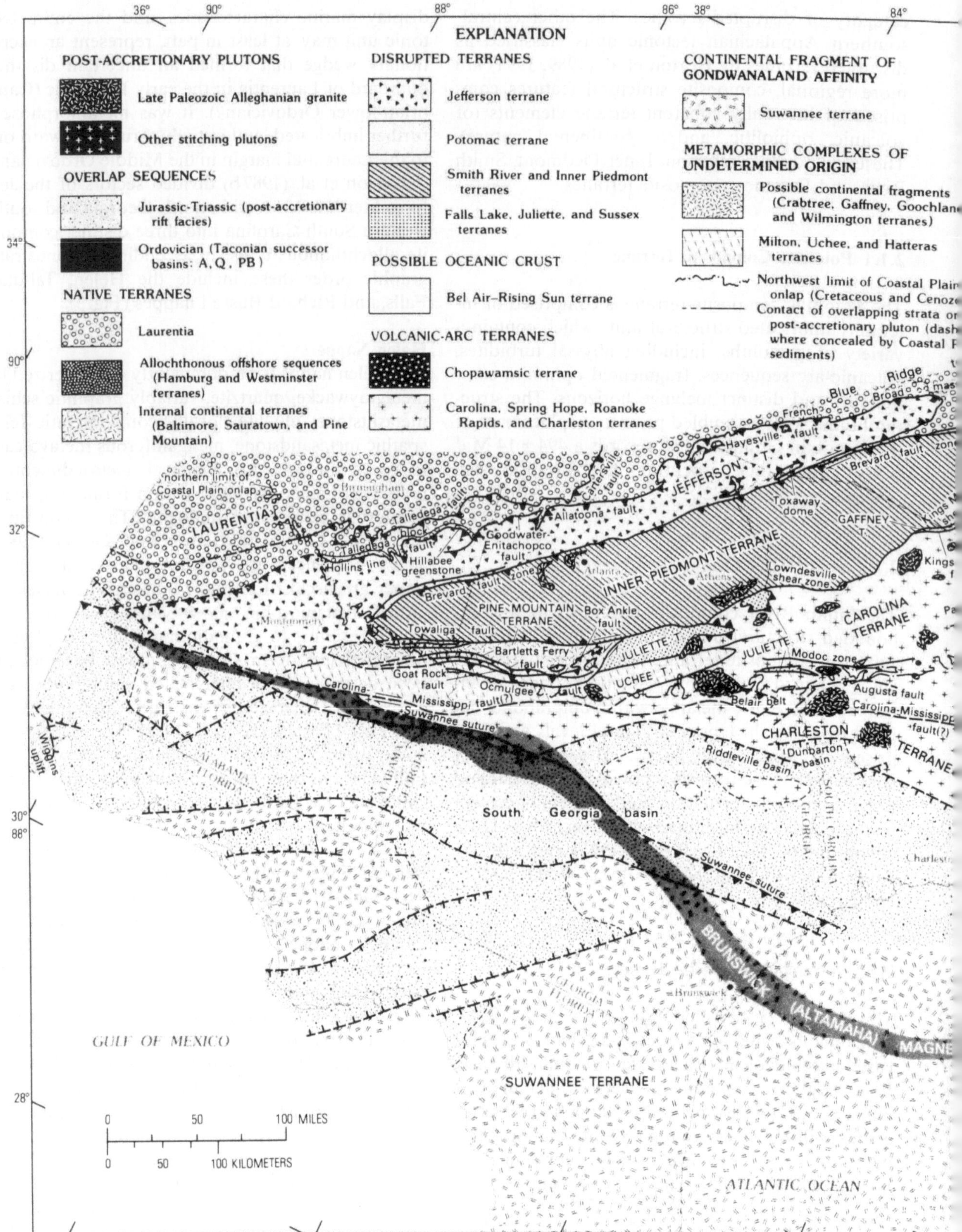

Fig. 2. Preliminary tectonostratigraphic terrane map of central and southern portions of the Appalachian orogen (Horton et al. 1989, 1991)

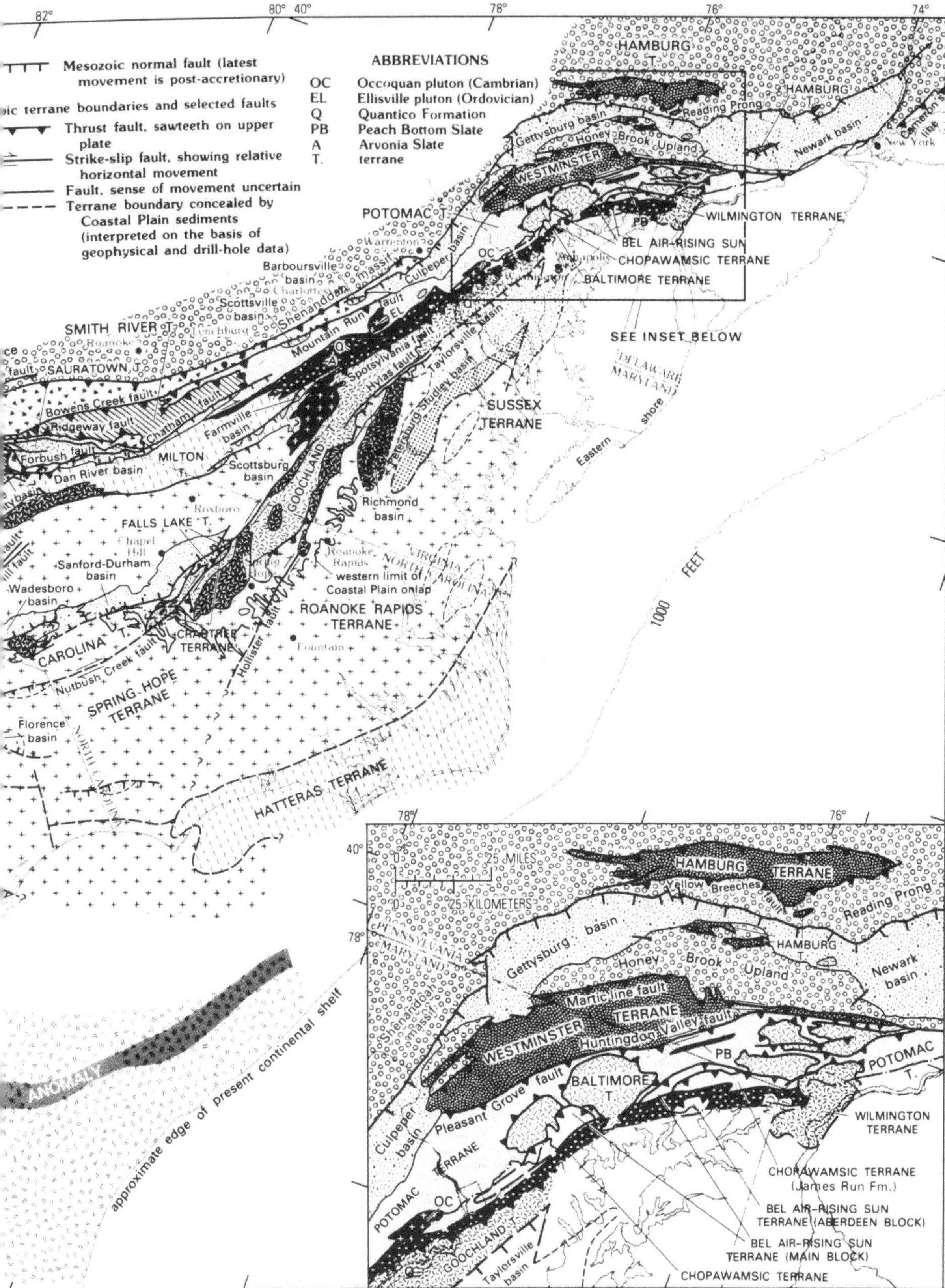

Mesozoic normal fault (latest movement is post-accretionary)
Thrust fault, sawteeth on upper plate
Strike-slip fault, showing relative horizontal movement
Fault, sense of movement uncertain
Terrane boundary concealed by Coastal Plain sediments (interpreted on the basis of geophysical and drill-hole data)
ABBREVIATIONS
OC Occoquan pluton (Cambrian)
EL Ellisville pluton (Ordovician)
Q Quantico Formation
PB Peach Bottom Slate
A Arvonia Slate
T. terrane
HAMBURG T.
Reading Prong
Gettysburg basin
Honey Brook Upland
Newark basin
Cameron's line
New York
WESTMINSTER T.
WILMINGTON TERRANE
BEL AIR-RISING SUN
CHOPAWAMSIC TERRANE
BALTIMORE TERRANE
POTOMAC T.
Culpeper basin
Barboursville basin
Scottsville basin
Shenandoah massif
Mountain Run fault
SMITH RIVER T.
SAURATOWN T.
Bowens Creek fault
Ridgeway fault
Chatham fault
Forbush fault
Dan River basin
MILTON T.
Farmville basin
Scottsburg basin
GOOCHLAND T.
Spotsylvania fault
Hylas fault
Taylorsville basin
Petersburg-Studley basin
Richmond basin
SUSSEX TERRANE
SEE INSET BELOW
DELAWARE
MARYLAND
Eastern shore
FALLS LAKE T.
Sanford-Durham basin
Wadesboro basin
CAROLINA T.
CRABTREE TERRANE
Nutbush Creek fault
Hollister fault
VIRGINIA
NORTH CAROLINA
western limit of Coastal Plain onlap
ROANOKE RAPIDS TERRANE
SPRING HOPE TERRANE
Florence basin
HATTERAS TERRANE
1000 FEET
ANOMALY
approximate edge of present continental shelf
82°
80°
40°
78°
76°
74°
38°
36°
72°
0 25 MILES
0 25 KILOMETERS
HAMBURG TERRANE
Yellow Breeches fault
PENNSYLVANIA
MARYLAND
Martic line fault
WESTMINSTER TERRANE
Huntingdon Valley fault
PB
BALTIMORE T.
Pleasant Grove fault
POTOMAC TERRANE
OC
GOOCHLAND T.
Taylorsville basin
CHOPAWAMSIC TERRANE (James Run Fm.)
BEL AIR-RISING SUN TERRANE (ABERDEEN BLOCK)
BEL AIR-RISING SUN TERRANE (MAIN BLOCK)
CHOPAWAMSIC TERRANE

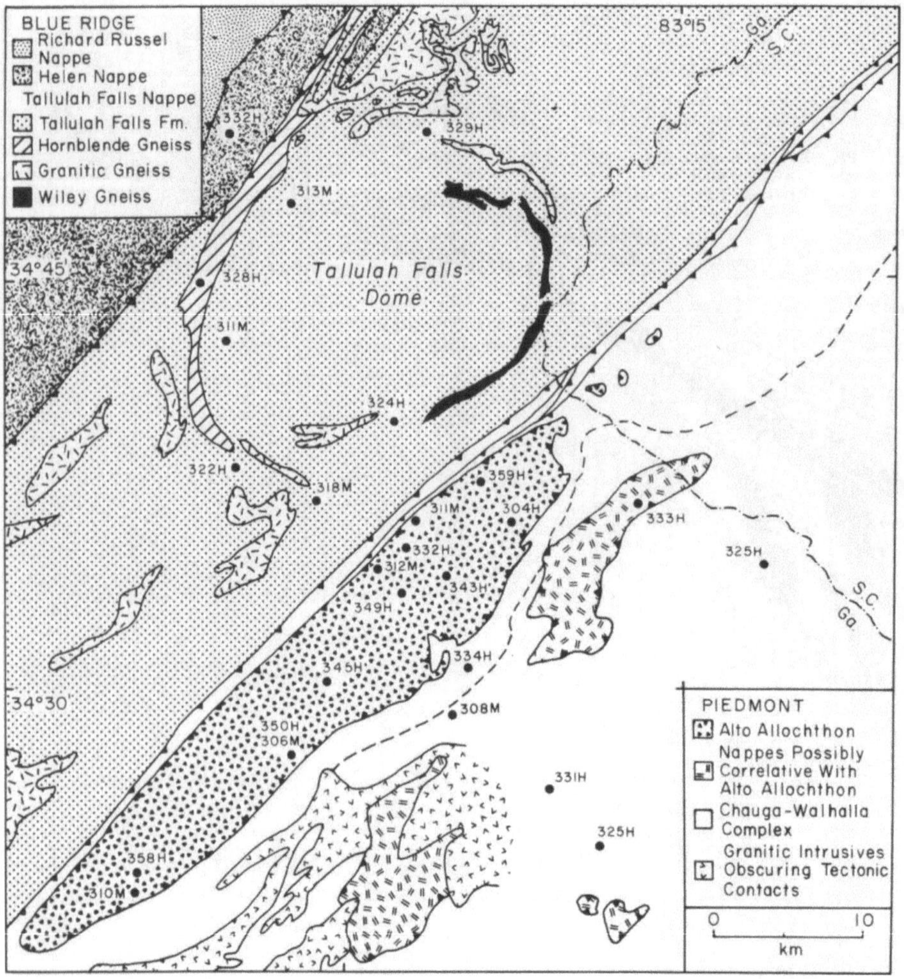

Fig. 3. Generalized geologic map of portions of the Jefferson and Inner Piedmont composite terranes in Georgia and South Carolina (Dallmeyer 1988). *Dashed line* approximately separates upper greenschist-lower amphibolite facies successions (west) from middle and upper amphibolite facies units (east) within the Chauga-Walhalla complex. Representative $^{40}Ar/^{39}Ar$ plateau ages (Dallmeyer 1988) are shown (in Ma) for hornblende *(H)* and muscovite *(M)*

(often represented by mylonitic orthogneisses) cross-cut basement and cover sequences. These include granite, granodiorite, quartz monzonite, and tonalite protolith compositions. Zircons from one of these gneiss units yields 410–420 Ma U-Pb zircon ages (Nelson 1989). Ultramafic schists composed of talc, anthophyllite and serpentine minerals occur locally in the Tallulah Falls nappe. These are exposed as a linear trend of discontinuous pods within enclosing metaclastic rocks, suggesting they comprise a tectonic melange.

Richard Russell Nappe

A large portion of the Richard Russell nappe is represented by rocks of the Richard Russell Formation (Nelson and Gillon 1985). These include biotite gneiss, variably argillaceous and feldspathic metasandstone and garnet-sillimanite-biotite schist. The Richard Russell Formation structurally overlies a more pelitic succession (Crooked Creek schist) which includes biotite-muscovite schist, quartz-feldspar gneiss, amphibolite, hornblende gneiss, calc-silicate, and variably argillaceous and feldspathic metasandstone and/or metagraywacke. Both successions are polydeformed and record lower to upper amphibolite facies regional metamorphic assemblages. Nelson (1989) reported a U-Pb zircon age of 1035 Ma from a granitic intrusive cross-cutting biotite gneiss of the Richard Russell Formation. Weiner and Merschat (1988) have described a paragneiss sequence in a Grenville-age terrane in North Carolina which is lithologically correlative with the Richard Russell Formation.

$^{40}Ar/^{39}Ar$ Geochronology

Dallmeyer (1988) reported $^{40}Ar/^{39}Ar$ plateau ages of 322-335 Ma for hornblende and 311-318 Ma for muscovite from nappe units of the Jefferson composite terrane exposed in Georgia and South Carolina (Fig. 3). These were interpreted to date post-metamorphic cooling through ca. 500° and 400 °C respectively. Dallmeyer (1986) reported slightly older plateau dates (360-365 Ma) for hornblende within variably mylonitic amphibolite developed from mafic-ultramafic sequences (interpreted as fragmented ophiolite by Shaw and Wasserburg 1984) located within a thin thrust sheet tectonically situated between the Jefferson composite terrane and structurally underlying Laurentian metamorphic sequences below the Hayesville thrust.

2.1.4 Inner Piedmont Composite Terrane

The Inner Piedmont composite terrane is an internally imbricated, composite thrust sheet which is composed of contrasting metamorphic sequences which are characterized by different tectonic affinites. Horton et al. (1989, 1991) suggested that it may represent an amalgamation of several disrupted terranes. A variety of protolith affinities are represented, including continental sequences (gneisses and charnockitic units) disrupted oceanic sequences (mafic-ultramafic complexes), together with deformed and metamorphosed melange complexes. In general, protolith age is not well constrained.

Within Georgia and South Carolina the Inner Piedmont composite terrane is largely characterized by polydeformed, generally high-grade aluminous and quartzofeldspathic metasedimentary rocks together with felsic metaplutonic and metavolcanic units. Although protolith age is not well defined, it is generally considered to be late Precambrian or earliest Paleozoic (e.g. Hatcher 1972; Odom et al. 1976; van Breemen and Dallmeyer 1984). Griffin (1977) suggested the Inner Piedmont composite terrane is represented by a high-grade core (sillimanite) which is in tectonic contact with lower grade (kyanite) flanks. Griffin (1975) defined three distinct thrust units (Walhalla, Six-Mile and Starr nappes) within this core-flank structural complex. More recent mapping (A. E. Nelson et al. 1985, 1987 a, b) in Georgia and South Carolina portions of the Inner Piedmont composite terrane has more clearly defined the extent of the Chauga-Walhalla complex and the Six-Mile nappe (Fig. 3).

The Chauga-Walhalla complex includes sequences of contrasting metamorphic grade. Low-grade units (middle greenschist to lower amphibolite facies) occur in the west and include variably graphitic phyllite, amphibolite, variably feldspathic quartzite and minor carbonate rocks (Chauga belt of Hatcher 1972), High-grade sequences (middle and upper amphibolite facies) occur in the east. These include various types of hornblende and biotite gneiss together with variably deformed, generally granitic plutons (Walhalla nappe of Griffin 1975). The contact between these contrasting sequences has been variably interpreted as a thrust (Griffin 1974) or a pronounced metamorhpic gradient (Hatcher 1978 a).

Southeast of the Brevard fault zone, a sequence of high-grade metasedimentary rocks and variably deformed granitic intrusions structurally overlies markedly lower grade rocks (Chauga belt) along the Toccoa Falls fault (e.g. Hatcher 1978 a; Hopson 1984; Hopson and Hatcher 1988). The high-grade sequence was termed the Alto allochthon by Hatcher (1978 a). It consists largely of pelitic schist (locally with sillimanite), biotite quartz feldspar gneiss, amphibolite, and granitic gneiss (mylonitic orothogneiss). Nelson et al. (1987 a, b) suggested that the Alto allochthon is one of several erosional remnants of high-grade rocks within the Six-Mile nappe (Fig. 3). Detailed mapping of northeastern portions of the Alto allochthon indicates that five episodes of deformation are common to the allochthon, Chauga belt, and eastern portions of the Jefferson composite terrane (Hopson 1984; Hopson and Hatcher 1988). The Alto allochthon appears to have been emplaced between D_2 and D_3 deformational events following attainment of maximum regional metamorphic conditions. The allochthon was subsequently affected by phases of both ductile and brittle faulting within the Brevard fault zone.

The timing of specific igneous, metamorphic, and deformational events recorded in the Inner Piedmont composite terrane of Georgia and South Carolina has been uncertain. Homogenization of Sr isotopes appears to have occurred at least locally between ca. 380-400 Ma (e.g. Higgins et al. 1980; van Breemen and Dallmeyer 1984). A suite of bimodal (granite to gabbroic), post-kinematic late Paleozoic plutons was emplaced within pre-deformed Inner Piedmont gneisses (Fig. 4). These do not display regionally penetrative strains and lack well-defined contact aureoles. Rb-Sr whole-rock isochron and/or U-Pb zircon crystallization ages between 330 and 320 Ma have been reported for several of the granitic plutons (Fullagar and Butler 1979; Atkins et al. 1980; Ross and Bickford 1980).

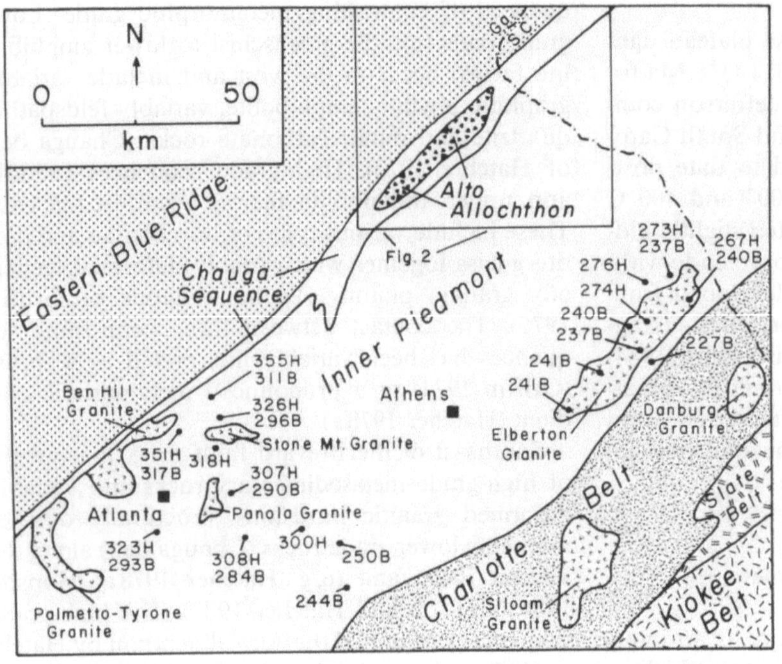

Fig. 4. Representative $^{40}Ar/^{39}Ar$ incremental-release plateau ages (in Ma) for hornblende *(H)* and biotite *(B)* from Georgia portions of the Inner Piedmont composite terrane (Dallmeyer 1978; Dallmeyer et al. 1981). Carboniferous (ca. 265-325 Ma) granitic plutons are named. The quarry investigated by van Breemen and Dallmeyer (1984) is located near Athens

$^{40}Ar/^{39}Ar$ ages were determined for hornblende and biotite along a traverse across Georgia portions of the Inner Piedmont composite terrane exposed near Atlanta (Dallmeyer 1978). Results (Fig. 4) are similar to previously reported K-Ar mineral ages, with a southeasterly younging trend in $^{40}Ar/^{39}Ar$ plateau dates recorded by hornblende (355-300 Ma) and biotite (317-244 Ma). Plateau dates recorded by biotite at each locality are ca. 25 Ma younger than those of coexisting hornblende. Dallmeyer et al. (1981) reported $^{40}Ar/^{39}Ar$ results from the Elberton granite and host Inner Piedmont gneisses (Fig. 4). Hornblende and biotite display internally concordant release spectra and each phase records similar ages within both the pluton and host gneisses. Plateau dates range between 276-267 Ma (hornblende) and 227-241 Ma (biotite), and are markedly younger than a ca. 320 Ma U-Pb zircon crystallization reported for the pluton (Ross and Bickford 1980). Muscovite from the pluton records a plateau age of 236 Ma.

Van Breemen and Dallmeyer (1984) reported Rb-Sr and $^{40}Ar/^{39}Ar$ dates from a quarry in Inner Piedmont gneisses near Athens, Georgia (Fig. 4). Results included Rb-Sr mineral dates of 285 Ma (muscovite) and 256 Ma (biotite) and a 252 Ma $^{40}Ar/^{39}Ar$ biotite age. A 317 ± 17 Ma Rb-Sr isochron was reported for 14 contiguous, 1 cm slabs cut from an individual gneiss sample. This was interpreted to date cooling through ca. 600 °C.

Dallmeyer (1988) reported 325-355 Ma $^{40}Ar/^{39}Ar$ plateau ages for hornblende (Fig. 3) from portions of the Inner Piedmont composite terrane exposed in Georgia and South Carolina (higher grade sequences of the Chauga-Walhalla complex and eastern portions of the Six-Mile nappe). Hornblende within lower grade portions of the Chauga-Walhalla complex record 335-340 Ma plateau ages. Hornblende within western portions of the Six-Mile nappe (Alto allochthon) records 335-360 Ma plateau dates. None of the hornblende spectra suggest a superposed metamorphic history, and the plateau ages were interpreted to date regionally diachronous cooling through ca. 500 °C following a Late Devonian or earlier high-grade metamorphism. Muscovite within all western structural units of the Inner Piedmont composite terrane record 305-315 Ma plateau ages interpreted to date post-metamorphic cooling through ca. 350 °C.

2.1.5 Brevard Fault Zone

Throughout much of its length the western boundary of the Inner Piedmont composite terrane is marked by the enigmatic Brevard fault zone (Figs. 1-3). The Brevard fault zone records a complex history involving polymetamorphism and phases of both ductile and brittle deformation (e.g. Roper and Dunn 1973; Hatcher 1978a, b). It is, in part, stratigraphically controlled and contains fault-bounded structural units of low-grade, platform-type carbonate rocks with bulk compositional affinities similar to North American Paleozoic sequences within the Valley and Ridge province (Hatcher 1971, 1978b). The overall style and sequence of metamorphic and folding events within the Brevard fault zone is generally similar to those recorded in immediately adjacent areas (e.g. Hopson 1984; Hopson and Hatcher 1988).

On the basis of Rb-Sr whole-rock ages, Odom and Fullagar (1973) suggested that mylonite formation within the Brevard fault zone was largely a Devonian event. Roper and Dunn (1973) concluded that widespread retrogressive metamorphism was associated with mylonite formation, requiring an even earlier age for initial prograde metamorphism. On the other hand, Sinha et al. (1986) interpreted a 270 Ma thin-slab isochron age for ultramylonite within the Brevard fault zone near Henderson, North Carolina, to closely date a phase of ductile deformation.

2.2 *Remnants of Oceanic Lithosphere*

Disrupted mafic and ultramafic complexes occur locally throughout the central-southern Appalachian orogen. However, the only extensive tract is within the Bel Air-Rising Sun terrane exposed in Maryland and Pennsylvania (Fig. 2). This may represent structural remnants of oceanic lithosphere, although isotopic characteristics are inconclusive (e.g. Shaw and Wasserburg 1984). On the other hand, the mafic and ultramafic complexes exposed at both Lake Chatuge and Chunky Gal, Georgia-North Carolina (associated with the Jefferson composite terrane) do display Sm-Nd and Rb-Sr isotopic signatures which are consistent with an oceanic crustal protolith derived from a depleted mantle source (Shaw and Wasserburg 1984).

2.3 *Volcanic Arc Terranes*

Structural units largely comprised of variably metamorphosed volcanic, volcaniclastic and sedimentary units are exposed throughout eastern portions of the central-southern Appalachian orogen. Horton et al. (1989, 1991) have termed these the Chopawamsic, Carolina, Spring-Hope and Roanoke Rapids terranes. In addition, Horton et al. (1989) suggested that the Charleston magnetic terrane (Higgins and Zietz 1983) beneath the Coastal Plains in North Carolina and Georgia is largerly underlain by a similar lithologic association. They termed this the Charleston terrane.

2.3.1 Chopawamsic Terrane

The Chopawamsic terrane is largely underlain by variably deformed and metamorphosed mafic and felsic volcanic units (protoliths including tuffaceous clastic rocks, breccias and flows) intercalated with nonvolcanic, clastic sedimentary sequences. Small felsic stocks with trondhjemitic to plagiogranitic composition occur locally. Discordant U-Pb zircon ages suggest an Early Cambrian age for most protoliths. The metamorphosed arc sequence is unconformably overlain by fossil-bearing, Middle-Upper Ordovician metasedimentary rocks. Horton et al. (1989, 1991) concluded that the Chopawamsic terrane formed in association with a west-dipping, Cambrian subduction zone, and was likely amalgamated with the Bel Air-Rising Sun and Potomac terranes prior to the Late Cambrian. They suggested that the resultant composite terrane was structurally emplaced onto the Baltimore terrane in the Early-Middle Ordovician.

2.3.2 Carolina Terrane

Secor et al. (1983) used the term Carolina terrane in reference to constrasting lithotectonic sequences previously ascribed to the Charlotte and Carolina slate belts of the central-southern Appalachians (Figs. 1, 2). Important chronologic and/or stratigraphic differences occur along strike within these belts, and the Carolina terrane is likely comprised of several distinct terrane elements.

2.3.2.1 Georgia and Central South Carolina Segments

Sequences of the Carolina terrane exposed in Georgia and South Carolina (Fig. 5) include low- to medium-grade, intermediate to felsic metavolcanic and epizonal metaplutonic rocks (lower levels) and metasedimentary rocks (upper levels). Metavolcanic rocks (Persimmon Fork Formation) occur predominantly in eastern sectors (Carolina slate belt), whereas epizonal metaplutonic rocks (Little Mountain pluton) occur mainly in western sectors (Charlotte belt). U-Pb zircon and Rb-Sr whole-rock ages

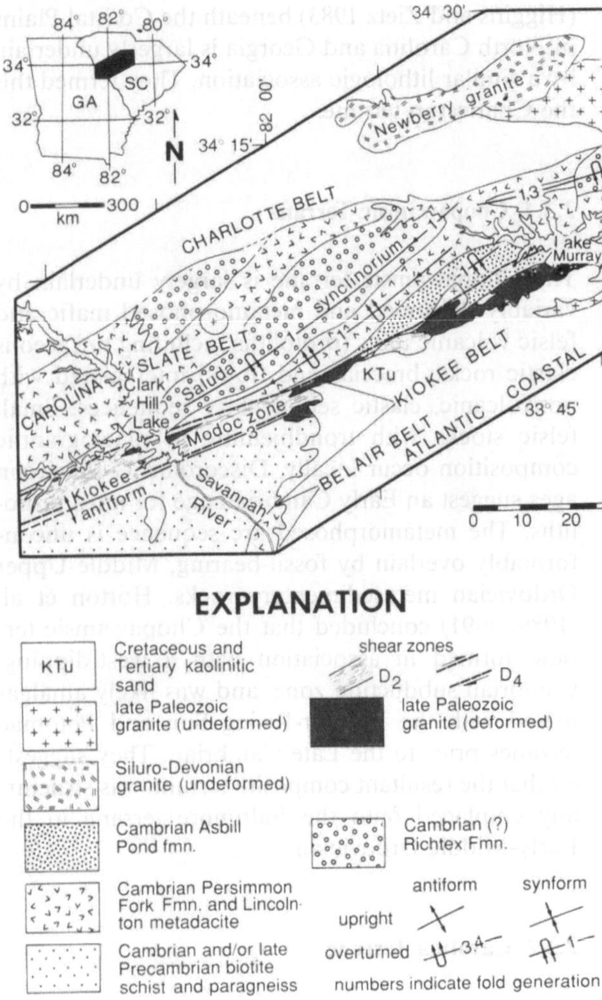

Fig. 5. Generalized geologic map of the Carolina terrane exposed in South Carolina and Georgia (after Overstreet and Bell 1965a, b; Wagener 1977; Secor et al. 1982; Halik 1983; Kirk 1985; and Secor et al. 1986a)

(Carpenter et al. 1982) from rocks interpreted to be correlative with the Persimmon Fork Formation indicate a 540–580 Ma crystallization age (Cambrian?). Textural, compositional, and stratigraphic evidence suggests that the Little Mountain Pluton (550 ± 4 Ma, U-Pb zircon crystallization age: Dallmeyer et al. 1986) is the intrusive equivalent of felsic metavolcanic rocks exposed in adjacent portions of the Carolina slate belt. The tectonic setting in which the volcanic and associated hypabyssal intrusive rocks originated is uncertain. Whitney et al. (1978) concluded that similar rocks exposed in the slate belt of east-central Georgia represent a subduction-related volcanic arc which developed on oceanic lithosphere. On the other hand, Rogers (1982) suggested a convergent continental margin setting.

Two different metasedimentary sequences overlie metavolcanic rocks in central South Carolina (Fig. 5). The Asbill Pond Formation (mudstone, quartz siltstone and quartz sandstone) occurs along the southeast side of the Carolina slate belt and in the adjacent Kiokee belt. The Asbill Pond Formation conformably overlies the Persimmon Fork Formation and is interpreted to have accumulated in a tidal-shelf depositional environment. Upper portions of the Asbill Pond Formation contain a mudstone member which has yield nine genera of Middle Cambrian trilobites with Acado-Baltic affinities (Samson 1984). These fossils indicate that the Asbill Pond and Persimmon Fork Formations are part of an exotic terrane (Carolina terrane of Secor et al. 1983) that accreted to North America subsequent to the Middle Cambrian.

The Richtex Formation (metawacke and thin-bedded metamudstone with subordinate amygdaloidal greenstone) overlies the Persimmon Fork Formation in central and northwestern parts of the Carolina slate belt. It is uncertain wheter the contact between the Richtex and Persimmon Fork Formations is depositional or tectonic.

Tight to isoclinal folding (D_1) with attendant de-

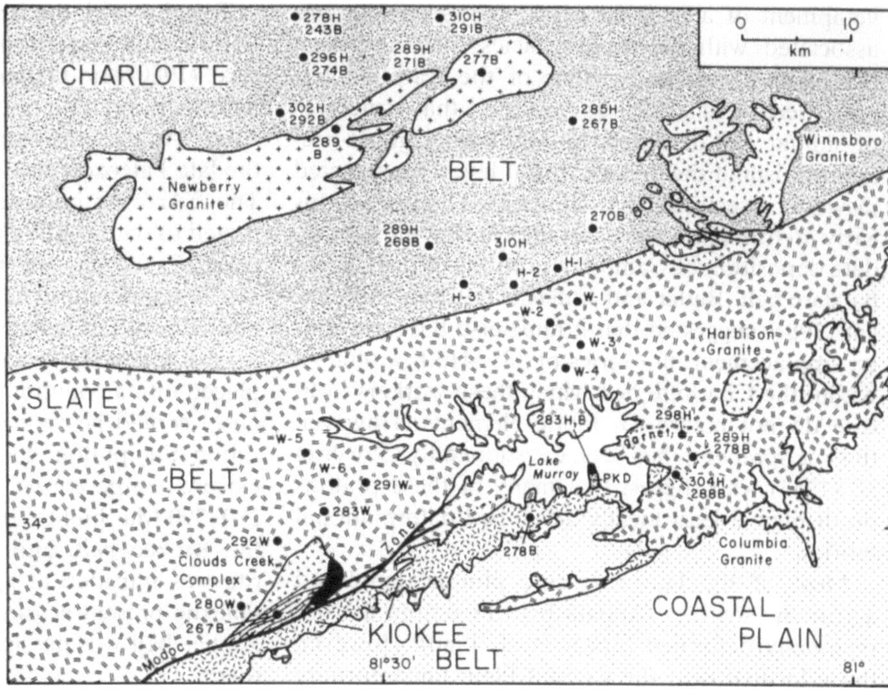

Fig. 6. Representative $^{40}Ar/^{39}Ar$ incremental-release plateau ages (in Ma) for hornblende *(H)*, biotite *(B)*, and whole-rock slate/phyllite *(W)* from the Carolina terrane in South Carolina (Dallmeyer et al. 1986). The Newberry Granite is not penetratively deformed and records a 415±9 Ma Rb-Sr whole-rock isochron crystallization age (Fullagar 1981). The Winnsboro, Harbinson, Columbia and Clouds Creek granites are not penetratively deformed, late Paleozoic plutons. Plutons of similar age within the Kiokee Belt (garnet = garnet isograd) are penetratively mylonitic (PKD = post-kinematic granodiorite within the Kiokee Belt): *H-1*- *H-3* are locations of samples within higher grade portions of the Carolina terrane displaying internally discordant hornblende $^{40}Ar/^{39}Ar$ age spectra; *W-1*- *W-6* are locations of whole rock slate/phyllite samples from lower grade segments of the Carolina terrane which record internally discordant $^{40}Ar/^{39}Ar$ age spectra. Significant ductile strain zones within the Clouds Creek granite are indicated

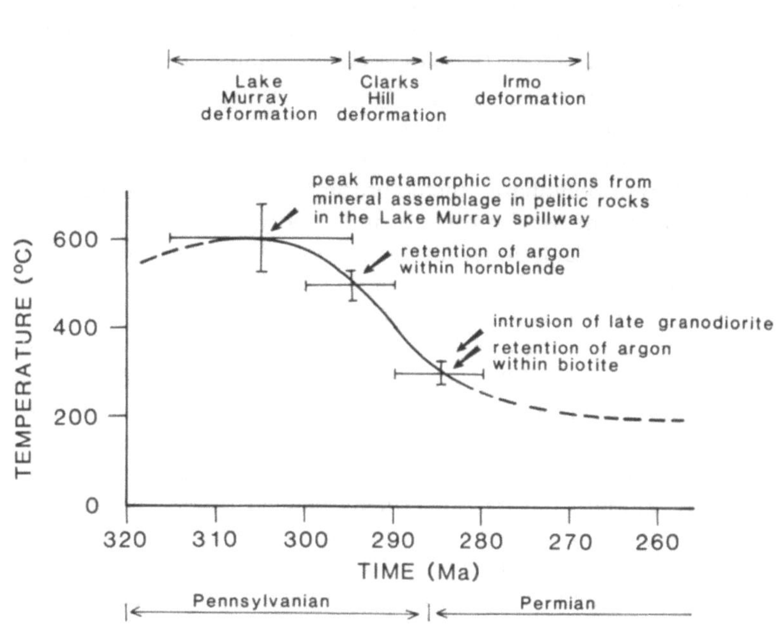

Fig. 7. Chronology of late Paleozoic (Alleghanian) tectonothermal evolution within the Carolina terrane exposed in South Carolina and Georgia (Dallmeyer et al. 1986)

velopment of a regional axial plane cleavage was associated with greenschist facies metamorphism (M_1) within eastern portions of the Carolina terrane. Secor et al. (1986a) suggested that structures of similar age developed during higher grade metamorphism at the deeper crustal levels represented in western sectors (Charlotte belt). The age of this early deformation and metamorphism must postdate deposition of Cambrian portions of the Carolina slate belt. Butler and Fullagar (1978) and Fullagar (1981) suggested that it must also have pre-dated emplacement of a series of 385-415 Ma plutons within the Charlotte belt which these workers indicate do not record D_1 fabric elements. However, this interpretation has been questioned by Glover et al. (1983) who suggest that many of these plutons do display at least locally developed deformational fabrics.

Most of the late Paleozoic plutons emplaced across the southern Appalachian Piedmont are not regionally deformed (Speer et al. 1980). However, those located near the Coastal Plain unconformity record evidence of penetrative, polyphase ductile strain; they constitute the major lithologic components of the Kiokee belt (Figs. 5, 6). Here a spectrum of pre-, syn- and post-kinematic plutons have been described, and their Rb-Sr whole-rock isochron and U-Pb zircon crystallization ages (Snoke et al. 1980; Dallmeyer et al. 1986) provide for calibration of the late Paleozoic tectonothermal record (Fig. 7). Lithologic units within the Carolina slate belt may locally be traced directly into higher grade equivalents within the Kiokee belt. In these areas the boundary between the two belts is arbitrarily placed along the garnet isograd (Fig. 6). Elsewhere, the boundary between the Kiokee and Carolina slate belts is marked by a polyphase, ductile-brittle fault collectively termed the Modoc zone. A similar polygenetic fault (Augusta fault zone) marks the boundary between the Kiokee belt and eastern, low-grade rocks of the Belair belt which are lithologically similar to sequences within the Carolina slate belt.

Dallmeyer et al. (1986) reported that hornblende from northeastern portions of the Kiokee belt displays internally concordant release spectra which define plateau dates between 278 and 285 Ma (Fig. 6). Biotite from this area yields plateau dates between 278 and 285 Ma. Similar ages of 278 and 277 Ma are recorded by biotite from variably mylonitic metagranite within the Kiokee belt. A late- to post-kinematic granodiorite stock intrudes mylonitic metagranite within the Kiokee belt. Both hornblende and biotite from this stock yield mutually identical plateau dates of 283 Ma.

Together with the Rb-Sr whole-rock isochron and U-Pb zircon crystallization ages reported by Snoke et al. (1980) and Dallmeyer et al. (1986) for variably deformed granitic rocks in the Kiokee belt, the $^{40}Ar/^{39}Ar$ mineral dates provide for calibration of the late Paleozoic tectonothermal evolution (Fig. 7). The $^{40}Ar/^{39}Ar$ hornblende plateau ages are interpreted to date cooling through argon retention temperatures of ca. 500 °C. Cooling through the ca. 300 °C temperatures required for argon retention in biotite occurred between 275 and 285 Ma. The late- to post-kinematic granodiorite was apparently emplaced into a country rock terrane which had previously cooled below ca. 500 °C. Following crystallization the pluton cooled quickly through biotite closure temperatures. This is consistent with the 280-290 Ma biotite dates recorded elsewhere in the Kiokee belt, and indicate that the granodiorite was intruded at approximately the same time that the present erosional level of the northeastern Kiokee belt was passing through the ca. 300 °C isotherm.

In an attempt to evaluate the regional extent of late Paleozoic metamorphism beyond the limits of the Kiokee belt, a series of $^{40}Ar/^{39}Ar$ whole-rock ages have been determined for slate/phyllite samples collected across the Carolina slate belt (Dallmeyer et al. 1986). Samples within ca. 5-6 km of the Kiokee belt record internally concordant release spectra which define plateau ages between ca. 280 and 290 Ma (Fig. 6). Samples collected farther from the Kiokee belt display internally discordant spectra in which apparent ages systematically increase over low- and intermediate-temperature increments to define high-temperature plateaux between ca. 340-360 Ma. The nature of this spectra discordancy suggests that the whole-rock systems initially recorded post-metamorphic cooling dates of 340-360 Ma and were variably overprinted during a mild geologic reheating at ca. 270-280 Ma. The lower age bracket is identical to that outlined for the tectonothermal record observed within the Kiokee belt, and indicates that at least slight late Paleozoic reheating affected the entire eastern Carolina slate belt. The $^{40}Ar/^{39}Ar$ results also suggest that original metamorphism (M_1) and deformation (D_1) within the Carolina slate belt likely occurred at ca. 340-360 Ma.

The Clouds Creek granite intrudes the eastern Carolina slate belt (Fig. 6). Fullagar and Butler (1979) reported a 319 ± 27 Ma Rb-Sr whole-rock isochron crystallization age for the pluton. Locally, it is affected by ductile shearing within the Modoc zone. Dynamically recrystallized biotite from two mylonitic metagranite samples record identical $^{40}Ar/^{39}Ar$ plateau ages of 267 Ma (Dallmeyer et al. 1986). This is slightly younger than that recorded by biotite within the Kiokee belt or reset whole-rock slate/phyllite systems in the eastern Carolina slate belt. The 267 Ma biotite dates likely reflect late ductile reactivation within the Modoc zone.

A series of $^{40}Ar/^{39}Ar$ ages have been determined for hornblende and biotite across western portions of the Carolina terrane (Charlotte belt: Dallmeyer et al. 1986). Hornblende from samples collected immediately adjacent to the slate belt display internally discordant spectra (Fig. 6). The nature of this discordancy suggest that a thermal overprint at ca. 270-280 Ma effected partial diffusive loss of radiogenic argon from systems which had previously cooled through argon retention temperatures at ca. 340-360 Ma. This in combination with results from the Carolina slate belt results suggests a Devonian age for initial metamorphism (M_1) and deformation (D_1) recorded within both areas. The remainder of the Charlotte belt hornblendes display internally concordant spectra yielding plateau dates which systematically decrease from 310 Ma in the southeast to 278 Ma in the northwest (Fig. 6). Biotite consistently records a slightly younger age than coexisting hornblende, but plateau dates show a similar trend (ranging from 288 Ma in the southeast to 251 Ma in the northwest). This includes 277 and 289 Ma biotite dates from the Newberry granite for which a 415 ± 9 Ma Rb-Sr whole-rock isochron crystallization age was presented by Fullager (1981).

Except for the discordant spectra displayed by three hornblende samples collected adjacent to the Carolina slate belt, the remainder of biotite and hornblende argon systems within western portions of the Carolina terrane show no record of post-crystallization thermal disturbance, although plateau dates are markedly younger than the ca. 340-360 Ma bracket estimated for initial D_1 tectonothermal activity. This suggests either that: (1) this portion of the Carolina terrane underwent diachronous late Paleozoic cooling after it had been maintained continuously above hornblende argon closure temperatures following initial M_1 metamorphism; or (2) previously closed hornblende and biotite argon systems within this section of the Carolina terrane were totally rejuvenated and then diachronously cooled following a distinct, late Paleozoic reheating. These alternative histories may be evaluated by considering the thermal evolution of generally undeformed, late Paleozoic, post-metamorphic plutons in the area.

Speer et al. (1980) summarized the field, geochemical, and isotopic characteristics of many of these plutons. Those within eastern portions of the Carolina terrane appear to have crystallized under lithostatic pressures of ca. 4.5-5.5 kb (15-18 km) and developed significant contact metamorphic aureoles (e. g. Clouds Creek, Liberty Hill, Winnsboro, and Pageland plutons). These underwent relatively rapid post-crystallization cooling, because in most instances K-Ar and $^{40}Ar/^{39}Ar$ ages of hornblende, biotite and muscovite are similar to Rb-Sr whole-rock isochron crystallization ages. Post-metamorphic plutons which intrude higher grade country rocks in the Inner Piedmont composite terrane and western Carolina terrane also appear to have been emplaced at depths of ca. 15-18 km; however, they do not display well-defined contact aureoles. In addition, K-Ar and $^{40}Ar/^{39}Ar$ mineral dates are consistently younger than Rb-Sr whole-rock isochron or U-Pb zircon crystallization ages. These differences suggest that markedly different temperatures were maintained at equivalent depths (reflecting contrasting geothermal gradients) within the various southern Appalachian terranes during emplacement of the 265-325 Ma plutons. These relationships together with the discordant $^{40}Ar/^{39}Ar$ age spectra displayed by hornblende from the initially higher crustal levels presently exposed along the Carolina slate belt contact combine to suggest that the observed variation in $^{40}Ar/^{39}Ar$ mineral ages across western segments of the Carolina terrane reflects diachronous cooling of a terrane which had been maintained above hornblende closure temperatures following a ca. 340-360 Ma regional metamorphic and deformational event.

Three distinct deformational events appear to have affected the Carolina terrane in Georgia and South Carolina between ca. 315 and 270 Ma. Together these comprise the Alleghanian orogeny (Secor et al. 1986a; Dallmeyer et al. 1986). The first event was associated with regional metamorphism of variable grade and the emplacement of felsic plutons at mid-crustal depths between ca. 315 and 295 Ma. A second Alleghanian event resulted in folding of early isothermal surfaces between ca. 295 and 285 Ma. The final phase of Alleghanian deformation led to development of dextral ductile shear zones between ca. 290 and 265 Ma.

Regional gravity and magnetic data (e. g. Hatcher and Zietz 1980; Cook and Oliver 1981) together with seismic results (e. g. Cook et al. 1979) indicate that all tectonic components of the southern Appalachian Piedmont are allochthonous, and were emplaced onto the North American margin along late

Paleozoic thrusts. Secor et al. (1986b) and Dallmeyer et al. (1986) suggest that regional variations in metamorphic grade and $^{40}Ar/^{39}Ar$ mineral cooling ages in the Carolina terrane exposed in Georgia and South Carolina are a result of the exposure of variable crustal levels which developed by regional flexure during late Paleozoic translation over thrust ramps. Their model suggests that Charlotte and Carolina slate belt segments of the Carolina terrane were initially contiguous and underwent a regional metamorphism sometime prior to ca. 340-360 Ma. Cooling following this metamorphism resulted in a crustal temperature distribution similar to that portrayed in Fig. 8. Upper crustal levels had cooled below hornblende, biotite, and whole-rock slate/phyllite argon closure temperatures while deeper crustal levels were continuously maintained above hornblende closure temperatures. Pre- to syn-tectonic granitic plutons were emplaced into eastern segments of the Carolina terrane between ca. 320 and 310 Ma. This heat influx resulted in widespread metamorphism and establishment of an amphibolite grade infrastructure (Kiokee belt) and a greenschist grade superstructure. In higher crustal levels the late Paleozoic thermal event variably reset intracrystalline argon systems which had previously cooled through closure temperatures. However, in deeper crustal levels ambient country rock temperatures had been continuously maintained above hornblende argon closure temperatures since initial Devonian metamorphism. This crustal section underwent westward transport onto the North American margin following emplacement of ca. 260-280 Ma late to post-kinematic plutons within the infrastructure. During translation, the high-grade intrastructure thickened as a result of stepping over a frontal ramp in the basal thrust (Fig. 8). This was accompanied by dextral ductile shearing and differential offset of the high-grade core and lower grade flanking zones along the polygenetic Modoc zone. The 267 Ma $^{40}Ar/^{39}Ar$ dates recorded by dynamically recrystallized biotite within the Modoc zone likely closely date final stages of this ductile activity. Depper crustal levels of the Carolina terrane (Charlotte belt) which had been maintained above hornblende argon closure temperatures since initial Devonian metamorphism were brought to higher crustal levels as a result of regional flexure over a western, higher level thrust ramp (Fig. 8). This model would interpret the discordant hornblende spectra from easternmost portions of the Charlotte belt to reflect exposure of relatively high crustal levels where cooling below ca. 500 °C had occurred prior to emplacement of late Paleozoic plutons in the infrastructural core. Similarly, whole-rock slate/phyllite systems within the Carolina Slate belt were progressively more affected by late Paleozoic heating in locations proximal to the infrastructural zone.

Final emplacement of Georgia and south Carolina segments of the allochthonous Carolina terrane into their present structural position over the Laurentian basement was clearly a late Paleozoic (Carboniferous) event. However, the chronology of initial accretion to Laurentia is uncertain. Paleomagnetic characteristics of plutons intruding the Carolina terrane are not consistent with significant post-Devonian latitudinal movement. This requires that the Carolina terrane was at least proximal to Laurentia prior to the Carboniferous (e.g. Ellwood 1982; Barton and Brown 1983; Dooley 1983). Sedimentological expression of outboard (eastern) tectonic instability is clearly documented in Silurian and Devonian successions deposited within sectors of the Laurentian miogeocline now exposed in Georgia and Alabama (e.g. Chowns 1989), and it is likely that at least some of the D_1 deformation and metamorphism recorded within the Carolina terrane developed during Devonian accretion to Laurentia. A tectonothermal record of this accretion also appears to be at least locally recorded in portions of the Jefferson and Inner Piedmont composite terranes. For example, the Alto allochthon appears to have been emplaced by west-directed thrusting from a Piedmont root zone following Late Devonian or earlier high-grade metamorphism (Hopson 1984; Hopson and Hatcher 1988; Dallmeyer 1988). This thermal chronology is consistent with work elsewhere in the Inner Piedmont composite terrane which suggests at least local homogenization of Sr isotopes between ca. 380-400 Ma (Higgins et al. 1980; van Breemen and Dallmeyer 1984). It is also consistent with ca. 355 Ma post-metamorphic $^{40}Ar/^{39}Ar$ cooling ages recorded by hornblende in westernmost portions of the Inner Piedmont composite terrane near Atlanta (Fig. 4).

Emplacement of the Alto allochthon occurred after Late Devonian or earlier high-grade metamorphism and prior to regional cooling through muscovite closure temperature (ca. 350 °C) at ca. 300-315 Ma. Upward transport to higher crustal levels during nappe transport is likely recorded by the diachronous cooling of the allochthon through ca. 500 °C between ca. 360 and 335 Ma. Generally similar times for ductile faulting are suggested by Rb-Sr whole-rock isochron ages reported for mylonitic rocks within thrusts bordering several of the southern Appalachian lithotectonic belts. These in-

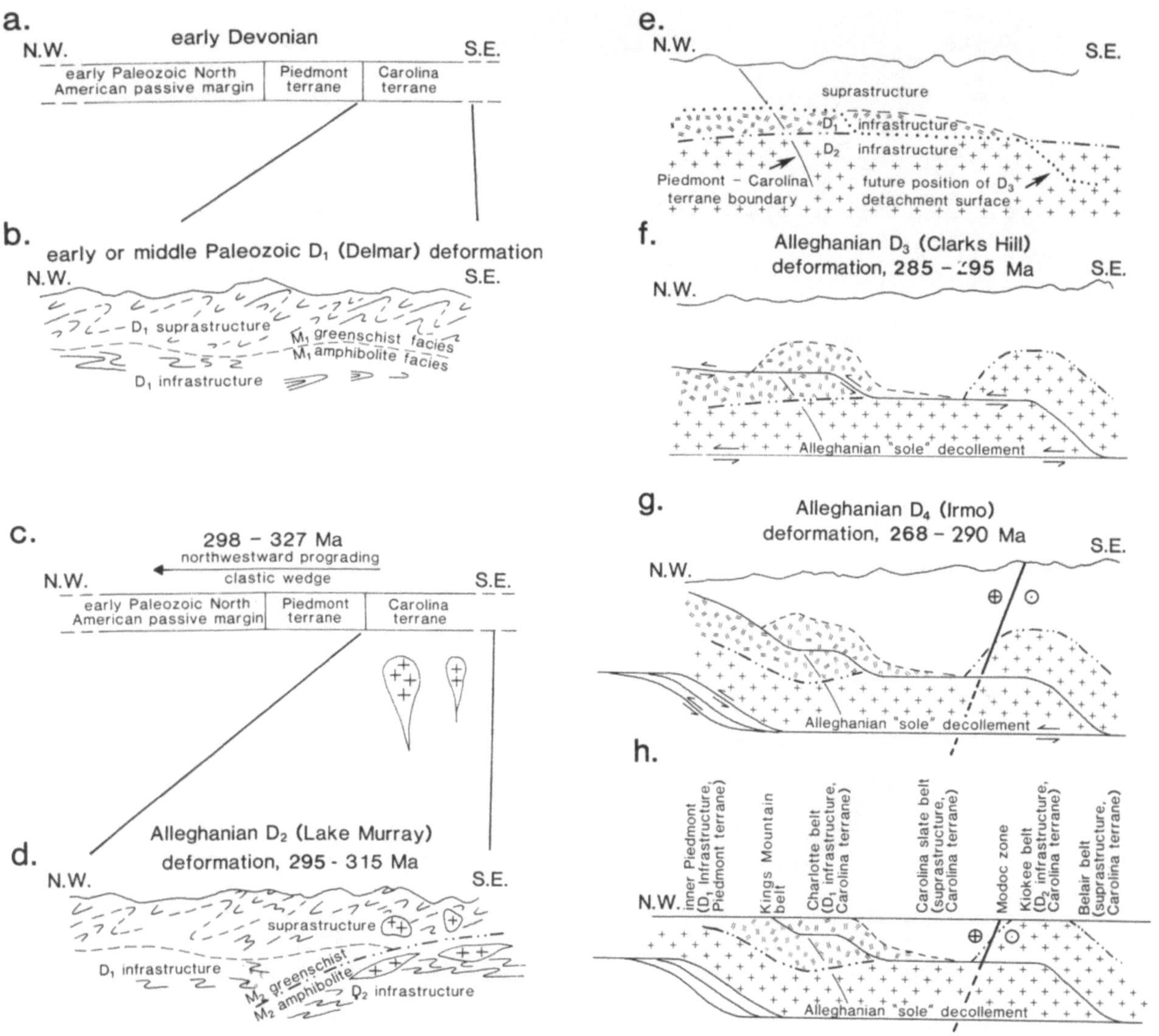

Fig. 8. Late Paleozoic tectonothermal evolution of the eastern Carolina terrane exposed in South Carolina and Georgia (Secor et al. 1986b). **a.** Crustal thermal setting following middle Paleozoic metamorphism and deformation (associated with accretion to Laurentia). **b., c.** Between 298 and 327 Ma intrusion of numerous granitic plutons resulted in high-grade metamorphism of outboard (eastern) terranes and formation of an infrastructural core. Regional thermal affects are only recorded in those mineral argon systems that had previously cooled through appropriate argon closure temperatures. Deeper crustal levels were continuously maintained above argon closure temperatures and therefore do not record a distinct late Paleozoic thermal disturbance. **d., e.** During 285-315 Ma the suprastructural and infrastructure is transported westward onto the Laurentian margin (after emplacement of the youngest post-kinematic plutons) during continental collision. Internal Alleghanian deformation occurs (D_2 and D_3). **f** 268-290 Ma the infrastructure oversteepens due to regional flexure over frontal ramps. This is accompanied by dextral shear within localized zones (D_4). **g., h.** At the present erosional level, exposure of alternating belts of low-, medium- and high-grade metamorphic rocks within the Carolina terrane are a consequence of D_3 Alleghanian folding and imbrication of paleoisothermal surfaces

clude: (1) the Brevard fault zone (356±28 Ma; Odon and Fullagar 1973); and (2) the Great Smoky fault (368±9 Ma; Hatcher and Odom 1980).

2.3.2.2 Northeastern North Carolina Segments

A sequence of late Precambrian to Early Cambrian (?) metavolcanic and metasedimentary rocks exceeding 20 km in thickness constitute the Carolina terrane exposed in the vicinity of Albemarle, North Carolina (Fig. 9). The oldest rocks comprise the Hy-

Fig. 9. Geologic map of the Carolina terrane exposed in northern North Carolina and southernmost Virginia (modified from Harris and Glover 1985)

co Formation which is a sequence of intermediate to felsic metavolcanic rocks and associated epizonal metaplutonic rocks that accumulated between ca. 600 and 680 Ma (Glover and Sinha 1973; McConnell and Glover 1982). The Uwharrie Formation unconformably overlies the Hyco Formation (Harris and Glover 1985) and is exposed in the axial region of the Troy anticlinorium (Fig. 9). The Uwharrie Formation is a bimodal sequence of mafic and felsic metavolcanic rocks. Wright and Seiders (1980) reported a 586 ± 10 Ma U-Pb zircon age for volcanic rocks within upper portions of te Uwharrie Formation.

The Uwharrie Formation is conformably overlain by predominantly metasedimentary rocks of the Albemarle Group which is comprised of an upward coarsening sequence of metamudstone, metasiltstone and metawacke with subordinate interlayered mafic to felsic metavolcanic rocks (Fig. 10). St. Jean (1973) described two fossils from a single block of metamudstone from within the Floyd Church Formation. Although St. Jean assigned these fossils to the trilobite genus *Paradoxides*, they were subsequently reassigned to the Ediacarian metazoan genus *Pteridinium* after several other examples of *Pteridinium* were found in the Floyd Church Formation (Gibson et al. 1984). The genus *Pteridinium* is characteristic of the Precambrian Ediacarian Period/System of Cloud and Glaessner (1982) and has never been found in Cambrian rocks. The occurrence of a fragment of *Pteridinium* at the top of the Floyd Church Formation (Gibson and Teeter 1984) indicates that the entire formation is of Precambrian age.

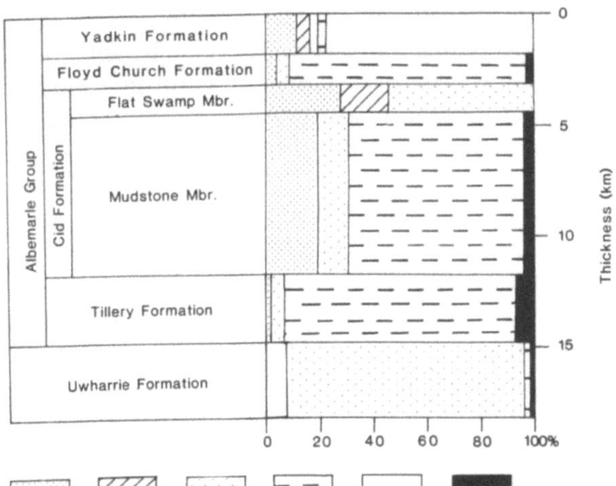

Fig. 10. Stratigraphic sequence of the Carolina terrane exposed in northern North Carolina indicating lithologic composition and thickness (modified from Seiders 1978)

The angular unconformity between the Hyco and overlying Uwharrie Formations has been interpreted to be a manifestation of late Precambrian deformation without accompanying metamorphism or foliation development (Glover and Sinha 1973). The effects of this Precambrian deformation have been more thoroughly documented in the vicinity of the Virginia-North Carolina border (Glover and Sinha 1973; Harris and Glover 1985). In the vicinity of Albemarle, rocks of the Uwharrie Formation and Albemarle Group experienced a *subsequent* regional metamorphism and folding with development of an attendant axial plane foliation (D_2). Map patterns of lithologic units are controlled by gently plunging major D_2 folds having wave lengths of 10-20 km and limb dips of 10°-30°. West of Albemarle, rocks of the Carolina slate belt are separated from the Charlotte belt by the Gold Hill and Silver Hill faults. Major D_2 folds in the slate belt are truncated by the Silver Hill fault. West of the Gold Hill fault (Charlotte belt) regional metamorphism is in the amphibolite facies, whereas east of the fault (Carolina slate belt) greenschist facies assemblages are dominant.

Geochronological studies indicate an early Paleozoic age for the D_2 event recorded within segments of the Carolina terrane exposed in northeastern North Carolina. In low-grade areas (Carolina slate belt), K-Ar whole-rock ages of phyllites suggest D_2 occurred at ca. 483 ± 15 Ma (Kish et al. 1979). Recent $^{40}Ar/^{39}Ar$ incremental release analyses of whole-rock slate from the Tillery Formation near Albemarle indicate that the D_2 cleavage developed at ca. 455-465 Ma (Noel et al. 1988). $^{40}Ar/^{39}Ar$ ages of metamorphic hornblende within higher grade portions of the Carolina terrane exposed in North Carolina (Charlotte belt) range between 425-430 Ma (Sutter et al. 1983, 1984), also suggesting an early Paleozoic age for D_2 in that area.

Paleomagnetic characteristics of rocks within the Albemarle Group indicate that the observed remnant magnetism was likely acquired immediately following D_2 folding at ca. 455-465 Ma (Noel et al. 1988). The results also suggest that the area was at that time located at a relatively low latitude (ca. 20 °C). These observations combine to suggest that the early Paleozoic (D_2) tectonothermal activity likely developed during accretion of northeastern segments of the Carolina terrane to Laurentia. Tectonothermal activity of similar age was documented by Farrar (1985) who demonstrated that low-grade portions of the Carolina terrane (Carolina slate belt) in northeastern North Carolina were thrust westward over middle Proterozoic basement rocks in the early Paleozoic. The paleomagnetic data of Noel et al. (1988) also suggest that northeastern segments of the Carolina terrane were not part of an extensive "Avalon terrane" which included sequences exposed in the northern Appalachian orogen [which have been shown, for example, by Van der Voo and Johnson (1985) to have been situated at relatively high paleolatitudes during the early Paleozoic].

Studies in higher grade elements of the Carolina terrane exposed immediately west of Albemarle (Butler and Fullagar 1978; Glover et al. 1983) indicate that the Salisbury group of plutons (385-415 Ma) are mildly deformed (D_3), whereas 285-325 Ma plutons are undeformed. The Carolina slate and Charlotte belts in the Albemarle area are separated by the Silver Hill and Gold Hill faults, and it is uncertain if the middle Paleozoic (D_3) deformation manifested in higher grade rocks of the Charlotte belt extends eastward across these faults into lower grade units of the slate belt.

No evidence has been documented for late Paleozoic (Alleghanian) deformation in the Albemarle area. However, slate belt rocks exposed at Albemarle are continuous with those exposed on the west side of the Raleigh belt in northeastern North Carolina where Farrar (1985) presented evidence for polyphase deformation together with upper greenschist to amphibolite facies regional metamorphism at ca. 285-315 Ma. The region of Alleghanian ductile deformation and regional metamorphism recorded in eastern segments of the Carolina terrane has been interpreted as an intraplate manifestation of late Paleozoic, oblique (dextral) collision with Gondwana and subsequent northwestward overthrusting of crystalline Piedmont rocks farther onto the North America plate (Harris et al. 1981; Gates et al. 1986; Secor et al. 1986b). Therefore, structural units comprising the Carolina terrane in the vicinity of Albemarle could have experienced late Paleozoic rotation and/or displacement relative to North America while not recording penetrative Alleghanian deformation.

2.3.2.3 South Carolina - North Carolina Transitional zone

Little detailed geologic mapping has been carried out in the Carolina terrane exposed in northeastern South Carolina. A generalized geologic map (Fig. 11) has been prepared from a data base including quadrangle mapping by Nystrom (1972) and Butler and Howell (1976), together with geological reconnaissance investigations of Overstreet and Bell (1965a, b), Bell et al. (1974), and Butler (1977).

Lower grade portions of the Carolina terrane exposed in northeastern South Carolina contain an extensive sequence of predominantly felsic metavolcanic rocks. The following characteristics suggest that these are lateral equivalents of the Persimmon Fork Formation exposed in central South Carolina:

1. They are lithologically similar and are, in part, exposed directly along strike from the Persimmon Fork Formation near Columbia, South Carolina.
2. A ca. 525 Ma Rb-Sr whole-rock age from felsic metavolcanic rocks in northeastern South Carolina (Butler and Fullagar 1975) suggests affinity with the Cambrian South Carolina sequence rather than with the late Precambrian sequence exposed in North Carolina.

Low-grade metasedimentary rocks within the Carolina terrane of northeastern South Carolina also appear to be lateral equivalents of the Cambrian South Carolina sequence. They lithologically resemble mudstones in upper portions of the Asbill Pond or Richtex formation, and, at one locality, Paleozoic sponge spicules have been described (Fig. 11; Bourland and Rigby 1982). These characteristics indicate that the northeast South Carolina metasedimentary sequence is, at least in part, too young to directly correlate with the Precambrian Albemarle Group.

Along the border between North and South Carolina, map patterns suggest a series of northeastward plunging folds, with the Cambrian(?) felsic metavolcanic rocks of the South Carolina sequence plunging northeastward beneath an upward facing Precambrian sequence within the Albemarle Group. These geometric relationships are incompatible with the inferred ages of the map units, and suggest, instead, that the North and South Carolina sequences are most likely tectonically juxtaposed.

The following lines of evidence suggest that a major fault, perhaps representing an internal terrane boundary within the Carolina terrane, is likely present in the eastern Piedmont in the vicinity of the North Carolina-South Carolina state line (Fig. 11):

1. The South Carolina sequence appears to be younger than ca. 565 Ma (Cambrian and Cambrian?), whereas the North Carolina sequence is entirely or almost entirely of Precambrian age.

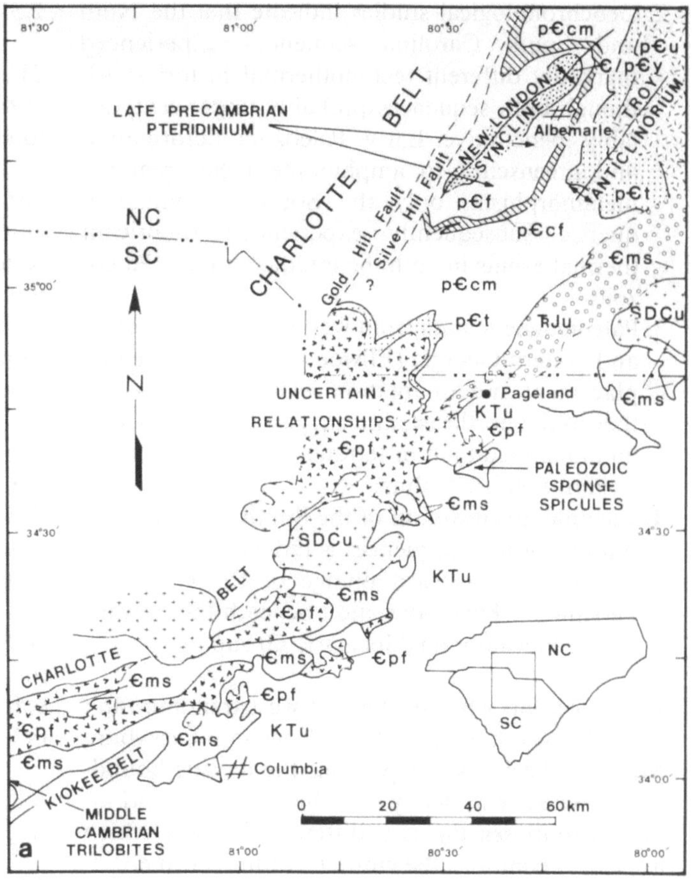

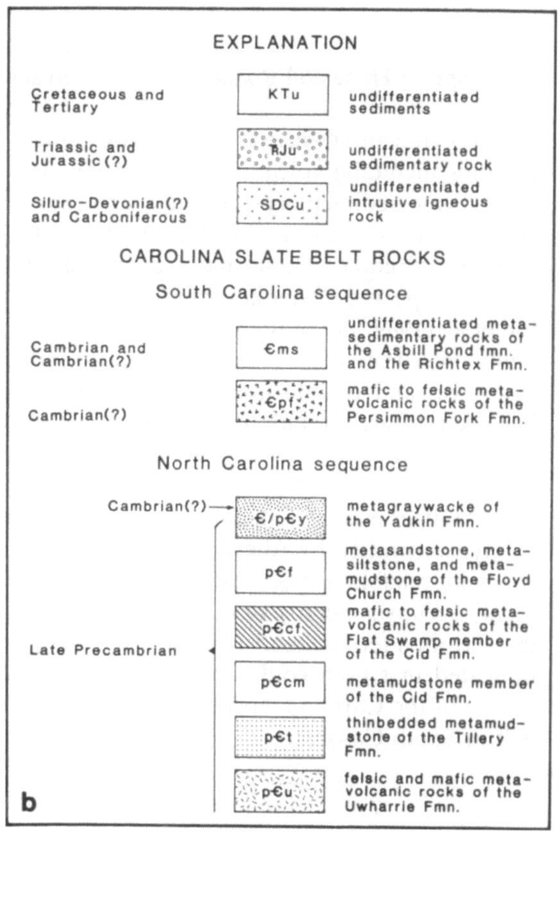

Fig. 11. Geologic map of the Carolina terrane exposed in southern North Carolina and northern South Carolina (complied in modified form from Overstreet and Bell 1965a, b; Nystrom 1972; Bell et al. 1974; Butler 1977; and Harris and Glover 1985)

2. Geochronological studies indicate that the Noth and South Carolina sequences experienced markedly different tectonothermal histories. Although both sequences probably underwent variably penetrative Early Paleozoic deformation and greenschist to amphibolite facies regional metamorphism, only the South Carolina sequence subsequently experienced significant thermal events in both the middle and late Paleozoic.
3. Paleomagnetic data from western South Carolina and eastern Georgia (Ellwood 1982) indicate displacement and/or rotation relative to areas farther to the northeast, and suggest the presence of an unmapped fault or terrane boundary in northeastern South Carolina.
4. Tectonic juxtaposition of the South Carolina and North Carolina sequences is indicated by the apparent northeastward plunge of the Cambrian Persimmon Fork Formation beneath the upward facing Precambrian Albemarle Group.

It is uncertain whether the east-west trending fault near the south end of the Wadesboro-Crowburg Mesozoic basin itself represents the terrane boundary or whether it merely cuts the boundary and locally juxtaposes the two different terranes which appear to comprise the entire Carolina terrane.

2.3.3 Spring Hope and Roanoke Rapids Terranes

Variably metamorphosed and polydeformed successions including volcanic, volcaniclastic and various clastic sedimentary protoliths are exposed in Nash County, North Carolina and in Halifax County, Virginia. These have been termed the Spring Hope and Roanoke Rapids terranes by Horton et al. (1989, 1991). Both sequences record a generally penetrative late Paleozoic tectonothermal overprint, and establishment of an earlier history has been difficult. Poorly-defined Rb-Sr whole-rock isochrons suggest lower Paleozoic (Cambrian?) crystallization ages for some extrusive sequences (e.g. Mauger et al. 1983), but most units are too severely overprinted by Alleghanian orogenic activity to yield reliable Rb-Sr results. Both terrane elements occupy generally similar structural settings as the Carolina terrane, and Horton et al. (1989, 1991) indicated that both could represent fault-detached blocks of the Carolina terrane (similar to the Belair belt in South Carolina).

2.3.4 Charleston Terrane

The Charleston terrane as defined by Horton et al. (1989, 1991) is generally equivalent to the "Charleston magnetic terrane" described by Higgins and Zietz (1983) or the "Brunswick terrane" of Williams and Hatcher (1982, 1983). This area is totally overlain by Coastal Plain sedimentary rocks (Fig. 2) and is locally covered by Mesozoic units within the South Georgia basin. Material recovered through drill penetrations displays characteristics generally similar to the low-grade metamorphosed volcanic, volcaniclastic and sedimentary units within the Carolina, Spring Hope and Roanoke Rapids terranes (Horton et al. 1989). Williams and Hatcher (1982, 1983) and Higgins and Zietz (1983) distinguished this area from the Carolina and Spring Hope terranes on the basis of characteristic, broadwavelength magnetic highs. These were interpreted to relate to mafic and granitic plutons. Sinha et al. (1980) reported U-Pb zircon ages of 372–377 Ma for various zircon size fractions separated from drill core recovered over one of the gravity lows (Springfield monzogranite of Speer 1982).

2.4 *Metamorphic Complexes with Uncertain Origins*

Several variably extensive, fault-bounded structural units containing metamorphic rocks within the central-southern Appalachians are poorly-known, and not easily classified as to terrane type. Horton et al. (1989, 1991) termed these "metamorphic complexes of undetermined affinity". They include the Milton, Gaffney, Uchee, Crabtree, Goochland, Wilmington and Hatteras terranes (Fig. 2).

2.4.1 Milton Terrane

The Milton terrane is exposed in North Carolina and Virginia (Fig. 2). It is composed of a heterogenous lithologic assemblage which includes intercalated mafic to felsic gneiss units and subordinate pelitic schist. Glover et al. (1971) reported discordant U-Pb zircon ages from a constituent felsic gneiss which defined an upper concordia intercept age of ca. 740 Ma. Several workers have suggested that the Milton terrane represents more penetratively deformed and higher grade metamorphic equivalents of successions within the Carolina terrane (e.g. Tobish and Glover 1969; Henika 1977). On the other hand, Kish (1983) suggested that units comprising the Milton terrane display characteristics more similar to rocks within the Inner Piedmont terrane.

2.4.2 Gaffney Terrane

The Gaffney terrane of Horton et al. (1989, 1991) is exposed along the South Carolina - North Carolina border (Fig. 2), and constitutes western segments of what has been classically referred to as the Kings Mountain belt. Predominant rock types include variably graphic pelitic units (schist and phyllite) together with intercalated amphibolite, quartzite, marble and calc-silicate intervals (Horton 1984). The age and tectonic setting of the Gaffney terrane are uncertain. Horton et al. (1989, 1991) suggested it could represent an internal continental terrane in which Laurentian crystalline basement is not exposed. Alternatively, the sequences could be metamorphic equivalents of rocks within the Carolina terrane.

2.4.3 Uchee Terrane

The Uchee terrane is exposed in Georgia and Alabama (Fig. 2), and is represented by a high-grade, generally migmatitic complex which includes paragneiss, amphibolite and granitic gneisses (e.g. Hanley 1986). Russell (1985) interpreted discordant U-Pb zircon results to suggest a late Proterozoic or early Paleozoic (Early Cambrian?) age. Stow et al. (1984) interpreted geochemical characteristics to suggest that constituent amphibolites were derived from tholeiitic basalt protoliths which formed in an oceanic rift environment.

2.4.4 Crabtree Terrane

A small, thrust-bounded sequence of quartzofeldspathic gneisses is exposed along the western flank of the Raleigh belt (Fig. 1) in North Carolina. This has been termed the Crabtree terrane by Horton et al. (1989). Stoddard et al. (1978) suggested protoliths were most likely of a felsic igneous and/or arkosic sedimentary nature. A poorly defined Rb-Sr whole-rock isochron (Kish and Campbell 1986) indicates that protoliths must pre-date the Devonian. Horton et al. (1989) tentatively interpreted the Crabtree terrane to represent a continental sequence with an uncertain age and unknown palinspastic affinities.

2.4.5 Goochland Terrane

Rocks ascribed to the Goochland terrane by Horton et al. (1989, 1991) are exposed in Virginia and North Carolina and may be traced in the subsurface beneath the Coastal Plain into Maryland (Fig. 2). Predominant rock types include amphibolite and high-grade gneissic units which locally preserve evidence of a granulite facies metamorphism. A small stock of anorthosite has also been mapped within this terrane (Farrar 1984). An poorly defined Rb-Sr whole-rock isochron age of 1031 ± 94 Ma was reported by Glover et al. (1978, 1982) for a gneissic unit within the Goochland terrane. Farrar (1984) and Glover et al. (1982) suggested that these sequences represent a fault-bounded fragment of Laurentia which was penetratively deformed during late Paleozoic orogenesis. On the other hand, Rankin et al. (1988) suggested that the Goochland terrane displays fundamental differences in both stratigraphy and structural style compared with proximal portions of Laurentia. They interpreted the Goochland terrane to be an exotic tectonic element.

2.4.6 Wilmington Terrane

A unique succession of very high-grade (granulite-facies) sedimentary, intrusive and volcanic rocks are exposed in Delaware. Horton et al. (1989, 1991) termed this sequence the Wilmington terrane (Fig. 2). Anorthositic gabbro has been locally described (Foland and Muessig 1978). A cross-cutting pluton appears to be lower Paleozoic (Cambrian?) in age based upon Rb-Sr whole-rock isochron results reported by Foland and Muessig (1978), implying that the host metastratified sequences must be older. Rao and Van der Voo (1980) and Brown and Van der Voo (1983) interpreted paleomagnetic data from the Wilmington terrane to indicate that it was removed from Laurentia in the Ordovician, and perhaps associated with "Armorican" continental fragments situated at relatively high latitudes at that time. However, as pointed out by Horton et al. (1989), this interpretation does not adequently consider potential modifications resulting from superposed deformational and/or metamorphic affects. Lithologic and tectonothermal characteristics are generally similar to those of the Goochland terrane, and the Wilmington could represent a fault-bounded extension of this continental terrane element.

2.4.7 Hatteras Terrane

Horton et al. (1989, 1991) used the term Hatteras terrane (Fig. 2) in reference to a subsurface sequence of medium-grade metamorphic rocks which host a series of plutonic rocks (ranging from granite to diorite) beneath the Coastal Plain in eastern North Carolina. Dennison et al. (1967) and Russell et al. (1981) reported Rb-Sr whole-rock isochron ages ranging between 583 and 685 Ma for the plutonic elements. These ages and the overall lithologic character suggests that the Hatteras terrane may represent higher grade metamorphic equivalents of either the Carolina and/or Roanoke Rapids terranes.

3 Basement Beneath the Gulf and Southeastern Atlantic Coastal Plain

3.1 Introduction

The nature of the pre-Mesozoic crystalline basement beneath the Atlantic and Gulf Coastal Plains of the southeastern United States has been revealed by penetrations associated with deep oil test drilling. Buried extensions of Appalachian tectonic elements (including the Valley and Ridge Province, Talledega slate belt and various Piedmont terranes) extend ca. 50–60 km southeast of the Coastal Plain unconformity (Fig. 12). These are bordered to the south by a series of fault-bounded basins containing Mesozoic, continental clastic rocks which are intruded by numerous diabase dikes. Three contrasting lithotectonic elements constitute the pre-Mesozoic crystalline basement which has been penetrated south of the Mesozoic basins. These include (Fig. 12): (1) a group of metamorphic rocks of

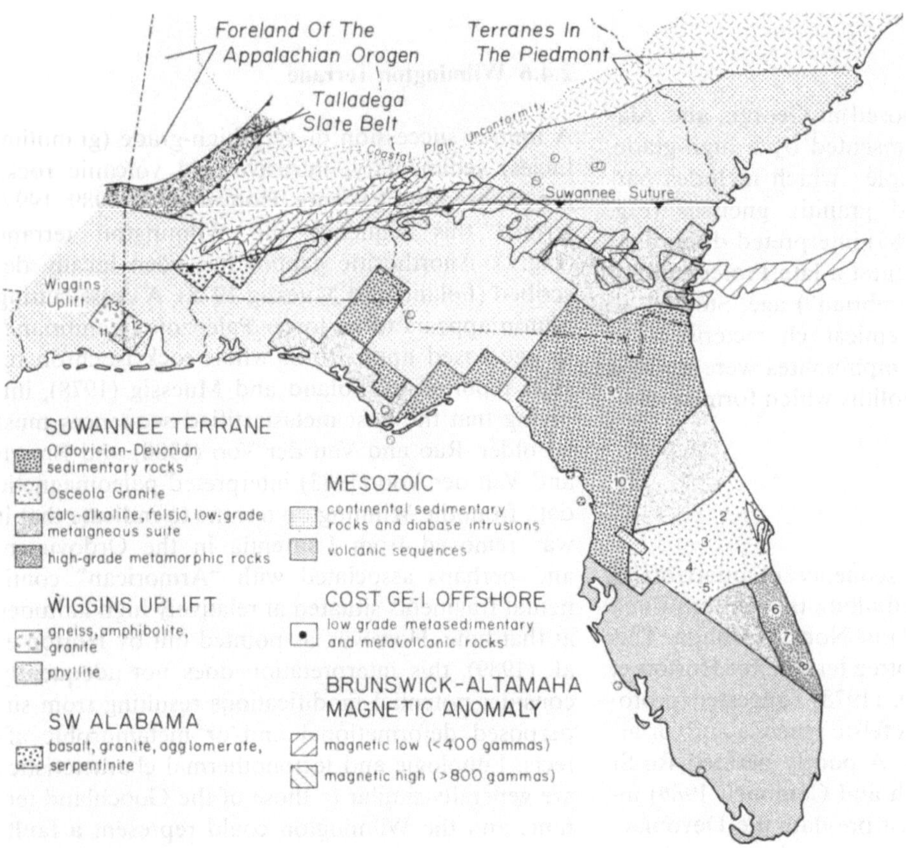

Fig. 12. Lithotectonic units within pre-Mesozoic crystalline basement of the Atlantic and Gulf coastal plains, southeastern United States (adapted from Chowns and Williams 1983; Thomas et al. 1989). All contacts with Mesozoic sequences shown as high-angle faults. Trace of Brunswick-Altamaha magnetic anomaly from Zeitz (1982). Suwannee suture from Thomas et al. (1989). The wells discussed in the text are located *(1-13)*

variable grade together with deformed and retrogressed granite in southwestern Alabama and southeastern Mississippi (Wiggins Uplift); (2) a suite of contrasting igneous rocks (granite, basalt and agglomerate) and serpentinite which occurs along the Brunswick-Altamaha magnetic anomaly in southwestern Alabama; and (3) an extensive, apparently coherent tectonic element comprised of undeformed granite, low-grade felsic metavolcanic rocks, a suite of high-grade metamorphic rocks (gneiss and amphibolite) and a succession of undeformed, Lower Ordovician-Middle Devonian sedimentary rocks. The latter association has been termed the Suwannee Terrane by Dallmeyer (1989a) et al. (1989, 1991). Thomas et al. (1989), and Horton.

3.2 Suwannee Terrane

3.2.1 Osceola Granite

An area of undeformed granite constitutes a large portion of the pre-Mesozoic crystalline basement of central peninsular Florida (Fig. 12). This has been termed the Osceola Granite by Chowns and Williams (1983). Dallmeyer et al. (1987) describe the pluton as heterogeneous, and composed dominantly of biotite granodiorite, leucocratic biotite quartz monzonite and biotite granite. Most of the samples examined by Dallmeyer et al. were composed dominantly of oligoclase, quartz, perthitic alkali feldspar and biotite. The feldspars typically display a magnetic character; plagioclase occurs as zoned crystals and alkali feldspar appears to have retained most of its original albite component in the form of perthitic lamellae. There is no textural evidence for significance subsolidus migration, exsolution, and/or recrystallization of felspar components. In view of these petrographic characteristics, Dallmeyer et al. suggested the pluton experienced relatively rapid post-magnetic cooling, and was likely emplaced at relatively shallow crustal levels.

Bass (1969) reported Rb-Sr analytical results from several density fractions of feldspar from two portions of a core from a well in Osceola County (well 1, Fig. 12). The data were extremely scattered and tentatively interpreted by Bass to reflect a crystallization age of ca. 530 Ma. Dallmeyer et al. (1987) reported $^{40}Ar/^{39}Ar$ incremental-release ages for five biotite concentrates from cuttings recovered from four wells penetrating the Osceola Granite (wells 2-5, Fig. 12). All the samples are unaltered, and the biotite concentrates display well defined plateau ages ranging between ca. 527 and 535 Ma. Dallmeyer et al. suggested that the biotite plateau ages likely closely date emplacement of the pluton in view of its high level character and apparently rapid post-magnetic cooling.

3.2.2 St. Lucie Metamorphic Complex

A suite of high-grade metamorphic and associated, variably deformed igneous rocks occurs southeast of the Osceola Granite. This has been termed the St. Lucie Metamorphic Complex by Dallmeyer (1986b) and Thomas et al. (1989). Predominant lithologies include amphibolite, biotite-muscovite schist and gneiss and quartz diorite. The complex has a distinctive aeromagnetic signature (Taylor et al. 1968; Klitgord et al. 1984), with marked northwest-trending magnetic lineations. Thomas et al. (1989) suggest these may reflect structural strike.

Bass (1969) reported isotopic ages for the high-grade complex, including a 503 Ma K-Ar date for a hornblende concentrate from amphibolite and a 530 Ma Rb-Sr model age for a biotite concentrate from interlayered gneiss within a well in St. Lucie County (well 6, Fig. 12). Hornblende concentrates prepared from amphibolite cuttings recovered from wells in St. Lucie and Martin Counties (wells 7 and 8, Fig. 12) record well defined $^{40}Ar/^{39}Ar$ plateau ages of ca. 511 and 513 Ma (Dallmeyer 1986b). These have been interpreted to date post-metamorphic cooling through appropriate argon retention temperatures.

3.2.3 Felsic Volcanic-Plutonic Complex

A felsic volcanic-plutonic complex has been penetrated in separated areas of the Coastal Plain pre-Mesozoic basement. Lithologic variants include felsic vitric tuff, felsic ashflow tuff, and tuffaceious arkose with subordinate andesite and basalt. Epizonal felsic plutons occur within some wells and likely represent subvolcanic equivalents of the volcanic sequences. Mueller and Porch (1983) presented geochemical analyses for representatives of the felsic complex which suggest calc-alkaline affinities. The complex is generally undeformed, however it nearly everywhere displays low-grade metamorphic assemblages.

The felsic igneous complex appears to be unconformably overlain by Lower Ordovician sandstone in a well located in central peninsular Florida. On the basis of these relationship, Chowns and Williams (1983) suggested a late Proterozoic-early Paleozoic age for the felsic igneous suite. This is consistent with stratigraphic relationship inferred from seismic characteristics in northwestern Florida by Arden (1974). Whole-rock, K-Ar ages ranging between ca. 165 and 480 Ma have been reported for many members of the felsic complex (summarized by Chowns and Williams 1983). A representative suite of 7 volcanic samples have been analyzed with whole-rock, $^{40}Ar/^{39}Ar$ incremental-release techniques (Dallmeyer, unpubl.). All samples display markedly discordant age spectra indicating widespread disturbance of initial intracrystalline argon systems. These results suggest that published K-Ar whole-rock ages may not be used to constrain either the time of magmatic or metamorphic events.

The COST GE1 well was drilled ca. 100 km east of northernmost Florida coast (Fig. 12) and penetrated ca. 600 m of low-grade metasedimentary rocks (argillite) overlying variably metamorphosed trachyte and sandstone (Scholle 1979). The relationship of this sequence to the mainland felsic igneous complex is uncertain. Whole-rock K-Ar ages of 374 and 346 Ma were reported for metasedimentary rocks recovered from the well (Simonis, in Scholle 1979). A slate sample from 11600' displays an internally discordant $^{40}Ar/^{39}Ar$ age spectrum defining a total-gas age of ca. 341 Ma (Dallmeyer, unpubl.). A felsic metavolcanic rock from 12350' also displays an internally discordant age spectrum; however, intermediate- and high-temperature increments correspond to an ca. 375 Ma plateau date. This is generally similar to a 363 ± 7 Ma Rb-Sr whole-rock isochron reported for seven samples from the COST well by Simonis (in Scholle 1979). These Devonian ages are more likely related to metamorphic overprinting than to initial magmatic events.

3.2.4 Paleozoic Sedimentary Rocks

A succession of generally undeformed sedimentary rocks occurs in several separate areas of the Coastal Plain crystalline basement. The base of the section is marked by Lower Ordovician littoral quartz sandstones (Carroll 1963). These are overlain with presumed conformity by Ordovician to Middle Devonian shales with locally significant horizons of siltstone and sandstone. Although a nearly continuous succession appears to be present, Cramer (1973) noted that the absence of Lower Silurian faunas may indicate a disconformity. Cold water, Gondwanan paleontological affinities are displayed by all fauna throughout the entire Paleozoic sequence (Whittington and Hughes 1972; Andress et al. 1969; Goldstein et al. 1969; Cramer 1971, 1973; Pojeta et al. 1976). Opdyke et al. (1987) reported ca. 1650-1800 Ma U-Pb zircon ages for detrital zircons within a core of Ordovician-Silurian sandstone recovered from a well penetrating the north Florida basin in Alachua, County, Florida (well 9, Fig. 12). Opdyke et al. also presented paleomagnetic results from the sandstone core which suggest a paleolatitude of ca. 40°. Dallmeyer (1987) reported a $^{40}Ar/^{39}Ar$ plateau age of ca. 504 Ma for detrital muscovite from a well penetrating Lower Ordovician sandstone in Marion County, Florida (well 10, Fig. 12).

3.2.5 Structure

The spacing of basement penetrations precludes reliable determination of the nature of the contacts between the various lithotectonic units which constitute the Suwannee Terràne. Applin (1951), Barnett (1975) and Chowns and Williams (1983) suggested that the Paleozoic sedimentary sequence occupies a regional synclinal structure (termed the North Florida Basin by Thomas et al. 1989). They suggested that later development of horsts and grabens during Mesozoic faulting significantly affected the subcrop distribution of basement units, particularly in south Georgia, southeastern Alabama and northwestern Florida (e. g. Smith 1983).

3.2.5 Relationship To Appalachian Elements

Elements of the pre-Cretaceaus crystalline basement benearth the Atlantic and Gulf Coastal Plains were initially correlated with Appalachian sequences in the Valley and Ridge province (Campbell 1939) and eastern Piedmont (Milton and Hurst 1965). However, the undeformed character and Gondwanan paleontological affinities of the Suwannee succession contrast markedly with Laurentian sequences of similar age exposed in the Valley and Ridge province. In addition, the age and character of the Osceola granite and bordering high-grade metamorphic sequence are unlike that of any structural elements within any exposed terranes in the central-southern Appalachians. Because of these inconsistances, most recent workers (e. g., Chowns and Williams 1983; Dallmeyer 1989a) have suggested that the pre-Cretaceous basement units are unrelated to exposed Appalachian tectonic elements.

3.3 *Wiggins Uplift*

The Wiggins Uplift is an elevated structural block of pre-Mesozoic crystalline rocks bordered by Mesozoic faults. It has been penetrated by several wells drilled in southwestern Alabama and southeastern Mississippi (e. g. Cagle and Khan 1983). Predominant rock types include phyllite, chlorite schist, metasandstone, amphibolite, gneiss and variably metamorphosed and deformed granite. K-Ar whole-rock ages ranging between ca. 275 and 300 Ma have been reported for various lithologies within the complex (Cagle and Khan 1983). $^{40}Ar/^{39}Ar$ incremental-release ages have been determined for several units within the Wiggins Uplift (Dallmeyer 1989c). These include: (1) a sample of deformed granite from a well in Jackson County, Mississippi (well 11, Fig. 12) which records a markedly discordant whole-rock age spectrum corresponding to a total-gas date of ca. 138 Ma; (2) phyllite from a well in Mobile County, Alabama (well 12, Fig. 12) which records a whole-rock plateau age of ca. 318 Ma; and (3) plateau ages of 310 and 305 Ma for hornblende and biotite concentrates from interlayered amphibolite and gneiss within core recovered from a well in Jackson County, Mississippi (well 13, Fig. 12).

3.4 *Southwestern Alabama Igneous Complex*

Several wells have penetrated various igneous rocks (including granite, basalt and volcanic agglomerate) along the trace of the Brunswick-Altamaha magnetic anomaly in southwestern Alabama (Neathery and Thomas 1975; Thomas et al. 1989). K-Ar whole-rock ages of ca. 335 and 267 Ma were reported by Neathery and Thomas (1975) for granite and basalt within the complex. Thomas et al. (1989) reported that massive serpentinite was penetrated in a well along the southern gradient of the anomaly.

3.5 *Regional Tectonic Relations*

Available basement penetrations allow demarkation of a boundary between Appalachian sequences and the Suwannee Terrane as shown in Fig. 12 (Chowns and Williams 1983; Dallmeyer 1989a; Thomas et al. 1989). This approximately coincides with the trace of the Altamaha-Brunswick magnetic anomaly in Alabama, and K. D. Nelson et al. (1985a, b) suggested that the anomaly everywhere marks a suture between Appalachian elements and the Suwannee Terrane. However, traced eastward across Georgia the anomaly and subsurface terrane boundary diverge (Fig. 12). On the basis of these relationships, Chowns and Williams (1983) suggested that although the anomaly may mark the deep crustal expression of a suture, shallower crustal levels may have been thrust northward, carrying the subcrop expression of the boundary over the deeper crustal interface.

The southern boundary of the Suwannee Terrane in peninsular Florida is defined by a major fault (the Jay Fault; Smith 1983) which is likely a projection of the Bahamas fracture zone (Klitgord et al. 1984). This may connect northwestward with the Pickens-Gilberton Fault System (Smith 1983) which can be traced into the midcontinent. A Mesozoic volcanic sequence occurs south and west of the Jay Fault. This succession probably developed in response to opening of the present Atlantic Ocean (Mueller and Porch 1983), but developed on older continental crust (Ross et al. 1986). Several subsurface fault-bounded blocks of crystalline basement with characteristics similar to that of the Suwannee Terrane appear to occur in southern Florida (Thomas et al. 1989). On the basis of geophysical characteristics, Klitgord and Popenoe (1984) have also suggested that several tracts of fault-bounded continental crust occur in the Gulf of Mexico west of Florida. In addition, continental crust with Pan African age affinities was penetrated in two DSDP holes drilled in the Gulf of Mexico northeast of Yucatan (Dallmeyer 1984).

An extensive series of northeast-trending Mesozoic grabens is developed along the boundary between the Suwannee Terrane and the various Appalachian elements. Higgins and Zietz (1983) suggested that these developed in response to initial phases of rifting of the present Atlantic Ocean. In northwestern Florida and southeastern Alabama a northwest-trending series of Mesozoic faults intersects the grabens, producing a complex series of smaller horst and graben structures (Smith 1983). The Wiggins Uplift appears to be localized within one of these horsts. The relationship of the Wiggins Uplift crystalline basement to that of the Suwannee Terrane is uncertain; however, it has clearly been extensively overprinted by late Paleozoic ductile strain and metamorphism. The relationship of the southwestern Alabama igneous suite to either the Suwannee Terrane or basement of the Wiggins Uplift is also uncertain.

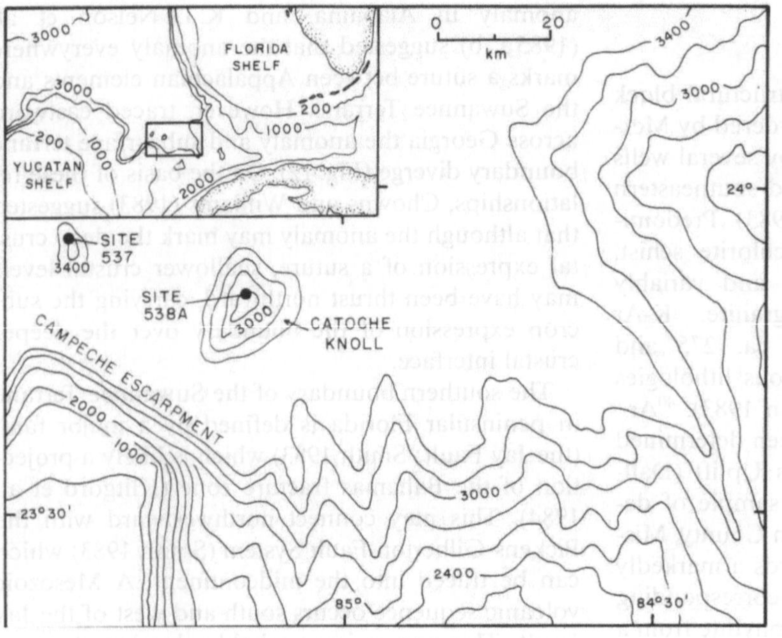

Fig. 13. Location of sites 537 and 538 A drilled during Leg 77 of the Deep Sea Drilling Project (Dallmeyer 1984): bathymetric contours in m

4 Pre-Mesozoic Basement in the Southeastern Gulf of Mexico

Drilling at Holes 537 and 538 A of the Deep Sea Drilling Project Leg 77 (Fig. 13) penetrated metamorphic rocks beneath abbreviated Mesozoic-Cenozoic sedimentary sequences capping relatively high-standing fault blocks in the southeastern Gulf of Mexico (Schlager et al. 1984; Buffler and Schlager 1984; Dallmeyer 1984). At Hole 538 A, located on Catoche Knoll, a foliated, regional metamorphic association of variably mylonitic felsic gneisses and interlayered amphibolite is intruded by post-tectonic diabase dikes. Hornblende from the amphibolite displays internally discordant $^{40}Ar/^{39}Ar$ age spectra (Dallmeyer 1984) suggesting initial post-metamorphic cooling at ca. 500 Ma followed by a mild thermal disturbance at ca. 200 Ma. A lower grade phyllitic metasedimentary sequence was penetrated at Hole 537. Whole-rock phyllite samples display internally discordant $^{40}Ar/^{39}Ar$ age spectra with plateau segments clearly documenting an early Paleozoic metamorphism at ca. 500 Ma.

$^{40}Ar/^{39}Ar$ results from Holes 537 and 538 A are mutually consistent and indicate that Mesozoic-Cenozoic sedimentary sections in the southeasternmost Gulf unconformably overlie a polymetamorphic terrane of variable grade. Initial metamorphism appears to have occurred at ca. 500 Ma, with markedly higher grade conditions maintained in that portion of the terrane penetrated in Hole 538 A. Variations in metamorphic grade between Holes 537 and 538 A may indicate that different crustal levels of the same fragmented basement complex were penetrated in the two fault blocks drilled on Leg 77. Such juxtaposition could reflect differential offset along normal faults that were active during initial continental extension preceeding late Mesozoic formation of the Gulf. The basement terrane at both Holes 537 and 538 A record the affects of a mild geologic reheating at ca. 200 Ma. This most likely resulted from elevation of temperatures caused by intrusion of diabase dikes during initial late Mesozoic crustal extension.

5 Correlations with West Africa

5.1 Introduction

Any attempt to correlate any Appalachian exotic terrane with tectonic elements comprising the West African orogens must first consider the overall crustal structure of West Africa. The Mauritanide orogen is bordered on the east by the West African craton and on the west by a thick section of Mesozoic and younger sedimentary sequences within the Senegal-Mauritania coastal basin (Fig. 14). The regional Bouguer gravity anomaly pattern suggests that the orogen extends a short distance westward beneath the basin (Lécorché et al. 1983; Roussel et

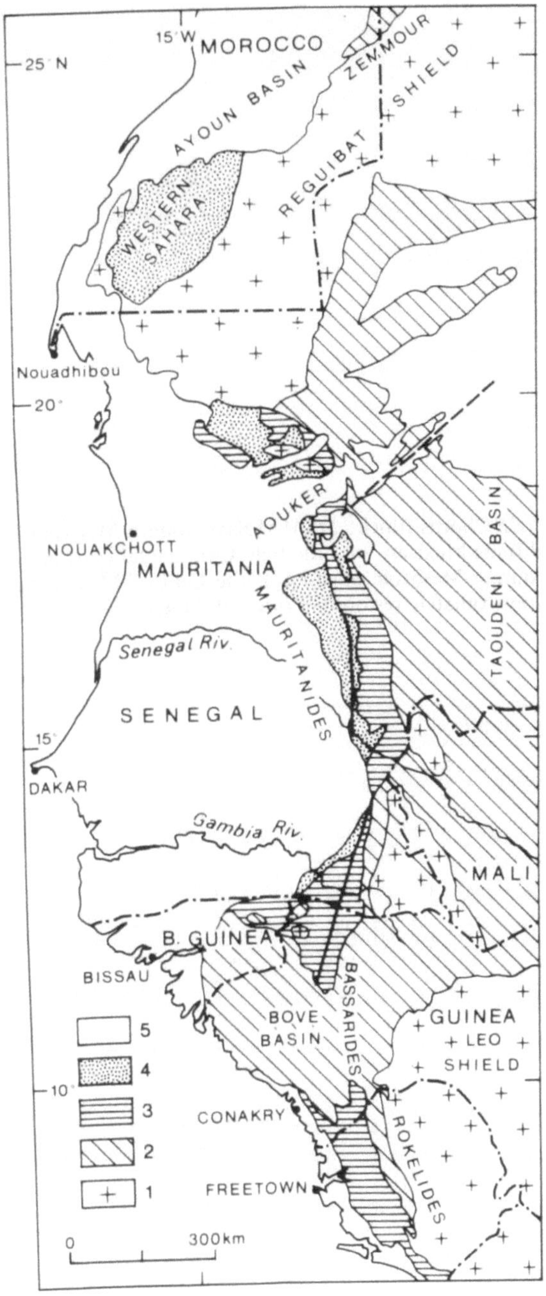

Fig. 14. Generalized geologic map of the West African orogens: *1* Crystalline basement (Archean-middle Proterozoic) of the West African Shield; *2* autochthonous sedimentary cover sequences (upper Proterozoic-late Paleozoic; *3* upper Proterozoic-lower Paleozoic (Pan-African) infrastructural orogenic terranes; *4* superstructural allochthonous complexes likely emplaced during late Paleozoic orogeny; *5* Mesozoic and younger sedimentary sequences

al. 1984; Roussel and Lesquer, this Vol.). Three distinct anomaly domains have been outlined: (1) a prominent, nearly continuous, NNW-SSE trending belt of positive anomalies (Mauritanian anomaly) which, although paralleling the axis of exposed portions of the orogen, is displaced slightly westward; (2) a broad regional negative anomaly east of the Mauritanian anomaly which is characterized by NE-SW gravity trends; and (3) a generally positive anomaly west of the Mauritanian anomaly. Lécorché et al. (1983) and Roussel et al. (1984) interpreted the regional gravity pattern east of the Mauritanian anomaly to largely reflect the signature of Precambrian basement beneath the West African craton. This distinctive signature may be traced westward beneath exposed portions of the orogen, clearly reflecting its allochthonous nature. The strong positive Mauritanian anomaly has been interpreted as an asymmetric mantle ridge overstep-pened eastward which has a crest at a depth of ca. 15 km. East of long. 15° 30'W the gravity character over the Mesozoic basin is very similar to that defined by Precambrian crystalline rocks beneath the West African craton. This led Ponsard et al. (1982), Lécorché et al. (1983), Roussel et al. (1984) and Ponsard (1984) to propose that a coastal structural block of continental crust similar in character to the West African Craton exists in the subsurface west of the Mauritanian anomaly. The two crustal blocks are therefore separated by the nearly continuous positive Mauritanian anomaly which likely marks a remnant, west-dipping suture zone.

Recent continental reconstructions (e.g. Ross et al. 1986; Rowley et al. 1986; Ross and Scotese 1988) suggest that final amalgamation of Laurentia and Gondwana resulted in a Pangea configuration similar to that portrayed in Fig. 15. This configuration suggests that most central and southern sectors of the Appalachian orogen were tectonically situated opposite the continental crustal block which presently underlies the Senegal-Mauritania coastal basin. These geometric relationships have significant bearing on potential Appalachian-West Africa correlations.

5.2 Central-Southern Appalachians and Eastern Coastal Plain Basement

The history of terrane accretion outlined by Horton et al. (1989) for central and southern portions of the Appalachian orogen is portrayed in Fig. 16. These workers suggest that essentially all of the western-

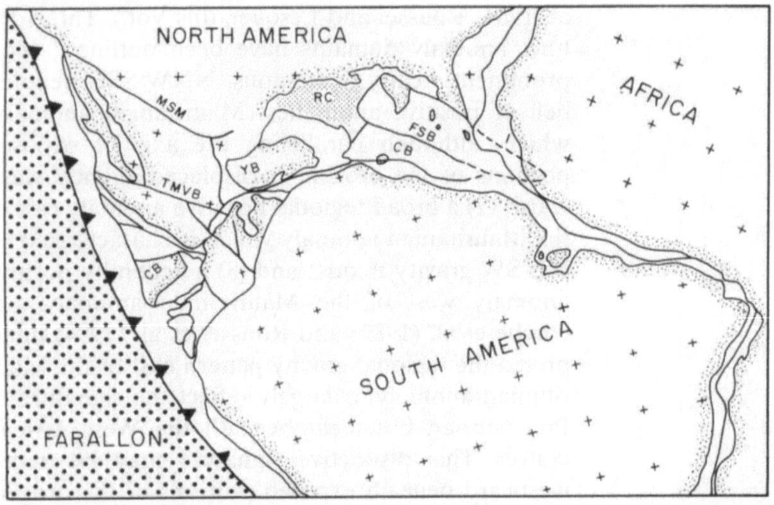

Fig. 15. Continental reconstruction prior to Mesozoic rifting and opening of the Gulf of Mexico and the Atlantic Ocean (Ross and Scotese 1988). Major tectonic elements include: *FSB* Florida Straits Block; *CB* Cuba Block; *YP* Yucatan Block; *CB* Chortis Block; *MSM* Mojave-Sonora Megashear; *TMVB* Trans-Mexican Volcanic Belt. Continental crust attenuated during Mesozoic opening of the Gulf of Mexico has been approximately restored to original dimensions *(RC)*

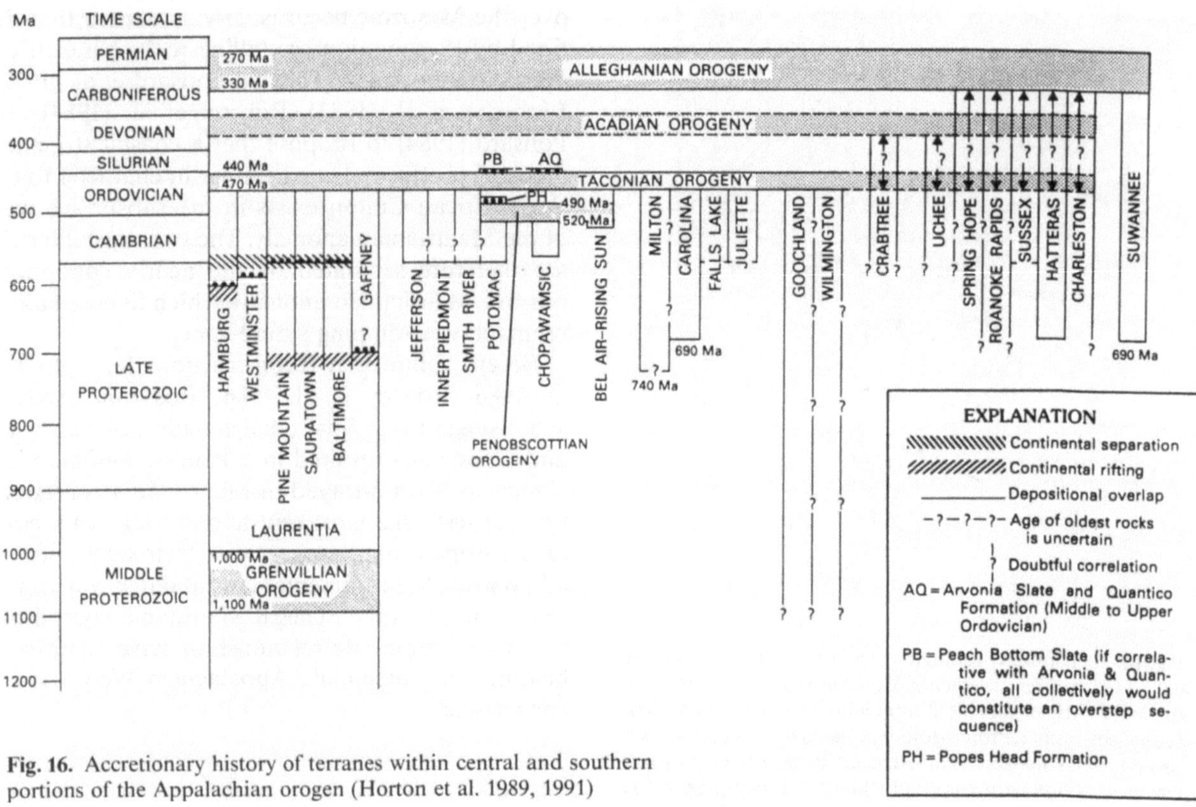

Fig. 16. Accretionary history of terranes within central and southern portions of the Appalachian orogen (Horton et al. 1989, 1991)

most Appalachian exotic terranes accreted to Laurentia in the early Paleozoic (Cambrian and Ordovician) at a time when most paleomagnetic continental reconstructions suggest considerable separation of Laurentia and Gondwana (e.g. Scotese et al. 1979; Scotese 1984). This and lithologic characteristics of the westernmost Appalachian exotic terranes preclude any protolith correlations with West African successions. On the other hand, most sequences within the Goochland terrane exposed in the central Appalachians are of continental affinity and therefore compatable with tectonic derivation from the West African coastal continental block. Most late Paleozoic continental reconstructions place this portion of the Appalachians opposite the Reguibat promentory of the West African craton (Fig. 15), and Lefort (1980) proposed that the promentory constituted a rigid "indentor" which played an important kinematic role during collision of Gandwana and Laurentia. This has been documented by Dallmeyer and Lécorché (1990) along southern portions of the Reguibat uplift in the northern Mauritanide orogen where supracrustal nappe complexes record extensive late Paleozoic thrust distances. A large, structurally similar supracrustal nappe complex tectonically overlies basement of the Reguibat uplift in the Western Sahara (Fig. 14: Lécorché et al., this Vol.). Structurally higher tectonic elements are, in part, comprised of granulite facies gneisses which are locally cut by anorthosite dikes. These lithologic characteristics are compareable with sequences of the Gouchland terrane, and perhaps both represent structural units derived from continental crust of the West African coastal black. They could have been thrust to anomalously high structural levels in the late Paleozoic as a result of the indenting affect of the Reguibat promentory.

Considerable confusion has centered on potential correlations of volcanic arc terranes within the central-southern Appalachians and other so-called "Avalonain" tectonic elements of the Circum-Atlantic realm. For example, units within the Carolina terrane are characterized by Middle Cambrian fauna with Acado-Baltic affinities which many workers have construed to indicate African connections. However, as noted by Palmer and Secor (1983), such inferences are not justified. Williams and Hatcher (1982, 1983) included the Carolina terrane in a single "Avalon superterrane" which they extended along the entire length of the orogen and suggested to have accreted as a single, previously-assembled structural unit. Rast and Skehan (1983) grouped the Avalon superterrane with sequences of similar age exposed in northwest Africa, western Europe and Britain to produce a microcontinent which they termed the "Avalonian plate". It is now clear that at least two contrasting tectonic elements comprise the Carolina terrane, and that these likely accreted to Laurentia at different times. Based upon paleomagnetic and geochronologic characteristics is appears that North Carolina segments of the Carolina terrane had interacted with Laurentia by the Early-Middle Ordovician. However, segments exposed in South Carolina and Georgia appear to not have been proximal to Laurentia until the Middle Devonian-Early Carboniferous. This tectonic interaction is recorded not only within sequences of the Carolina terrane, but also within western exotic terranes (e. g. Inner Piedmont and Jefferson composite terranes) as well as in autochthonous sedimentary sequences within the southern Appalachian miogeocline. Final transport of the Carolina (and other) southern Appalachian terranes into their present, allochthonous structural positions on the Laurentian margin occurred during terminal stages of collision of Laurentia and Gondwana (Fig. 17). No direct lithologic, faunal and/or structural correlations can be established between either components of the Carolina terrane and tectonic elements within any of the West African orogens.

5.3 Basement Beneath the Gulf and Southeastern Atlantic Coastal Plain

The crystalline basement rocks penetrated beneath the southeastern Atlantic and Gulf Coastal Plains were initially correlated with successions in either the Valley and Ridge or Piedmont Provinces of the Appalachians (e. g. Campbell 1939; Milton and Hurst 1965). However, on the basis of the Gondwanan paleontological affinities of the Paleozoic sedimentary succession, correlations with West African sequences have been suggested by most recent workers (e. g. Wilson 1966; Rodgers 1970). Recent collaborative field and geochronologic studies in the Mauritanide, Bassaride and Rokelide orogens of West Africa have helped resolve the tectonothermal evolution of these areas, thereby permitting direct correlation with counterparts comprising the Suwannee Terrane beneath the Coastal Plain of the southeastern United States. These include (Fig. 18):

1. Correlation of the subsurface Osceola Granite and the post-tectonic Coya Granite exposed in the northern Rokelide orogen (Guinea). Both record ca. 530 Ma crystallization ages (Dallmeyer et al. 1987) and display similar petrographic characteristics. Dallmeyer et al. proposed that the two plutons were initially part of a sequence of plutons emplaced along the northwestern margin of Gondwana following a ca. 550 Ma Pan African II tectonothermal event.

2. Correlation of the subsurface Paleozoic sequence in the North Florida Basin with sequences of similar age in the Bové Basin (Senegal and Guinea: Chowns and Williams 1983; Villeneuve 1984). This is suggested by similarities in fauna and stratigraphic successions. In addition, the ca. 505 Ma

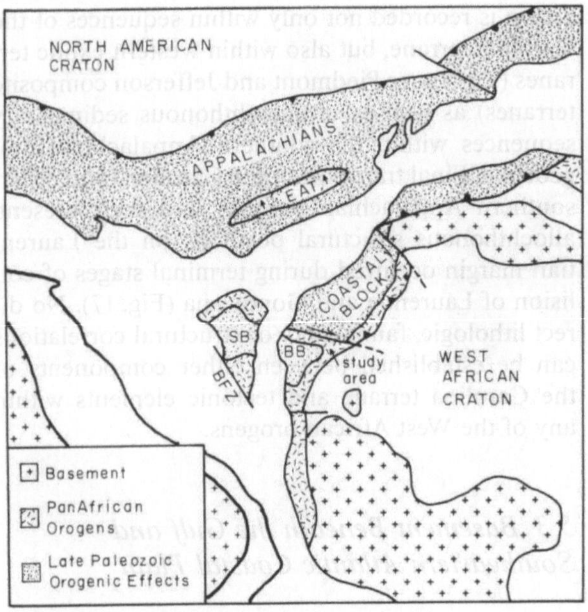

Fig. 17. Schematic reconstruction illustrating the proposed tectonic role of the West African coastal structural block during late Paleozoic amalgamation of Pangea: *EAT* Exotic Appalachian terranes which had earlier (Ordovician-Devonian) accreted to Laurentia and underwent thrust transport onto the North American margin during Late Carboniferous collision of Gondwana and Laurentia; *SB* Suwannee Basin (Florida subsurface): *BB* Bové Basin; *BFZ* Bahamas Fracture Zone (Mesozoic feature)

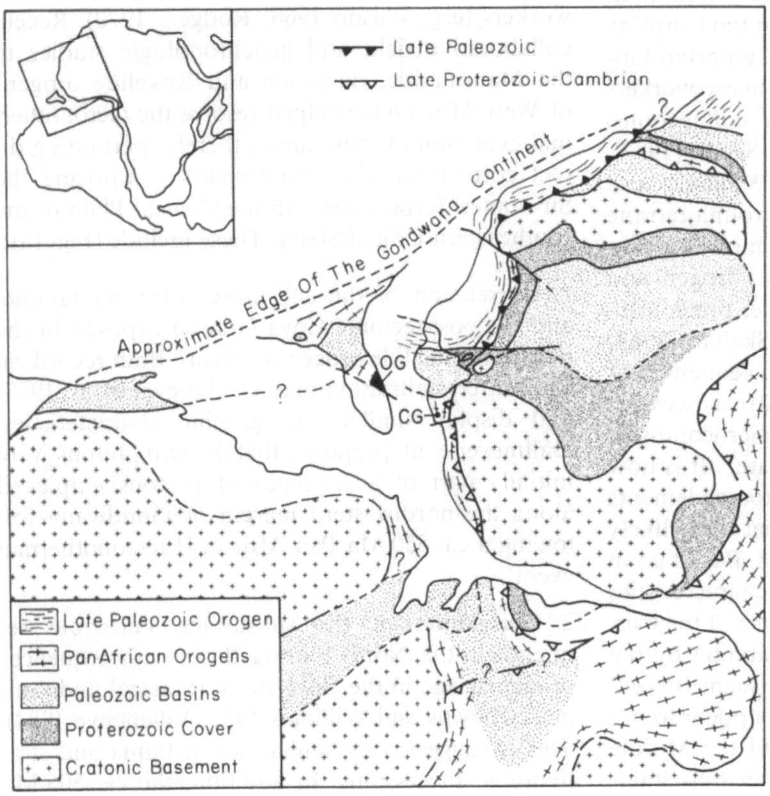

Fig. 18. Schematic reconstruction of the northwestern margin of Gondwana illustrating the relationship of pre-Cretaceous lithotectonic units in the subsurface of the southeastern United States to correlative sequences in West Africa and northeastern South America (Dallmeyer et al. 1987); OG Osceola Granite; CG Coya Granite

$^{40}Ar/^{39}Ar$ plateau age recorded by detrital muscovite within Ordovician sandstone in the Florida subsurface suggests a metamorphic source similar in age to that within the Bassaride and Rokelide orogens. The ca. 1650–1800 Ma U-Pb ages reported by Opdyke et al. (1987) for detrital zircons within Ordovician-Silurian sandstone in the Florida subsurface suggest derivation from a source similar in age to the basement of the West African Craton (e.g. Bessoles 1977). Paleomagnetic results from the sandstone core suggest a paleolatitude of ca. 49° which is in marked contrast to the ca. 28° paleolatitude suggested for Laurentia in the Ordovician-Silurian. It clearly supports a Gondwana linkage.

3. Correlation of the subsurface St. Lucie Metamorphic Complex and portions of the Rokelide Orogen (Chowns and Williams 1983). The effects of a penetrative ca. 550 Ma Pan African II tectonothermal event are recorded throughout the Rokelide orogen. Here components of a western exotic gneissic succession are imbricated with cover sequences and penetratively mylonitic and retrogressed structural units of West African Shield basement (Allen 1967, 1969; Williams 1978; Thorman 1976). In northern parts of the orogen, hornblende records K-Ar and $^{40}Ar/^{39}Ar$ post-metamor-

phic cooling ages ranging between ca. 550 and 580 Ma (Allen et al. 1967; Beckinsale et al. 1981; Dallmeyer 1989b). Post-metamorphic cooling appears to have been younger in the southern Rokelides where hornblende records K-Ar dates between ca. 485 and 530 Ma (Hurley et al. 1971; Hedge et al. 1975). The ca. 510-515 Ma $^{40}Ar/^{39}Ar$ plateau ages recorded by hornblende within the St. Lucie metamorphic complex clearly support a linkage with central portions of the Rokelide orogen.

4. Correlation of the subsurface felsic igneous complex with a calc-alkaline, variably deformed and metamorphosed igneous sequence that occurs along western portions of the Mauritanide, Bassaride and northernmost Rokelide orogens (Dallmeyer and Villeneuve 1987; Dallmeyer et al. 1987). This sequence includes felsic volcaniclastic units together with associated, hypabyssal subvolcanic plutons. Radiometric ages suggest that the calc-alkaline igneous sequence developed between ca. 650 and 700 Ma (Lille) 1969; Bassot and Caen-Vachette 1983; Dallmeyer and Villeneuve 1987).

These various correlations suggest that at least portions of the pre-Cretaceous basement beneath the Atlantic and Gulf Coastal Plains represent a fragment of Gondwana which, following amalgamation of Pangea, was stranded during Mesozoic rifting associated with development of the present Atlantic ocean. The Late Carboniferous suture between Laurentia and the Gondwanian sequenes (Fig. 12) appears to be approximately marked by the trace of the Brunswick magnetic anomaly (e.g. K. D. Nelson et al. 1985a, b). Concentration of tectonothermal activity in close proximity to the suture likely explains the 300-310 Ma rejuvenation of hornblende and biotite argon systems within proximal portions of the Gondwanan sequences.

5.3.1 Implications For Terrane Accretion In The Southern Appalachian Orogen

Geophysical, lithologic and/or faunal characteristics suggest that all southern Appalachian lithotectonic units east of the Hayesville thrust fault (Figs. 1, 2) are allochthonous, non-Laurentian terranes which presently structurally overlie autochthonous or parautochthonous North American successions (e.g. Cook et al. 1979; Harris et al. 1981; Secor et al. 1983; Secor et al. 1986b; Horton and Drake 1986). Paleomagnetic characteristics of plutons intruding exotic terrane elements of the southern Appalachians are not consistent with significant post-Devonian latitudinal movement. This requires that the various contrasting terranes were at least proximal to North America before the Carboniferous (e.g. Ellwood 1982; Barton and Brown 1983; Dooley 1983). Sedimentological expression of outboard (eastern) tectonic instability is clearly documented in Silurian and Devonian successions deposited within the Laurentian miogeocline (e.g. Chowns 1989), and it is likely that at least some of the metamorphism and ductile thrusting recorded within the crystalline southern Appalachians is a result of initial terrane accretion to North America (e.g. metamorphism and transport of the Alto allochthon; pre-Carboniferous metamorphism and folding of the Carolina terrane). Geochronological results within the Carolina terrane suggests that contrasting tectonic elements may have accreted to Laurentia at different times between the Ordovician and the Devonian.

Westward transport of previously accreted southern Appalachian terranes into their present structural positions on the North American margin occurred during the Alleghanian orogeny (Fig. 17) which resulted from the collision of Laurentia and Gondwana (e.g. Secor et al. 1986b; Dallmeyer 1986, 1988). Initial phases of Alleghanian tectonothermal activity occurred between ca. 315 and 295 Ma, and involved folding, metamorphism and emplacement of felsic plutons at middle crustal levels. The second episode of Alleghanian activity was associated with crustal uplift and resultant rapid post-metamorphic cooling between ca. 295 and 285 Ma. This was accompanied by westward-vergent folding as crystalline nappes moved over ramps during thrust transport. Regional post-metamorphic cooling appears to have occurred slightly earlier (between ca. 335 and 305 Ma) in the Jefferson and Inner Piedmont composite terranes. These sequences were likely maintained at elevated temperatures following a Late Devonian or earlier metamorphism which probably accompanied their initial accretion to Laurentia. Final cooling is interpreted to have occurred during transport to higher crustal levels as they were thrust onto the North American margin. The final phase of Alleghanian deformation resulted in development of dextral shear zones in easternmost portions of the Carolina terrane between ca. 290 and 268 Ma. This strain has been interpreted to have developed as a result of relative rotation between Gondwana and Laurentia during final stages of Pangea amalgamation (Secor et al. 1986b) Eastward transport of the West African coastal block also occurred at this time, and was accompanied by development of ductile strain zones along its borders (Dallmeyer and Villeneuve 1987). Recent continental reconstructions (e. g. Ross et al. 1986; Rowley et al. 1986; Ross and Scotese 1988) suggest that final amalgamation of Laurentia and Gondwana resulted in a Pangea configuration similar to that portrayed in Fig. 15. Fragments of Gondwana continental crust were stranded during Mesozoic opening of the Gulf of Mexico and the Atlantic ocean (basement in the southeastern Gulf and beneath the Atlantic and Gulf Coastal Plains).

Acknowledgments.
Various phases of the work summarized here were supported by grants from the US National Science Foundation (EAR-8020469; EAR-8514013) and the Petroleum Research Foundation of the American Chemical Society (PRF 13920-AC2).

References

Allen PM (1967) The geology of part of an orogenic belt in Sierra Leone. Ph D Thesis, Univ Leeds, UK, 313 pp

Allen PM (1969) Geology of part of an orogenic belt in western Sierra Leone. Geol Rundsch 58: 588-619

Allen PM, Snelling NJ, Rex D (1967) Age determinations in western Sierra Leone. In: Variations in isotopic abundance of strontium, calcium and argon and related topics. Mass Inst Technol 1381-15, Rept US Atomic Energy Com, pp 17-22

Andress NE, Cramer FH, Goldstein RF (1969) Ordovician chitizoans from Florida well samples. Gulf Coast Assoc Geol Soc Trans 19: 369-375

Applin PL (1951) Preliminary report on buried pre-Mesozoic rocks in Florida and adjacent states. US Geol Surv Circ 91: 28

Arden DD Jr (1974) Geology of the Suwannee basin interpreted from geophysical profiles. Gulf Coast Association of Geological Societies Transactions 24: 223-230

Atkins RL, Higgins MW, Gottfried D (1980) Geochemical data bearing on the origin of the Stone Mountain granite, Panola granite, and Lithonia Gneiss near Atlanta Georgia. Geol Soc Am Progr Abstr 12 (4): 170

Barnett RS (1975) Basement structure of Florida and its tectonic implications. Gulf Coast Assoc Geol Soc Trans 25: 122-142

Barton C, Brown L (1983) Paleomagnetism of Carboniferous intrusions in North Carolina. J Geophys Res 88: 2327-2335

Bass MN (1969) Petrography and ages of crystalline of basement rocks of Florida - some extrapolations. Am Assoc Petrol Geol Mem 11: 283-310

Bassot JP, Caen-Vachette M (1983) Données nouvelles sur l'âge du massif de granitoîde du Niokolo-Koba (Sénégal oriental); implication sur l'âge du stade précoce do la chaîne des Mauritanides: J Afr Earth Sci 1: 159-165

Beckinsale RD, Pankhurst RJ, Snelling NJ (1981) The geochronology of Sierra Leone (Appendix). Overseas Mem Inst Geol Soc Lond 7: 16 pp

Bell H III, Butler JR, Howell DE, Wheeler WH (1974) Geology of the Piedmont and Coastal Plain near Pageland, South Carolina and Wadesboro, North Carolina. Columbia, S C Geol Surv Carol Geol Soc Guidebook 1974, 23 pp

Bessoles B (1977) Géologie de l'Afrique, I. Le craton Ouest africain. Mém Bur Rech Géol Min Paris 88: 402

Bourland WC, Rigby JK (1982) Sponge spicules from early Paleozoic rocks of the Carolina slate belt. In: Bearce DN, Black WW, Kish SA, Tull JF (eds) Tectonic studies in the Talladega and Carolina slate belts, southern Appalachian orogen. Geol Soc Am Spec Pap 191: 125-132

Brown PM, Van der Voo R (1983) A paleomagnetic study of Piedmont metamorphic rocks from northern Delaware. Geol Soc Am Bull 94: 815-822

Buffler RT, Schlager W (1984) Initial Reports of the Deep Sea Drilling Project, V. LXXVII, US Gov printing Office, Washington

Butler JR (1977) Generalized map of the Spartanburg quadrangle, SC and NC. In: Heffner JD, Ferguson RB (eds) Savannah river laboratory hydrogeochemical and stream sediment reconnaissance; preliminary raw data release, Spartanburg 1° × 2° NTMS area, North Carolina and South Carolina, National Uranium Resource Evaluation program. Aiken, S C Savannah River Lab, p c-28

Butler JR (1980) Review of potential host rocks for radioactive waste disposal in the Piedmont province of North Carolina. Aiken, S C du Pont de Nemours & Co, DP-1562; 47

Butler JR, Fullagar PD (1975) Lilesville and Pageland plutons and associated metarhyolites, eastern Carolina slate belt: Geol Soc Am Progr Abstr 7: 475

Butler JR, Fullagar PD (1978) Petrochemical and geochronological studies of plutonic rocks in the southern Appalachians: III. Leucocratic adamellites of the Charlotte belt near Salisbury, North Carolina. Geol Soc Am Bull 89: 460-466

Butler JR, Howell DE (1976) Geology of the Taxahaw quadrangle, Lancaster County, South Carolina. Columbia, S C Geol Surv Geol Notes 20: 133-149

Cagle JW, Khan MA (1983) Smackover-Norphlet stratigraphy south Wiggins Arch, Mississippi and Alabama. Trans Gulf Coast Assoc Geol Soc 33: 23-29

Campbell RB (1939) Paleozoic under Florida? Am Assoc Petrol Geol Bull 23: 1712-1713

Carpenter RH, Odom AL, Hartley ME III (1982) Geochronological investigation of the Lincolnton metadacite, Georgia and South Carolina. In: Bearce DN; Black WW, Kish SA, Tull JF (eds) Tectonic studies in the Talladega and Carolina slate belts, southern Appalachian orogen. Geol Soc Am Spec Pap 191: 145-152

Carroll D (1963) Petrography of some sandstones and shales of Paleozoic age from borings in Florida. US Geol Surv Prof Pap 454A: 15

Chowns TM (1989) Record of Paleozoic tectonism in the Valley and Ridge province of Georgia and Alabama. In: Dallmeyer RD (ed) Tectonostratigraphic expression of terrane accretion in the circum-Atlantik, Paleozoic orogens. Progr Abstr 4, IGCP Proj 233, Athens, Georgia, p 17

Chowns TM, Williams CT (1983) Pre-Cretaceous rocks beneath the Georgia Coastal Plain-regional implications. In: Gohn GS (ed) Studies related to the Charleston, South Carolina, earthquake of 1886-tectonics and seismicity. US Geol Surv Prof Pap 1313: L1-L41

Cloud P, Glaessner MF (1982) The Ediacarian Period and System: Metazoa inherit the Earth. Science 218: 783-792

Coney PJ, Jones DL, Monger JWH (1980) Cordilleran suspect terranes. Nature (Lond) 288: 329-333

Cook FA, Oliver JE (1981) The late Precambrian-early Paleozoic continental edge in the Appalachian orogen. Am J Sci 281: 993-1008

Cook FA, Albaugh DS, Brown LD, Kaufman S, Oliver JE; Hatcher RD Jr (1979) Thin-skinned tectonics in the crystalline southern Appalachians; COCORP seismic-reflection profiling of the Blue Ridge and Piedmont. Geology 7: 563-567

Cramer FH (1971) Position of the north Florida lower Paleozoic block in Silurian time - Phytoplankton evidence. J Geophys Res 76: 4754-4757

Cramer FH (1973) Middle and upper Silurian chitinozoan succession in Florida subsurface. J Paleontol 47: 279-288

Dainty AM, Frazier JE (1984) Bouguer gravity in northeastern Georgia: A buried suture, a surface suture, and granites. Geol Soc Am Bull 95: 1168-1175

Dallmeyer RD (1978) $^{40}Ar/^{39}Ar$ incremental-release ages of hornblende and biotite across the Georgia inner Piedmont: their bearing on late Paleozoic - early Mesozoic tectonothermal history. Am J Sci 278: 124-149

Dallmeyer RD (1984) $^{40}Ar/^{39}Ar$ ages from a pre-Mesozoic crystalline basement penetrated at Holes 537 and 538A of the Deep Sea Drilling Project Leg 77, southeastern Gulf of Mexica: Tectonic implications. In: Buffler RT, Schlager W (eds) Initial Reports of the Deep Sea Drilling Project. Washington DC LXXVII: 496-504

Dallmeyer RD (1986) Polyphase terrane accretion in the southern Appalachians. Geol Soc Am Progr Abstr 18 (6): 579

Dallmeyer RD (1987) $^{40}Ar/^{39}Ar$ age of detrital muscovite within Lower Ordovician sandstone in the Coastal Plain basement of Florida: Implications for West African terrane linkages. Geology 15: 998-1001

Dallmeyer RD (1988) Late Paleozoic tectonothermal evolution of the western Piedmont and eastern Blue Ridge, Georgia: Controls on the chronology of terrane accretion and transport in the southern Appalachian orogen. Geol Soc Am Bull 100: 702-713

Dallmeyer RD (1989a) Contrasting accreted terranes in the southern Appalachian orogen and Atlantic-Gulf Coastal Plains and their correlations with West African sequences. In: Dallmeyer RD (ed) Terranes in the Circum-Atlantic paleozoic orogens. Geol Soc Am Spec Pap 230: 247-267

Dallmeyer RD (1989b) A tectonic linkage between the Rokelide orogen (Sierra Leone) and the St. Lucie metamorphic complex in the Florida subsurface. J Geol 89: 183-195

Dallmeyer RD (1989c) $^{40}Ar/^{39}Ar$ ages from subsurface crystalline basement of the Wiggins uplift and southwesternmost Appalachian Piedmont: Implications for late Paleozoic terrane accretion during assembly of Pangea. Am J Sci 289: 812-828

Dallmeyer RD, Lécorché JP (1990) $^{40}Ar/^{39}Ar$ polyorogenic mineral age record within the northern Mauritanide orogen, West Africa. Tectonophysics 177: 81-107

Dallmeyer RD, Villeneuve M (1987) $^{40}Ar/^{39}Ar$ mineral age record of polyphase tectonothermal evolution in the southern Mauritanide orogen, southeastern Senegal. Geol Soc Am Bull 98: 602-611

Dallmeyer RD, Hess JR, Whitney JA (1981) Post-magmatic cooling of the Elberton Granite: bearing on the late Paleozoic tectonothermal history of the Georgia Inner Piedmont. J Geol 89: 585-600

Dallmeyer RD, Wright JE, Secor DT, Snoke AW (1986) Character of the Alleghanian orogeny in the southern Appalachians: Part II. Geochronological constraints on the tectonothermal evolution of the eastern Piedmont in South Carolina. Bull Geol Soc Am 97: 1329-1344

Dallmeyer RD, Caen-Vachette M, Villeneuve M (1987) Emplacement age of post-tectonic granites in southern Guinea (West Africa) and the peninsular Florida subsurface: implications for origins of southern Appalachian exotic terranes. Geol Soc Am Bull 99: 87-93

Dennison RE, Raveling HP, Rouse JT (1967) Age and descriptions of subsurface basement rocks, Pamlico and Albemarle Sound area, North Carolina. Am Assoc Petrol Geol Bull 51: 268-272

Dooley RE (1983) Paleomagnetism of some mafic intrusions in the South Carolina Piedmont. I. Magnetic systems with single characteristic directions. Phys Earth Planet Interiors 31: 241-268

Ellwood BB (1982) Paleomagnetic evidence for the continuity and independent movement of a distinct major crustal block in the southern Appalachians. J Geophys Res 87: 5339-5350

Farrar SS (1984) The Goochland terrane: remobilized Grenville basement in the eastern Virginia Piedmont. Geol Soc Am Spec Pap 194: 215-227

Farrar SS (1985) Tectonic evolution of the easternmost Piedmont, North Carolina. Geol Soc Am Bull 96: 362-380

Foland KA, Muessig KW (1978) A Paleozoic age for some charnockitic-anorthositic rocks. Geology 6 (3): 143-146

Fullagar PD (1981) Summary of Rb-Sr whole rock ages for South Carolina, Columbia. S C Geol Surv S C Geol 25: 29-32

Fullagar PD, Butler JR (1979) 325 to 265 m.y.-old granitic plutons in the Piedmont of the Southern Appalachians. Am J Sci 279: 161-185

Fullagar PD, Hatcher RD Jr, Merschat CE (1979) 1200 m.y.-old gneisses in the Blue Ridge province of North and South Carolina. S E Geol 20: 69-77

Gates AE, Simpson C, Glover L III (1986) Appalachian Carboniferous dextral strike-slip faults: An example from Brookneal, Virgina. Tectonics 5: 119-133

Gibson GG, Teeter SA (1984) A stratigrapher's view of the Carolina slate belt, southcentral North Carolina. Charlotte, Dep Geogr Earth Sci, Univ N C Charlotte, 43 pp

Gibson GG, Teeter SA, Fedonkin MA (1984) Ediacarian fossils from the Carolina slate belt, Stanley County, North Carolina. Geology 12: 387-390

Glover L III, Sinha AK (1973) The Virginina deformation, A late Precambrian to Early Cambrian (?) orogenic event in the central Piedmont of Virginia and North Carolina. Am J Sci 273-A: 234-251

Glover L III, Sinha AK, Higgins MW, Kirk WS (1971) U-Pb dating of Carolina slate belt and Charlotte belt rocks, Virgilina district, Virginia and North Carolina. Geol Soc Am Prog Abstr 3: 313

Glover L III, Mose DG, Poland FB, Bobyarchick AR, Bourland WC (1978) Grenville basement in the eastern Piedmont of Virginia: implications for orogenic models. Geol Soc Am Progr Abstr 10: 169

Glover L III, Mose DG, Costain JK, Poland FB, Reilly JM (1982) Grenville basement in the Eastern Piedmont of Virgina: a progress report. Geol Soc Am Progr Abstr 14: 20

Glover L III, Speer JA, Russell GS, Farrar SS (1983) Ages of regional metamorphism and ductile deformation in the central and southern Appalachians. Lithos 16: 223-245

Goldstein RF, Cramer FH, Andress NE (1969) Silurian chitinozoans from Florida well samples. Gulf Coast Assoc Geol Soc Trans 19: 377-384

Griffin VS Jr (1971) Inner Piedmont belt of the southern crystalline Appalachians. Geol Soc Am Bull 82: 1885-1898

Griffin VS Jr (1972) Inner Piedmont belt of the southern crys-

talline appalachians: Reply. Geol Soc Am Bull 83: 1897-1900
Griffin VS Jr (1974) Analysis of the Piedmont in northwest South Carolina: Geol Soc Am Bull 85: 1123-1138
Griffin VS Jr (1975) Geology of the Seneca, Westminster, and portions of the Avalon and Oakway Quadrangles, S. C. S C State Dev Board, Div Geol, Geol Notes, 18: 13-23
Griffin V Jr (1977) Preliminary geologic map of the South Carolina Inner Piedmont Belt. S C State Dev Board Div Geol, Geol Notes 21 (4): 198-204
Halik RS (1983) Characterization of the Charlotte/Carolina slate belt boundary in the Silverstreet and Denny quadrangles, South Carolina: A stratigraphic, structural and metamorphic study. M. S. Thesis, Columbia, University S C, 83 pp
Hanley TB (1986) Petrography and structural geology of Uchee Belt rocks in Columbus, Georgia, and Phenix City, Alabama. In: Neathery TL (ed) Centinnial Field Guide Volume 6, Southeastern Section of the Geological Society of America. Boulder, Colorado, Geol Soc Am, pp 297-300
Harris CW, Glover L III (1985) The Virgilina deformation: Implications of stratigraphic correlation in the Carolina Slate Belt. Charlotteville, Virginia Div Miner Resour, Carol Geol Soc Field Trip Guidebook 1985, 58 pp
Harris LD, Harris AG, de Witt Wallace Jr (1981) Evaluation of the southeastern overthrust belt beneath the Blue Ridge-Piedmont thrust. Am Assoc Petrol Geol Bull 65 (12): 9
Hatcher RD, Jr (1971) The geology of Rabun and Habersham Counties, Georgia. Georgia Geol Surv Bull 83: 48
Hatcher RD, Jr (1972) Developmental model for the southern Appalachians. Geol Soc Am Bull 83: 2735-2760
Hatcher RD, Jr (1976) Introduction to the geology of the eastern Blue Ridge of the Carolinas and nearby Georgia. Carol Geol Soc Guidebook, 53 pp
Hatcher RD, Jr (1978a) The Alto allochthon: A major tectonic feature of the Piedmont of northeast Georgia. Georgia Geol Surv Bull 93: 83-86
Hatcher RD Jr (1978b) Tectonics of the western Piedmont and Blue Ridge, southern Appalachians: Review and speculations. Am J Sci 278: 276-304
Hatcher RD Jr (1987) Tectonics of the southern and central Appalachian internides. Ann Rev Earth Planet Sci 15: 337-362
Hatcher RD Jr, Odom AL (1980) Timing of thrusting in the southern Appalachians, U. S. A.: Model for orogeny?. Q J Geol Soc Lond 137: 321-327
Hatcher RD Jr, Zietz I (1978) Thin crystalline thrust sheets in the Southern Appalachian Inner Piedmont and Blue Ridge: Interpretation based upon regional aeromagnetic data. Geol Soc Am Progr Abstr 70: 417
Hatcher RD Jr, Zietz I (1980) Tectonic implications of regional aeromagnetic and gravity data from the southern Appalachians. In: Wones DR (ed) Proc Caledonides USA, Blacksburg, Dep Geol Sci, Virginia Polytectonic Inst State Univ Mem 2: 235-244
Hedge CE, Marvin RF, Naser CW (1975) Age provinces in the basement rocks of Liberia. J Res, US Geol Surv 3: 425-429
Henika WS (1977) Geology of the Blairs, Mount Hermon, Danville and Ringgold quadrangles, Virginia, with section on the Triassic system by P. A. Thayer. Virginia Div Miner Resour Publ 2: 45
Higgins MW, Zietz I (1983) Geologic interpretation of geophysical maps of the pre-Cretaceous "basement" beneath the Coastal Plain of the southeastern United States. Geol Soc Am Mem 158: 125-130
Higgins MW, Atkins RL, Dooley RE (1980) Structure and stratigraphy of the Atlanta area, Georgia. Geol Soc Am Progr Abstr 12: 180
Higgins MW, Atkins RL, Crawford TJ, Crawford RF, Brooks R, Cook RB (1986) The structure, stratigraphy, tectonostratigraphy, and evolution of the southernmost part of the Appalachian orogen, Georgia and Alabama. US Geol Surv Open-File Rep 86-372: 162
Higgins MW, Atkins RL, Crawford TJ, Crawford RF, Brooks R, Cook RB (1988) The structure, stratigraphy, tectonostratigraphy, and evolution of the southernmost part of the Appalachian orogen. US Geol Surv Prof Pap 1475: 173
Hopson JL (1984) Stratigraphy and structure of the Alto allochthon, Ayersville Quadrangle, Georgia. Ms Thesis, Univ S C, 151 pp (unpubl)
Hopson JL (1988) Petrogenesis of the Lake Burton mafic-ultra-mafic complex (abs.). Geol Soc Am Progr Abst 20 (4): 271
Hopson JL, Hatcher RD Jr (1988) Tectonothermal evolution of the Alto allochthon of the Southern Appalachian orogen. Bull Geol Soc Am 100: 339-350
Horton JW (1984) Stratigraphic nomenclature in the Kings Mountain belt, North Carolina and South Carolina. In: Stratigraphic Notes, 1983: US Geol Surv Bull 1537-A: A59-A67
Horton JW, Drake AA (1986) Tectonostratigraphic terranes and their boundaries in the central and southern Appalachians. Geol Soc Am Progr Abstr 18 (6): 640
Horton JW, Drake AA, Rankin DW (1989) Tectonostratigraphic terranes and their Paleozoic boundaries in the central and southern Appalachians. In: Dallmeyer RD (ed) Terranes In The Circum-Atlantic Paleozoic Orogens. Geol Soc Am Spec Pap 230: 213-246
Horton JW, Rankin DW, Drake AA, Dallmeyer RD (1991) Preliminary tectonostratigraphic terrane map of the central and southern Appalachians (1:2,000,000) U. S. Geological survey, Miscellaneous Investigation Series, Map I2163
Howell DG, Jones DL (1984) Tectonostratigraphic terrane analysis and some terrane vernacular. In: Howell DG, Jones DL, Cox A, Nur A (eds) Proc Circum-Pacific Terrane Conf Stanford Univ Publ Geol Sci 18: 6-9
Howell DG, Jones DL, Schermer ER (1985) Tectonostratigraphic terranes in the circum-Pacific region. In: Howell DG (ed) Tectonostratigraphic terranes of the Circum-Pacific region. Circum-Pacific Counc Energy Miner Resour, Houston, Earth Sci Ser 1: 3-30
Hurley PM, Leo GW, White RW, Fairbairn HW (1971) Liberian Age province (about 2700 Ma) and adjacent provinces in Liberia and Sierra Leone. Geol Soc Am Bull 82: 1004-1005
Jones DL, Howell DG, Coney PJ, Monger JWH (1983a) Recognition, character, and analysis of tectonostratigraphic terranes in western North America. In: Hashimoto M, Uyeda S (eds) Advances in earth and planetary sciences. Terra Sci, Tokyo, pp 31-35
Jones DL, Howell DG, Coney PJ, Monger JWH (1983b) Recognition, character, and analysis of tectonostratigraphic terranes in western North America. J Geol Educ 31: 295-303
King PB (1955) A geologic section across the southern Appalachians: An outline of the geology in the segment in Tennessee, North Carolina and South Carolina. In: Russell RJ (ed) Guides to southeastern geology. Geol Soc Am, pp 332-373
Kirk PD (1985) Geology of the southern half of the Prosperi-

ty 7½ quadrangle and the northern half of the Delmar 7½′ quadrangle, South Carolina. MS thesis, Univ S C Columbia, 30 pp

Kish SA (1983) A geochronological study of deformation and metamorphism in the Blue Ridge and Piedmont of the Carolinas. PhD dissert, Chapel Hill, Univ N C, 220 pp

Kish SA, Campbell SK (1986) A middle Paleozoic plutonic terrane in the eastern Piedmont of North Carolina. Geol Soc Am Progr Abstr 18 (6): 658

Kish SA, Butler JR, Fullagar PD (1979) The timing of metamorphism and deformation in the central and eastern Piedmont of North Carolina. Geol Soc Am Progr Abstr 11: 184-185

Klitgord KD, Popenoe P (1984) Florida: A Jurassic transform plate boundary. J Geophys Res 89: 7753-7772

Klitgord KD, Schouten H (1981) Mesozoic evolution of the Atlantic Caribbean and Gulf of Mexico. In: Pilger RH (ed) The Origin of the Gulf of Mexico and the Early Opening of the Central North Atlantic Ocean. Proc Symp Feb 1981, Houston Geol Soc, Houston, pp 100-101

Klitgord KD, Dillon WP, Popenoe P (1983) Mesozoic tectonics of the southeastern United States Coastal Plain and continental margin. In: Gohn GS (ed) Studies related to the Charleston, South Carolina, earthquake of 1886 - tectonics and Seismicity. US Geol Surv Prof Pap 1313: P1-P15

Klitgord KD, Popenoe P, Schouten H (1984) Florida: a Jurassic transform plate boundary. J Geophy Res 89: 7753-7772

Lécorché JP, Roussel J, Sougy J, Guetat Z (1983) An interpretation of the geology of the Mauritanides orogenic belt (West Africa) in the light of geophysical data. In: Hatcher RD, Williams H, Zietz I (eds) Contributions To The Tectonics And Geophysics Of Mountain Chains. Geol Soc Am Mem 158: 131-147

Lefort JP (1980) Un "fit" structural de l'Atlantique Nord: arguments geologiques pour correler les marqueurs geophysiques reconnus sur les marges. Mar Geol 37: 355-369

Lille R (1967) Etude géologique du Guidimakha (Mauritanie). Essai de résolution structurale d'une série épimetamorphique. Mém Bur Rech Géol Min Paris, Fr, 55: 397

Lille R (1969) Précambrien et Cambro-Ordovician du Guidimaka (Mauritanie orientale). Bull Soc Géol Fr 7, XI, 2: 257-267

Mauger RL, Spruill RK, Christopher MT, Shafiquallah M (1983) Petrology and geochemistry of peralkalic metagranite and metarhyolite dikes, Fountain quarry, Pitt County, North Carolina. S E Geol 24 (2): 67-89

McConnell KI, Glover L III (1982) Emplacement of the Flat River complex, an Eocambrian sub-volcanic pluton near Durham, North Carolina. In- Bearce DN, Black WW, Kish SA, Tull JF (eds) Tectonic studies in the Talladega and Carolina slate belts, southern Appalachian orogen. Geol Soc Am Spec Pap 191: 133-143

Milton C, Hurst VJ (1965) Subsurface "basement" rocks of Georgia. Georgia Geol Surv Bull 76: 56

Mose DG, Nagel MS (1982) Plutonic events in the Piedmont of Virginia. S E Geol 23: 25-39

Mueller PA, Porch JW (1983) Tectonic implications of Paleozoic and Mesozoic igneous rocks in the subsurface of Peninsular Florida. Trans Gulf Coast Assoc Geol Soc 33: 169-173

Neathery TL, Thomas WA (1975) Pre-Mesozoic basement rocks of the Alabama Coastal Plain. Gulf Coast Assoc Geol Soc Trans 25: 86-99

Nelson AE (1989) Structure and tectonostratigraphy of the western and eastern Blue Ridge allochthonous complexes, Georgia. In: Dallmeyer RD (ed) Tectonostratigraphic expression of terrane accretion in the southern Appalachian orogen: a geotraverse excursion. IGCP Project 233 Guidebook Series, University of Georgia, Athens 4: 7-1-7-21

Nelson AE, Gillon KA (1985) Stratigraphic nomenclature in the Richard Russell and Helen thrust sheets, North Georgia. In: Contributions to Stratigraphy, stratigraphic notes, 1984. US Geol Surv Bull 1605-A: A59-A62

Nelson AE, Horton JW Jr, Clarke JW (1985c) Stacked thrust sheets of the Blue Ridge and Piedmont Provinces, northeast Georgia and western South Carolina. Geol Soc Am Progr Abst 17 (7): 674

Nelson AE, Horton JW Jr, Clarke JW (1987a) Generalized tectonic map of the Greenville 1° by 2° quadrangle, Georgia, South Coralina and North Carolina. US Geol Surv, Map MF-1898, 1:250000

Nelson AE, Horton JW Jr, Clarke JW (1987b) Tectonic map showing distribution of thrust sheets in the Greenville two-degree quadrangle, Georgia, South Carolina, and North Carolina. US Geol Surv Misc Field Stud Map 1898, scale 1:250000

Nelson KD, Arnow JA, McBride JH et al. (1985a) New COCORP profiling in the southeastern United States. Part I. Late Paleozoic suture and Mesozoic rift system. Geology 13: 714-718

Nelson KD, McBride JH, Arnow JA, Oliver JE, Brown LD, Kaufman S (1985b) New COCORP profiling in the southeastern United States. Part II. Brunswick and east coast magnetic anomalies, opening of the north-central Atlantic Ocean. Geology 13: 718-721

Noel JR, Spariosu DJ, Dallmeyer RD (1988) Paleomagnetism and $^{40}Ar/^{39}Ar$ ages from the Carolina Slate Belt, Albemarle, N.C.: Implications for terrane history. Geology 16: 64-68

Nystrom PG (1972) Geology of the Catarrh NW quadrangle, South Carolina. Univ S Carol Masters thesis, Columbia, 49 pp

Odom AL, Fullagar PD (1973) Geochronologic and tectonic relationships between the Inner Piedmont, Brevard zone, and Blue Ridge belts, North Carolina. Am J Sci 273-A: 133-149

Odom AL, Russell GS (1975) The time of regional metamorphism of the Inner Piedmont and Smith River allochthon: inferences from whole rock ages. Geol Soc Am Progr Abstr 7 (4): 522-523

Odom AL, Russell GS, Russell CW (1976) Distribution and age of Precambrian basement in the southern Appalachians. Geol Soc Am Progr Abstr 8: 238

Opdyke ND, Jones DS, MacFadden BJ, Smith DL, Mueller PA, Shuster RD (1987) Florida as an exotic terrane: Paleomagnetic and geochronologic investigation of lower Paleozoic rocks from the subsurface of Florida. Geology 15: 900-903

Overstreet WC, Bell H III (1965a) Geologic map of the crystalline rocks of South Carolina. Washington DC, US Geol Surv Misc Geol Invest Map I-413, 1:250000

Overstreet WC, Bell H III (1965b) The crystalline rocks of South Carolina. Washington DC, US Geol Surv Bull 1183: 126

Palmer AR, Secor DT (1983) Comment: Do they really mean Africa? Geol Soc Am Bull 94: 1380

Pojeta J Jr, Kriz J, Berdan JM (1976) Silurian-Devonian pelecypods and Paleozoic stratigraphy of subsurface rocks in

Florida and Georgia and related Silurian pelecypods from Bolivia and Turkey. US Geol Surv Prof Pap 879: 32

Ponsard JF (1984) La marge du craton Ouest africain du Sénégal a la Sierra Léone: Interprétation géophysique de la chaine panafricaine et des bassins du Protérozoique a l'Actuel. Thèse Univ Aix-Marseille III, Fr, 198 pp

Ponsard JF, Lesquer A, Villeneuve M (1982) Une suture panafricaine sur la bordure occidentale du carton ouest-africain? Compt Rend Acad Sci, Paris 295: 1161-1164

Rankin DW (1975) The continental margin of eastern North America in the southern Appalachians. The opening and closing of the Proto-Atlantic Ocean. Am J Sci 275-A: 298-336

Rankin DW, Drake AA Jr, Glover L III et al. (1988) Pre-orogenic terranes. In: Hatcher RD Jr, Thomas WA, Viele GW (eds) The Appalachian-Ouachita orogen in the United States. Geol Soc Am Geol N Am Boulder, Colorado, F-2 (in press)

Rao KV, Van der Voo R (1980) Paleomagnetism of a Paleozoic anorthosite from the Appalachian Piedmont, northern Delaware: Possible tectonic implications. Earth Planet Sci Lett 47: 113-120

Rast N, Skehan JW (1983) The evolution of the Avalonian plate. Tectonophysics 100: 257-286

Rodgers J (1970) The tectonics of the Appalachians. Wiley, N Y 271 pp

Rogers JJW (1982) Criteria for recognizing environments of formation of volcanic suites: application of these criteria to volcanic suites in the Carolina Slate Belt. In: Bearce DN, Black WW, Kish SA, Tull JF (eds) Tectonic studies in the Talladega and Carolina Slate Belts, Southern Appalachian Orogen. Geol Soc Am Spec Pap 191: 97-107

Roper PJ, Dunn DE (1973) Superposed deformation and polymetamorphism in the Brevard zone, South Carolina. Geol Soc Am Bull 84: 3373-3386

Ross CR, Bickford ME (1980) The U-Pb age of zircons from the Elberton granite, Piedmont of Georgia. In: Stormer JC, Whitney JA (eds) Geological, geochemical and geophysical studies of the Elberton Batholith, Eastern Georgia. GA Dep Nat Resour Guidebook, Atlanta, 19: 52-62

Ross MI, Scotese CR (1988) A hierachical tectonic model of the Gulf of Mexico and Caribbean region. Tectonophysics 155: 139-168

Ross MI, Scotese CR, Mann P (1986) Computer animation of Caribbean tectonostratigraphic development. Geol Soc Am Progr Abstr 18 (6): 734

Roussel J, Dia O, Lecorche JP, Ponsard JF, Sougy J, Villeneuve M (1984) Panafrican to Hercynian deformations in the Mauritanides and tectonic significance of gravity anomalies. Tectonophysics 109: 41-59

Rowley DB, Pindell J, Lottes AL, Ziergler AM (1986) Phanerozoic reconstructions of northern South America, west Africa. North America, and the Caribbean Region. Geol Soc Am Progr Abstr 18 (69): 735

Russell GS (1985) Reconnaissance geochronological investigations in the Phenix City Gneiss and Bartletts Ferry mylonite zone. In: Kish SA, Hanley TB, Schamel S (eds) Geology of the southwestern Piedmont of Georgia, Geological Society of America 1985 Annual Meeting, Orlando, Florida. Florida State Univ, Tallahassee, Dep Geol pp 9-11

Russell GS, Russell CW, Speer JA, Glover L III (1981) Rb-Sr evidence of latest Precambrian to Cambrian and Alleghanian plutonism along the eastern margin of the sub-coastal plain Appalachians, North Carolina and Virginia (abs). Geol Soc Am Progr Abstr 13 (7): 543

Samson SL (1984) Middle Cambrian fauna of the Carolina slate belt, central South Carolina. MS thesis, Univ S Carol, Columbia, 50 pp

Schlager W et al (1984) Deep Sea Drilling Project, Leg 77, Southeastern Gulf Of Mexico. Bull Geol Soc Am 95: 226-236

Scholle PA (ed) (1979) Geological studies of the COST Ge-1 well, United States South Atlantic outer continental shelf area. US Geol Serv Chirc 800: 113

Scotese CR (1984) Paleozoic paleomagnetism in the assembly of Pangea. In: Van der Voo R, Scotese R, Bonhommet N (eds) Plate reconstruction from Paleozoic magnetism. Am Geophys Union Geodyn Ser 2: 1-10

Scotese CR, Bambach RK, Barton C, Van der Voo R, Ziegler AM (1979) Paleozoic base maps. J Geol 87: 217-277

Secor DT Jr, Peck LS, Pitcher DM et al. (1982) Geology of the area of induced seismic activity at Monticello Reservoir, South Carolina. J Geophys Res 87: 6945-6957

Secor DT Jr, Samson SL, Snoke AW, Palmer AR (1983) Confirmation of the Carolina slate belt as an exotic terrane. Science 221: 649-651

Secor DT Jr, Snoke AW, Bramlett KW, Costello OP, Kimbrell OP (1986a) Character of the Alleghanian orogeny in the southern Appalachians. Part I. Alleghanian deformation in the eastern Piedmont of South Carolina. Geol Soc Am 97: 1319-1328

Secor DT Jr, Snoke AW, Dallmeyer RD (1986b) Character of the Alleghanian orogeny in the southern Appalachians. Part III. Regional tectonic relations. Geol Soc Am Bull 97: 1345-1353

Seiders VM (1978) A chemically bimodal, calc-alkalic suite of volcanic rocks, Carolina state belt, central north Carolina. S E Geol 19: 241-265

Shaw HF, Wasserburg G (1984) Isotopic constraints on the origin of Appalachian mafic complexes. Am J Sci 284: 319-349

Sinha AK, Costain JK, Glover L III (1980) Distribution and analysis of 300 m.y. old granites as a potential geothermal resource. Rep VPI&SU-LASL-2, Los Alamos Sci Lab Contr N28-7750G-1

Sinha AK, Hewitt DA, Rimstidt JD (1986) Radiometric ages and metamorphic geochemistry of mylonites. Terra Cognita 6 (2): 253

Smith DL (1983) Basement model for the panhandle of Florida. Gulf Coast Assoc Geol Soc Trans 23: 203-208

Snoke AW, Kish SA, Secor DT Jr (1980) Deformed Hercynian granitic rocks from the Piedmont of South Carolina. Am J Sci 280: 1018-1034

Speer JA (1982) Descriptions of the granitoid rocks associated with two gravity minima in Aiken and Barnwell Counties, South Carolina. S Carol Geol 26: 15-24

Speer JA, Becker SW, Farrar SS (1980) Field relations and petrology of the postmetamorphic coarse-grained granitoids and associated rocks of the southern Appalachian Piedmont. In: Wones DR (ed) Proc Caledonides USA Dep Geol Sci Blacksburg, Virginia Polytechn Inst State Univ, Mem 2: 137-148

Steiger RH, Jäger E (1977) Subcommission on geochronology: convention on the use of decay constants in geo- and cosmochronology. Earth Planet Sci Lett 36: 359-362

St Jean J (1973) A new Cambrian trilobite from the Piedmont of North Carolina. Am J Sci 273A: 196-216

Stoddard EF, Cavaroc VV, McDaniel RD (1978) Status of geologic research in the Raleigh belt. In: Snoke AW (ed) Geological investigations of the eastern Piedmont, southern Appalachians (Carolina Geological Society field trip guidebook 1978). S Carol Geol Surv Columbia, pp 8-12

Stow SH, Neilson MJ, Neathery TL (1984) Petrography, geochemistry, and tectonic significance of the amphibolites of the Alabama Piedmont. In: Misra KC, McSween HY Jr (eds) Mafic and ultramafic rocks of the Appalachian orogen. Am J Sci 284: 416-436

Sutter JF, Milton DJ, Kunk MJ (1983) $^{40}Ar/^{39}Ar$ age spectrum dating of gabbro plutons and surrounding rocks in the Charlotte belt of North Carolina. Geol Soc Am Progr Abstr 15: 110

Sutter JF, Horton JW Jr, Kunk MJ (1984) Timing of Alleghanian metamorphism in the Kings Mountain belt of North Carolina and South Carolina. Geol Soc Am Progr Abstr 16: 201

Taylor PT, Zietz I, Dennis LS (1968) Geologic implications of aeromagnetic data for the eastern continental margin of the United States. Geophysics 33: 755-780

Thomas WA, Chowns TM, Daniels DL, Neathery TL, Glover L, Geason RJ (1989) (in press) The subsurface Appalachians beneath the Atlantic and Gulf Coastal Plains. Geol Soc Am DNAG Ser, Appalachian-Ouachita Vol

Thorman CH (1976) Implications of klippen and a new sedimentary unit at Gibi mountains (Liberia, West Africa), in the problem of Pan african - Liberian age province boundary. Geol Soc Am Bull 87: 251-268

Tobish OT, Glover L (1969) Metamorphic changes across part of the Carolina Slate Belt-Charlotte Belt boundary, North Carolina and Virgina. US Geol Surv Prof Pap 650-C: C1-C7

Van Breemer O, Dallmeyer RD (1984) The scale of Sr isotopic diffusion during post-metamorphic cooling of gneisses in the Inner Piedmont of Georgia, southern Appalachians. Earth Planet Sci Lett 68: 141-150

Van der Voo R, Johnson RJE (1985) Paleomagnetism of the Dunn Point Formation (Nova Scotia): High paleolatitudes for the Avalon terrane in the late Ordovician. Geophys Res Lett 12: 337-340

Villeneuve M (1984) Etude géologique sur la bordure sud-ouest du craton ouest-Africain: La suture Panfricaine et l'évolution des bassins sedimentares protérozoiques et palérozoiques de la marge nw du continent de Gondwana. Thèse Doct Etat, Univ Aix-Marseille III, 552 pp

Villeneuve M, Dallmeyer RD (1987) Geodynamic evolution of the Mauritanide, Bassaride, and Rokelide Orogens (West Africa). Precambrian Res 37: 19-28

Wagener HD (1977) The granitic stone resources of South Carolina. S Carol Geol Surv, Columbia, Miner Resour Ser 5: 65

Weiner SW, Merschat CE (1988) Summary of geology between the Great Smoky fault at Parksville, Tennessee and basement rocks of the Blue Ridge at Glade Gap, North Carolina. Tenn Div Geol Rep Invest 37, Field trips S Appalachians, Milici, RC 86 pp

Whitney JA, Paris TA, Carpenter RH, Hartley ME III (1978) Volcanic evolution of the southern slate belt of Georgia and South Carolina: A primative oceanic island arc. J Geol 86: 173-192

Whittington HB, Hughes CP (1972) Ordovician geography and faunal provinces deduced from trilobite distribution. R Soc Lond Philos Trans Ser B, 263: 235-278

Williams H, Hatcher RD (1982) Suspect terranes and accretionary history of the Appalachian orogen. Geology 10: 530-536

Williams H, Hatcher JD (1983) Appalachian suspect terranes. In: Hatcher RD Jr, Williams H, Zietz I (eds) Contributions to the tectonics and geophysics of mountain chains. Geol Soc Am Mem 158: 33-53

Williams HR (1978) The Archean geology of Sierra Leone. Precambrian Res 6: 251-268

Williams HR (1979) An archean suture in Sierra Leone. Nature (Lond) 286: 608-609

Wilson JT (1966) Did the Atlantic close and then re-open? Nature (Lond) 211: 676-681

Wright JE, Seiders VM (1980) Age of zircon from volcanic rocks of the central North Carolina Piedmont and tectonic implications for the Carolina volcanic slate belt. Geol Soc Am Bull I 91: 287-294

Zeitz I (1982) Composite magnetic anomaly map of the United States. US Geol Surv Map GP-954A, scale 1: 2500000

Zen E-an (1981) An alternative model for the development of the allochthonous southern Appalachian Piedmont. Am J Sci 281: 1153-1163

Crustal Evolution and the Brasiliano Orogeny in Northeast Brazil

R. Caby[1], A. N. Sial[2], M. Arthaud[3], and A. Vauchez[4]

1 Introduction

Northeast Brazil comprises two structural provinces defined by Almeida et al. (1981). To the north the Parnaiba province includes the São Luis craton as a probable remnant of the West Africa craton in Brazil (Hurley 1967) and the Gurupi belt (Hasui et al. 1984). Most of northeast Brazil belongs to the Borborema province (Fig. 1) which extends for about 400 000 km^2. The major tectonic feature of this province is a system of sinuous and branched, anastomosing shear zones, which divide the province into elongate domains that often differ in lithology, metamorphic grade and structure, suggesting large-scale relative displacements (Schobbenhaus et al. 1981).

Routine geochronology, by Rb-Sr and K-Ar methods, provided a few ages determinations, ranging from > 2.7 Ga to 450 Ma. More precise data obtained in the last decade suggest the existence of three major tectonothermal events at about 2.6 Ga (Archean), 2 Ga (Early Proterozoic or Trans-Amazonian), and 700 to 550 Ma, the age of the Brasiliano event during which the province was cratonized.

A major Late Proterozoic tectonothermal event in this province was proposed by Almeida et al. (1981). Rb-Sr and K-Ar mineral ages usually lie in the range of 650–450 Ma, as well as whole-rock Rb-Sr isochron dates obtained on syn- and late-orogenic granites. Intensity of reactivation of pre-Brasiliano basement rocks is, however, variable and distinction of poly-from monocyclic units, as proposed by Ferreira (1967), is still debatable. The distinction based solely on grade of metamorphism is not very successful, as it varies significantly along and across the belts. Unconformity beyond dispute, between older, probably reworked Archean gneisses, and Proterozoic sediments, although rare, has been locally observed. The basement-cover relationships are often obscured by deformation or migmatization. Post-orogenic molassic units are represented by late-Brasiliano reddish sandstones to conglomerates preserved in down-faulted blocks adjacent to Sobral-Pedro II fault zone (Brito Neves, 1975; Costa et al. 1979) and along the northern margin of the São Francisco craton (Brito Neves 1975).

The Borborema Province is subdivided here into three structural domains (Médio Coreaú, Central and Sergipano domains), and this review will focus mainly on their geology and petrostructural evolution.

2 Central domain

2.1 Reworked Archean

Inliers of older gneisses of assumed or demonstrated Archean age are widely exposed in the Borborema province. They form domes or half domes bounded by shear zones, and also klippen recorded in the Ceará state (Caby and Arthaud 1986). In the Seridó region these gneisses have been termed the Caicó complex (Meunier 1964; Brito Neves et al. 1975).

Unmigmatized Archean gneisses are mostly plagioclase-hornblende-biotite gray gneisses derived from a predominant tonalite-trondhjemite-granodiorite (TTG) plutonic association (Caby 1989). These gneisses usually display a prominent banding formed by ductile deformation, as proposed elsewhere by Myers (1978). Locally, a progressive tran-

[1] Centre Géologique et Géophysique, UPR 361 CNRS Université des Sciences et Techniques du Languedoc, 34095, Montpellier cedex 5 France
[2] Université Féderale de Pernambuco, Dep. de Geologia, Recife, PE, 50000 Brazil
[3] Universidade de Fortaleza, Dept. de Geologia, C. P. 1258, 60000, Fortaleza, CE, Brazil
[4] Laboratoire de Tectonophysique, Université des Sciences et Techniques du Languedoc 34095 Montpellier Cedex 5 France, et Laboratoire de Géologie dynamique, Faculté des Sciences Saint-Jéôme, F-13307 Marseille Cedex 18 France

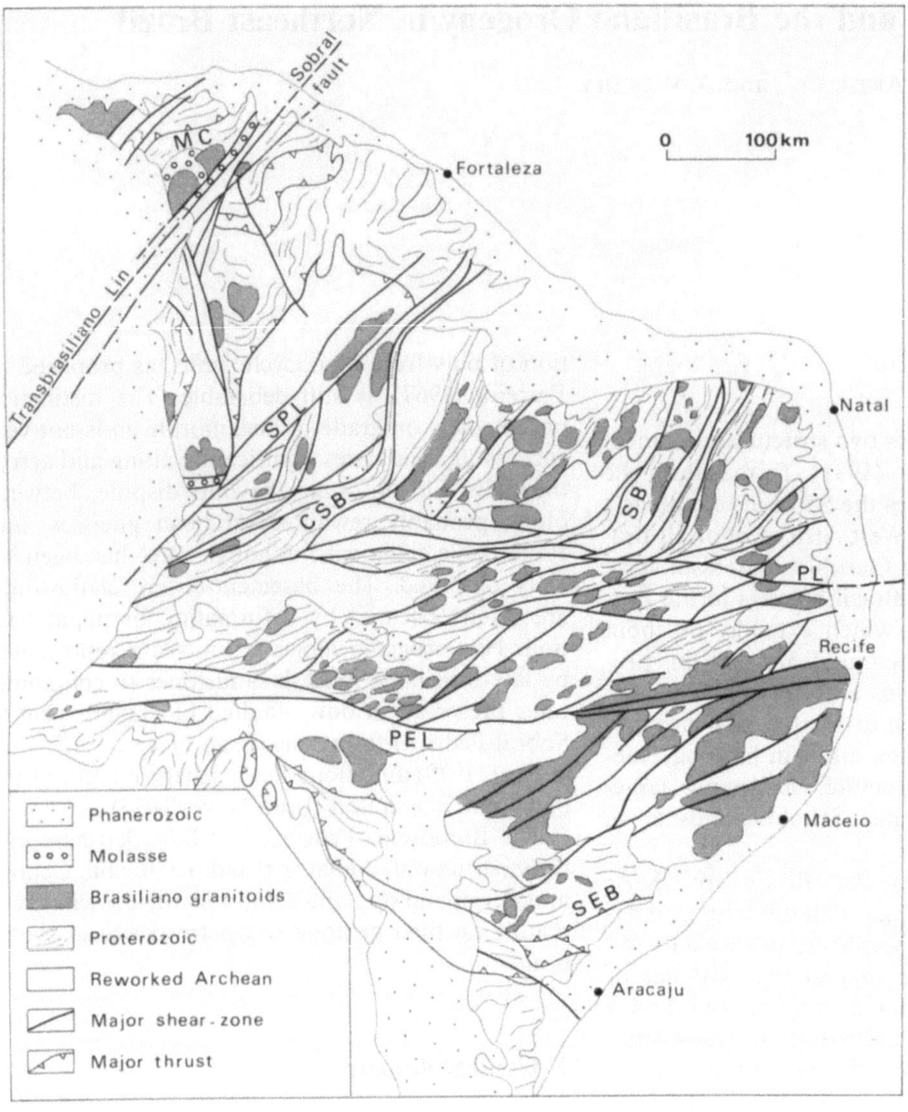

Fig. 1. Map of northeast Brazil (modified from Santos et al. 1984)

sition from undeformed plutonic rocks, in which angulous mafic enclaves are occasionally present, to strongly deformed banded gneiss, can be observed (Fig. 2A).

Mineralogy of gray gneiss essentialy comprises blastic plagioclase associated with epidote granules, which testify to its recrystallization of more calcic plagioclase, elongate quartz, often greenish biotite, green hornblende, sphene, ilmenite and minor amounts of K-feldspar + myrmekite.

Amphibolites constitute layered units conformable with gneisses, and may derive from Mg-rich bodies which may predate the emplacement of the TTG suite, since they are petrographically similar to the enclaves.

Porphyritic granites grading into augen-gneiss by increasing deformation are also part of this igneous association (G_1 granites, Jardim de Sá et al. 1981b) or perhaps younger.

Greenstones and metasediments have been proved locally to be the country rocks of gneissified Archean granitoids (Caby and Arthaud 1986). They include mafic, Mg-rich amphibolite with possible remnants of pillow structure and volcanic breccias, tremolite schists, impure carbonates, siliceous rocks and Al-rich metapelitic schists.

A more complete greenstone association including Cr-rich ultramafics, metagabbros, pyroxenite, amphibolite and tremolite schist, and pillowed metabasalts, may be represented in the Tróia massif of Brito Neves (1975). Associated metasediments comprise calcareous metaquartzite, Fe-rich cummingtonite, sulphide-rich and Mn-rich metasediments,

Fig. 2 **A** Porphyritic granodiorite with dioritic inclusion. The foliated rock progressively grades downward to a streaky veined gneiss *(below)* and to a wavy gneiss cut by a successively sheared leucosome *(above)*. **B** Sharp gradation between a veined tonalitic gneiss into a wavy migmatite to the *left*. **C** Inclusion of a mafic tonalite gneiss in a more feldspathitic homogeneous migmatitic gneiss grading downward into a migmatitic granodiotite

marbles and Al-rich graphitic schist (Pessoa and Archanjo 1984).

Migmatitic gneisses derived from grey gneiss crop out widely in the whole Borborema province (Almeida et al. 1981). Migmatization progressively or abruptly appears in the form of plagioclase-rich streaky leucosome, followed by injection of several crosscutting generations of anatectic mobilizates (Fig. 2B). Nebulitic gneiss sometimes includes schollen of non-migmatized, already deformed orthogneiss (Fig. 2c).

The highest metamorphic grade in anatectic gneisses is defined by the two pyroxenes mineral association typical of granulite facies (see Sect. 2.2).

Younger crosscutting veins of Brasiliano age, sometimes foliated, granodiorite to tonalite, are considered as anatectic products. They are abundant in and restricted to migmatitic areas.

Mylonitic gneisses resulted from the younger stage of reworking of the gray gneisses. They display mylonitic banding and include lenses and boudins of mafic rocks. The mylonitic foliation is defined by newly recrystallized tiny biotite and polycrystalline quartz and plagioclase ribbons, with clasts of amphibole and sphene. Mylonite gneisses occurs either along the steep shear zones or underline major low-angle thrusts.

In the state of Ceará, strongly migmatitic gneisses locally constitute a flat-lying large klippe overlying Proterozoic metasediments (Caby and Arthaud 1986). Mylonitic gneisses form imbricated units bounded by moderately-dipping shear zones considered as lateral ramps of nappes. The less reactivated, non-migmatitic TTG suite and the derived gray gneisses crop out as the lowermost unit exposed south of Independência (Fig. 2).

Polycyclic gray gneisses display the same foliation as that developed in adjacent meta-plutonic rocks. Polyphase folding, however, affected aplitic-pegmatitic veins before development of the regional foliation, suggesting that these rocks derived from pre-Brasiliano material converted into gneiss before (or during?) emplacement of the TTG igneous suite.

Geochronology. Due to the high degree of reworking which is always higher than amphibolite facies, Rb-Sr data on this pre-Brasiliano basement have given scattered ages (mostly errorchrons) in the range 2–2.7 Ga (Santos and Brito Neves 1984). It seems conclusive, however, that more recent data obtained on typical non- migmatitic gneiss have given a well-defined whole-rock Rb-Sr isochron at 2540 ± 60 Ma (Pessoa et al. 1986) with low 87Sr/86Sr initial ratio < 0.702. K-Ar and Rb-Sr mineral ages have invariably yielded Brasiliano ages (Santos and Brito Neves 1984).

The well-preserved, locally non-foliated plutonic rocks of the TTG suite, as well as the preserved pillow basalts and volcanic breccias, allow one to stress that Archean metamorphism did not exceed greenschist facies as in many Archean gneiss-greenstone associations elsewhere in the world. The TTG association is thus considered a Late Archean large-scale igneous suite of probably juvenile character. Post-Archean amphibolitized dikes with a superimposed Brasiliano foliation have been encountered, but their distinction from syn-orogenic mafic intrusives of Brasiliano age is not easy in the field.

2.2 *Proterozoic Sequences*

The distinction of monocyclic, Proterozoic metasedimentary sequences from pre-Brasiliano basement is widely accepted (Santos and Brito Neves 1984). However, divergent interpretations about stratigraphy, mono- or polycyclic evolution of these metasediments have been proposed (Jardim de Sá et al. 1987 and references therein).

The Seridó region (Seridó Fold Belt) is the classical area where two lithostratigraphic units (Fig. 1) were first distinguished (Meunier 1964; Ferreira 1967; Ebert 1970; Jardim de Sá 1984; Santos and Brito Neves 1984). Proterozoic sequences crop out as a series of schist belts in between Archean basement.

The Jucurutu Group (early Jucurutu Formation of Ferreira 1967) includes the quartzitic Equador Formation (Ebert 1970), which locally appears to be unconformable on, and undetached from, the Archean basement. This widely distributed basal unit comprises both micaceous and feldspathic metaquartzites with preserved quartz pebbles (gravels locally present) and oblique bedding. Associated rocks include calc-magnesian, siliceous and Fe-rich metasediments. The rest of the Jucurutu Group includes epidote-rich paragneisses and a thick pile of calc-silicates and impure marbles interlayered with both mafic and felsic metavolcanics (Jardim de Sá 1984). This assemblage represents the host-rock of the widespread scheelite mineralizations encountered in this region.

Conformable sheets of foliated granites intruded this group prior to deformation and metamorphism. This family of intrusions (G_2, Jardim de Sá et al. 1981b) has a banded appearance, interpreted as a deformed magmatic layering, defined by porphyritic or non-porphyritic layers, an abundance of Fe-Mg minerals and sometimes thin bands enriched in accessory minerals. Whole rock Rb-Sr dating of G_2 orthogneisses yielded poorly defined alignments around 2 Ga. Ages of 2086 ± 84 Ma have been proposed (Jardim de Sá et al. 1987). The same authors have, however, considered this G_2 family as syn-orogenic granites related to the Trans-Amazonian orogenic cycle. In contrast, Caby (1989) considers these layered orthogneisses as anorogenic intrusions with superimposed Brasiliano tectono-metamorphic imprint only.

The Seridó Group is here regarded as a distinct, probably unconformable unit, with basal polymictic conglomerates (Santos and Brito Neves 1984; Neves et al. 1984; Archanjo and Salim 1986). It is a monotonous semi-pelitic and pelitic unit possibly up to several kilometers in thickness, with some quartzite, calc-silicate and amphibolitic layers. When only affected by greenschist facies, bedding with a rhythmic layering is often preserved, suggesting turbiditic deposits. Ca-Fe-Mg-carbonate nodules, and lenses are conspicuous in this unit. By increasing metamorphism, it grades into two-mica schists, often with giant Al silicates porphyroblasts (andalusite, cordierite, staurolite), whereas deformation is clearly polyphase (Jardim de Sá 1978).

Whole rock Rb-Sr determinations on Seridó schists yielded poorly defined alignments at ca. 560, 580 and 600 Ma (quoted in Santos and Brito Neves 1984).

The Cachoeirinha-Salgueiro Fold Belt (same as Piancó-Alto Brigida, Brito Neves 1975). The Cachoeirinha Group (Barbosa 1970) is commonly considered as a possible equivalent of the Seridó Group. It crops out between the Patos and the Pernambuco lineaments, as a monotonous flysh-type greenschist facies unit, which consists of green metasiltites, metapelites and fine-grained meta-arenites, locally with hematite-rich phyllites (e.g. São José do Belmonte mine, Pernambuco state, Silva and Gentil 1969). Load structure, slumps, convolute and graded bedding, brittle beds and incomplete Bouma's sequence have been observed, suggesting that these pelitic sediments may represent distal turbidites. Metagreywackes, matrix-supported conglomerates and a bimodal volcanic suite (spilitic metabasalt and felsic volcanics with associated calcareous and volcanoclastic deposits) are also part of this succession (Munis and Santos 1980). Pre-metamorphic intrusives are represented by diorite, tonalite and granodiorite small stocks, as well as slices of serpentinite. Increasing metamorphism may have resulted in a gradual passage into the Salgueiro two-mica schists ± garnet ± staurolite, synkinematic fibrolite being restricted around the thermal aureoles of some Brasiliano granites.

The Ceará region. Proterozoic sequences form linear schist belts that follow major NNE-SSW trend-

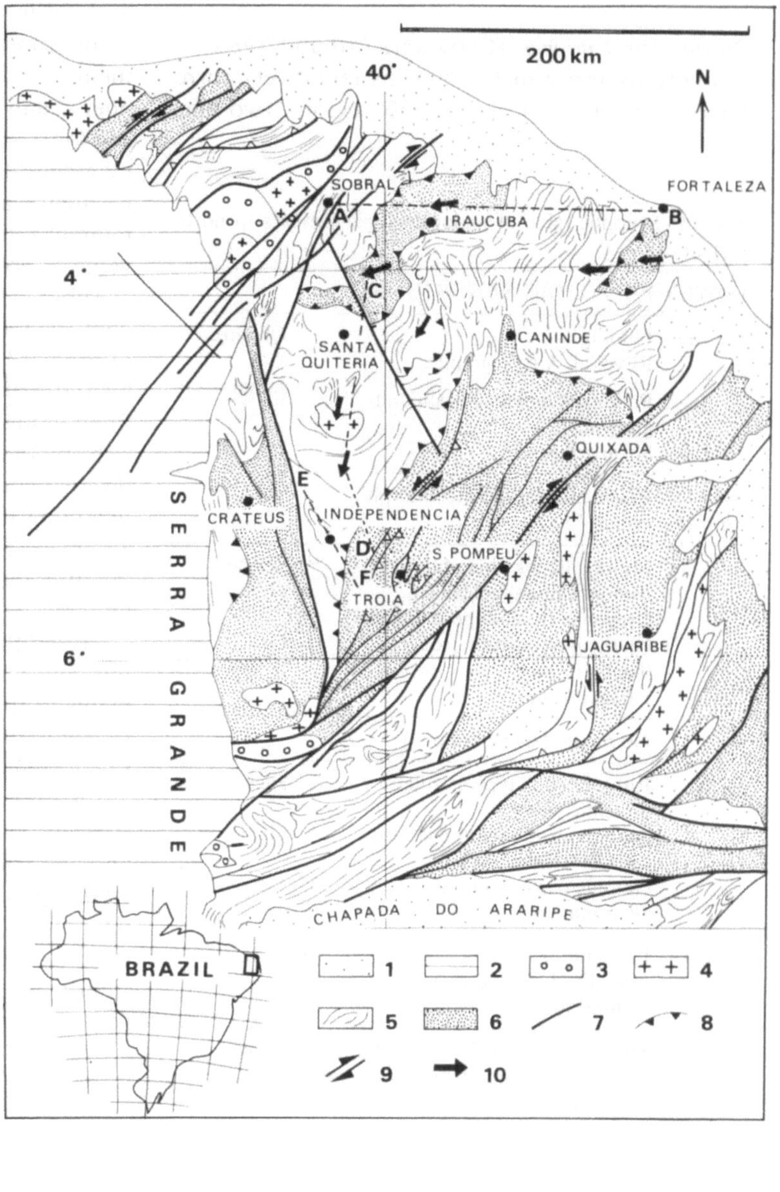

Fig. 3. Ceará state and adjacent areas (after Caby and Arthaud 1986). *1* Mesozoic and Tertiary; *2* Paleozoic Maranhão basin; *3* Late Brasiliano molasse; *4* major Brasiliano granites; *5* Ceará Group (Proterozoic) *6* Archean basement rocks; *7* major shear zones; *8* major thrust faults; *9* major strike-slip faults; *10* direction of movement of nappes. Letters *A*-*B* and *C*-*D* indicate lines of sections in Fig. 2A, B respectively

ing shear zones, but to north they are also involved in large-scale horizontal tectonics (Caby and Arthaud 1986; Fig. 3).

The Orós unit represents the lowest grade unit preserved along two sinuous major shear zones (Fig. 2). It comprises black schists with interlayered marble lenses, which overlie quartzites with an overall parallel bedding suggestive of a marine environment. These metasediments contain thick sheets of layered orthogneiss in which it is possible to recognize metarhyolite, metaporphyry and K-rich porphyritic granite (Martins Sá et al., in press). These intrusives predate regional deformation and metamorphism, here characterized by almandine and staurolite in metapelites. A preliminary whole rock Rb-Sr isochron in orthogneiss and on metarhyolites (Macedo et al. 1988) yielded an age of ca. 1.7 Ga.

North Ceará nappes. (Caby and Arthaud 1986) essentially comprise metasediments tentatively correlated with those of the Seridó region. Two-mica schists and paragneisses are the major constituents. Metamorphic foliation of upper amphibolite facies has thoroughly swept out the bedding, which, however, can be present in metaquartzites. A large volume of marbles, tectonically accumulated in part, also crop out in central Ceará, together with calc-silicates, calcareous micaschists and quartzites. The overall boudinage at all scales is conspicuous in these units, which often display a gently dipping foliation. Boudins of amphibolite and of metaquartzite, as well as rootless folds, are frequently found in metapelites. A conspicuous lithology is represented by garnet-calcite-scapolite-bearing mafic amphibolites which may be derived from impure, Fe-Mg-rich carbonates deposited in an evaporitic environment.

Pre-metamorphic intrusives in these units comprise mafic sills later converted into amphibolites, and felsic sub-alkalic to syenitic intrusives. They form conformable sheets in between metasediments, mainly in the carbonate units. Syenitic gneiss and meta-syenite form a large massif (< 150 km^2) south of Santa Quitéria. The contorted magmatic layering defined by variable abundance of augite and amphibole is partly obliterated by migmatization, and is cut by pegmatitic syenite veins of possible anatectic origin.

Sub-alkalic gneisses form also interlayered sheets conformable with impure marbles and micaschists of the flat-lying nappe north of Independência (Caby and Arthaud 1986). The magmatic layering is defined in these gneisses by more or less abundant Fe-rich biotite and hastingsitic amphibole, and by discrete concentrations of accessory minerals often including metamict allanite, sphene, magnetite and zircon. K-rich and strongly Na-enriched varieties are present, and fluorite is a frequent accessory mineral.

From lithostratigraphic data, an evolution involving deposition of early Proterozoic shelf-type sediments of the Jucurutu Group in marine and evaporitic environments followed by anorogenic, possibly rift-related intrusions of 2 to 1.7 Ga age is suggested. A similar evolution characterizes both the Pan-African domains of West Africa (Caby 1987, 1989) and the stable cratonic domains of central Africa (Bonhomme et al. 1982). The paraconformable or unconformable, younger Seridó Group is interpreted as a flysch-type unit which may have been deposited in an orogenic environment. Table 1 summarizes the proposed geological evolution.

Table 1. Synoptic table of major geological events in northeast Brazil

LOWER PALEOZOIC SANDSTONES		
530 Ma	Folding Rhyolite, A type granites Pull-apart molassic basin and graben Shear zones	Ubajara & Jaibaras Gr
580 Ma 620 Ma	Brasiliano tectono-metamorphic events	
700? Ma	Minor calc-alkaline plutonism Minor bimodal volcanism Orogenic, flysch-type sediments Opening of ensialic/trough (back-arc?) Paraconformity	Serido Gr
	No record (stable craton?)	
1.7 Ga	Rift-related A-type granites, tholeiitic sills Aluminous quartzites: schists, carbonates etc. cratonic to peri-cratonic environment	Jucurutu Gr Ceara Gr
2.0 Ga	Late granitoids Long-lived, progressive cratonization	
> 2.6 Ga	Trondjhemite-tonalite-granodiorite association, minor greenstones and sediments	Caico

2.3 Tectono-Metamorphic Evolution

The structure of the Borborema province is dominated by a system of sinuous and branched, often steep, continental-scale shear zones, which partly match with those of West Africa (Caby 1989). This system of shear zones has been considered an early feature, and several stages of movement have been documented (Santos and Brito Neves 1984). Some of these zones have clearly controlled the structural evolution, the metamorphic zonation and the emplacement of synorogenic granites. These relationships suggest that they represent first-order, synorogenic shear zones.

On the other hand, dominant ductile thrust and nappe tectonics have been previously described in the Sergipano structural domain (Brito Neves et al. 1977; Davison and Santos, 1989, Fig. 1), south of the Borborema province, and in Ceará (Caby and Arthaud 1986). In these areas, strike-slip faulting is associated with or subsequent to syn-to late-metamorphic nappes. It becomes more evident that several structural subdomains should be distinguished on the basis of the overall tectonic regime dominant during the Brasiliano orogeny. The Central Structural Domain can be subdivided into four subdomains: the Ceará and the Seridó to the north, the Cachoeirinha-Salgueiro and the Riacho do Pontal (Sial 1987) to the south. K-Ar and Rb-Sr mineral ages systematically in the range of 620–480 Ma, provide evidence that the Brasiliano tectonothermal event is strongly expressed in the entire province. Pre-Brasiliano mineral ages are restricted to the São Luis and São Francisco cratons, which have not recorded the Brasiliano event at all (Fig. 1).

Central Ceará. Major nappes involving reworked Archean basement and Proterozoic metasediments have been described in this area by Caby and Arthaud (1986). Geometrical and structural relationships allow one to distinguish highly reactivated and anatectic Archean basement overlying Proterozoic units to the north, from nappes of high-grade Proterozoic rocks with an apparent reverse metamorphism overthrusted onto mildly reactivated Archean rocks to the south (Fig. 3).

Archean plutonic rocks were mostly transformed into banded gneiss sometimes displaying anatectic mobilization. Lenses of various size of undeformed or slightly deformed plutonic rocks have been preserved in strongly foliated gneiss. These rocks have clearly undergone a single tectonothermal episode involving the development of banding under upper amphibolite facies conditions, and several subsequent phases of folding. Intrusion of leucotonalitic to trondhjemitic and granodioritic veins (sometimes displaying *en échelon* pattern) along the axial surface of folds strongly suggest that folding occurred at still high temperature (Fig. 4A). Mineral and stretching lineations are generally well expressed, especially by the alignment of pseudomorphic aggregates after igneous Fe-Mg-minerals (biotite, amphibole, ilmenite), and by xenoliths and quartz grains elongation. Poikilitic minerals, triple point grain boundaries and quartz overgrowth indicate post-kinematic annealing. Plagioclase, amphibole and sphene microclasts enclosed in overgrown quartz have retained an asymmetric shape and suggest an earlier porphyroclastic texture. Undeformed crosscutting hornblende-ilmenite-bearing anatectic mobilizates attest that high-temperature conditions required for collection of liquids of leuco-tonalitic to trondhjemitic compositions ($T > 700\,°C$) were maintained in these rocks after the main deformation.

Granulite facies conditions were locally reached along a linear belt parallel to the Senador Pompeu shear zone around 6° Lat S (Fig. 3). Quartz + plagioclase + opx + garnet + ilmenite primary association has been observed in banded grey gneiss. Orthopyroxene appears as poikilitic new grains nucleated along quartz-plagioclase boundaries. The local occurrence of granulite facies conditions in gray gneisses may be tentatively related to intrusion of syntectonic mafic tonalite and two-pyroxenes pyriclasite veins (Fig. 4B).

Late metamorphic mylonitic fabrics, in contrast, characterize Archean gneisses involved in nappes, imbricates and shear zones around Tróia. In this domain the mylonitic foliation is associated with extensive boudinage of mafic rocks, aplites and pegmatites. The texture is porphyroclastic with quartz ribbons. NE-trending stretching lineation is outlined by truncated and elongated prophyroclasts and streaks of minute secondary biotite. High temperature isotropic fabric was only preserved in the core of mafic lenses.

Proterozoic units display a pervasive flat-lying metamorphic foliation usually parallel to lithologic boundaries. In the pelitic gneisses beneath the Archean migmatitic rocks thrust sheet of northern Ceará, a subhorizontal foliation defined by the orientation of muscovite, biotite and sillimanite is crosscut by aplite, pegmatite and granite veins. Some garnet-rich layers and lenses have retained a quartz + plagioclase + garnet + kyanite + biotite +

Fig. 4. A Highly foliated tonalite gneiss with folded leucocratic mobilizates. Note the axial disposition of the sheared leucocratic late veins. Horizontal section parallel to XZ. **B** Banded granodiorite gneiss with a conformable mafic band with two-pyroxene-garnet granulite facies paragenesis. The gneiss is cut by a mafic pyriclasitic vein. Note the partial assimilation of the dyke by the leucocratic gneiss, consistent with a high temperature emplacement. **C** Syn-kinematic diorite with more mafic enclaves emplaced in the Seridó schists. Note the asymmetric figure consistent with a dextral movement. Nearly horizontal section parallel to XZ

rutile + graphite earlier high-pressure assemblage; incipient fibrolitic sillimanite developed from biotite and kyanite. The foliation is defined by elongated monocrystalline quartz ribbons, biotite films and elongated, boudinaged and bent kyanite prisms. These metapelites ($d \simeq 2.9$ g/cm^2) are regarded as the deepest exposed rocks with monocyclic evolution of Brasiliano age.

In the southern part of Ceará, Proterozoic units of nappes display a refolded initially gently dipping foliation. Metapelitic schists and gneisses are muscovite- and sillimanite-bearing with incipient melting. Marble-calc-silicate alternances display coarse-grained subisotropic fabrics formed during post-kinematic annealing and have experienced boudinage at all scales.

Plurifacial, syn-kinematic metamorphism has been observed in metapelitic rocks involved in large-scale nappe tectonics around Independência (Fig. 5A). The metamorphic zonation from top to the sole of the allochthon is (Fig. 5B):

1. Two-mica schists with both prismatic and fibrolitic sillimanite + plagioclase + almandine; K-feldspar is restricted to migmatitic rocks. Concordant and crosscutting Al-rich leucosomes have been emplaced together with veins and sheets of leucocratic, garnet-muscovite leuco granite with a random fabric. Fibrolite is often a late phase superimposed along discrete shear planes crosscutting even leucocratic anatectic mobilizates.

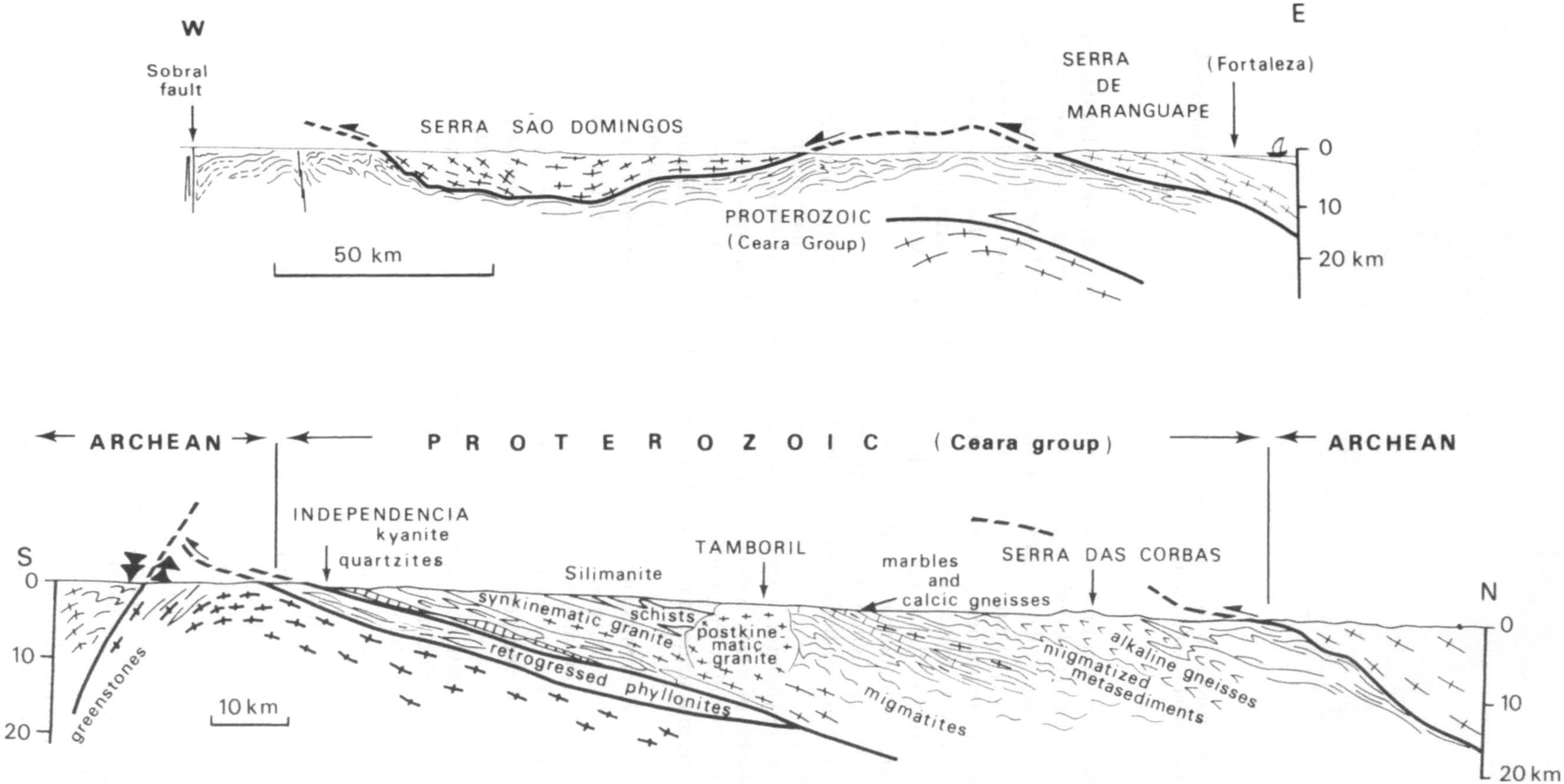

Fig. 5. A Interpretative cross-section (A-B in Fig. 3) between Fortaleza and Sobral. Archean rocks to east include gray gneisses of plutonic origin cut by amphibolitized dykes, and some granitic orthogneisses of subalkaline affinity. Those from klippe are mostly biotite-hornblende orthogneisses of tonalitic to granodioritic composition, grading downward into their migmatitic equivalents. The Proterozoic Ceará Group is mainly composed of slightly migmatized metasediments with kyanite or sillimanite (after Caby and Arthaud 1986). **B** Interpretative cross-section (C-D in Fig. 3) across Proterozoic of Independência, interpreted as a duplex about 25 km thick between two Archean units; sole thrusts of phylonites derived from unmigmatized kyanite gneisses underneath. These metasediments are overlain by migmatite, various orthogneisses and granitoids cut by post-kinematic granites. Map *symbols* as in Fig. 3 (After Caby and Arthaud 1986)

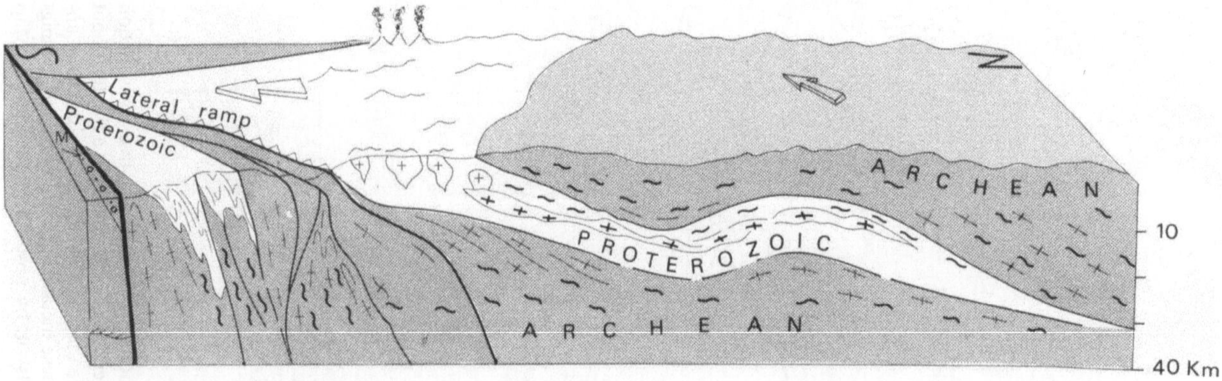

Fig. 6. Interpretative cross-section of Ceará nappes northeast Brazil, prior to present-day topography (modified after Caby and Arthaud 1986). Horizontal = vertical scale

2. Two-mica schists with a primary quartz + plagioclase + biotite + garnet + kyanite + rutile metamorphic assemblage. Only few garnetiferous leucosome veins are present. Late fibrolite has grown at grain boundaries and partly replaced kyanite. Downwards, muscovite replaced kyanite during reactivation of the metamorphic cleavage under retrogressive conditions, giving a porphyroclastic texture (e. g. sigmoidal muscovite clasts, fragmented garnet, crystallization of secondary minute biotite).
3. Retrogressive phyllonites (100 m thick) with abundant matrix of newly formed biotite + muscovite, include boudins (pegmatite, calcsilicates) of various size, and scattered clasts (feldspar, garnet).
4. Ultramylonite (<10 m thick) with minute biotite, needle-like garnet aggregates, and long quartz ribbons, in which sub-spherical coarser grained rocks may be present (scapolite-bearing garnet amphibolite, calc-silicate, pegmatite).

This metamorphic zonation is consistent with a late-metamorphic emplacement of the nappe (Fig. 6). The overall N-S trend of stretching lineations, the gross geometry and the reverse metamorphic zonation are consistent with southward emplacement of this thrust sheet. Trend of isoclinal folds in metaquartzites is also parallel to stretching lineations, and sheath folds have been observed locally. The sigmoidal shape of clasts and the development of north-dipping shear bands, however, points out to a last top to north incremental shear component related to an extensional regime, as is nowadays widely reported elsewhere in crystalline nappes (Platt 1986).

Muscovite-free, kyanite-bearing metamorphic assemblages in this tectonic pile are thus exposed through large-scale late-metamorphic overthrusting and compare well with the Himalayan Tibetean slab (Caby et al. 1983). Thus, these belong to the lowermost exposed crustal level of the Central Structural Domain, possibly detached from underlying granulite facies rocks such as those of the Médico-Coreaú, Granja kinzigites (see Sect. 3.2.).

Seridó low-pressure high-temperature belt. The Seridó subdomain is limited to the south by a major shear zone (Patos shear zone, Fig. 4). Approaching this shear zone, the trend of metamorphic cleavage and of gross lithology displays an arcuate pattern. Pre-Brasiliano basement crops out as elongated domes and half-domes, and the Proterozoic cover forms synformal structures associated with sinuous shear zones that converge to the Patos lineament to the south.

A similar trend of mineral and stretching lineation has been recorded in the basement and the supracrustal rocks, independent of the attitude of the foliation. Archean gneisses display a flat-lying foliation parallel with the S_1 foliation observed in overlying Proterozoic rocks (Fig. 7). Upward in the succession, upright folding with sub-horizontal axis and axial-plane cleavage characterize the Seridó two-mica schists. Refolding of the S_1 primary cleavage by F_2 folds with various axis orientation is frequent. Dextal steeply dipping shear zones coeval with injection of granodiorite veins can be observed. In domains deformed under higher temperature, low-pressure metamorphism, large euhedral poikiloblasts of andalusite, cordierite, staurolite and biotite connected to the syn-kinematic emplacement of granitoids, grew in black metapelites which are affected by open folds with steep axial surface. These hornfels-like rocks escaped any subsequent

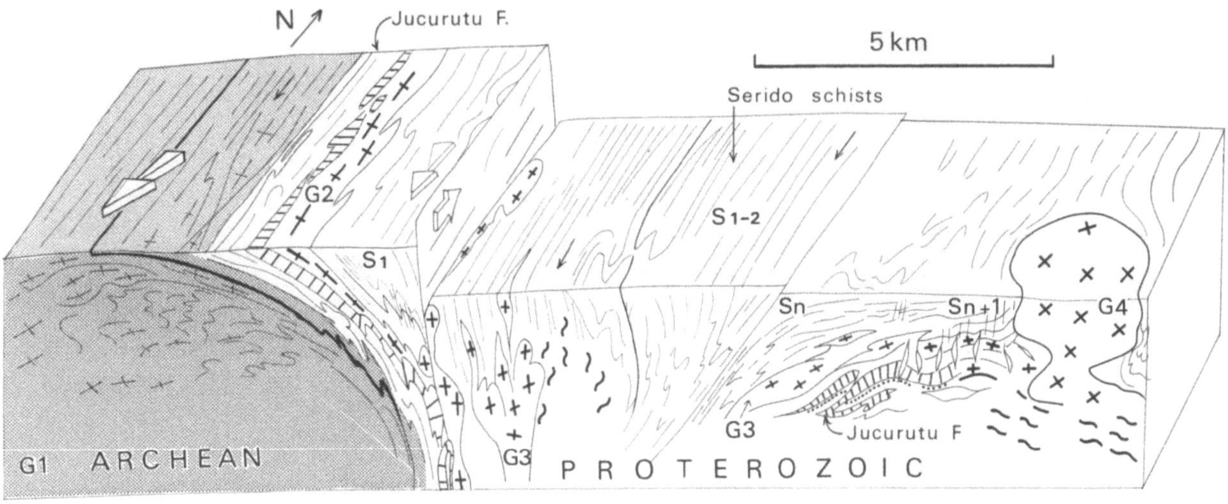

Fig. 7. Geometric relationships between horizontal tectonics and strike-slip tectonics in the Seridó region, northeast Brazil. G_2 to G_4 are the different generations of granites of Jardim de Sá et al. (1987). Note the parallelism of stretching lineations in Archean domes and in Proterozoic synform with S_{1-2} cleavage and low-pressure-type metamorphism. Horizontal = vertical scale

retrogression. Asymmetry of shear sense indicators suggest dextral movement.

In the central part of the Seriodó belt, the zone with recumbent cleavage is about 15 km wide. Finite strain computation in metaconglomerates suggest a highly constrictional deformation (Archanjo 1988). Horizontal extension up to 500% and shortening normal to the foliation plane of about 50 to 70% have been estimated. In the Acari synform, refolding of S_1 cleavage, during horizontal tectonics took place in the sillimanite stability field. A possible pre-Brasiliano age for the main metamorphism and polyphase deformation in the Seridó subdomain has been continuously challenged by Jardim de Sá (e.g. 1984) and Jardim de Sá et al. (1987), based on relationships between granitic intrusions and successive foliations besides stratigraphical informations from outside this subdomain.

Within the Seridó subdomain, some NE-SW trending, mostly dextral strike-slip faults developed in a parallel direction with the folds. These shear zones are some tens to some hundreds of metres wide, steeply-dipping; they display retrogressive greenschist facies mylonitic or phyllonitic fabric. Brasiliano syntectonic granites are deformed by ductile shear zones, suggesting shearing represents a late stage in the deformation history (Fig. 4C). However, in the syn- to late-kinematic Acari granite of Brasiliano age (Fig. 8), the pattern of the magmatic fabric, as well as shear criteria in the magmatic fluidality, strongly suggest that a dextral transcurrent tectonic regime was active during the emplacement of this plutonic body (Jardim de Sá et al. 1987). Whether the synmetamorphic horizontal stretching, parallel to fold axes observed in metaconglomerates, was associated with a transcurrent regime is not clearly understood and needs complementary investigations.

Cachoeirinha-Salgueiro subdomain (same as the Piancó-Alto Brigida, Brito Neves 1975) is delimited by the EW-trending Patos and Pernambuco shear zones. This fold belt extends for 20000 km^2 and has an overall NE trend; it is composed of the Cachoeirinha and the Salgueiro Groups, which could represent different crustal levels of the same orogen. The shape of this belt is sinuous and show a succession of NE- and EW-trending segments; the later ones are, at least locally, due to the proximity of major EW shear zones.

The overall NE-SW structural orientation of the belt seems to be mainly a result of the late stages of the tectonic evolution. In the northern part of the Cachoeirinha belt, asymmetric folds are associated with a steeply dipping incipient slaty cleavage. Southward a deeper crustal level is exposed; it is characterized by the development of a gently dipping foliation associated with southward-verging, E-W-trending isoclinal folds, both contemporaneous with prograde metamorphism whose culmination is marked by crystallization of garnet and staurolite. Although the metamorphic transformations and the deformation, the rocks have frequently retained sedimentary structures.

The Salgueiro Group displays a more complex tectonometamorphic evolution. It is considered as a

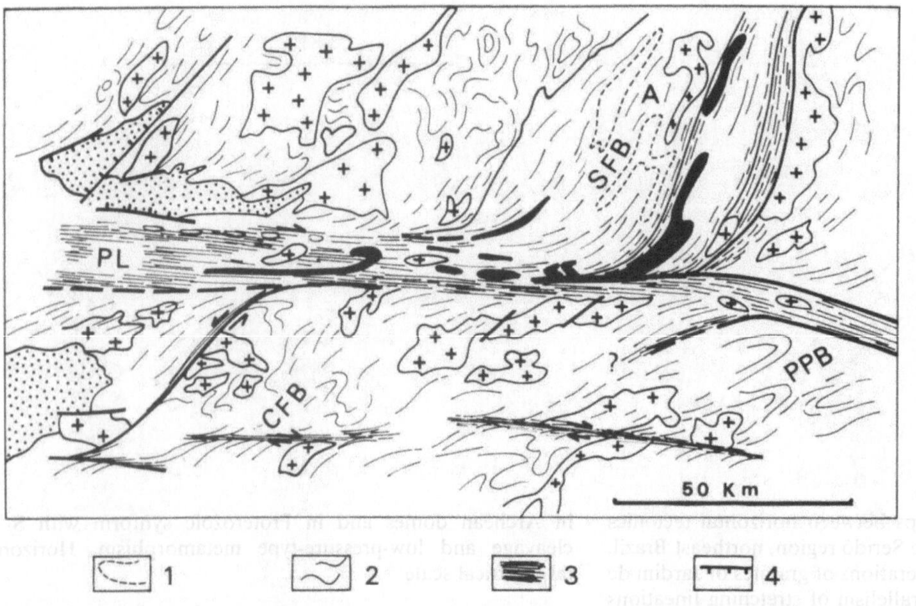

Fig. 8. The structural trend within the Patos shear zone and within the Seridó belt (modified from Archanjo, 1988). *P* Patos; *A* Acari granite; *I* Itaporanga granite; *CFB* Cachoeirinha Fold Belt; *PPB* Pajeu Paraiba Belt; *SFB* Serido Fold Belt. *1* Lithologic limit; *2* structural trend; *3* shear zone; *4* normal fault; *in black* metaquartzite

lateral, higher-grade equivalent to the Cachoeirinha, although Barbosa (1970) and Silva Filho (1985) have mapped it as a separate unit. Near Salgueiro town, two-mica schist with garnet ± staurolite is the country rocks of Brasiliano trondhjemitic granitoids (see Sect. 2.5). The steeply-dipping, refolded foliation often displays steeply plunging folds and stretching lineations parallel to biotite streaks. Incipient anatectic mobilization of leucocratic material can also be observed. Earlier magmatic phases represented by muscovite granite veins that cut the Salgueiro metasediments display the same steeply-plunging lineation, defined there by muscovite streaks, suggesting that at least part of the deformation of the schists took place during emplacement of Brasiliano granites.

2.4 Major Shear Zones

The most striking tectonic structures of the Borborema province probably are the EW-trending Pernambuco and Patos shear zones, both transecting the entire northeastern part of Brazil (Fig. 1). These structures were recognized long ago as major fault zones [e. g. see Ebert (1970) or Santos (1971) for the Pernambuco lineament, and Kegel (1965) for the Patos lineament]. Furthermore, interpretation of remote sensing imagery (Braun 1982) gave a more accurate picture of these fault zones, and emphasized their importance in the building of the Borborema tectonic province.

Improved mapping together with geochronological and tectonic studies progressively suggested a more complex network of shear zones. NE-trending ductile shear zones, most of them several kilometers wide, have been described. They are especially well-developed in the northwestern part of the Ceará province (Caby and Arthaud 1986; Caby 1989; Sá et al. in press) and between the Patos and Pernambuco lineaments (Brito Neves 1983). Some NE-SW subsidiary shear zones have been also described in the Seridó fold belt (e. g. Jardim de Sà et al. 1987).

2.4.1 Shear zones of Ceará

Shear zones of Ceará (Fig. 3) appear in the field as thick belts up to several kilometers wide, and displaying a vertical or steeply dipping, variably less developed mylonitic foliation with general horizontal mineral stretching lineations. Mylonite to ultramylonite bands, 10 to 100 m wide, includes remnants of less deformed rocks. Several shear zones crosscut adjacent lithologic units and the nappes; others are parallel or at low-angles with the regional foliation and may converge to sole-thrusts. Anastomosing steep shear zones define large-scale sigmoidal wedges suggesting they represent strike-slip duplexes.

Regional descriptions of some studied shear zones are presented below.

The Sobral-Pedro II shear zone is about 2 km wide. It is underlined by banded ultramylonites that include lenses, sometimes cigar-shaped or subspherical boudins of coarse-grained non-mylonitic rocks. In ultramylonite, foliation show a complex geometry that fits with the shape of enclosed, non-mylonitic rocks. The linear fabric is defined by elongation of quartz and by streaks of minute biotite. Syn-mylonitic folds display the geometry of sheath folds that are symptomatic of a progressive deformation. A marked sigmoidal shape of clasts and asymmetric pressure shadows has been observed in rocks with a planar protomylonitic fabric, but displacement inferred from shear criteria close to boudins are inconsistent.

In thin section, ultramylonites display mineral assemblages recrystallized under upper greenschist facies conditions; minute biotite is a widespread newly formed phase, whereas mixed clasts with a frequent subspherical shape may persist in many rocks. Out of the shear zone, to the southeast, open folding of the subhorizontal foliation, progressively disappears in a few kilometers. Zones of subsequent brittle deformation may portray a normal fault component linked to the down-faulted Jaibaras graben to the northwest.

Initiation of the Sobral shear zone system which is part of the Trans-Brasiliano lineament (TBL; Caby 1989) thus appears to be an essentially late-orogenic feature post-dating horizontal tectonics under sillimanite-grade conditions. High- to low-angle retrogressive shear zones from the Médio Coreaú domain are also part of this system (Abreu et al., in press; Gama Junior et al., 1988). A large dextral displacement was probably accomodated along the TBL, since immediate correlation of lithostratigraphic units and P/T regimes on both sides is not possible.

The Senador Pompeu shear zone system is underlined by elongated batholiths of the Quixadá Quixeramobim complex. These mainly calcalkalic granites of Brasiliano age intruded reworked Archean gneisses during an early stage of strike-slip movement. This is demonstrated by the parallelism between the vertical magmatic planar fabric of undeformed granite with globular quartz near the center of the batholiths and the tectonic fabric defined by flattened and horizontally stretched xenoliths and phenocrysts developed along their margins, and progressively turn into mylonitic foliation within the shear zone. A belt of steeply-dipping reworked Archean, high-temperature metamorphic rocks (including granulites, cf. Sect. 2.3) underlines the shear zone system to the north. These rocks display a tectonic banding only present within the shear zone. High-temperature syn-kinematic fabrics were entirely overprinted by post-kinematic annealing at still high temperature. This evolution suggests thermal softening of the crust at the time of granite emplacement, followed by hardening during cooling of the batholith and other associated intrusives. Lower temperature reactivation of this shear zone system is localized in narrow (< dm) faults with negligible chlorite infill, although it is parallel to lower temperature phyllonites of the shear zone to the northwest. To the southeast, however, significant superimposed movement (transtension) took place during differential uplift and infill of a pull-apart molassic graben.

The Orós shear zone system displays a rather curved pattern. Two steep shear zones, that to the north parallel the Senador Pompeu shear zone, delimit a narrow N-S-trending strip of mylonitised metasediments to the north.

Along the virgation to the south at least three parallel northwest dipping shear zones have been recognized (Sá et al., 1988). The regional metamorphic cleavage in metasediments is defined by tiny white micas and is axial-plane of tight curviplanar folds. Linear fabric defined by quartz ribbons and acicular minerals are sub-horizontal. Successive growths of euhedral almandine + stauro lite + biotite ± andalusite took place in schists, in possible connection with emplacement of garnetiferous granite sheets dated at 665 ± 40 Ma by whole-rock Rb-Sr isochron (Macedo et al. 1988). Dextral strike-slip faulting along the NW-dipping shear zone took place in retrogressive conditions, and resulted in the development of phyllonites and ultramylonites and in tight folding of the earlier metamorphic cleavage. Gently dipping shear zones, and lineations at high angle to the strike suggest combined wrenching and thrusting towards the northeast. Sá et al. (1988) interpreted the Orós structure as a flower structure (Sanderson and Marchini 1984), assuming that initiation of the regional cleavage and strike-slip movement along shear zones have been contemporaneous.

2.4.2 East-West Shear Zones of the Pernambuco-Patos System

A major problem concerning E-W lineaments is to determine whether they represent lithospheric faults developed during a late stage of the tectonic evolution or permanent fault zones inherited from a pre-Brasiliano rifting stage and active all along the Brasiliano orogeny. Some recent preliminary data on the mylonitic fabric of rocks cropping out along the Patos shear zone give an insight into the evolution of this structure.

The Patos lineament is nearly 50 km wide (Fig. 8). Many large-scale lenses of non-mylonitic rocks considered to belong to the Seridó belt are embedded in mylonitic to ultramylonitic rocks. The mylonitic foliation is steeply dipping and bears horizontal mineral stretching lineation. Shear sense criteria are frequent features; they include S/C fabric, asymmetric pressure shadows, asymmetric recrystallized tails around feldspars, S-shaped micas, rotated porphyroclasts (mostly feldspar or staurolite) and synthetic shear bands. Most of the inferred shear senses are dextral, few are sinistral and no evidence of movement perpendicular to the lineation has been found.

Field and thin section observations suggest that shearing was long lasting and occurred at different crustal levels during uplift of the area. Coarse-grained protomylonites formed under amphibolite facies metamorphism (probable with T not less than 500 °C) is supported by dislocation creep evidences in K-feldspars, myrmekite-rich recrystallized mantle around K-feldspar, and activation of prismatic slip systems in quartz. They have been reworked during a later deformational event under retrograde greenschist facies conditions producing fine-grained mylonites and ultramylonites. Finally, cataclasites developed from mylonites when the evolution of metamorphic conditions allow a transition from crystal plasticity to brittle behavior. Shear senses are consistently dextral in all kinds of fault rocks. These preliminary data would suggest that the Patos shear zone is a syn-orogenic structure active during the main metamorphic event and reactivated several times during the uplift of the area. The importance of post-Cretaceous reactivation is, however, still debatable.

The Pernambuco lineament displays features similar to the Patos one (Brito Neves et al. 1982). Low-temperature ultramylonite zones have been seen to affect even high-level, late orogenic granite c. a. 60 km west of Recife. Although detailed kinematic studies are lacking, geologic, magnetic and gravimetric data strongly suggest an overall dextral strike-slip movement along the Pernambuco shear zone, the data also suggest dextral strike-slip faulting along the Bom Conselho-Maragoji lineament, south of the Pernambuco shear zone (Brito Neves et al. 1982).

2.4.3 Kinematics and Timing

Unlike the simple kinematic model of conjugate fault zones suggested by the geometrical pattern of shear zones, preliminary kinematic data require a more complex history. Motions along the E-W shear zones are consistently dextral strike-slip. On the contrary, dextral or sinistral strike-slip movements have been reported from the NE-SW shear zones.

A Brasiliano age for the Ceará shear zones is well documented. The major movements took place there in a late stage of the Brasiliano tectonothermal events. This episode of strike-slip motions along NE-SW-trending shear zones in Ceará was contemporary with emplacement of large nappes (Caby and Arthaud 1986), since flat-lying sole thrusts, low- to high-angle shear zones are marked by the same retrogressive phyllonites and ultramylonites with the same movement pattern. Younger reactivation of the Sobral and Senador Pompeu shear zones under transtension resulted in infill and closure of molassic graben.

The Brasiliano age of structures in the Seridó belt is still debatable (Jardim de Sá et al. 1988) but major movement along the NE-SW shear zones of the same area as well as along the E-W Patos and Pernambuco lineaments are considered to be of Brasiliano age.

Finally, studies of the successive generation of granites emplaced during the Brasiliano orogeny emphasize that the distribution of granitic bodies is very likely partly controlled by shear zones, suggesting both genetic relationships between granites and shear zones and an early syn-orogenic initiation of shear zones (see Sect. 2.5).

2.5 Granites

The enormous volume of syn- to late-tectonic granites of Brasiliano age in the Borborema province attests that the Brasiliano orogeny was a major tectonothermal event in this area. The distribution of this igneous bodies is partly controlled by shear zones. Most granitoids are Proterozoic in age, although some are Middle Cambrian (e.g. Meruoca, Mocambo; Sial et al. 1981) as well as a few dike swarms, including acidic members (e.g. Sobral,

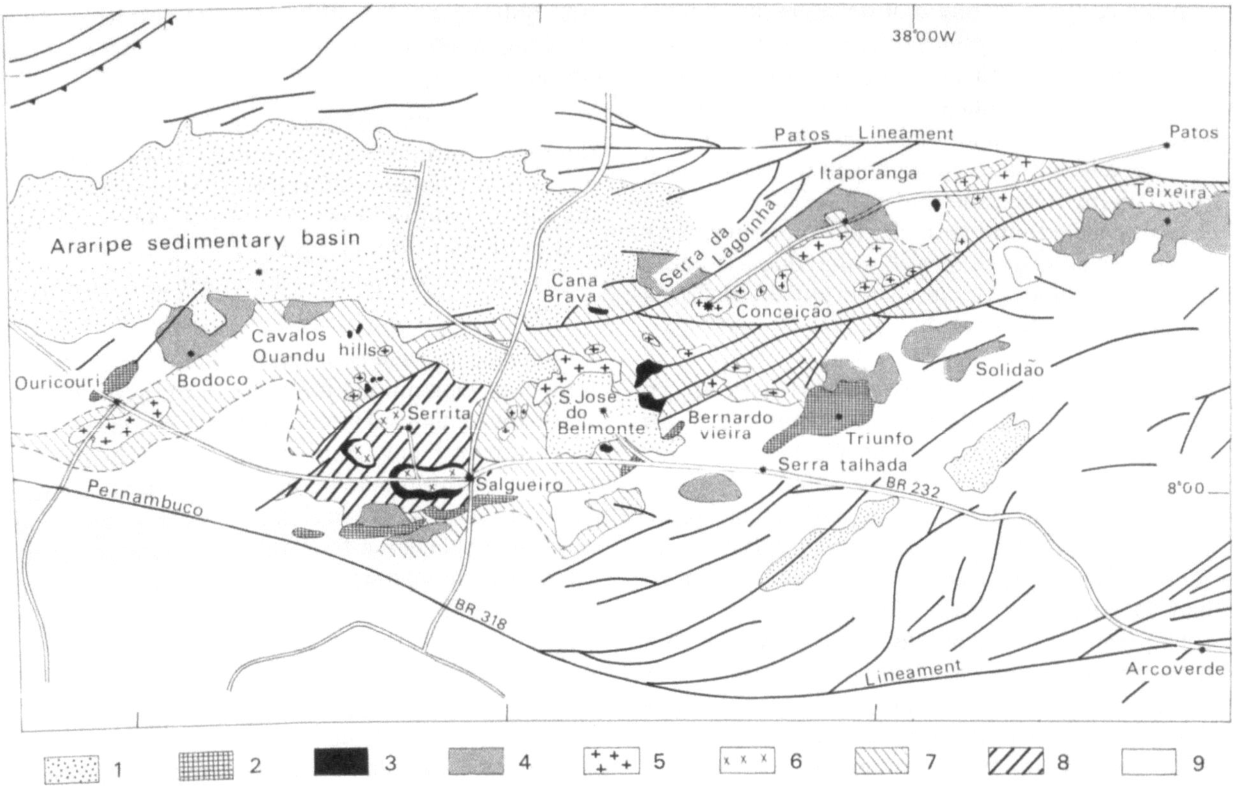

Fig. 9. Brasiliano granites of the Cachoeirinha-Salgueiro belt (after Sial and Ferreira 1988). *1* Cretaceous sedimentary over; *2* saturated peralkaline plutons (Triunfo type); *3* oversaturated peralkaline plutons (Catingueira type); *4* shoshonitic granitoids (Solidão, Teixeira, part of the Salgueiro batholith) and potassic calc-alkaline granitoids (Itaporanga type); *5* epidote-bearing calc-alkaline granitoids (Conceição type); *6* granitoids with trondhjemitic affinities (Serrita type); *7* Cachoeirinha group (schists, phyllites); *8* Salgueiro group (schists, gneisses); *9* basement rocks (mainly gneisses and migmatites)

Tauá, Independência, in the Ceará state). These early Paleozoic granitoids and dykes are commonly peralkalic or with peralkalic affinities, which are typical for the time when the Brasiliano cycle was coming to a close in this part of Brazil. Granitic plutons are mostly elongated in a SW-NE direction and, from a structural viewpoint, they are best known in the Seridó subdomain.

Syn- to late-kinematic intrusions (G_3, G_4, according to Jardim de Sá et al. 1981b) appear as large diapiric complexes or as minor intrusions. Many G_3 granitoids in the Seridó belt are associated with the latest deformational event and they are the most voluminous intrusive type in this fold belt. They are *dent-de-cheval* porphyritic granites and quartz-monzonites with hornblende and biotite, and they display a magmatic preferred orientation. The G_4 intrusive rocks are late- to post-kinematic; they are stocks of biotite granite, granodiorite, or muscovite-biotite-leucogranites. They are always devoid of amphibole, and show magmatic flow textures and preferential orientation of xenoliths.

Granitoids in the Cachoeirinha-Salgueiro subdomain (Fig. 9) are less well-known, from the structural viewpoint, than those in the Seridó subdomain, but they are better known geochemically. According to Almeida et al. (1967), the syn-kinematic bodies were emplaced at 650 ± 30 Ma, late tectonic bodies at 540 ± 25, followed by 500 Ma-old post-tectonic plutons. These granitoids can be divided into four distinct groups: 1) potassic-calc-alkalic, 2) calc-alkalic, 3) granitoids with trondhjemitic affinities, and 4) shoshonitic to ultrapotassic plutons (Sial 1986, 1987; Sial and Ferreira, 1988). Each group has a distinct oxygen isotope and REE signature and has a clearly defined geographic distribution.

The potassic-calc-alkalic granitoids (Itaporanga type) represent a potassic diorite-quartz monzonite association. They are emplaced along the northern boundary of the Cachoeirinha-Salgueiro subdomain (e.g. Itaporanga, Serra de Lagoinha, Bodocó batholith, Fig. 9) and appear to be contemporary with the regional F_2 phase of deformation. Chemical and textural evidence suggest that these rocks are examples of mixing of magmas (potassic diorite and a more felsic, potassic magma). They appear to be diapirically emplaced and are commonly exposed at a level that corresponds to the neck part of the "balloon" (Mariano and Sial, 1988). Comparable plutons are found in the Seridó subdomain (e.g. the syn-tectonic Acari massif, a polydiapiric batholith intruded under a transcurrent ductile deformation; Jardim de Sá et al. 1987).

The calc-alkalic granitoids (Conceição-type) are epidote-bearing granitoids to tonalites emplaced within the area of outcrop of the Cachoeirinha low grade metasediments (Sial 1986, 1987). Typically, they carry amphibolite inclusions that may represent fragments of the source rocks, and they are ^{18}O-enriched. Whole-rock ^{18}O values are greater than 11 permil, in distinct contrast to the +7 to +8.5 permil of the Itaporanga-type granitoids (Sial 1986). Preliminary estimates for pressure of crystallization of two of these bodies, using the Al in amphibole as a geobarometer (Hammarstrom and Zen 1986) yielded numbers between 6.5 and 8 Kbar. Moreover, the Conceição-type granitoids contain primary (?) epidote (Sial 1986), implying crystallization below 6-8 Kbar pressure (Zen 1985). Because they intruded low-to intermediate grade metamorphic rocks, these granites may therefore represent an important clue to an interpretation of regional tectonics and a further correlation of Northeast Brazil and West Africa.

Igneous hornblende (hbl) and biotite (bi) from the Conceição-type pluton of the type locality yield internally discordant $^{40}Ar/^{39}Ar$ age spectra suggesting that initial post-magmatic cooling through appropriate argon retention temperatures occurred between 625 (hbl) and 605 (bi) Ma. This was followed by a distinct low-temperature overprint ca. 580 Ma (Dallmeyer et al. 1987). Data from the Itaporanga-type pluton at the type locality corroborate these two thermal events; a Rb-Sr whole-rock isochron age yielded 620±22 Ma (R=0.7058±0.0003; Sial 1986) and internally discordant $^{40}Ar/^{39}Ar$ age spectra from hornblende and biotite from this pluton, provide plateau ages of 580 and 540 Ma respectively (Dallmeyer et al. 1987).

Rb-Sr ages for the Itaporanga-type plutons (e.g. Acari, Jardim de Sá et al. 1987; Bodocó, J. McMurry, oral commun 1988; Monte das Gameleiras, McMurry 1982) should be considered with caution, as these rocks resulted from the mixing, in various proportions of two magmas which likely had different initial $^{87}Sr/^{86}Sr$ ratio, and the resulting Rb-Sr systematics are in usually pseudo isochrons or errorchrons.

Granitoids with trondhjemitic chemical affinities (Serrita-type) intruded amphibolite facies metasediments of the Salgueiro Group. At Serrita, they form the topographically low core of two ring-structures that grade into topographically high aegirine-bearing syenites at the margins (Sial 1986 and references herein).

Shoshonitic to peralkalic granitoids of the Central Structural Domain are the most thoroughly studied of the four granitoid types (Sial and Ferreira, 1988). As shown in Fig. 9, emplacement of several sodic-potassic to ultrapotassic, aegirine-bearing syenites was controlled by active shear zones (e.g. the syenitoid line emplaced at the southern boundary of the Cachoeirinha-Salgueiro subdomain, Ferreira and Sial 1986). These rocks may represent alkali-feldspar crystal mushes suspended in highly envolved alkalic silica-saturated liquids. They show oval, streaky to globular pyroxenite inclusions and are associated with syn-plutonic dikes of the same composition in a pyroxenite-syenite association. Northward, the main syenitoid line changes to quartz-monzonite with shoshonitic affinities (e.g., Teixeira, Solidão; Sial and Ferreira 1988). Crosscutting relationships in some localities indicate that shoshonitic plutons are slightly older than the ultrapotassic ones, although elsewhere (e.g. Salgueiro, Pernambuco) a gradation between the two can be observed.

The shoshonitic to peralkalic granitoids are high to extremely high in K, Ba, Sr, P, LREE, and low in Nb and Ti (Ferreira 1986; Ferreira and Sial 1986; Sial 1987). A 660 Ma Rb-Sr age (Brito Neves 1982) was reported for the Triunfo batholith, the largest batholith in the main syenitoid line. Rocks of the gabbro-peralkalic syenite association at Ouricuri, along the northern boundary of the Cachoeirinha-Salgueiro domain, show a flat-lying foliation and were probably emplaced concomitantly to the Triunfo batholith. Other peralkalic plutons (silica-saturated to oversaturated) are unfoliated and crosscut ultrapotassic plutons in the main syenitoid line (e.g. Terra Nova dikes in Pernambuco) or intruded Cachoeirinha metasediments (e.g. Quandu, Riacho dos Cavalos, Serra da Cana Brava stocks; Ferreira 1986).

Further geochronological studies by the U-Pb method and structural data are required to constrain the age of emplacement and the age of deformation and metamorphism for these foliated plutons. Indeed peralkalic syenites of the northern part of the São Francisco craton have yielded Rb-Sr, whole-rock ages in the range of 1.8 to 2.2 Ga (Figueiredo 1976; Bartels et al. 1977). Sub-alkalic gneisses and associated metalbitites totally reworked during the Brasiliano orogeny also constitute voluminous "anorogenic" intrusions (e. g. Lagoa Real, Bahia state) possibly emplaced in a rift-type environment with U-Pb zircon ages of ca. 1.7 Ga (Turpin et al. 1988).

The most-studied examples of the ultrapotassic province in the Cachoeirinha-Salgueiro subdomain provide an unique opportunity to examin deeper crustal levels of ultrapotassic peralkalic activity emplaced through pull-apart processes along sinuous shear zones.

Preliminary data suggest that the ultrapotassic rocks may extend beyond this region (Sial and Ferreira, 1988). At the Riacho do Pontal segment of the Central Structural Domain, for example, a biotite pyroxenite-syenite association is found next to Afranio, Pernambuco. The Serra de Santa Luzia elongate peralkalic pluton was emplaced along a shear zone in the Seridó Belt, Paraiba state (Sial and Ferreira, 1988).

Whole-rock ^{18}O, range between 6.5 to 8.3 permil (Ferreira et al. 1988). Such values are comparable to those of the ultrapotassic Roman province in Italy for which an LILE-enriched mantle source has been proposed (Taylor et al. 1984). The ultrapotassic rocks of the Cachoeirinha-Salgueiro Fold Belt exhibit a similar LILE-enrichment, suggesting that there may have been at different times of the Proterozoic evolution an LILE-enriched mantle underneath northeast Brazil (Sial and Ferreira 1988). This must have been a rather shallow feature, within the lithospheric mantle, otherwiese it would have been destroyed by convection. Detailed isotope investigations including Pb, Sr and O isotopes could help to constrain the physical extent and duration of such an anomaly in the mantle.

3 Médio-Coreaú Domain

This domain is delimited to the southeast by the Sobral fault which is considered as part of the shield-scale Trans-Brasiliano lineament (TBL, Fig. 3). It is divided into sub-domains by major, often steep faults underlined by late-metamorphic mylonites showing horizontal lineations and thus interpreted as strike-slip faults. However, the occurrence of molassic graben attests that some of them also had a late orogenic normal fault component.

3.1 Lithostratigraphic Units

Units are listed in order of increasing age:

1. The Jaibaras Group (3000 m thick) is a continental molassic unit (Costa et al. 1979) with interlayered volcanics. It is preserved in a graben, and paraconformably overlies the Ubajara Group. A whole-rock Rb-Sr isochron at 535 ± 27 Ma (Novais et al. 1979) may refer to the low-grade, prehnite-pumpelleyite facies (Jardim de Sá et al. 1979).
2. The Ubajara Group (4000 m thick) includes pelite-psammite, arenite-greywacke and limestone units (Costa et al. 1979). Both groups were intruded by hypovolcanic alkalic granites (Gorayeb et al. in press) with a Rb-Sr whole-rock isochron age of 550 ± 30 Ma (Sial et al. 1981).
3. The Chaval sheared synorogenic plutonic suite of Brasiliano age.
4. The Martinópole phyllites preserved in a down-faulted block. They include also impure carbonates, meta-tuffs and mylonitic quartzites, all with lower greenschist facies, often refolded recumbent cleavage.
5. The São Joaquim Group includes aluminous metaquartzites (> 1500 m thick) and various interlayered gneisses both of sedimentary and igneous origin.
6. The Granja granulites include pyriclasites, pyrigarnites, kinzigites and crosscutting veins of gneissic charnockite.
7. Amphibole-biotite veined gneisses of dioritic, tonalitic and granodioritic composition may represent reworked pre-Brasiliano basement (Archean?), possibly unconformably overlain by metaquartzite of the São Joaquim Group.

Lithostratigraphic correlation with the central domain are speculative; they will be discussed in sect. 2.2 in the light of petro-structural evidence, with reference to the external unitis of the Pan-African belt of Togo-Benin.

3.2 Petro-Structural Evolution of Premolassic Units

Martinópole phyllites (Brito Neves 1975 and references therein) display a penetrative greenschist facies slaty cleavage (S_1) defined by chlorite and sericite orientation, and by mylonitic quartz ribbons. S_1 has been refolded by ENE-trending compressional crenulation and folds, associated with a variably dipping cleavage. Both mineral and stretching lineations within S_1 are defined by tiny micas and pressure shadows around opaque minerals; they have a main ENE to EW trend. Large EW elongation of detrital quartz grains within the S_1 horizontal cleavage, which cuts bedding at a low angle, has been observed in sericite-rich quartzites. In plane perpendicular to the lineation, mylonitic quartzites display refolded isoclinical folds with eye-shaped sections. They could represent sheath folds formed during a progressive deformation which also resulted in a complete reorientation of early ENE-trending linear fabrics. This initial orientation has been retained in some fold hinges where extremely elongated quartz ribbons parallel to the S_0/S_1 intersection have been observed. These elements suggest that an early L-S fabric subsequently evolved into a constrictional regime. A late westward displacement along flat-lying shear zones is evidenced in metaquartzite by *en échelon* quartz veins and, in thin section, by S-shaped micas and C/S planar fabric.

São Joaquim metaquartzites display a penetrative flat-lying foliation. A well-defined linear fabric usually parallel to the bedding-S_1 intersection is marked by the preferred orientation of kyanite in often muscovite-free aluminous metaquartzites and locally by highly elongated quartz gravels. Although only small-scale isoclinal folds have been observed, larger-scale "a"-folds are suspected in this area. The coarse-grained, high-temperature fabric of quartz is characterized by large blastic grains up to several centimeters long. Previously microboudinaged kyanite crystals, frequently rimed by statically grown staurolite, hematite, tourmaline and rutile are included in quartz; this early, syn-metamorphic, fabric is fairly well preserved. It is noticeable that superimposed, low-temperature recrystallization of quartz and replacement of pyrophyllite after kyanite were negligible and restricted to the vicinity of the major tectonic contacts. To the southeast, fibrolitic and prismatic sillimanite (as a possible pseudomorph after kyanite) is present without primary muscovite. Associated metasediments include diopside-bearing calc-silicates, biotite- and biotite-muscovite-garnet-gneiss, cut by pegmatities and anatectic mobilizates of various composition. Mineral lineations have similar trend in kyanite and sillimanite-grade assemblages. Sigmoidal micas in paragneisses with northward gently-dipping C/S planes indicate westward late-metamorphic displacements.

Tonalitic gneisses display a refolded compositional layering of tectonic origin parallel to metamorphic foliation with conformable early anatectic mobilizates. Post-foliation open to tight folds with steep ENE-trending axial-planes occurred at still high temperatures, as evidenced by axial-plane disposition of secondary, anatectic mobilizates. A linear fabric parallel to F_2 folds is defined by green hornblende, ilmenite and biotite streaks, and by elongated feldspar and quartz. Migmatitic rocks with nebulitic texture, including refolded mafic lenses, all with an isotropic igneous-like microstructure, and with up to 40% leuco-tonalitic to trondhjemitic anatectic mobilizates are also present.

The poikilitic amphibole, the systematic presence of epidote in oligoclase, and of myrmekite attests to a complete recrystallization of protoliths of generally tonalitic composition. Later mylonitic reworking of the high-temperature foliation associated with crystallization of minute biotite and chlorite streaks is common approaching shear zones that delimit this unit.

Granja granulites display a refolded catazonal foliation mimetic with a pre-metamorphic layering defined by contrasted lithologies. A roadcut immediately north of Granja exposes layered kinzigites and alternating boudinaged leptynite and metabasic rocks. Banding of kinzigites is defined by compositional variations from inherited sedimentary layering, but also by lenticular bands (khondalite) considered as early anatectic mobilizates. Coarse-grained, typomorphic mineral association of kinzigites comprises quartz + mesoperthite + plagioclase + garnet + biotite + prismatic sillimanite + graphite + rutile as major phases. Early kyanite preserved only in garnet, attests to earlier higher pressure conditions. Early anatectic mobilizates with quartz + mesoperthite + plagioclase + garnet contain minor amounts of rutile and biotite. Cordierite and green spinel have been observed in mesocratic graphite-biotite-garnet enriched layers. Metabasic rocks present as conformable bands and boudins are two pyroxene garnetiferous pyriclasites, with kelyphitic garnet + symplectite of secondary orthopyroxene + pargasitic amphibole. Ilmenite and sphene instead of rutile are present. Other garnet-free, two pyroxenes + pargasitic brown amphibole pyriclasites, with a random, triple point, microstructure, probably derived from static recrystallization of noritic gabbro, apatite needles of igneous origin being still preserved in labradorite. This suggests that at least some mafic rocks may represent syn-metamorphic intrusives.

Some gneissic charnockite veins intruded during granulite facies conditions crosscut banding and veining of leptynites. The coarsegrained primary igneous mineralogy comprises two feldspars, hypersthene, elongated quartz, ilmenite, euhedral apatite and zircon.

Upright folds with subhorizontal, ENE-trending axes, affected the granulitic banding. This folding took place at still high temperature, as suggested by younger, crosscutting, gneissic leucocratic pegmatoids with bluish quartz, mesoperthite and ilmenite. A late, pervasive deformation resulting in tightening of the ENE folds under dextral shear regime occurred after annealing in the sillimanite stability field of the granulite facies. It produced a steep protomylonitic foliation, parallel to the axial surface of ENE-trending folds, which was recorded in all kind of rocks except pyriclasites and garnet-rich granulites. This deformation event is responsible for the flattened fabric of quartz, the deformation of kelyphitic minerals, and for the kink bands which affected pyroxene cleavage. Discrete, minor shear zones underlined by tiny secondary biotite are associated with this event. The constant asymmetry of sigmoidal feldspar clasts observed in leptynites and anatectic mobilizates, indicates a dextral shear regime. The fine-grained mosaic of quartz within protomylonitic rocks (although minor retrogression of high-temperature minerals and replacement of early biotite by newly formed biotite has occurred), is consistent with the existence of low-temperature, dry conditions during this late penetrative deformation associated with dextral movements along ENE-trending shear zones. Metastability of high-temperature phases up to brittle conditions suggest one single stage of evolution ending with excavation of granulites of Brasiliano ages. A similar evolution has been described in the Pan-African granulites of Benin (Caby 1989).

The Chaval synorogenic plutonic suite comprises diversely foliated mylonitized tonalite, adamellite granodiorite and granite. The mainly subvertical magmatic layering with a NE trend is defined by variation in grain size, biotite-enriched layers and by aplite-pegmatite leucocratic veins. A post-magmatic deformation affected most of this composite massif. It produced a variable NE-trending protomylonitic foliation outlined by newly formed minute biotite, and numerous associated anastomosed shear zones. Stretching lineations are nearly horizontal. Shear sense indicators (C/S planes, sigmoidal feldspars) suggest that an homogeneous dextral shear regime affected the entire massif during and after its crystallization.

The Pedra do Sal amphibole-biotite granite which crops out along the shore farther northwest is also affected by sinuous, usually steeply dipping NE-trending shear zones. Kinematic indicators are also consistent with a dextral movement under greenschist facies conditions, postdating granite emplacement (Gama Junior et al. 1988). Within the shear zone, ultramylonitic bands some centimeters thick with grain-size < 5 microns were formed and subsequently transformed in cataclasites of possible Phanerozoic age.

The similar trend of stretching primary mineral lineations within the different metamorphic units of the Médio-Córeaú domain suggests that these units represent portion of the upper and lower crust affected by the same orogenic event. It is noteworthy that retrogression of both kyanite-grade and granulitic assemblages is negligible, though superimposition of a regionally developed protomylonitic, NE-trending foliation and associated mylonitic shear zones. This deformation which post-dates regional metamorphism and synorogenic plutonism consists in a dextral shear regime of regional scale ending under greenschist facies conditions.

3.3 Post-Metamorphic-Granitoids

The post-metamorphic Brasiliano granitoids in the state of Ceará are characterized by high-level plutons (Meruoca, Mocambo, Morrinhos, São Paulo) some of them in ring-structures (Taperuaba), and as several dike sets (e. g. Sobral, Independência, Tauá). The meruoca and Mocambo plutons at Sobral (Fig. 3), Ceará, emplaced next to Sobral-Pedro II lineament. The northernmost part of the Meruoca is composed of fayalite-bearing and fayalite-free granites, quartz syenites and alkali-feldspar granites, emplaced probably through a pull apart process (Sial et al. 1981). The internal distribution of the rock types suggests a structure of a ring complex, partially obscured by a set of faults which surrounds the batholith. Similar intrusions exist near Sobral (Taperuaba ring complex, Haddad and Leonardos 1980) and next to Tauá which bears some peralkalic affinities and have associated granophyres and quartz-porphyry.

The Meruoca and Mocambo plutons were emplaced in Cambrian time. Rb-Sr whole-rock analysis yielded 550 ± 30 Ma (Sial et al. 1981) und 512 ± 10 Ma (Nascimento et al. 1981).

3.4 *Post-Molassic Tectonics*

Post-molassic tectonics have generated upright folds with NE-SW often steep axis, associated with an incipient slaty cleavage only in the Ubajara Group. Horizontal stylolites surfaces in limestones of the Ubajara Group indicate a moderate vertical shortening. This high crustal level, apparently inhomogeneous deformation may be explained by local compressive stress fields in an overall extensional regime (Abreu et al. 1988). Adjacent high-grade metamorphics were uplifted in Cambrian time while intramontane basins were filled (>9 km) and then closed. Post-Paleozoic reactivation of the Sobral fault is evidenced to the southwest by offset of the unconformable flat-lying Early Paleozoic sandstones.

4 Sergipano Belt

The Sergipano belt has been regarded for long time as a thrust- and -fold-belt displaced southward onto the São Francisco craton (Santos and Brito Neves 1984; Jardim de Sá et al. 1986). The belt displays a convex, arcuate pattern marked by major thrusts and shear zones (Fig. 10A and B). Davison and Santos (1989), have given a rather new stratigraphic and tectonic interpretation that completes the conclusion of Santos and Brito neves (1984).

The northern part of the belt includes high-grade gneisses similar to reworked pre-Brasiliano basement of the Borborema province described above. To the south, a large area exists where flat-lying foliation and polyphase folds occurred under amphibolite facies metamorphism; migmatites and late-kinematic intrusions are also present. The southernmost allochtonous domain includes carbonate to clastic-shelf-slope metasediments subjected to monophase folding under greenschist facies grade. It was thrust southwestward onto the São Francisco craton and its slightly deformed Proterozoic cover, with anchizonal metamorphism dated at 490± 11 Ma by Rb-Sr whole rock and fine clay fraction (Brito Neves et al. 1977).

Kinematic data from Davison and Santos (1989) support southward to southwestward displacements of the allochtonous units, followed by sinistral displacement along superimposed, steeply dipping, arcuate faults showing transpressive characteristics (Fig. 10B). Lithostratigraphic correlations between the different zones are impossible, suggesting displacement over large, but still unknown distances. Suspect terranes are also present in the central part of the Belt (Figs. 7, 8): 1) The Canindé subdomain which includes gabbros, ultramafic schists, felsic lavas and metasediments among which graphitic schists are present, and 2) the Macururé subdomain composed mainly of metarhyolite, metadacite, amphibolites, metaconglomerates derived from volcanics, a quartzite-carbonate metapelitic sequence, and serpentinized ultramafic bodies with podiform chromitites, a feature highly suggestive of oceanic mantle peridotite. The occurrence of an ophiolitic assemblage previously proposed by Silva Filho (1979) and of a possible island-arc type terrane (Jardim de Sá et al. 1986) suggest that the two subdomains described above may represent pieces of an outboard, suspect terrane that may have docked along the northern paleo-continent prior to collision during the Brasiliano orogeny with the São Francisco craton and related passive continental margin assemblage (Davison and Santos, 1989). Dating of some late kinematic granites in this belt has yielded Rb-Sr whole-rock isochron of 600 Ma (Davison and Santos, 1989).

Unconformable outliers of red-type sediments preserved in graben related to E-W major fault at the northern edge of the São Francisco craton are interpreted to represent an external molasse, lithologically similar to the Jaibaras Group (cf. 3.1; Mello 1977).

5 Summary and Conclusions

The structural framework of northeast Brazil is dominated by a complex network of anastomozing and sinuous shear zones. Archean basement is present in several areas in the form of highly reworked gray gneisses, with ages around 2.6 Ga (Rb-Sr, whole rock isochrons). In Ceará State, Proterozoic sediments were deposited onto this old basement prior to 1.7 Ga, age of anorogenic, possibly rift-related, igneous complexes well characterized in Ceará as well as in the São Francisco craton. It is likely that a similar evolution also took place in the Seridó region where similar Early Proterozoic sediments have been intruded by granites, now orthogneisses with Rb-Sr whole rock isochron ages around 2 Ga. Another point of view has been developed by Jardim de Sá and co-workers (Jardim de Sá et al., in press) who consider the 2 Ga-old orthogneisses as derived from syntectonic granites, and thus ascribe a trans-Amazonian age to the ma-

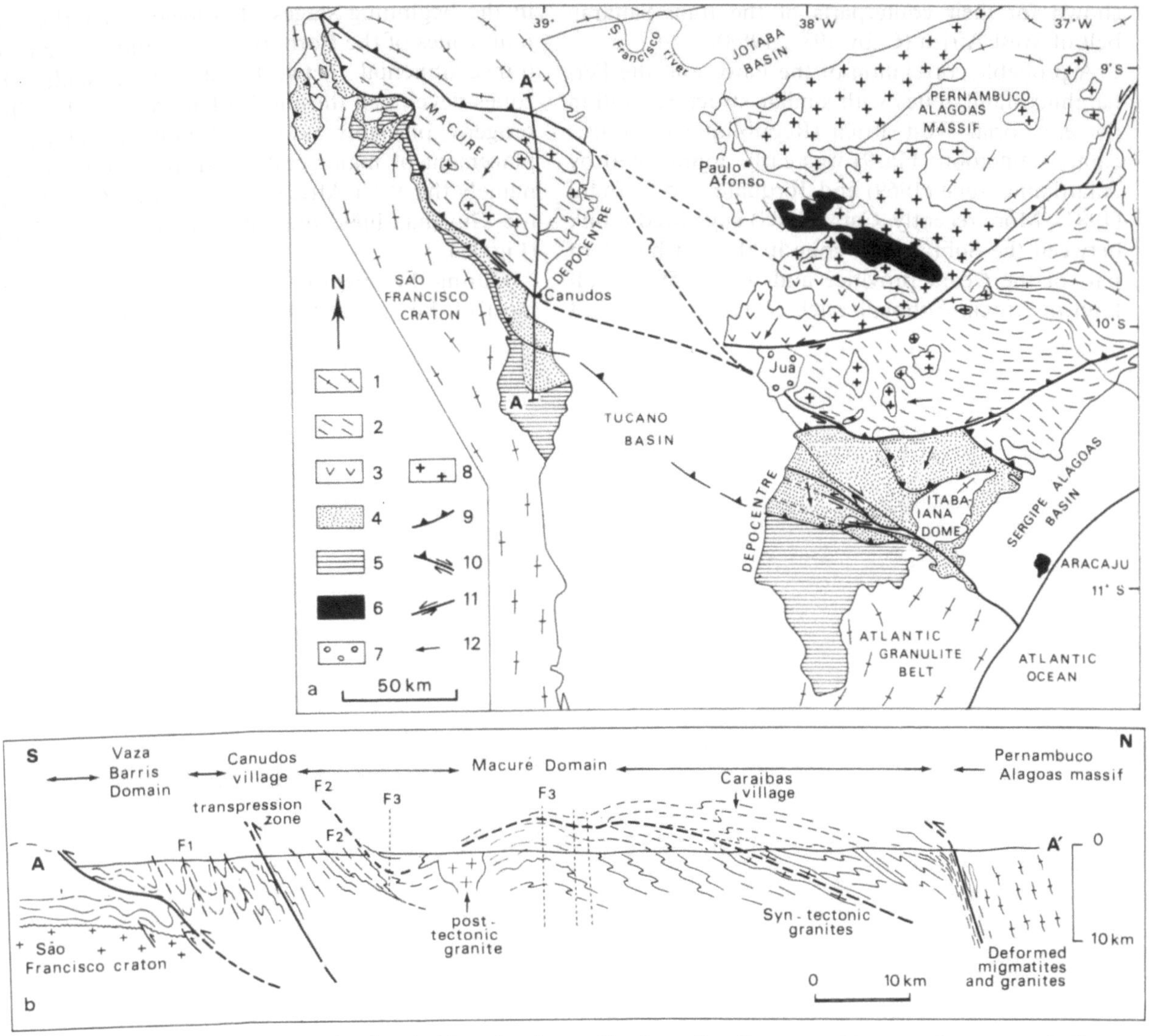

Fig. 10 A. The Sergipano Fold Belt showing the Domains described in this paper. Modified from Silva Filho (1979). *1* High-grade basement gneisses; *2* Macururé Domain, *lines* indicate S_2 foliation trends; *3* Maranco Domain; *4* Vaza Barris Domain; *5* Estância Group; *6* Canindé Complex, *7* Juá Graben; *8* Brasiliano age plutons; *9* major thrust faults; *10* major strike-slip zones with sense of movement indicated; *11* transpression zones with a thrust and transcurrent component of movement; *12* major tectonic transport direction. *Line AA'* denotes cross-section in **B. B** The Macurure Domain (*AA'* in **A**) (Modified from Davison and Santos, 1989)

jor tectonothermal event in the Seridó region. Moreover, Jardim de Sá et al. (in press) state that southward, similar metamorphic assemblages (i. e. the Salgueiro schists, cf. Sect. 2.3) can be traced south of the Pernambuco lineament and are unconformably overlain by > 1.7 Ga-old quartzites of the Espinhaço Group.

Further petrostructural and geochronological investigations are required to solve this apparent divergence. It is also possible that the north-trending schists exposed south of the Pernambuco lineament are already part of the > 1.8 Ga-old São Francisco craton, as suggested by Davison et al. (1988). Flysch-type units (i. e. Cachoeirinha Group) preserved between the Patos and Pernambuco lineaments have not been proved to be intruded by 1.7 or 2 Ga-old granites. They may represent a distinct younger unit suggestive of an "orogenic" environment and are interpreted as the younger deposits involved in the Brasiliano orogeny (Caby 1989).

The amount of lateral displacement along the major shear zones is still unknown, but it is assumed to have been considerable, as it has been shown for their conterparts of the trans-Saharan belt of West Africa (Caby 1987, 1988).

A probable correlation of the Patos and the Pernambuco fault zones with similar structures within the equatorial West africa (Republique Centrafricaine, Cameroon; Fig. 9) was early argumented by Allard and Hurst (1969) and Torquato and Cordani (1981). More recently, Caby (1988) suggested a link between the Sobral fault zone in the northwestern part of the Ceará province and the 4° 50′ Kandi fault of Africa. Although such correlations are just in the beginning stages, it appears that the major fault zones of the Borborema Province are part of a transcontinental network of crustal-scale shear zones developed during the Pan-African-Brasiliano orogeny, probably associated with the relative displacement of major cratons surrounding this area, namely the West African Craton, the Zaire Craton, the Guyana shield and the São Francisco Craton (Fig. 11).

During the opening of the South Atlantic ocean, most of the crustal anisotropies inherited from the

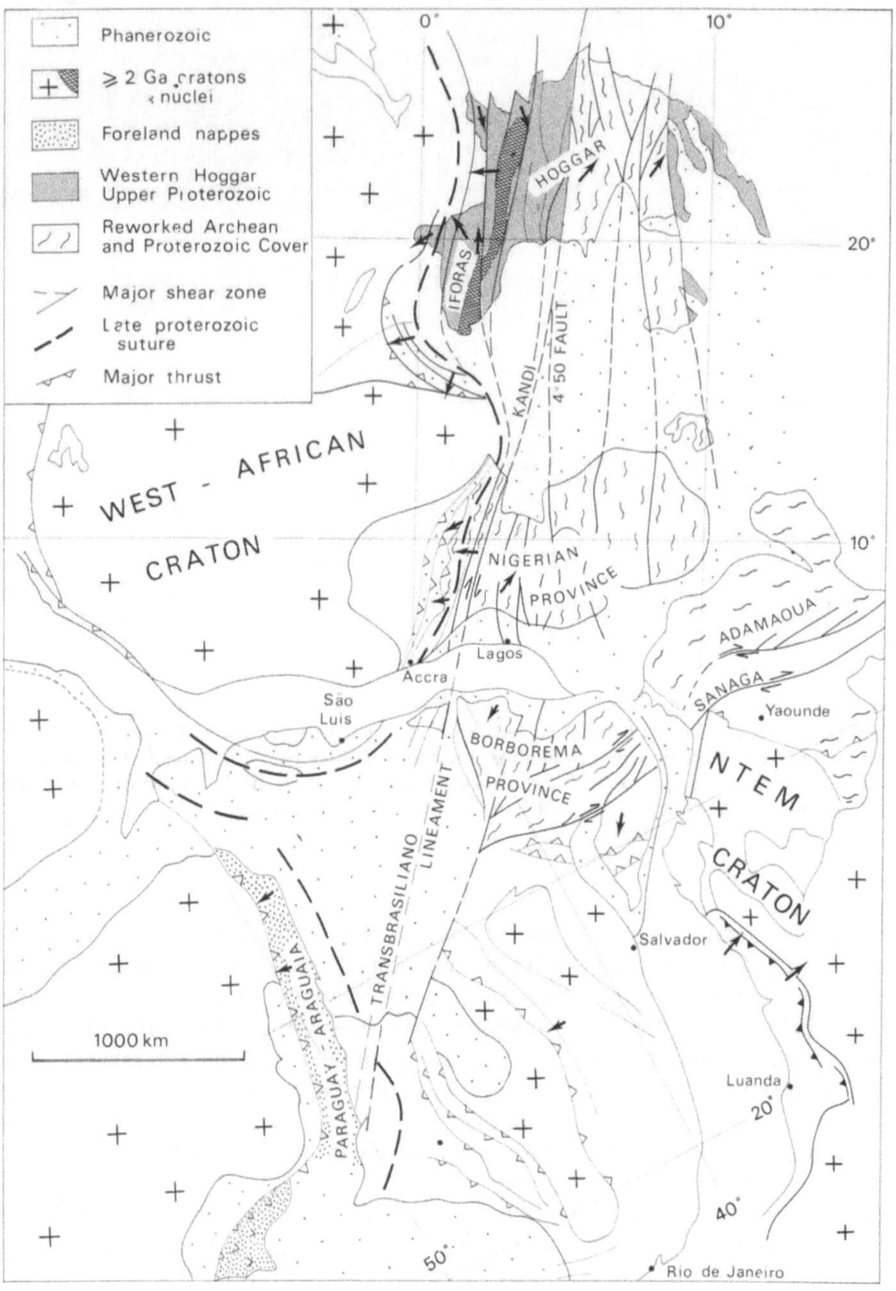

Fig. 11. The Trans-Saharan and Nigerian segments of the Pan-African belt and the Brasiliano belt in a pre-Mesozoic drift reconstruction. *Arrows* indicate direction of movement in nappes and along shear zones (after Caby, 1989)

Brasiliano orogeny were reactivated and controlled the formation of Cenozoic sedimentary basins (Castro 1987; Popoff 1988). These basins are preferentially located along the fault zones, where they display brecciated boundaries. In those from southern Ceará the dip of Jurassic-Cretaceous strata toward the fault is conspicuous, and a post Cretaceous (?) age for cataclasites and pseudotachylites is probable. Preliminary sedimentologic data strongly suggest that the faults have acted, mainly as normal fault, contemporaneously with sediments deposit (Senant and Popoff, pers. commun. 1987).

Offshore data along the northern margin of the Borborema province suggest that the margin results from the reactivation of an inherited E-W-trending shear zone, very similar to the Patos and the Pernambuco lineaments, during the disruption of the Gondwana continent.

Present-day seismicity in the Borborema province is closely related to local reactivation of inherited fault zones (Ferreira and Assunção 1983). Most earthquake epicenters are located along the Brasiliano lineaments, especially the Pernambuco and Patos lineaments, the Seridó eastern boundary fault system and the Orós fault system. Focal mechanism exhibited reverse and strike-slip faulting on steeply-dipping planes (Assunção et al. 1985), suggesting an E-W direction of shortening related to the South Atlantic expansion (Mendiguri and Richter 1978).

Acknowledgements

We are grateful to CNPq, EEC (project CI 1 - 0320-F-CD) and PADCT/FINEP for financial support.

References

Abreu FAM, Gama TG Jr, Gorayeb PSS, Hasui Y (1988) O cinturão de cisalhamento noroeste do Ceará. VII Congr Latino-Americ de Geol Belém, Brasil, 1: 20-34

Allard GO, Hurst VJ (1969) Brazil-Gabon geologic link supports for continental drift. Science 163: 528-532

Almeida FFM de, Leonardos OH, Valença J (1967) Review on granitic rocks of Northeast South America. IUGS/UNESCO Symp, Recife, p 41

Almeida FFM de, Hasui Y, Brito Neves BB de, Fuck R (1981) Brazilian structural provinces. Earth Sci Rev 17: 1-29

Archanjo CJ, Salim J (1986) Posicao da forma ao Seridó no contexto estratigrafico regional (RN-PB) Atas XII Simp Proc Geol Nordeste, pp 270-281

Archanjo CJ (1988) A deformação constrictional nos conglomerados da faixa Seridó. Anais XXV Congr Bras Geol Belém, Brasil, 5: 2240-2247

Assunção M, Suarez G, Veloso JA (1985) Fault plane solutions of intraplate earthquakes in Brazil: some constraints on the regional stress field. Tectonophysics 113: 283-293

Barbosa O (1970) Geologia econômica de parte da região do mèdio São Francisco, Nordeste do Brasil. Dept Nac Prod Miner (DNPM), Div Fom Bol 140: 97

Bartels RL, Torquato JRF, Oliveira MAFT (1977) Geochronological investigations in granulitic terrain Northeast of Feira de Santana, Bahia state. Proc VIII Simp Geol Nord Soc Bras Geol Bol Esp Natal pp 18-19

Bonhomme MG, Gauthier-Lafaye F, Weber F (1982) An example of Lower Proterozoic sediments: the Fancevillian in Gabon. Precambrian Res 18: 87-102

Braun OPG (1982) A structural synthesis of Brazil, based on the study of major lineaments derived from remote sensing imagery interpretation. Photogrametria 37: 77-108

Brito Neves BB de (1975) Regionalização geotectônica do Precambriano Nordestino. Thesis, Univ São Paulo, São Paulo, p 198

Brito Neves BB de (1982) Sintese da geocronologia. In: Mapa geológico do estado da Paraiba. Texto explicativo. Secret Energ Rec Miner (CDRM), João Pessoa, pp 55-69

Brito Neves BB de (1983) O mapa geológico do Nordeste oriental do Brasil, escala 1/1000000. Thesis, Univ São Paulo, São Paulo, p 177

Brito Neves BB de, Kawashita K, Pessoa RR (1975) A posição estratigráfica do complexo Caicó. Proc VII Simp Geol Nord Soc Bras Geol, pp 289-297

Brito Neves BB de, Sial AN, Patrocinio JP (1977) Vergência centrifuga residual no sistema de dobramentos Sergipano. Rev Bras Geoc 7: 102-114

Brito Neves BB de, Sial AN, Rand HM, Manso VAV (1982) The Pernambuco-Alagoas massif, Northeast Brazil. Rev Bras Geoc 12: 240-250

Caby R (1987) The pan-African belt of West Africa from the Sahara Desert to the Gulf of Benin. In: Schaer JP, Rodgers, J (eds) Anatomy of mountain ranges. Princeton Univ Press, Princeton, pp 129-170

Caby R (1989) Precambrian terranes of Benin-Nigeria and Northeast Brazil and the late Proterozoic South Atlantic fit. Geol Soc Am Spec Pap 230; R. D. Dallmeyer, Ed.: 145-158

Caby R, Arthaud M (1986) Major Precambrian nappes of the Brazilian belt, Ceará, northeast Brazil. Geology 14: 871-874

Caby R, Pêcher A, Lefort P (1983) Le MCT himalayen: nouvelles données sur le métamorphisme inverse à la base de la dalle du Tibet. Rev Geogr Phys Geol Dyn 24 (2): 89-100

Castro ACM (1987) The northeastern Brazil and Gabon Basins: a double rifting system associated with multiple crustal detachment surfaces. Tectonics 6: 727-758

Costa MJ, França JB, Lins CAC, Bacchiegga IF, Habekost OR, Cruz WB (1979) Geologia da bacia Jaibaras, Ceará, Piaui, Maranhão. Minist Minas Energ, Dep Nac Prod Mineral (DNPM), Brasilia, p 108 (Geol Bàsica 11)

Dallmeyer RD, Sial AN, Long LE, McMurry J (1987) New evidence for polyphase tectonothermal evolution of the Brasiliano orogen, North-eastern Brazil. Geol Soc Am Ann Meet Abstr Progr, Phoenix, p 634

Davison I, Santos RA (1989) Sergipano fold belt, NE Brazil: tectonic evolution during the Brasiliano orogeny. Precambrian Res 45: 319-342

Davison I, Teixeira JGB, Silva MG, Neto MBR, Matos FMV

(1988) The Rio Itapicuru greenstone belt, Bahia, Brazil: Structure and stratigraphical outline. Precambrian Res 42: 1-17

Ebert H (1962) Baustil und Regionalmetamorphose im präkambrischen Grundgebirge Brasiliens. Tscherm Mineral Petrogr Mitt 8: 49-81

Ebert H (1970) The Precambrian geology of the Borborema belt, (State of Paraiba and Rio Grande do Norte, northeastern Brazil) and the origin of its mineral resources. Geol Rundsch 59: 1299-1326

Ferreira JAM (1967) Considerações sobre uma nova estratigrafia do Seridó. Rev Eng Miner Metal, Rio de Janeiro, 45 (265): 25-28

Ferreira JAM, Assunção M (1983) Sismicidade do Nordeste do Brasil. Rev Bras Geofis 1: 67-88

Ferreira VP (1986) Petrologia e geoquímica de rochas peralcalinas do cinturão de dobramentos. Cachoeirinha-Salgueiro, Nordeste do Brasil. Univ Federal Pernambuco, Poŝ-graduação em Geociências, Centr Tecnol, MS thesis, Recife, 177 pp (unpubl)

Ferreira VP, Sial AN (1986) The peralkaline magmatism in the Precambrian Cachoeirinha-Salgueiro Fold Belt, Northeast Brazil: geochemical aspects. Rev Bras Geoc 16: 78-85

Ferreira VP, Sial AN, Whitney JA, Wenner DB (1988) Petrology and oxygen isotope geochemistry of Brasiliano-age shoshonitic to ultra-potassic plutons of Northeast Brazil. Geol Soc Am Ann Meet, Denver, (abstr) A74

Figueiredo MCH (1976) Geologia e geoquímica da região de Poço de Fora, vale do Rio Curaçá, Bahia. Thesis, Univ Federal da Bahia, Salvador, p 87

Gama Junior T, Gorayeb PSS, Abreu FAM (1988) O granito Pedra do Sal e suas feições de cisalhamento. Rev Bras Geoc 18 (4): 424-432

Gorayeb PS, Abreu FAM, Correia JAM, Moura CAV (1988) Relações estratigràficas entre o granito Meruoca e a sequência Ubajara-Jaibaras. VII Congr Latino-Amer de Geol

Haddad RC, Leonardos OH (1980) Granitos anelares de Taperuaba (Ceará) e processos metassomàticos associados. Proc XXXI Congr Bras Geol, Soc Bras Geol, Camboriú, pp 2626-2633

Hammarstrom JM, Zen E-An (1986) Aluminum in hornblende: an empirical igneous geobarometer. Am Mineral 71: 1297-1313

Hasui Y, Costa JBS, Abreu FAM (1984) Provìncia Tocantins, Setor Setentrional. In: Almeida FFM, Hasui Y (eds) O Pré-Cambriano do Brasil. Blücher, São Paulo, pp 187-204

Hurley PM (1967) Test of continental drift by comparison of radiometric ages. Science 157 (3788): 495-500

Jardim de Sá EF (1978) Revisao preliminar sobre a "Faixa Dobrasa de Seridó" e eventuais correlatos no Nordeste. Rev Ciência, UFRN, Natal 1: 77-83

Jardim de Sá EF (1984) A evolução Proterozóica da Provìncia Borborema. Proc XI Simp Geol Nord, Soc Bras Geol, pp 297-315

Jardim de Sá EF et al. (1979) Observações sobre a estratigrafia e sequências ìgneas da Bacia de Jaibaras, noroeste do Cearà. Proc XI Simp Geol Nord, Soc Bras Geol, Natal, pp 28-38

Jardim de Sá EF, Legrand JM, Hackspacher PC (1981a) Contraste de estilos estruturais e metamorficas na Faixa Dobrada Sergipana. Rev Bras Geoc 11: 128-137

Jardim de Sá EF, Legrand JM, McReath I (1981b) Estratigrafia de rochas granitóides na região do Seridó (RN-PB), com base em critérios estruturais. Rev Bras Geoc 11: 50-57

Jardim de Sá EF, Moraes JAC, D'El-Rey Silva LJH (1986) Tectonica tengencial na Faixa Sergipana. An XXXIV Congr Bras Geol 3: 1246-1259

Jardim de Sá EF, Macedo MH, Legrand JM, McReath I, Galindo AC, Sá JM (1987) Proterozoic granitoids in a polycyclic setting: the Seridó region, NE Brazil. Proc Int Symp Gran Assoc Mineral, Salvador, Brazil, pp 103-110

Jardim de Sá EF, Macedo MHF, Torres HHF, Kawashita K (1988) Geochronology of metaplutonics and the evolution of supracrustal belts in the Borborema Province, NE Brazil. VII Congr. Latino-Americ de Geol, Belém, Brasil, 1: 49-62

Kegel W (1965) A estrutura geològica do Nordeste do Brasil. Div Geol Mineral, Minist Minas Energ, 227: 47

Macedo MHF, Jardim de Sá EF, Sá JM (1984) Datação Rb-Sr em orthogneisses e a idade do Grupo Seridó. Proc XI Simp Geol Nord, Soc Bras Geol, pp 253-257

Macedo MHF, Martins Sá J, Kawashita K (1988) A idade da Faixa Oroś: dados preliminares. Rev Bras Geoc 18 (3): 362-367

Mariano G, Sial AN (1988) Magma mixing: evidence in the Itaporanga batholith, Northeastern Brazil. RC Soc Ital Mineral Petrol I 43: 555-567

Sà MS, Bezerra FHR, Macedo MHF, Pereira R (1988) Middle Proterozoic supracrustals and Brasiliano orogeny in the southeast Ceara stata: a monocyclic evolution. VII Congr Latino-Americ de Geol, Belém, Brasil, 1: 35-48

McMurry J (1982) Petrology and Rb-Sr geochemistry of the Monte das Gameleiras and Dona Inês plutons, Northeastern Brazil. Thesis, Univ Texas, Austin, p 180

Mello ZF (1977) Considerações tectono-petrográficas em seqüências molássicas do Nordeste Oriental do Brasil. Recife, Dissertação de Mestrado Universidade Federal de Pernambuco, 254 pp

Mendiguri JA, Richter F (1978) On the origin of compressional intraplate stresses in South America. Phys Earth Planet Int 16: 318-326

Menot RP, Seddoh KF (1981) Le massif basique stratifié précambrien de Djabatore-Soutouboua: pétrographie et évolution métamorphique. Bull Bur Rech Geol Min Paris 4 (2): 319-337

Meunier A (1964) Succession stratigraphique et passages latéraux dus au métamorphisme de la série Cearà, Antécambrien du Nord-Est brésilien. CR Acad Sci Paris 259: 3796-3799

Munis MS, Santos EJ (1980) Prospecto cobre, chumbo e zinco na faixa Cachoeirinha-Salgueiro- relatório técnico-científico, Companhia. Rec Miner Recife, Spec publ, p 51

Myers JS (1978) Formation of banded gneisses by deformation of igneous rocks. Precambrian Res 6: 43-64

Nascimento DA, Gava A, Pires JL, Teixeira W (1981) Projeto RADAM Brasil. 21, folha SA-24 (Fortaleza) Minist Minas Energ Dep Nac Prod Miner Geol, p 483

Neves SP (1987) Granitic rocks with trondhjemitic affinities and associated peralkaline granites in the Serrita stocks, Northeast Brazil. Proc Int Symp Gran Assoc Mineral, Salvador, Brazil, pp 169-171

Neves SP, Ferreira VP, Farias VP, Albuquerque AC, Melo EB (1984) Petrologia e análise estrutural de uma área a sul de Santa Luzia, Paraìba. Proc XXXIII Congr Bras Geol, Soc Bras Geol, Rio de Janeiro, pp 2537-2548

Novais F, Brito Neves BB de, Kawashita K (1979) Reconhecimento cronoestratigráfico na região Nordeste do Ceará. Proc IX Simp Geol Nord Natal, pp 33

Pessoa RR, Archanjo CJ (1984) Tectônica de empurrões na região de Tróia, Ceará. Proc XXXIII Congr Bras Geol, Soc Bras Geol, Rio de Janeiro, pp 1721-1728

Pessoa RR, Brito Neves BB, Kawashita K, Pessoa DR, Fuck AR (1986) Contribuição ao estudo da evolução geocronològica do maciço de Tróia, CE. Proc XII Simp Geol Nord, Soc Bras Geol, João Pessoa, pp 75-93

Platt JP (1986) Dynamics of orogenic wedges and the uplift of high-pressure metamorphic rocks. Geol Soc Am Bull 97: 1037-1053

Popoff M (1988) Du Gondwana à l'Atlantique Sud: les connexions du fossé de la Bénoué avec les bassins du NE-brésilien jusqu'à l'ouverture du golf de Guinée au Crétacé inférieur. J Afr Earth Sci 7: 409-431

Sanderson DJ, Marchini WRD (1984) Transpression. J Struc Geol 6: 449-458

Santos EJ dos (1971) As feicoes estruturais da falha Arcoverde, Pernambuco e o mecanismo dos falhamentos do "zona tranversal". Rev Min Met, Rio de Janeiro, 53 (313): 35-40

Santos EJ dos, Brito Neves BB de (1984) Provincia Borborema. In: Almeida FFM, Hasui Y (eds) O pré-Cambriano do Brazil. Blücher, São Paulo (Brazil) pp 123-186

Santos EJ dos, Coutinho MGN, Costa MPA, Ramalho R (1984) A direção de dobramentos Nordeste e a Bacia do Parnaiba, incluindo o cráton de São Luiz e as bacias marginais. In: Shobbenhaus C, Campos DA, Derze DGR, Asmus He (eds) Geologia do Brasil. Dep Nac Prod Miner, pp 131-189

Schobbenhaus C, Campos DA, Derze GR, Asmus HE (1981) Geologic map of Brazil and adjoining ocean floor including mineral deposits, scale 1/2500000. Minist Minas Energ, Brasilia, Brazil

Sial AN (1986) Granite types in Northeast Brazil: current knowledge. Rev Bras Geoc 16 (1): 54-72

Sial AN (1987) Granitic rocks of northeast Brazil. Proc Int Symp Gran Assoc Mineral, Salvador, Brazil, pp 103-110

Sial AN, Ferreira VP (1988) Brasiliano-age peralkaline plutonic rocks of the central structural domain, Northeast Brazil. RC Soc Ital Mineral Petrol 43 (2): 307-342

Sial AN, Figueiredo MCH, Long LE (1981) Rare-earth element geochemistry of the Meruoca and Mocambo plutons, Ceará, Northeast Brazil. Chem Geol 31: 281-273

Silva GB, Gentil J (1969) Geologia da jazida de ferro de São José do Belmonte. Univ Federal de Pernambuco, Inst Geol, Recife, Spec Iss, pp 1-20

Silva Filho MA (coord) (1979) Geologia da geossinclinal Sergipana e do seu embasamento. Proj Baixo São Francisco/ Vaza Barris, Dep Nac Prod Miner, Brasilia, p 131, Sér Geol 13, Seção Geol Básica 10

Silva Filho MA da (coord) (1985) Projeto Cachoeirinha. Dep Nac Prod Miner, Recife, 1: 128

Taylor HP, Turi B, Cundari A (1984) $^{18}O/^{16}O$ and chemical relationships in K-rich volcanics rocks from Australia, East Africa, Antarctica and San Venanzo-Cupaello, Italy. Earth Planet Sci Lett 69: 263-276

Torquato JR, Cordani UG (1981) Brazil-Africa geological links. Earth Sci Rev 17: 155-176

Turpin L, Maruejol P, Cuney M (1988) U-Pb, Rb-Sr and Sm-Nd chronology in a polycyclic area: the Lagõa Real granite and uranium-bearing albitites (south Bahia, Brazil). Contrib Mineral Petrol 98: 139-147

Zen E-An (1985) Implications of magmatic epidote-bearing plutons on crustal evolution in the accreted terranes of northwestern North America. Geology 13 (4): 265-266

Subject Index

MIX
Papier aus verantwortungsvollen Quellen
Paper from responsible sources
FSC® C105338

If you have any concerns about our products,
you can contact us on
ProductSafety@springernature.com

In case Publisher is established outside the EU,
the EU authorized representative is:
Springer Nature Customer Service Center GmbH
Europaplatz 3, 69115 Heidelberg, Germany

Printed by Libri Plureos GmbH
in Hamburg, Germany